Susan
Mc Iver

ANNUAL REVIEW OF ENTOMOLOGY

ANNUAL REVIEW OF ENTOMOLOGY

VOLUME 32, 1987

THOMAS E. MITTLER, *Editor*

University of California, Berkeley

FRANK J. RADOVSKY, *Associate Editor*

California Academy of Sciences, San Francisco

VINCENT H. RESH, *Associate Editor*

University of California, Berkeley

ANNUAL REVIEWS INC.   4139 EL CAMINO WAY   P.O. BOX 10139   PALO ALTO, CALIFORNIA 94303-0897

International Standard Serial Number: 0066–4170
International Standard Book Number: 0–8243–0132-3
Library of Congress Catalog Card Number: A56–5750

Typesetting by Kachina Typesetting Inc., Tempe, Arizona; John Olson, President Typesetting coordinator, Janis Hoffman

PRINTED AND BOUND IN THE UNITED STATES OF AMERICA

PREFACE

The editorial committee thought the readership might be interested in the process involved in the planning of an *Annual Review of Entomology* volume.

Each December, ten to fifteen scientists, chosen to represent the breadth of subdisciplines of our science, meet to determine topic priorities for the volume that will be published two years later. Each committee member strives to make sure that his or her areas of interest in entomology are adequately represented; yet each also recognizes that for the good of the overall volume the final choice of topics must represent the diverse interests of our readership. Most committee members develop an approach of friendly persuasion based on give and take.

Some 200–300 potential topics (suggested by our readership, past authors, foreign correspondents, meeting guests, and members of the committee) are discussed each year, but only 30–40 invitations can be sent out for each volume. Clearly, only the fittest topics survive the full committee's scrutiny.

The exercise of planning an *Annual Review* volume is entertaining, broadening, and enriching for participants. After one and a half days of discussions that span current topics of entomological interest from biochemistry to paleoentomology and from genetics to applied agricultural entomology, each member's breadth of knowledge has been increased substantially.

Among the most valuable benefits of attending the annual editorial committee meetings are the friendship and camaraderie that develop during each member's five-year tenure. After having fought, teased, and cajoled each other, we know each other better and share a certain pride in the volumes we helped design. Considering the uniqueness of this experience, it is no wonder that most who have served on the editorial committee of the *Annual Review of Entomology* remain for life avid supporters of this important institution of service to entomology, worldwide.

We thank Ms. Andrea Perlis, our Production Editor, for the outstanding manner in which she has handled the many demanding tasks leading to the publication of the volume. We are also pleased to acknowledge the high quality of work by our compositor and printer.

THE EDITORIAL COMMITTEE

OTHER REVIEWS OF ENTOMOLOGICAL INTEREST

From the *Annual Review of Ecology and Systematics,* Volume 17 (1986)

The Ecology of Tropical Arthropod-Borne Viruses, Thomas M. Yuill
Phenetic Taxonomy: Theory and Methods, Robert R. Sokal
Parasite Mediation in Ecological Interactions, Peter W. Price, Mark Westoby,
   Barbara Rice, Peter R. Atsatt, Robert S. Fritz, John N. Thompson, and Kristine
   Mobley
Genetic Polymorphism in Heterogeneous Environments: A Decade Later, Philip W.
   Hedrick

From the *Annual Review of Genetics,* Volume 20 (1986)

Neurogenetics of Membrane Excitability in Drosophila, B. Ganetzky and C.-F. Wu

From the *Annual Review of Microbiology,* Volume 40 (1986)

Natural History of Rickettsia rickettsii, Joseph E. McDade and Verne F. Newhouse
The Natural History of Japanese Encephalitis Virus, Leon Rosen
The Molecular Biology of Parasporal Crystal Body Formation in Bacillus thurin-
   giensis, H. R. Whiteley and H. Ernest Schnepf

From the *Annual Review of Neuroscience,* Volume 10 (1987)

Neurosecretion: Beginnings and New Directions in Neuropeptide Research, Berta
   Scharrer

From the *Annual Review of Phytopathology,* Volume 24 (1986)

Remote Sensing of Biotic and Abiotic Plant Stress, Ray D. Jackson
Biological Control of Plant-Parasitic Nematodes, P. Jatala

Annual Review of Entomology
Volume 32, 1987

CONTENTS

INSECTS AS MODELS IN NEUROENDOCRINE RESEARCH, *Berta Scharrer*   1

CHEMOSYSTEMATICS AND EVOLUTION OF BEETLE CHEMICAL
DEFENSES, *Konrad Dettner*   17

INSECT HYPERPARASITISM, *Daniel J. Sullivan*   49

CHITIN BIOCHEMISTRY: SYNTHESIS AND INHIBITION, *E. Cohen*   71

BIOGEOGRAPHY OF THE MONTANE ENTOMOFAUNA OF MEXICO AND
CENTRAL AMERICA, *Gonzalo Halffter*   95

THE BIOLOGY OF DACINE FRUIT FLIES, *B. S. Fletcher*   115

IMPROVED DETECTION OF INSECTICIDE RESISTANCE THROUGH
CONVENTIONAL AND MOLECULAR TECHNIQUES, *Thomas M. Brown
and William G. Brogdon*   145

ARTHROPODS OF ALPINE AEOLIAN ECOSYSTEMS, *John S. Edwards*   163

CULTURAL ENTOMOLOGY, *Charles L. Hogue*   181

BIOLOGY OF LIRIOMYZA, *Michael P. Parrella*   201

ECOLOGICAL CONSIDERATIONS FOR THE USE OF ENTOMOPATHOGENS
IN IPM, *J. R. Fuxa*   225

BIOLOGY OF RIFFLE BEETLES, *H. P. Brown*   253

SCORPION BIONOMICS, *S. C. Williams*   275

VISUAL ECOLOGY OF BITING FLIES, *Sandra A. Allan, Jonathan
F. Day, and John D. Edman*   297

FACTORS AFFECTING INSECT POPULATION DYNAMICS: Differences
Between Outbreak and Non-Outbreak Species, *W. E. Wallner*   317

INSECT PESTS OF SUGAR BEET, *W. Harry Lange*   341

ECOLOGICAL GENETICS OF INSECTICIDE AND ACARICIDE RESISTANCE,
Richard T. Roush and John A. McKenzie   361

BIOSYNTHESIS OF ARTHROPOD EXOCRINE COMPOUNDS, *Murray
S. Blum*   381

COMPUTER-ASSISTED DECISION-MAKING AS APPLIED TO
ENTOMOLOGY, *Robert N. Coulson and Michael C. Saunders*   415

viii   CONTENTS   *(continued)*

PHYSIOLOGY OF OSMOREGULATION IN MOSQUITOES, *T. J. Bradley*   439

ROLE OF SALIVA IN BLOOD-FEEDING BY ARTHROPODS,
  J. M. C. Ribeiro   463

ADVANCES IN MOSQUITO-BORNE ARBOVIRUS/VECTOR RESEARCH,
  G. R. DeFoliart, P. R. Grimstad, and D. M. Watts   479

ECOLOGY AND MANAGEMENT OF SOYBEAN ARTHROPODS, *Marcos
  Kogan and Samuel G. Turnipseed*   507

INDEXES

Subject Index   539

Cumulative Index of Contributing Authors, Volumes 23–32   550

Cumulative Index of Chapter Titles, Volumes 23–32   553

Ann. Rev. Entomol. 1987. 32:1–16

INSECTS AS MODELS IN NEUROENDOCRINE RESEARCH

Berta Scharrer

Department of Anatomy and Structural Biology and Department of Neuroscience, Albert Einstein College of Medicine, Bronx, New York 10461

INTRODUCTION

The use of model systems for the study of basic biological and biomedical phenomena has a long history, during which models have largely been restricted to certain mammalian species. Because of a century-old preoccupation with homology (a concept introduced by Owen in 1843, which emphasizes relatedness based on the common evolutionary ancestry of organ structure), the functional correspondence of analogous biological systems was given little attention (51). In recent years, however, suitable animal models have also been sought among invertebrates as well as nonmammalian vertebrates, and now the principle of analogy is of primary concern. This change of orientation is reflected in a comprehensive report prepared by the Committee on Models for Biomedical Research at the request of the Commission on Life Sciences of the National Research Council (7a).

The current focus is on lower forms for several reasons, including relative simplicity, greater accessibility, and applicability for certain experimental procedures. But first and foremost, the relevance of information gained from the use of such models has been established by the demonstration of remarkable structural and functional parallels between the two animal phyla. These analogies apply from the molecular to the organismic level of organizational complexity. Moreover, the rewards gained from experimentation with lower organisms, including unicellular organisms, reach beyond the detection of commonalities judged to be practically useful for modeling in biomedical research (see 47); broadly based comparative studies provide information on the evolutionary history of basic biological phenomena and quite frequently

1

lead to the discovery of new principles that would not easily be found through mammalian studies alone.

This article addresses a specific sector in this area, i.e. the analysis of insect systems to elucidate how the neuroendocrine apparatus maintains integrative control over the body's functions. This regulatory mechanism is as important to these highly differentiated invertebrates as it is to mammals (including humans), and its operation is certainly no less complex. Therefore, the parallels discussed here in some detail between selected regulatory phenomena in insects and in mammals should demonstrate the usefulness of insects as experimental models.

THE NEUROENDOCRINE SYSTEM OF INSECTS

The processes in an insect's life cycle that require precisely coordinated control are embryonic and postembryonic development, reproductive activity, and changing metabolic and behavioral patterns. The episodic events, including the molting cycle and metamorphic transformations that lead to the emergence of adult insects, are programed with greater precision than the developmental steps leading to maturity in most vertebrates. The cyclicity in the reproductive activity of the females of certain insect species resembles that of mammals. It is not surprising, therefore, that unlike lower invertebrates, insects share with vertebrates the use of two integrative control systems, the neural and the endocrine.

It cannot be denied that insights gained from the study of insects and vertebrates contributed equally to the evolvement of our current concepts on the interaction of these two regulatory systems, which culminated in the emergence of the new discipline of neuroendocrinology. The remarkable discovery by the Polish biologist Kopeć (25, 26) that the brain of the lepidopteran insect *Lymantria* furnishes a "pupation hormone" was not only the first demonstration of an endocrine activity in any invertebrate, but was also the first indication anywhere in the animal kingdom that the nervous tissue is capable of producing hormones. Much time had to pass before this discovery, which was so much ahead of its time, could be fitted into the general framework of neuroendocrine control systems.

It is now firmly established that in order to operate effectively, the two systems of integration must be in constant communication with each other. Thus mechanisms are required to enable endocrine centers to "understand the language" of the nervous system and vice versa (see 52).

Good evidence indicates that at least certain neurons have receptors that enable them to receive afferent hormonal signals arriving via circulatory channels. Efferent neural directives aimed at endocrine organs can reach their destinations in more than one way. The most effective of these is the dispatch by certain neurons of hormonal messengers addressing endocrine cells in their own language. Cytological evidence suggested that these special neuronal elements are related to the "pupation hormone" of insects, but the functional significance of this class of neurons turned out to be much more complex than anticipated (48).

The possible role of these neurons in providing a channel of communication with the endocrine system was suggested as early as 1928 by Ernst Scharrer (50) when he discovered the cell bodies that give rise to the hypothalamo-hypophysial tract in the preoptic nucleus of a teleost fish, *Phoxinus laevis*. Because these cells produce proteinaceous messenger substances that, because of their dispatch via the circulation, are produced in much greater quantity than conventional transmitters, the cells resemble gland cells as much as impulse-conducting nerve cells. They are therefore referred to as "neurosecretory neurons."

Analogous groups of glandular nerve cells were detected in the nervous systems of many invertebrates. The first observation, in the opisthobranch snail *Aplysia* (39), was soon followed by comparable observations in annelids and arthropods (40). One of the insects that has provided much information on neuroendocrine functions, the blattid *Leucophaea maderae,* is repeatedly used as an example in the following sections.

The peptidergic granules of neurosecretory neurons can be stained selectively, e.g. by the aldehyde fuchsin method originally developed by Gomori for the demonstration of β-cells in the mammalian pancreas. Electron micrographs reveal that these secretory elements (identified by light microscopy) are accumulations of smaller ($\sim$ 200–300 nm diameter) membrane-bounded electron-dense granules. At both levels of magnification, these cytoplasmic inclusions serve as convenient markers for tracing the course of axons that deliver the material to its site of release. The fact that this transport occurs by axoplasmic flow was demonstrated by severance of neurosecretory nerves in insects (41) as well as in mammals (21). In both instances, this operation resulted in the accumulation of the secretory product proximal and its depletion distal to the plane of section.

The cerebral neurosecretory centers in which the active material is synthesized represent an essential component of a neuroendocrine organ complex that is remarkably similar, in both structure and function, in insects and vertebrates. The second component of this complex is an extracerebral neurohemal structure where the secretory material is stored and released. The third is a nonneural gland of internal secretion, which is in close spatial relationship with the second.

The paired groups of neurosecretory cells in the protocerebrum of insects are analogous to the hypothalamic neurosecretory centers of vertebrates. Their axon bundles (nervi corporis cardiaci) enter the corpus cardiacum which, in its neurohemal capacity, corresponds to the posterior lobe of the pituitary gland. In addition to neurohormones of cerebral origin, the corpus cardiacum delivers products of its intrinsic neuroglandular cells into the hemolymph. Some axonal projections leave the corpus cardiacum to enter the adjacent corpus allatum, i.e. the third component of the complex. This important nonneural endocrine insect gland can be considered the analog of the adenohypophysis of vertebrates.

In both the brain-cardiacum-allatum system and the hypothalamic-hypophysial system the dispatch of peptide hormones synthesized in the brain results in two types of signals. Released from neurohemal centers into the circulation, the signals, may (a) address terminal effector sites ("target cells") directly or (b) control a given physiological activity indirectly by giving instructions to an intervening endocrine gland (corpus allatum or adenohypophysis, respectively).

A number of direct (first-order) neurosecretory functions, paralleling those of the so-called posterior lobe hormones of vertebrates, have been demonstrated in insects. In lower invertebrates these are the only neurosecretory functions. Indirect (second-order) mechanisms, exemplified in vertebrates by the operation of the hypophysiotropic factors of the hypothalamus, have a counterpart in the neurosecretory control over the production and release of juvenile hormones by the corpus allatum. The same is true for a second endocrine organ, the prothoracic gland, which is active during the postembryonic development of insects.

Detailed studies on the neural regulation of the corpus allatum (for review, see 64) revealed that it closely parallels the regulation of the anterior lobe of the hypophysis. In both cases, the directives may be either stimulatory or inhibitory. Moreover, the regulating factors may reach the gland not only via a circulatory route (hemolymph and hypophysial portal circulation, respectively) but may deliver their messages at close range. The latter type of operation is illustrated by the existence in the corpus allatum (44) as well as in the mammalian pars intermedia (3) of synapse-like (synaptoid) terminals of neurosecretory neurons. A variant of spatial proximity is the separation of the release site of a neuropeptide from its putative site of action by a narrow interstitium of extracellular matrix.

A neurohormone that stimulates corpus allatum activity (allatotropin) has been demonstrated by various experimental procedures, including brain implantation and cauterization of selected brain areas, to originate in the pars intercerebralis of the protocerebrum (e.g. 14, 71). Effective signals, both stimulatory and inhibitory, are known to be conveyed to the gland periodically by the nervi corporis cardiaci. In a number of species, severance of the nervi corporis cardiaci results in activation of the corpora allata, reflected by characteristic structural parameters as well as by elevation of the rate of juvenile hormone synthesis (see 64).

The corpus allatum shows a distinctive structural plasticity in conjunction with changing functional states in *Leucophaea maderae*. Active glands are significantly larger and have an increased number of cells. In addition, cellular as well as nuclear diameters are increased as compared with those in inactive glands, and the abundant cytoplasm contains more numerous, better-differentiated organelles (41, 43). Similar structure-function relationships have been reported in the viviparous cockroach *Diploptera punctata* (24).

At certain developmental stages (i.e. in late-instar nymphs), denervation of the corpora allata does not lead to full activation of the glands. This can be taken as a sign that, in addition to close-range neural signals, blood-borne inhibitory signals (allatostatin/allatinhibin, comparable to vertebrate somatostatin) may also be in operation (59). However, it seems that in this interplay of directives neural inhibition takes precedence over neurohormonal inhibition.

As is the case in the adenohypophysis, the type of directive dispatched to the corpus allatum by a neuropeptide depends on the total interoceptive and exteroceptive signals received by the brain (see 52). Among the latter signals, photoperiodic cues are prevalent throughout the animal kingdom. Important interoceptive directives are delivered to the brain by hormones, as illustrated by the operation of a three-step feedback mechanism controlling the dispatch of tropic hormones by the corpus allatum as well as by the adenohypophysis. Hormonal messages representing the afferent link in the self-regulatory process of an endocrine gland must detour via the brain because centralization of possibly conflicting signals is needed to ensure the receipt of integrated, effective commands by the endocrine system (42).

Peptidergic neurons that represent the "final common pathway" to the glands of internal secretion are electrically excitable and capable of receiving instructions via excitatory and inhibitory synaptic signals. In insects as well as vertebrates, these peptidergic neurons are known to generate action potentials (27, 31), which effect the release of the peptidergic neuroregulators (see 69). The exteriorization generally occurs via exocytosis (see e.g. 49).

It should be mentioned here that a number of peptidergic neurons have been identified outside the neuroendocrine axis. As demonstrated by electron microscopy, somatic elements, e.g. salivary gland cells of the blattarian insect *Byrsotria fumigata* (45), may receive strictly localized peptidergic signals. Moreover, junctional complexes have been observed between neurosecretory terminals and other neurons of either the conventional or the neurosecretory type (1). In the corpus allatum of the orthopteran *Arphia pseudonietana*, peptidergic fibers of the same ultrastructural type have been found in an intriguing symmetrical (mirror-image type) arrangement. This observation, which suggests a reciprocal exchange of "information" that is presumed to be related to the control of the corpus allatum (45), presents a challenge for future exploration.

CHEMISTRY OF INSECT NEUROPEPTIDES

The structural and functional analogies in the neuroendocrine systems of insects and vertebrates discussed in the preceding section have a counterpart in the chemical similarity of peptidic messenger substances used by the nervous system for signaling to the endocrine apparatus or to terminal effector cells. In recent years, extensive use of immunocytochemical methods, including work at the ultrastructural level (see 58), has contributed much to the elucidation of the chemical nature, the sites of origin, and the sites of release of these neurosecretory substances (22).

An important finding has been that substances antigenically related to known vertebrate neuropeptides occur in the neuroendocrine system of many insects. By the same token, peptides first identified in invertebrates turned out to be present also in vertebrates. For example, representatives of the FMRF-amide family of cardioexcitatory neuropeptides discovered in molluscs occur in insects as well as in vertebrates (15, 28, 29, 65). A growth-promoting principle, the "head activator" discovered in the coelenterate *Hydra* and sequenced by Schaller & Bodenmüller (38), has subsequently been demonstrated in insects as well as mammals, including humans.

The peptidergic neurons of *Leucophaea* corresponding to those of the hypothalamic-hypophysial system in vertebrates contain material closely related to oxytocin, vasopressin, somatostatin, and a number of additional peptides (17, 18). Reaction products have been demonstrated in axon bundles (nervi corporis cardiaci) within the corpora cardiaca and in some axons that enter the corpora allata, where they end in close contact with the glands' parenchymal cells.

Hansen et al (18), using region-specific immunocytochemistry, demonstrated that the neuroendocrine complex of *Leucophaea* contains a peptide material with antigenic determinants that recognize at least part of a known vertebrate neuropeptide, the (1–3) and the (11–17) regions of the ACTH (1–24) molecule. This observation supports the view that this substance, as well as related peptide molecules, has a long evolutionary history.

Comprehensive studies on the differential distribution of neuropeptides in a number of other insect species (e.g. 2, 9, 10, 13, 34, 76) revealed peptide material related to Substance P, somatostatin, hypothalamic growth hormone–releasing factor (GRF), glucagon, insulin, gastrin/cholecystokinin, vasoactive intestinal peptide (VIP), pancreatic polypeptide (PP), secretin, and endogenous opioids. Much of this material is located in an area of the corpus cardiacum that bespeaks its extrinsic (cerebral) origin, but some neuropeptides, among them the adipokinetic insect hormone, can be localized in the organ's intrinsic neuroglandular cells and their processes (54).

The neuropeptides traced thus far to the corpora allata [material closely resembling Substance P, β-endorphin, somatostatin, luteinizing hormone–releasing factor, VIP, insulin, and α-melanocyte-stimulating hormone (αMSH)] all seem to be derived from extrinsic cells in the brain.

In insects and vertebrates, a variety of recently identified neuropeptides have been localized not only within the nervous system but also in nonneural organs such as the gut. The differential distribution of axon terminals containing an immunoreactive material resembling the insect neurohormone proctolin suggests that this pentapeptide may function also as a neurotransmitter or neuromodulator in the nervous system of *Periplaneta americana* (1) and in the digestive system of *Leptinotarsa decemlineata* (72). The same function can be attributed to vertebrate-type neuropeptides including pancreatic polypeptide, somatostatin, and β-endorphin in neuronal and "diffuse endocrine cells" of the digestive system of insects and mammals (30).

The identification of rising numbers of neuropeptides occurred hand in hand with the recognition of specific receptors for these substances. The search for such receptors in insects is still in its infancy. However, first results strongly suggest yet another analogy with mammals and other vertebrates. These studies revealed the existence in the brain and midgut of *Leucophaea* of specific high-affinity binding sites for a synthetic analog of the opioid peptide Met-enkephalin (56, 57). Binding was shown to be monophasic, saturable with respect to the concentration of the radioligand used, and stereospecific. The binding-site density for this peptide was reduced by sodium and lithium and increased by manganese. In these and other respects, the opioid receptors presumed to operate in these insect organs meet the criteria established by corresponding incubation experiments carried out in mammalian systems (75).

The discovery of such signal-receptor interactions in invertebrates led Stefano (55) to suggest an interesting explanation for the conservation of neuropeptides throughout their long evolutionary history. He called attention to the complexity of the gene-controlled establishment of a peptidergic intercellular signaling device in which several components (signal molecule, stereospecific receptor molecule, specific enzymes controlling biosynthesis and inactivation) must evolve simultaneously to make the system operational. It is this need for precise conformational matching of molecules that is thought to be the determining factor in keeping the signal molecules relatively intact during the development of increasingly sophisticated systems of integration.

Another intriguing subject addressed in current neuropeptide research is the functional significance of the coexistence of more than one regulatory substance within a single neuron, especially when conventional, i.e. nonpeptide, transmitters are combined with peptidergic neuroregulators. This phenomenon is widespread among vertebrates (see 7, 23) and also occurs in insects, as demonstrated by the immunocytochemical identification of dopamine and an α-endorphin-like substance in certain neurons of the silkworm, *Bombyx mori* (62). Here too, insect studies may contribute to the solution of an as yet little understood phenomenon.

NEUROENDOCRINE CONTROL OF REPRODUCTIVE ACTIVITY

Juvenile hormone owes its name to the fact that during postembryonic development it enables the growing insect to retain its immature characteristics until the insect has reached the size appropriate for the adult stage. After metamorphosis, however, the same hormone takes over certain roles in the control of reproductive activity. Aside from its influence on the proper function of the accessory sex glands, the presence of juvenile hormone ensures ovarian competence. This dependency of the female gonad on the activity of the corpus allatum has been demonstrated in a number of insect species. Allatectomy and starvation (during which the gland remains inactive) both prevent the maturation of the oocytes.

According to current information, juvenile hormone, in cooperation with a second endocrine factor, ecdysone, enables the ovary to deposit yolk (see 32, 64). The mechanism by which this is accomplished is remarkably similar to the process of vitellogenesis in vertebrates. Both activities are under neuroendocrine control. In both instances, vitellogenins, i.e. precursors of yolk protein, are synthesized in analogous organs outside of the ovary, the liver in vertebrates and the fat body in insects. The protein reaches the ovary by way of the circulation, and is taken up by the developing oocytes to be packaged in the form of yolk granules, as first demonstrated by Roth & Porter (36) in the mosquito *Aedes aegypti*. In some insects, e.g. *Drosophila,* an additional site of vitellogenin synthesis has been identified, i.e. the ovarian follicle cell. Juvenile hormone controls the biosynthetic process in both locations as well as the uptake of the material by the oocytes (12). The influence of ecdysone seems to be restricted to the fat body.

A detailed analysis (see 32) has shown that vitellogenin synthesis in insects is gene-encoded. The fact that the process takes place in a sex-specific fashion is thought to be related to a difference in the number or nature of hormone receptors in females versus males.

The subcellular events responsible for the selective endocytotic uptake of three female-specific insect vitellogenins by the ovary parallel those characteristic of receptor-mediated endocytosis in many vertebrate systems, including that of the incorporation of vitellogenin by amphibian oocytes (74).

Of special interest as instructive model systems are insects whose reproductive cyclicity compares to that of mammals, i.e. viviparous and ovoviviparous blattarian species. In *Leucophaea,* for example, periods of oocyte growth and maturation are followed by longer periods of ovarian dormancy during which the embryos develop in the mother's brood pouch. As in mammals, the timing of these events depends on the concomitant cyclic activity of the neuroendocrine apparatus: During the period of yolk deposition by the oocytes the corpus allatum is active, apparently because the inhibitory influence of the neurosecretory elements of the brain is at that time minimal or absent. After the fully grown oocytes, laden with yolk, enter the brood pouch (ovulation), the biosynthesis of juvenile hormone ceases, and yolk formation in the next batch of terminal oocytes is halted throughout the period of "pregnancy." This holding pattern is dictated by afferent signals to the brain attributable to the presence of an ootheca in the brood pouch.

The periods of reproductive quiescence can be terminated prematurely by experimental interventions that remove the programed restraint on corpus allatum activity, i.e. by denervation of the gland or forced removal of the ootheca from the brood pouch. As expected, both procedures lead to accelerated reactivation of the corpora allata, which is expressed not only by renewed biosynthesis of juvenile hormone, but also by a return of the ultrastructural features characteristic of the active gland as discussed above for *Leucophaea*.

The ovarian dormancy in *Leucophaea* is comparable to the suppression of embryonic development in marsupials when a suckling sibling is present in the mother's pouch. Here too, diapause can be experimentally broken by removal of the pouch young, denervation of the mammary gland, or hypophysectomy (70).

In insects as well as vertebrates, afferent gonadal (feedback) signals are known to contribute information determining the types of gonadotropic directives dispatched by the neuroendocrine axis. Stimuli conveyed by an afferent regulatory factor (ecdysteroid) released by the insect ovary may reach the corpus allatum either directly or indirectly via the brain (33). Removal of the ovary leads to activation of the corpora allata, which in the case of *Leucophaea* may result in a level of activity greater than that of normal glands at the height of their activity. Ultrastructurally, this hyperactivity expresses itself in supranormal cellular dimensions and in a striking abundance of smooth-surfaced endoplasmic reticulum (46). Since this organelle is the site of the final steps in the biosynthesis of juvenile hormone (63), its prominence indicates functional hyperactivity of the glands. This result may perhaps be explained by the constancy of the demand on the corpora allata of castrates. In the intact animal the brain turns off juvenile hormone activity in response to afferent signals during the lengthy periods of gestation; in catrates these afferent signals no longer exist. This interpretation is in line with that which accounts for the appearance of "castration cells" in the anterior pituitary of gonadectomized mammals.

Additional insights regarding the neural control over the cyclicity of the corpus allatum and the ovary are based on the following considerations. The search for neuropeptide receptors in the brain of *Leucophaea* revealed that the content of binding sites for the synthetic enkephalin analog DALA (D-Ala^2-Met-enkephalinamide) per unit protein was 30% higher in reproducing adult females than in adult males, old females, and nymphs of either sex (6, 56). It is reasonable to propose that the receptors in question bind to the endogenous enkephalin-like neuropeptide(s) demonstrated immunocytochemically in neurons of the brain (34) and in fibers of the retrocerebral neuroendocrine complex (17, 73). Rémy & Dubois (34) suggested that because of their close spatial relationship with the fiber tracts known to control corpus allatum activity, the enkephalinergic neurons in question modulate the dispatch of allatotropic neuropeptide. Such a role would seem to be particularly called for during the period of reproductive cyclicity in females, and may well be reflected by the higher density of neuropeptide receptors presumed to mediate modulatory control during that time.

This conclusion is of particular interest since it matches a report by Hammer (16) concerning the sexual dimorphism of the medial preoptic area of the rat hypothalamus. He found a higher concentration of opiate receptors in this area in adult female rats during estrus and diestrus than in males. The dependency of this sex difference on the hormonal milieu was demonstrated by the fact that the higher concentration was abolished in females receiving testosterone injections.

Adaptive mechanisms that enable the offspring to develop under favorable environmental conditions are known in insects as well as vertebrates. In both instances, neuroendocrine controlling factors have been identified. In *Bombyx mori* diapausing eggs are produced under the influence of a diapause hormone extractable from the subesophageal ganglion (19). The same or closely related neuropeptides, present in the corpora cardiaca, corpora allata, and subesophageal ganglia of other insect species, e.g. *Locusta migratoria,* are also capable of inducing egg diapause in *Bombyx* females deprived of their own subesophageal ganglia (61).

By the same token, the activity of the hypothalamic-adenohypophysial system is responsible for embryonic diapause, i.e. delayed implantation of fertilized eggs, which is well known among mammalian species. The actual signals for obligatory seasonal quiescence recorded in certain kangaroo species, for example, seem to be provided by photoperiod and cyclic changes in pineal function (70).

ADDITIONAL PHENOMENA UNDER NEUROENDOCRINE CONTROL

Several additional regulatory phenomena under neuroendocrine control lend themselves to comparative studies, including those related to metabolism, various behavior patterns, and biological rhythms. No attempt is made here to report in detail on these systems in insects.

Significant analogies between metabolic functions in insects and those in higher organisms are based on the fact that the fundamental mechanisms involved appear to have arisen early in the course of evolution. A family of neuropeptide hormones that control fat, carbohydrate, and water metabolism has been identified. An adipokinetic decapeptide extractable from the intrinsic cells of the corpora cardiaca of several insect species promotes the release of diacyl glycerol from the fat body (see 53). The same or a closely related hyperglycemic (hypertrehalosemic) factor furnished by the corpora cardiaca controls the glycogen content of the fat body and the sugar level of the hemolymph (e.g. 20, 37, 77). An opposite (hypoglycemic) action can be attributed to a neuropeptide originating in protocerebral neurosecretory cells (see 8). Similarities of these regulatory substances with glucagon and insulin are being examined.

Considerable efforts have been made in recent years to search, across a broad taxonomic range, for analogous structures and neuroendocrine processes involved in certain behaviors. Here too, insects have been found to provide useful systems for study (68). A rather complicated physiological event, ecdysis (eclosion), may demonstrate this point. According to Truman & Copenhaver (67), the terminal molt (as well as preceding larval molts) in the moth *Manduca sexta* is triggered by the eclosion hormone, a neuropeptide originating in the brain and released by the corpus cardiacum. This principle acts on neurons, inducing an abrupt shift from the slow pupal behavior to the stereotyped motor pattern characteristic of the terminal molt. These movements are preprogramed in the abdominal ganglia. The motor neurons involved are subsequently turned off, and the abdominal muscles regress. In the adult insect, juvenile hormones produce several behavioral effects, among them aggression, dominance, and female sexual responses.

A further topic of current interest is the capacity of many organisms for endogenous control over the temporal organization of physiological processes. Biological (or circadian) clocks are known to influence a variety of cellular and organismic processes, including those under neuroendocrine control. Much effort is being directed toward the identification of the cellular sites (not presumed to be exclusively neural) and the mode of operation of these pacemakers (oscillators).

The study of insects has yielded significant insights into the transduction of environmental, e.g. photoperiodic, stimuli into hormonal signals via neurosecretory pathways. An example is the regulation of pupal diapause in *Manduca,* which is determined by the photoperiod that the insect has experienced at an earlier (i.e. larval) stage of its development (4, 11). Another example is the control over the activity of the prothoracic gland and the rise in ecdysteroid titer leading to a molt. Here two components of a circadian system, one light-sensitive and located within the brain, the other temperature-sensitive and located outside the brain, interact to transmit the appropriate signal by way of an identified neuron (66). In this context it is of interest that in *Periplaneta americana* identified cells in the circadian pacemaker loci of the optic lobe, which is comparable to the suprachiasmatic nucleus of mammals, react with antibodies to several neuropeptides, e.g. pancreatic polypeptide, gastrin/cholecystokinin, and β-endorphin (60).

CONCLUSIONS

With due consideration for the apparent diversity of neuroendocrine control systems, a strong case can be made for the value of comparative studies on a wide front, with the aim of exploring common denominators. An intertaxonomic approach has principally yielded new insights into the functional capacities of the two systems of integration, the nervous system (see 5) and the endocrine system (see 35). The discovery of the brain's competency for multiple modes of chemical signaling, ranging from the use of locally acting neuroregulators to the use of neurohormones, has removed the sharp borderline formerly separating the two systems. Because of their wide distribution and versatility in neurochemical signaling, the neuropeptides have now moved to center stage. Comparative studies of these regulatory substances have elucidated their long evolutionary history and their interaction with specific receptor molecules. The analysis of insect systems also promises to bear fruit in future programs aimed at the elucidation of integrative phenomena.

ACKNOWLEDGMENTS

The author's research referred to in this chapter has been supported by grants from the National Science Foundation and the United States Public Health Service.

Literature Cited

1. Agricola, H., Eckert, M., Ude, J., Birkenbeil, H., Penzlin, H. 1985. The distribution of a proctolin-like immunoreactive material in the terminal ganglion of the cockroach, *Periplaneta americana* L. *Cell Tissue Res.* 239:203–9
2. Andriès, J. C., Belemtougri, G., Tramu, G. 1984. Immunohistochemical identification of growth hormone–releasing material in the nervous system of an insect, *Aeshna cyanea* (Odonata). *Neuropeptides* 4:519–28
3. Bargmann, W., Lindner, E., Andres, K. H. 1967. Über Synapsen an endokrinen Epithelzellen und die Definition sekretorischer Neurone. Untersuchungen am Zwischenlappen der Katzenhypophyse. *Z. Zellforsch. Mikrosk. Anat.* 77:282–98
4. Bowen, M. F., Saunders, D. S., Bollenbacher, W. E., Gilbert, L. I. 1984. *In vitro* programming of the photoperiodic clock in an insect brain-retrocerebral complex. *Proc. Natl. Acad. Sci. USA* 81:5881–84

5. Bullock, T. H. 1984. Comparative neuroscience holds promise for quiet revolutions. *Science* 225:473–78
6. Burrowes, W. R., Assanah, P., Chapman, A., Iannone, B., Martin, R., et al. 1984. Increase in enkephalin levels and decrease in high-affinity opioid receptor density in invertebrate neural tissues during aging. *Ann. NY Acad. Sci.* 435:245–47
7. Chan-Palay, V., Jonsson, G., Palay, S. L. 1978. Serotonin and substance P coexist in neurons of the rat's central nervous system. *Proc. Natl. Acad. Sci. USA* 75:1582–86
7a. Committee on Models for Biomedical Research. 1985. *Models for Biomedical Research: A New Perspective.* Washington, DC: Natl. Acad. Press
8. Duve, H. 1978. The presence of a hypoglucemic and hypotrehalocemic hormone in the neurosecretory system of the blowfly *Calliphora erythrocephala*. *Gen. Comp. Endocrinol.* 36:102–10
9. Duve, H., Thorpe, A. 1981. Gastrin/cholecystokinin (CCK)-like immunoreactive neurones in the brain of the blowfly, *Calliphora erythrocephala* (Diptera). *Gen. Comp. Endocrinol.* 43:381–91
10. El-Salhy, M., Falkmer, S., Kramer, K. J., Speirs, R. D. 1983. Immunohistochemical investigations of neuropeptides in the brain, corpora cardiaca, and corpora allata of an adult lepidopteran insect, *Manduca sexta*. *Cell Tissue Res.* 232:295–317
11. Gilbert, L. I., Bollenbacher, W. E., Goodman, W., Smith, S. L., Agui, S., et al. 1980. Hormones controlling insect metamorphosis. *Recent Prog. Horm. Res.* 36:401–49
12. Giorgi, F. 1979. *In vitro* induced pinocytotic activity by a juvenile hormone analogue in oocytes of *Drosophila melanogaster*. *Cell Tissue Res.* 203:241–47
13. Girardie, A., Girardie, J., Lavenseau, L., Proux, J., Rémy, C., Vieillemaringe, J. 1985. Insect neurosecretion: Similarities and differences with vertebrate neurosecretion. In *Neurosecretion and the Biology of Neuropeptides*, ed. H. Kobayashi, H. A. Bern, A. Urano, pp. 392–400. Tokyo/Berlin: Jpn. Sci. Soc. Press/Springer-Verlag
14. Granger, N. A., Borg, T. K. 1976. The allatotropic activity of the larval brain of *Galleria mellonella* cultured *in vitro*. *Gen. Comp. Endocrinol.* 29:349–59
15. Greenberg, M. J., Price, D. A., Lehman, H. K. 1985. FMRF amide-like peptides of molluscs and vertebrates:

Distribution and evidence of function. See Ref. 13, pp. 370–76
16. Hammer, R. P. 1984. The sex hormone–dependent ontogeny of opiate receptors in the rat medial preoptic area. *Anat. Rec.* 208:69A
17. Hansen, B. L., Hansen, G. N., Scharrer, B. 1982. Immunoreactive material resembling vertebrate neuropeptides in the corpus cardiacum and corpus allatum of the insect *Leucophaea maderae*. *Cell Tissue Res.* 255:319–29
18. Hansen, B. L., Hansen, G. N., Scharrer, B. 1986. Immunocytochemical demonstration of material resembling vertebrate ACTH and MSH in the corpus cardiacum–corpus allatum complex of the insect *Leucophaea maderae*. In *Handbook of Comparative Opioid and Related Neuropeptide Mechanisms*, ed. G. B. Stefano. Boca Raton: CRC. 1:213–22
19. Hasegawa, K. 1957. The diapause hormone of the silkworm, *Bombyx mori*. *Nature* 176:1300–1
20. Hayes, T. K., Keeley, L. L. 1985. Properties of an *in vitro* bioassay for hypertrehalosemic hormone of *Blaberus discoidalis* cockroaches. *Gen. Comp. Endocrinol.* 57:246–56
21. Hild, W. 1951. Experimentell-morphologische Untersuchungen über das Verhalten der "Neurosekretorischen Bahn" nach Hypophysenstieldurchtrennungen, Eingriffen in den Wasserhaushalt und Belastung der Osmoregulation. *Virchows Arch.* 319:526–46
22. Hökfelt, T., Elde, R., Fuxe, K., Johansson, O., Ljungdahl, A., et al. 1978. Aminergic and peptidergic pathways in the nervous system with special reference to the hypothalamus. In *The Hypothalamus*, ed. S. Reichlin, R. J. Baldessarini, J. B. Martin, pp. 69–135. New York: Raven
23. Hökfelt, T., Johansson, O., Goldstein, M. 1984. Chemical anatomy of the brain. *Science* 225:1326–34
24. Johnson, G. D., Stay, B., Rankin, S. M. 1985. Ultrastructure of corpora allata of known activity during the vitellogenic cycle in the cockroach *Diploptera punctata*. *Cell Tissue Res.* 239:317–27
25. Kopeć, S. 1917. Experiments on metamorphosis of insects. *Bull. Acad. Sci. Cracovie Sci. Math. Nat. Sér. B* 1917:57–60
26. Kopeć, S. 1922. Studies on the necessity of the brain for the inception of insect metamorphosis. *Biol. Bull. Woods Hole Mass.* 42:323–42
27. Krauthamer, V. 1985. Morphology of identified neurosecretory and non-neuro-

secretory cells in the cockroach pars intercerebralis. *J. Exp. Zool.* 234: 221–30

28. Moore, R. Y., Gustafson, E. L., Card, J. P. 1984. Identical immunoreactivity of afferents to the rat suprachiasmatic nucleus with antisera against avian pancreatic polypeptide, molluscan cardioexcitatory peptide and neuropeptide Y. *Cell Tissue Res.* 236:41–46

29. Myers, C. M., Evans, P. D. 1985. An FMRF-amide antiserum differentiates between populations of antigens in the central nervous system of the locust, *Schistocerca gregaria. Cell Tissue Res.* 242:109–14

30. Nishiisutsuji-Uwo, J., Endo, Y., Takeda, M. 1985. Insect brain-midgut endocrine system: with special reference to paraneurons. See Ref. 13, pp. 410–17

31. Orchard, I., Finlayson, L. H. 1977. Electrical properties of identified neurosecretory cells in the stick insect. *Comp. Biochem. Physiol.* 58A:87–91

32. Postlethwait, J. H., Kunert, C. J. 1985. Endocrine and genetic regulation of vitellogenesis in *Drosophila.* In *Comparative Endocrinology: Developments and Directions,* ed. C. L. Ralph, pp. 33–52. New York: Liss

33. Rankin, S. M., Stay, B. 1985. Ovarian inhibition of juvenile hormone synthesis in the viviparous cockroach, *Diploptera punctata. Gen. Comp. Endocrinol.* 59: 230–37

34. Rémy, C., Dubois, M. P. 1981. Immunohistological evidence of methionine enkephalin-like material in the brain of the migratory locust. *Cell Tissue Res.* 218:271–78

35. Roth, J., Le Roith, D., Shiloach, J., Rubinovitz, C. 1983. Intercellular communication: An attempt at a unifying hypothesis. *Clin. Res.* 31:354–63

36. Roth, T. F., Porter, K. R. 1964. Yolk protein uptake in the oocyte of the mosquito *Aedes aegypti* (L.). *J. Cell Biol.* 20:313–32

37. Scarborough, R. M., Jamieson, G. C., Kalish, F., Kramer, S. J., McEnroe, G. A., et al. 1984. Isolation and primary structure of two peptides with cardioacceleratory and hyperglycemic activity from the corpora cardiaca of *Periplaneta americana. Proc. Natl. Acad. Sci. USA* 81:5575–79

38. Schaller, H. C., Bodenmüller, H. 1985. Role of the neuropeptide head activator for nerve function and development. *Biol. Chem. Hoppe-Seyler* 336:1003–7

39. Scharrer, B. 1935. Über das Hanström-sche Organ X bei Opisthobranchiern. *Pubbl. Stn. Zool. Napoli* 15:132–42

40. Scharrer, B. 1937. Über sekretorisch tätige Nervenzellen bei wirbellosen Tieren. *Naturwissenschaften* 25:131–38

41. Scharrer, B. 1952. Neurosecretion. XI. The effects of nerve section on the intercerebralis-cardiacum-allatum system of the insect *Leucophaea maderae. Biol. Bull. Woods Hole Mass.* 102:261–72

42. Scharrer, B. 1959. The role of neurosecretion in neuroendocrine integration. In *Comparative Endocrinology,* ed. A. Gorbman, pp. 134–48. New York: Wiley

43. Scharrer, B. 1964. Histophysiological studies on the corpus allatum of *Leucophaea maderae.* IV. Ultrastructure during normal activity cycle. *Z. Zellforsch. Mikrosk. Anat.* 62:125–48

44. Scharrer, B. 1972. Neuroendocrine communication (neurohormonal, neurohumoral, and intermediate). In *Progress in Brain Research,* ed. J. Ariëns Kappers, J. P. Schadé, 38:7–18. Amsterdam/London/New York: Elsevier

45. Scharrer, B. 1974. New trends in invertebrate neurosecretion. In *Neurosecretion—The Final Neuroendocrine Pathway,* ed. F. Knowles, L. Vollrath, pp. 285–87. Berlin/Heidelberg/New York: Springer-Verlag

46. Scharrer, B. 1978. Histophysiological studies on the corpus allatum of *Leucophaea maderae.* VI. Ultrastructural characteristics in gonadectomized females. *Cell Tissue Res.* 194:533–45

47. Scharrer, B. 1985. Invertebrate neurosecretion: A round table discussion. See Ref. 13, pp. 435–41

48. Scharrer, B. 1985. Neurosecretion: The development of a concept. In *Current Trends in Comparative Endocrinology,* ed. B. Lofts, W. N. Holmes, 1:23–27. *Proc. 9th Int. Symp. Comp. Endocrinol. Hong Kong, 1981.* Hong Kong: Univ. Press

49. Scharrer, B., Wurzelmann, S. 1978. Neurosecretion. XVII. Experimentally induced release of neurosecretory material by exocytosis in the insect *Leucophaea maderae. Cell Tissue Res.* 190:173–80

50. Scharrer, E. 1928. Die Lichtempfindlichkeit blinder Elritzen (Untersuchungen über das Zwischenhirn der Fische). *Z. Vgl. Physiol.* 7:1–38

51. Scharrer, E. 1946. Anatomy and the concept of analogy. *Science* 103:578–79

52. Scharrer, E., Scharrer, B. 1963. *Neuroendocrinology.* New York/London: Columbia Univ. Press. 289 pp.

53. Schooneveld, H., Romberg-Privee, H. M., Veenstra, J. A. 1985. Adipokinetic hormone-immunoreactive peptide in the endocrine and central nervous system of several insect species: A comparative immunocytochemical approach. *Gen. Comp. Endocrinol.* 57:184–94

54. Schooneveld, H., Romberg-Privee, H. M., Veenstra, J. A. 1986. Immunocytochemical differentiation between adipokinetic hormone (AKH)-like peptides in neurons and glandular cells in the corpus cardiacum of *Locusta migratoria* and *Periplaneta americana* with C-terminal and N-terminal specific antisera to AKH. *Cell Tissue Res.* 243:9–14

55. Stefano, G. B. 1986. Conformational matching: A possible evolutionary force in the evolvement of signal systems. See Ref. 18, 2:271–77

56. Stefano, G. B., Scharrer, B. 1981. High affinity binding of an enkephalin analog in the cerebral ganglion of the insect *Leucophaea maderae* (Blattaria). *Brain Res.* 225:107–14

57. Stefano, G. B., Scharrer, B., Assanah, P. 1982. Demonstration, characterization and localization of opioid binding sites in the midgut of the insect *Leucophaea maderae* (Blattaria). *Brain Res.* 253:205–12

58. Sternberger, L. A. 1986. *Immunocytochemistry.* New York: Wiley. 3rd ed. 524 pp.

59. Szibbo, C. M., Tobe, S. S. 1983. Nervous humoral inhibition of C_{16} juvenile hormone synthesis in last instar females of the viviparous cockroach, *Diploptera punctata*. *Gen. Comp. Endocrinol.* 49:437–45

60. Takeda, M., Endo, Y., Saito, H., Nishimura, M., Nishiitsutsuji-Uwo, J. 1985. Neuropeptide and monoamine immunoreactivity of the circadian pacemaker in *Periplaneta*. *Biomed. Res.* 6:395–406

61. Takeda, S., Girardie, A. 1985. An active principle from the corpora cardiaca, corpora allata and suboesophageal ganglion of the locust, inducing egg diapause in *Bombyx mori*. *J. Insect Physiol.* 31:761–66

62. Takeda, S., Vieillemaringe, J., Geffard, M., Rémy, C. 1986. Immunohistological evidence of dopamine cells in the cephalic nervous system of the silkworm *Bombyx mori*. Coexistence of dopamine and α-endorphin-like substance in neurosecretory cells of the suboesophageal ganglion. *Cell Tissue Res.* 243:125–28

63. Tobe, S. S., Saleuddin, A. S. M. 1977. Ultrastructural localization of juvenile hormone biosynthesis by insect corpora allata. *Cell Tissue Res.* 183:25–32

64. Tobe, S. S., Stay, B. 1985. Structure and regulation of the corpus allatum. In *Advances in Insect Physiology*, ed. M. J. Berridge, J. E. Treherne, V. B. Wigglesworth, 18:305–432. London: Academic

65. Triepel, J., Grimmelikhuijzen, C. J. P. 1984. Mapping of neurons in the central nervous system of the guinea pig by use of antisera specific to the molluscan neuropeptide FMRFamide. *Cell Tissue Res.* 237:575–86

66. Truman, J. W. 1984. Physiological aspects of the two oscillators that regulate the timing of eclosion in moths. In *Photoperiodic Regulation of Insect and Molluscan Hormones*, ed. R. Porter, C. M. Collins. pp. 221–39. Newark, NJ: Ciba Pharm. Med. Ed. Div.

67. Truman, J. W., Copenhaver, P. F. 1985. Towards a cellular analysis of the regulation of eclosion hormone release in the hawkmoth *Manduca sexta*. See Ref. 13, pp. 179–85

68. Truman, J. W., Weeks, J. C., Levine, R. B. 1985. Developmental plasticity during the metamorphosis of an insect nervous system. In *Comparative Neurobiology: Modes of Communication in the Nervous System*, ed. M. J. Cohen, F. Strumwasser, pp. 25–44. New York: Wiley

69. Tublitz, N. J., Truman, J. W. 1985. Intracellular stimulation of an identified neuron evokes cardioacceleratory peptide release. *Science* 228:1013–15

70. Tyndale-Biscoe, C. H., 1986. Embryonic diapause in a marsupial: Roles of the corpus luteum and pituitary in its control. See Ref. 32, pp. 137–55

71. Ulrich, G. M., Schlagintweit, B., Eder, J., Rembold, H. 1985. Elimination of the allatotropic activity in locusts by microsurgical and immunological methods: Evidence for humoral control of the corpora allata, hemolymph proteins, and ovary development. *Gen. Comp. Endocrinol.* 59:120–29

72. Veenstra, J. A., Romberg-Privee, H. M., Schooneveld, H. 1985. A proctolin-like peptide and its immunocytochemical localization in the Colorado potato beetle, *Leptinotarsa decemlineata*. *Cell Tissue Res.* 240:535–40

73. Verhaert, P., De Loof, A. 1985. Immunocytochemical localization of a methionine-enkephalin-resembling neuropeptide in the central nervous system of

the American cockroach, *Periplaneta americana* L. *J. Comp. Neurol.* 239:54–61

74. Wall, D. A., Meleka, I. 1985. An unusual lysosome compartment involved in vitellogenin endocytosis by *Xenopus* oocytes. *J. Cell Biol.* 101:1651–64

75. Way, E. L., ed. 1980. *Endogenous and Exogenous Opiate Agonists and Antagonists.* New York: Pergamon

76. Yui, R., Fujita, T., Ito, S. 1980. Insulin-, gastrin-, pancreatic polypeptide-like immunoreactive neurons in the brain of the silkworm, *Bombyx mori*. *Biomed. Res.* 1:42–46

77. Ziegler, R. 1979. Hyperglycaemic factor from the corpora cardiaca of *Manduca sexta* (L.) (Lepidoptera: Sphingidae). *Gen. Comp. Endocrinol.* 39:350–57

Ann. Rev. Entomol. 1987. 32:17–48

CHEMOSYSTEMATICS AND EVOLUTION OF BEETLE CHEMICAL DEFENSES

Konrad Dettner

Institut für Biologie II (Zoologie), RWTH Aachen, Kopernikusstrasse 16, D-5100 Aachen, Federal Republic of Germany

INTRODUCTION

Though mechanically well protected by an exoskeleton, Coleoptera, which comprise the largest group of organisms at the order level, have evolved a variety of defense mechanisms (21, 39, 156). At each developmental stage beetles are protected by various structural modifications, types of behavior, mimetic, cryptic, or aposematic appearances, and, in particular, chemically based defenses (21, 39).

As a rule these defensive substances are multifunctional. As repellents, toxicants, insecticides, or antimicrobics they are directed against a large array of potential target organisms; they may also function as surfactants (10, 38). Although these allomones contribute to the fitness of the producers, it is often experimentally difficult or impossible to determine their adaptive value (9, 38).

Usually Coleoptera biosynthesize and store their defensive compounds either in complex glands or in the hemolymph (10, 156). In the latter case, toxicants are liberated by reflex bleeding; species that employ this defense often exhibit warning colorations or produce warning odors and may be models for mimicry (10, 21). Some beetles profit from exogenous plant- and animal-derived toxicants, which are tolerated and sequestered (10).

Since the last review on comparative aspects of beetle chemical defense in 1978 (156), in which 330 species from 161 genera were considered, a wealth of information has emerged. To date, approximately 900 species from 380

17

genera have been studied chemically. It is now possible to assess both phylogenetic and ecological influences on the chemical defense systems of beetles. The qualitative composition of defensive chemicals that are not derived from exogenous sources seems to be determined primarily by the taxonomic framework and the requirement for certain physicochemical properties (30, 38, 149). Specific life habits and other ecological conditions of the allomone producers are of secondary importance (29, 32, 149). I apply Hennig's principle (68) to trace the phylogenetic influences on beetle chemical defense systems and evolutionary trends.

BEETLE CHEMICAL DEFENSE AND CLADISTICS

The general importance of chemosystematics with regard to defense systems of various insect orders has previously been stressed (129). To interpret glandular morphological and chemical data on a cladistic basis, information on homologization and polarization of character state transformations is first required, since only derived characters (apomorphies) but not primitive characters (plesiomorphies) are used to construct a cladogram (68).

Beetle chemical defenses evolved independently and at different times within several taxa, as illustrated in Figure 1. The defense glands of all adephagan beetles evolved in the Triassic, whereas members of the Staphylinidae acquired their gland systems at different times and much later (20, 21, 35). Chemosystematic deductions are restricted only to beetle taxa with morphologically homologous glands. In a few cases, convergent evolution of a defense system seems probable if the body parts concerned are exposed (28). In one case, convergent evolution of defensive quinones from different precursors has been proved (10). If highly sensitive techniques are applied, defensive secretions that are not derived exogenously can be shown to be qualitatively constant (10). Only a few tenebrionids show intraspecific differences in their gland constituents (149). Quantitative variations usually found in beetle allomones are dependent on sex (10, 74), age (10, 29, 74), and season (10, 29), or have a genetic basis (46).

Polarizations of morphological glandular characteristics are evident if glands of primitive species are compared with highly advanced taxa and the origin of the insect glands is taken into account (28). Coleopteran exocrine glands originate from epidermal glandular cells, which may be localized on either sclerites or intersegmental membranes. Originally these invaginated membranes were small and unpaired, and contained few secretions. The glands tend to be extremely enlarged and lateralized in advanced species (28, 37, 150, 151). Gland cells that were originally scattered on the reservoir

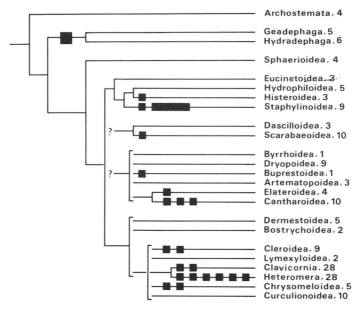

Figure 1 Phylogeny of Coleoptera (20, 21, 94, 95), with numbers of families *(right)* and independent evolution of chemical defense systems at the family level *(squares)*. *Large square* for Adephaga comprises 11 families with homologous glands; *rectangle* symbolizes independent evolution of chemical defense systems within the single family Staphylinidae (Geadephaga: Rhysodidae, Paussidae, Trachypachidae, Cicindelidae, Carabidae; Hydradephaga: Haliplidae, Gyrinidae, Amphizoidae, Hygrobiidae, Noteridae, Dytiscidae; Histeroidea: Histeridae; Staphylinoidea: Silphidae, Staphylinidae; Scarabaeoidea: Scarabaeidae; Buprestoidea: Buprestidae; Elateroidea: Elateridae; Cantharoidea: Lycidae, Lampyridae, Cantharidae; Cleroidea: Trogositidae, Melyridae; Clavicornia: Endomychidae, Coccinellidae; Heteromera: Anthicidae, Meloidae, Oedemeridae, Pyrochroidae, Tenebrionidae, Lagriidae; Chrysomeloidea: Cerambycidae, Chrysomelidae).

surface tend to be aggregated in specific areas and have gradually become isolated from the reservoir surface, with which they are connected by collecting ducts (85, 150). Apart from lateralization, which is also found in gland aggregations scattered on sclerites (25), in a further trend segmental glands have become reduced and are confined to exposed parts of the body (34). Together with such advanced behavior as headstanding, stridulation, spraying, and crepitation, these morphological adaptations optimize lateral delivery of secretions and the capability to react to multiple molestations. Further, the evolution of collecting ducts may prevent the secretions from backing up into the sensitive secretory structures during discharge (54, 147).

Chemically simple *p*-quinones are certainly plesiomorphic characteristics, since *o*-quinones are synthesized by epidermal cells for the tanning of the

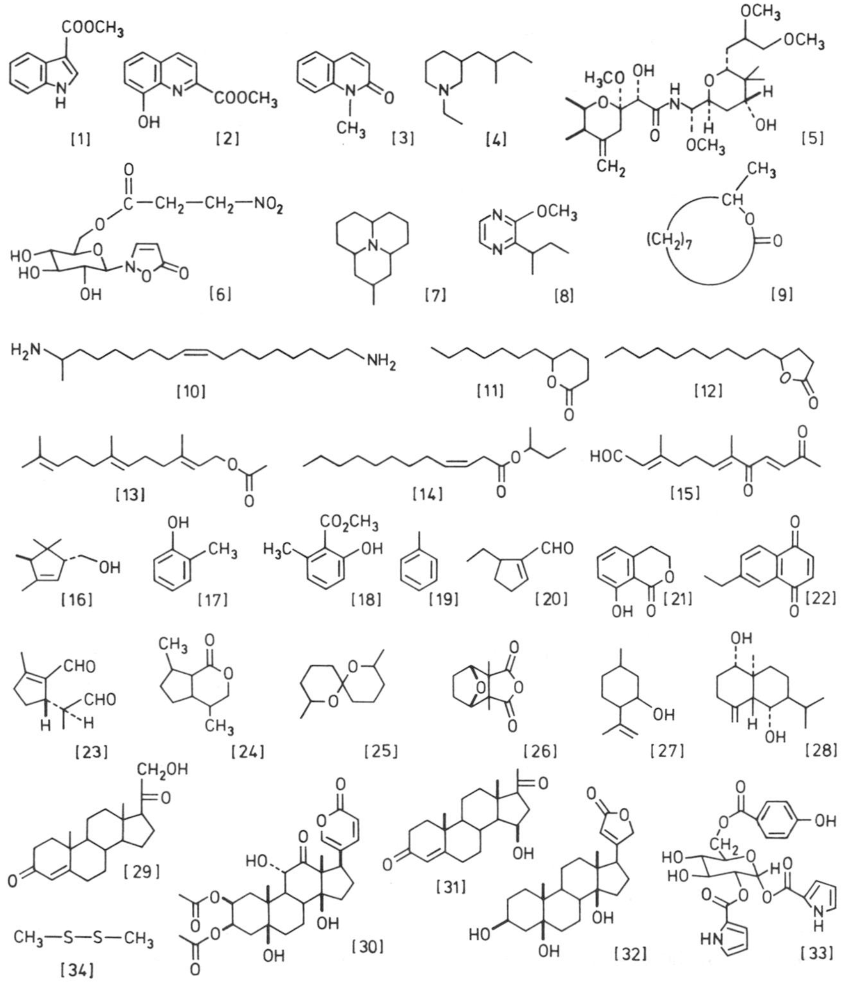

COOCH₃

[1]   [2]   [3]   [4]   [5]

[6]   [7]   [8]   [9]

[10]   [11]   [12]

[13]   [14]   [15]

[16]   [17]   [18]   [19]   [20]   [21]   [22]

[23]   [24]   [25]   [26]   [27]   [28]

[29]   [30]   [31]   [32]   [33]

CH₃—S—S—CH₃
[34]

Figure 2 Structurally complex defensive compounds from Coleoptera, which are of diagnostic use (derived characters) in the defined taxa. (*1*) methylindole-3-carboxylate: Amphizoidae (K. Dettner, unpublished); (*2*) methyl 8-hydroxyquinoline-2-carboxylate: *Ilybius* (Dytiscidae) (136); (*3*) 1-methyl-2-quinolone: *Metriorrhynchus* (Lycidae) (111); (*4*) stenusin: Steninae (Staphylinidae) (135); (*5*) pederin: *Paederus* (Staphylinidae) (10, 123); (*6*) 2-(6'-(3''-nitropropanoyl)-β-D-glucopyranosyl)-3-isoxazolin-5-one: Chrysomelinae (Chrysomelidae) (119); (*7*) myrrhine: *Myrrha* (Coccinellidae) (152); (*8*) 2-methoxy-3-*sec*-butylpyrazine: *Metriorrhynchus* (Lycidae) (111); (*9*) phoracantholide: *Phoracantha* (Cerambycidae) (108); (*10*) (Z)-1,17-diaminooctadec-9-ene (Coccinellidae) (16); (*11*) δ-dodecalactone: Oxytelinae (Staphylinidae) (32, 36); (*12*) γ-tetradecalactone: Oxytelinae (Staphylinidae) (32, 36); (*13*) farnesylacetate: *Oxytelus* (Staphylinidae) (32, 36); (*14*) *sec*-butyl-(Z)-3-dodecenoate: *Deleaster* (Staphylinidae) (35); (*15*) gyrinidal (Gyrinidae) (133); (*16*) α-necrodol: *Necrodes* (Silphidae) (44); (*17*) o-cresol: Cerambycidae,

cuticle. Primitive tenebrionids, for example, use quinones in an aqueous phase as a defensive secretion (39, 150), whereas advanced taxa also synthesize various solvents (149). Figure 2 illustrates chemically complex and thus derived defensive compounds. These constituents require evolution of considerably more enzymes for the numerous biogenetic steps involved, and their convergent evolution seems highly improbable (10, 63). Such derived micromolecules, which are taxonomically important, are confined to certain beetle taxa (see trivial names in Figure 2) or have a very limited distribution outside the coleopteran order (10).

In order to interpret the micromolecular gap between primitive and highly derived compounds it is necessary to analyze the biogenetic principles observed in arthropods (62, 63).

Defensive compounds are either synthesized via a biogenetic chain A,B,C,D, or several compounds B,C,D are each produced from the same precursor, A. In the first case, if the biogenetic pathway is expanded, D should be derived rather than C. Benzoic acid (D) or actinidin (D), for example, should be derived rather than the possible precursors phenylacetic acid [C; Hydradephaga (29)] or iridodial [C; Staphylinidae (26)], respectively. If biogenetic reductions can be excluded, derived species should be characterized by the presence of A,B,C, and D or just C and D (with A and B absent or present only in trace amounts).

In the case of a single precursor, as observed in Oxytelinae (Staphylinidae) (30; see Figure 3), A gives rise to either B,C,D, or E. For example, 3-enoic acids (A) represent common precursors of isopropyl esters (B), *sec*-butyl esters (C), 1-alkenes (D), or γ-lactone (E). Schildknecht and coworkers (138) described evolution as a process of successive invention of new constituents. Sometimes the actual succession may be determined by efficiency tests (assuming anagenesis and known target) and by chemotaxonomic patterns using morphologically based phylogenetic data (30, 35).

Defensive secretions of advanced beetles are generally more complex and

---

Tenebrionidae (106, 98); (*18*) methyl 2-hydroxy-6-methylbenzoate: *Dyschirius* (Carabidae) (110); (*19*) toluene: Cerambycidae (106); (*20*) phoracanthal: *Phoracantha* (Cerambycidae) (107); (*21*) 3,4-dihydro-8-hydroxyisocoumarin: *Apsena* (Tenebrionidae) (98); (*22*) 6-ethyl-1,4-naphthoquinone: *Argoporis* (Tenebrionidae) (149); (*23*) chrysomelidial: Chrysomelinae (Chrysomelidae) (14, 122); (*24*) dihydronepetalactone: *Creophilus* (Staphylinidae) (78); (*25*) (*E,E*)-2,8-dimethyl-1,7-dioxaspiro[5.5]undecane: *Ontholestes* (Staphylinidae) (33); (*26*) cantharidin (Meloidae, Oedemeridae) (10, 93); (*27*) isopulegol: *Xantholinus* (Staphylinidae) (33); (*28*) platambin: *Platambus* (Dytiscidae) (133); (*29*) cortexone (Dytiscidae) (132); (*30*) 12-oxo-2β,3β-di-O-acetyl-5β,11α-dihydrobufalin: Lampyridae (10, 61); (*31*) 15β-hydroxyprogesterone: *Silpha* (Silphidae) (102); (*32*) periplogenin: Chrysomelinae (Chrysomelidae) (119, 122); (*33*) buprestin B: Buprestidae (18, 112); (*34*) dimethyl disulfide: Amphizoidae (K. Dettner, unpublished).

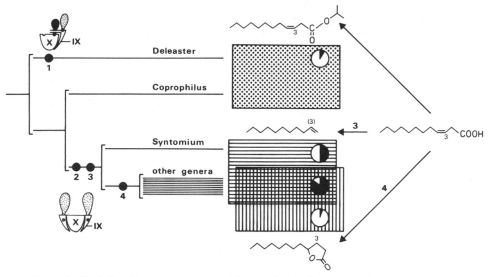

Figure 3 Evolution of solvent types *(center)* in the quinoid defensive secretions of Oxytelinae (Staphylinidae) (30, 35). The cladogram of Oxytelinae *(left;* points indicate derived characters 1–4) is based on chemistry and gland morphology. The biogenetic scheme *(right)* symbolizes the biogenetic steps of isopropyl esters *(above),* 1-alkenes *(center;* character 3), and γ-lactones *(below;* character 4) from a 3-enoic acid precursor. *Solid sectors* (in *circles*) indicate % mortalities of *Lucilia* larvae (n = 20) 3 hr after topical treatment with 1 μl quinone-saturated synthetic defensive secretion (30).

contain different compounds, which often arise from several biogenetic pathways (10, 30, 129). Selective pressure toward secretion diversity could provide defense against a greater number of predators (38, 121). Moreover, predators could not so easily evolve resistance to the chemical complexity of a secretion (38, 121). Unlike primitive Oxytelinae (Staphylinidae), which have physicochemically similar solvents (27, 30, 35), derived species of this subfamily mix different solvents. By this means they vary and optimize the physicochemical parameters of their secretions (volatility, wettability, or lipophility), producing variable repellencies or toxicities (30, 32, 36). When these species are able to synthesize and mix definite volumes of single solvents, they succeed in increasing the defensive potential of their secretions by quasisynergisms (30). At the same time they save toxic quinones by improving the formulation and reduce the volatility of the secretion by adding fixatives (30, 36). Therefore the physicochemical properties of the secretion as a whole may be as important as the specific chemical nature of the individual compounds, which seems taxonomically determined by preexisting biogenetic pathways (149).

COMPARATIVE ANALYSIS OF BEETLE CHEMICAL DEFENSES

Geadephaga

All terrestrial adephagan beetles (Rhysodidae, Paussidae, Trachypachidae, Cicindelidae, Carabidae) possess a pair of pygidial defensive glands whose reservoirs open onto the membranous cuticle behind the eighth abdominal tergite. These glands have a collecting canal of variable length, which connects the secretory tissue with the gland reservoir (53, 54). Primitive species allow the secretion to ooze over the cuticle of their hind segments, and a few taxa apotypically eject their secretions by exerting strong muscular pressure on the reservoirs. As a refinement, both Brachininae (Carabidae) and Paussidae possess a spontaneous and exothermic crepitation mechanism, which utilizes a two-chamber system in conjunction with fingerlike secretory lobes and extremely prolonged collecting canals (54). Quinoles plus hydrogen peroxide, water, and hydrocarbons from the gland reservoir enter a reaction chamber, where the exothermic reactions that produce both reactive quinones and oxygen are catalyzed (132). It is not clear whether the specialized gland morphology and crepitation mechanisms are a consequence of evolutionary convergence in Paussidae (Paussinae, Ozaeninae, Metriinae) and Brachininae or whether they have evolved only once (42). Within bombardier beetles, the paussid *Homopterus arrowi* fortifies its quinoic mixture by adding 2-methyl-3-methoxy-1,4-benzoquinone (42). The diagnostic elytral flanges of Paussinae lie just in front of the two gland openings, and have been shown to serve as launching guides for anteriorly directed discharges, since Paussinae do not aim their spray by revolving their abdominal tip (40).

Pygidial glands of the tiger beetles (Cicindelidae) morphologically resemble hydradephagan glands in both their lengthened secretory lobes and their gland cell structure (53). Five *Megacephala* species synthesize benzaldehyde, which is derived from the unstable cyanogen mandelonitrile and is produced together with hydrogen cyanide (12). The morphology of the defense glands of the primitive Rhysodidae and Trachypachidae (no chemical data available) resembles that of the Carabidae (54).

The results of a chemical study of 325 Carabidae species from 133 genera and the glandular morphological data reveal only a few evolutionary trends, but illustrate a considerable parallelism in evolution (23, 82, 105, 113, 138). Chemically, carabid secretions are composed of highly reactive polar constituents such as acids, phenols, aldehydes, or quinones. These sometimes occur in admixture with nonpolar minor components such as hydrocarbons, ketones, or esters, which possibly facilitate the penetration of epicuticles by the toxic chemicals. Weakly acid, slightly toxic, branched, saturated aliphatic acids (e.g. 2-methylpropanoic acid, 3-methylbutanoic acid) derived from amino acids (e.g. valine, leucine) represent the most primitive carabid allomones; these are found in primitive Loricerini, Elaphrini, or Omophronini (138). Independently, the more toxic, wetting, unsaturated acids (e.g. methacrylic and tiglic acid) evolved within several taxa; other ground beetles from several unrelated taxa may secrete the toxic formic acid in admixture with hydrocarbons or esters (105, 113, 138). Schildknecht et al (138) have suggested that among Carabidae defensive chemicals apparently evolve through successive production of new compounds. Gradual optimization of the mixtures may result in loss of the original compounds, or they may be present only as trace constituents. The presence of unexpected minor components in defensive chemicals of several taxa underscores this concept (81). Chemically, Pterostichini are divided into three groups, which secrete either formic acid + alkanes, methacrylic + tiglic acid, or salicylaldehyde (23). A species of the relict genus *Abacomorphus,* however, synthesizes each of these acids (113). *Platynus dorsalis* synthesizes both formic acid + alkanes and methyl salicylate (138). *Harpalus capito* secretes formic acid (as do other Harpalini) and additionally stores traces of unsaturated acids (81). Carabini species are characterized by either unsaturated acids or salicylaldehyde in their gland reservoirs, but four representatives of *Calosoma,* a *Callistenus* species, and a *Campalita* species produce both types of chemicals (23). Below the tribal level and within several Carabidae, distinct, exclusive chemical patterns are discernible: Notiophilini produce saturated or unsaturated acids, Scaritini produce quinones, unsaturated acids, or iridodial + ketones + a substituted methyl benzoate, Bembidiini produce either salicylaldehyde + valeric acid or various other acids, and Chlaeniini produce quinones or *m*-cresol, which occurs with hydrocarbons (23, 105).

A few ground beetles synthesize unique compounds such as ethylphenol (Panagaeini) or straight-chain formates and acetates, which are confined to Panagaeini and a few Pterostichini, Helluonini, and Dryptini (23). *Dyschirius wilsoni* (Scaritini) is unique in producing methyl 2-hydroxy-6-methylbenzoate (Figure 2–18) together with iridodial and ketones (110). Only *Carabus yaconinus* (Carabini) contains isocrotonic acid, which is admixed with various other acids in its pygidial glands (81). Below the genus level, chemical characteristics of Carabidae become very uniform and every chemical discontinuity may provide the basis for generic revision (23). Only few trends are obvious in the morphology of Geadephaga defense glands. Species that produce unsaturated acids tend to have smaller reservoirs relative to body volume than those that secrete formic acid (54). Moreover, formic acid–producing species can discharge a fine spray in any preferred direction owing to their strong muscular reservoir (54). The shape of the secretory lobes seems to be correlated to some extent with the main chemical components; prolonged and thickened lobes are found in *m*-cresol- or quinone-producing species, but production of acids always seems to be associated with spherical lobes (54, 82). In several tribes various accessory glands have evolved independently (53, 54). Mimicry seems to be confined to Lebiini; models are provided by members of the chrysomelid subfamily Alticinae (115, 146). Despite their chemical weapons, which are sometimes associated with stridulation, carabids have a wide array of predators, ranging from spiders to birds and mammals (146).

An exceptional intraspecific utilization of pygidial gland secretion has been reported for *Pterostichus lucublandus:* Unreceptive females use it to repel males (86). Larvae of the carabid genus *Chlaenius* are unique in possessing eversible gland sacs located on the metathoracic epimeral sclerites (67).

Hydradephaga

Aquatic adephagan beetles (Haliplidae, Gyrinidae, Amphizoidae, Hygrobiidae, Noteridae, Dytiscidae) are characterized by pygidial glands that are homologous to the equivalent structures in Geadephaga (52, 53). The glands obviously evolved only once since the Triassic (20, 21). The secretory lobes of the Hydradephaga, apart from those of the Haliplidae, which are oval, are usually prolonged and sometimes branched (29). Collecting canals may be absent (*Enhydrus,* Gyrinidae) or distinctly prolonged (*Haliplus,* Haliplidae). The muscle layer covering the gland reservoir is thin, unlike the thick layer in Geadephaga. Contrary to all remaining Adephaga, Dytiscidae are characterized by two distinctly different types of secretory cells (52); accessory glands have been found only in Hydroporinae (Dytiscidae).

As typical antimicrobics and fungicides, hydradephagan pygidial gland constituents primarily prevent the attachment of microorganisms on the water beetle chitin surfaces. The beetles leave the water to coat their surfaces with pygidial gland secretion, using their legs as brushes (29, 132). Especially Gyrinidae but also Amphizoidae and several Dytiscidae may deplete their pygidial gland reservoirs when molested by arthropods or vertebrates. Stinking or sweetish odors are also often perceivable, and in a few species an intensely colored drop appears at the abdominal tip when the beetle is disturbed (29). The good wetting properties of hydradephagan gland constituents and the cleaning behavior of temporarily dried water beetles indicate that these chemicals have another function, namely to increase the wettability of the beetle's hydrophobic surface after a stay on land so it can dive through the water surface (29, 31).

Gyrinidae, Hygrobiidae, and Amphizoidae exhibit characteristic gland chemistry and are differentiated from the remaining Adephaga. When molested, whirligig beetles (Gyrinidae) immediately emit from their pygidial glands a stinking milky droplet that may contain 3-methylbutanal-1 and the corresponding alcohol (140). Chemically this taxon is isolated by the co-occurring toxic sesquiterpenes gyrinidone, gyrinidione, gyrinidal (Figure 2–15), and isogyrinidal, which are probably derived from farnesal and have been found in eight species of the genera *Dineutus* and *Gyrinus* (29, 133).

Hygrobia tarda (Hygrobiidae) produces a series of unusual 2-hydroxy acids such as 2-hydroxyhexanoic and S-methyl-2-hydroxy mercaptobutanoic acids, which may form diverse lactides (29). Traces of benzoic acid and *p*-hydroxybenzaldehyde (K. Dettner, unpublished) indicate a relationship to Dytiscidae.

On molestation, representatives of the third relict family, Amphizoidae, ooze a malodorous yellow drop from the pygidial glands (29) which apotypically contains dimethyl disulfide (Figure 2–34) and the probably antimicrobic methylindole-3-carboxylate (Figure 2–1; K. Dettner, unpublished). The co-occurrence of methyl *p*-hydroxybenzoate and the yellow pigment marginalin (4',5-dihydroxybenzalisocumaranone) (137) in Amphizoidae demonstrates a possible relationship to Dytiscidae (K. Dettner, unpublished).

The remaining Hydradephaga, comprising 88 chemically tested species from 33 genera, produce aromatic compounds and are subdivided into a phenylacetic acid group (Haliplidae, Noteridae, Hydroporinae), a benzoic acid–hydroxybenzaldehyde group (Dytiscinae, Colymbetinae), and the Laccophilinae, which synthesize various fungicidal β-hydroxy acids and methyl 3,4-dihydroxyphenylacetate (29, 141). The first two groups contain traces of the constituents typical of the neighboring groups. The primitive genus *Copelatus* (Colymbetinae) is characterized by main compounds from both aromatic groups; this indicates that compounds from one aromatic group represent primitive characters that can always accompany the derived constituents. The presence of phenylacetic acid within unrelated groups also indicates that this aromatic acid was originally present in the pygidial glands of Dytiscidae and neighboring groups, and that it represents a plesiomorphic character. This is more marked in the case of expansion of the biogenetic chain of phenylalanine, phenylpyruvic acid, phenylacetic acid, and benzoic acid. It is interesting that phenylacetic acid was found in the most primitive Hydroporinae genera, *Laccornis* and *Hydrovatus* (159; K. Dettner, unpublished). The presence of this acid in small species does not seem to be correlated with aquatic life habits, since the first terrestrial "water" beetle, *Geodessus,* from Himalayan litter, produced phenylacetic acid, as do its aquatic relatives (29). Erratically distributed compounds within certain water beetle taxa are indole-3-acetic acid (Noteridae, Hydroporinae) (29, 31), phenylpropionic acid *(Hydaticus)* (31), methyl 2,5-dihydroxyphenylacetate *(Dytiscus)* (137), ethyl 3,4-dihydroxybenzoate *(Cybister)* (132), tiglic acid *(Ilybius)* (31), marginalin *(Dytiscus, Agabus)* (29, 137), and octenoic acid *(Laccophilus, Platambus)* (29, 141).

Prothoracic defensive glands are only present in Dytiscidae and consist of two pocketlike reservoirs covered with gland cells (52). These reservoirs open behind the anterior prothoracic margin. However, the glands of Hygrobiidae open more posteriolaterally, and a muscle covers the reservoir; thus homologization is questionable (53). The prothoracic gland reservoirs contain both toxic anaesthetic and odorous substances, which the water beetles deplete when molested.

Prothoracic defensive agents primarily act on vertebrate predators such as fish, amphibians, and sometimes small mammals (29). Amphibians have been observed to regurgitate live or dead water beetles (132). A minnow bioassay has been developed to measure concentrations, anaesthetic activities, and toxicities of prothoracic gland constituents (103). Schildknecht identified a large series of defensive compounds from the prothoracic glands of Dytiscinae and Colymbetinae, and chemical data are now available for 21 species of 8 genera (10, 132). The presence of huge amounts of steroids, some identical to well-known vertebrate hormones, is impressive. Pregnenes (e.g. cortexone, Figure 2–29) and pregnadienes (e.g. cybisterone, cybisterol) dominate in the secretions of most species. Representatives of *Ilybius* also secrete considerable amounts of vertebrate hormones such as testosterone, estradiol, or estrone (10, 103). Water beetles cannot biosynthesize the steroid skeleton de novo; thus precursors such as cholesterol must be obtained from exogenous sources.

Prothoracic defensive glands of *Colymbetes fuscus* contain the nucleoproteid colymbetin, which lowers blood pressure in mammals (139). Together with various steroids, *Ilybius fenestratus* oozes the unique and powerful antiseptic methyl 8-hydroxyquinoline-2-carboxylate (Figure 2–2; 136), which originates from tryptophane and produces clonic spasms in mice. In *Platambus* the unique sesquiterpene platambin (Figure 2–28; 133) is the main constituent of the secretion.

Odorous substances from the prothoracic defensive glands of *Ilybius fenestratus* have been identified (134) as sesquiterpenes (e.g. δ-cadinene, various isomeric muurolenes, and cubebenes). Blum (10) has suggested that these odorants act either as intraspecific alarm pheromones or as early warning signals for predators.

Histeroidea

Histeridae may exude minute defensive droplets from the pores of the lateral edges of the body (Abraeinae, Dendrophilinae, Histerinae) (96). The yellowish secretion of an *Atholus* species contains octadecyl- and eicosylacetate (K. Dettner & S. Tietz, manuscript in preparation).

Staphylinoidea

On molestation, Silphidae (carrion beetles) show thanatosis and ooze or spray a malodorous fluid from the anal region (44, 102, 142). In both Silphinae and

Necrophorinae the secretion originates from a secretory diverticulum that opens into a blind sac near the rectum (44, 102). The defensive secretion of *Silpha americana* contains considerable amounts of interesting 20-ketopregnanes such as 15β-hydroxyprogesterone (102; Figure 2–31). *Necrodes surinamensis* synthesizes caprylic, capric, and two decenoic acids together with the terpene alcohol lavandulol and two unique necrodols (44; Figure 2–16). The alkaline anal effluent of *Oeceoptoma*, *Silpha*, and *Phosphuga* species is a 4.5% aqueous solution of ammonia (142), which is probably produced in the mammillated glands of the hindgut (73). Remarkably, anal effluents of carrion beetles not only defend against arthropods and vertebrates, but may also be used aggressively; *Ablattaria laevigata* attacks snails by discharging its blind sac over the body of the snail (72).

The evolutionary success of Staphylinidae (rove beetles) may in part be ascribed to their lengthened body, reduced elytra, and freely movable abdomen, which allow them to colonize interstices (26, 35). Owing to the unprotected abdomen in several taxa of Staphylinidae, a wide variety of abdominal defensive glands evolved independently (4). With their numerous exocrine glands, rove beetles are unique among the Coleoptera. The phylogenetic characterization of low Staphylinidae taxa on the basis of gland systems reveals that these chemical defense systems must have originated a short time ago. Systems that evolved independently are found in Paederinae, Omaliinae + Proteininae, Steninae, Aleocharinae, Staphylininae + Xantholininae, and Oxytelinae + Pseudopsinae (4, 35, 116).

In Paederinae, representatives of the genus *Paederus* usually show a black-orange warning coloration. Three vesicant compounds, pederin (Figure 2–5), pederon, and pseudopederin, have been isolated from their hemolymph. Chemically, these are the most complicated beetle toxins known. They are derived via polyketid biosynthesis, and exhibit antitumor activity (10, 123). *Paederus* shows no reflex bleeding, and must be crushed before the toxic hemolymph amides (0.025% of beetle wet weight) can cause skin and eye lesions (pederosis), as is well known in the tropics (2).

Both Omaliinae and Proteininae possess an unpaired abdominal defensive gland, composed of a sac-like reservoir and two types of gland cells, which is situated above the proximal margin of sternite 8 (4, 89, 90). Comparative gross morphology reveals a distinct trend toward a lateralization of this unpaired gland reservoir, which may provide advantageous reservoir enlargement and specific defense against lateral attacks (89). The secretion of four *Eusphalerum* species contains (Z)-2-hexenal and 3-methylbutanoic acid, and repels ants, spiders, and bugs that are found together in blossoms (89, 90).

Representatives of Steninae possess a pair of eversible abdominal gland reservoirs that are covered by secretory cells and open beside the rectum (79). Beetles of the genus *Stenus* usually occur in wet habitats, especially on the banks of pools and rivers. Occasionally they walk on the water surface, where they sometimes employ a remarkable escape mechanism: They evert their glands and discharge their gland secretion, which spreads on the water surface and thereby propels the beetle forward (135). The exudate of *S. comma* is primarily composed of the monoterpenes isopiperitenole, 1,8-cineole, and 6-methyl-5-hepten-2-one; the main spreading agent is represented by the weakly toxic stenusine [*N*-ethyl-3-(2-methylbutyl)piperidin] (Figure 2–4; 135). The secretion may also act against microorganisms and may repel arthropods.

Representatives of Aleocharinae are characterized by an unpaired tergal gland situated between tergites 6 and 7 (4). Two types of gland cells secrete a defensive mixture, which is stored in an apically divided gland reservoir (4). This tergal gland is reduced in both termitophile and myrmecophile Aleocharinae (121). Six of eight species of Schistogenini, Aleocharini, and Zyrasini studied contained toxic quinones (benzoquinone, toluquinone and ethyl quinone, 2-methoxy-3-methyl-1,4-benzoquinone) and corresponding hydroquinones (11, 17, 57, 125). Quinone solvents are represented either by hydrocarbons and methyl- and ethylesters of octanoic, decanoic, and dodecanoic acids (Schistogenini) (57); by hydrocarbons (Zyrasini) (11); or by hydrocarbons and aldehydes (Zyrasini, Aleocharini) (17, 125). Myrmecophile representatives of *Zyras* are unique in producing either isovaleric acid (*Z. humeralis*) (92) or citronellal (*Z. japonicus, Z. comes*) (88); the latter compound is also an alarm pheromone of host ants (88). When disturbed, Aleocharines emit droplets of the repellent tergal gland secretion from their elevated abdomen. Intraspecifically, high concentrations of the secretion of *Aleochara curtula* inhibit male copulatory response, although in small amounts the secretion acts as a mating stimulant and acts synergistically with aphrodisiac female sex pheromones from the epicuticular lipids (124). Larvae of some genera of Aleocharinae (*Oligota, Alianta, Lomechusa, Bryothinusa, Halomaeusa, Bolitochara, Gyrophaena, Phytosus, Thectura, Diaulota, Liparocephalus, Phloeopora, Microglotta*) possess an osmeterium in the eighth abdominal segment (114). This structure probably represents a derived character; the clear separation of genera that do and do not possess the gland seems to be a first step toward a natural classification of Aleocharinae (114) and away from the artificial system used hitherto. The process on the eighth abdominal tergum generates no silk (5), but gives off a defensive secretion that contains toluquinone, 2-methoxy-3-methyl-1,4-benzoquinone dissolved in ethyl hexadecanoate (*Bolitochara*) (K. Dettner, unpublished).

Rove beetles of the subtribes Philonthina and Staphylinina are characterized by paired defensive gland reservoirs between tergites 8 and 9, which are partially everted (4, 26, 33). Gland cells lie in a band directly on the reservoir surfaces (4, 26). In Staphylinina, tubules of gland cells project into the gland reservoir and may increase evaporation in the partly everted reservoirs (78). Gland reservoirs in Philonthina are more differentiated owing to separated gland cell tissues and an unusual evaporation tissue that prevents complete eversion of the reservoir and excessive loss of secretion (26).

Staphylinina (4 species) produce iridodial as the main secretion component, but in the Philonthina secretions (18 species) the alkaloid actinidine dominates (26). Each taxon produces trace amounts of the constituents typical of its sister group. Unique Staphylinina defensive compounds were found in *Ontholestes* [2-heptanone, 6-methyl-2-heptanone, and (*E,E*)-2,8-dimethyl-1,7-dioxaspiro[5.5]undecane (Figure 2–25)] (33), *Creophilus* [dihydronepetalactone (Figure 2–24), isoamylacetate, and (*E*)-8-oxocitronellylacetate] (78), and *Staphylinus* (4-methylhexan-3-one) (51). While Staphylinina use terpenoids as solvents for iridodial (33, 51, 78), Philonthina generally manufacture a series of acetates and hydrocarbons as solvents for actinidine (7, 26). Philonthina are more advanced, as shown by the separated glands and the presence of actinidine, which is synthesized from the precursor iridodial (26).

Xantholininae are characterized by an unpaired, noneversible gland reservoir that opens near the anus and an elongated duct that connects the gland tissue with the reservoir (55). Iridoid compounds such as iridodial *(Thyreocephalus)* (7, 55) and actinidine *(Xantholinus)* (33) are typical constituents of the Xantholininae secretion (4 species) and are dissolved in terpenoid aldehydes and ketones (7, 26, 33). Unique constituents have been identified in *Thyreocephalus* (isovaleraldehyde) (7) and *Xantholinus* [limonen and isopulegol (Figure 2–27)] (33). In spite of distinct morphological differences between defensive glands of Staphylininae and Xantholininae, the common presence of iridoids in both structures indicates a common origin (26). Quediini (actually integrated into Staphylininae) hold a unique position because they either lack abdominal defensive glands, as in *Quedius* (26), or possess noneversible abdominal defensive glands, as in *Algon grandicollis* (83). This beetle produces hexanoic acid, hexanal, and (*E*)-2-hexenal, and is unique among rove beetles because it is capable of spraying its defensive secretions (83).

The 1700 representatives of Oxytelinae known worldwide (69) possess paired, noneversible defensive gland reservoirs, which are associated with the subdivided ninth abdominal tergites and are always connected with glandular tissues by an efferent duct (3, 4, 66). By comparing gland morphology and chemistry of both phylogenetically advanced and extremely primitive Oxytelinae species, it has been possible to trace clear evolutionary trends (27, 30, 32, 35, 36). On molestation, all Oxytelinae bend their abdominal tip dorsally to emit a reddish toluquinone-containing droplet. In primitive genera *(Deleaster, Coprophilus)*, defensive gland reservoirs open in the soft region at the anterior border of the ninth tergites, which are still connected by a bridge (Figure 3; 27, 35, 36). As species progress toward more derivation, the gland openings tend to be moved posteriorly from the soft intersegmental membrane to the center of the completely divided, sclerotized ninth tergites (Figure 3, character 2) (36). The basis for the shift from sensitive to thickened cuticle regions is possibly the gradual increase in repellency and toxicity of the secretions and an increased need for self-protection. Additional evolutionary trends observed are (a) enlargement of the gland reservoir volume relative to body volume, (b) extreme lengthening of efferent ducts, and (c) increasing ramification of cuticular ductules within secretory tissues (35, 36). Enlargement of gland reservoirs seems to be advantageous in the case of repeated molestations, since regeneration of secretory constituents is often time-consuming (10).

All Oxytelinae secrete the toxic p-toluquinone as the active principle of their defensive secretions, which are always quinone-saturated (30, 36). However, during evolution the solvents for this quinone changed drastically for improved secretion repellency (30, 35). Primitive species *(Deleaster, Coprophilus)* use homologous series of isopropyl esters and *sec*-butyl esters (Figure 3, Figure 2–14; 27, 35). The main esters, which only exhibit a minor defensive activity (Figure 3), have a double bond at the 3-position and are synthesized from a 3-enoic acid (35). Members of the primitive genus *Syntomium* replace esters by 1-alkenes in order to dissolve quinone; this increases the efficiency of the secretion (Figure 3, character 3). The same 3-enoic acid is decarboxylated to produce the 1-alkene; otherwise another position of the double bond would be realized (36). Nineteen highly advanced Oxytelinae species of the genera *Trogophloeus, Ancyrophorus, Bledius, Aploderus, Platystethus,* and *Oxytelus* utilize mixtures of diverse solvents composed of a γ-lactone (Figure 2–12; Figure 3, character 4; 32, 36, 157) and a corresponding 1-alkene shortened by one carbon atom (Figure 3, character 3); this formulation exhibits maximal defensive activity. Both the ineffectual lactone and the effective 1-alkene are probably formed from a 3-enoic acid, the former by decarboxylation and the latter by addition across the double bond (30, 35). Moreover, phylogenetically derived Oxytelinae manufacture lactone and alkene solvents (γ-dodecalactone and 1-undecene) in a volume ratio of 1:5, which provides maximum repellency when the secretion is saturated with quinone (30). The correlation of this ratio with the maximal penetration rate of the toxic quinone through the integument of arthropods represents a case of quasisynergism (30).

Although the derived species studied were taken from different habitats (seashore, shores of rivers and brooks, dung, rotting plants) and prefer different foods (ranging from algae to dung and arthropods) (36, 64), they nonetheless have uniform chemical secretions with corresponding pairs of γ-lactones and 1-alkenes (Figure 3). In addition, several advanced species contain δ-dodecalactone (Figure 2–11; 32) or citral (32, 36). Only *Oxytelus piceus* contains decyl-, undecyl-, and dodecylacetates together with farnesyl-acetate (Figure 2–13; 32), whereas *Oxytelus (Anotylus) sculpturatus* produces hexyldecanoate (36).

Deleaster is distinct from other Oxytelinae (Figure 3, character 1) because its unique paired glands are associated with the tenth tergite and are proximal to its second gland system, which is typical of Oxytelinae. On molestation, *Deleaster* partially depletes all four gland reservoirs simultaneously (35) in order to mix its quinoic defensive secretion (gland system associated with the ninth tergite) with four iridodial isomers (gland system associated with the tenth tergite). These dialdehydes polymerize on exposure to air, forming an adhesive that deters small predatory arthropods. This sticky gland is unique within Coleoptera (Figure 3, character 1; 35, 36).

Both chemotaxonomical data and efficiency tests reveal that isopropyl esters and *sec*-butyl esters represent the primitive solvents in the evolution of Oxytelinae. 1-Alkenes (Figure 3, character 3) must have evolved before lactones (Figure 3, character 4) because quinone-saturated lactones are virtually ineffective if administered alone and the primitive genus *Syntomium* exclusively produces alkene solvents (30, 36). In accordance with Herman's classification (69), *Syntomium* therefore represents a sister group of the derived genera (Figure 3), and certainly shows no affinities with *Deleaster* (117). Within phylogenetically advanced species, the main advantage is the precise mixing of solvents that are physicochemically different (e.g. lactone, alkene), with the possibility of optimizing secretion parameters such as wettability, volatility, or repellency (30). If a series of homologous, physicochemically similar solvents (such as isopropyl esters of primitive species) are mixed and saturated with quinone, no variability in the above secretion parameters is observed (30, 36).

Representatives of the genus *Pseudopsis* possess variously developed gland reservoirs at the anterior margin of the still-undivided ninth tergite (70), and thus exhibit the predicted primitive condition. Therefore the glands of *Pseudopsis* and Oxytelinae are not convergent (71, 116), and *Pseudopsis* may be a sister group of Oxytelinae, as was suggested earlier (70). *Nanobius, Zalobius,* and *Asemobius* (Pseudopsinae) primarily or secondarily lack appropriate glands. However, it seems probable that the inflections in the anterior margins of the ninth tergite in the three genera (116) indicate the presence of small primitive defensive glands.

Scarabaeoidea

Within Scarabaeidae, chemical defense systems have evolved separately in Cetoniinae, Trichiinae, and Scarabaeinae. *Cremastocheilus* (Cetoniinae) repels ants with a defensive fluid that originates from propygidial spiracles (87, 91). The source and significance of *Osmoderma* (Trichiinae) secretions remain unknown. In 15 Scarabaeinae genera, both sexes produce highly odorous substances in the sometimes voluminous pygidial glands (128). The most elaborate gland systems are found in taxa exhibiting especially complicated nidification and sexual behavior (128). The distribution and development of scarab beetle pygidial glands have recently been used to redefine Scarabaeinae taxa (128).

Buprestoidea

Jewel beetles (Buprestidae), which are often brightly colored, metallic, and able to mimic aposematic lycids, contain bitter tasting buprestins (1% wet weight; 18, 112) in their hemolymph. Buprestins A and B (Figure 2–33) [β-D-glucose-1,2,6-triesters with pyrrol-2-carboxylic acid as the main esterifying moiety (18)] are derived family characters. Although buprestids show no reflex bleeding, they are only occasionally attacked by arthropods and birds. Buprestins have been found in the hemolymph of 26 Australian species (9 tribes), irrespective of their bright coloration or their ability to mimic lycids (112).

Elateroidea

On molestation, both sexes of *Lacon (Adelocera) murinus* (Elateridae) do not click, but show thanatosis and exhibit two gland reservoirs between the last abdominal tergites. Nothing is known of the chemistry of this secretion, which is probably defensive (K. Dettner, personal observation).

Cantharoidea

An astonishing variety of chemical defensive principles is found in Lycidae, Lampyridae, and Cantharidae. Both Lycidae and Lampyridae (fireflies) are distasteful, often aposematically colored, and capable of reflex bleeding when roughly handled (13, 43, 111). Members of both families are rejected by both arthropods and vertebrates, and Lycidae in particular often serve as models for mimicry complexes (21, 43, 97). Even phytophagous mimetic Cerambycidae may feed on lycid body fluids and probably profit from toxic compounds (43). The characteristic warning odor of the lycid *Metriorrhynchus rhipidus* is due to unique methoxyalkylpyrazines (Figure 2–8; 111). Chemically interesting bitter principles from the hemolymph are 3-phenylpropanamide and 1-methyl-2-quinolone (Figure 2–3), which are probably derived from phenylalanine. The co-occurring 1.8-cineole and especially the acetylenic lipids are presumed to function as antifeedants (111).

Nine novel deterrent bufadienolides (lucibufagines) (Figure 2–30; 10, 45, 61), together with fluorescent pterines, were found in three *Photinus* species

(Lampyridae). *Photuris* females, which are incapable of synthesizing lucibu-fagines, lure males of *Photinus* with imitated flash signals in order to acquire these toxic steroid pyrones from the diet (40). Lampyridae larvae may release resin- and peppermint-scented defensive secretions from segmental eversible gland sacs *(Luciola)* (118) or may show thanatosis *(Lampyris, Phausis)*.

Larvae, pupae, and adults of Cantharidae possess paired segmentally arranged defensive gland reservoirs, which were originally present on the thorax and abdomen and which may be secondarily reduced (144). Adults of Cantharinae may produce minute defensive droplets from two glands on the apex of the abdomen (144; K. Dettner, personal observation). Aposematically colored Chauliognathini emit droplets of antifeedant (Z)-8-dihydromatricaria acid from segmental glands (41, 101). This acetylenic compound is probably of dietary origin, since *Chauliognathus* feeds on Compositae and also sequesters this acid in its hemolymph (41). Unexpectedly, *Chauliognathus pulchellus* synthesizes typical ladybird alkaloids in its defensive glands (109). Although unoxygenated bases dominate in cantharid glands (in Coccinellidae hemolymph water-soluble N-oxides dominate) the question of whether these unique natural products are synthesized or acquired from the diet should be investigated (109).

Cleroidea

Trogositid larvae possess paired tubular defensive glands on the nine abdominal tergites (94). Adults of Melyridae, which are often brightly colored, possess colored membranes at the sides of the prothorax and metathorax; these are everted on molestation, but they do not represent defensive gland reservoirs (47). Originally these membranes were arranged segmentally (Carphurinae) (47). Cantharinae larvae possess dorsal and lateral paired defensive glands in the thorax and abdomen; polysaccharides and SH groups were detected histochemically in their secretions (50).

Clavicornia

Only two families of Clavicornia (Endomychidae, Coccinellidae) have been known to use chemical defense systems.

Some Endomychidae show reflex bleeding (K. Dettner, personal observation), and three species of this family (Eumorphinae) are attracted to and feed on cantharidin (Figure 2–26) as do certain heteromeran beetles (161).

Coccinellidae (ladybird beetles) may show thanatosis and often have bright yellow, red, and orange coloring with black markings. Aposematic Coccinellidae show reflex bleeding from the tibio-femoral joints of the legs (65). In larvae, reflex bleeding may occur from bleeding pores (84), specially modified areas of integument, or broken spines of the body surface. Owing to the bitter taste of their coagulating and alkaloid-enriched hemolymph, ladybirds are usually well protected against ants and birds (65). Their peculiar smell seems to be a warning odor and might be due to alkylpyrazines (111), which are also present in Lycidae and Cerambycidae.

The chemically unique alkaloids from coccinellid eggs, larvae, and adults have been named coccinellines and have evolved as repellents or deterrents against a variety of invertebrate and vertebrate predators (80, 120). One species of Epilachninae and 27 species of Coccinellinae (mainly Hippodamiini and Coccinellini) produce these interesting alkaloids, which may occur as free bases and corresponding N-oxides (80, 120). The bicyclic ketoamine adaline is restricted to the genus *Adalia;* hippocasine and hippocasine oxide are only known in a *Hippodamia* species (6); myrrhine (Figure 2–7) occurs in a *Myrrha* species (152); propyleine is exclusively found in the hemolymph of *Propylaea* (80, 120), but n-octylamine could be isolated from *Hippodamia* (16). The unique (Z)-1,17-diaminooctadec-9-ene (Figure 2–10; 16) and the co-occurring pairs precoccinelline + coccinelline and hippodamine + convergine have a more widespread distribution among Coccinellidae taxa (120).

Biosynthesis of coccinelline alkaloids occurs via the polyketide pathway (10, 152). The origin and presence of these unique natural compounds in Cantharidae defensive glands remain a mystery (109). Pasteels et al (120) studied, in addition to aphidiophagous Coccinellidae, a few plant- and fungus-eating species to correlate the presence of nitrogen-containing hemolymph constituents with nutrition biology. Remarkably, brown species of the genera *Pullus* (Scymnini) and *Rhizobius* (Coccidulini) are devoid of hemolymph alkaloids; this is also especially evident in *Aphidecta obliterata,* the only species of Hippodamiini that lacks nitrogen-containing hemolymph constituents (120). Toxins from exogenous sources are reported in *Coccinella undecimpunctata.* This species apparently sequesters cardenolides, which are derived from its oleander-feeding prey *Aphis nerii* (131).

Heteromera

Hemolymph toxins and defense glands have an important role in Anthicidae, Meloidae, Oedemeridae, Pyrochroidae, and Tenebrionidae. Both Oedemeridae (false blister beetles) and Meloidae (blister beetles) are thought to be phylogenetically related on the grounds of the common presence of the hemolymph toxin cantharidin (Figure 3–26). This chemically unique monoterpene anhydride, which is also a drug, is a powerful vesicant, insecticide, and feeding deterrent (2, 19, 93). In Meloidae only males (via accessory glands) and larvae are capable of biosynthesizing cantharidin, which is transferred to females by copulation (126, 161). The often-aposematic blister beetles liberate cantharidin-containing hemolymph by reflex bleeding (2, 19). Both blister beetle subfamilies, Meloinae and Nemognathinae, have been found to contain cantharidin, with the possible exception of the tribe Horiini (Nemognathini), which may have lost the unique toxin secondarily (1).

Pyrochroidae (Pyrochroinae, Pedilinae), Anthicidae, and representatives of other insect orders positively orient themselves toward cantharidin bait or the beetles that produce it (49, 160, 161). They damage these Coleoptera by chewing their elytra. It is supposed that they primarily profit from the toxic cantharidin thus acquired by diet. On the other hand, cantharidin producers could provide aggregation sites for mating. The reason for the dominance of attracted male Pyrochroidae and Anthicidae remains unclear. Certain male anthicids are characterized by glandular elytral cavities and the production of cantharidinlike substances in these has been postulated (1, 49). The occurrence and biological significance of cantharidin indicate that Pyrochroidae, Anthicidae, Oedemeridae, and Meloidae are closely related and that this unique compound was present in an ancestral heteromeran stock. In Oedemeridae and Meloidae a defensive function has been maintained; in Anthicidae and Pyrochroidae this compound could partially serve as a sex pheromone (1). Similar phenomena are known from the utilization of pyrrolizidine alkaloids by certain Lepidoptera (143). Important exocrine glands are present in Pyrochroidae and Meloidae. Both sexes of red *Pyrochroa* and *Schizotus* species possess paired abdominal defensive glands between the visible sternites 5 and 6, which are absent in both the related Anthicidae and Pythidae (28). Molested specimens emit an unpleasant volatile secretion of unknown chemistry. Pyrochroidae defensive glands are morphologically homologous with corresponding gland systems in tenebrionids; it is possible, however, that both glands evolved by convergence (28).

Except in the genera *Spastonyx* and *Hornia,* mesothoracic odoriferous glands are found in both sexes of most Meloidae (Meloini, Eupomphini, Pyrotini, Lyttini, Epicautini, Tetraomycini, Nemognathini, Zonitini) (8).

The often dull black or brown Tenebrionidae (darkling beetles) are usually nocturnal as adults and feed on a variety of materials, mostly of plant origin. Most species of Tenebrionidae (including Lagriidae, Alleculidae, and Nilionidae) have abdominal defensive glands (37, 94, 155), but in some taxa prothoracic defensive glands are also present. The absence of abdominal glands as in Tentyriinae, Zolodinini, and Belopini probably represents the primary and plesiomorphic state; these taxa are therefore clearly separated from other Tenebrionidae (37, 85, 151). Independently, the intersegmental membranes between sternites 7 and 8 or between sternites 8 and 9 have become glandular; sometimes these glands have been lost secondarily (37, 151).

In Adeliini a pair of long tapering reservoirs supplied with dorsally scattered defensive gland cells has evolved between sternites 8 and 9 (151). When disturbed, *Adelium* stridulates and everts its gland reservoir, emitting a quinoid defensive secretion (40). The remaining Tenebrionidae, apart from primary glandless species, evolved abdominal glands between sternites 7 and 8. Whether these glands originated at the same time independently in Pycnocerini, Lagriinae, and the remaining taxa has been questioned (28, 37). The small medial glandular pouch in Pycnocerini and in the genus *Phrenapates,* which exhibits randomly scattered gland cells on the dorsal surface of the reservoir, is the most primitive system (37, 85, 150, 151). Within several taxa, evolution has proceeded independently by lateralization, separation and enlargement of reservoirs, concentration of secretory tubules, and formation of elongated collecting ducts (37, 151). Apart from their muscular reservoir surfaces, Lagriinae glands are morphologically and chemically similar to the common tenebrionid glands. Paired common tenebrionid glands, which are also present in Alleculinae and Nilioninae, may be primitive, as in *Tenebrio,* or highly differentiated, as in Coelometopini; sometimes glands exhibit a mosaic of primitive and advanced features. Secondary loss of defensive glands occurred in Goniaderini and probably in Cossyphini, manifested by the persisting externalized intersegmental membranes. Chemically defended heteromeran beetles are always characterized by external membranes between sternites 5 and 7; the membranes are rapidly internalized if glands are lost, since otherwise these soft structures could be pierced by predators (37).

Tenebrionid secretions are fundamentally quinoic mixtures (149). In each of the 193 hitherto analyzed species from 87 genera, ethyl quinone and toluquinone are produced together with small amounts of *p*-benzoquinone. In primitive species small amounts of quinones are in equilibrium with an aqueous phase, whereas advanced species synthesize huge amounts of quinones in admixture with diverse alkenes (mainly 1-tridecene) and chemicals of other biogenetic origin (149, 150). Several taxa have been found to contain distinctive compounds and can often be identified by the characteristic odors of their defensive exudates. Scaurini are extremely heterogenous owing to the occurrence of naphthoquinones (Figure 2–22) in the genus *Argoporis* (149) and the presence of antibiotic, substituted 8-hydroxyisocoumarins (Figure 2–21); the latter occur together with *o*-cresol (Figure 2–17) in *Apsena pubescens* (98). Pedinini genera and *Eleodes obsoleta* (Eleodini) are unique in producing propylquinone (149). The strong-smelling *Uloma tenebrionides* stores, in addition to quinones and hydrocarbons, unusual ketones such as nonadec-10-en-2-one, heptadec-10-en-2-one, or methoxyphenol (60). In Amarygmini, *Amarygmus tristis* is differentiated by the presence of 4-methyl-hex-1-en-3-one and its saturated analog (59). *Eleodes* species produce either aldehydes, methoxyquinones, or octanoic acid (100, 148). Four *Artystona* species (Tenebrionini) secrete α-pinene and limonene, and should probably be excluded from this heterogenous tribe (58). The proposal that Misolampini are polyphyletic is supported by the unique defensive chemicals of a *Chrysopeplus* species (methyl-6-methylsalicylate, methyl-6-ethylsalicylate) (56).

Depletion of defensive chemicals is often correlated with stridulatory, thanatotic, or headstanding behaviors, which are also found within glandless tenebrionids (147). As a rule, headstanding species spray their secretions, are larger, and always secrete considerable amounts of hydrocarbons to dissolve their quinones. If an aqueous quinone solution is mixed with hydrocarbons in a headstanding species, the originally subnatant quinonic repellent phase becomes supernatant and is preferably expelled. A selective pressure for headstanding species to produce considerable amounts of hydrocarbons therefore seems conceivable (147).

Although unpalatability of some tenebrionids has been demonstrated, glands afford no absolute protection (85, 149). The scent of abdominal glands of the genus *Blaps* may cause aggregation or disaggregation, depending on the concentration (145).

Several Tenebrioninae, Nilioninae, and Alleculinae taxa possess prothoracic defensive glands (85, 149). Prothoracic gland cells of *Zophobas* secrete an emulsion of phenol, *m*-cresol, and *m*-ethylphenol (74, 149), whereas those of *Tribolium* secrete quinones identical to those from the abdominal glands. Unexpectedly, alkylated 1.4-benzoquinones of the *Eleodes* abdominal defensive glands are biosynthesized from aliphatic acids, whereas 1.4-benzoquinone is generated from an aromatic amino acid (10).

Concentrations of abdominal gland constituents are age-dependent, showing either a maximum early in adult life *(Zophobas)* (74) or a continuous increase until 30–50 days after eclosion *(Tribolium)* (158).

Chrysomeloidea

Both Cerambycidae and Chrysomelidae are characterized by effective chemical defensive systems.

Paired mandibular and paired metasternal glands have been detected in several Cerambycidae taxa; the contents of both may be discharged on molestation. Mandibular glands were found in *Stenocentrus ostricilla* and *Syllites grammicus* (106). They are characterized by reservoirs extending almost to the abdominal base. The strong smell, which is like that of carbolic acid, is due to the presence of *o*-cresol (Figure 2–17) and toluene, which is probably biogenetically related to *o*-cresol (Figure 2–19). A large fraction of Cerambycidae exhibit metasternal defensive glands containing aromatic secretions whose odors often resemble those of musk or attar of roses (97). The paired metasternal glands lie in the thorax, and associated reservoirs open near the articulated coxa of the hind legs. *Aromia moschata* synthesizes stereoisomers of rose oxide and iridodial (153), and representatives of the eucalyptus-feeding Australian cerambycid genus *Phoracantha* use 2-hydroxy-6-methylbenzaldehyde in admixture with macrocyclic lactones [e.g. *P. syn-*

onyma synthesizes phoracantholide (Figure 2–9; 108)] or unique cyclopentyl compounds such as phoracanthal (Figure 2–20) and phoracanthol [*P. semipunctata* (107)]. The strong, pineapplelike odor of the *P. synonyma* secretion is due to methyl and ethyl esters of 2-methylbutyric and related acids (108). Feeding experiments with radioactive compounds revealed that adults of *P. semipunctata* may not synthesize the cyclopentyl compounds de novo, but may retain precursors from the wood-feeding larval stage (107). Many Cerambycidae are brightly colored and mimic other beetles, especially Lycids, Cantharids, or Chrysomelids (97). It is fascinating that Cerambycids mimicking the lycid *Metriorrhynchus rhipidus* contain alkylpyrazines in their metasternal glands, which are known to be typical lycid warning odors (111). On the other hand, four brightly colored *Tetraopes* species (*T. tetrophthalmus, T. basilis, T. femoratus,* and *T. oregonensis*), which feed on Asclepiadaceae and Apocynaceae, are known to sequester small amounts of cardiac glycosides from the food plants within their hemolymph (10, 77).

The phytophagous leaf beetles (Chrysomelidae) have evolved a fascinating diversity of chemical defense mechanisms; these are especially evident in larvae and adults (122). Larval stages of Galerucinae exhibit reflex bleeding when disturbed (154); Alticinae larvae are protected by chemicals secreted at the tips of spined tubercles of the body surface (127); and most Chrysomelinae larvae have evolved three types of defensive gland systems that probably originated from segmentally arranged eversible glandular vesicles supplied with a few voluminous gland cells (34). Apart from Timarchini, in which absence of larval glands appears to represent the primary condition (34), Phyllodectina and Chrysomelina are characterized by paired defensive glands located dorsolaterally on the mesothorax and metathorax and on the first seven abdominal segments. When molested, these larvae cling to the leaf surface and discharge droplets of a defensive fluid, which may later be withdrawn. The defensive secretions are not only directed against predatory arthropods, but may also act as repellents against disturbance by approaching larvae and adults of the same species (130). In other taxa the segmentally arranged glands have probably been completely reduced and replaced by a single pair of defensive vesicles at the abdominal apex situated either between tergites 7 and 8, as in Paropsina and Gonioctenina, or between tergites 8 and 9, as in Doryphorina and Chrysolinina (34). Larvae with apically situated gland reservoirs must move the hindpart of the body dorsally to emit a highly volatile droplet to repel predators.

A comparative chemical analysis of Chrysomelina and Phyllodectina larval secretions revealed a fascinating variability of production of allomones de novo versus use of defensive compounds from food plants (122). Originally, larvae fed on herbs and synthesized unique cyclopentanoid monoterpenoids such as chrysomelidial (Figure 2–23), plagiodial, or plagiolactone, which are sometimes mixed with alkyl and alkenyl acetates. The acetates and especially the monoterpenoids are characterized by a pronounced repellency against ants (14, 122). Some larvae of Chrysomelina and Phyllodectina have shifted from feeding on herbs to feeding on certain trees (Salicaceae). Consequently these larvae use salicylaldehyde as a repellent (sometimes in admixture with benzaldehyde); they form salicylaldehyde from salicin, a phenylglucoside present in the leaves of their food plants. The primitive, herb-feeding Chrysomelina and Phyllodectina larva may have switched to feeding on Salicaceae in part for this cheap, presynthesized defensive compound and for the energy available from its glucose moiety (122). *Gastrolina depressa* larvae incorporate juglone (from the *Juglans* food plant) in their defensive scent (99). These chrysomelid larvae may show a further shift toward food plants devoid of salicin, and may synthesize phenylethyl esters. These esters probably take the place of the terpenoids, because the insects had lost the ability to produce terpenoids while feeding on salicin-containing trees.

Fascinating interrelations between defensive molecules from plants and larval allomones synthesized de novo may probably also be found in larvae that possess one pair of defensive glands at the apex of the abdomen. Chemical data, however, are only available from Paropsina and Gonioctenina. Larvae of the first group have been found to secrete hydrogen cyanide, benzaldehyde, and glucose (104). (*R*)-Mandelonitrile and the cyanogenic glucoside prunasin have been established as the source of hydrogen cyanide in all stages of *Paropsis atomaria* (114a). Cyanogenic compounds are not sequestered from the *Eucalyptus* food plant and have been found in all body parts of all *P. atomaria* developmental stages (114a). The defensive secretion of *Gonioctena viminalis* completely differs from the Paropsina allomones in the multifarious occurrence of compounds that show biogenetic affinities for typical allomones of Chrysomelina and Phyllodectina (34). Phenylethanol is present as phenylethyl esters in other chrysomelid larvae that possess segmental glands (10); 6-methyl-5-hepten-2-one and the corresponding alcohol may be biogenetically derived from citral, a major precursor of cyclopentanoid monoterpenes, which are typically found in most Chrysomelina and Phyllodectina (10). Larvae of two *Gonioctena* subgenera have obviously lost their glands secondarily, which indicates that the advantage of chemical defense systems may sometimes be limited (34).

Whether the mode of pupation of larvae with segmental glands is correlated with the secretion chemistry remains unknown. *Chrysomela* pupae, which hang from the leaf surface, retain the larval integument with the salicylaldehyde-rich defensive glands. When the pupae are molested the odorous aldehyde may be released (99). Freshly emerged adults bathe in the larval salicylaldehyde, which protects them as long as they are still soft (75). When pupation takes place in the ground or on the leaf surface (without utilization of allomones from the last larval skin), only cyclopentanoid monoterpenoids are present (99). It has now been shown that even egg clusters may be chemically protected by the ant-deterrent oleic acid in *Gastrophysa* species (75).

Although adults of Chrysomelidae are mechanically protected against smaller predators by their hard cuticle, various defensive strategies have developed, such as jumping mechanisms (Alticinae), cryptic coloration (Cassidinae), reflex bleeding (Chrysomelinae, Galerucinae), chitinous expansions (Hispinae), falling behavior, or aposematic coloration (21, 24). In addition, several species possess clusters of glands that open into grooves on the surface of the pronotum and elytra (24). These glands evolved monophyletically in Criocerinae, Chrysomelinae, Galerucinae, and Alticinae; their secondary loss has been suggested for some species of the latter subfamilies (25). Glandular defensive secretions of adults have been chemically studied in the Chrysomelinae tribes Doryphorina, Chrysolinina, Chrysomelina, and Phyllodectina. Both Doryphorina and Chrysolinina representatives use amino acid derivatives and hydrocarbons in their defensive exudates. This similarity, together with the identical larval glands, suggests a close relationship between the two taxa. Numerous species of the genera *Chrysolina* and *Oreina* secrete unique cardenolides, which they probably synthesize from a plant sterol precursor (22, 119, 122). The concentrated, bitter secretion (100–200 μg cardenolides μl^{-1}) may contain sarmentoginin, bipindogenin, periplogenin (Figure 2–32) and their corresponding xylosides; this protects beetles from vertebrate predators. All immature stages were found to contain hemolymph cardenolides in concentrations that increase with age, although no reflex bleeding was found. Interestingly, species that feed on chemically protected plants do not produce cardenolides. In lieu of cardenolides, Chrysomelina and Phyllodectina beetles secrete isoxazolin-5-one glucosides (Figure 2–6), which may be esterified with β-nitropropionic acid; their physiological effects remain to be investigated (119). Owing to the different chemical defensive mechanisms of both larvae and adults, the tribe Phyllodectina has no relationship to Gonioctenina, contrary to previous suggestion.

Larvae of *Pyrrhalta luteola* (Galerucinae) (76) may derive protection against predators from hemolymph anthraquinones and anthrones, which are not derived from Ulmaceae food plants. *P. luteola* exhibits no reflex bleeding, and the storage of highly reactive naphthoquinones in the hemolymph may also protect the beetle larva against diseases or parasites.

Various species of leaf beetles sequester secondary compounds from food plants in their hemolymph. *Diabrotica* species (Galerucinae) (48) sequester bitter cucurbitacins, and males of *Gabonia* (Alticinae) (15) are attracted to and feed on pyrrolizidine alkaloid–containing plants, as do certain Lepidoptera (143). *Chrysolina brunsvicensis* (Chrysolinina) sequesters the toxic hypericin, and both *Chrysochus cobaltinus* (Eumolpinae) and *Labidomera clivicollis* (Doryphorina) may contain small amounts of plant-derived cardenolides (10). Neurotoxic hemolymph proteins have been found in several *Leptinotarsa* species (Doryphorina) (156) and in larvae of the genus *Diamphidia* (Alticinae) (115), which are used by African bushmen as a source of efficient arrow poison. Certain *Lebistina* ground beetles that feed on leaf beetle larvae mimic *Diamphidia* and are also used as a source for arrow poison (115, 146).

ACKNOWLEDGMENTS

I wish to thank Dr. M. S. Blum (Georgia) and Dr. H. Schildknecht (Heidelberg) for the valuable discussions we have had, and Dr. H. Cooper-Schlüter, R. Djenanian, R. Grümmer, C. Gürzenich, M. Hennes, and H. Winkens for their help in preparing the manuscript. I am especially grateful to G. Schwinger (Hohenheim) for his cooperation and to the Deutsche Forschungsgemeinschaft (Bonn) for generous support.

Literature Cited

1. Abdullah, M. 1964. *Protomeloe argentinensis,* a new genus and species of Meloidae (Coleoptera), with remarks on the significance of cantharidin and the phylogeny of the families Pyrochroidae, Anthicidae, Meloidae and Cephaloidae. *Ann. Mag. Nat. Hist.* 7:247–54

2. Alexander, J. O. 1984. *Arthropods and Human Skin.* Berlin: Springer. 422 pp.

3. Araujo, J. 1973. Morphologie et histologie de la glande pygidiale défensive de *Bledius spectabilis* Kr. (Staphylinidae Oxytelinae). *C. R. Acad. Sci. Paris* 276D:2713–16

4. Araujo, J. 1978. Anatomie comparée des systèmes glandulaires de défense chimique des Staphylinidae. *Arch. Biol.* 89:217–50

5. Ashe, J. S. 1981. Construction of pupal cells by larvae of Aleocharinae (Coleoptera: Staphylinidae). *Coleopt. Bull.* 35:341–43

6. Ayer, W. A., Bennett, M. J., Browne, L. T., Purdham, J. T. 1976. Defensive substances of *Coccinella transversoguttata* and *Hippodamia caseyi,* ladybugs indigenous to western Canada. *Can. J. Chem.* 54:1807–13

7. Bellas, T. E., Brown, W. V., Moore, B. P. 1974. The alkaloid actinidine and plausible precursors in defensive secretions of rove beetles. *J. Insect Physiol.* 20:227–80

8. Berrios-Ortiz, H. 1985. The presence of mesothoracic glands in *Epicauta segmenta* (Gay) (Coleoptera: Meloidae) and other blister beetles. *J. Kans. Entomol. Soc.* 58:179–81

9. Blum, M. S. 1980. Arthropods and ecomones: better fitness through ecological chemistry. In *Animals and Environmental Fitness,* ed. R. Gilles, pp. 207–22. Oxford: Pergamon. 613 pp.

10. Blum, M. S. 1981. *Chemical Defenses of Arthropods.* New York: Academic. 562 pp.

11. Blum, M. S., Crewe, R. M., Pasteels, J. M. 1971. Defensive secretion of *Lomechusa strumosa,* a myrmecophilous beetle. *Ann. Entomol. Soc. Am.* 64:975–76

12. Blum, M. S., Jones, T. H., House, G. J., Tschinkel, W. R. 1981. Defensive secretions of tiger beetles: Cyanogenetic basis. *Comp. Biochem. Physiol.* 69B: 903–4

13. Blum, M. S., Sannasi, A. 1974. Reflex bleeding in the lampyrid *Photinus pyralis:* Defensive function. *J. Insect Physiol.* 20:451–60

14. Blum, M. S., Wallace, J. B., Duffield, R. M., Brand, J. M., Fales, H. M., Sokoloski, E. A. 1978. Chrysomelidial in the defensive secretion of the leaf beetle *Gastrophysa cyanea* Melsheimer. *J. Chem. Ecol.* 4:47–53

15. Boppré, M., Scherer, G. 1981. A new species of flea beetle (Alticinae) showing male-biased feeding at withered *Heliotropium* plants. *Syst. Entomol.* 6: 347–54

16. Braconnier, M. F., Braekman, J. C.,

Daloze, D., Pasteels, J. M. 1985. (Z)-1,17-Diaminooctadec-9-ene, a novel aliphatic diamine from Coccinellidae. *Experientia* 41:519–20

17. Brand, J. M., Blum, M. S., Fales, H. M., Pasteels, J. M. 1973. The chemistry of the defensive secretion of the beetle, *Drusilla canaliculata*. *J. Insect Physiol.* 19:369–82

18. Brown, W. V., Jones, A. J., Lacey, M. L., Moore, B. P. 1985. The chemistry of buprestins A and B. Bitter principles of jewel beetles (Coleoptera: Buprestidae). *Aust. J. Chem.* 38:197–206

19. Carrel, J. E., Eisner, T. 1974. Cantharidin: Potent feeding deterrent to insects. *Science* 183:755–57

20. Crowson, R. A. 1975. The evolutionary history of Coleoptera, as documented by fossil and comparative evidence. *Atti Congr. Naz. Ital. Entomol.* 10:47–90

21. Crowson, R. A. 1981. *The Biology of the Coleoptera*. London: Academic. 802 pp.

22. Daloze, D., Pasteels, J. M. 1979. Production of cardiac glycosides by chrysomelid beetles and larvae. *J. Chem. Ecol.* 5:63–77

23. Dazzini-Valcurone, M., Pavan, M. 1980. Glandole pigidiali e secrezioni difensive dei Carabidae (Insecta Coleoptera). *Pubbl. Ist. Entomol. Univ. Pavia* 1980(12):1–36

24. Deroe, C., Pasteels, J. M. 1977. Defensive mechanisms against predation in the Colorado beetle (*Leptinotarsa decemlineata*, Gay). *Arch. Biol.* 88:289–304

25. Deroe, C., Pasteels, J. M. 1982. Distribution of adult defense glands in chrysomelids (Coleoptera: Chrysomelidae) and its significance in the evolution of defense mechanisms within the family. *J. Chem. Ecol.* 8:67–82

26. Dettner, K. 1983. Vergleichende Untersuchungen zur Wehrchemie und Drüsenmorphologie abdominaler Abwehrdrüsen von Kurzflüglern aus dem Subtribus Philonthina (Coleoptera, Staphylinidae). *Z. Naturforsch.* 38c:319–28

27. Dettner, K. 1984. Isopropylesters as wetting agents from the defensive secretion of the rove beetle *Coprophilus striatulus* F. (Coleoptera, Staphylinidae). *Insect Biochem.* 14:383–90

28. Dettner, K. 1984. Description of defensive glands from cardinal beetles (Coleoptera, Pyrochroidae)—Their phylogenetic significance as compared with other heteromeran defensive glands. *Entomol. Basiliensia* 9:204–15

29. Dettner, K. 1985. Ecological and phylogenetic significance of defensive compounds from pygidial glands of Hydradephaga (Coleoptera). *Proc. Acad. Nat. Sci. Philadelphia* 137:156–71

30. Dettner, K., Grümmer, R. 1986. Quasisynergism as evolutionary advance to increase repellency of beetle defensive secretions. *Z. Naturforsch.* 41c:493–96

31. Dettner, K., Schwinger, G. 1980. Defensive substances from pygidial glands of water beetles. *Biochem. Syst. Ecol.* 8:89–95

32. Dettner, K., Schwinger, G. 1982. Defensive secretions of three Oxytelinae rove beetles (Coleoptera, Staphylinidae). *J. Chem. Ecol.* 8:1411–20

33. Dettner, K., Schwinger, G. 1986. Volatiles from the defensive secretions of two rove beetle species (Coleoptera: Staphylinidae). *Z. Naturforsch.* 41c:366–68

34. Dettner, K., Schwinger, G. 1986. Chemical defence in the larvae of the leaf beetle *Gonioctena viminalis* L. (Coleoptera: Chrysomelidae). *Experientia* In press

35. Dettner, K., Schwinger, G., Wunderle, P. 1985. Sticky secretion from two pairs of defensive glands of rove beetle *Deleaster dichrous* (Grav.) (Coleoptera: Staphylinidae). Gland morphology, chemical constituents, defensive functions, and chemotaxonomy. *J. Chem. Ecol.* 11:859–83

36. Dettner, K., Wunderle, P., Schwinger, G. 1987. Comparative chemical and morphological analysis of the chemical defensive system within the rove beetle subfamily Oxytelinae (Coleoptera: Staphylinidae). *Z. Zool. Syst. Evolutionsforsch.* In press

37. Doyen, J. T., Tschinkel, W. R. 1982. Phenetic and cladistic relationships among tenebrionid beetles (Coleoptera). *Syst. Entomol.* 7:127–83

38. Duffey, S. S. 1976. Arthropod allomones: Chemical effronteries and antogonists. *Proc. 15th Int. Congr. Entomol.*, *Washington*, pp. 323–94.

39. Eisner, T. 1970. Chemical defense against predation in arthropods. In *Chemical Ecology*, ed. E. Sondheimer, J. B. Simeone, pp. 157–217. New York: Academic. 336 pp.

40. Eisner, T. 1980. Chemistry, defense, and survival: Case studies and selected topics. In *Insect Biology in the Future*, ed. M. Locke, D. S. Smith, pp. 847–78. New York: Academic

41. Eisner, T., Hill, D., Goetz, M., Jain, S., Alsop, D., et al. 1981. Antifeedant action of Z-hihydromatricaria acid from

44   DETTNER

soldier beetles (*Chauliognathus* spp.). *J. Chem. Ecol.* 7:1149–58
42. Eisner, T., Jones, T. H., Aneshansley, J., Tschinkel, W. R., Silberglied, R. E., Meinwald, J. 1977. Chemistry of defensive secretions of bombardier beetles (Brachinini, Metriini, Ozaenini, Paussini). *J. Insect Physiol.* 23:1383–86
43. Eisner, T., Kafatos, F. C., Linsley, E. G. 1962. Lycid predation by mimetic adult Cerambycidae (Coleoptera). *Evolution* 16:316–24
44. Eisner, T., Meinwald, J. 1982. Defensive spray mechanism of a silphid beetle *(Necrodes surinamensis)*. *Psyche* 89:357–67
45. Eisner, T., Wiemer, D. F., Haynes, L. W., Meinwald, J. 1978. Lucibufagins: Defensive steroids from the fireflies *Photinus ignitus* and *P. marginellus* (Coleoptera: Lampyridae). *Proc. Natl. Acad. Sci. USA* 75:905–8
46. Engelhardt, M., Rapoport, H., Sokoloff, A. 1965. Odorous secretion of normal and mutant *Tribolium confusum*. *Science* 150:632–33
47. Evers, A. M. J. 1968. Carphurinae oder Carphuridae? *Entomol. Bl.* 64:17–27
48. Ferguson, J. E., Metcalf, R. L. 1985. Cucurbitacins plant-derived defense compounds for diabroticites (Coleoptera: Chrysomelidae). *J. Chem. Ecol.* 11:311–18
49. Fey, F. 1954. Beiträge zur Biologie der canthariphilen Insekten. *Beitr. Entomol.* 4:180–87
50. Fiori, G. 1960. Le glandole tegumentali segmentali del *Malachius sardous* Er. *Studi Sassar. Sez. 3* 8:90–102
51. Fish, L. J., Pattenden, G. 1975. Iridodial, and a new alkanone, 4-methylhexan-3-one, in the defensive secretion of the beetle, *Staphylinus olens*. *J. Insect Physiol.* 21:741–44
52. Forsyth, D. J. 1968. The structure of the defence glands in the Dytiscidae, Noteridae, Haliplidae and Gyrinidae (Coleoptera). *Trans. R. Entomol. Soc. London* 120:159–81
53. Forsyth, D. J. 1970. The structure of the defence glands of the Cicindelidae, Amphizoidae, and Hygrobiidae (Insecta: Coleoptera). *J. Zool.* 160:51–69
54. Forsyth, D. J. 1972. The structure of the pygidial defence glands of Carabidae (Coleoptera). *Trans. Zool. Soc. London* 32:249–309
55. Gnanasunderam, C., Butcher, C. F., Hutchins, R. F. N. 1981. Chemistry of the defensive secretions of some New Zealand rove beetles (Coleoptera: Staphylinidae) *Insect Biochem.* 11:411–16

56. Gnanasunderam, C., Young, H., Blum, M. H. 1984. Defensive secretions of New Zealand tenebrionids. III. The identification of methyl esters of 6-methyl and 6-ethylsalicylic acid in *Chrysopeplus expolitus* (Coleoptera: Tenebrionidae). *Insect Biochem.* 14:159–61
57. Gnanasunderam, C., Young, H., Butcher, C. F., Hutchins, R. F. N. 1981. Ethyl decanoate as a major component in the defensive secretion of two New Zealand Aleocharinae (Staphylinidae) beetles—*Tramiathaea cornigera* (Broun) and *Thamiaraea fuscicornis* (Broun). *J. Chem. Ecol.* 7:197–202
58. Gnanasunderam, C., Young, H., Hutchins, R. F. N. 1981. Defensive secretions of New Zealand tenebrionids. I. Presence of monoterpene hydrocarbons in the genus *Artystona* (Coleoptera, Tenebrionidae). *J. Chem. Ecol.* 7:889–94
59. Gnanasunderam, C., Young, H., Hutchins, R. F. N. 1982. Defensive secretions of New Zealand tenebrionids. II. Presence of the unsaturated ketone 4-methylhex-1-en-3-one in *Amarygmus tristis* (Coleoptera: Tenebrionidae). *Insect Biochem.* 12:221–24
60. Gnanasunderam, C., Young, H., Hutchins, R. F. N. 1985. Defensive secretions of New Zealand tenebrionids. V. Presence of methyl ketones in *Uloma tenebrionides* (Coleoptera: Tenebrionidae). *J. Chem. Ecol.* 11:465–72
61. Goetz, M. A., Meinwald, J., Eisner, T. 1981. Lucibufagins. IV. New defensive steroids and a pterin from the firefly *Photinus pyralis* (Coleoptera: Lampyridae). *Experientia* 37:679–80
62. Gottlieb, O. R. 1980. Micromolecular systematics: Principles and practice. In *Chemosystematics: Principles and Practice*, ed. F. A. Bisby, J. G. Vaughan, C. A. Wright, pp. 329–52. New York: Academic. 450 pp.
63. Gottlieb, O. R. 1982. *Micromolecular Evolution, Systematics and Ecology*. Berlin: Springer. 170 pp.
64. Hammond, P. M. 1976. A review of the genus *Anotylus* C. G. Thomson (Coleoptera: Staphylinidae). *Bull. Br. Mus. Nat. Hist. Entomol.* 33:137–87
65. Happ, G. M., Eisner, T. 1961. Hemorrhage in a coccinellid beetle and its repellent effect on ants. *Science* 134:329–31
66. Happ, G. M., Happ, C. M. 1973. Fine structure of the pygidial glands of *Bledius mandibularis* (Coleoptera: Staphylinidae). *Tissue Cell* 5:215–31
67. Hayes, W. P., Chu, H. F. 1947. An undescribed eversible gland in the larvae

of *Chlaenius* (Coleoptera Carabidae). *J. Kans. Entomol. Soc.* 20:142–45
68. Hennig, W. 1982. *Phylogenetische Systematik.* Berlin: Parey. 246 pp.
69. Herman, L. H. 1970. Phylogeny and reclassification of the genera of the rove-beetle subfamily Oxytelinae of the world (Coleoptera, Staphylinidae). *Bull. Am. Mus. Nat. Hist.* 142:343–454
70. Herman, L. H. 1975. Revision and phylogeny of the monogeneric subfamily Pseudopsinae for the world (Staphylinidae, Coleoptera). *Bull. Am. Mus. Nat. Hist.* 155:242–317
71. Herman, L. H. 1977. Revision and phylogeny of *Zalobius, Asemobius,* and *Nanobius,* new genus (Coleoptera, Staphylinidae, Piestinae). *Bull. Am. Mus. Nat. Hist.* 159:45–86
72. Heymons, R., von Lengerken, H. 1932. Studien über die Lebenserscheinungen der Silphini (Coleopt.). VIII. *Ablattaria laevigata* F. *Z. Morphol. Oekol. Tiere* 24:259–87
73. Heymons, R., von Lengerken, H., Bayer, M. 1927. Studien über die Lebenserscheinungen der Silphini (Coleopt.). II. *Phosphuga atrata* L. *Z. Morphol. Oekol. Tiere* 9:271–312
74. Hill, C. S., Tschinkel, W. R. 1985. Defensive secretion production in the tenebrionid beetle, *Zophobas atratus:* Effects of age, sex and milking frequency. *J. Chem. Ecol.* 11:1083–91
75. Howard, D. F., Blum, M. S., Jones, T. H., Phillips, D. W. 1982. Defensive adaptations of eggs and adults of *Gastrophysa cyanea* (Coleoptera: Chrysomelidae). *J. Chem. Ecol.* 8:453–62
76. Howard, D. F., Phillips, D. W., Jones, T. H., Blum, M. S. 1982. Anthraquinones and anthrones: occurrence and defensive function in a chrysomelid beetle. *Naturwissenschaften* 69:91–92
77. Isman, M. B., Duffey, S. S., Scudder, G. G. E. 1977. Cardenolide content of some leaf- and stem-feeding insects on temperate North American milkweeds (*Asclepias* spp.). *Can. J. Zool.* 55:1024–28
78. Jefson, M., Meinwald, J., Nowicki, S., Hicks, K., Eisner, T. 1983. Chemical defense of a rove beetle *(Creophilus maxillosus). J. Chem. Ecol.* 9:159–80
79. Jenkins, M. F. 1957. The morphology and anatomy of the pygidial glands of *Dianous coerulescens* Gyllenhal (Coleoptera: Staphylinidae). *Proc. R. Entomol. Soc. London A* 32:159–67
80. Jones, T. H., Blum, M. S. 1983. Arthropod alkaloids: distribution, functions, and chemistry. In *Alkaloids: Chemical and Biological Perspectives,*

ed. S. W. Pelletier, 1:33–84. New York: Wiley
81. Kanehisa, K., Kawazu, K. 1982. Fatty acid components of the defensive substances in acid-secreting carabid beetles. *Appl. Entomol. Zool.* 17:460–66
82. Kanehisa, K., Murase, M. 1977. Comparative study of the pygidial defensive systems of carabid beetles. *Appl. Entomol. Zool.* 12:225–35
83. Kanehisa, K., Shiraga, T., Kawazu, K. 1984. Defensive secretory organs of the rove beetles (Coleoptera: Staphylinidae). *Nogaku Kenkyu* 60(3):111–21
84. Kendall, D. A. 1971. A note on reflex bleeding in the larvae of the beetle *Exochomus quadripustulatus* (L.) (Col.: Coccinellidae). *Entomologist London* 104:233–35
85. Kendall, D. A. 1974. The structure of defence glands in some Tenebrionidae and Nilionidae (Coleoptera). *Trans. R. Entomol. Soc. London* 125:437–87
86. Kirk, V. A., Dupraz, B. J. 1972. Discharge by a female ground beetle *Pterostichus lucublandus* (Coleoptera: Carabidae), used as a defense against males. *Ann. Entomol. Soc. Am.* 65:513
87. Kistner, D. H. 1982. The social insects' bestiary. In *Social Insects,* ed. H. R. Hermann, 3:1–244. New York: Academic
88. Kistner, D. H., Blum, M. S. 1971. Alarm pheromone of *Lasius (Dendrolasius) spathebus* (Hymenoptera: Formicidae) and its possible mimcry by two species of *Pella* (Coleoptera: Staphylinidae). *Ann. Entomol. Soc. Am.* 64:589–94
89. Klinger, R. 1979. *Eine Sternaldrüse bei Kurzflügelkäfern. Systematische Verbreitung sowie Bau, Inhaltsstoffe und Funktion bei Eusphalerum minutum* (L.) (Coleoptera: Staphylinidae). PhD thesis. Univ. Frankfurt. 17 pp.
90. Klinger, R., Maschwitz, U. 1977. The defensive gland of Omaliinae (Coleoptera: Staphylinidae). *J. Chem. Ecol.* 3:401–10
91. Kloft, W. J., Woodruff, R. E., Kloft, E. S. 1979. *Formica integra* (Hymenoptera: Formicidae). IV. Exchange of food and trichome secretions between worker ants and the inquiline beetle *Cremastocheilus castaneus* (Coleoptera: Scarabaeidae). *Tijdschr. Entomol.* 122:47–57
92. Kolbe, W., Proske, M. G. 1973. Iso-Valeriansäure im Abwehrsekret von *Zyras humeralis* Grav. (Coleoptera, Staphylinidae). *Entomol. Bl.* 69:57–60
93. Kurosa, K., Watanabe, H. 1958. On the toxic substance of *Xanthochroa*

waterhousei (Coleoptera: Oedemeridae). *Jpn. J. Sanit. Zool.* 9:200–1
94. Lawrence, J. F. 1982. Coleoptera. In *Synopsis and Classification of Living Organisms,* ed. S. P. Parker, 2:482–553. New York: McGraw-Hill. 1232 pp.
95. Lawrence, J. F., Newton, A. F. Jr. 1982. Evolution and classification of beetles. *Ann. Rev. Ecol. Syst.* 13:261–90
96. Lindner, W. 1967. Ökologie und Larvalbiologie einheimischer Histeriden. *Z. Morphol. Oekol. Tiere* 59:341–80
97. Linsley, E. G. 1959. Ecology of Cerambycidae. *Ann. Rev. Entomol.* 4:99–138
98. Lloyd, H. A., Evans, S. L., Khan, A. H., Tschinkel, W. R., Blum, M. S. 1978. 8-Hydroxyisocoumarin and 3,4-dihydro-8-hydroxyisocoumarin in the defensive secretion of the tenebrionid beetle, *Apsena pubescens. Insect Biochem.* 8:333–36
99. Matsuda, K., Sugawara, F. 1980. Defensive secretion of chrysomelid larvae *Chrysomela vigintipunctata costella* (Marseul), *C. populi* L. and *Gastrolina depressa* Baly (Coleoptera: Chrysomelidae). *Appl. Entomol. Zool.* 15:316–20
100. Meinwald, Y. C., Eisner, T. 1964. Defense mechanisms of arthropods. XIV. Caprylic acid: An accessory component of the secretion of *Eleodes longicollis. Ann. Entomol. Soc. Am.* 57:513–14
101. Meinwald, J., Meinwald, Y. C., Chalmers, A. M., Eisner, T. 1968. Dihydromatricaria acid: acetylinic acid secreted by soldier beetle. *Science* 160:890–92
102. Meinwald, J., Roach, B., Hicks, K., Eisner, T. 1985. Defensive steroids from a carrion beetle *(Silpha americana). Experientia* 41:516–19
103. Miller, J. R., Mumma, R. O. 1976. Physiological activity of water beetle defensive agents. I. Toxicity and anesthetic activity of steroids and norsesquiterpenes administered in solution to the minnow *Pimephales promelas* Raf. *J. Chem. Ecol.* 2:115–30
104. Moore, B. P. 1967. Hydrogen cyanide in the defensive secretion of larval Paropsini (Coleoptera: Chrysomelidae). *J. Aust. Entomol. Soc.* 6:36–38
105. Moore, B. P. 1979. Chemical defense in carabids and its bearing on phylogeny. In *Carabid Beetles: Their Evolution, Natural History and Classification,* ed. T. L. Erwin, G. E. Ball, D. R. Witehead, pp. 193–203. The Hague: Junk
106. Moore, B. P., Brown, W. V. 1971. Chemical defence in longhorn beetles of the genera *Stenocentrus* and *Syllitus* (Coleoptera: Cerambycidae). *J. Aust. Entomol. Soc.* 10:230–32

107. Moore, B. P., Brown, W. V. 1972. The chemistry of the metasternal gland secretion of the common eucalypt longicorn, *Phoracantha semipunctata* (Coleoptera: Cerambycidae). *Aust. J. Chem.* 25:591–98
108. Moore, B. P., Brown, W. V. 1976. The chemistry of the metasternal gland secretion of the eucalypt longicorn, *Phoracantha synonyma* (Coleoptera: Cerambycidae). *Aust. J. Chem.* 29:1365–74
109. Moore, B. P., Brown, W. V. 1978. Precoccinelline and related alkaloids in the Australian soldier beetle *Chauliognathus pulchellus* (Coleoptera: Cantharidae). *Insect Biochem.* 8:393–95
110. Moore, B. P., Brown, W. V. 1979. Chemical composition of the defensive secretion in *Dyschirius bonelli* (Coleoptera: Carabidae: Scaritinae) and its taxonomic significance. *J. Aust. Entomol. Soc.* 18:123–25
111. Moore, B. P., Brown, W. V. 1981. Identification of warning odour components, bitter principles and antifeedants in an aposemantic beetle: *Metriorrhynchus rhipidius* (Coleoptera: Lycidae). *Insect Biochem.* 11:493–99
112. Moore, B. P., Brown, W. V. 1985. The buprestins: Bitter principles of jewel beetles (Coleoptera: Buprestidae). *J. Aust. Entomol. Soc.* 24:81–5
113. Moore, B. P., Wallbank, B. E. 1968. Chemical composition of the defensive secretion in carabid beetles and its importance as a taxonomic character. *Proc. R. Entomol. Soc. London B* 37:62–72
114. Moore, I. 1978. Usefulness of larval osmeteria in determining natural classification in Aleocharinae (Coleoptera: Staphylinidae). *Entomol. News* 89:245–46
114a. Nahrstedt, A., Davis, R. H. 1986. *(R)*-Mandelonitrile and prunasin, the sources of hydrogen cyanide in all stages of *Paropsis atomaria* (Coleoptera: Chrysomelidae). *Z. Naturforsch.* 41c:In press
115. Neuwinger, H. D., Scherer, G. 1976. Das Larven-Pfeilgift der Buschmänner. *Biol. Zeit* 6:75–82
116. Newton, A. F. 1982. Redefinition, revised phylogeny, and relationships of Pseudopsinae (Coleoptera, Staphylinidae). *Am. Mus. Novit.* 2743:1–13
117. Newton, A. F. 1982. A new genus and species of Oxytelinae from Australia, with a description of its larva, systematic position, and phylogenetic relationships (Coleoptera, Staphylinidae). *Am. Mus. Novit.* 2744:1–24
118. Okada, Y. K. 1928. Two Japanese aquatic glowworms. *Trans. R. Entomol. Soc. London* 1928:101–7

119. Pasteels, J. M., Braekman, J. C., Daloze, D., Ottinger, R. 1982. Chemical defence in chrysomelid larvae and adults. *Tetrahedron* 38:1891–97

120. Pasteels, J. M., Deroe, C., Tursch, B., Braekman, J. C., Daloze, D., Hootele, C. 1973: Distribution et activités des alcaloides défensifs des Coccinellidae. *J. Insect Physiol.* 19:1771–84

121. Pasteels, J. M., Grégoire, J.-C., Rowell-Rahier, M. 1983. The chemical ecology of defense in arthropods. *Ann. Rev. Entomol.* 28:263–89

122. Pasteels, J. M., Rowell-Rahier, M., Braekman, J. C., Daloze, D. 1984. Chemical defences in leaf beetles and their larvae: The ecological, evolutionary and taxonomic significance. *Biochem. Syst. Ecol.* 12:395–406

123. Pavan, M. 1975. Sunto delle attuali conoscenze sulla pederina. *Pubbl. Ist. Entomol. Univ. Pavia* 1975(1):1–35

124. Peschke, K. 1983. Defensive and pheromonal secretion of the tergal gland of *Aleochara curtula*. II. Release and inhibition of male copulatory behavior. *J. Chem. Ecol.* 9:13–31

125. Peschke, K., Metzler, M. 1982. Defensive and pheromonal secretion of the tergal gland of *Aleochara curtula*. I. The chemical composition. *J. Chem. Ecol.* 8:773–83

126. Peter, M. G., Woggon, W. D., Schmid, H. 1977. Identifizierung von Farnesol als Zwischenstufe in der Biosynthese des Cantharidins aus Mevalonsäurelacton. *Helv. Chim. Acta* 60:2756–62

127. Phillips, W. M. 1977. Observations on the biology and ecology of the chrysomelid genus *Haltica* Geoff. in Britain. *Ecol. Entomol.* 2:205–16

128. Pluot-Sigwalt, D. 1983. Les glandes tégumentaires des Coléoptères Scarabaeidae: répartition des glandes sternales et pygidiales dans la famille. *Bull. Soc. Entomol. Fr.* 88:597–602

129. Prestwich, G. D. 1983. Chemical systematics of termite exocrine secretions. *Ann. Rev. Ecol. Syst.* 14:287–311

130. Renner, K. 1970. Über die ausstülpbaren Hautblasen der Larven von *Gastroidea viridula* de Geer und ihre ökologische Bedeutung. *Beitr. Entomol.* 20:527–33

131. Rothschild, M. 1972. Secondary plant substances and warning colouration in insects. In *Insect Plant Relationships, Symp. R. Entomol. Soc. London* 6:59–83

132. Schildknecht, H. 1970. Die Wehrchemie von Land- und Wasserkäfern. *Angew. Chem.* 82:17–25

133. Schildknecht, H. 1976. Chemische Öko-logie—Ein Kapitel moderner Naturstoffchemie. *Angew. Chem.* 88:235–43

134. Schildknecht, H. 1977. Protective substances of arthropods and plants. *Pontif. Acad. Sci. Scripta Varia* 41:59–107

135. Schildknecht, H., Berger, D., Krauss, D., Connert, J., Gehlhaus, J., Essenbreis, H. 1976. Defense chemistry of *Stenus comma* (Coleoptera: Staphylinidae). LXI. *J. Chem. Ecol.* 2:1–11

136. Schildknecht, H., Birringer, H., Krauss, D. 1969. Aufklärung des gelben Prothorakalwehrdrüsen-Farbstoffes von *Ilybius fenestratus*. *Z. Naturforsch.* 24b:38–47

137. Schildknecht, H., Körnig, W., Siewerdt, S., Krauss, D. 1970. Aufklärung des gelben Pygidialwehrdrüsen-Farbstoffes des Gelbrandkäfers *(Dytiscus marginalis)*. *Liebigs Ann. Chem.* 734:116–25

138. Schildknecht, H., Maschwitz, U., Winkler, H. 1968. Zur Evolution der Carabiden-Wehrdrüsensekrete. *Naturwissenschaften* 55:112–17

139. Schildknecht, H., Tacheci, H. 1971. Colymbetin, a new defensive substance of the water beetle, *Colymbetes fuscus*, that lowers blood pressure. LII. *J. Insect Physiol.* 17:1889–96

140. Schildknecht, H., Tauscher, B., Krauss, D. 1972. Der Duftstoff des Taumelkäfers *Gyrinus natator* L. *Chem. Ztg.* 96:33–35

141. Schildknecht, H., Weber, B., Dettner, K. 1983. Über Arthropedenabwehrstoffe. LXV. Die chemische Ökologie des Grundschwimmers *Laccophilus minutus*. *Z. Naturforsch.* 38b:1678–85

142. Schildknecht, H., Weis, K. H. 1962. Über die chemische Abwehr der Aaskäfer. *Z. Naturforsch.* 176:452–55

143. Schneider, D. 1987. The strange fate of pyrrolizidine alkaloids. In *Perspectives in Chemoreception and Behavior*, ed. J. G. Stoffolano, E. A. Bernays, R. F. Chapman. Berlin: Springer. In press

144. Sulc, K. 1949. On the repellent stink glands in the beetles of the genus *Cantharis*, Coleoptera. *Rozpr. II. Třidy Česk. Akad.* 59(8):1–22 (In Czech)

145. Tannert, W., Hien, B. C. 1973. Nachweis und Funktion eines "Versammlungsduftstoffes" und eines "Alarmduftstoffes" bei *Blaps mucronata* Latr., 1804 (Coleopt.—Tenebrionidae). *Biol. Zentralbl.* 92:601–12

146. Thiele, H. U. 1977. *Carabid Beetles in Their Environments*. Berlin: Springer. 369 pp.

147. Tschinkel, W. R. 1975. A comparative study of the chemical defensive system of tenebrionid beetles. Defensive be-

havior and ancillary features. *Ann. Entomol. Soc. Am.* 68:439–53

148. Tschinkel, W. R. 1975. Unusual occurrence of aldehydes and ketones in the defensive secretion of the tenebrionid beetle, *Eleodes beameri. J. Insect Physiol.* 21:659–71

149. Tschinkel, W. R. 1975. A comparative study of the chemical defensive system of tenebrionid beetles: Chemistry of the secretions. *J. Insect Physiol.* 21:753–83

150. Tschinkel, W. R. 1975. A comparative study of the chemical defensive system of tenebrionid beetles. III. Morphology of the glands. *J. Morphol.* 145:355–70

151. Tschinkel, W. R., Doyen, J. T. 1980. Comparative anatomy of the defensive glands, ovipositors and female genital tubes of tenebrionid beetles (Coleoptera). *Int. J. Insect Morphol. Embryol.* 9:321–68

152. Tursch, B., Daloze, D., Braekman, J. C., Hootele, C., Pasteels, J. M. 1975. Chemical ecology of arthropods. X. The structure of myrrhine and the biosynthesis of coccinelline. *Tetrahedron* 31:1541–43

153. Vidari, G., Bernardi, M., Pavan, M., Ragozzino, L. 1973. Rose oxide and iridodial from *Aromia moschata* L. (Coleoptera: Cerambycidae). *Tetrahedron Lett.* 41:4065–68

154. Wallace, J. B., Blum, M. S. 1971. Reflex bleeding: A highly refined defensive mechanism in *Diabrotica* larvae (Coleoptera: Chrysomelidae). *Ann. Entomol. Soc. Am.* 64:1021–24

155. Watt, J. C. 1974. A revised subfamily classification of Tenebrionidae (Coleoptera). *NZ J. Zool.* 1:381–452

156. Weatherston, J., Percy, J. E. 1978. Venoms of Coleoptera. In *Arthropod Venoms*, ed. S. Bettini, pp. 511–54. Berlin: Springer. 977 pp.

157. Wheeler, J. W., Happ, G. M., Araujo, J., Pasteels, J. M. 1972. γ-Dodecalactone from rove beetles. *Tetrahedron Lett.* 46:4635–38

158. Wirtz, R. A., Taylor, S. L., Semey, H. G. 1978. Concentrations of substituted *p*-benzoquinones and 1-pentadecene in the flour beetles *Tribolium confusum* J. du Val and *Tribolium castaneum* (Herbst). *Comp. Biochem. Physiol.* 61B:25–28

159. Wolfe, G. W. 1985. A phylogenetic analysis of pleisiotypic hydroporine lineages with an emphasis on *Laccornis* Des Gozis (Coleoptera: Dytiscidae). *Proc. Acad. Nat. Sci. Philadelphia* 137:132–55

160. Young, D. K. 1984. Cantharidin and insects: An historical review. *Great Lakes Entomol.* 17:187–94

161. Young, D. K. 1984. Field records and observations of insects associated with cantharidin. *Great Lakes Entomol.* 17:195–99

Ann. Rev. Entomol. 1987. 32:49–70

INSECT HYPERPARASITISM

Daniel J. Sullivan

Department of Biological Sciences, Fordham University, Bronx, New York 10458

PERSPECTIVES AND OVERVIEW

Insect hyperparasitism is a highly evolved behavior that is restricted to three orders: the Hymenoptera, Diptera, and Coleoptera. It involves the development of a secondary parasitoid, or hyperparasitoid, at the expense of a primary parasitoid (3, 18, 21, 52, 63–65, 134). Hence, an insect hyperparasitoid attacks another insect that is itself parasitic on a host insect. The host is usually phytophagous, but could also be a predator or a scavenger. The terminology, taxonomy and evolution, bionomics and behavior of selected examples, ecology and host specificity, food web and mathematical models, and finally the impact on biological control of these insect hyperparasitoids are reviewed here. The fundamental theme running through this presentation of insect hyperparasitism is the ecological concept of the "food web" and community structure (2, 33, 54, 78, 98, 99). There exists in nature a complex of interlocking food chains consisting of the host plant, then phytophagous insects, and finally several levels of entomophagous insects that form a two-dimensional ecological community or biocenosis with both inter- and intraspecific components.

Terminology

Obligate hyperparasitoids are always secondary parasitoids; their progeny can develop only in or on a primary parasitoid. There are also *facultative* hyperparasitoids; their progeny can develop as either primary or secondary parasitoids. Another classification of hyperparasitoids is based on their feeding behavior: *endophagous* hyperparasitoids have larvae that feed inside the host, while the larvae of *ectophagous* species feed externally. Finally, *direct* hyperparasitoids attack the primary parasitoid directly by ovipositing in or on it. *Indirect* secondary parasitoids, on the other hand, attack the primary

49

0066-4170/87/0101-0049$02.00

parasitoid's phytophagous host and thus only attack the parasitoid itself indirectly. In this case, the female hyperparasitoid oviposits into the phytophagous host whether it is parasitized or not (31).

TAXONOMY AND EVOLUTION

Only three insect orders (Hymenoptera, Diptera, and Coleoptera) have evolved hyperparasitic behavior. The taxonomic survey by Gordh (43) is summarized below.

Hymenoptera

Most insect hyperparasitism is found in Hymenoptera, especially in the following six superfamilies.

CHALCIDOIDEA   Because so many of the species in this superfamily are parasitoids, it is not surprising that 11 of 17 families display hyperparasitism: Pteromalidae, Encyrtidae, Chalcididae, Aphelinidae, Eulophidae, Eupelmidae, Signiphoridae, Torymidae, Eurytomidae, Elasmidae, and Perilampidae (8, 11, 42, 44, 90, 120, 125, 129, 130, 138).

ICHNEUMONOIDEA   All species are parasitic, but only the Ichneumonidae behave as hyperparasitoids. Hyperparastism is restricted to four subfamilies: Ephaltinae, Gelinae, Mesochorinae, and Tryphoninae. Often their primary parasitoid hosts are other ichneumonids and braconids on Lepidoptera (128).

CYNIPOIDEA   Most members of the family Cynipidae are phytophagous, but the subfamily Alloxystinae has three genera (Alloxysta, Phaenoglyphis, Lytoxysta) that are all hyperparasitic on the primary parasitoids that attack aphids (1, 25, 100).

CERAPHRONOIDEA   Hyperparasitism occurs in two families. Some Ceraphronidae are facultative hyperparasitoids on ichneumonids, braconids, bethylids, and dryinids. Some species of Megaspilidae (Dendrocerus) have evolved a more specialized host range and are hyperparasitic on the primary parasitoids of aphids (19, 30).

PROCTOTRUPOIDEA   In the family Diapriidae there is only one reported case of hyperparasitism, in the genus Ismarus (13, 76).

TRIGONALOIDEA   In this small superfamily, only the family Trigonalidae has hyperparasitoids, which parasitize ichneumonids and tachinids attacking Lepidoptera (17, 127).

Diptera

Although this is also a large order with a number of families that are parasitic, hyperparasitism seems restricted to only two families, Bombyliidae (22a) and Conopidae (102).

Coleoptera

This largest order has only a few examples of hyperparasitism in two families, Rhipiphoridae and Cleridae (89).

Evolution of Hyperparasitism in Hymenoptera

The hymenopteran suborder Symphyta ("Phytophaga" such as the sawflies and horntails) is the most primitive in structure and behavior, and is considered closest to the ancestors of the Hymenoptera. The other suborder, Apocrita ("Heterophaga"), is usually divided into the Parasitica (Terebrantia) and the Aculeata (bees, wasps, and ants).

Concerning primary parasitism, Malyshev (81) held that the Parasitica were derived from the original Symphyta, while the Aculeata originated from primitive Parasitica or together with them from common ancestors. It is not known how phytophagous feeding evolved into entomophagous behavior, but Telenga (126) suggested that the phytophagous sawflies first evolved predation, and later parasitism of coleopteran larvae also living in tree trunks. These primary parasitoids were at first ectophagous, but endophagous parasitism appeared very early in the Parasitica.

Hyperparasitism (except in the Aculeata) evolved from among early Parasitica, probably independently and several times in different taxa. Brues (10) noted that many insect hosts exhibit defense reactions against the parasitic larva, and the greater the taxonomic and physiological distance between parasitoid and host, the stronger this defense reaction. Therefore, if a hymenopteran parasitoid accidentally oviposited in a host of the same order, the primary parasitic larva would find the hymenopteran a very suitable host. Telenga (126) agreed, and added that facultative secondary parasitism provides a transitional stage to obligatory hyperparasitism.

Gordh (43) concluded his taxonomic survey by pointing out that although the host spectrum of hyperparasitoids is broader at the species level than that of primary parasitoids, it is mainly restricted to immature hymenopteran hosts that are natural enemies of phytophagous insects, mainly in three orders: Homoptera, Lepidoptera, and the hymenopteran suborder Symphyta. On the other hand, hyperparasitoids rarely attack the egg and adult stages of primary parasitoids. It is also interesting that some families of insects that are well known for their parasitic behavior have no hyperparasitoids. Such is the case in the hymenopteran families Aphidiidae, Braconidae, Trichogrammatidae,

Mymaridae, Tetracampidae, and Eucharitidae, and in almost the entire super-family Proctotrupoidea. Similarly, in the order Diptera, hyperparasitoids are completely absent in the parasitic families Tachinidae, Acroceridae, Pipunculidae, and Nemestrinidae.

APHID HYPERPARASITOIDS

The most intensive studies of hyperparasitism have been conducted on the Hymenoptera that attack the Homoptera, and in particular the superfamily Aphidoidea (132). That which holds true for aphid hyperparasitoids, however, has relevance for hyperparasitoids that affect primary parasitoids of other insect groups. Hence, aphids are given special coverage in this review.

Taxonomy

Not all of the superfamilies and families listed above in the general taxonomic survey include species of aphid hyperparasitoids. Instead, aphid hyperparasitism is restricted to three hymenopteran superfamilies (52, 87, 119): Chalcidoidea [Pteromalidae: *Asaphes, Pachyneuron, Coruna* (44, 68, 108, 111); Encyrtidae: *Aphidencyrtus* (60–62); and Eulophidae: *Tetrastichus* (8, 90)], Ceraphronoidea [Megaspilidae: *Dendrocerus* (19–21, 30, 113, 121, 142)], and Cynipoidea [Cynipidae (subfamily Alloxystinae): *Alloxysta, Phaenoglyphis,* and *Lytoxysta* (1, 25–29, 100)]. Within some of these five families and nine genera, there are also species that are not aphid hyperparasitoids.

Primary Parasitoid Development

To understand the behavior of aphid hyperparasitoids, a knowledge of the development of the primary parasitoids of aphids is necessary. The latter are classified both taxonomically and behaviorally into only two families, the Aphidiidae (Ichneumonoidea) and the Aphelinidae (Chalcidoidea) (52, 80, 112). A well-studied species is the aphidiid wasp *Aphidius smithi* introduced into North America to control the pea aphid, *Acyrthosiphon pisum,* an exotic pest. The female wasp oviposits into the aphid, and over a period of approximately 8 days the parasitic larva gradually devours the aphid internally and kills it. The fourth instar larva spins a cocoon inside the dead aphid, whose exoskeleton becomes hard and changes color from green to light brown (this is referred to as a "mummy"). The larva then pupates, and approximately 4 days later (or about 12 days after the original oviposition), the new adult primary parasitoid cuts a circular emergence hole in the dorsum of the mummy and pulls itself out.

Hyperparasitoid Development

Sullivan (115–117), Matejko & Sullivan (83), and others (21, 112) divide aphid hyperparasitoids into two categories based on adult ovipositional and larval feeding behaviors. (*a*) The female wasp of endophagous species deposits her egg inside the primary parasitoid larva while it is still developing inside the live aphid, before the aphid is mummified. The egg does not hatch until after the mummy is formed, and then the hyperparasitic larva feeds internally on the primary larval host. (*b*) The female wasp of ectophagous species deposits her egg on the surface of the primary parasitoid larva after the aphid is killed and mummified. Then the hyperparasitic larva feeds externally on the primary larval host while both are still within the mummy. Based on these behavioral criteria, the nine genera listed taxonomically above can be arranged as follows: Endophagous hyperparasitoid species in the genera *Alloxysta, Phaenoglyphis, Lytoxysta,* and *Tetrastichus;* Ectophagous hyperparasitoids in the genera *Asaphes, Dendrocerus, Pachyneuron,* and *Coruna;* and *Aphidencyrtus,* a special case in which the larva is essentially endoparasitic but the adult can manifest either ovipositional behavior.

Comparative Attack and Ovipositional Behavior

The behaviors of representative species from four genera of hyperparasitoids are described.

ALLOXYSTA   *Alloxysta* (=*Charips*) *victrix,* in the family Cynipidae (subfamily Alloxystinae), is an example of an endophagous hyperparasitoid (9, 50, 82, 115–117). The female approaches a live, already parasitized aphid and rapidly antennates its surface. She mounts the dorsum of the aphid and assumes a squatting position with her abdomen slightly bent (Figure 1*a*). The female then inserts her ovipositor through the thin exoskeleton of the aphid and deposits her egg inside the primary larva, which is still feeding. The *Alloxysta* egg does not hatch until after the primary larva has completely devoured the aphid internally and killed it in the usual manner as if it had not been hyperparasitized. Only after the dead aphid is mummified does the hyperparasitoid larva hatch from the egg within the primary parasitoid larva. The secondary larva feeds endophagously until it kills and completely consumes the primary larva. Then it metamorphoses into a pupa and emerges from the mummy as an adult approximately 19 days after the original oviposition.

ASAPHES   The first example of an ectophagous hyperparasitoid is *Asaphes californicus* or *A. lucens,* species with similar behavior, in the family Pteromalidae (7, 70, 77, 115–117). The ovipositional behavior differs from that of *Alloxysta* in that the primary larva is not attacked until after the aphid is killed and the mummy is formed. The female mounts the mummy, drills a hole with her ovipositor (Figure 1*b*), and injects a venom into the primary parasitoid larva developing inside the mummy; this results in paralysis and

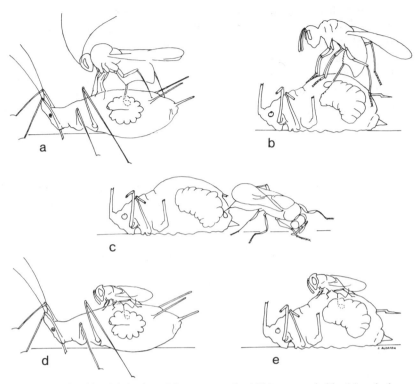

Figure 1   Ovipositional behavior of four genera of aphid hyperparasitoids: (*a*) endophagous
Alloxysta victrix; (*b*) ectophagous *Asaphes lucens;* (*c*) ectophagous *Dendrocerus carpenteri;* (*d*)
endophagous *Aphidencyrtus aphidivorus* ovipositing in primary parasitoid larva inside live aphid;
and (*e*) *A. aphidivorus* ovipositing in primary parasitoid larva inside dead mummy. From Sullivan
(117) with permission from Elsevier Science Publishers. See text for description.

termination of its development (7). An egg is then laid on the surface of the
primary larva, which gradually deteriorates into a soft, blackened mass. In
spite of the decay of the host, the newly hatched *Asaphes* larva continues to
feed ectophagously. The primary parasitoid is devoured, and after metamorphosis an *Asaphes* adult emerges approximately 21 days after egg deposition.

DENDROCERUS   The species of *Dendrocerus,* family Megaspilidae, are also
ectophagous hyperparasitoids, but the ovipositional behavior and venom are
quite different from those of *Asaphes* (5, 83, 116, 117). Instead of mounting
the top of the mummy, the female *Dendrocerus carpenteri* turns around 180°,
backs into the side or rear of the mummy, and drills a hole (Figure 1*c*). An
egg is deposited on the surface of the primary larva within the mummy.
Venom is injected but, whereas the venom of *Asaphes* caused blackening and

decay, the primary larva retains its bright yellow color during feeding by the *Dendrocerus* larva, as clearly shown in the color plate of Bocchino & Sullivan (7). The *Dendrocerus* larva also becomes yellow. Development from egg to adult takes approximately 16 days.

APHIDENCYRTUS   *Aphidencyrtus aphidivorus* in the family Encyrtidae is a special case; it has "dual" ovipositional behavior (69, 116, 117). Although the larva is endophagous, the female hyperparasitoid can attack the primary parasitoid larva either while the aphid is still alive (Figure 1*d*), in the manner of *Alloxysta,* or after the mummy has been formed (Figure 1*e*), like *Asaphes* and *Dendrocerus.* Choice experiments indicate a preference to attack through the mummy. In both cases, however, the egg of *Aphidencyrtus* is laid inside the primary parasite larva, where it feeds endophagously.

Tertiary Hyperparasitism in Aphids

At the next higher trophic level, aphid hyperparasitoids attack each other (Figure 2). Although difficult to prove in the field, it has been demonstrated in the laboratory that both intraspecific tertiary parasitism (or autohyperparasitism) (5, 77) and interspecific tertiary hyperparasitism (or allohyperparasitism) (83, 115) can occur. Success in the competition between hyperparasitic larvae depends on the developmental age of the hyperparasitoid larva already inside the mummy at the time of oviposition by the second hyperparasitoid.

HOST SPECIFICITY

This topic has received greater attention at the level of primary parasitoids (139, 140), for it was thought that hyperparasitoids tended toward polyphagy with little host specificity. Contrary evidence from field and laboratory research, especially that on the well-studied ecosystems in which aphids are the insect pests, was reviewed by van den Bosch (132). He pointed out that feeding behavior involves a continuum, and that "host specificity" can range from monophagy to some level of oligophagy, as shown in the five aphid complexes discussed below. There is indeed some host specificity among the endophagous aphid hyperparasitoids, but much less, if any, in the ectophagous genera.

Cabbage Aphid Complex

Hafez (51) studied the seasonal population fluctuations in the relatively stable ecosystem of a brussels sprout–cabbage aphid *(Brevicoryne brassicae)* habitat in the Netherlands. There was strong host specificity and a bimodal phenological synchronization between the endophagous hyperparasitoid, *Alloxysta* (=*Charips*) *ancylocera,* and the primary parasitoid, *Diaeretiella rapae.* On

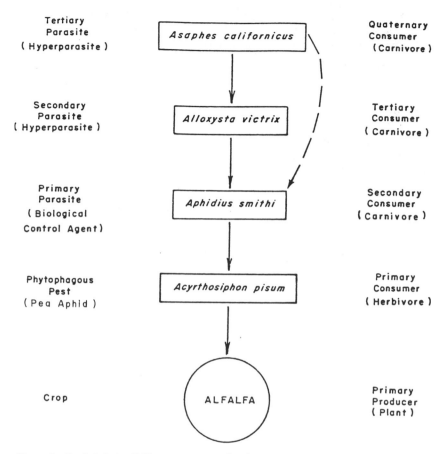

Figure 2 Food chain in alfalfa agroecosystem showing pea aphid, primary parasitoid, and two hyperparasitoids at the secondary, tertiary, and quaternary trophic levels. After van den Bosch et al (134), with permission from Plenum Press.

the other hand, neither host specificity nor phenological synchrony were found between the two ectophagous hyperparasitoids, the pteromalids *Asaphes* and *Pachyneuron,* and the primary parasitoid. Hence, the endophagous *Alloxysta* was the dominant hyperparasitoid. Evenhuis (29) and Chua (14) consider this an example of *Alloxysta*'s strong temporal, behavioral, and biological adaptation for monophagous specificity to a primary host. Chemical cues for habitat selection may also be involved (101, 136).

Two Aphids on Alfalfa

Using a different species, *Alloxysta (=Charips) victrix,* Gutierrez (45–48) with van den Bosch (49, 50) further substantiated the presence of host specificity in this endophagous genus. They studied alfalfa fields in California where two different species of aphids coexisted (Figure 3): the pea aphid *(Acyrthosiphon pisum)* and the spotted alfalfa aphid *(Therioaphis trifolii).* Not only did *Alloxysta* show a preference for potential hosts in the pea aphid over those in the spotted alfalfa aphid, but it even discriminated among nine primary parasitoids, ovipositing in *Aphidius* most frequently. In laboratory experiments Gutierrez demonstrated that *Alloxysta* could distinguish between two primary parasitoids that had parasitized the same species (the preferred pea aphids), and that only *Aphidius* was suitable for the development of the hyperparasitoid's larva (Figure 4).

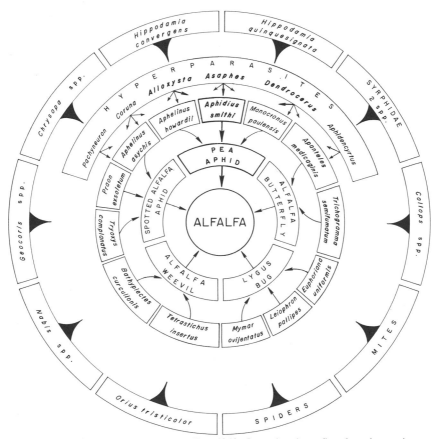

Figure 3   Diagram of food web surrounding alfalfa. Inner ring shows five phytophagous insect pests. Second ring shows entomophagous primary parasitoids and predators, followed by partial ring of hyperparasitoids that attack only aphidophagous parasitoids. In outer ring are other predaceous arthropods associated with this agroecosystem. Redrawn from van den Bosch et al (134), with permission from Plenum Press.

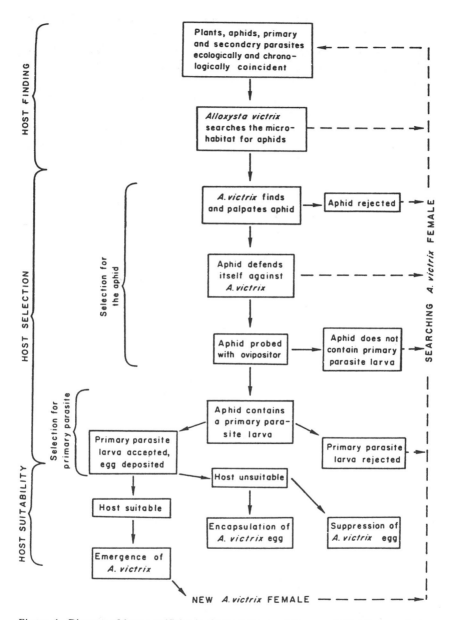

Figure 4 Diagram of host specificity in the endophagous hyperparasitoid *Alloxysta victrix*, showing the interaction of ecological and physiological factors. After the initial host-habitat finding, the sequence of behaviors includes host finding, host selection, and determination of host suitability. From van den Bosch et al (134), with permission from Plenum Press.

Three Apple-Infesting Aphids

Evenhuis (24–29) reported that three different species of endophagous hyperparasitoids had each selected specific primary parasitoid hosts that in turn were specific on each of three different aphids: *Alloxysta quateri–Binodoxys–Aphis pomi, Alloxysta sp.–Ephedrus–Dysaphis plantaginea, Phaenoglyphis–Monoctonus–Rhopalosiphum insertum*. In contrast, however, the three ectophagous hyperparasitoids present *(Dendrocerus, Asaphes,* and *Pachyneuron)* all shared the three primary parasitoid hosts. As a result of this difference in host specificity, the alloxystines were the dominant hyperparasitoids.

Potato Aphid Complex

Sullivan & van den Bosch (119) in California reported a similar pattern of host specificity in two species of the endophagous *Alloxysta* that were hyperparasitic on potato aphids *(Macrosiphum euphorbiae)*. Among six species of hyperparasitoids, *Alloxysta victrix* was the most abundant and showed the strongest preference for *Aphidius,* which in turn was the dominant primary parasitoid. Another *Alloxysta* species showed host specificity for a different primary parasitoid, *Aphelinus.* Perhaps because of this competition, the ectophagous *Asaphes* showed some host specificity by concentrating on the third primary parasitoid, *Ephedrus,* and was the most numerous of the six hyperparasitoids that attacked it. The remaining hyperparasitoids *(Pachyneuron, Dendrocerus,* and *Aphidencyrtus)* showed little host specificity and were of minor importance.

Walnut Aphid Complex

Frazer & van den Bosch (38) and van den Bosch et al (133) reported on the biological control program against the walnut aphid *(Chromaphis juglandicola)* in California, in which the imported primary parasitoid was *Trioxys pallidus.* In the Palearctic region, where both the walnut aphid and *Trioxys* are native, the dominant hyperparasitoids are the endophagous alloxystines; this follows the pattern demonstrated in the previous examples. In California, however, where the walnut aphid is an invader and the primary parasitoid was introduced, the alloxystines are insignificant in the walnut aphid ecosystem. Instead, this ecological niche is filled by another hyperparasitoid, *Aphidencyrtus aphidivorus,* which has become dominant.

FACULTATIVE HYPERPARASITISM

Some hyperparasitoids behave both as typical secondary and primary parasitoids (23, 36, 37, 88, 96, 114, 145, 146). Kfir et al (71–75) studied an interesting case of facultative hyperparasitism in the pteromalid *Pachyneuron concolor,* a cosmopolitan and polyphagous secondary parasitoid that attacks encyrtid primary parasitoids in soft scale insects, mealybugs, aphids, and coccinellid larvae. *P. concolor* can also develop as a tertiary parasitoid on members of its own species or on various other chalcidoids that have developed as secondary parasitoids on an encyrtid host (74).

Of special interest here, however, is the proof that *P. concolor* is truly a facultative hyperparasitoid, since it also develops as a primary parasitoid of fly puparia (104). Host selection appears to be based on locating a soft-bodied host within a hard, dry shell, independent of whether the host is a dipterous pupa in its puparium or a primary parasitoid in its mummified host. Unfortunately, when *P. concolor* behaves as a primary parasitoid, the dipterous host that it attacks happens to be a beneficial aphidophagous fly. Hence, *P. concolor* could be detrimental to a biological control program.

ADELPHOPARASITISM IN THE APHELINIDAE

"Adelphoparasitism" is unique to the Aphelinidae; this unusual autoparasitic behavior involves a hyperparasitic male larva attacking a primary female larva of its own species. Most Aphelinidae are primary parasitoids of Homoptera (Aphidoidea, Aleyrodoidea, and Coccoidea); Viggiani (138) reviewed the bionomics of this interesting chalcidoid family. Although some species of Aphelinidae exhibit hyperparasitism, it is not widely distributed (only 11 of 42 genera). The females of three of these genera *(Marietta, Ablerus,* and *Azotus)* attack primary parasitoids of Homoptera, and the hyperparasitic development of both sexes is quite normal. Abnormal behavior or deviant male ontogeny is associated with sex differentiation in the host relations of some species in the eight remaining genera *(Aneristus, Coccophagus, Coccophagoides, Euxanthellus, Encarsia, Lounsburyia, Physcus,* and *Prococcophagus).* The female larvae always develop as primary endoparasitoids in Homoptera (mealybugs, scale insects, and whiteflies), while the male ontogeny is deviant.

Adelphoparasitism as an example of deviant male ontogeny is better understood in light of the categories of sex differentiation in primary-parasitoid hosts (137, 144, 147, 149, 150): (*a*) In some species, the male is a primary parasitoid of the same host as the female, but the haploid egg has been laid externally (not internally as with the female) and the male larva feeds ectophagously. (*b*) In other species, the male is again a primary parasitoid, but in a different host from the female; the haploid egg has been laid internally, so the male larva feeds endophagously as does the female larva. (*c*) Finally, in some species the male larva develops as an external or internal secondary parasitoid of preimaginal stages of a primary parasitoid. This has been well studied in species of *Coccophagus* (34, 35). Behavior in which a male hyperparasitoid attacks a female larva of its own species is termed autoparasitism; in this special case of a male hyperparasitoid it is termed adelphoparasitism. Examples of adelphoparasitoids are *Coccophagoides similis* and *C. utilis*. This sex differentiation in host relations and deviant male ontogeny (especially adelphoparasitism) must be taken into consideration when using Aphelinidae in biological control programs (32). It may be necessary to release unmated females after mated females to produce males at the correct time.

FOOD WEBS AND MATHEMATICAL MODELS

In the last 20 years there has been an increasing number of field studies of phytophagous insects (usually economic pests) that include not merely primary parasitoids, but also hyperparasitoids in the food web (14, 22, 24, 36, 37, 39, 41, 49, 51, 66, 67, 87, 91–95, 97, 103, 105, 107, 109, 110, 119, 122–124, 131, 136, 142, 143, 148). Two food webs are used as examples.

Alfalfa Agroecosystem

This food web (Figure 3) is composed of a two-dimensional, multilevel ecological community. Horizontally, at the same trophic level, there is both inter- and intraspecific competition; vertically, at different trophic levels, there are several food chains (134). Alfalfa, as the central plant, is the producer at the first trophic level, with phytophagous insects such as the pea aphid and the spotted alfalfa aphid among the herbivorous consumers at the second trophic level (Figure 2). The aphid food chain continues with the carnivorous consumers or entomophagous insects at the third level, e.g. aphidophagous predators and primary parasitoids. These in turn are fed upon by secondary parasitoids or hyperparasitoids (*Alloxysta, Asaphes, Dendrocerus, Aphidencyrtus,* etc.) at the fourth trophic level. Tertiary parasitism may also exist (83), making a fifth trophic level. Yet this aphid food chain constitutes just one component of the food web in this alfalfa agroecosystem.

Cassava Mealybug Complex

Another interesting food web is associated with the cassava plant *(Manihot esculenta),* which was brought to Africa from South America by the Portuguese about 300 years ago. In 1973, the cassava mealybug, *Phenacoccus manihoti,* was discovered in Zaire as an accidental import from South America. Before the cassava mealybug arrived in Africa, the only insects commonly found in large numbers on cassava were grasshoppers (*Zonocerus* spp.).

As the mealybug spread across the African cassava belt, so did the diversity of the associated insect fauna. This became especially noticeable in 1981, when the encyrtid wasp *Epidinocarsis lopezi* was introduced from South America as a primary parasitoid of the cassava mealybug. *E. lopezi* is now the only common primary parasitoid of the cassava mealybug in Africa (91), despite defensive behavior by the mealybug involving encapsulation and melanization of the wasp's eggs and larvae; the wasp has coevolved to overcome this defense by laying extra eggs (118). About 12 other species of primary parasitoids that attack *P. manihoti* are indigenous to Africa, but most are adapted to other species of mealybugs and therefore have not been effective biological control agents of the cassava mealybug. Although exotic hyperparasitoids were carefully excluded during importation of the natural enemies from South America, four species of indigenous hyperparasitoids followed their normal primary parasitoid hosts into the cassava mealybug ecosystem and began attacking *E. lopezi* as well. In addition, the cassava complex includes several introduced and native coccinellid predators with their parasitoids and hyperparasitoids. The complex food web associated with the cassava mealybug and its introduced primary parasitoid now consists of about 110 species of insects.

Mathematical Models

There is a commonly held thesis in population ecology that increasing the complexity of the food web automatically leads to increased community stability. Based on mathematical models, however, May (84) expressed some reservations, for in some cases the reverse may be true. The history, evolution, and equations of mathematical models should be consulted (15, 40, 54–59, 85, 98, 135, 151). Many models do not explicitly include hyperparasitoids, but several groups of researchers have indeed included them. Luck et al (79) based their model on the Hassell & Varley (57) modification of the Nicholson-Bailey equation, using Q_h as the quest constant or "area of discovery" of an obligate hyperparasitoid. Hassell & Waage (58) reviewed the multispecies interactions of primary parasitoids and hyperparasitoids, and noted that there have been relatively few attempts to assess the ecological impact of obligate hyperparasitoids (16, 49, 71, 119) and to analyze this mathematically (4, 53, 54, 85).

The mathematical models indicate that although an obligate hyperparasitoid contributes to the complexity of the food web and may add stability, it also raises the host's equilibrium and increases pest abundance. Hence, obligate hyperparasitoids are usually detrimental to biological control programs.

IMPACT ON BIOLOGICAL CONTROL

Although hyperparasitoids have traditionally been considered harmful to the beneficial primary parasitoids of insect pests, obligate hyperparasitoids should be distinguished from facultative, and exotic from indigenous.

Obligate Hyperparasitoids

It has been the policy in biological control projects to exclude exotic hyper-parasitoids when introducing foreign natural enemies. Bennett (6) discussed the problems inherent in trying to exclude exotic, obligate hyperparasitoids at each of the standard steps (64, 134) in importation for biological control. Once introduced and colonized, however, the primary parasitoid faces the danger of indigenous hyperparasitoids in its new habitat (12, 106). After the primary parasitoid has been introduced, there is a gradual transition from the colonizing stage to permanent establishment. The role of the hyperparasitoids also changes, so that sometimes what was expected to be a major threat by secondary parasitoids has not in fact impeded the success of the primary ones. Moreover, the diversity of species as well as the numbers of hyperparasitoids can vary greatly from place to place and from year to year, making their long-term influence on the primary parasitoids difficult to predict. Although there is a need for more field data and mathematical modeling, Bennett (6), Luck et al (79), Hassell & Waage (58, 141), and other researchers have concurred that exotic obligate hyperparasitoids should continue to be ex-cluded during the foreign importation part of a biological control program. Little can be done about indigenous hyperparasitoids, since they antedate the exotic primary parasitoid and move easily into the food web.

Facultative Hyperparasitoids

Bennett (6) and others (23, 146) also considered the more complex situation of importing facultative hyperparasitoids (which behave both as primary and also as secondary parasitoids). Species are evaluated according to whether the hyperparasitic behavior is predominant or only occasional. Species that are predominantly hyperparasitic are treated as if they were obligatory hyperpar-asitoids and hence are excluded. The problem is with those species that are usually primary but only occasionally secondary parasitoids. In a serious pest situation where there are no normal primary parasitoids available for biologi-cal control, perhaps a calculated risk should be taken as a last recourse.

SUMMARY AND CONCLUSIONS

Based on this limited review, a "laundry list" for future research projects might include the following. More research is needed on the components of the behavior of hyperparasitoids that result in host specificity (host-habitat finding, host location, host acceptance, and determination of host suitability), in order to see how widespread oligophagy is among endophagous and perhaps even ectophagous hyperparasitoids. Related to this are morphological and physiological studies on chemoreception of pheromones and kairomones.

Host suitability also involves defense mechanisms on the part of the host, such as encapsulation and melanization; yet reports of primary parasitoids using this phenomenon against hyperparasitoids are rare. There is also a dearth of ethological studies on courtship and mating behavior of hyperparasitoids that might lead to control measures. Comparative biochemical analyses of the venoms of endophagous and ectophagous hyperparasitoids could reveal useful components.

Finally, the consensus remains that exotic obligate hyperparasitoids should be excluded as part of biological control programs. Whether or not exotic facultative hyperparasitoids should be imported must be evaluated separately for each candidate species depending on the availability of conventional natural enemies and the seriousness of the insect pest problem. Indigenous hyperparasitoids already form part of the existing food web and may or may not interfere significantly with exotic primary parasitoids introduced in a biological control program. Realistically, they cannot be eliminated from the ecosystem, but monitoring of their impact on the primary parasitoids is important and must continue. Indigenous hyperparasitoids add complexity to the food web, but more field studies to complement mathematical modeling are needed to analyze their impact on stability. If the influence is positive, and the extreme oscillations of primary parasitoids are dampened, then some insect hyperparasitoids might even be considered beneficial.

ACKNOWLEDGMENTS

The author was supported by a US Government Fulbright Research Fellowship (1984–1985) at the International Institute of Tropical Agriculture (IITA) in Ibadan, Nigeria, where the laboratories of the "Africa-wide Biological Control Programme" were used. Thanks are due to Dr. Julia Alzofon for her illustrations (Figure 1) and redrawings (Figures 2–4), and to Dr. Gerard Iwantsch for his critical review of the manuscript. Mr. William Carew kindly assisted in the proofreading.

Literature Cited

1. Andrews, F. G. 1978. Taxonomy and host specificity of Nearctic Alloxystinae with a catalog of the world species (Hymenoptera: Cynipidae). *State Calif. Dep. Food Agric. Occas. Pap. Entomol.* 25:1–128
2. Askew, R. R. 1961. On the biology of the inhabitants of oak galls of Cynipidae (Hymenoptera) in Britain. *Trans. R. Entomol. Soc. London* 14:237–68
3. Askew, R. R. 1971. *Parasitic Insects.* London: Heinemann. 316 pp.
4. Beddington, J. R., Hammond, P. S. 1977. On the dynamics of host-parasite-hyperparasite interactions. *J. Anim. Ecol.* 46:811–21
5. Bennett, A. W., Sullivan, D. J. 1978. Defensive behavior against tertiary parasitism by the larva of *Dendrocerus carpenteri*, an aphid hyperparasitoid. *J. NY Entomol. Soc.* 86:153–60
6. Bennett, F. D. 1981. Hyperparasitism in the practice of biological control. In *The Role of Hyperparasitism in Biological Control: A Symposium. Priced Publ. 4103*, ed. D. Rosen, pp. 43–49. Berkeley, Calif: *Div. Agric. Sci. Univ. Calif.*

7. Bocchino, F. J., Sullivan, D. J. 1981. Effects of venoms from two aphid hyperparasitoids, *Asaphes lucens* and *Dendrocerus carpenteri* (Hymenoptera: Pteromalidae and Megaspilidae), on larvae of *Aphidius smithi* (Hymenoptera: Aphidiidae). *Can. Entomol.* 113:887–89

8. Bouček, Z., Askew, R. R. 1968. *Index of Entomophagous Insects. Palearctic Eulophidae (exc. Tetrastichinae) (Hym. Chalcidoidea).* Paris: Le Francois. 260 pp.

9. Broussal, G. 1964. Comparaison des fécondités de *Charips ancylocera* (Hymenoptère: Cynipidae), hyperparasite et d'*Aphidius brassicae* (Hymenoptère: Aphidiidae), parasite primaire de *Brevicoryne brassicae* (Homoptère: Aphididae). *Ann. Univ. Reims Assoc. Rég. Étud. Réch. Sci. (ARERS)* 2:135–37

10. Brues, C. T. 1921. Correlation of taxonomic affinities with food habits in Hymenoptera with special reference to parasitism. *Am. Nat.* 55:134–64, 636–38

11. Burks, B. D. 1979. Family Pteromalidae. In *Catalog of Hymenoptera in America North of Mexico,* Vol. 1, ed. K. V. Krombein, P. D. Hurd, D. R. Smith, B. D. Burks, pp. 768–835. Washington, DC: Smithsonian Inst. 1198 pp.

12. Burton, R. L., Starks, K. L. 1977. Control of a primary parasite of the greenbug with a secondary parasite in greenhouse screening for plant resistance. *J. Econ. Entomol.* 70:219–20

13. Chambers, V. H. 1955. Some hosts of *Anteon* spp. (Hym., Dryinidae) and a hyperparasite *Ismarus* (Hym., Belytidae). *Entomol. Mon. Mag.* 91:114–15

14. Chua, T. H. 1977. Population studies of *Brevicoryne brassicae,* its parasites and hyperparasites in England. *Res. Popul. Ecol.* 19:125–39

15. Chua, T. H. 1978. A model of an aphid-parasite-hyperparasite system, with reference to timing of attack. *J. Malays. Agric.* 51:375–86

16. Chua, T. H. 1979. A comparative study of the searching efficiencies of a parasite and a hyperparasite. *Res. Popul. Ecol.* 20:178–87

17. Cooper, K. W. 1954. Biology of eumenine wasps. IV. A trigonalid wasp parasitic on *Rygchium rugosum* (Saussure). (Hymenoptera: Trigonalidae). *Proc. Entomol. Soc. Wash.* 56:280–88

18. DeBach, P., ed. 1964. *Biological Control of Insect Pests and Weeds.* London: Chapman & Hall. 844 pp.

19. Dessart, P. 1972. Révision des éspèces Européennes du genre *Dendrocerus* Ratzeburg, 1852 (Hymenoptera: Ceraphronoidea). *Mém. Soc. R. Belge Entomol.* 32:1–312

20. Dessart, P. 1974. Compléments à l'étude des *Dendrocerus* Européens (Hymenoptera: Ceraphronoidea, Megaspilidae). *Bull. Ann. Soc. R. Belge Entomol.* 110:69–84

21. Dessart, P. 1985. A propos des Hyménoptères parasites. *Nat. Belg.* 66:97–120

22. des Vignes, W. G. 1977. Seasonal distribution of *Diatraea* spp., their parasites and hyperparasites on sugar cane and grasses. *Caroni Research Station, Annual Report,* 1977:234–37. Carapichaima, Trinidad: Caroni Res. Stn.

22a. Du Merle, P. 1975. Les Hôtes et les stades pré-imaginaux des diptères Bombyliidae: Révue bibliographique annottée. *Bull. Sect. Rég. Ouest Paléarct. (SROP)* 289 pp.

23. Ehler, L. E. 1979. Utility of facultative secondary parasites in biological control. *Environ. Entomol.* 8:829–32

24. Evenhuis, H. H. 1964. The interrelations between apple aphids and their parasites and hyperparasites. *Entomophaga* 9:227–31

25. Evenhuis, H. H. 1971. Studies on Cynipidae Alloxystinae. 1. The identity of *Alloxysta rubriceps* (Kieffer, 1902), with some general remarks on the subfamily. *Entomol. Ber. Amsterdam* 31: 93–100

26. Evenhuis, H. H. 1972. Studies on Cynipidae Alloxystinae. 2. The identity of some species associated with aphids of economic importance. *Entomol. Ber.* 32:210–17

27. Evenhuis, H. H. 1973. Studies on Cynipidae Alloxystinae. 3. The identity of *Phaenoglyphis ruficornis* (Foster, 1869). *Entomol. Ber.* 33:218–19

28. Evenhuis, H. H. 1974. Studies on Cynipidae Alloxystinae. 4. *Alloxysta macrophadna* (Hartig, 1841) and *Alloxysta brassicae* (Ashmead, 1887). *Entomol. Ber.* 34:165–68

29. Evenhuis, H. H. 1976. Studies on Cynipidae Alloxystinae. 5. *Alloxysta citripes* (Thompson) and *Alloxysta ligustri* n. sp., with remarks on host specificity in the subfamily. *Entomol. Ber.* 36:140–44

30. Fergusson, N. D. M. 1980. A revision of the British species of *Dendrocerus* Ratzeburg (Hymenoptera: Ceraphronoidea) with a review of their biology as aphid hyperparasites. *Bull. Br. Mus. Nat. Hist. Entomol. Ser.* 41:255–314

31. Flanders, S. E. 1943. Indirect hyper-

parasitism and observations on three species of indirect hyperparasites. *J. Econ. Entomol.* 36:921–26

32. Flanders, S. E. 1959. Differential host relations of the sexes in the parasitic Hymenoptera. *Entomol. Exp. Appl.* 2: 125–42

33. Flanders, S. E. 1963. Hyperparasitism, a mutualistic phenomenon. *Can. Entomol.* 95:716–20

34. Flanders, S. E. 1964. Dual ontogeny of the male *Coccophagus gurneyi* Comp. (Hymenoptera: Aphelinidae): a phenotypic phenomenon. *Nature* 204:944–46

35. Flanders, S. E. 1967. Deviate ontogenies in the aphelinid male (Hym.) associated with the ovipositional behavior of the parental female. *Entomophaga* 12: 415–27

36. Force, D. C. 1970. Competition among four hymenopterous parasites of an endemic host. *Ann. Entomol. Soc. Am.* 63:1675–88

37. Force, D. C. 1974. Ecology of insect host-parasitoid communities. *Science* 184:624–32

38. Frazer, B. D., van den Bosch, R. 1973. Biological control of the walnut aphid in California: the interrelationship of the aphid and its parasite. *Environ. Entomol.* 2:561–68

39. Gambino, P., Sullivan, D. J. 1982. Phenology of emergence of the spotted tentiform leafminer, *Phyllonorycter crataegella* (Lepidoptera: Gracillariidae), and its parasitoids in New York. *J. NY Entomol. Soc.* 90:229–36

40. Getz, W. M., Gutierrez, A. P. 1982. A perspective on systems analysis in crop production and insect pest management. *Ann. Rev. Entomol.* 27:447–66

41. Gonzales, D., Mizakoki, M., White, W., Takada, H., Dickson, R., Hall, J. 1979. Geographical distribution of *Acyrthosiphon kondoi* Shinji (Homoptera: Aphididae) and some of its parasites and hyperparasites in Japan. *Kontyu* 47:1–7

42. Gordh, G. 1979. Family Encyrtidae. See Ref. 11, pp. 890–967

43. Gordh, G. 1981. The phenomenon of insect hyperparasitism and its taxonomic occurrence in the Insecta. See Ref. 6, pp. 10–18

44. Graham, M. W. R. de V. 1969. The Pteromalidae of north-western Europe (Hymenoptera: Chalcidoidea). *Bull. Br. Mus. Nat. Hist. Entomol. Suppl.* 16:1–908

45. Gutierrez, A. P. 1970. Studies on host selection and host specificity of the aphid hyperparasite *Charips victrix* (Hymenoptera: Cynipidae). 3. Host suitability studies. *Ann. Entomol. Soc. Am.* 63:1485–91

46. Gutierrez, A. P. 1970. Studies on host selection and host specificity of the aphid hyperparasite *Charips victrix* (Hymenoptera: Cynipidae). 4. The effect of age of host on host selection. *Ann. Entomol. Soc. Am.* 63:1491–94

47. Gutierrez, A. P. 1970. Studies on host selection and host specificity of the aphid hyperparasite *Charips victrix* (Hymenoptera: Cynipidae). 5. Host selection. *Ann. Entomol. Soc. Am.* 63: 1495–98

48. Gutierrez, A. P. 1970. Studies on host selection and host specificity of the aphid hyperparasite *Charips victrix* (Hymenoptera: Cynipidae). 6. Description of sensory structures and a synopsis of host selection and host specificity. *Ann. Entomol. Soc. Am.* 63:1705–9

49. Gutierrez, A. P., van den Bosch, R. 1970. Studies on host selection and host specificity of the aphid hyperparasite *Charips victrix* (Hymenoptera: Cynipidae). 1. Review of hyperparasitism and the field ecology of *Charips victrix*. *Ann. Entomol. Soc. Am.* 63:1345–54

50. Gutierrez, A. P., van den Bosch, R. 1970. Studies on host selection and host specificity of the aphid hyperparasite *Charips victrix* (Hymenoptera: Cynipidae). 2. The bionomics of *Charips victrix*. *Ann. Entomol. Soc. Am.* 63:1355–60

51. Hafez, M. 1961. Seasonal fluctuations of population density of the cabbage aphid, *Brevicoryne brassicae* (L.) in the Netherlands, and the role of its parasite *Aphidius (Diaeretiella) rapae* (Curtis). *Tijdschr. Plantenziekten* 67:445–548

52. Hagen, K. S., van den Bosch, R. 1968. Impact of pathogens, parasites, and predators on aphids. *Ann. Rev. Entomol.* 13:325–84

53. Hassell, M. P. 1978. *The Dynamics of Arthropod Predator-Prey Systems.* Princeton, NJ: Princeton Univ. Press. 237 pp.

54. Hassell, M. P. 1979. The dynamics of predator prey interactions, polyphagous predators, competing predators and hyperparasitoids. *Br. Ecol. Soc. Symp.* 20:283–306

55. Hassell, M. P., May, R. M. 1973. Stability in insect host-parasite models. *J. Anim. Ecol.* 42:693–726

56. Hassell, M. P., May, R. M. 1974. Aggregation of predators and insect

parasites and its effect on stability. *J. Anim. Ecol.* 43:567–94

57. Hassell, M. P., Varley, G. C. 1969. New inductive population model for insect parasites and its bearing on biological control. *Nature* 223:1133–37

58. Hassell, M. P., Waage, J. K. 1984. Host-parasitoid population interactions. *Ann. Rev. Entomol.* 29:89–114

59. Hassell, M. P., Waage, J. K., May, R. M. 1983. Variable parasitoid sex ratios and their effect on host parasitoid dynamics. *J. Anim. Ecol.* 52:889–904

60. Hoffer, A. 1970. Erster Beitrag zur Taxonomie der Palaearktischen Arten der Gattung *Aphidencyrtus* Ashm. (Hymenoptera: Chalcidoidea, Encyrtidae). *Stud. Entomol. For.* 1:25–42

61. Hoffer, A. 1970. Zweiter Beitrag zur Taxonomie der Palaearktischen Arten der Gattung *Aphidencyrtus* Ashm. (Hymenoptera: Chalcidoidea, Encyrtidae). *Stud. Entomol. For.* 1:65–80

62. Hoffer, A., Starý, P. 1970. Zur Biologie der Palaearktischen Arten der Gattung *Aphidencyrtus* Ashm. (Hymenoptera: Chalcidoidea, Encyrtidae). *Stud. Entomol. For.* 1:81–95

63. Huffaker, C. B., ed. 1980. *New Technology of Pest Control.* New York: Wiley-Interscience. 500 pp.

64. Huffaker, C. B., Messenger, P. S., eds. 1976. *Theory and Practice of Biological Control.* New York: Academic. 788 pp.

65. Huffaker, C. B., Rabb, R. L., eds. 1984. *Ecological Entomology.* New York: Wiley-Interscience. 844 pp.

66. Hughes, G. Hammond, P. S., des Vignes, W. G. 1982. Population cycles of the small moth-borers of sugar cane, *Diatraea* spp., and their primary and secondary parasitoids, in Trinidad, West Indies. *Agro-Ecosyst.* 8:13–25

67. Humble, L. M. 1985. Final-instar larvae of native pupal parasites and hyperparasites of *Operophtera* spp. (Lepidoptera: Geometridae) on southern Vancouver Island. *Can. Entomol.* 117:525–34

68. Kamijo, K., Takada, H. 1973. Studies on aphid hyperparasites of Japan. II. Aphid hyperparasites of the Pteromalidae occurring in Japan (Hymenoptera). *Insecta Matsummurana* 2:39–76

69. Kanuck, M. 1981. *The biology and host preference behavior of* Aphidencyrtus aphidivorus *(Mayr), an aphid hyperparasitoid (Hymenoptera: Encyrtidae).* PhD dissertation. Fordham Univ., New York, NY. 146 pp.

70. Keller, L. J., Sullivan, D. J. 1976. Oviposition behavior and host feeding of *Asaphes lucens,* an aphid hyperparasi-toid (Hymenoptera: Pteromalidae). *J. NY Entomol. Soc.* 84:206–11

71. Kfir, R., Podoler, H., Rosen, D. 1976. The area of discovery and searching strategy of a primary parasite and two hyperparasites. *Ecol. Entomol.* 1:287–95

72. Kfir, R., Rosen, D. 1981. Biology of the hyperparasite *Cheiloneurus paralia* (Walker) (Hymenoptera: Encyrtidae) reared on *Microterys flavus* (Howard) in brown soft scale. *J. Entomol. Soc. South. Afr.* 44:131–39

73. Kfir, R., Rosen, D. 1981. Biology of the hyperparasite *Marietta javensis* (Howard) (Hymenoptera: Aphelinidae) reared on *Microterys flavus* (Howard) in brown soft scale. *J. Entomol. Soc. South. Afr.* 44:141–50

74. Kfir, R., Rosen, D. 1981. Biology of the hyperparasite *Pachyneuron concolor* (Forster) (Hymenoptera: Pteromalidae) reared on *Microterys flavus* (Howard) in brown soft scale. *J. Entomol. Soc. South. Afr.* 44:151–63

75. Kfir, R., Rosen, D., Podoler, H. 1983. Laboratory studies of competition among three species of hymenopterous hyperparasites. *Entomol. Exp. Appl.* 33:320–28

76. Kozlov, M. A. 1971. Proctotrupoids (Hymenoptera: Proctotrupoidea) of the USSR. *Tr. Vses. Entomol. Ova.* 54:3–67 (In Russian)

77. Levine, L., Sullivan, D. J. 1983. Intraspecific tertiary parasitoidism in *Asaphes lucens* (Hymenoptera: Pteromalidae), an aphid hyperparasitoid. *Can. Entomol.* 115:1653–58

78. Liss, W. J., Gut, L. J., Westigard, P. H., Warren, C. E. 1986. Perspectives on arthropod community structure, organization, and development in agricultural crops. *Ann. Rev. Entomol.* 31:455–78

79. Luck, R., Messenger, P. S., Barbieri, J. F. 1981. The influence of hyperparasitism on the performance of biological control agents. See Ref. 6, pp. 34–42

80. Mackauer, M., Finlayson, T. 1967. The hymenopterous parasites (Hymenoptera: Aphidiidae et Aphelinidae) of the pea aphid in eastern North America. *Can. Entomol.* 99:1051–82

81. Malyshev, S. I. 1966. *Genesis of the Hymenoptera and the Phases of Their Evolution.* London: Methuen. 319 pp.

82. Matejko, I., Sullivan, D. J. 1979. Bionomics and behavior of *Alloxysta megourae,* an aphid hyperparasitoid (Hymenoptera: Cynipidae). *J. NY Entomol. Soc.* 87:275–82

83. Matejko, I., Sullivan, D. J. 1984. In-

terspecific tertiary parasitoidism between two aphid hyperparasitoids: *Dendrocerus carpenteri* and *Alloxysta megourae* (Hymenoptera: Megaspilidae and Cynipidae). *J. Wash. Acad. Sci.* 74:31–38

84. May, R. M. 1973. *Stability and Complexity in Model Ecosystems*. Princeton, NJ: Princeton Univ. Press. 235 pp.

85. May, R. M., Hassell, M. P. 1981. The dynamics of multiparasitoid-host interactions. *Am. Nat.* 117:234–61

86. Deleted in proof

87. Mertins, J. W. 1985. Hyperparasitoids from pea aphid mummies, *Acyrthosiphon pisum* (Homoptera: Aphididae), in North America. *Ann. Entomol. Soc. Am.* 78:186–97

88. Muesebeck, C. F. W. 1931. *Monodontomerus aereus* Walker, both a primary and secondary parasite of the brown-tail moth and the gypsy moth. *J. Agric. Res.* 43:445–60

89. Muesebeck, C. F. W., Dohanian, S. M. 1927. A study in hyperparasitism, with particular reference to the parasites of *Apanteles melanoscelus* (Ratzeburg). *US Dep. Agric. Bull.* 1487:1–36

90. Nealis, V. G. 1983. *Tetrastichus galactopus* (Hym.: Eulophidae) a hyperparasite of *Apanteles rubecula* and *Apanteles glomeratus* (Hym.: Braconidae) in North America. *J. Entomol. Soc. BC* 80:25–28

91. Neuenschwander, P., Hennessey, R. D., Herren, H. R. 1985. Food web of insects associated with the cassava mealybug. In *IITA Annual Report for 1984*, pp. 130–33. Ibadan, Nigeria: Int. Inst. Trop. Agric. 220 pp.

92. Oatman, E. R. 1973. Parasitization of natural enemies attacking the cabbage aphid on cabbage in southern California. *Environ. Entomol.* 2:365–67

93. Paetzold, D., Vater, G. 1967. Populationsdynamische Untersuchungen an den Parasiten und Hyperparasiten von *Brevicoryne brassicae* (L.) (Homoptera: Aphididae). *Acta Entomol. Bohemoslov.* 64:83–90

94. Paetzold, D., Vater, G. 1968. Zur Teratologie der Primär-Hyperparasiten von *Brevicoryne brassicae*. *Dtsch. Entomol. Z.* 15:409–26

95. Paetzold, D., Vater, G. 1969. Untersuchungen zum Einfluss der Hyperparasiten auf die Populationsdynamik von *Diaeretiella rapae* (McIntosh) (Hymenoptera: Aphididae). *Ber. 10te Wanderversamml. Dtsch. Entomol., Dresden, 1965* 80:365–75. Berlin, GDR: Dtsch. Akad. Landwirtschaftswiss.

96. Patnaik, N. C., Satpathy, J. M. 1984. Facultative hyperparasitism/predation on *Platygaster oryzae*, an egg-larval parasite of the rice gall midge, *Orseolia oryzae*. *J. Entomol. Res. New Delhi* 8:106–8

97. Polgar, L. 1984. Some new records of parasites, predators and hyperparasites of aphids and a leaf-miner fly, *Cerodonta incisa* (Diptera: Antomyiidae), living in maize ecosystems in Hungary. *Folia Entomol. Hung.* 45:191–94

98. Price, P. W. 1984. *Insect Ecology*. New York: Wiley-Interscience. 607 pp. 2nd ed.

99. Price, P. W., Bouton, C. E., Gross, P., McPheron, B. A., Thompson, J. N., Weis, A. E. 1980. Interactions among three trophic levels: influence of plants on interactions between insect herbivores and natural enemies. *Ann. Rev. Ecol. Syst.* 11:41–65

100. Quinlan, J., Evenhuis, H. H. 1980. Status of the subfamily names Charipinae and Alloxystinae (Hymenoptera: Cynipidae). *Syst. Entomol.* 5:427–30

101. Read, D. P., Feeny, P. P., Root, R. B. 1970. Habitat selection by the aphid parasite *Diaeretiella rapae* (Hymenoptera: Braconidae) and the hyperparasite *Charips brassicae* (Hymenoptera: Cynipidae). *Can. Entomol.* 102:1567–78

102. Rettenmeyer, C. W. 1961. Observations on the biology and taxonomy of flies found over swarm raids of army ants (Diptera: Tachinidae, Conopidae). *Univ. Kans. Sci. Bull.* 52:993–1066

103. Rosen, D. 1967. The hymenopterous parasites and hyperparasites of aphids on citrus in Israel. *Ann. Entomol. Soc. Am.* 60:394–99

104. Rosen, D., Kfir, R. 1983. A hyperparasite of coccids develops as a primary parasite of fly puparia. *Entomophaga* 28:83–88

105. Santas, L. A. 1979. Distribution of aphids on citrus and cotton and their parasites in Greece. *Biol. Gallo-Hell.* 9:315–19

106. Schlinger, E. I. 1960. Diapause and secondary parasites nullify the effectiveness of rose-aphid parasites in Riverside, California, 1957–1958. *J. Econ. Entomol.* 53:151–54

107. Sedlag, U. 1964. Zur Biologie und Bedeutung von *Diaeretiella rapae* (McIntosh) als Parasit der Kohlblattlaus (*Brevicoryne brassicae* L.). *Nachrichtenbl. Dtsch. Pflanzenschutzdienst Berlin* 18:81–86

108. Sekhar, P. S. 1958. Studies on *Asaphes fletcheri* (Crawford), a hyperparasite of *Aphidius testaceipes* (Cresson) and

Praon aguti (Smith), primary parasites of aphids. *Ann. Entomol. Soc. Am.* 51:1–7

109. Shands, W. A., Simpson, G. W., Muesebeck, C. F. W., Wave, H. E. 1965. Parasites of potato-infesting aphids in northeastern Maine. *Maine Agric. Exp. Stn. Tech. Bull.* T19:1–77

110. Soteres, K. M., Berberet, R. C., McNew, R. W. 1984. Parasitic insects associated with lepidopterous herbivores on alfalfa in Oklahoma. *Environ. Entomol.* 13:787–93

111. Specht, H. B. 1969. Hyperparasitism of the pea aphid parasite *Aphelinus semiflavus* by *Asaphes vulgaris* in a greenhouse. *Ann. Entomol. Soc. Am.* 62:1207

112. Starý, P. 1970. *Biology of Aphid Parasites (Hymenoptera: Aphidiidae) with Respect to Integrated Control.* The Hague: Junk. 643 pp.

113. Starý, P. 1977. *Dendrocerus* hyperparasites of aphids in Czechoslovakia (Hymenoptera: Ceraphronidae). *Acta Entomol. Bohemoslov.* 74:1–9

114. Strand, M. R., Vinson, S. B. 1984. Facultative hyperparasitism by the egg parasitoid *Trichogramma pretiosum* (Hym.: Trichogrammatidae). *Ann. Entomol. Soc. Am.* 77:679–86

115. Sullivan, D. J. 1972. Comparative behavior and competition between two aphid hyperparasites: *Alloxysta victrix* and *Asaphes californicus* (Hymenoptera: Cynipidae; Pteromalidae). *Environ. Entomol.* 1:234–44

116. Sullivan, D. J. 1986. Aphid hyperparasites: taxonomy and ovipositional behavior. In *Ecology of Aphidophaga,* ed. I. Hodek, pp. 511–18. Prague: Academia

117. Sullivan, D. J. 1987. Aphid hyperparasites. In *Aphids, Their Biology, Natural Enemies and Control,* ed. P. Harrewijn, A. K. Minks. Amsterdam: Elsevier. In press

118. Sullivan, D. J., Neuenschwander, P. 1985. Melanization: the mealybug defends itself. See Ref. 91, pp. 127–29

119. Sullivan, D. J., van den Bosch, R. 1971. Field ecology of the primary parasites and hyperparasites of the potato aphid, *Macrosiphum euphorbiae,* in the East San Francisco Bay Area (Homoptera: Aphididae). *Ann. Entomol. Soc. Am.* 64:389–94

120. Tachikawa, T. 1974. Hosts of the Encyrtidae (Hymenoptera: Chalcidoidea). *Mem. Coll. Agric. Ehime Univ.* 19:185–204

121. Takada, H. 1973. Studies on aphid hyperparasites of Japan. I. Aphid hyperparasites of the genus *Dendrocerus* Ratzeburg occurring in Japan (Hymenoptera: Ceraphronidae). *Insecta Matsumurana* 2:1–37

122. Takada, H. 1976. Studies of aphids and their parasites on cruciferous crops and potatoes. I. Parasite complex of aphids. *Kontyu* 44:234–53 (In Japanese with English summary)

123. Takada, H. 1976. Studies of aphids and their parasites on cruciferous crops and potatoes. II. Life-cycle. *Kontyu* 44:366–84 (In Japanese with English summary)

124. Takada, H., Takenaka, Y. 1982. Parasite complex of *Myzus persicae* on tobacco (Japan). *Kontyu* 50:556–68

125. Tanton, M. T., Epila, J. S. O. 1984. Description of the planidium of *Perilampus tasmanicus* (Hymen.: Chalcidoidea), a hyperparasitoid of larvae of *Paropsis atomaria* (Coleop.: Chrysomelidae). *J. Aust. Entomol. Soc.* 23:149–52

126. Telenga, N. A. 1952. *Origin and Evolution of Parasitism in Hymenoptera Parasitica and Development of Their Fauna in the USSR.* Acad. Sci. Ukr. SSR, Inst. Entomol. Phytopathol., Kiev. Transl. Isr. Prog. Sci. Transl. Jerusalem, 1969. 112 pp.

127. Townes, H. 1956. The Nearctic species of trigonalid wasps. *US Natl. Mus. Proc.* 106:295–304

128. Townes, H. 1969. The genera of Ichneumonidae, Pt. 1. *Mem. Am. Entomol. Inst.* Ann Arbor 11:1–300

129. Trjapitzin, V. A. 1973. Classification of parasitic Hymenoptera of the family Encyrtidae (Hymenoptera: Chalcidoidea), Pt. 1. Survey of the systems of classification. The subfamily Tetracneminae Howard, 1892. *Entomol. Rev.* 52:118–25

130. Trjapitzin, V. A. 1973. Classification of parasitic Hymenoptera of the family Encyrtidae (Chalcidoidea), Pt. 2. Subfamily Encyrtinae Walker, 1837. *Entomol. Rev.* 52:287–95

131. Valentine, E. W. 1975. Additions and corrections to Hymenoptera hyperparasitic on aphids in New Zealand. *NZ Entomol.* 6:59–61

132. van den Bosch, R. 1981. Specificity of hyperparasites. See Ref. 6, pp. 27–33

133. van den Bosch, R., Hom, R. R., Matteson, P., Frazer, B. D., Messenger, P. S., Davis, C. S. 1979. Biological control of the walnut aphid in California: impact of the parasite, *Trioxus pallidus.* *Hilgardia* 47:1–13

134. van den Bosch, R., Messenger, P. S.,

Gutierrez, A. P. 1982. *An Introduction to Biological Control.* New York: Plenum. 247 pp.

135. Varley, G. C., Gradwell, G. R., Hassell, M. P. 1973. *Insect Population Ecology.* Oxford: Blackwell. 212 pp.

136. Vater, G. 1971. Dispersal and orientation of *Diaeretiella rapae* with references to the hyperparasites of *Brevicoryne brassicae. Z. Angew. Entomol.* 68: 187–225

137. Viggiani, G. 1981. Hyperparasitism and sex differentiation in the Aphelinidae. See Ref. 6, pp. 19–26

138. Viggiani, G. 1984. Bionomics of the Aphelinidae. *Ann. Rev. Entomol.* 29: 257–76

139. Vinson, S. B. 1976. Host selection by insect parasitoids. *Ann. Rev. Entomol.* 21:109–33

140. Vinson, S. B., Iwantsch, G. F. 1980. Host suitability for insect parasitoids. *Ann. Rev. Entomol.* 25:397–419

141. Waage, J. K., Hassell, M. P. 1982. Parasitoids as biological control agents—a fundamental approach. *Parasitology* 84:241–68

142. Walker, G. P., Cameron, P. J. 1981. Biology of *Dendrocerus carpenteri,* parasite of *Aphidius* spp., and field observations of *Dendrocerus* spp. as hyperparasites of *Acyrthosiphon* sp. *NZ J. Zool.* 8:531–38

143. Walker, G. P., Nault, L. R., Simonet, D. E. 1984. Natural mortality factors acting on potato aphid *(Macrosiphum euphorbiae)* populations in processing tomato fields in Ohio. *Environ. Entomol.* 13:724–32

144. Walter, G. H. 1983. Divergent male ontogenies in Aphelinidae (Hymenoptera: Chalcidoidea): a simplified classification and a suggested evolutionary sequence. *Biol. J. Linn. Soc.* 19:63–82

145. Weseloh, R. M. 1969. Biology of *Cheiloneurus noxius,* with emphasis on host relationships and oviposition behavior. *Ann. Entomol. Soc. Am.* 62: 299–305

146. Weseloh, R. M., Wallner, W. E., Hoy, M. 1979. Possible deleterious effects of releasing *Anastatus kashmirensis,* a facultative hyperparasite of the gypsy moth. *Environ. Entomol.* 8:174–77

147. Williams, J. R. 1977. Some features of sex-linked hyperparasitism in Aphelinidae (Hymenoptera). *Entomophaga* 22: 345–50

148. Wilson, C. G., Swincer, D. E. 1984. Hyperparasitism of *Therioaphis trifolii* f. *maculata* (Hom.: Aphididae) in South Australia. *J. Aust. Entomol. Soc.* 23:47–50

149. Zinna, G. 1961. Ricerche sugli insetti entomofagi. II. Specializzazione entomoparassitica negli Aphelinidae: studio morfologico, etologico e fisiologico del *Coccophagus bivittatus,* nuovo parassita del *Coccus hesperidum* L. per l'Italia. *Boll. Lab. Entomol. Agrar. F. Silvestri Portici* 19:301–58

150. Zinna, G. 1962. Ricerche sugli insetti entomofagi. III. Specializzazione entomoparassitica negli Aphelinidae: interdipendenze biocenotiche tra due specie associate. Studio morfologico, etologico e fisiologico del *Coccophagoides similes* (Masi) e *Azotus matritensis* Mercet. *Boll. Lab. Entomol. Agrar. F. Silvestri Portici* 20:73–184

151. Zwölfer, H. 1971. The structure and effect of parasitoid complexes attacking phytophagous host insects. In *Dynamics of Populations,* ed. P. J. den Boer, G. R. Gradwell, pp. 405–18. Wageningen: Cent. Agric. Publ. Doc. 611 pp.

Ann. Rev. Entomol. 1987. 32:71–93
Copyright © 1987 by Annual Reviews Inc. All rights reserved

CHITIN BIOCHEMISTRY: SYNTHESIS AND INHIBITION

E. Cohen

Department of Entomology, Hebrew University of Jerusalem, Faculty of Agriculture, Rehovot 76 100, Israel

INTRODUCTION

Chitin is an amino-sugar polysaccharide that serves as a supporting element in extracellular structures, notably in exoskeletons of arthropods (91) and cell walls of various fungi (88). Chitin is prevalent in invertebrates (43, 105) and is found in certain diatom algae (45). In insects, this biopolymer is a major carbohydrate component of chito-protein complexes such as the cuticle and the peritrophic membrane (1, 92, 100, 105).

Taking into consideration the global biomass of arthropods, particularly the zooplankton, polymerization of *N*-acetyl-D-glucosamine into chitin is a major synthetic event, second only to cellulose production. Synthesis of chitin involves concerted multifaceted cellular activities starting from biotransformations of simple metabolites and culminating in the emergence of a polymer to be extruded outside cell membranes. Many enzymatic and nonenzymatic parts of this intricate process are still poorly described. Information regarding the mechanism of polymerization and its regulation in insect (18, 60, 65, 66) and fungal (10, 35, 37, 88, 106, 120) systems has only recently started to emerge. Future studies on the role of insect hormones coupled with in-depth investigations using various types of inhibitors that interfere with chitin synthesis will shed light on these inadequately understood events. In particular, inhibitors will be instrumental in chemical microsurgery for elucidation of the polymerization step. Information related to subsequent steps such as extrusion of nascent polysaccharides, crystallization, orientation of microfibrils, and attachment of microfibrils to extracellular proteins is incomplete or lacking. Thorough studies should not only address academic curiosity about this fundamental biochemical phenomenon, but should also

71

have considerable applied aspects, since chitin has been viewed for a long time as an attractive target for selective pesticides.

CHITIN

Chitin is a high–molecular weight, unbranched helical homopolymer consisting of $\beta(1-4)$-linked N-acetyl-D-glucosamine (GlcNAc) units. This polysaccharide, which is insoluble in most solvents, is deacetylated in hot concentrated alkali to chitosan; deacetylated and hydrolyzed in hot concentrated acids yielding glucosamine; and enzymatically degraded by chitinases from various sources. The helicoidal structure of the polymer chain is stabilized by intramolecular hydrogen bonds. In a crystalline state, stabilized by hydrogen bonding between adjacent polymers, the polysaccharide chains are assembled into microfibrils, which constitute the fibrous component of insect exoskeletons. X-ray diffraction studies (105) of various chitin crystallites have revealed that an antiparallel arrangement of polymers (α-chitin) is the most abundant form. The rare parallel disposition (β-chitin) is found in cocoons of certain beetles. The structure of an uncommon third polymorphic form (γ-chitin), in which the arrangement of chains evidently alternates so that for every three polymers two are parallel, has not yet been confirmed. For detailed accounts regarding physical and chemical aspects of chitin, the reader is referred to several excellent reviews (43, 60, 92, 105).

In insects, chitin microfibrils are intimately associated with various proteins in structures such as the cuticle and the peritrophic membrane (43, 60, 91, 100). This fibrous component confers the needed tensility to these structures.

Chitin Synthesis and Deposition

The process of chitin formation involves an orderly sequence of complicated cellular events (Figure 1). Active catalytic units assembled in cell membranes polymerize GlcNAc molecules into extracellular chitin chains. The immediate substrate for polymerization, 5'-uridine diphospho-N-acetyl-D-glucosamine (UDP-GlcNAc), is an end metabolite of a cascade of cytoplasmic biochemical transformations that start from the disaccharide trehalose or from glucose. The sequence of formation of metabolites includes hydrolysis of trehalose, phosphorylation of glucose, transmutation to form phosphorylated fructose, amination, acetylation, and conversion to a nucleotide phosphate–acetylated amino sugar. Precursors listed in Figure 1 are also chanelled to alternative biochemical pathways. However, amination of fructose-6-P to glucosamine-6-P is a turning point for incorporation of metabolites into amino sugar–containing biopolymers such as glycoproteins, mucopolysaccharides, and chitin.

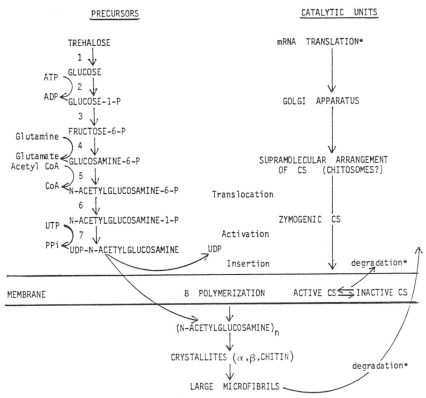

Figure 1   Schematic illustration of chitin formation. *1:* trehalase, *2:* hexokinase, *3:* glucose-6-P isomerase, *4:* glutamine fructose–6-P aminotransferase, *5:* glucosamine-6-*N*-acetyltransferase, *6:* phosphoacetylglucosamine mutase, *7:* UDP-*N*-acetylglucosamine pyrophosphorylase, *8:* chitin synthetase. *Hormonally regulated.

The various metabolites and enzymes related to chitin formation that have been identified in chitin-containing yeast and filamentous fungi (10, 35, 37) have been also reported in insects. Candy & Kilby (12) detected activities of enzymes involved in biotransformation of glucose to UDP-GlcNAc in the desert locust, *Schistocerca gregaria.* Fristrom (32) and Silvert & Fristrom (112), studying hexosamine metabolism in imaginal disks of the fruit fly *Drosophila melanogaster,* found various intermediates of the chitin formation pathway. Benson & Friedman (8) studied the enzyme glucosamine phosphate isomerase in adult house flies. Enzymes involved in the chitin synthesis pathway have also been shown in epidermal cells of crustaceans (39, 98). Chitin polymerization in a cell-free system extracted from the fungus *Neurospora crassa* was demonstrated by Glaser & Brown (34) in 1957. Since then, many chitin-polymerizing enzymes have been extracted from yeast and filamentous fungi (10, 37, 88). Chitin polymerization by cell-free systems has recently been reported in a number of insect species (19, 21, 58, 72).

The epithelium engaged in chitin production is a monolayer of polarized cells in which clusters of chitin-polymerizing units should be nonrandomly distributed in cell membranes at the apical area. It is conceivable that for bundles of chitin polymers to undergo further crystallization and fibrillogenesis, the polymerizing enzymes must be essentially immobile within the membranes. Such rigidity can be accomplished by clustering of the catalytic units or by their attachment to cytoskeletal elements. Electron microscopy studies carried out by Locke & Huie (67) with epithelial cells of the larger cana leafroller, *Calpodes ethlius,* suggest that clusters of the chitin synthesis apparatus are restricted to specific domains on the tips of fingerlike projecting microvilli. Moreover, chitin synthesis by imaginal disks of the Indian meal moth, *Plodia interpunctella,* was severely disrupted by compounds such as colcemid and vinblastine, which are known to interfere with formation of microtubular proteins (95).

Insect Chitin Synthetase

The membrane-bound chitin synthetase (CS) [UDP-2-acetamido-2-deoxy-D-glucose: chitin 4-β-acetamidodeoxyglucosyltransferase (EC 2.4.1.16)] is the key enzyme in chitin formation. Its activity has been demonstrated in a number of insect cell-free systems (19, 21, 58, 72). The enzyme preparations are still crude, obtained from either integumental (21, 58, 118) or gut (19) tissues. In the case of the stable fly, *Stomoxys calcitrans* (72), the tissue source of the CS is unidentified. Suspension of 10^4 g (mitochondrial) or 10^5 g (microsomal) pellets in appropriate buffers served as a source of the polymerizing enzyme. The radioassay (at pH values close to the neutral range), using labeled ^{3}H- or ^{14}C-UDP-GlcNAc, included magnesium ions and in certain cases GlcNAc as activators. The reaction was carried out at 30°C for *S. calcitrans* (72) and the sheep blowfly, *Lucilia cuprina* (118), and at 37°C for the brine shrimp, *Artemia salina* (50), CS enzymes. The optimum temperature for the gut enzyme prepared from larvae of the flour beetle, *Tribolium castaneum,* was only 22°C (19). The enzymatic reaction required magnesium (10–30 mM) for all but the stable fly CS, for which it was inhibitory. However, with the *T. castaneum* enzyme magnesium could be partially replaced by manganese or cobalt cations. Activity of CS in several cell-free preparations (19, 21) was greatly stimulated in the presence of GlcNAc (about 17 mM), which presumably serves along with the substrate as an allosteric effector. The degree of in vivo CS activity is possibly regulated by availability of UDP-GlcNAc, magnesium, and perhaps GlcNAc at the polymerization site. To further advance research on insect CS, solubilization and subsequent purification are certainly required. Attempts to solubilize the *T. castaneum*

(22) and *S. calcitrans* (72) particulate CS enzymes have been unsuccessful. Activity of the *T. castaneum* polymerase was inhibited by nonionized detergents as well as by high levels of digitonin (22), which was helpful in solubilizing fungal enzymes (30, 36).

Inhibition of the insect CS activity by the potent nucleoside peptide antibiotic polyoxin-D (see Figure 3, *I*) served as a standard indicator for chitin polymerization. Criteria used to verify the radioactive product as chitin were (*a*) immobility on thin layer chromatoplates, (*b*) insolubility and resistance to degradation by hot alkali, (*c*) degradation by chitinase, and (*d*) hydrolysis after prolonged treatment with hot concentrated hydrochloric acid, yielding glucosamine.

Based on studies with fungal enzymes (10, 30, 107), it has been proposed that the insect polymerizing enzyme is initially zymogenic and is proteolytically activated. When trypsin was used as a replacement for the presumed natural protease (19, 50, 72), it became evident that the extent of enzymatic stimulation (30–70%) was relatively low compared with that of the zymogenic fungal enzymes. The slight enhancement of CS activity was attributed to unmasking of active or allosteric sites of the enzyme (19). A similar effect was proposed for the increase in *T. castaneum* CS activity with dimethylsulfoxide and low levels of the detergent digitonin (22).

Polymerization of GlcNAc units in fungi and insects is independent of previous formation of a lipid intermediate (22, 30). Such an intermediate, recently detected in a crustacean species (51), is required for synthesis of sugar moieties of various glycoproteins. Tunicamycin, which is an effective inhibitor of dolichyl phosphate GlcNAc (42), and which suppressed incorporation of glucosamine into chitin in epidermal tissues of the bloodsucking bug *Triatoma infestans* (103), had no effect on insect-derived CS activity (22, 73). This inhibitor also did not affect chitin polymerization by CS of the yeast *Saccharomyces cerevisiae* (30). A need for a primer as a template for chitin polymerization has not been demonstrated in insects (17, 22). Apparently the crude enzyme preparations contain enough of these primers. It is noteworthy that solubilized and purified yeast CS polymerized GlcNAc into chitin without any included primer (11). It has been suggested that the enzyme itself served as an initiator of chitin chains.

Uridine 5'-diphosphate (UDP) (see Figure 3, *III*), released upon chitin polymerization, is a strong competitive inhibitor of the insect CS (see Table 1; 20, 21). Activity of UDPase shown in the fungus *Coprinus cinereus* (27) supposedly decreases inhibitory levels of UDP at the site of chitin polymerization. It has been proposed that this hydrolytic enzyme has a similar role in relation to the mode of action of insecticidal benzoylphenyl ureas (22).

It has been demonstrated by the use of protein cross-linking reagents such as glutaraldehyde that the membrane-bound CS in the yeast *S. cerevisiae* faces the cytoplasm (29). The catalytic center of the insect enzyme conceivably has the same location. Nevertheless, Mitsui and coworkers (81) claimed that the CS in midgut cells of larvae of the cabbage armyworm, *Mamestra brassicae,* faces the outer surface of the cell membrane. Consequently, they proposed that the high–molecular weight UDP-GlcNAc is translocated across cell membranes by a special active transport system. Meanwhile, there is no solid evidence for the location of the enzyme or for any transport system for its substrate.

Crystallization and Fibrillogenesis

Formation of chitin microfibrils to be integrated into matrices of cuticle and peritrophic membrane is dependent on crystallization of nascent poly-GlcNAc and fusion of bundles of chitin polymers into larger microfibrillar entities (Figure 1). The packed arrangement of CS units as supramolecular structures in cell membranes is most likely indispensible for formation and orientation of chitin microfibrils. The apparent nearness of the polymers formed enables fast crystallization by presumably physicochemical forces. Unlike the mechanism of the parallel polymer arrangement in β-chitin, that of the antiparallel orientation of the more common α-chitin is still unresolved. Speculated antiparallel arrangements include two adjacent sterically different enzymes polymerizing GlcNAc in opposite directions and chitin polymers folding on themselves (92). Elorza et al (31) interfered with chitin formation in the yeasts *S. cerevisiae* and *Candida albicans* by using Calcofluor white, a whitening reagent known to interact with hydrogen bonds. They maintain that the polymerized chitin chains progressively and spontaneously crystallize. Nevertheless, using the same reagent, Herth (44) concluded that polymerization and crystallization are separable processes. Questions pertaining to the roles of cell surfaces or cytoskeletal elements in fibrillogenesis and the orientation of microfibrils are still open.

Cohen (15) demonstrated in vitro microfibril formation by the *T. castaneum* cell-free preparation. Under standard assay conditions (19) a precipitate containing chitin microfibrils was detected after 10 min of incubation. Negative staining, which followed mild alkali treatment of the precipitate to remove proteins, revealed a considerable network of large microfibrils (Figure 2, *top left;* 15). It appears that in the absence of organized cell membranes, the chitin polymers crystallized and coalesced to form longitudinally oriented microfibrils (Figure 2, *upper* and *lower left*). The large diameter of the microfibrils formed in vitro might be due to the artificial conditions of the assay, which did not impose the restrictions that apparently exist in intact living systems.

Electron micrographs of the negatively stained preparation revealed spheric structures resembling chitosomes, from which fibrous filaments emerged

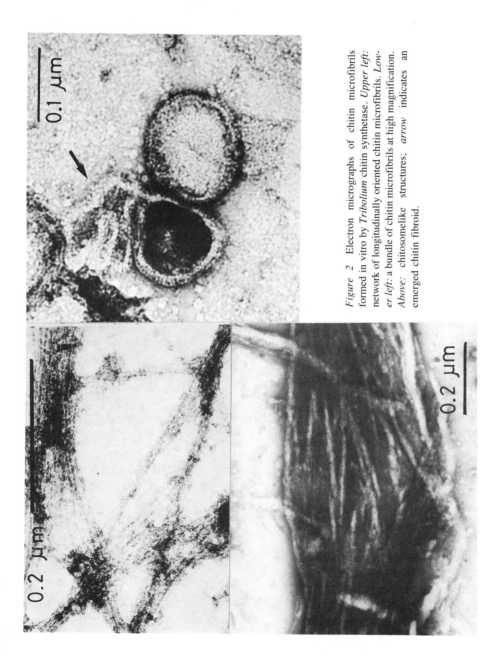

Figure 2 Electron micrographs of chitin microfibrils formed in vitro by *Tribolium* chitin synthetase. *Upper left:* network of longitudinally oriented chitin microfibrils. *Lower left:* a bundle of chitin microfibrils at high magnification. *Above:* chitosomelike structures; *arrow* indicates an emerged chitin fibroid.

(Figure 2, *right;* 15). Chitosomes found in various fungal CS preparations (3, 107, 108) were claimed to be true cellular organelles composed of clusters of zymogenic CS units being translocated to their proper location in the cell membrane. If insect chitosomes exist, the mechanism that guides them to apical areas of epithelial cells is obscure. It is still disputed whether chitosomes occur naturally or are merely an artifact related to the intense physical forces imposed by the high-speed centrifugation used in preparing the CS source. Incidentally, recent electron microscopy studies have shown cytoplasmic chitosomelike bodies in taxonomically diverse fungi (3) and in a mite (84).

Since the CS is an integrated membrane enzyme, the nascent chitin chains must be extruded outside the cell membrane. The mechanism involved in extrusion and the membrane components possibly linked to this process are completely unknown.

Control of Chitin Synthesis

The discontinuous synthesis of cuticular components obviously requires some mechanism of regulation and coordination. Kimura (57) has suggested that the molting hormone is directly involved in chitin synthesis and degradation. The use of insect integumental organ cultures has greatly facilitated research on cuticle biochemistry, including research on chitin and effects of insect growth hormones (18). In long-term organ cultures (5, 58, 70, 94, 114) ecdysterone was normally required, and chitin formation was detected about 24 hr after exposure to the hormone. In imaginal disks of *P. interpunctella* (94), enhanced chitin biosynthesis was detected following increased levels of exogenously applied ecdysterone. In short-term cultures where chitin synthesis is readily demonstrable, no addition of exogenous molting hormone was needed (41, 74, 82, 104, 116, 117, 123). It is believed that in these cases integumental cells had already been triggered by the endogenous hormone.

There is some evidence that recruitment of sugar precursors for chitin synthesis at periodic critical stages of cuticle formation is under hormonal control. It is still unclear whether the synthesis or activity of the polymerizing enzymes are under similar control. Some indications came from the study with CS of the giant silk moth, *Hyalophora cecropia* (21). In this investigation no chitin-polymerizing activity was detected in wing tissues of the diapausing pupae. Ecdysterone injected into the insects markedly stimulated CS activity. It is still unclear whether preexistent inactive enzyme was affected or de novo synthesis of CS units was stimulated. A dynamic interchange between active and inactive states of insect CS within the cell membranes has been proposed (Figure 1; 18). Oberlander and associates (93), using imaginal disks of *P. interpunctella,* observed a block of chitin synthesis after treatment with either RNA or protein-synthesis inhibitors. This may indicate de novo synthesis of CS units, but also could reflect a negative feedback mechanism involving synthesis of chitin and cuticular proteins.

Of great importance are dynamic aspects related to CS synthesis, translocation, and insertion into appropriate locations in the apical cell membranes, as well as its possible activation and turnover. The studies of Locke (66) and Locke & Huie (67) using *C. ethlius* epidermal cells indicate hormonal control of the turnover of the catalytic units. Plasma membrane plaques in apical areas of epithelial cells that apparently engage in chitin synthesis are periodically degraded and formed de novo. Such a dynamic turnover occurs at apolysis and coincides with cyclic changes in titers of the molting hormone. Actinomycin-D inhibited chitin synthesis in crayfish only during the premolt period of each molting cycle (115). This again points to periodic synthesis of CS complexes triggered by hormonal release.

It appears reasonable that recruitment of metabolites for chitin synthesis is also affected by the molting hormone. This does not necessarily imply that a precise biochemical event is regulated. Perhaps chitin formation and deposition are part of well-aligned cellular processes associated with growth, molting, and metamorphosis, which are controlled by insect hormones.

Unlike integumental tissues, the alimentary canal is continuously engaged in chitin synthesis for the chito-protein structure of the peritrophic membrane (100). This membrane is formed during active feeding stages of insects and is thus presumably under some physiological control.

It is quite remarkable that the pathway of chitin synthesis (which includes synthesis of metabolites and CS complexes) as well as properties and requirements of the polymerizing enzymes are similar in insects (18) and fungi (37, 88). This indicates that the chitin synthesis process is evolutionarily conservative. However, the physiological control of this composite process is likely to vary in two groups as taxonomically distant as insects and fungi.

INHIBITION OF CHITIN FORMATION

All currently known inhibitors of chitin synthesis act on the CS or the polymerization step. This is not surprising, since only at the junction of polymerization is the pathway outlined in Figure 1 "committed" to produce chitin. Inhibitors of chitin formation include unrelated, diverse groups of compounds. Some inhibitors, such as the nucleoside peptides, act directly at the catalytic site, whereas others interfere with essential sulfhydryl-sensitive sites of the enzyme. Compounds of great potential for disrupting chitin synthesis, such as the insecticidal benzoylphenyl ureas, act in an as yet unidentified way on polymerization or on events tightly associated with it.

Inhibitors Acting at the Catalytic Site

PEPTIDYL NUCLEOSIDE ANTIBIOTICS   Potent inhibitors of the chitin-polymerizing enzyme in insect and fungal systems are natural compounds extracted from actinomycetes (88). They are largely pyrimidine nucleoside peptides that structurally resemble the CS substrate (UDP-GlcNAc) and act as powerful competitive inhibitors (35, 47, 59, 88). Polyoxins and nikkomycins are the major groups of *Streptomyces*-derived inhibitors. Representative compounds of the former group have been commercially applied for control of fungal phytopathogens that cause diseases in rice, vegetables, and fruit trees (46, 78, 79).

Japanese scientists isolated polyoxins (A–M) from cultures of *Streptomyces cacaoi* var. *asoensis* more than a decade ago (49). Essentially, polyoxins consist of a uridyl-ribose moiety attached to a dipeptide. Kinetic studies carried out with various polyoxins revealed that the nucleoside moiety of the inhibitor binds to a specific site, whereas the peptidyl part most likely interacts with the catalytic site of the fungal CS (47, 48, 79). Nikkomycins (Figure 3, *II*), structurally related to polyoxins, were recently extracted from *Streptomyces tendae* cultures (25, 59). Like polyoxins, these inhibitors are dipeptidyl compounds; they have a uridyl but also an imidazoyl ribose moiety and a pyridine ring attached at the dipeptidyl end (25, 59).

The antibiotics are effective fungicides but poor insect-control agents (4, 9, 23, 59, 87, 88), evidently because the polar nature of the inhibitors drastically restricts their penetration through the hydrophobic insect epicuticle. When ingested, these compounds are likely to be degraded inside the insect alimentary canal. Alternatively, their relatively high molecular weight prevents their translocation across membranes of midgut cells. Insects apparently lack the active transport system for di- and tripeptides that translocates polyoxins and nikkomycins across fungal cell membranes (76, 99). Despite the poor toxicity of peptidyl nucleosides to insects, polyoxin-D and nikkomycin were toxic when fed to larvae of the sheep blowfly, *Lucilia cuprina* (117), and the Mexican bean beetle, *Epilachna varivestis* (110), respectively. Polyoxin-D drastically affected peritrophic membrane formation in the blowfly *Calliphora erythrocephala* (6); the degree of chitin synthesis was related to levels of polyoxin-D applied. Nikkomycin interfered with molting of the twospotted spider mite, *Tetranychus urticae* (85). Histological observations revealed disrupted arrangement of procuticular layers and disorder in muscle attachment to the cuticle (85, 110). Polyoxin-A administered by injection was toxic to immature stages of the migratory grasshopper, *Melanoplus sanguinipes* (124); mortality was due to disruption of chitin synthesis. In vitro studies with integumental tissues supported the in vivo observation (123). Malformed endocuticular layers were reported after injection of polyoxin-D

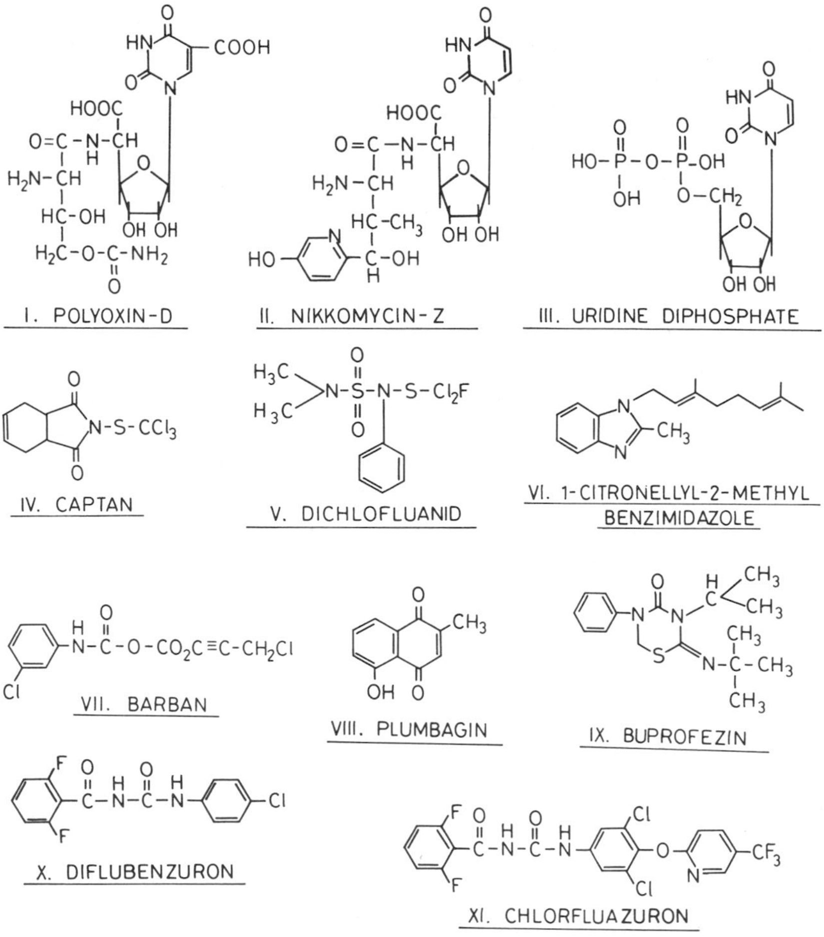

Figure 3 Chemical structures of selected chitin-synthesis inhibitors.

into larvae of the large cabbage white, *Pieris brassicae* (33). Similar observations were reported in larvae of *L. cuprina* fed with the same inhibitor (7). Using labeled precursors such as glucose, glucosamine, and *N*-acetylglucosamine, it has been shown that polyoxin-D is a powerful inhibitor (I_{50} values ranged from 1.9×10^{-5} M to 6×10^{-7} M) of chitin synthesis in isolated integumental tissues obtained from orthopteran, hemipteran, dipteran, and lepidopteran species (41, 58, 74, 114, 117).

The peptidyl nucleoside antibiotics are established competitive inhibitors of fungal polymerizing enzymes (37, 88). Polyoxin-D action was assayed using the recently available insect CS preparations. The chitin-polymerizing enzymes extracted from *T. castaneum* (20), from the cabbage looper, *Tricho-*

plusia ni (21), and from the Asiatic rice borer, *Chilo suppressalis* (58), were highly sensitive to the drug (I_{50} values of 4, 14, and 50 μM, respectively). CS preparations obtained from *H. cecropia* (21), *S. calcitrans* (72), and the brine shrimp, *A. salina* (50), were inhibited by relatively high levels of polyoxin-D. This insensitivity might be related to the crudeness of the CS preparations and also to a possible incorporation of the substrate into nonchitinous polymers. Nikkomycin (mixture of Z and X compounds) was the most efficient CS inhibitor; I_{50} values were only 0.02 and 0.06 μM for enzymes prepared from *T. castaneum* (20) and *T. ni* (21), respectively.

NUCLEOSIDE PHOSPHATES   UDP released upon chitin polymerization is an effective inhibitor of insect (20) and fungal (37, 88) CS enzymes. In vivo accumulation of UDP reaching inhibitory levels at the polymerization site might drastically reduce chitin formation. Hydrolysis of the nucleotide by UDPase, studied in the fungus *C. cinereus,* was suggested as a possible regulatory influence on chitin synthesis (27). UDP and uridine 5'-triphosphate (UTP) are potent inhibitors of *Tribolium* and *Trichoplusia* CS enzymes (Table 1; 20). The stable fly polymerizing enzyme is considerably less sensitive to these nucleotides (72). The similar slopes of the inhibition curves for polyoxin-D and UDP (20) indicate that both compounds interact with the same site on the enzyme. Other pyrimidine ribose di- and triphosphates, and in particular UTP, showed inhibitory effects on *T. castaneum* gut enzyme (Table 1). The ribosyl purine di- and trinucleotides as well as cyclic AMP did not affect the polymerizing-enzyme activity (20).

Inhibitors of Polymerization

SULFHYDRYL REACTIVE REAGENTS   The fungicide captan (Figure 3, *IV*) is known to interact with sensitive thiol-containing molecules in fungal cells (111). This toxicant was reported to inhibit cuticle formation in a cockroach-leg regenerate system (71) and to severely disrupt the formation of peritrophic membrane in isolated gut from adults of the blowfly, *Calliphora erythrocephala* (5). However, the fungicide did not inhibit cuticle formation by cultured integuments obtained from *Chilo suppressalis* (58). Captan included in cell-free systems prepared from larvae of *Tribolium castaneum* (20) and *Trichoplusia ni* (21) and from pupae of *H. cecropia* (21) strongly inhibited chitin polymerization. The fungicide was similar in potency to polyoxin-D, but the different slopes of their inhibition curves suggest distinct modes of action (20). Involvement of sensitive thiol groups associated with the insect CS was illustrated by the protective effect of the sulfhydryl-containing compound dithiothreitol included in the reaction mixture with captan (20). The action of captan can be mimicked by related sulphenimide

Table 1 Inhibition of insect chitin synthetase

Compound	Enzyme source	Inhibition (%)	Concentration (μM)	Ref.
Nucleoside peptide antibiotics				
polyoxin-D	*L. cuprina*	48	0.5	(118)
polyoxin-D	*T. castaneum*	50	4	(20)
polyoxin-D	*T. ni*	50	14	(21)
polyoxin-D	*C. suppressalis*	50	50	(58)
polyoxin-D	*A. salina*[a]	17	200	(50)
polyoxin-D	*H. cecropia*	14	300	(21)
polyoxin-D	*S. calcitrans*	40	1000	(72)
polyoxin-B	*T. ni*	50	0.5	(21)
nikkomycin	*T. castaneum*	50	0.02	(20)
nikkomycin	*T. ni*	50	0.06	(21)
Nucleoside phosphates[b]				
UDP	*T. ni*	50	70	(21)
UDP	*T. castaneum*	50	100	(20)
UDP	*S. calcitrans*	38	1000	(72)
UTP	*T. ni*	50	70	(21)
UTP	*T. castaneum*	61	170	(20)
UTP	*S. calcitrans*	36	1000	(72)
CDP	*T. castaneum*	64	170	(20)
CTP	*T. castaneum*	57	170	(20)
Sulphenimides				
captan	*T. castaneum*	50	6	(20)
captan	*T. ni*	99	30	(21)
captan	*H. cecropia*	50	40	(21)
captafol	*T. castaneum*	>95	60	(20)
dichlofluanid	*T. castaneum*	>95	60	(20)
Terpenoyl benzimidazoles				
1-geranyl-2-methylbenzimi-dazole	*T. castaneum*	50	50	(24)
1-citronellyl-benzimidazole	*T. castaneum*	50	68	(24)
Phenyl carbamates				
barban	*T. castaneum*	50	160	(20)
H-24108	*T. castaneum*	38	400	(20)
Quinones				
plumbagin	*T. ni*	71	300	(61)

[a]A crustacean species.
[b]UDP, UTP, CDP, and CTP are uridine di- and triphosphate and cytidine di- and triphosphate, respectively.

fungicides such as captafol and folpet or by the reactive dichlofluanid (Figure 3, *V*) (20). It is noteworthy that CS enzymes from the yeast *S. cerevisiae* (20) and the soilborne phytopathogen *Sclerotium rolfsii* (23) were insensitive to captan; this suggests that they differ significantly from insect enzymes in their response to sulfhydryl groups.

INSECT GROWTH REGULATORS   Compounds that interfere in any way with chitin biosynthesis exert their toxic effects at the time of molting, when the damaged newly formed cuticle cannot withstand hemolymph pressure and muscular traction. Therefore, unlike conventional neuroactive insecticides, chitin-synthesis inhibitors elicit symptoms of poisoning a few days after treatment. Compared with the nucleoside peptide antibiotics, certain insect growth regulators exhibit moderate to low inhibitory effects on the insect CS. The economically important benzoylphenyl-urea insecticides are completely ineffective as inhibitors of insect (17, 20, 72) or fungal (23, 122) cell-free chitin-polymerization systems. The exact biochemical lesion caused by these chitin synthesis–inhibiting insect growth regulators is still unknown.

Benzimidazoles   1-Geranyl-2-methyl benzimidazole was reported to disrupt larval molting of the silkmoth *Bombyx mori* (62), which displayed poisoning symptoms similar to those produced by benzoylphenyl ureas. Since a large number of terpenoyl benzimidazoles and related compounds were available, their inhibitory action on chitin synthesis was assayed using the efficient and convenient *T. castaneum* CS probe (24). The range of benzimidazole compounds permitted the first structure-activity relationship (SAR) study using an enzyme assay for chitin-synthesis inhibitors. Benzimidazoles with neryl, geranyl, or citronellyl (Figure 3, *VI*) moieties were the best inhibitors. Although activity of the most potent benzimidazoles was about one order of magnitude less than that of polyoxin-D, structure optimization could yield considerably more active CS inhibitors.

Buprofezin   This interesting insecticidal insect growth regulator (Figure 3, *IX*) is very effective against hemipteran pests (128). Symptoms of poisoning, which occur at molting, resemble those induced by benzoylphenyl ureas (2). Buprofezin disrupted cuticle deposition and inhibited in vivo and in vitro chitin synthesis in the brown rice planthopper, *Nilaparvata lugens* (119). Mortality of insects was positively correlated with the level of chitin synthesis inhibition (55). Protein synthesis was insensitive to buprofezin, and synthesis and formation of nucleic acids was only slightly decreased (55). The potency of buprofezin as an insect CS inhibitor has not been reported.

Phenyl carbamates   Based on the growth-retarding effect of the phenyl carabamate H-24108 (109), several related compounds were assayed for inhibition of *T. castaneum* CS. Except for the herbicide barban (Figure 3, *VII*), which had moderate inhibitory effects (Table 1), a series of H-24108 structural analogs displayed weak activity (20).

Plumbagin   This natural quinone (Figure 3, *VIII*) isolated from roots of the shrub *Plumbago capensis* affected molting of *B. mori* as well as other lepidopteran species (61). When assayed with *T. ni* integumental CS, plumbagin displayed low inhibition potency.

Benzoylphenyl ureas   The renewed interest in cuticle biochemistry and particularly in chitin synthesis stems from the unexpected discovery of bioactive benzoylphenyl ureas more than a decade ago. Scientists at Philips-Duphar (The Netherlands), initially looking for potent herbicides, combined dichlobenil and diuron, two established weed-control agents (121, 125, 126). The benzoylphenyl urea formed, DU-19111, exhibited poor herbicidal activity but surprising insecticidal activity upon a number of insect pests that ingested treated leaves (86, 125). Poisoning symptoms observed a few days after treatment were characterized by difficulties in molting (125). The unexpected biological activity of the new compounds prompted an intensive synthesizing effort; the optimal molecule of this benzoylphenyl urea series was diflubenzuron (TH-6040) (Figure 3, *X*), also known as the commercial product Dimilin® (125). Later, other bioactive compounds based on the original urea molecules, such as BAY SIR 8514 (131) and IKI-7899 (chlorfluazuron, Figure 3, *XI*) (40), were synthesized and successfully commercialized. A thorough quantitative structure-activity relationship (QSAR) study was recently conducted for a large number of benzoylphenyl ureas, with larvicidal activity and inhibition of cuticle formation in cultured integuments used as probing parameters (89, 90).

In early biological studies using these novel active insecticides, typical lesions in cuticle (125) and disrupted production of peritrophic membrane were observed (5, 14, 113). Histological investigations revealed malformation of new cuticle and disruption of endocuticular layers (33, 38, 64, 86). Later research focused on inhibition of chitin synthesis, since benzoylphenyl ureas affected neither the formation of cuticular proteins (52, 56, 86, 122, 125, 132) nor the sclerotization process (52, 56). The various biotransformations of sugars involved in the chitin synthesis pathway (Figure 1) were unaffected by the urea compounds (125), yet the incorporation of labeled glucose and amino sugars into chitin was blocked in a very short time (28, 41, 102). A large number of in vitro studies using various integumental tissues confirmed the involvement of insecticidal urea compounds in inhibition of chitin synthesis (41, 58, 74, 82, 114, 117). The recently available chitin-forming cell line derived from embryonic tissues of the tobacco hornworm,

Manduca sexta, has emerged as a powerful tool for investigation of chitin synthesis and inhibition (68). Using these continuously growing cultured cells, Marks et al (68) reported inhibition of chitin synthesis by diflubenzuron. Other studies demonstrated that the urea inhibitors and polyoxin-D caused similar symptoms (33, 122, 124). In several cases, UDP-GlcNAc was accumulated, which suggests that biochemical lesions caused by benzoylphenyl ureas involve the polymerization step (41, 101, 122, 125). Toxicological and SAR studies using isolated abdomens of the large milkweed bug, *Oncopeltus fasciatus,* illustrated that insecticidal activities of benzoylphenyl urea derivatives were positively correlated with the potency of their inhibition of chitin synthesis (41). At this stage of the research the chitin polymerization enzyme was considered as the likely target for the new insecticides.

When insect cell-free systems capable of chitin synthesis became available, the benzoylphenyl ureas were immediately assayed. In nearly all cases, the insect CS was insensitive to the compounds (16, 20, 21, 58, 72). Also, in vitro chitin synthesis and fibrillogenesis proceeded unaffected in the presence of high levels of diflubenzuron (15). The CS extracted from integumental tissues of *L. cuprina* was reported to be sensitive to diflubenzuron (118). However, the data are difficult to interpret, since the inhibition did not exceed 50% and the substrate was apparently also incorporated into nonchitinous molecules. The insensitivity of the insect CS to insecticidal benzoylphenyl ureas leaves the nature of their precise biochemical lesion unresolved. These compounds arrested proliferation of imaginal histoblasts in *S. calcitrans* pupae (77) and were implicated in inhibition of DNA synthesis (26, 80) and decreased uptake of nucleosides by cultured melanoma cells (75). The ovicidal and chemosterilizing properties of benzoylphenyl ureas (13, 38, 96, 127) might be related in part to these effects. Other effects, such as enhancement of chitinase and phenoloxidase (53) and interference with the insect endocrine system (129, 130), are insubstantial or are regarded as secondary (22, 28, 41, 97).

A large body of evidence points to inhibition of chitin synthesis as the major target for the action of benzoylphenyl ureas. A number of possibilities for their mode of action have been raised (22). Some are considered unlikely, while others cannot be evaluated until a suitable experimental system is developed. A direct effect of the insecticidal ureas could involve the generation of an outstanding active metabolite at the CS target. Nevertheless, this appears unlikely owing to the slow metabolism of benzoylphenyl ureas in insects (20, 54) and the rapid inhibition of chitin synthesis (28, 41). The suggestion that the chitin-synthesis inhibitors affect a putative protease, activating a zymogenic CS (63, 69), is repudiated by the blockage of an already operative polymerizing enzyme and by the speed with which this effect becomes evident. Other speculated modes of action include inhibition of UDPase (22), which would generate inhibitory levels of UDP at the polymerization site, and inhibition of the access of the substrate UDP-GlcNAc or the allosteric effector GlcNAc to the catalytic sites (22). Mitsui and coworkers (81), who suggested that the CS catalytic site is located at the outer surface of epithelial cells, maintained that diflubenzuron inhibits translocation of UDP-GlcNAc across membranes of *M. brassicae* midgut epithelial cells (83). Nevertheless, the investigators were unable to demonstrate any inhibitory effect of diflubenzuron on Na^+-K^+ or $Ca^{2+}-Mg^{2+}$ ATPase-associated active-transport systems (81).

Pertinent to the mode of action of benzoylphenyl ureas is the in vivo insensitivity of fungal systems toward these compounds. This possibly indicates that inhibitors act on regulatory mechanisms present in insects but not in fungi. Other hypotheses are attractive but are difficult to prove because it is not really understood how polymerization and extrusion of newly formed chitin chains occur. At present it is merely a speculation that cytoskeletal elements or membrane components linked to the supramolecular arrangement of CS units are involved with polymerization and extrusion. The mechanism of direct or indirect regulation of dynamic processes such as catalysis, extrusion of polymers, and fibrillogenesis, and of the possibly associated membrane components is unknown. The biochemical events related to chitin synthesis and deposition are undoubtedly highly integrated and coordinated. Insecticidal benzoylphenyl ureas could interfere at sensitive sites of these complex processes, disrupting either regulation or membrane elements. Discovery of the mode of action of the urea compounds might be key to the elucidation of obscure and intricate aspects of chitin biochemistry.

CONCLUDING REMARKS

Despite some progress toward an understanding of the properties and inhibition of insect chitin synthetase and fibrillogenesis, many black boxes still exist. The unknown aspects include (*a*) the regulatory mechanisms of discontinuous chitin biosynthesis by cuticle-producing cells, which differ from those of continuous synthesis by gut cells engaged in the formation of peritrophic membrane; (*b*) the mechanisms involved in the directed insertion and in the clustering of enzyme units in plasma membranes at the apical region of epithelial cells; (*c*) the zymogenicity of the insect polymerizing enzyme, and the mechanism and possible regulation of its proteolytic activation; (*d*) intrinsic factors modulating levels of substrate and activators at the polymerization site; (*e*) in situ polymerization and formation of polymorphic chitin crystallites; (*f*) cellular and membrane components associated with extrusion and orientation of nascent chitin polymers, and the exact mech-

88   COHEN

anisms of these processes; and (g) association of the microfibrillar components with various proteins in cuticle and peritrophic membrane structures. Definition of the mode of action of chitin-synthesis inhibitors, notably the benzoylphenyl ureas, should certainly help us understand at least some aspects of chitin biochemistry.

Literature Cited

1. Andersen, S. O. 1979. Biochemistry of insect cuticle. *Ann. Rev. Entomol.* 24: 24–61
2. Asai, T., Fukada, M., Maekawa, S., Ikeda, K., Kanno, H. 1983. Studies on the mode of action of buprofezin. I. Nymphicidal and larvicidal activities on the brown rice planthopper, *Nilaparvata lugens* Stål (Homoptera: Delphacidae). *Appl. Entomol. Zool.* 18:550–52
3. Bartnicki-Garcia, S., Bracker, C. E., Reyes, E., Ruiz-Herrera, J. 1978. Isolation of chitosomes from taxonomically diverse fungi and synthesis of chitin microfibrils *in-vitro. Exp. Mycol.* 2: 173–92
4. Bartnicki-Garcia, S., Lippman, E. 1972. Inhibition of *Mucor rouxii* by polyoxin-D: Effects on chitin synthesis and morphological development. *J. Gen. Microbiol.* 71:301–9
5. Becker, B. 1978. Effects of 20-hydroxyecdysone, juvenile hormone, Dimilin, and captan on *in vitro* synthesis of peritrophic membranes in *Calliphora erythrocephala. Insect Biochem.* 24:699–705
6. Becker, B. 1980. Effects of polyoxin-D on *in vitro* synthesis of peritrophic membranes in *Calliphora erythrocephala. Insect Biochem.* 10:101–6
7. Bennington, K. C. 1985. Ultrastructural changes in the cuticle of the sheep blowfly, *Lucilia,* induced by certain insecticides and biological inhibitors. *Tissue Cell* 17:131–40
8. Benson, R. L., Friedman, S. 1970. Allosteric control of glucosamine phosphate isomerase from the adult housefly and its role in the synthesis of glucosamine 6-phosphate. *J. Biol. Chem.* 245: 2219–28
9. Brillinger, G. U. 1979. Metabolic products of microorganisms. 181. Chitin synthase from fungi, a test model for substances with insecticidal properties. *Arch. Microbiol.* 121:71–74
10. Cabib, E. 1981. Chitin: Structure, metabolism, and regulation of biosynthesis. In *Encyclopedia of Plant Physiology,* Vol. 13, *Plant Carbohydrates. II.*

Extracellular Carbohydrates, ed. W. Tanner, F. A. Loewus, pp. 395–415. Berlin: Springer-Verlag
11. Cabib, E., Roberts, R., Bowers, B. 1982. Synthesis of the yeast cell wall and its regulation. *Ann. Rev. Biochem.* 51:763–93
12. Candy, D. J., Kilby, B. A. 1962. Studies on chitin synthesis in the desert locust. *J. Exp. Biol.* 39:129–40
13. Chang, S. C. 1979. Laboratory evaluation of diflubenzuron, penfluron, and BAY SIR 8514 as female sterilants against the house fly. *J. Econ. Entomol.* 72:479–81
14. Clarke, L., Temple, G. H. R., Vincent, J. F. V. 1977. The effects of a chitin inhibitor—Dimilin—on the production of peritrophic membrane in the locust, *Locusta migratoria. J. Insect Physiol.* 23:241–46
15. Cohen, E. 1982. In vitro chitin synthesis in an insect: formation and structure of microfibrils. *Eur. J. Cell Biol.* 16:284–94
16. Cohen, E. 1985. Chitin synthetase activity and inhibition in different insect microsomal preparations. *Experientia* 41: 470–72
17. Cohen, E. 1986. Inhibition of chitin synthesis in insect systems. In *Chitin in Nature and Technology,* ed. R. A. A. Muzzarelli, C. Jeuniaux, G. W. Gooday. New York: Plenum. In press
18. Cohen, E. 1986. Interference with chitin biosynthesis in insects. In *Chitin and Benzoylphenyl Ureas,* ed. J. Wright, A. Retnakaran. The Hague: Junk. In press
19. Cohen, E., Casida, J. E. 1980. Properties of *Tribolium* gut chitin synthetase. *Pestic. Biochem. Physiol.* 13:121–28
20. Cohen, E., Casida, J. E. 1980. Inhibition of *Tribolium* gut chitin synthetase. *Pestic. Biochem. Physiol.* 13:129–36
21. Cohen, E., Casida, J. E. 1982. Properties and inhibition of insect integumental chitin synthetase. *Pestic. Biochem. Physiol.* 17:301–6
22. Cohen, E., Casida, J. E. 1983. Insect chitin synthetase as a biochemical probe for insecticidal compounds. In *Pesticide*

Chemistry. Human Welfare and the Environment, ed. J. Miyamoto, P. C. Kearney, pp. 25–32. Oxford: Pergamon
23. Cohen, E., Elster, I., Chet, I. 1986. Properties and inhibition of Sclerotium rolfsii chitin synthetase. Pestic. Sci. 17:175–82
24. Cohen, E., Kuwano, E., Eto, M. 1984. The use of Tribolium chitin synthetase assay in studying effects of benzimidazoles with a terpene moiety and related compounds. Agric. Biol. Chem. 48:1617–20
25. Dähn, U., Hagenmaier, H., Höhne, H., König, W. A., Wolf, W. A., et al. 1976. Stoffwechselprodukte von Mikroorganismen. 154. Mitteilung. Nikkomycin, ein neuer Hemmstoff der Chitinsynthese bei Pilzen. Arch. Microbiol. 107:143–60
26. DeLoach, J. R., Meola, S. M., Mayer, R. T., Thompson, J. M. 1981. Inhibition of DNA synthesis by diflubenzuron in pupae of the stable fly Stomoxys calcitrans (L.). Pestic. Biochem. Physiol. 15:172–80
27. de Rousset-Hall, A., Gooday, G. W. 1975. A kinetic study of a solubilized chitin synthetase preparation from Coprinus cinereus. J. Gen. Microbiol. 89: 146–54
28. Deul, D. J., deJong, B. J., Kortenbach, J. A. M. 1978. Inhibition of chitin synthesis by two 1-(2,6-disubstituted benzoyl)-3-phenylurea insecticides. II. Pestic. Biochem. Physiol. 8:98–105
29. Duran, A., Bowers, B., Cabib, E. 1975. Chitin synthetase zymogen is attached to the yeast plasma membrane. Proc. Natl. Acad. Sci. USA 72:3952–55
30. Duran, A., Cabib, E. 1978. Solubilization and partial purification of yeast chitin synthetase. Confirmation of the zymogenic nature of the enzyme. J. Biol. Chem. 253:4419–25
31. Elorza, M. V., Rico, H., Sentandreu, R. 1983. Calcofluor white alters the assembly of chitin fibrils in Saccharomyces cerevisiae and Candida albicans cells. J. Gen. Microbiol. 129:1577–82
32. Fristrom, J. W. 1968. Hexosamine metabolism in imaginal disks of Drosophila melanogaster. J. Insect Physiol. 14:729–40
33. Gijwijt, M. J., Deul, D. H., deJong, B. J. 1979. Inhibition of chitin synthesis by benzoyl-phenylurea insecticides. III. Similarity in action in Pieris brassicae (L.) with polyoxin-D. Pestic. Biochem. Physiol. 12:87–94
34. Glaser, L., Brown, D. H. 1957. The synthesis of chitin in cell free extracts of

Neurospora crassa. J. Biol. Chem. 228:729–42
35. Gooday, G. W. 1979. Chitin synthesis and differentiation in Coprinus cinereus. In Fungal Wall and Hyphal Growth, ed. J. H. Burnett, A. P. J. Trinci, pp. 203–23. Cambridge: Cambridge Univ. Press
36. Gooday, G. W., de Rousset-Hall, A. 1975. Properties of chitin synthetase from Coprinus cinereus. J. Gen. Microbiol. 89:137–45
37. Gooday, G. W., Trinci, A. P. J. 1980. Wall structure and biosynthesis in fungi. In Eukaryotic Microbial Cell. Soc. Gen. Microbiol. Symp., Cambridge, 30:207–52. Cambridge, UK: Cambridge Univ. Press
38. Grosscurt, A. C. 1978. Diflubenzuron: Some aspects of its ovicidal and larvicidal mode of action and an evaluation of its practical possibilities. Pestic. Sci. 9: 373–86
39. Gwinn, J. F., Stevenson, J. R. 1973. Role of acetylglucosamine in chitin synthesis in crayfish. II. Enzymes in the epidermis for incorporation of acetylglucosamine into UDP-acetylglucosamine. Comp. Biochem. Physiol. 45B:777–85
40. Haga, T., Toki, T., Koyanagi, T., Nishiyama, R. 1982. Structure-activity relationships of a series of benzoylpyridyloxyphenyl urea derivatives. Abstr. 5th. Int. Congr. Pestic. Chem., Kyoto, Jpn. IId-7
41. Hajjar, N. P., Casida, J. E. 1979. Structure-activity relationships of benzoylphenyl ureas as toxicants and chitin synthesis inhibitors in Oncopeltus fasciatus. Pestic. Biochem. Physiol. 11:33–45
42. Heifetz, A., Keenan, R. W., Elbein, A. D. 1979. Mechanism of action of tunicamycin on the UDP-GlcNAc: Dolichylphosphate GlcNAc-1-phosphate transferase. Biochemistry 18:2186–92
43. Hepburn, H. E. 1985. Structure of the integument. In Comprehensive Insect Physiology, Biochemistry and Pharmacology, ed. G. A. Kerkut, L. I. Gilbert. 3:1–58. Oxford: Pergamon
44. Herth, W. 1980. Calcofluor white and Congo red inhibit chitin microfibril assembly of Poterioochromonas: Evidence for a gap between polymerization and microfibril formation. J. Cell Biol. 87:442–50
45. Herth, W., Zugenmaier, P. 1977. Ultrastructure of the chitin fibrils of the centric diatom Cyclotella cryptica. J. Ultrastruct. Res. 61:230–39
46. Hori, M., Eguchi, J., Kakiki, K., Misato, T. 1974. Studies on the mode of

action of polyoxins. VI. Effect of polyoxin B on chitin synthesis in polyoxin-sensitive and resistant strains of *Alternaria kikuchiana*. *J. Antibiot.* 27:260–66

47. Hori, M., Kakiki, K., Misato, T. 1974. Further study on the relation of polyoxin structure to chitin synthetase inhibition. *Agric. Biol. Chem.* 38:691–98

48. Hori, M., Kakiki, K., Misato, T. 1974. Interaction between polyoxin and active center of chitin synthetase. *Agric. Biol. Chem.* 38:699–705

49. Hori, M., Kakiki, K., Suzuki, S., Misato, T. 1971. Studies on the mode of action of polyoxins. Part III. Relation of polyoxin structure to chitin synthetase inhibition. *Agric. Biol. Chem.* 35:1280–91

50. Horst, M. N. 1981. The biosynthesis of crustacean chitin by a microsomal enzyme from larval brine shrimp. *J. Biol. Chem.* 256:1412–19

51. Horst, M. N. 1983. The biosynthesis of crustacean chitin. Isolation and characterization of polyprenol-linked intermediates from brine shrimp microsomes. *Arch. Biochem. Biophys.* 223:254–63

52. Hunter, E., Vincent, J. F. V. 1974. The effects of a novel insecticide on insect cuticle. *Experientia* 30:1432–33

53. Ishaaya, I., Casida, J. E. 1974. Dietary TH 6040 alters composition and enzyme activity of housefly larval cuticle. *Pestic. Biochem. Physiol.* 4:484–90

54. Ivie, G. W., Wright, J. E. 1978. Fate of diflubenzuron in the stable fly and house fly. *J. Agric. Food Chem.* 26:90–94

55. Izawa, Y., Uchida, M., Sugimoto, T., Asai, T. 1985. Inhibition of chitin biosynthesis by buprofezin analogs in relation to their activity controlling *Nilaparvata lugens* Stål. *Pestic. Biochem. Physiol.* 24:343–47

56. Ker, R. F. 1977. Investigation of locust cuticle using the insecticide diflubenzuron. *J. Insect Physiol.* 23:39–48

57. Kimura, S. 1973. The control of chitin deposition by ecdysterone in larvae of *Bombyx mori*. *J. Insect Physiol.* 19:2177–81

58. Kitahara, K., Nakagawa, Y., Nishioka, T., Fujita, T. 1983. Cultured integument of *Chilo suppressalis* as a bioassay system of insect growth regulators. *Agric. Biol. Chem.* 47:1583–89

59. Kobinata, K., Uramoto, M., Nishii, M., Kusakabe, H., Nakamura, G., et al. 1980. Neopolyoxins A, B, and C, new chitin synthetase inhibitors. *Agric. Biol. Chem.* 44:1709–11

60. Kramer, K. J., Dziadik-Turner, C., Koga, D. 1985. Chitin metabolism in insects. See Ref. 43, 3:75–115

61. Kubo, I., Uchida, M., Klocke, J. A. 1983. An insect ecdysis inhibitor from the African medicinal plant, *Plumbago capensis* (Plumbaginaceae); a naturally occurring chitin synthetase inhibitor. *Agric. Biol. Chem.* 47:911–13

62. Kuwano, E., Sato, N., Eto, M. 1982. Insecticidal benzimidazoles with a terpenoid moiety. *Agric. Biol. Chem.* 46:1715–16

63. Leighton, T., Marks, E., Leighton, F. 1981. Pesticides: Insecticides and fungicides as chitin synthesis inhibitors. *Science* 213:905–7

64. Lim, S. J., Lee, S. S. 1982. The toxicity of diflubenzuron to *Oxya japonica* (Willemse) and its effect on moulting. *Pestic. Sci.* 13:537–44

65. Locke, M. 1984. Epidermal cells. In *Biology of the Integument*, ed. J. Bereiter-Hahn, A. G. Matoltsy, K. S. Richards, 1:502–22. Berlin: Springer-Verlag

66. Locke, M. 1985. A structural analysis of post-embryonic development. See Ref. 43, 2:87–149

67. Locke, M., Huie, P. 1979. Apolysis and the turnover of plasma membrane plaques during cuticle formation in an insect. *Tissue Cell* 11:277–91

68. Marks, E. P., Balke, J., Klosterman, H. 1984. Evidence for chitin synthesis in an insect cell line. *Arch. Insect Biochem. Physiol.* 1:225–30

69. Marks, E. P., Leighton, T., Leighton, F. 1982. Mode of action of chitin synthesis inhibitors. In *Insecticide Mode of Action*, ed. J. R. Coats, pp. 281–313. New York: Academic

70. Marks, E. P., Leopold, R. A. 1970. Cockroach leg regeneration: Effect of ecdysterone in vitro. *Science* 167:61–62

71. Marks, E. P., Sowa, B. A. 1976. Cuticle formation in vitro. In *The Insect Integument*, ed. H. R. Hepburn, pp. 339–57. Amsterdam: Elsevier

72. Mayer, R. T., Chen, A. C., DeLoach, J. R. 1980. Characterization of a chitin synthase from the stable fly, *Stomoxys calcitrans* (L.). *Insect Biochem.* 10:549–56

73. Mayer, R. T., Chen, A. C., DeLoach, J. R. 1981. Chitin synthesis inhibiting insect growth regulators do not inhibit chitin synthase. *Experientia* 37:337–38

74. Mayer, R. T., Meola, S. M., Coppage, D. L., DeLoach, J. R. 1980. Utilization of imaginal tissues from pupae of the stable fly for the study of chitin synthesis and screening of chitin synthesis

inhibitors. *J. Econ. Entomol.* 73:76–80

75. Mayer, R. T., Netter, K. J., Leising, H. B., Schachtschabel, D. O. 1984. Inhibition of the uptake of nucleosides in cultured Harding-Passey melanoma cells by diflubenzuron. *Toxicology* 30:1–6

76. McCarthy, P. J., Troke, P. F., Gull, K. 1985. Mechanism of action of nikkomycin and the peptide transport system of *Candida albicans. J. Gen. Microbiol.* 131:775–80

77. Meola, S. M., Mayer, R. T. 1980. Inhibition of cellular proliferation of imaginal epidermal cells by diflubenzuron in pupae of the stable fly. *Science* 207:985–87

78. Misato, T. 1982. Present status and future aspects of agricultural antibiotics. *J. Pestic. Sci.* 7:301–5

79. Misato, T., Kakiki, K., Hori, M. 1979. Chitin as a target for pesticide action: Progress and prospect. In *Advances in Pesticide Science*, ed. H. Geissbuhler, G. T. Brooks, P. C. Kearney, pp. 458–64. Oxford: Pergamon

80. Mitlin, N., Wiygul, G., Haynes, J. W. 1977. Inhibition of DNA synthesis in boll weevils (*Anthonomus grandis* Boheman) sterilized by Dimilin. *Pestic. Biochem. Physiol.* 7:559–63

81. Mitsui, T., Nobusawa, C., Fukami, J.-I. 1984. Mode of inhibition of chitin synthesis by diflubenzuron in the cabbage armyworm, *Mamestra brassicae* L. *J. Pestic. Sci.* 9:19–26

82. Mitsui, T., Nobusawa, C., Fukami, J.-I., Colins, J., Riddiford, L. M. 1980. Inhibition of chitin synthesis by diflubenzuron in *Manduca* larvae. *J. Pestic. Sci.* 5:335–41

83. Mitsui, T., Tada, M., Nobusawa, C., Yamaguchi, I. 1985. Inhibition of UDP-N-acetylglucosamine transport by diflubenzuron across biomembranes of the midgut epithelial cells in the cabbage armyworm, *Mamestra brassicae* L. *J. Pestic. Sci.* 10:55–60

84. Mothes, U., Seitz, K. A. 1981. A possible pathway of chitin synthesis as revealed by electron microscopy in *Tetranychus urticae* (Acari, Tetranychidae). *Cell Tissue Res.* 214:443–48

85. Mothes, U., Seitz, K. A. 1982. Action of the microbial metabolite and chitin synthesis inhibitor nikkomycin on the mite *Tetranychus urticae;* an electron microscopy study. *Pestic. Sci.* 13:426–41

86. Mulder, R., Gijswijt, M. J. 1973. The laboratory evaluation of two promising new insecticides which interfere with cuticle deposition. *Pestic. Sci.* 4:737–45

87. Müller, H., Furter, R., Zähner, H., Rast, D. M. 1981. Metabolic products of microorganisms. 203. Inhibition of chitosomal chitin synthetase and growth of *Mucor rouxii* by nikkomycin Z, nikkomycin X, and polyoxin A: A comparison. *Arch. Microbiol.* 130:195–97

88. Muzzarelli, R. A. A. 1977. *Chitin.* Oxford: Pergamon

89. Nakagawa, Y., Iwamura, H., Fujita, T. 1985. Quantitative structure-activity studies of benzoylphenylurea larvicides. II. Effect of benzyloxy substituents at aniline moiety against *Chilo suppressalis* Walker. *Pestic. Biochem. Physiol.* 23:7–12

90. Nakagawa, Y., Kitahara, K., Nishioka, T., Iwamura, H., Fujita, T. 1984. Quantitative structure-activity studies of benzoylphenylurea larvicides. I. Effects of substituents at aniline moiety against *Chilo suppressalis* Walker. *Pestic. Biochem. Physiol.* 21:309–25

91. Neville, A. C. 1975. *Biology of Arthropod Cuticle.* New York: Springer-Verlag

92. Neville, A. C. 1984. Cuticle organization. See Ref. 65, 1:611–25

93. Oberlander, H., Ferkovich, S., Leach, E., Van Essen, F. 1980. Inhibition of chitin biosynthesis in cultured imaginal disks: Effects of alpha-amanitin, actinomycin-D, cycloheximide and puromycin. *Wilhelm Roux Arch. Dev. Biol.* 188:81–86

94. Oberlander, H., Ferkovich, S. M., Van Essen, F., Leach, C. E. 1978. Chitin biosynthesis in imaginal disks cultured in vitro. *Wilhem Roux Arch. Dev. Biol.* 185:95–98

95. Oberlander, H., Lynn, D. E., Leach, C. E. 1983. Inhibition of cuticle production in imaginal discs of *Plodia interpunctella* (cultured in vitro): Effects of colcemid and vinblastine. *J. Insect Physiol.* 29:47–53

96. Oliver, J. E., DeMilo, A. B., Brown, R. T., McHaffey, D. G. 1977. AI3-63223: A highly effective boll weevil sterilant. *J. Econ. Entomol.* 70:286–88

97. O'Neill, M. P., Holman, G. M., Wright, J. E. 1977. β-Ecdysone levels in pharate pupae of the stable fly, *Stomoxys calcitrans* and interaction with the chitin inhibitor diflubenzuron. *J. Insect Physiol.* 23:1243–44

98. Pahlic, M., Stevenson, J. R. 1978. Glucosamine-6-phosphate synthesis in the crayfish epidermis. *Comp. Biochem. Physiol.* 60B:281–85

99. Payne, J. W., Shallow, D. A. 1985.

Studies on drug targeting in the pathogenic fungus *Candida albicans:* Peptide transport mutants resistant to polyoxins, nikkomycins and bacilysin. *FEMS Microbiol. Lett.* 28:55–60

100. Peters, W. 1976. Investigations on the peritrophic membranes of Diptera. See Ref. 71, pp. 515–43

101. Post, L. C., deJong, B. J., Vincent, W. R. 1974. 1-(2,6-disubstituted benzoyl)-3-phenylurea insecticides: Inhibitors of chitin synthesis. *Pestic. Biochem. Physiol.* 4:473–83

102. Post, L. C., Vincent, W. R. 1973. A new insecticide inhibits chitin. *Naturwissenschaften* 60:431–32

103. Quesada-Allue, L. A. 1982. The inhibition of insect chitin synthesis by tunicamycin. *Biochem. Biophys. Res. Commun.* 105:312–19

104. Retnakaran, A., Hackman, R. H. 1985. Synthesis and deposition of chitin in larvae of the Australian sheep blowfly, *Lucilia cuprina. Arch. Insect Biochem. Physiol.* 2:251–63

105. Rudall, K. M., Kenchington, W. 1973. The chitin system. *Biol. Rev.* 49:597–636

106. Ruiz-Herrera, J., Bartnicki-Garcia, S. 1976. Proteolytic activation and inactivation of chitin synthetase from *Mucor rouxii. J. Gen. Microbiol.* 97:241–49

107. Ruiz-Herrera, J., Lopez-Romero, E., Bartnicki-Garcia, S. 1977. Properties of chitin synthetase in isolated chitosomes from yeast cells of *Mucor rouxii. J. Biol. Chem.* 252:3338–43

108. Ruiz-Herrera, J., Sing, V. O., Van Der Woude, W. J., Bartnicki-Garcia, S. 1975. Microfibril assembly by granules of chitin synthetase. *Proc. Natl. Acad. Sci. USA* 72:2706–10

109. Schaefer, C. H., Wilder, W. H., Mulligan, F. S. III, Dupras, E. F. Jr. 1974. Insect development inhibitors: Effects of Altosid®, TH6040 and H24108 against mosquitoes (Diptera: Culicidae). *Calif. Mosq. Control Assoc.* 42:137–39

110. Schlüter, U. 1982. Ultrastructural evidence for inhibition of chitin synthesis by nikkomycin. *Wilhelm Roux Arch. Dev. Biol.* 191:205–7

111. Siegel, M. R. 1970. Reactions of certain trichloromethyl sulfenyl fungicides with low molecular weight thiols. *In vitro* studies with glutathione. *J. Agric. Food Chem.* 18:819–22

112. Silvert, D. J., Fristrom, J. W. 1980. Biochemistry of imaginal discs: Retrospect and prospect. *Insect Biochem.* 10:341–55

113. Soltani, N. 1984. Effects of ingested diflubenzuron on the longevity and the peritrophic membrane of adult mealworms (*Tenebrio molitor* L.). *Pestic. Sci.* 15:221–25

114. Sowa, B. A., Marks, E. P. 1975. An *in vitro* system for the quantitative measurement of chitin synthesis in the cockroach: Inhibition by TH 6040 and polyoxin D. *Insect Biochem.* 5:855–59

115. Stevenson, J. R., Tung, D. A. 1971. Inhibition by actinomycin D of the initiation of chitin biosynthesis in the crayfish. *Comp. Biochem. Physiol.* 39B:559–67

116. Surholt, B. 1975. Studies *in vivo* and *in vitro* on chitin synthesis during the larval-adult moulting cycle of the migratory locust, *Locusta migratoria* L. *J. Comp. Physiol.* 102:135–47

117. Turnbull, I. F., Howells, A. J. 1982. Effects of several larvicidal compounds on chitin biosynthesis by isolated larval integuments of the sheep blowfly *Lucilia cuprina. Aust. J. Biol. Sci.* 35:491–503

118. Turnbull, I. F., Howells, A. J. 1983. Integumental chitin synthase activity in cell-free extracts of larvae of the Australian sheep blowfly, *Lucilia cuprina,* and two other species of Diptera. *Aust. J. Biol. Sci.* 36:251–62

119. Uchida, M., Asai, T., Sugimoto, T. 1985. Inhibition of cuticle deposition and chitin biosynthesis by a new growth regulator, buprofezin, in *Nilaparvata lugens* Stål. *Agric. Biol. Chem.* 49:1233–34

120. Ulane, R. E., Cabib, E. 1974. The activating system of chitin synthetase from *Saccharomyces cerevisiae.* Purification and properties of an inhibitor of the activating factor. *J. Biol. Chem.* 249:3418–22

121. van Daalen, J. J., Meltzer, J., Mulder, R., Wellinga, K. 1972. A selective insecticide with a novel mode of action. *Naturwissenschaften* 59:312–13

122. van Eck, W. H. 1979. Mode of action of two benzoylphenyl ureas as inhibitors of chitin synthesis in insects. *Insect Biochem.* 9:295–300

123. Vardanis, A. 1976. An *in vitro* assay system for chitin synthesis in insect tissue. *Life Sci.* 19:1949–56

124. Vardanis, A. 1978. Polyoxin fungicides: Demonstration of insecticidal activity due to inhibition of chitin synthesis. *Experientia* 34:228–29

125. Verloop, A., Ferrell, C. D. 1977. Benzoylphenyl ureas—A new group of larvicides interfering with chitin deposition. In *Pesticide Chemistry in the 20th*

Century, ACS Symp. Ser., ed. J. R. Plimmer, 37:237–70. Washington, DC: Am. Chem. Soc.

126. Wellinga, K., Mulder, R., van Daalen, J. J. 1973. Synthesis and laboratory evaluation of 1-(2,6-disubstituted benzoyl)-3-phenylureas, a new class of insecticides. I. 1-(2,6-dichlorobenzoyl)-3-phenylureas. *J. Agric. Food Chem.* 21: 348–54

127. Wright, J. E., Harris, R. L. 1976. Ovicidal activity of Thompson-Hayward TH 6040 in the stable fly and horn fly after surface contact by adults. *J. Econ. Entomol.* 69:728–30

128. Yasui, M., Fukada, M., Maekawa, S. 1985. Effects of buprofezin on different developmental stages of the greenhouse whitefly, *Trialeurodes vaporariorum* (Westwood) (Homoptera: Aleyrodidae). *Appl. Entomol. Zool.* 20:340–47

129. Yu, S. J., Terriere, L. C. 1975. Activities of hormone metabolizing enzymes in house flies treated with some substituted urea growth regulators. *Life Sci.* 17:619–26

130. Yu, S. J., Terriere, L. C. 1977. Ecdysone metabolism by soluble enzymes from three species of Diptera and its inhibition by the insect growth regulator TH-6040. *Pestic. Biochem. Physiol.* 7: 48–55

131. Zoebelein, G., Hamman, I., Sirrenberg, W. 1980. BAY SIR 8514, a new chitin synthesis inhibitor. *Z. Angew. Entomol.* 89:289–97

132. Zomer, E., Lipke, H. 1981. Tyrosine metabolism in *Aedes aegypti*. II. Arrest of sclerotization by MON 0585 and diflubenzuron. *Pestic. Biochem. Physiol.* 16:28–37

Ann. Rev. Entomol. 1987. 32:95–114

BIOGEOGRAPHY OF THE MONTANE ENTOMOFAUNA OF MEXICO AND CENTRAL AMERICA

Gonzalo Halffter

Instituto de Ecología, Apartado Postal 18-845, Delegación Miguel Hidalgo, 11 800 México, D. F.

PERSPECTIVES AND OVERVIEW

Faced with the challenge of writing a review on the distribution of insects in the mountains of Mexico and Central America (the Mexican Transition Zone), I found two possible alternatives: (*a*) to synthesize the latest publications or (*b*) to attempt to formulate a coherent theory that, although omitting some information and relevant references because of obvious space restrictions, would have the advantage of its consistency. I have followed the second alternative. I present a synthesis of information and hypotheses that may serve as a base to explain the distribution of montane insects in the Transition Zone. This is an exposition of essential features, not a summation of individual cases.

I have defined (32) as the Mexican Transition Zone a complex and varied area in which the Neotropical and the Nearctic faunas overlap. This area includes part of the southwestern United States, all of Mexico, and a large part of Central America extending to the Nicaraguan lowlands. The isthmus south of Lake Nicaragua (Costa Rica and Panama), although rich in endemisms, presents marked South American affinities and a Nearctic penetration similar to that of the northern part of South America. For these reasons I believe it is preferable not to consider this area as part of the Transition Zone.

The physiography of the large Transition Zone is varied and complex as a result of the eventful geologic history that has prevailed since the Middle Cenozoic. The area's flora and fauna are exceptionally rich, owing on the one hand to the great variety of environments and ecological refuges available in

95

0066-4170/87/0101-0095$02.00

the zone, and on the other to the adequate routes for dispersion of faunas of different origin, ranging from cold-temperate mountains to humid tropical corridors. The north-south direction of these corridors has been extremely important, as this disposition has facilitated the displacement of faunas and floras during great climatic changes. Equally significant in the Transition Zone are the enormous possibilities of allopatric differentiation that derive from an extremely complex orography in a tropical region.

It is difficult to understand the biological composition of the Transition Zone without taking into account the fact that communication with North America has existed since the Mesozoic, despite a few temporary barriers, particularly in the region of the Isthmus of Tehuantepec. In contrast, communication with South America, and therefore the possibility of the introduction of South American elements that at present form an important part of the biota in the zone, has been subjected to extremely variable conditions in the course of time.

There are various studies of what we may call "statistical biogeography," in which the zone is divided into provinces and subprovinces (28, 72). The distribution of plants has been studied with this approach (68). Amphibians and reptiles have been the object of reports encompassing either specific areas or the entire zone; many of these reports involved a dynamic and evolutionary approach distinct from descriptive, static biogeography (16, 17, 69, 71). However, this approach has not been extensively applied to insects; few studies attempt a broad understanding of the biogeographic problem. Ball (3), in an excellent and not very well-known paper, analyzed the penetration of Holarctic carabids in Mexico. His hypotheses and conclusions can be extended to all the groups of Holarctic origin. Whitehead (75) discussed the origin and evolution of the insect fauna of Middle America and also included a good bibliographic review. Whitehead (74) and Ball (4) analyzed the expansion of taxa of South American origin in the Transition Zone and in North America. Halffter (27–33) has developed a set of hypotheses to explain comprehensively the distribution of insects in the Transition Zone.

Much information, some of it very relevant, on insect distribution in the Transition Zone is dispersed as biogeographic comments in taxonomic studies or in local analyses. Due to space limitations only some of these publications are included in this review. The present paper is thus the first attempt to analyze comprehensively the distribution of insects in the mountains of the Transition Zone.

In my work I frequently use the concept of pattern (30–32), which corresponds to a synthesis of the essential features of the distribution of a set of coexisting organisms that originated or became integrated in a given area and time, are subjected to the same macroecological pressures for a prolonged period, live under the same physiographic conditions, and have a common biogeographic history. The concept of pattern is a generalization. It is intended for reference to help us analyze and compare differences in the distribution of each taxon.

There are some excellent studies on biogeography of montane insects in other parts of the world (23, 59), including a book that discusses the subject in general (48). However, the biogeography of montane insects is covered by many fewer studies than that of island insects.

Insect biogeography has been the subject of general bibliographic reviews that cover the literature of recent decades through 1972 (25, 26, 57). Ross's studies on the insect biogeography of North America (67) help elucidate the Transition Zone. Reichardt's study of South American fauna is similarly helpful (63). Other authors (20, 38, 58) have reviewed concepts and methodology.

THE MOUNTAINS OF THE TRANSITION ZONE

The distribution of a montane fauna depends, comparably only to that of an insular fauna, on the present structure and on the history of its geographic substrate. The ecologically unfavorable conditions of the lowlands constrain the opportunity for dispersal from one mountain to the next. The historically changing and ecologically discountinous spacial structure of mountains are factors that lead to isolation and speciation.

Seventy percent of the Transition Zone is above 1000 m; this percentage gives an idea of the importance of mountains in this region. The major mountain systems are shown in Figure 1.

History

Toward the end of the Cretaceous (Maestrichtian) the Laramidian orogeny started and determined the main physiographic features of the mountains in Mexico and northern Central America, with the exception of the Transverse Volcanic Belt. The altitude of the Sierra de Chiapas, the Sierra Madre del Sur, and the Eastern Sierra Madre, and the initial elevation of the Central Plateau and northern Central America give an idea of the magnitude of this revolution.

Probably toward the Middle Miocene the Central Plateau had reached its maximum elevation. The topography of the modern plateau is the result of a Miocene/Pliocene process. During the Pliocene and the Pleistocene, deposits up to 2000 m thick formed, and only the remains of some of the big ranges emerged in the form of smaller sierras.

The Transverse Volcanic Belt (or Neovolcanic Axis) started to develop during the Oligocene, although it reached its present form later. This system

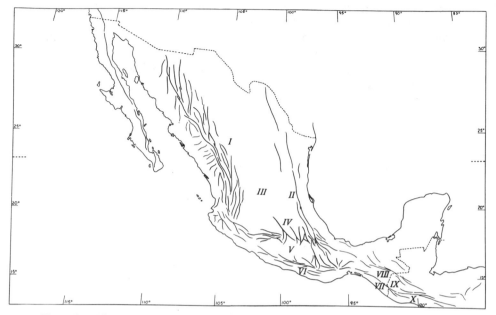

Figure 1   Main orographic systems of Mexico and northern Central America. *I*, Western Sierra Madre; *II*, Eastern Sierra Madre; *III*, Mexican High Plateau; *IV*, Transverse Volcanic Belt; *V*, Balsas Depression; *VI*, Southern Sierra Madre; *VII*, Sierra Madre de Chiapas; *VIII*, Chiapas Central Massif; *IX* and *X*, Central American Nucleus.

forms an immense dam across the middle of Mexico, stopping the drainage of the central part, which would otherwise flow toward the Pacific, in the Balsas Depression. This is the origin of the series of lakes in the south of the Mexican Plateau. The Transverse Volcanic Belt constitutes the most important east-west mountain range on the American Continent. It is of enormous importance for evolutionary processes, particularly for allopatric speciation and vicariance. It practically cuts Mexico in two, favoring fragmentation more than dispersion.

Major Systems

WESTERN SIERRA MADRE   This mountain system is the longest and most continuous in the Transition Zone. It runs parallel to the Pacific coast from the United States–Mexico border to the boundaries between the states of Nayarit and Jalisco, where it converges with the Transverse Volcanic Belt. Its total length is 1400 km, and its average width is 200 km. It can be considered as part of the orographic system that runs from Alaska to middle Central America, although a significant discontinuity occurs at the latitude of the United States–Mexico border, where the Sonoran Desert intervenes.

EASTERN SIERRA MADRE   This mountain range consists of a series of elongated folds with a general NNW-SSE direction. It ends toward the north on the Texas platform, and toward the southeast it is interrupted by the igneous flows of the Transverse Volcanic Belt. South of this point it continues in a southeast direction under the name Sierra de Juárez, and it ends in the Isthmus of Tehuantepec. Its length is approximately 600 km with an average width of 80 km.

Although the continental slope of the Eastern Sierra Madre is dry, the external slope facing the Gulf of Mexico is wet. This climatic dichotomy, together with the existence of elongated valleys attributed to synclines or sinking blocks, facilitates the penetration of many tropical elements at relatively high elevations, while many mountain elements descend into the coastal plain.

TRANSVERSE VOLCANIC BELT (OR NEOVOLCANIC AXIS)   This system is situated between the 19° and 20° parallels, along a band 950 km long and 50–150 km wide. It is of recent origin and overlays older structures. The general east-west orientation is not valid for individual ranges; many are integrated (often as a result of Pliocene-Recent vulcanism) and have a northwest-southeast orientation. An extreme and disjunct extension of the Volcanic Belt is the Sierra de los Tuxtlas, in the Gulf of Mexico to the south of the coastal plain, which has a northwest-southeast direction. The Transverse Volcanic Belt started to form during the Oligocene, although it underwent more intense development later. The Michoacán area, among others, experienced intense volcanic activity during the Pleistocene and the Recent.

SOUTHERN SIERRA MADRE   This mountain system runs parallel to the Pacific Coast from the State of Jalisco southeast to the western part of the Isthmus of Tehuantepec. Between the Southern Sierra Madre and the Transverse Volcanic Belt lies the Balsas Depression; its lower elevations are between 300 and 500 m and have a warm, dry climate. This depression constitutes an old and efficient barrier for mountain fauna. North of Oaxaca and southeast of Puebla, the Sierra and the Volcanic Belt are linked by mountains. The length of the Southern Sierra Madre is 1100 km and the average width is 120 km.

The Southern Sierra Madre is not tectonically related to the Sierra de Chiapas, as they are separated by the depression of the Isthmus of Tehuantepec. Nevertheless, both systems are of Laramidian origin, and are over 70 million years old.

MOUNTAIN SYSTEMS OF SOUTHEAST MEXICO AND CENTRAL AMERICA   The two mountain systems of southeastern Mexico, the Central Massif and the Sierra Madre de Chiapas, constitute the northern projections of the Central American mountain system.

In both historic and present contexts, two mountain systems can be clearly distinguished in Central America (from the Isthmus of Tehuantepec to Panama). Toward the north lies the Central American Nucleus, which includes 80% of the Central American highlands (above 600 m altitude) (71). Toward the south a narrow range, the Sierra de Talamanca, is situated between Costa Rica and Western Panama. The Nicaraguan Depression, lowlands 200 km long, lies between the two mountain systems. The Central American Nucleus includes the ranges and highlands of Chiapas, central and southern Guatemala, Honduras, El Salvador, and northern Nicaragua. This system emerged during the Laramidian revolution toward the end of the Cretaceous (62, 71). During the Pliocene the relief of the area was increased by the massive outpouring of igneous materials. Toward the end of the Pliocene this area attained its present altitude of approximately 2000 m, although a few crests were probably higher.

In the mountains of Cuchumatán, northwest of Guatemala, there is evidence of glaciation near the 3600 m level, probably of Wisconsin age (1, 34, 46). These glaciations induced the descent of montane vegetation to lower elevations. The mountain cloud forest, for example, was 1000 m below its present altitude (73).

Climate and Ecological Conditions

From the Transverse Volcanic Belt southward, the Mexican Transition Zone experienced intense vulcanism during the Middle and Upper Cenozoic and also during the Recent. The possibility that this volcanic activity could have caused massive extinction should not be exaggerated. Even during periods of generalized vulcanism, an extensive area in its entirety was never simultaneously affected in biological time. Vulcanism in the Volcanic Belt and in Central America occurred in relatively brief periods on a geologic scale but not on a biological one. Recolonization of an area, even of an extensive one, can occur in a few thousand years. Furthermore, some plants and animals can recolonize an area in a few centuries. Lava flows and cinder deposits only a few thousand years old, which are common in the Volcanic Belt, illustrate the speed with which flora and fauna can be reestablished. Thus, as eruptions were not synchronous on a biological scale, there is no reason to suppose that catastrophic extinctions occurred. Possibly the most important effect of volcanic eruptions was climatic change (induced by the interception of solar radiation by ash suspended in the atmosphere). The effect and magnitude of climatic changes are not fully known.

The microregional effect of vulcanism, on the other hand, may have been very important. The fragmentation of distribution areas favored allopatric evolution. An eruption could also induce new environmental conditions, such as vacant niches established by the sequence of colonization. Under these conditions, vulcanism was more a factor in fragmentation and diversification than a cause of extensive extinctions.

Most of the studies on Pleistocene climate in the mountains of Mexico are based on glaciers. At present (45) glaciers remain only on Pico de Orizaba, Popocatépetl, and Ixtaccíhuatl, although other high mountains in Central Mexico show evidence of former glacial activity.

The last glaciation contributed to increasing aridity in the Northern Hemisphere and induced a dry period in Mexico (36). Comparing climatic changes in the mountains of Central Mexico with those elsewhere, Klaus (40) concluded that during the glacial maximum, precipitation was strongly reduced. This explains the absence of glaciers in Mexico during that period. After the glacial maximum, the increase in world temperatures brought higher rainfall to the area. The MII (12,100 years ago) and MIII (9000–10,000 years ago) glacial formations on Mexican volcanoes occurred during these warmer periods. Tropical mountain glaciers were formed in a period between the glacial maximum and the interglacial maximum, which were both dry periods. The effects of climatic changes during the Pleistocene on North American insects have been reviewed by Howden (37) and Coope (14).

MONTANE INSECT FAUNA IN THE MEXICAN TRANSITION ZONE

In no other part of the world is the insect fauna of high mountains (i.e. above 2000 m altitude) as different from that of the contiguous lowlands as in the Mexican Transition Zone north of the Isthmus of Tehuantepec. This is not only because insularity and refuge effects are brought together in mountains separating the two great zoogeographic regions of the American Continent, but particularly because these mountains extend the Nearctic Region into the Transition Zone and form a penetration corridor for Nearctic fauna. It is in these mountains that the largest number of Paleo-American species (i.e. species of old north temperate origin) are conserved. The Mexican Plateau, on the other hand, reaches elevations above 2000 m on its southern extreme but is basically colonized by elements of very early South American origin that have evolved in situ. Lastly, the tropical lowlands are colonized by South American elements of a much more recent origin, which show much less differentiation.

The montane insects from the Transition Zone north of Tehuantepec have a predominantly northern affinity. The more characteristic portion of this insect fauna follows the Nearctic distribution pattern; the fauna above 3000 m is constituted almost exclusively by elements that follow this pattern. (The

Nearctic pattern, as well as the Meso-American and the Paleo-American patterns, are defined later.) There is also a strong contribution of elements from Paleo-American lines that, at the level of species or group of species, are exclusive to these mountains. The penetration of some elements corresponding to the Mexican Plateau distribution pattern can be observed on the internal slopes, sometimes reaching elevations above 2000 m. Additionally, some vicariant species from Meso-American lines can be found, particularly in the mountains of Oaxaca and Guerrero, the Transverse Volcanic Belt, and the Eastern Sierra Madre.

The resulting fauna is relatively homogeneous (the proportion of elements of South American origin is much lower than in the Mexican Plateau), with well-defined altitudinal zonation. In general, elements from the Mexican Plateau or of Meso-American origin are not found above 2200–2300 m. A mixed, Nearctic–Paleo-American fauna dominates between 2000 and 3000 m. Above 3000 m insects are nearly exclusively Nearctic.

South of the Isthmus of Tehuantepec the composition of the insect fauna is markedly different. From the mountains of Chiapas to those of Nicaragua the Meso-American element dominates. There are few Paleo-American species and these, like the less numerous Nearctic ones, show vicariance with forms north of the isthmus. The decrease in the number of species and in the diversity of lineages of northern origin shows that in numerous cases the isthmus has been an effective barrier.

The Transition Zone ends in the Nicaraguan lakes. Meso-American and some very rare Nearctic elements (21) can be found in the Sierra de Talamanca, but in the highest parts of this range the insect fauna is extremely poor. The impression is that these mountains are being colonized, without much success, by the surrounding Neotropical fauna, although some elements have been cited as showing Andean affinities.

In spite of the Central American bridge that was gradually established during the Miocene-Pliocene and definitely consolidated 5.7 million years ago (62), very few animals of Andean origin can be found in the mountains north of the Isthmus of Tehuantepec and very few Nearctic elements have reached the Andes. Some Paleo-American elements do occur in the Andes (8), but very few. The number of Andean elements in the whole Mexican Transition Zone is also very low. This contrasts strongly with the distribution of plants (68).

The Central American Nucleus has received a strong southern (but not Andean) contribution, and the elements that have evolved there (Meso-American pattern) have expanded in turn toward South America and toward the north of the Isthmus of Tehuantepec.

The study of the distribution of insects in the mountains north of the Isthmus of Tehuantepec involves the analysis of recent processes of speciation, subspeciation, and incipient separation of populations under Pleistocene and Recent conditions. It also involves the analysis of an early dispersal process, equivalent to "taxon pulse" as used by Erwin (18, 19), that preceded the present mountain physiography, and of a more recent one in which these mountains served as corridors. The expansion of Nearctic fauna toward the south is not only recent but is ongoing. Dispersal, vicariance, and consequent changes in the distributional areas are part of a dynamic process in which the biota evolves under the influence of a changing geographic scenario. This scenario entails alternately isolation or communication, expansion or contraction of distribution areas, and increased or decreased extinction of species. This dynamism suggests the sterility of many confrontations over vicariance and dispersion. As Brundin (11) stated, "if appearance of barriers instigates vicariance, then disappearance of barriers instigates dispersal."

Some aspects of the distribution of insects in the mountains of the Transition Zone have not been fully explained. Among these is the interruption in the distribution of numerous Paleo-American and Nearctic species that are typical of the mountains of Durango (Western Sierra Madre) and are found again toward the center of the Transverse Volcanic Belt (Michoacán and the state of México), but are apparently missing in Nayarit, Jalisco, and the western part of Michoacán. The interruption corresponds to the contact area between the Western Sierra Madre and the Transverse Volcanic Belt, where the mountains decrease in elevation and the 2000 m contour line is interrupted. This topography, in turn, creates discontinuities in ecologically adequate sites. Additionally, access to this region has been difficult until recently, and remains so even today, so that collecting has been limited.

A more significant problem is a difference in affinities of fauna north and south of the Isthmus of Tehuantepec. As previously stated, the montane insect fauna of the north has strong Nearctic affinities. In the south, Meso-American elements dominate. Why have so few mountain forms of Nearctic origin crossed to the Central American Nucleus? The Isthmus of Tehuantepec seems to have constituted a much more important barrier than it has generally been considered. Additionally, and perhaps very importantly, the climatic and ecological conditions of the mountains in the Central American Nucleus were and are more favorable to groups adapted to tropical humid mountains. The drier conditions of the Western and Eastern Sierras Madres (particularly the western system, which has been the main north-south migration corridor) and the Transverse Volcanic Belt have clearly hindered the migration of these groups. That is, few elements adapted to warm, humid climates were able to travel through these mountains.

The previous hypothesis is further supported by the principal expansion route of the Meso-American elements toward the north, the mountains of Oaxaca-Guerrero, the eastern part of the Volcanic Belt, and the southern part of the Eastern Sierra Madre, which are all more humid than the Western Sierra Madre and the western part of the Volcanic Axis. In the rare instances in which a Meso-American line has reached the Western Sierra Madre, it has been found in the southern part on the Pacific slope, i.e. in the most humid and warm part of the range.

An interesting and rather general characteristic of the Nearctic species and subspecies in the Transverse Volcanic Belt is that they are very recent and, in many well-studied cases, apomorphic. A higher rate of speciation in the mountains has been noted. Kavanaugh (39), referring to carabids, stated: "Average differentiation rates in the lowland groups appear to be much slower than rates suggested for the montane groups." [Coope (14) reported that taxonomic change in insects was markedly slow in the Cenozoic.]

In the Nearctic lines, expansion and differentiation to the species or subspecies level (vicariance) occurred recently. Ecologically, these lines are adapted to cold or cold-temperate conditions. This adaptation is expressed by phyletic relationships that are more latitudinal than altitudinal (28, 32, 39).

In contrast with the composition of flora and fauna in other mountain regions (southern Central America, South America, and high mountains in tropical Africa) (23, 35), that in the Transition Zone north of the Isthmus of Tehuantepec is more the result of dispersal under favorable conditions and later fragmentation under unfavorable conditions than of in situ evolution from the surrounding biota.

Following an opposite trend, the fauna in the tropical mountains south of the isthmus derived from elements of the adjacent lowlands. This phenomenon is noticeable in the Central American Nucleus, but it is even more marked in the Sierra de Talamanca in Costa Rica and Panama.

The distribution patterns of insects in the Mexican Transition Zone have been defined by Halffter (27–33).

Nearctic Distribution Pattern

The Nearctic distribution pattern is followed by Holarctic and Nearctic genera of comparatively recent (Pliocene-Pleistocene) penetration into the Mexican Transition Zone (3, 6, 7, 22, 28, 32, 42, 60, 77). Within the zone these genera are restricted to the mountain systems of Mexico and northern Central America; very few go beyond the Isthmus of Tehuantepec. In the mountains of the Transition Zone they are found at high elevations (usually above 1700 m) in temperate conifer forests and high-altitude grasslands. This pattern includes nearly all the species that are found in the highest regions (boreal forests, high-altitude prairies, and alpine tundra). For these insects, the orographic systems of the Transition Zone have served as a north-south penetration route (particularly the Western Sierra Madre) and also as diversification centers. This second phenomenon is very noticeable in the Transverse Volcanic Belt.

Species or groups of species within the Paleo-American lines can also follow the Nearctic distribution model (78). These species are of northern origin but penetrated into the Transition Zone earlier; they are completely adapted to the high mountains.

ORIGIN AND PENETRATION INTO AMERICA   The penetration route into America for many insects that follow the Nearctic distribution pattern, as for Paleo-American elements, seems to have been the Bering Bridge. However, groups of insects that follow the Nearctic pattern are adapted to cold-temperate, cold, or even Arctic conditions; the above supports the possibility that the Bering Bridge was crossed during the Upper Cenozoic, and situates the ecologic-geographic distribution of the Nearctic group in America and away from the Palearctic Region.

VARIATIONS WITHIN THE PATTERN   Within the Nearctic distribution pattern two variations can be found, sometimes clearly defined and at other times difficult or even impossible to distinguish: (a) lines with exclusively Nearctic affinities and (b) Holarctic lines.

The first group of lines (2) corresponds to an older distribution. These lines mainly have affinities with elements of the Alleghenian fauna. They are insects of temperate to cold-temperate climate. In general, in the mountains of the Mexican Transition Zone the species with strictly Nearctic affinities are found at a lower altitude than Holarctic species. Plant species with an equivalent distribution exist in the eastern United States and appear as far south as Chiapas and Guatemala (68).

The Holarctic elements (3, 6, 10) have a more or less continuous distribution in the northern regions of the great continental masses of the Northern Hemisphere. Ecologically, they are adapted to cold or cold-temperate ecosystems; even Arctic species are included. This explains the fact that in Mexico Holarctic elements are found above 1700 m; almost all the truly high-altitude species belong to this group, while the rest are derived from Paleo-American groups.

The Holarctic elements have penetrated toward the south following the mountain ranges of western North America, and exhibit relatively few affinities with the forms of the eastern and southeastern United States. Once more, everything seems to indicate that in the United States the division between eastern and western faunas was an early phenomenon. The elements of the Mexican mountains are more similar to the western than to the eastern fauna because of the continuity of the western mountain systems.

The typical Holarctic and strictly boreo-montane elements in the Transition Zone expanded southward, taking advantage of the cold, humid conditions of the four glaciations and particularly of the periods immediately following them. The fact that the great glaciations of the Northern Hemisphere were not simultaneous with the montane glaciations in Mexico (discussed above) helped the expansion of these groups.

The continuity of the orographic systems has been an extremely important factor in the Nearctic distribution. Series of mountains and highland plateaus occur more or less continuously from Alaska to northern Central America, interrupted only by a few barriers. The main barriers are the low xeric area between the mountains of the western United States and the Western Sierra Madre, approximately at the latitude of the Mexico–United States border; the Balsas Depression, which restricts the contact between the Transverse Volcanic Belt and the mountains of Guerrero-Oaxaca to a reduced area in the east where the Volcanic System meets the Eastern Sierra Madre; and most importantly, the hiatus represented by the tropical lowlands of the Isthmus of Tehuantepec.

PENETRATION INTO THE TRANSITION ZONE   Based on the distribution of carabid beetles, Ball (3) made an excellent analysis of the way in which the Holarctic boreo-montane groups extended southward, following mountain systems. To support the idea of a "modern" penetration, Ball pointed out the close similarity and the allopatric distribution of related species, which suggest recent speciation. In carabids, as in other insects, it is difficult to distinguish in certain cases between these recent penetrations and older ones; often the same taxonomic group followed the same orographic corridors during the Upper Cenozoic.

Undoubtedly, the barriers that interrupt the continuity of the orographic systems and hinder the penetration of northern boreo-montane elements also favor the speciation and survival of those elements that cross them; the barriers stop the migration of new forms and subsequent competition. On the other hand, the effect of barriers has not always been the same. Cold climates provided conditions more favorable than present ones for crossing these barriers, particularly the deserts and grasslands that separate the mountain ranges of the western United States from the Western Sierra Madre. During dry, warm periods (interglacials and interpluvials) the boreo-montane fauna has been concentrated in the high, cold mountains. During these periods barriers have been extremely effective.

The effect of the particular ecological conditions of the western mountains must be added to that of the barriers. Rivers and lakes are scarce in the mountains of the southwestern United States and in the Mexican Sierras. This has hindered the southward displacement of hygrophilous North American species.

The north-south disposition, the width, and the continuity of the Western Sierra Madre make it the main Nearctic penetration corridor. The Transverse Volcanic Belt, with an east-west disposition and subdivision into more or less isolated units, is the most important diversification area within the orographic systems of the Transition Zone. In these mountains there is a considerable degree of vicariance, which originated in the low number of forms that arrived through the Sierra Madre (50).

One of Ball's main conclusions (3) was that in general the boreo-montane Holarctic fauna has not contributed significantly to the Mexican fauna as a whole. The hygrophilous groups have not penetrated into the Mexican mountains, while the mesophilous and xerophilous ones have moved southward with difficulty. I believe that species from Paleo-American groups that have adapted to high-mountain conditions partly compensate for these limitations. According to Ball, the mountains are not a good route for penetration into the tropics. Thus, the tropics have been colonized not by groups that have penetrated along mountain ranges, but by northern groups that have developed forms adapted to relatively hot conditions in the north.

Between 15,000 and 12,000 years ago the climate of the Wisconsin Glacial Period started to change toward more temperate conditions. This interval is known as the Pluvial Postglacial Period. During this time some reptiles associated with humid conditions were found nearly continuously throughout most of the Chihuahuan Desert, owing to the existence of a continuous network of subhumid forest throughout what is now the desert. Martin (49) postulated that this network was the Pleistocene connection route between the two Sierras Madres throughout the Highland Plateau. During the Xerothermic Period, which followed the Pluvial, arid conditions advanced toward the north and the east, and the subhumid corridors of the Highland Plateau were interrupted and transformed into their present condition.

Meso-American Montane Pattern

The Meso-American montane pattern is represented by elements that evolved in the Central American Nucleus (4, 5, 24, 41, 43, 53–55, 75, 76, 78). [Comparison between the distribution of the insects (75) and that of the herpetofauna (16, 17, 69, 71) in Middle America is particularly interesting; Stuart (71) includes a description of the area.] The affinities are mainly of early South American origin, but this pattern also includes some elements from the region of Mexico north of the Isthmus of Tehuantepec. The latter entered the Central American Nucleus during the Cenozoic, evolved in this area, and later expanded jointly with the lines of South American origin. Access between the southern extreme of North America (approximately what is now Oaxaca) and the Central American Nucleus was intermittent but relatively easy, so that as elements of the Nucleus could migrate to the north, North American elements adapted to tropical conditions could reach the Nucleus and evolve there.

Ecologically, the elements that exhibit this pattern are very closely linked to mountain temperate and tropical forests and to cloud forests, penetrating in some cases to the more humid pine-oak forests. Moisture, which is basic to the structure of tropical mountain forests, seems to have a great importance in the development of this pattern. Humid conditions are also fundamental in the limited expansion of insect groups to the north and southeast of the Central American Nucleus. The main barriers for such expansion have been the Isthmus of Tehuantepec and the Nicaraguan depression, which both are covered by tropical forests but have drier conditions than the highlands.

The Central American Nucleus was elevated toward the end of the Cretaceous. The first important South American elements, however, must have arrived during the Oligocene-Miocene. Until the Miocene, the exchange between the Central American Nucleus and South America was very difficult and was only possible for those groups that were capable of passing across ocean barriers ("island hoppers") (9, 12, 15, 44, 47, 52). Even between the Miocene and the Middle Pliocene the faunistic interchange was difficult and only included a small portion of the South American fauna. The Central American Nucleus, on the other hand, had good communication with areas in Mexico north of the Isthmus of Tehuantepec. When the Balsas gate was closed during the Upper Cretaceous the only barrier that remained was the isthmus itself, which was a marine barrier during the Miocene and a climatic one throughout the Cenozoic and at present. Humid-mountain groups probably crossed this barrier most easily during the wetter episodes of the Pleistocene (16). The most important factor in the expansion to the north is the orography and its influence in the distribution of humid mountain forests. North of the Isthmus of Tehuantepec the species of Meso-American affiliation are most abundant in the mountains of Oaxaca-Guerrero and in the Southern Sierra Madre, on its external and more humid slopes.

Present distribution, as well as degree of taxonomic divergence, allows us to speculate about when northward expansion of a given group occurred. If its distribution follows the contour of the mountains of southern Mexico, particularly of the Transverse Volcanic Belt and the Eastern Sierra Madre, we can hypothesize that a group expanded when these orographic conditions already existed. This places the origin of most distributional patterns between the Pliocene and the Recent.

The reestablishment of the Central American bridge during the Pliocene allowed the expansion of Meso-American elements toward the Isthmus of Panama and South America. This expansion, however, was somewhat restricted in comparison to the northern expansion. The principal reason for this is ecological. The only adequate environmental conditions for montane Meso-American elements are found in the mountains of Talamanca, which are separated from the Central American Nucleus by the lowlands of the Nicaraguan depression, an important barrier for humid-mountain fauna. Some

Meso-American lines have given rise to elements adapted to the lowlands of southern Central and even South America. A good part of the montane insect fauna of the Talamanca range derives secondarily from these elements. Another part had its origin in typical South American elements. Thus the Talamanca range was colonized mainly by insects from tropical lowlands.

The distribution of the tribe Proculini of the family Passalidae illustrates the evolution of the Meso-American pattern (13, 46, 61, 64, 65). The Proculini are a predominantly montane group. At the generic level they are totally montane; no characteristically lowland genus exists. At the species level, two thirds are montane and around 30% have reduced wings. Among the forms that Pedro Reyes-Castillo (personal communication) calls "insular montane," the proportion of taxa with reduced wings reaches 45%. Adaptation to mountain environments has restricted dispersal and favored endemism, which is very marked in the mountains of the Central American Nucleus as well as in those situated north of the Isthmus of Tehuantepec. In Costa Rica–Panama and in the South American Andes there are fewer montane species of Passalidae, and they belong to a predominantly lowland tribe, the Passalini. There are good reasons (65) to suppose that the ancestral form from which the Proculini originated was an island hopper (passalids can be good flyers and colonize oceanic islands) that came to the Central American Nucleus from South America in the Oligocene-Miocene. The present South American Proculini belong to genera that show a southward expansion from the Central American Nucleus through the Isthmus of Panama. This expansion occurred after the Pliocene.

The northward expansion of the Proculini, although intermittently interrupted in the Isthmus of Tehuantepec, has occurred from the first stages of the group's formation, i.e. since the Oligocene-Miocene. As is the case with other Meso-American groups, the Proculini radiated in a secondary area of diversification formed by the Mexican mountains north of the Isthmus of Tehuantepec, where numerous genera of wide distribution in the Central American Nucleus have originated. The Costa Rican mountain system has been less important as a secondary center of diversification.

The Proculini also include excellent examples of disjunct distributions of the same species (without noticeable taxonomic changes) on different volcanoes that represent biogeographic islands. MacVean & Schuster (46) pointed out that "the passalid beetle fauna of seven Guatemalan volcanoes is apparently quite uniform. A given species has a similar altitudinal range on different volcanoes. . . . Dispersal of these species among the volcanoes probably occurred during the Pleistocene glacial periods when temperature depressions caused drops in the altitudinal limits of the montane wet forest and cloud forests allowing them to coalesce from one volcano to the next." These volcanoes correspond to what Dengo (15) calls the Volcanic Chain and are relatively near each other.

Paleo-American Pattern

The Paleo-American is the last dispersal pattern that contributes significantly to the montane insect fauna within the Mexican Transition Zone (28, 32, 51, 56). I have already indicated that species or groups of species of Paleo-American lines may have distributional characteristics similar to those of taxa arriving later through the Nearctic pattern. The fundamental difference is that in the case of elements corresponding to Paleo-American lines, the derivation has occurred within the Transition Zone.

The Paleo-American pattern results from an early immigration to the American continent (32). The lines that follow this pattern are more taxonomically and ecologically diverse in the Old World than in America. Even though in America a secondary expansion is frequently observed, this has occurred from a reduced number of ancestral species (often only one), clearly derived from lines that originated in the Old World. In some cases it is difficult to determine if the place of origin of a given group is the Old World or America. It is possible that various elements that we consider as belonging to the Paleo-American pattern originated in North America and then extended on the one hand to the Transition Zone and on the other to Eurasia via the Behring Strait. All the evidence, however, seems to indicate that a majority of the Paleo-American elements originated in the Old World, particularly in the tropics.

The groups that gave rise to Paleo-American elements exhibit wide ecologic diversification both in America and in the Old World, although they clearly tend to predominate in warm and warm-temperate climates. There is a marked contrast, both in America and elsewhere, between the Paleo-American elements and the cold or cold-temperate forms that follow the Nearctic Pattern.

The lines that follow the Paleo-American pattern have existed for a prolonged period in the Mexican Transition Zone; the majority of species are restricted to this zone. In these lines the effect of ecological heterogeneity is conspicuous (32). (It is also conspicuous, but to a smaller degree, in the other group of early origin, the South American species following the high plateau distribution pattern.) A given line or genus can include species adapted to arid conditions, grasslands, mountains, and even tropical rain forests. This is a phenomenon of ecological vicariance that occurs in lines of wide and ancient dispersion; series of populations (later species or groups of species) are fragmented and isolated by geographic and ecologically changing conditions. In the mountains this phenomenon is particularly important. The ranges north of the Isthmus of Tehuantepec contain many Paleo-American elements, which decrease in number south of the isthmus but are still important. This interpretation of the evolution of the Paleo-American pattern coincides with Sharp's (70) analysis of the Mexican flora.

SUMMARY

The montane insects of the Mexican Transition Zone form two well-defined groups. The first occupies the ranges north of the Isthmus of Tehuantepec. The second occupies the mountain systems south of the isthmus, the Sierra Madre de Chiapas and the mountains of Central America extending to the Nicaraguan depression.

A high proportion of the montane insects north of the Isthmus of Tehuantepec are clearly of northern origin. The mountain ranges have been the route for expansion of these insects toward the south, particularly for the most modern insects in the Transition Zone. But these ranges (especially the Transverse Volcanic Belt and the mountains of Oaxaca and Guerrero) have also been areas of frequent isolation and vicariance.

The northern elements fit into the Nearctic or Paleo-American distribution patterns, according to the time of their entrance into the Transition Zone. Additionally, the mountains north of the Isthmus of Tehuantepec include a certain number of Meso-American elements. Insects from the tropical lowlands and from the highland plateau are exceptional.

The mountains north of the isthmus have been, in general, a poor route for the expansion of the more modern northern fauna (Nearctic pattern), particularly the hygrophilous elements. Expansion is hampered by the lack of lakes and rivers in the Mexican ranges, particularly in the Western Sierra Madre.

Most of the insects of the mountains south of the isthmus evolved in the Central American Nucleus. This region received an ancient and very important contribution from South America (but not from the Andes), and also acquired some elements of northern origin. The fauna that evolved in the area has expanded toward South America and toward the north through the Isthmus of Tehuantepec; this expansion corresponds to the Meso-American Distribution Pattern.

Immediately south of the Transition Zone, in the Sierra de Talamanca (Costa Rica–Panama), the mountains were colonized in large part by lines from the tropical lowlands. This is an exceptional phenomenon in the Transition Zone.

ACKNOWLEDGMENTS

During the last few years I have discussed different aspects of the biogeography of Mexico with Gustavo Aguirre, Miguel Angel Morón, and Pedro Reyes-Castillo, researchers of the Institute of Ecology (Mexico); they have provided very valuable information.

Pedro Reyes-Castillo has been collaborating with me for many years. Our collaboration has been close and fruitful. I am very grateful for his careful comments about this paper.

Exequiel Ezcurra, also from the Institute of Ecology, translated the paper from Spanish to English. I thank him very much for the meticulousness and care with which he accomplished this task.

Literature Cited

1. Anderson, T. H. 1969. *Geology of the San Sebastian Huehuetenango Quadrangle, Guatemala, Central America.* PhD thesis. Univ. Texas, Austin
2. Ball, G. E. 1959. A taxonomic study of the North American Licinini with notes on the Old World species of the genus *Diplocheila* Brullé. *Mem. Am. Entomol. Soc.* 16:1–258
3. Ball, G. E. 1968 (1970). Barriers and southward dispersal of the holarctic boreomontane element of the family Carabidae in the mountains of Mexico. *An. Esc. Nac. Cienc. Biol. Méx.* 17:91–112
4. Ball, G. E. 1975. Pericaline Lebiini: Notes on classification, a synopsis of the New World genera, and a revision of the genus *Phloeoxena* Chaudoir (Coleoptera: Carabidae). *Quaest. Entomol.* 11:143–242
5. Ball, G. E. 1978. The species of the Neotropical genus *Trichopselaphus* Chaudoir (Coleoptera: Carabidae): Classification, phylogeny and zoogeography. *Quaest. Entomol.* 14:447–89
6. Ball, G. E., Erwin, T. L. 1969. A taxonomic synopsis of the tribe Loricerini (Coleoptera: Carabidae). *Can. J. Zool.* 47:877–907
7. Ball, G. E., Negre, J. 1972. The taxonomy of the Nearctic species of the genus *Calathus* Bonelli (Coleoptera: Carabidae). *Trans. Am. Entomol. Soc. Philadelphia* 98:412–533
8. Bell, R. T. 1979. Zoogeography of Rhysodini—Do beetles travel on driftwood? In *Carabid Beetles: Their Evolution, Natural History, and Classification,* ed. T. L. Erwin, G. E. Ball, D. R. Whitehead, pp. 331–42. The Hague: Junk. 635 pp.
9. Bohnenberger, O. 1977. Plate tectonics hypothesis as related to Central America. *Bol. Inst. Geol. Méx.* 101:33–46
10. Bolívar y Pieltain, C., Hendrichs, J. 1972. Distribución en Norteamérica del género holártico *Pteroloma* Gyllenkal, y estudio de tres nuevas formas mexicanas (Col., Silph.). *Ciencia Méx.* 27(6):207–16
11. Brundin, L. Z. 1981. Croizat's panbiogeography versus phylogenetic biogeography. In *Vicariance Biogeography: A Critique,* ed. G. Nelson, D. E. Rosen, pp. 94–138. New York: Columbia Univ. Press. 593 pp.
12. Burke, K., Fox, P. J., Segnor, A. M. C. 1978. Buoyant ocean floor and the evolution of the Caribbean. *J. Geophys. Res.* 83:3949–54
13. Castillo, C., Reyes-Castillo, P. 1984. Biosistemática del género *Petrejoides* Kuwert (Coleoptera: Passalidae). *Acta Zool. Mex.* NS 4:1–84
14. Coope, G. R. 1979. Late Cenozoic fossil Coleoptera: Evolution, biogeography, and ecology. *Ann. Rev. Ecol. Syst.* 10:247–67
15. Dengo, G. 1968. *Estructura Geológica, Historia Tectónica y Morfología de América Central.* México/Buenos Aires: Cent. Reg. Ayuda Téc. (AID). 52 pp.
16. Duellman, W. E. 1966. The Central American herpetofauna: an ecological perspective. *Copeia* 4:700–19
17. Duellman, W. E. 1970. *The Hylid Frogs of Middle America.* Lawrence, Kansas: Mus. Nat. Hist. Univ. Kansas. 753 pp.
18. Erwin, T. L. 1979. Thought on the evolutionary history of ground beetles: hypotheses generated from comparative faunal analyses of lowland forest sites in temperate and tropical regions. See Ref. 8, pp. 539–92
19. Erwin, T. L. 1979. The American connection, past and present, as a model blending dispersal and vicariance in the study of biogeography. See Ref. 8, pp. 335–67
20. Erwin, T. L. 1981. Taxon pulses, vicariance, and dispersal: an evolutionary synthesis illustrated by carabid beetles. See Ref. 11, pp. 169–83
21. Erwin, T. L. 1982. Small terrestrial ground-beetles of Central America (Carabidae). *Proc. Calif. Acad. Sci.* 42(19): 455–96
22. Evans, H. E. 1966. A revision of the Mexican and Central American spider wasps of the subfamily Pompilinae (Hymenoptera: Pompilidae). *Mem. Am. Entomol. Soc.* 20:1–442
23. Fittkau, E. J., Illies, J., Klinge, H., Schwabe, G. H., Sioli, H., eds. 1969. *Biogeography and Ecology in South America.* The Hague: Junk
24. Goulet, H. 1974. Classification of the North and Middle American species of

the genus *Pelmatellus* Bates (Coleoptera: Carabidae). *Quaest. Entomol.* 10:80–102

25. Gressitt, J. L. 1958. Zoogeography of insects. *Ann. Rev. Entomol.* 3:207–30
26. Gressitt, J. L. 1974. Insect biogeography. *Ann. Rev. Entomol.* 19:293–343
27. Halffter, G. 1962. Explicación preliminar de la distribución geográfica de los Scarabaeinae mexicanos. *Acta Zool. Mex.* 5(4–5):1–17
28. Halffter, G. 1964. La entomofauna americana, ideas acerca de su origen y distribución. *Folia Entomol. Mex.* 6:1–108
29. Halffter, G. 1964. Las regiones Neártica y Neotropical desde el punto de vista de su entomofauna. *An. 2ndo Congr. Lat.-Am. Zool., São Paulo* 1:51–61
30. Halffter, G. 1972. Eléments anciens de l'entomofaune Neotropicale: ses implications biogéographiques. In *Biogeographie et Liasons Intercontinentales au Cours du Mésozoique, 17me Congr. Int. Zool., Monte-Carlo* 1:1–40
31. Halffter, G. 1974. Eléments anciens de l'entomofaune Neotropicale: ses implications biogéographiques. *Quaest. Entomol.* 10:223–62
32. Halffter, G. 1976. Distribución de los insectos en la Zona de Transición Mexicana. Relaciones con la entomofauna de Norteamérica. *Folia Entomol. Mex.* 35:1–64
33. Halffter, G. 1978. Un nuevo patrón de dispersión en la Zona de Transición Mexicana: el Mesoamericano de montaña. *Folia Entomol. Mex.* 39–40:219–22
34. Hastenrath, V. S. 1974. Spuren pleistozaner Vereisung in den Altos de Cuchumatanes, Guatemala. *Eiszeitalter Ggw.* 25:25–34
35. Hedberg, O. 1969. Evolution and speciation in a tropical high mountain flora. *Biol. J. Linn. Soc.* 1:135–48
36. Heine, K. 1973. Variaciones más importantes del clima durante los últimos 40,000 años en México. *Comun. Proyecto Puebla-Tlaxcala (Fund. Alemana Invest. Cient.)* 7:51–8
37. Howden, H. F. 1969. Effects of the Pleistocene on North American insects. *Ann. Rev. Entomol.* 14:39–56
38. Illies, J. 1983. Changing concepts in biogeography. *Ann. Rev. Entomol.* 28:391–406
39. Kavanaugh, D. H. 1979. Rates of taxonomically significant differentiation in relation to geographical isolation and habitat: examples from a study of the Nearctic *Nebria* fauna. See Ref. 8, pp. 35–57
40. Klaus, D. 1973. Las fluctuaciones del

clima en el valle Puebla-Tlaxcala. *Comun. Proyecto Puebla-Tlaxcala (Fund. Alemana Invest. Cient.)* 7:59–62
41. Kohlmann, B. 1984. Biosistemática de las especies norteamericanas del género *Ateuchus* (Coleoptera: Scarabaeidae). *Folia Entomol. Mex.* 60:3–81
42. Linsley, E. G. 1963. Behring arc relationship of Cerambycidae and their host plants. In *Symposium: Pacific Basin Biogeography,* ed. J. L. Gressitt, pp. 159–78. Honolulu: Bishop Mus. Press
43. Llorente-Bousquets, J. 1983. Sinopsis sistemática y biogeográfica de los Dismorphiinae de México con especial referencia al género *Enantia* (Lepidoptera: Pieridae). *Folia Entomol. Mex.* 58:1–207
44. Lloyd, J. J. 1963. Tectonic history of the South Central American Orogen. *Am. Assoc. Pet. Geol. Mem.* 2:88–100
45. Lorenzo, J. L. 1964. *Los Glaciares de México.* México City: Inst. Geofís. UNAM. 123 pp.
46. MacVean, C., Schuster, J. C. 1981. Altitudinal distribution of passalid beetles (Coleoptera: Passalidae) and Pleistocene dispersal on the volcanic chain of northern Central America. *Biotropica* 13(1):29–38
47. Malfait, B. T., Dinkelmann, M. G. 1972. Circun-Caribbean tectonic and igneous activity and the evolution of the Caribbean Plate. *Geol. Soc. Am. Bull.* 83:251–72
48. Mani, M. 1968. *Ecology and Biogeography of High Altitude Insects.* The Hague: Junk. 527 pp.
49. Martin, P. S. 1958. A biogeography of reptiles and amphibians in the Gómez Farías region, Tamaulipas, México. *Misc. Publ. Mus. Zool. Univ. Mich.* 101:1–102
50. Mateu, J. 1974. Sobre algunos linajes de carábidos boreo-montanos de México y sus relaciones con el poblamiento entomológico del Sistema Volcánico Transversal. *Rev. Soc. Mex. Hist. Nat.* 35:181–224
51. Mateu, J. 1974. Sur les *Microlestes* Schmidt-Goebel du Mexique (Coleoptera: Carabidae). *Entomol. Arb. Mus. G. Frey Tutzing Muenchen* 25:261–75
52. Minster, J. B., Thomas, H. J. 1978. Present-day plate motions. *J. Geophys. Res.* 83:5331–56
53. Morón, M. A. 1981. Descripción de dos nuevas especies de *Plusiotis* Burm. y discusión de algunos aspectos zoogeográficos del grupo de especies "*costata*" (Coleoptera: Melolonthidae). *Folia Entomol. Mex.* 49:49–69
54. Morón, M. A. 1983. A revision of the

subtribe Heterosternina (Coleoptera: Melolonthidae: Rutelinae). *Folia Entomol. Mex.* 55:31–101

55. Morón, M. A. 1986. Análisis biogeográfico preliminar del género *Plusiotis* Burmeister (Coleoptera: Melolonthidae). In *Memorias Simposio Biogeografía de Mesoamérica*, ed. L. D. Gómez. San José, Costa Rica: Mus. Nacl. Costa Rica/Mesoamerican Ecology Inst. In press

56. Morón, M. A. 1986. *El Género Phyllophaga en México. Morfología, Distribución y Sistemática Supraespecífica (Coleoptera).* México City: Inst. Ecol. 341 pp.

57. Munroe, E. 1965. Zoogeography of insects and allied groups. *Ann. Rev. Entomol.* 10:325–44

58. Noonan, G. R. 1979. The science of biogeography with relation to carabids. See Ref. 8, pp. 295–317

59. Noonan, G. R. 1981. South American species of the subgenus *Anisotarsus* Chaudoir (Coleoptera: Carabidae). II. Evolution and biogeography. *Milwaukee Public Mus. Contrib. Biol. Geol.* 45:1–117

60. Peck, S. B., Anderson, R. S. 1985. Taxonomy, phylogeny and biogeography of the carrion beetles of Latin America (Coleoptera: Silphidae). *Quaest. Entomol.* 21(3):247–317

61. Quintero, G., Reyes-Castillo, P. 1983. Monografía del género *Oileus* Kaup (Coleoptera: Passalidae). *Folia Entomol. Mex.* 54:1–50

62. Raven, P. H., Axelrod, D. I. 1975. History of the flora and fauna of Latin America. *Am. Sci.* 63:420–29

63. Reichardt, H. 1979. The South American carabid fauna: endemic tribes and tribes with African relationships. See Ref. 8, pp. 319–25

64. Reyes-Castillo, P. 1978. Revisión monográfica del género *Spurius* Kaup (Coleoptera: Passalidae). *Stud. Entomol.* 20(1–4):269–90

65. Reyes-Castillo, P., Halffter, G. 1978. Análisis de la distribución geográfica de la tribu Proculini (Coleoptera: Passalidae). *Folia Entomol. Mex.* 39–40:222–26

66. Reyes-Castillo, P., Quintero, G. 1977. The species of *Oileus* Kaup and their distribution (Coleoptera: Passalidae). *Folia Entomol. Mex.* 37:31–41

67. Ross, H. H. 1967. The evolution and past dispersal of the Trichoptera. *Ann. Rev. Entomol.* 12:169–206

68. Rzedowski, J. 1978. *Vegetación de México.* México City: Limusa. 432 pp.

69. Savage, J. M. 1966. The origins and history of the Central American herpetofauna. *Copeia* 4:719–66

70. Sharp, A. J. 1966. Some aspects of Mexican phytogeography. *Ciencia México City* 24(5–6):229–32

71. Stuart, L. C. 1966. The environment of the Central American cold-blooded vertebrate fauna. *Copeia* 4:684–99

72. Udvardy, M. D. F. 1975. A classification of the biogeographical provinces of the world. *Int. Union Conserv. Nat. Occas. Pap. 18* Morges, Switzerland: IUCN

73. Wake, D. B., Lynch, J. F. 1982. Evolutionary relationships among Central American salamanders of the *Bolitoglossa franklini* group, with a description of a new species from Guatemala. *Herpetologica.* 38(2):257–72

74. Whitehead, D. R. 1972. Classification, phylogeny, and zoogeography of *Schizogenius* Putzeys (Coleoptera: Carabidae). *Quaest. Entomol.* 8:131–348

75. Whitehead, D. R. 1976. Classification and evolution of *Rhinochenus* Lucas (Coleoptera: Curculionidae), and Quaternary Middle American zoogeography. *Quaest. Entomol.* 12(2):118–201

76. Whitehead, D. R., Ball, G. E. 1975. Classification of the Middle American genus *Cyrtolaus* Bates (Coleoptera: Carabidae). *Quaest. Entomol.* 11:591–619

77. Zunino, M. 1984. Sistematica generica dei Geotrupinae (Coleoptera, Scarabaeoidea: Geotrupidae), filogenesi della sottofamiglia e considerazioni biogeografiche. *Boll. Mus. Reg. Sci. Nat. Torino* 2(1):9–162

78. Zunino, M., Halffter, G. 1986. Análisis taxonómico, ecológico y biogeográfico de un grupo americano de *Onthophagus* (Coleoptera: Scarabaeidae). *Boll. Mus. Reg. Sci. Nat. Torino.* In press

Ann. Rev. Entomol. 1987. 32:115–44

THE BIOLOGY OF DACINE FRUIT FLIES

B. S. Fletcher

CSIRO, Division of Entomology, 55 Hastings Road, Warrawee, New South Wales 2074, Australia

INTRODUCTION AND OVERVIEW

Dacine fruit flies, one of the major subfamilies of the Tephritidae, are a biologically interesting and economically important group of Diptera. The Dacinae have traditionally been divided into two main genera, *Dacus* and *Callantra*.[1] Certain *Dacus* species, e.g. *D. cucurbitae, D. dorsalis, D. oleae, D. tryoni,* and *D. zonatus,* together with the other major pest tephritids belonging to the genera *Anastrepha, Rhagoletis,* and *Ceratitis* of the subfamilies Trypetinae and Ceratitinae, comprise one of the most important global groups of pests attacking fruit and vegetable crops (12, 22).

Many aspects of dacine biology and ecology have been covered in earlier reviews on the family Tephritidae (10, 27, 163), and this review concentrates on those aspects that are of special interest, or areas in which major advances have been made in recent years. Owing to space restrictions, economic aspects are not covered.

There are approximately 700 known species of dacine fruit flies, and the rate of discovery of new species suggests there may be upwards of a thousand species in total. Apart from *D. oleae,* which occurs in southern Europe, they are endemic to Africa, Asia, Australia, and the South Pacific (1, 41, 91, 92, 106, 151), and occur predominantly in the moist rain forests of the tropical and subtropical zones. They are absent from the Americas, although three species have been introduced into Hawaii in the past century (21).

---

[1]In a recent monograph, Munro (151) raised the Dacinae to family rank as the Dacidae, and divided it into approximately 50 genera. As it is uncertain if this higher taxonomic classification will be generally adopted, the original generic names have been retained for species mentioned in this review.

0066-4170/87/0101-0115$02.00

Dacines can be divided into two major groups based on whether abdominal tergites 3–5 are free or fused. Of the 182 species recorded from Africa and Madagascar, all but six (including the recently introduced *D. cucurbitae*) have fused tergites, whereas all but about 25 of the species from other areas have free tergites. Both in Africa and elsewhere, the majority of species with fused tergites utilize plants belonging to the families Asclepiadaceae and Cucurbitaceae as larval hosts (41, 151).

Although a few of the African species breed in stems or leaves of asclepiads and some species, including *D. curcurbitae,* infest the male flowers of cucurbits (151), the vast majority of dacines are frugivorous and during the larval stages feed on and develop in fruits or seed pods. Most are monophagous or stenophagous, and the majority of the others are oligophagous; Of the 83 Australian species for which there are host records, 80% have been recorded from a single host. Polyphagous species are relatively uncommon; only two of the Australian species, *D. tryoni* and *D. neohumeralis,* fit into this category (54).

Owing to their frugivorous habits, a number of species have become major pests of fruits and vegetables. It is significant that with the exception of *D. oleae,* which is restricted to olives, the most serious pest species are both polyphagous and multivoltine, e.g. *D. dorsalis, D. cucurbitae, D. tryoni,* and *D. zonatus.* Much of the interest in dacine fruit flies has centered on the development of effective control and/or eradication procedures against these polyphagous tropical species and *D. oleae.* Measures include area suppression with bait sprays or traps, male annihilation with synthetic lures, and release of sterile insects (22, 41, 173). In recent years the application of these techniques and the development of associated technologies such as mass rearing, radiation biology, timing of spray programs, establishment of economic thresholds, and population monitoring have greatly increased the success of suppression and eradication programs.

Nearly all known dacines have a similar basic life cycle. The females deposit their eggs into ripening host fruit. Larvae pass through three instars before puparium formation, which normally takes place in the ground, although in a few species it occurs inside the host fruit. After emergence, the adults have a maturation period of several days before becoming sexually active, which coincides in females with the completion of vitellogenesis. Males can mate frequently, but after mating females become sexually unreceptive for several weeks (12, 46, 196). The adults need to feed regularly on carbohydrates and water to survive, and the females require proteinaceous material for eggs to mature (10, 27).

All the polyphagous and oligophagous species are multivoltine, as are the majority of stenophagous and monophagous species. Some of the monophagous species have restricted breeding periods owing to the limited seasonal availability of their hosts. A few species, e.g. *D. opiliae,* are essentially univoltine (52, 53). No known dacines have a true diapausing stage, but the adults of some species are able to pass unfavorable periods of the year in a facultative reproductive "diapause" during which they aggregate in suitable refuges and remain in or revert to a sexually immature state (52, 88, 182).

FUNCTIONAL MORPHOLOGY AND ANATOMY

Morphology and anatomy of the different life stages, particularly characters useful for taxonomic purposes, have been described in detail elsewhere (41, 151). Only those aspects relevant to an understanding of the group's biology are considered here.

The adult body coloration of different species varies from black through various shades of brown to orange or yellow. Yellow marks, particularly on the thorax, give many species a somewhat wasplike appearance. This resemblance is particularly pronounced in certain *Dacus* subgenera and *Callantra* spp., which have petiolate abdomens, heavily fuscated costal stripes on the wings, and a jerky, wasplike walk (41, 151).

The paired antennae each consist of three segments. Scanning electron microscope studies on *D. tryoni* (78) and *D. oleae* (86) indicate that the outer segment is covered with long cuticular spines interspersed with large numbers of chemosensilla of several distinct morphological types. The functions of the chemosensilla are unknown, although electroantennogram (EAG) studies on *D. oleae* (197) indicate that both sexes have sensilla that respond to components of the female sex pheromone, and behavioral studies suggest that there are sensilla that respond to a wide range of volatiles, including fruit and food odors (13, 38, 75).

Male dacines, except those of some groups such as *Gymnodacus,* typically have a pair of combs (or pectens) comprised of stiff curved bristles on the lateral hind margins of the third abdominal tergite. These combs function as stridulatory organs during courtship (41, 151). Both sexes have a pair of tergal glands (ceromae) that open onto the surface of tergite 5. These are apparently absent from other tephritids. They consist of dense groups of minute alveolae that secrete a waxy substance, which is spread onto the body and wings during preening (151).

In females, abdominal segments 7–9 form the ovipositor, which is usually smooth and pointed but is serrated in some species. The ovipositor enables females to pierce the skin of their hosts. The apical segment has a number of chemosensilla; the most prominent are the preapical setae that arise from lateral grooves on either side of the segment (91). These presumably play an important role in fruit discrimination.

The general anatomy of dacine fruit flies is fairly typical of cyclorrhaphan Diptera. The structure of the brain of *D. oleae* (178) and the nervous and neuroendocrine systems of *D. tryoni,* including the distribution of neurosecretory cells, have been studied in some detail (5, 32).

In dacines and many other tephritids, a diverticulum leads from the esophagus anterior to the brain, and terminates in a blind sac (the esophageal bulb) (79). In *D. oleae,* this sac has been implicated in the storage, development, and release into the alimentary canal of symbiotic bacteria (79, 138, 159). Female *D. oleae* also have about 25 bacteria-containing diverticula (smearing organs), which open into the rectal tube (138, 159).

In the males of all species examined, the ventral posterior part of the rectum includes a modified glandular region, which in the subgenus *Bactrocera* consists of a reservoir and secretory sac (59). This region has been shown in *D. tryoni, D. dorsalis,* and *D. cucurbitae* to be the site of production and storage of a sex attractant pheromone (59, 119). In *D. oleae,* a species in which the female produces the principal sex attractant, both sexes have glandular complexes (43, 135).

The female reproductive system consists of paired ovaries containing a number of polytrophic ovarioles, two spermathecae used for sperm storage, and paired collateral glands. During oogenesis, follicles containing 16 cystocytes, of which one becomes the oocyte and the others become nurse cells, are budded off from the germarium in sequence (72, 95, 125). Ovarian maturation in dacines is asynchronous; terminal follicles at different stages of development occur together in the same ovary.

In unfavorable conditions, the contents of developing follicles may be resorbed. This has been observed in *D. tryoni* females during winter months when mean temperatures fall below the threshold for development. Resorption occurs in *D. oleae* during the summer months, owing to high temperatures, low humidity, and lack of suitable fruit (62, 72, 147).

The number of ovarioles per ovary varies intraspecifically depending on body size. There are also interspecific differences. A survey of some Australian species revealed a broad correlation between ovariole number and host range (54). Two highly polyphagous species (*D. tryoni* and *D. neohumeralis*) had more than 35 ovarioles per ovary; nine oligophagous species (including *D. halfordiae, D. cucumis,* and *D. kraussi*) had 20–29 ovarioles; and six monophagous or stenophagous species (including *D. cacuminatus, D. aglaia,* and *D. visendus*) had 8–25 ovarioles. The major exception was the banana fly, *D. musae,* which, although stenophagous, had 38 ovarioles per ovary.

Egg size and structure show some variation from species to species, but there is no obvious correlation between egg size and body size or ovariole number (54). Certain of the cucurbit-infesting species, e.g. *D. cucurbitae* and

D. cucumis, produce the largest eggs; these have the shortest hatching time (54, 198).

The larvae of dacines are typical acephalic cyclorrhaphan maggots with an involuted head, three thoracic segments, and eight abdominal segments. The most important diagnostic features are the mouth hooks and anterior and posterior spiracles, which change during each instar (2). Larvae of *D. oleae* have four spherical-shaped mycetomes that arise as diverticula just posterior to the proventriculus and that contain bacteria (159).

BACTERIAL ASSOCIATIONS

One of the most interesting aspects of dacine biology is the role of bacteria in the nutrition and survival of larvae and adults. In his pioneering study, Petri (159) concluded that *Pseudomonas savastanoi,* a bacterial pathogen of olive trees, had a symbiotic relationship with *D. oleae.* Recent studies have confirmed the presence of symbiotic bacteria in the alimentary canal of *D. oleae,* and the ultrastructure and mode of transmission of these bacteria have been described in detail (79, 138). They multiply in the esophageal bulb and are released as compact masses, which pass down the gut and break down at the beginning of the hind gut. Electron microscopy has indicated that the bacteria subsequently collect and multiply in the diverticula that open into the rectum, and are transferred from there to the eggs during oviposition; some enter the micropyle and others remain on the egg surface (138). The bacteria enter the larvae at the time of hatching and multiply in the four mycetomes. Small numbers remain in the pupal stage and reinvade the esophageal bulb when it forms.

The identity of the bacteria in *D. oleae* remains uncertain, although current evidence suggests that they are not *P. savastanoi.* It has been postulated that they are a specific symbiotic bacterium difficult to culture in vitro (81). However, a wide variety of bacteria have now been isolated from the esophageal bulb and alimentary canal of both laboratory-reared and wild *D. oleae* (187, 205); these bacteria include many species also present on olive leaves. Furthermore, the species of bacteria both in wild flies and on leaves change from season to season. *P. savastanoi* was not detected in isolates from wild olive flies, although it occurs commonly on olive leaves.

A single *Pseudomonas*-type bacterium was identified from eggs, larvae, and adults of *D. cucurbitae* (85), but diverse bacterial species were found associated with different life stages in a number of other dacines including 15 species from *D. tryoni* and 14 from *D. jarvisi* (58). The three most common bacterial species in the crop and stomach of adult *D. tryoni, D. neohumeralis, D. cacuminatus,* and *D. musae* collected in Queensland were *Erwinia herbicola, Enterobacter cloacae,* and *Klebsiella oxytoca* (42). The last two, interestingly, have also been identified as the most common bacteria associated with *Rhagoletis* spp. in North America (94). The same three bacteria were also the most common species in *D. tryoni* oviposition punctures and larval-induced fruit rot. Female *D. tryoni* may introduce the bacteria into the fruit in a different manner than *D. oleae*. They apparently regurgitate fluid containing bacteria onto the fruit surface and oviposit through it (42).

Petri (159) suggested that the bacteria of *D. oleae* might help in the breakdown and digestion of fruit tissues. More recently it has been shown that when mature larvae or adults of *D. oleae* or their oviposition sites in newly infested fruit are treated with antibiotics, the larval development of offspring is suppressed (122, 195). When adults were treated, larval offspring were not able to develop in green olives, and they died in the first instar. However, they were able to develop in ripe olives, which have a higher amino acid content. *D. jarvisi* larvae were unable to complete development in artificial medium containing unhydrolyzed proteins in the absence of bacteria, but developed normally when *E. cloacae* were present (58). Bacteria that released proteases to hydrolyze the proteins in situ were detrimental to the larvae. Together, these results suggest that the bacteria are ingested and digested to provide amino acids and possibly other growth factors. The bacteria may also have other functions, including detoxification of plant defense chemicals and suppression of pathogenic fruit-rot microorganisms (94).

The significance of bacteria in the biology of adult flies is still conjectural. It has been proposed that leaf-surface bacteria are a major source of dietary protein for fruit flies in tropical areas (30, 38). However, recent studies by Drew & Lloyd (42) indicated that colonization of fruit surfaces by the three types of bacteria found in the guts of *D. tryoni* increases significantly after adult flies start to arrive in host trees. This finding led the researchers to suggest that the flies inoculate the surface of the fruit with bacteria by regurgitation, and utilize the resulting bacterial colonies as a convenient protein source for egg development (42).

BEHAVIOR

Adult Activities

In general, dacine fruit flies are diurnal insects and at night rest on the undersides of leaves of host plants or other trees. Their daytime activities can be divided into five main functional categories: feeding, mating, ovipositing, dispersing, and resting or sheltering. The time spent in each type of activity depends upon many factors including age, sex, availability of mates and hosts, and short- and long-term climatic conditions.

DAILY RHYTHMS   Certain activities are restricted to fairly specific times of day owing to the interaction of internal circadian rhythms and external factors such as temperature and light intensity (112, 177). In certain species mature males and mature virgin females show peaks of locomotory activity and other sexually associated behaviors, including male stridulation, pheromone release, and response to sex pheromones, at dusk (68, 177, 193). Studies on *D. tryoni* indicate that the synchronization of the various components of sexual behavior with dusk involves the effect of decreasing light intensity on an underlying rhythm of readiness to mate, which has a genetic basis (176, 193). This rhythm is under circadian control; it persisted for at least four days, with a periodicity of 28 hr, under a constant low light intensity of 10 lux (191, 193).

DETECTION OF HOST PLANTS   Feeding, mating, and ovipositing all involve an initial period of foraging for specific resources (168). Mating and feeding in dacines do not appear to be as closely confined to host plants as in many other tephritids. In some of the cucurbit-infesting species, feeding and mating take place almost exclusively on nonhost plants (10, 134, 179a). Nevertheless, in the majority of species host trees seem to be important sites for foraging for food and mates, as well as for oviposition.

Because they are highly mobile, the polyphagous species might be expected to have evolved efficient mechanisms for detecting hosts, but information on this is limited. All the dacines studied responded strongly to flat yellow squares or rectangles with peak reflectance close to that of green leaves (550 nm), and were less attracted to colored boards with peak reflective wavelengths above or below this value (12, 93). In *D. tryoni*, response was increased with checkerboard patterns, which gave a contrast effect; yellow and green and yellow and red induced the maximum response (144). Visual cues with the spectral wavelength of green leaves appear to act as supranormal foliage stimuli in tephritids and in numerous other herbivorous insects (167), and in combination with certain shapes act as token tree stimuli. *D. oleae*, however, whose sole host, the olive tree, has leaves with a characteristic undersurface color that reflects maximally between 380 and 520 nm, showed no specific response either to the undersurface of olive leaves or to yellow tints with the same spectral reflection (165).

The available data suggest that host detection may be relatively short-range and may involve important olfactory as well as visual cues. Although adult dacines respond to a number of plant extracts, many are from nonhosts or are of general occurrence (84, 114), and there is no evidence at present that flies can locate host trees solely on the basis of visual or olfactory cues from the foliage. There is an increasing amount of evidence that fruit volatiles are important in enabling flies not only to locate fruit, but also to discriminate between hosts and nonhosts and among fruits at different stages of ripeness (52, 56, 75).

OVIPOSITION   Studies on *D. tryoni* using artificial fruit models containing pieces of fruit (75) indicated that attraction to the fruit, exploratory behavior after arrival, piercing of the fruit, and egg-laying are all influenced by olfactory stimuli, although color and shape also have a role.

Which of the large number of volatiles that emanate from fruit induce these responses is not known, although 2-chloroethanol *(D. tryoni)* (73) and *trans*-6-nonen-1-ol acetate *(D. cucurbitae)* (116) attracted flies and stimulated probing behavior in laboratory bioassays. Polyphagous species such as *D. tryoni* and *D. dorsalis* may respond to a wide range of fruit volatile combinations, whereas monophagous species are likely to respond much more specifically. In laboratory studies monophagous species were able to discriminate between host and nonhost fruits even before landing on them (52, 54).

D. oleae females can recognize olive fruits purely by vision, on the basis of shape, size and color (90). A preference for fruit models 35 mm in diameter and for normal-sized (18 mm) yellow models reflecting maximally between 560 and 610 nm suggests that the response can be influenced by supranormal stimuli (109, 110). Mature olives do, however, release volatiles that are attractive to females (47). Some compounds that stimulate oviposition, including certain derivatives of oleuropeine, have been extracted from mature olives (82, 127).

Once a potentially suitable fruit has been located, the female explores it before attempting to oviposit. Many species, including *D. cucurbitae, D. dorsalis,* and *D. tryoni* (155, 160, 166) oviposit in recent oviposition stings of other females or in breaks in the skin caused by other agents. Females of *D. tryoni* (160) and *D. ciliatus* (183) choose shaded rather than sunny positions on fruit for oviposition.

Different species have different clutch sizes, ranging from 1 egg per oviposition in *D. oleae* and *D. latifrons* (28, 199) to 30–40 as observed on some occasions in *D. cucurbitae* (155, 166). Clutch size is correlated to some extent with the size of host fruits; species that infest small hosts normally lay the fewest eggs per clutch (54). Females of some species, e.g. *D. tryoni* and *D. dorsalis,* adjust clutch size depending on the size of the fruit, laying fewer eggs in smaller fruit. Length of host deprivation, suitability of the host for larval development, and even the effort required to puncture the fruit skin can also influence clutch size (134, 174a; B. S. Fletcher, unpublished).

Preferences for particular hosts appear to be genetically fixed rather than determined by conditioning in many polyphagous species (52, 75). Although there is often a positive correlation between host preferences and host suitability for larval development (96), females sometimes show preference for and lay large numbers of eggs in fruit varieties that inhibit larval development or cause high larval mortality (19, 182).

Time spent ovipositing varies, but *D. tryoni* takes about 1–3 min per oviposition and *D. jarvisi,* which lays more eggs per clutch, takes about 4–6 min (55). If other females land on the fruit, ovipositing females will interrupt their activity and attempt to drive them off (160).

There is no evidence that any dacine deposits an epideictic oviposition-deterring pheromone on the fruit after egg-laying (12, 55, 166) as many other tephritids do (164). However, after oviposition female *D. oleae* suck up the olive juice that exudes from the oviposition puncture and spread it on the surface of the fruit (28) to deter other females from ovipositing. A number of compounds in olive juice, including dihydroxyphenyl ethanol (201a), pyrocatechol, benzaldehyde and acetophenone (83), have been found to have deterrent properties.

Presence of larvae in fruit has a deterrent effect on ovipositing females of *D. oleae* (28), *D. cucurbitae* (166), and *D. tryoni* and *D. jarvisi* (55). Fitt (55) found that the effect was detectable as soon as the larvae hatched and started feeding, and that in *D. tryoni* and *D. jarvisi* the effect was interspecific. The stimulus was shown to be present in partly worked fruit tissue from which larvae had been removed. In *D. oleae,* a deterrent effect was noted in fruit containing second- and third-stage larvae, and it has been suggested that this was due to liposoluble volatile substances released from olive tissues attacked by the feeding larvae (83).

SEXUAL BEHAVIOR   The majority of dacine fruit flies, including most pest species, mate at dusk under low light intensity (less than 1000 lux), although some species, e.g. *D. oleae,* start mating in the late afternoon at a somewhat higher light intensity (4, 124, 193, 170, 181). In contrast, there are a few species, including *D. neohumeralis, D. expandens, D. tenuifascia,* and *D. tsuneonsis,* that mate during the day at high light intensities (48, 97, 177).

Mating behavior has rarely been observed in the field. It seems, however, that mating occurs predominantly on the foliage of host plants. Even so, matings on nonhosts have been observed in *D. tryoni* (10), and many of the cucurbit-infesting species mate exclusively on nonhosts (134, 179a). Field-cage studies indicate that males of *D. tryoni* (192) and *D. cucurbitae* (121) engage in lekking behavior. As light intensity drops near dusk, they aggregate on specific parts of a tree and take up individual territories on leaves, which they aggressively defend from incursions by other males. Lekking behavior was not observed in *D. neohumeralis,* which is a day-mating species (192).

Sexually active males of *D. tryoni, D. dorsalis,* and *D. cucurbitae* produce a high-pitched buzzing sound by rapidly drawing their wings across their stridulatory organs, and simultaneously release pheromone from their rectal pheromone glands (77, 119, 121). The role of stridulation in dacines has not been determined, but in some other tephritids it is known to stimulate male sexual activity and heighten the response of females (202). Studies in laboratory bioassays and field cages indicate that the male-produced pheromone attracts females from several meters and evokes copulatory responses at close range (7, 15, 49). The courtship sequence in *D. oleae* is not known but presumably differs from that of the other dacines studied, because the females release the major attractant pheromone (135, 136).

A number of compounds have been identified in the rectal pheromone gland secretions and cold-trapped volatiles from dacines, but relatively little is known about their functions in most species.

D. oleae females release an airborne sex pheromone that is a blend of four compounds, 1,7-dioxaspiro(5.5)undecane, α-pinene, nonanol, and ethyl dodecanoate (6, 136); this attracts males from a distance. The major component, both in quantity and in biological activity, is the spiroketal, which alone can attract males to traps in the field. *E*-6-Nonen-1-ol and *p*-cymene have also been identified in the cold-trapped effluent of sexually active females (76), but the role of these compounds as pheromone components remains uncertain (6, 136).

Wild males of *D. oleae* produce the same spiroketal as females, plus diethyl-5-oxononadioate, in their rectal glands (137). Recent results indicate that the spiroketal is released as a racemate. Males respond only to the R$-(-)-$ enantiomer, which functions as a long-range sex attractant, and females respond only to the S$-(+)-$ enantiomer, which may function as a short-range arrestant and aphrodisiac (89).

The major components (quantitatively) in the secretion from the rectal pheromone glands of *D. tryoni* were identified as six amides. Synthetic samples of these amides acted as a short-range aphrodisiac for females in laboratory bioassays, but lacked the longer-range attractancy of crude extracts. This suggests that other unidentified compounds form part of the natural pheromone blend (15).

Analysis of volatiles from the rectal pheromone glands of male *D. cucurbitae* (7) yielded three amides, three pyrazine derivatives, and 2-ethoxybenzoic acid. One of the amides (also found in *D. tryoni*) elicited increased flight activity in females. The roles of the other components have not yet been determined.

The "smoke" emitted by males of *D. cucurbitae* and *D. dorsalis* was found to consist mainly of trisodium and potassium phosphates, which may act as carriers for the volatile pheromone components. The major volatiles associated with the smoke in *D. cucurbitae,* mainly saturated hydrocarbons (none of which showed any activity in bioassays) and an unsaturated lactone (which was not tested), were quite different from the pheromone gland secretions. In *D. dorsalis* the major organic compound was an amide (present also in *D. tryoni* but not in *D. cucurbitae*), which was marginally attractive to females in laboratory bioassays (156).

Recent chemical studies on some dacine species from the Australasian region have indicated that the major components of the male rectal glands frequently fall into two categories (118). They may be either amides, as reported for *D. tryoni, D. neohumeralis,* and *D. aquilonis,* or spiroketals, as found in *D. cucumis, D. halfordiae,* and *D. cacuminatus.* However, both amides and spiroketals may be present, as in *D. dorsalis* and *D. opiliae.* In *D. occipitalis* the major component of the gland secretion is 6-oxononan-1-ol, but an unusual even-numbered spiroketal that also occurs in *D. dorsalis* and *D. latifrons* is also present. The roles of these compounds have not yet been determined.

MALE LURES   Among the most remarkable unexplained phenomena involving dacine fruit flies is their response to the synthetic lures methyl eugenol, cue-lure [4-(*p*-acetoxyphenyl)-2-butanone], and the hydroxy derivative of cue-lure (Willison's lure) (23). Of the major pest species, *D. cucurbitae* and *D. tryoni* respond to cue-lure, and *D. dorsalis* and *D. zonatus* respond to methyl eugenol (23, 172). Interestingly, none of the endemic African species appear to respond to methyl eugenol, although many species respond to cue-lure. Males of *D. vertebratus,* which do not respond to either major lure, respond to methyl-*p*-hydroxybenzoate (vert-lure), a compound that, like methyl eugenol and cue-lure, consists of a benzene ring with acetoxy or methoxy groups attached (88).

In a survey carried out in northern Australia and the South Pacific (36), males of 79 species of dacines were trapped with lures; 56 responded to cue-lure or Willison's lure and 23 to methyl eugenol. In a later study carried out in Australia (39), 39 species responded to cue-lure and 16 to methyl eugenol; 23 species known to occur in the area did not respond to either lure. There are no substantiated records of a species responding to both lures.

The response of males to lures is influenced by their age, their physiological maturity, and the time of day (26, 50). The development of the response is normally correlated with sexual maturation (50, 61), and the level of response is very low during winter when adults are sexually inactive. Males of dusk-mating species respond to lures during the daylight hours when they are not sexually active (50). In contrast, mature virgin females isolated from males respond to male lures at the same time of day and under the same lighting conditions that invite their response to the male sex pheromone (48).

It has been suggested that "male" lures act as rendezvous stimuli that in nature would normally bring the sexes together in proximity to suitable host plants (149). However, these compounds are mostly recorded from nonhost plants (65), and they appear to function mainly as secondary defensive chemicals to prevent attack by phytophagous insects. A few tropical plants do appear to use these or closely related compounds to attract male fruit flies as pollinating agents (98, 113). However, it is not known if the males obtain in return any nutritional or other advantage from visiting such flowers.

FEEDING   Dacines require a diet rich in amino acids, vitamins, minerals, carbohydrates, and water to survive and reproduce (186). Newly emerged adults contain some reserves carried over from the larval stage, which enable them to survive for 1–2 days after emergence if other food is not available. *D. oleae* carries over enough protein to mature a few eggs (71, 133). Feeding, however, is normally a daily activity that entails foraging for food on both host and nonhost trees (38, 108). Peak feeding normally occurs in the morning, but some feeding may take place at other times (183).

Laboratory studies on *D. tryoni* maintained at 25°C and allowed to feed ad lib indicated that protein ingestion rose from 0.7 mg per female just after emergence to 1.2 mg per female on days 7–8, which coincided with the late stages of vitellogenesis. Sugar ingestion was about 2.7 mg on the first day, and subsequently varied from 1 to 2 mg. Protein and sugar intake of males was generally lower but showed the same overall trends. In temperature regimes in which daily day-degrees were below the threshold for development, consumption of both protein and sugar fell to low levels (117, 126).

Dacines have been observed both on host and nonhost trees feeding on honeydew, plant exudates, extrafloral nectaries, pollen, fruit juice, ripe fruits, microorganisms, and bird droppings (10, 38, 108, 185). However, very little information is available about the quantitative aspects of adult nutrition in the field or about the factors that influence food-foraging behavior. Both sexes of many *Dacus* species respond to protein hydrolysate baits. Much of the attractancy is due to the volatiles released as a result of amino acid degradation by microorganisms and nonbiotic chemical reactions; ammonia appears to be one of the important compounds involved (13).

Recently, studies in Australia have shown that *D. tryoni* and some other species are attracted to and feed on cultures of the Enterobacteriaceae found on leaf and fruit surfaces. Feeding experiments indicated that females could mature eggs when fed on these bacteria. It has been suggested that the bacteria might be the major source of amino acids, vitamins, and other growth factors for adult fruit flies in tropical regions, where honeydew is scarce or absent (30, 38). Subsequent studies have indicated that adult flies inoculate the surface of host fruit with bacteria from the gut (see section on bacterial associations), possibly to provide a readily available protein source (42).

MOVEMENT   The polyphagous, multivoltine tropical and subtropical species, e.g. *D. cucurbitae*, *D. dorsalis*, *D. tryoni*, and *D. zonatus*, are strong fliers and are highly mobile. They engage in extensive dispersive movements during the postteneral period prior to host-seeking and mating, and mature flies leave locations where hosts are dwindling in search of new hosts (14, 24, 25, 60, 87, 101, 111, 152, 171). During these periods some individuals may move large distances in a few weeks. Maximum distances recorded for marked individuals are 200 km for *D. cucurbitae* (150), 65 km for *D. dorsalis* (180), 90 km for *D. tryoni* (130), and 40 km for *D. zonatus* (171).

When hosts are plentiful in an area, mature flies tend to restrict their movements to foraging flights in search of food, water, and oviposition sites; the overall movements are nondispersive (10). Flies may show a daily pattern of movement between hosts and surrounding vegetation. This is particularly evident in cucurbit-infesting species such as *D. cucurbitae, D. ciliatus, D. frontalis, D. vertebratus,* and *D. vivittatus,* which enter plots containing cucurbit fruits at fairly specific times of day to oviposit and then return to nearby vegetation to feed and rest (10, 134, 179a).

Adults of *D. oleae* often emerge beneath fruiting olive trees and do not have an adaptive postteneral dispersive phase. They disperse when suitable fruit becomes scarce and at the end of the season, when because of the biennial fruit cycle large numbers of adults emerge in locations where there is no new fruit crop available (44, 67, 69). Generally, *D. oleae* movements seem to be much less extensive than those of the polyphagous tropical species, but movements of up to 10 km have been recorded (20, 44).

Very little is known about the movements of the nonpest species. Trap catches indicate that adults of *D. opiliae* move extensively in search of hosts at the beginning of the wet season when they leave their dry season refuges (53).

RESTING AND SHELTERING   Adults normally rest on the undersides of leaves of both hosts and nonhosts at night and in the daytime during periods of inactivity. When daytime temperatures have risen above 35°C, adults of *D. zonatus* (182) and *D. oleae* (B. S. Fletcher, personal observation) have been observed to leave the trees and congregate on low bushy undergrowth.

During cooler periods of the year when conditions are unsuitable for breeding, individuals of a number of species seek out refuges where they remain until the return of warmer weather. One interesting aspect of this behavior, observed both in Africa (88) and Asia (182), is the establishment of communal roosts, usually on the undersurfaces of large leaves. Large numbers of both sexes of several species may aggregate in these roosts and remain there for several weeks. When weather permits they often leave the roost during the day to feed but return before nightfall. In Africa, species that exhibit this behavior include *D. vertebratus, D. amphoratus, D. binotatus, D. umbeluzinus, D. brevis, D. serratus,* and *D. butianus;* in Pakistan, *D. cucurbitae, D. hageni, D. scutellaris,* and *D. diversus.* Overwintering *D. tryoni* also aggregate in sheltered refuges with broad-leaved evergreen trees, but large aggregations on individual leaves have not been observed (62, 63).

Larval Behavior and Nutrition

Fruit tissues are low in protein and it is generally thought that larvae have to rely on bacteria to provide certain essential amino acids and other growth factors. The chemical and nutritional components of olive fruit and the major biochemical groups (i.e. lipids, proteins, amino acids, and peptide acids) in mature larvae and newly emerged adults of *D. oleae* (132, 133) have been determined. Synthetic diets have been developed for the major pest species, but many aspects of larval nutrition remain to be investigated.

Larvae of many of the specialist species are able to develop in commercial fruit that they do not utilize as hosts in the field (56a, 194), but development in nonhost wild fruits is often inhibited. This suggests that monophagous and stenophagous species have evolved specialized mechanisms that enable them to exploit their particular hosts, and that host selection is maintained largely by female oviposition preferences (56).

During development the larvae tunnel in the fruit, macerate the tissues, and ingest the broken-down tissues and associated bacteria. In larger fruit they move toward the center, which may offer some protection from parasites and certain predators. When mature, larvae of most species leave the fruit and burrow several centimeters into the soil, where they pupate (51, 154). Mature larvae of all species studied have the ability to "hop," which appears to be a defense against ground-dwelling insect predators, particularly ants. *D. oleae* pupates in fruit during summer and early autumn and in the ground in the winter and spring (104). Prior to pupation in the fruit, the mature larva tunnels to the fruit surface and eats away all but the thin outer membrane of the pericarp so that the emerging adult can escape from the fruit.

The two most pronounced rhythmical behaviors during development are larval exit from the fruit prior to pupation and adult eclosion; both peak around dawn and are controlled by light and temperature cycles (3, 123, 131, 177).

POPULATION ECOLOGY

In recent years a number of intensive studies have been carried out on the population ecology of pest dacines (9, 14, 16–18, 33–35, 40, 45, 52, 53, 60–63, 99, 100, 102–105, 120, 128, 129, 154, 184, 200, 201, 206, 207). Most of them, however, have been performed in ecosystems disturbed by agricultural practices, and there is still little information available about the dynamics of dacine populations in endemic rain forests or other natural habitats. Space here only permits consideration of the major factors determining the spatial and temporal abundance of particular species and the life history strategies they have evolved.

Determinants of Abundance

Many factors, both biological and environmental, can influence the endemic distribution and demography of fruit fly populations by directly or indirectly affecting survival and development rates of the different life stages and the fecundity of females. The most important factors appear to be temperature, moisture, and availability of hosts; natural enemies and competition may also be important in some circumstances.

TEMPERATURE AND MOISTURE   These two climatic factors not only have a direct effect on the demography of a species, but also an indirect effect through their influence on its hosts and natural enemies.

One major influence of temperature on the multivoltine species is in determination of development times and thus the number of generations per year. In the highly polyphagous species like *D. tryoni* and *D. dorsalis,* the number of generations per year may vary from three to eight in different parts of their range (142, 174). *D. oleae* has four to six generations per year in the Mediterranean basin, with a break during the summer period when high temperatures and/or lack of fruit curtail breeding (33–35, 104).

Studies on the different life stages of *D. cucurbitae* (148, 157), *D. dorsalis* (148, 174), *D. oleae* (31, 70, 188, 189), and *D. tryoni* (162) at a series of constant temperatures have indicated that the temperature–development rate curve has the same general shape in all stages and species. Above a low-temperature development threshold, which for the immature stages lies between 6° and 9°C, the relationship is sigmoidal up to a maximum between 26° and 30°C, above which development rates start to decrease again.

Temperature–development rate relationships for the different life stages based either on the linear summation of day-degrees above the lower development threshold or on some nonlinear development-rate model have been calculated for the major pest species (31, 70, 80, 148, 153, 157, 162, 174, 188, 189). Development times are fairly similar for all species; eggs take 1–2 days, larvae 7–8 days, and pupae 10–11 days to complete development at 25°C. Comparison of predictions with field observations in *D. tryoni* and *D. oleae* suggest that generation times in the field are largely determined by temperature (66, 158). However, other factors, including moisture content and ripeness of fruit in the case of eggs (175, 190), and ripeness, fruit variety, and degree of crowding in the case of larvae (21, 96, 153), can have a significant influence on development rates.

Ovarian maturation is similarly influenced by temperature, except that the lower threshold is about 12–13°C (72, 161). At 25°C, ovarian maturation takes 6–8 days. Day-degrees needed to complete maturation in the laboratory or in field cages have been determined for *D. dorsalis* (157), *D. cucurbitae* (174), *D. oleae* (72), and *D. tryoni* (161). Other factors that affect ovarian maturation include availability of hosts, water relationships, availability of dietary protein, and presence of males (71, 72, 161).

Temperature also acts as a major mortality factor. Outside the optimum temperature range of approximately 18–27°C mortality increases, and there are upper and lower lethal thresholds beyond which no individuals survive long enough to complete development (31, 70, 145). At fluctuating temperatures the relationship between temperature and mortality is complex, and is influenced by moisture levels and the ability of some life stages to acclimate (141). Some development can occur at temperatures that can be lethal if prolonged, both at the upper and lower ends of the temperature scale. Repeated exposures of a *D. tryoni* population to an extreme temperature result in increased mortality, and exposure to suboptimal conditions during one life stage reduces the chances of survival of later stages (145).

Although the immature stages of dacines can survive short periods of high temperature (>30°C) or low temperature (<5°C), adults are normally the best able to survive prolonged periods at such temperatures. Therefore, most dacines pass the unfavorable winter and summer months in the adult stage. There are some exceptions, however; most individuals of *D. oleae* pass the winter months as pupae (104), and *D. zonatus* overwinters in the larval or pupal stage (183). Overwintering survival is enhanced by acclimation that enables the adults to reduce their cold-torpor level and increase their cold-survival time in response to decreasing temperatures as winter approaches (74, 139–141). The torpor threshold can drop to as low as 2°C in winter from 7°C in summer. Acclimation can also occur during the late larval stage and pharate adult stage, so that adults emerging close to winter are fully acclimated (143). Although flight threshold is also lowered by cold acclimation, flight propensity is reduced. In *D. tryoni,* the mating threshold is not changed by acclimation (146).

In *D. oleae,* the upper threshold for maturation is lower than the lethal threshold. Thus during the hot summer months flies resorb their follicles and remain sexually inactive. The upper threshold for maturation is influenced by moisture and fruit availability (71, 72).

Temperature affects the water relationships of the different stages, and this can have a significant impact upon dacine populations. In the species so far studied, tolerance of a particular maximum temperature is greater at high humidity, which reduces the rate of water loss. The stress caused by suboptimal humidity can also prolong development rates of the immature stages (90) and inhibit the maturation of adults (72). Highly significant correlations between rainfall and population levels have been noted in a number of species, including *D. tryoni* (9) and *D. dorsalis* (128), owing to the direct influence of moisture on survival, fecundity, and movement and its indirect effects on host-fruit availability.

Species of *Dacus* that utilize hosts that fruit in drier areas or during the dry season in tropical areas have evolved mechanisms that make them more resistant to desiccation. Newly emerged adults of *D. oleae* can survive for longer periods without water than those of *D. tryoni* (133). The tendency of *D. oleae* to pupate within the fruit during the hot summer period may be an adaptation to reduce mortality when hot, dry soil conditions prevail (169). The pupae of *D. tenuifascia,* which breeds in a dry season host and frequently pupates in hot, dry soil, are more resistant to desiccation than those of another northern Australian dacine, *D. opiliae,* which attacks a wet season host and pupates in damp soil (51). One dacine, *D. newmani* (41), has been able to adapt totally to the hot, dry, almost desertlike conditions that occur in central Australia, but unfortunately nothing is known about its ecology.

An examination of the bioclimatic potential of *D. tryoni* in Australia (142) indicated that the southern limits of its current distribution are set by thermal restrictions on development rates, which determine the number of generations per year; the western limits are set by insufficient summer rainfall; and altitude limits in the cooler southern parts of its range are set by lethally low minimum winter temperatures.

LARVAL AND ADULT FOOD   The abundance of larval hosts is one of the major factors regulating dacine populations (35, 104, 184, 200). In the polyphagous species, populations are usually at their lowest level at the end of the winter owing to the slowdown or cessation of breeding (63), and it may take several generations for them to build up to peak levels (60, 128, 171). In the more specialized species that utilize hosts that only fruit for limited periods each year, minimum population levels generally occur just prior to the infestation of a new crop (53, 104). In these specialized species the presence of hosts appears to strongly influence egg production; some species do not mature eggs in the absence of hosts (56).

It has been suggested that in the tropics proteinaceous food sources may be relatively scarce, and that the flies depend on leaf-surface bacteria that may be intermittent in supply (30). At present, however, there are no data to indicate that fecundity and survival of wild flies are limited by food availability.

NATURAL ENEMIES   The larvae of most dacines are attacked by hymenopterous parasites, particularly of the family Braconidae (107, 203). Sometimes natural parasitism can be high, e.g. up to 57% by braconid wasps on *D. dorsalis* in Malaysia (201) and up to 80–95% by eupelmid wasps on *D. oleae* in parts of Greece and Italy in late summer (33–35, 129). Generally, however, parasitism rates are fairly low: not exceeding 30% in seven of eight generations of *D. oleae* in Corfu in 1976–1978 (105); 0–11% for *D. opiliae* at 15 sites in northern Australia (52), and a maximum of 18% for *D. cacumina-*

tus in Queensland (37). The larvae of species that breed in commercial hosts normally escape parasitism, as the wasps are not able to reach them in large fruit (9). Egg and pupal parasites have been recorded from some species, but they normally occur in small numbers and have little impact on endemic populations. However, after its introduction into Hawaii *Opius oophilus* caused high egg mortality in *D. dorsalis* (155, 204).

Other insects act as predators. The most important are ants, which remove larvae and pupae from fruit as well as from the soil (12, 17, 105). Earwigs have been observed removing larvae of *D. ciliatus* (183) and *D. musae* (175) from fruit. Staphylinid and carabid beetles prey on larvae and pupae in the soil (12), and spiders capture some adults (63).

Vertebrates may also cause high mortality of the immature stages. Frugivorous birds consume *D. oleae*–infested olives in Crete (18), and fruit-eating pigeons are a major cause of mortality of *D. cacuminatus* larvae in *Solanum mauritianum* fruits in Queensland (37). In rain forests, ground-living mammals and large birds such as the cassowary eat much of the fallen fruit containing dacine larvae, while some rodents break open fruits of *Planchonella australis* and remove the larvae of *D. halfordiae* (37).

Microbial pathogens, including fungi and bacteria, have often been found associated with dead larvae and pupae, but it is not always possible to determine if infection was the cause of death. In Hawaii, *Mucor* was identified as one of the major factors in mortality of *D. dorsalis* larvae in decaying fruit (155). Significant mortality caused by microorganisms has also been reported in other species, including *D. opiliae* larvae (52) and *D. zonatus* pupae (182).

INTRA- AND INTERSPECIFIC COMPETITION   There is some evidence that intraspecific competition may limit or depress population levels when a species becomes abundant in relation to its resources. The most obvious competitive interactions in dacines occur between females on fruit (160). Their aggression may reduce overall fecundity by restricting egg-laying and possibly by encouraging females to disperse, so that more energy is diverted into metabolic activities than into oogenesis. Interactions between females, however, are relatively infrequent except at very high population levels. Competition among larvae in fruit appears to be more frequent and important.

In *D. oleae*, larval competition is of the "contest" type; only one individual normally completes development, irrespective of the number of eggs laid per fruit. This limits the increase of field populations when there is a relative shortage of hosts and fruits receive more than one egg (45). In most other dacines, "scramble" competition between larvae occurs to some extent, so that as larval density per fruit increases the resulting pupae and adults become smaller. Small females have fewer ovarioles, lower fecundity, and a reduced chance of survival, particularly in suboptimum conditions (54, 57, 131).

In the polyphagous species that infest large fruits, scramble competition may operate over a wider range of larval densities than in the monophagous species that breed in small hosts (e.g. *D. cacuminatus* and *D. opiliae,* whose larvae appear to suffer high overall mortality even at moderate levels of overcrowding) (40, 52). Scramble competition only operates between larvae of a single age class; older larvae have a highly detrimental effect on younger larvae in the same fruit, owing to an interference of unknown mechanism (57).

There is some evidence that *D. dorsalis* has displaced the Mediterranean fruit fly, *Ceratitis capitata,* in some parts of Hawaii, possibly owing to larval interactions (115). However, there is very little evidence that interspecific competition occurs between sympatric dacines, even when several species breed in the same fruits (57).

Demography

Data on development rates, age-specific mortality, and fecundity have been used to construct life tables and to calculate the innate capacities for increase (r_m) of laboratory populations of a number of species including *D. cucurbitae* (198), *D. dorsalis* (198), *D. tryoni* (8, 54), *D. jarvisi* (54), and *D. latifrons* (199). The polyphagous species have the highest r_m, and *D. jarvisi* and *D. latifrons* the lowest. Demographic data on wild populations are limited, however, although partial life tables have been constructed for *D. oleae* (16–18, 105) and some demographic data have been published for *D. tryoni* (9, 14, 63).

The major mortality factors acting on *D. oleae* are quite similar in much of the Mediterranean region. High temperatures during summer cause high mortality of eggs and young larvae in fruit (33, 105, 169, 179) and of pupae in the soil in the first generation (105). Parasitism is the major factor in mortality of second and third stage larvae in the first and second generations (18, 105, 129, 169, 179). Cold, wet soil conditions are a major cause of pupal mortality in the third generation (16, 105, 154), and frugivorous birds can cause significant mortality of larvae in fruit (18). Dry soil and in some cases predation on pupae are the main mortality factors in the spring generation (17, 105, 154).

Although parasites and other mortality factors have some impact on population size, key-factor analysis of the Corfu data indicate that natality, largely determined by host availability, is the most important regulatory factor in *D. oleae* (105). This conclusion is supported by other field studies (33). Simulation studies also indicate that number of fruit is the major determinant of population size in most circumstances (29, 66).

Bateman (9) reported a highly significant correlation between summer rainfall and peak insect numbers in a *D. tryoni* population toward the southern end of its range. Further north, in the rain forests that are the endemic home of many Australian dacines, consumption of fruit by frugivorous birds and mammals, both on trees and on the ground, may be a more important factor in population regulation (37).

Life History Strategies

The two most important parameters influencing the life history strategies of dacine fruit flies are (*a*) favorableness of the environment for reproduction and survival and (*b*) host availability in time and space (64).

The life history characteristics of the polyphagous species are best suited for exploiting resources that occur intermittently throughout most of the year but are unpredictable in time and space. The adults have high mobility, relatively long life span (often more than three months), high potential fecundity (>1000 eggs per female) (54, 198), scramble type competition in the larval stages, several generations per year, and the ability to pass unfavorable periods of the year in a facultative reproductive diapause when necessary.

The life history characteristics of the oligophagous species are similar in many respects to those of the polyphagous species. However, the majority of oligophagous species have a lower potential fecundity (400–600 eggs) (54, 64) and their rate of egg production is more directly influenced by the availability of hosts (56).

The life history strategies of the monophagous and stenophagous species vary depending upon the fruiting characteristics of their hosts. In general, they have fewer ovarioles and lower potential fecundity than the other groups. The potential fecundity of *D. oleae* and *D. latifrons* (199) is about 300 eggs per female. *D. musae* females, however, have about the same number of ovarioles as polyphagous females, and therefore possibly have the ability to lay a large number of eggs in a relatively short period when hosts are abundant.

D. opiliae is univoltine and spends most of the year as an adult, although the actual reproductive period is quite short. Adults of *D. oleae* can survive several months when reproductively inactive, but during periods of intense oviposition, particularly in late summer, their longevity is markedly reduced (104). During much of the year adults probably live a maximum of 1–2 months. Unlike most dacines, *D. oleae* lays only one egg per fruit and then marks the fruit with fruit juice. It is the only dacine known in which contest-type competition normally prevents the development of more than one larva per fruit. A number of other monophagous species that infest small fruits, e.g. *D. cacuminatus* and *D. opiliae,* do not use marking compounds; overcrowding due to multiple ovipositions can result in high mortality of larvae (40, 52), because the type of competition switches from scramble toward contest at relatively low larval densities per fruit.

The importance of dispersal also varies among monophagous species. Those that infest hosts that fruit erratically throughout the year, e.g. *D. cacuminatus,* appear to be highly mobile, and move around frequently in search of hosts. *D. opiliae,* which infests a host with a limited fruiting period, moves mainly from emergence site to dry-season refuge, and then in search of fruiting host plants at the beginning of the subsequent wet season. In *D. oleae* the dispersive movements are determined largely by host availability in the areas where they emerge.

The polyphagous dacines are the most strongly r-selected, and the monophagous species, such as *D. oleae,* are the least strongly r-selected, although they still fall nearer the r than the K end of the spectrum compared with most other tephritids (11, 64).

CONCLUSIONS

Most of the information about the biology of dacine fruit flies has been derived from studies on the economically important pest species. Most of these, apart from *D. oleae,* are highly polyphagous, which is not typical of dacines as a group. Very little is known about the biology of the many monophagous and oligophagous species that breed only in native hosts, which are mostly restricted to tropical and subtropical rain forests.

There are still many important gaps in our knowledge of the biology of dacines, even in the case of the major pest species. Much of the information that we do have is based on laboratory or field-cage observations.

Physiological and behavioral aspects that should be particularly rewarding for further study include larval and adult nutrition, especially the role and significance of bacteria; various aspects of sexual behavior, including the roles of the various chemical components of the rectal pheromone gland secretions used during courtship and mating; the sites of assembly in nature and the significance of lekking behavior; and host-detection mechanisms, general foraging strategies, and the ways they are influenced by host availability and other environmental variables.

Further information is also required on the following aspects of dacine ecology in both disturbed agricultural systems and indigenous habitats: adult movement, mortality factors of immature stages and adults, fecundity, and the influence of host availability, population density, and climatic factors on the preceding aspects. In addition, studies on reproductive traits and other life history parameters in a greater variety of dacines would significantly increase our knowledge of the diversity of life history strategies within the group. Such knowledge could possibly provide a significant contribution to general life history theory.

Literature Cited

1. Agarwal, M. L. 1986. Zoogeography of Indian Dacinae (Diptera: Tephritidae). *J. Bombay Nat. Hist. Soc.* In press
2. Anderson, D. T. 1963. The larval development of *Dacus tryoni* (Frogg.) (Diptera: Trypetidae). I. Larval instars, imaginal discs, and haemocytes. *Aust. J. Zool.* 11:202–18
3. Arai, T. 1976. Effects of temperature and light-dark cycles on the diel rhythm of emergence in the oriental fruit fly, *Dacus dorsalis* Hendel (Diptera: Trypetidae). *Jpn. J. Appl. Entomol. Zool.* 20:69–76 (In Japanese with English summary)
4. Arakaki, N., Kuba, H., Soemori, H. 1984. Mating behavior of the oriental fruit fly, *Dacus dorsalis* Hendel (Diptera: Tephritidae). *Appl. Entomol. Zool.* 19:42–51
5. Armati, P. 1976. The anatomy of the nervous system in the Queensland fruit fly *Dacus tryoni* (Froggatt). *J. Aust. Entomol. Soc.* 14:429–39
6. Baker, R., Herbert, R., Howse, P. E., Jones, O. T. 1980. Identification and synthesis of the major sex pheromone of the olive fly *(Dacus oleae)*. *J. Chem. Soc. Chem. Commun.* 1:52–53
7. Baker, R., Herbert, R. H., Lomer, R. A. 1982. Chemical components of the rectal gland secretions of male *Dacus cucurbitae*, the melon fly. *Experientia* 38:232–33
8. Bateman, M. A. 1967. Adaptations to temperature in geographic races of the Queensland fruit fly, *Dacus (Strumeta) tryoni* (Froggatt). *Aust. J. Zool.* 15:1141–61
9. Bateman, M. A. 1968. Determinants of abundance in a population of the Queensland fruit fly. *Symp. R. Entomol. Soc. London* 4:119–31
10. Bateman, M. A. 1972. The ecology of fruit flies. *Ann. Rev. Entomol.* 17:493–518
11. Bateman, M. A. 1979. Dispersal and species interaction as factors in the establishment and success of tropical fruit flies in new areas. *Proc. Ecol. Soc. Aust.* 10:106–12
12. Bateman, M. A., Boller, E. F., Bush, G. L., Chambers, D. L., Economopoulos, A. P., Fletcher, B. S., et al. 1976. Fruit flies. In *Studies in Biological Control*, ed. V. L. Delucchi, 1:11–49. Cambridge: Cambridge Univ. Press. 304 pp.
13. Bateman, M. A., Morton, T. C. 1981. The importance of ammonia in proteinaceous attractants of fruit flies (family: Tephritidae) *Aust. J. Agric. Res.* 32: 883–903
14. Bateman, M. A., Sonleitner, F. J. 1967. The ecology of a natural population of the Queensland fruit fly *Dacus tryoni*. I. The parameters of the pupae and adult populations during a single season. *Aust. J. Zool.* 15:303–35
15. Bellas, T. E., Fletcher, B. S. 1979. Identification of the major components in the secretion from the rectal pheromone glands of the Queensland fruit flies *Dacus tryoni* and *Dacus neohumeralis* (Diptera: Tephritidae). *J. Chem. Ecol.* 5:795–803
16. Bigler, F. 1980. Einfluss der Wirspflanze auf die postembryonalen Stadien der Olivenfliege, *Dacus oleae* Gmel. (Dipt. Trypetidae). *Mitt. Schweiz. Entomol. Ges.* 53:347–63
17. Bigler, V. F. 1982. Die postlarvale Mortalität der Olivenfliege, *Dacus oleae* Gmel. (Dipt., Tephritidae), in Oleastergebieten von Westkreta. *Z. Angew. Entomol.* 93:76–89
18. Bigler, V. F., Delucchi, V. 1981. Wichtigste Mortälitatsfaktoren während der präpupalen Entwicklung der Olivenfliege, *Dacus oleae* Gmel. (Dipt., Tephritidae) auf Oleastern und Kultivierten Oliven in Westkreta, Griechenland. *Z. Angew. Entomol.* 92:343–63
19. Bower, C. C. 1977. Inhibition of larval growth of the Queensland fruit fly, *Dacus tryoni* (Diptera: Tephritidae) in apples. *Ann. Entomol. Soc. Am.* 70:97–100
20. Brnetic, D. 1981. Biological control of olive fly by means of sterile male technique and by *Opius concolor*. *Rep. Proj. YO-ARS-9-JB5*, Inst. Jadranske Kult. I Melior. KRSA, Split, Yugoslavia. 90 pp.
21. Carey, R. J., Harris, E. J., McInnis, D. O. 1985. Demography of a native strain of the melon fly, *Dacus cucurbitae*, from Hawaii. *Entomol. Exp. Appl.* 38: 195–99
22. Cavalloro, R., ed. 1983. *Fruit Flies of Economic Importance, CEC/IOBC Symp., Athens, 1982.* Rotterdam: Balkema. 642 pp.
23. Chambers, D. L. 1977. Attractants for fruit fly survey and control. In *Chemical Control of Insect Behavior*, ed. H. H. Shorey, J. J. McKelvey, 19:327–44. New York: Wiley. 414 pp.
24. Chapman, M. G. 1983. Experimental analysis of the propensity to flight of the Queensland fruit fly, *Dacus tryoni*. *Entomol. Exp. Appl.* 33:163–70
25. Chiu, H. 1983. Movements of oriental fruit flies in the field. *Chin. J. Entomol.* 3:93–102

26. Chiu, H. 1984. Influence of environmental factors to the attractiveness of oriental fruit flies by using of methyl eugenol. *Plant Prot. Bull. Taiwan ROC* 26:355–64

27. Christenson, L. D., Foote, R. H. 1960. Biology of fruit flies. *Ann. Rev. Entomol.* 5:171–92

28. Cirio, U. 1971. Reperti sul meccanismo stimolo-risposta nell'ovideposizione del *Dacus oleae* Gmelin (Diptera, Trypetidae). *Redia* 52:577–600

29. Comins, H. N., Fletcher, B. S. 1986. Simulation of fruit fly population dynamics with particular reference to the olive fruit fly, *Dacus oleae. Ecol. Model.* In press

30. Courtice, A. C., Drew, R. A. I. 1984. Bacterial regulation of abundance in tropical fruit flies (Diptera: Tephritidae) *Aust. Zool.* 21:251–66

31. Crovetti, A., Quaglia, F., Loi, G., Rossi, E., Malafatti, P., et al. 1982. Influenza di temperatura e umidita sullo sviluppo degli stadi preimmaginali di *Dacus oleae* (Gmelin). *Frustula Entomol.* (NS) 5:133–66

32. Dankers-Williems, G. M. 1976. *The neurosecretory system and control of ovarian maturation in the Queensland fruit fly,* Dacus tryoni *(Frogg.).* PhD thesis. Univ. Sydney, Australia. 205 pp.

33. Delrio, G. 1978. Fattori di regolazione delle popolazioni di *Dacus oleae* Gmelin nella Sardegna nord-occidentale. *Not. Mal. Piante* 98–99:27–45

34. Delrio, G., Cavalloro, R. 1977. Reperti sul ciclo biologico e sulla dinamica di popolazione del *Dacus oleae* Gmelin in Liguria. *Redia* 60:221–53

35. Delrio, G., Prota, R. 1975–1976. Osservazioni eco-etologiche sul *Dacus oleae* Gmelin nella Sardegna nord-occidentale. *Boll. Zool. Agrar. Bachic.* (Ser. II) 13:49–118

36. Drew, R. A. I. 1974. The responses of fruit fly species (Diptera: Tephritidae) in the South Pacific area to male attractants. *J. Aust. Entomol. Soc.* 13:267–70

37. Drew, R. A. I. 1986. Reduction in fruit fly populations in their endemic rainforest habitat by frugivorous vertebrates. *Aust. J. Zool.* In press

38. Drew, R. A. I., Courtice, A. C., Teakle, D. S. 1983. Bacteria as a natural source of food for adult fruit flies (Diptera: Tephritidae). *Oecologia* 60:279–84

39. Drew, R. A. I., Hooper, G. H. S. 1981. Responses of Australian fruit fly species to attractants. *J. Aust. Entomol. Soc.* 20:201–5

40. Drew, R. A. I., Hooper, G. H. S. 1983. Population studies of fruit flies (Diptera: Tephritidae) in South-East Queensland. *Oecologia* 56:153–59

41. Drew, R. A. I., Hooper, G. H. S., Bateman, M. A. 1982. *Economic Fruit Flies of the South Pacific Region.* Brisbane, Australia: Qld Dept. Primary Ind. 139 pp. 2nd ed.

42. Drew, R. A. I., Lloyd, A. C. 1987. Bacteria associated with fruit flies and their host plants. See Ref. 173. In press

43. Economopoulos, A. P., Giannakakis, A., Tzanakakis, M. E., Voyadjoglou, A. V. 1971. Reproductive behavior and physiology of the olive fruit fly. I. Anatomy of the adult rectum and odors emitted by adults. *Ann. Entomol. Soc. Am.* 64:1112–16

44. Economopoulos, A. P., Haniotakis, G. E., Mathioudis, J., Missis, N., Kinigakis, P. 1978. Long distance flight of wild and artificially-reared *Dacus oleae* (Gmelin) (Diptera, Tephritidae). *Z. Angew Entomol.* 87:101–8

45. Economopoulos, A. P., Haniotakis, G. E., Michelakis, S., Tsiropoulos, G. J., Zervas, J. A., et al. 1982. Population studies on the olive fly, *Dacus oleae* (Gmel.) (Dipt. Tephritidae) in Western Crete. *Z. Angew. Entomol.* 93:463–76

46. Fay, H. A. C., Meats, A. 1983. The influence of age, ambient temperature, thermal history and mating history on mating frequency in males of the Queensland fruit fly, *Dacus tryoni. Entomol. Exp. Appl.* 34:273–76

47. Fiestas Ros de Ursinos, J. A., Canstante, E. G., Duran, R. M., Roncero, A. V. 1972. Étude d'un attractif naturel pour *Dacus oleae. Ann. Soc. Entomol. Fr.* 8:179–88

48. Fitt, G. P. 1981. Responses by female Dacinae to "male" lures and their relationship to patterns of mating behaviour and pheromone response. *Entomol. Exp. Appl.* 29:87–97

49. Fitt, G. P. 1981. Inter- and intraspecific responses to sex pheromone in laboratory bioassays by females of three species of tephritid fruit flies from northern Australia. *Entomol. Exp. Appl.* 30:40–44

50. Fitt, G. P. 1981. The influence of age, nutrition and time of day on the responsiveness of male *Dacus opiliae* to the synthetic lure, methyl eugenol. *Entomol. Exp. Appl.* 30:83–90

51. Fitt, G. P. 1981. Pupal survival of two northern Australian tephritid fruit fly species. *J. Aust. Entomol. Soc.* 20:139–44

52. Fitt, G. P. 1981. The ecology of north-

ern Australian Dacinae (Diptera: Tephritidae). I. Host phenology and utilization of *Opilia amentacea* Roxb. (Opiliaceae) by *Dacus (Bactrocera) opiliae* Drew and Hardy, with notes on some other species. *Aust. J. Zool.* 29:691–705

53. Fitt, G. P. 1981. The ecology of northern Australian Dacinae. II. Seasonal fluctuations in trap catches of *Dacus opiliae* Drew and Hardy and *D. tenuifascia* (May) and their relationship to host phenology and climatic factors. *Aust. J. Zool.* 29:885–94

54. Fitt, G. P. 1983. *Factors limiting host range of tephritid fruit flies, with particular emphasis on the influence of* Dacus tryoni *on the distribution and abundance of* Dacus jarvisi. PhD thesis. Univ. Sydney, Australia. 291 pp.

55. Fitt, G. P. 1984. Oviposition behaviour of two tephritid fruit flies, *Dacus tryoni* and *Dacus jarvisi,* as influenced by the presence of larvae in the host fruit. *Oecologia* 62:37–46

56. Fitt, G. P. 1986. The influence of a shortage of hosts on the specificity of oviposition behaviour in species of *Dacus* (Diptera: Tephritidae). *Physiol. Entomol.* In press

56a. Fitt, G. P. 1986. The roles of adult and larval specialisations in limiting the occurrence of five species of *Dacus* (Diptera: Tephritidae) in cultivated fruits. *Oecologia* 69:101–9

57. Fitt, G. P. 1987. The importance of interspecific interactions among tephritids. See Ref. 173. In press

58. Fitt, G. P., O'Brien, R. W. 1985. Bacteria associated with four species of *Dacus* (Diptera: Tephritidae) and their role in the nutrition of the larvae. *Oecologia* 85:447–54

59. Fletcher, B. S. 1969. The structure and function of the sex pheromone glands of the male Queensland fruit fly, *Dacus tryoni. J. Insect Physiol.* 15:1309–22

60. Fletcher, B. S. 1973. The ecology of a natural population of the Queensland fruit fly, *Dacus tryoni.* IV. The immigration and emigration of adults. *Aust. J. Zool.* 21:541–65

61. Fletcher, B. S. 1974. The ecology of a natural population of the Queensland fruit fly *Dacus tryoni.* V. The dispersal of adults. *Aust. J. Zool.* 22:189–202

62. Fletcher, B. S. 1975. Temperature-regulated changes in the ovaries of overwintering females of the Queensland fruit fly, *Dacus tryoni. Aust. J. Zool.* 23:91–102

63. Fletcher, B. S., 1979. The over-wintering survival of adults of the Queensland fruit fly, *Dacus tryoni,* under natural conditions. *Aust. J. Zool.* 27:403–11

64. Fletcher, B. S., 1987. Life history strategies of tephritid fruit flies. See Ref. 173. In press

65. Fletcher, B. S., Bateman, M. A., Hart, N. K., Lamberton, J. A. 1975. Identification of a fruit fly attractant in an Australian plant, *Zieria smithii,* as *o*-methyl eugenol. *J. Econ. Entomol.* 68:815–16

66. Fletcher, B. S., Comins, H. 1985. The development and use of a computer simulation model to study the population dynamics of *Dacus oleae* and other fruit flies. *Atti XIV Congr. Naz. Ital. Entomol.,* pp. 561–75

67. Fletcher, B. S., Economopoulos, A. P. 1976. Dispersal of normal and irradiated laboratory strains and wild strains of the olive fly *Dacus oleae* in an olive grove. *Entomol. Exp. Appl.* 20:183–94

68. Fletcher, B. S., Giannakakis, A. 1973. Factors limiting the response of females of the Queensland fruit fly, *Dacus tryoni,* to the sex pheromone of the male. *J. Insect Physiol.* 19:1147–55

69. Fletcher, B. S., Kapatos, E. T. 1981. Dispersal of the olive fly, *Dacus oleae,* during the summer period on Corfu. *Entomol. Exp. Appl.* 29:1–8

70. Fletcher, B. S., Kapatos, E. T. 1983. An evaluation of different temperature development rate models for predicting the phenology of the olive fly, *Dacus oleae.* See Ref. 22, pp. 321–29

71. Fletcher, B. S., Kapatos, E. T. 1983. The influence of temperature, diet and olive fruits on the maturation rates of female olive flies at different times of the year. *Entomol. Exp. Appl.* 33:244–52

72. Fletcher, B. S., Pappas, S., Kapatos, E. 1978. Changes in the ovaries of olive flies during the summer and their relationship to temperature, humidity and fruit availability. *Ecol. Entomol.* 3:99–107

73. Fletcher, B. S., Watson, C. A. 1973. The ovipositional response of the tephritid fruit fly *Dacus tryoni* to 2-chloroethanol in laboratory bioassays. *Ann. Entomol. Soc. Am.* 67:21–23

74. Fletcher, B. S., Zervas, G. 1977. Acclimation of different strains of the olive fly *Dacus oleae* to low temperatures. *J. Insect Physiol.* 231:649–53

75. Fowler, A. J. 1977. *Host-selection in wild and laboratory-culture* Dacus try-

oni. Honours thesis. Univ. Sydney, Australia. 137 pp.

76. Gariboldi, P., Jommi, G., Rossi, R., Vita, G. 1982. Studies on the chemical constitution and sex pheromone activity of volatile substances emitted by *Dacus oleae*. *Experientia* 38:441–44

77. Giannakakis, A. M. 1976. *Behavioural and physiological studies on the Queensland fruit fly, Dacus tryoni, in relation to the male sex pheromone*. PhD thesis. Univ. Sydney, Australia. 212 pp.

78. Giannakakis, A. M., Fletcher, B. S. 1985. Morphology and distribution of antennal sensilla of *Dacus tryoni*, (Froggatt) (Diptera: Tephritidae). *J. Aust. Entomol. Soc.* 24:31–35

79. Girolami, V. 1973. Reperti morfoistologici sulle batterio simbiosi del *Dacus oleae* Gmelin e di altri ditteri Tripetidi, in natura e negli allevamenti su substrati artificiali. *Redia* 54:269–94

80. Girolami, V. 1979. Studi biologici e demoecologici sul *Dacus oleae* (Gmelin). I. Infuenza dei fattori ambientali abiotici sull'adulto e sugli stadi preimmaginali. *Redia* 62:147–91

81. Girolami, V., Cavalloro, R. 1972. Aspetti della simbiosi batterica di *Dacus oleae* Gmelin in natura e negli allevamenti di laboratorio. *Ann. Soc. Entomol. Fr.* 8:561–71

82. Girolami, V., Stapazzon, A., De Gerloni, P. F. 1983. Insect/plant relationship in olive flies: general aspects and new findings. See Ref. 22, pp. 258–67

83. Girolami, V., Vianello, A., Stapazzon, A., Ragazzi, E., Veronese, G. 1981. Ovipositional deterrents in *Dacus oleae*. *Entomol. Exp. Appl.* 29:177–88

84. Guerin, P. M., Katsoyannos, B. I., Delrio, G., Remund, U., Boller, E. F. 1983. Fruit fly electroantennogram and behaviour responses to some generally occurring fruit volatiles. See Ref. 22, pp. 248–51

85. Gupta, M., Pant, N. C., Lal, B. S. 1982. Symbiotes of *Dacus cucurbitae* (Coquillett). II. Cultivation and identification. *Indian J. Entomol.* 44:331–36

86. Hallberg, E., Van der Perrs, J. N. C., Haniotakis, G. E. 1984. Funicular sensilla of *Dacus oleae*: fine structural characteristics. *Entomol. Hell.* 2:41–46

87. Hamada, R. 1980. Studies on the dispersal behavior of melon flies, *Dacus cucurbitae* Coquillett (Diptera: Tephritidae) and the influence of gamma-irradiation on dispersal. *Appl. Entomol. Zool.* 15:363–71

88. Hancock, D. L. 1985. New species and records of African Dacinae (Diptera: Tephritidae). *Arnoldia Zimbabwe* 9: 299–314

89. Haniotakis, G., Francke, W., Mori, K., Redlich, R., Schurig, V. 1986. Sex specific activity of R−(−)− and S−(+)−1, 7-dioxaspiro [5.5] undecane, the major pheromone of *Dacus oleae*. *J. Chem. Ecol.* 12:In press

90. Haniotakis, G., Voyadjoglou, A. 1978. Oviposition regulation in *Dacus oleae* by various olive fruit characters. *Entomol. Exp. Appl.* 24:387–92

91. Hardy, D. E. 1969. Taxonomy and distribution of the oriental fruit fly and related species (Tephritidae-Diptera). *Proc. Hawaii. Entomol. Soc.* 20:395–428

92. Hardy, D. E. 1983. The fruit flies of the genus *Dacus* Fabricius of Java, Sumatra and Lombock, Indonesia (Diptera: Tephritidae). *Treubia* 29:1–45

93. Hill, A. R., Hooper, G. H. S. 1984. Attractiveness of various colours to Australian tephritid fruit flies in the field. *Entomol. Exp. Appl.* 35:119–28

94. Howard, D. J., Bush, G. L., Breznak, J. A. 1985. The evolutionary significance of bacteria associated with *Rhagoletis*. *Evolution* 39:405–17

95. Hwang, S., Lee, W., Ho, K. 1979. Comparison of the histological structure of the spermathecae between normal and irradiated oriental fruit fly, *Dacus dorsalis*. *Bull. Inst. Zool. Acad. Sin.* 18:59–70

96. Ibrahim, A. G., Rahman, M. D. A. 1982. Laboratory studies on the effects of selected tropical fruits on the larvae of *Dacus dorsalis*, Hendel. *Pertanika* 5: 90–94

97. Ichinohe, F., Mizobuchi, M., Kosei, I. 1980. Notes on the biology of *Dacus expandens* Walker (Diptera: Tephritidae) with morphological description of the immature stages of *D. expandens* and *D. dorsalis*. *Res. Bull. Plant Prot. Serv. Jpn.* 16:35–40

98. Ichinohe, F., Tanaka, K., Ishikawa, K., Mizobuchi, M. 1983. Observations on attraction of the male melon flies, *Dacus cucurbitae*, to the blossoms of a common orchid, *Dendrobium superbum*. *Res. Bull. Plant Prot. Serv. Jpn.* 19:1–8

99. Ichinohe, F., Tanaka, K., Miura, M., Ito, Y. 1978. An estimation of the winter population density of *Dacus cucurbitae* Coq. in Southern Okinawa and effect of gamma radiation on the released male flies. *Appl. Entomol. Zool.* 13:316–18

100. Ishizuka, Y., Okumura, O., Yoshida, T. 1984. Studies on the adaptability of the melon fly to low temperatures. *Res.*

Bull. Plant Prot. Serv. Jpn. 20:55–61 (In Japanese with English summary)

101. Iwahashi, O. 1972. Movement of the oriental fruit fly adults among islets of the Ogasawara Islands. *Environ. Entomol.* 1:176–79

102. Kapatos, E. T., Fletcher, B. S. 1983. Seasonal changes in the efficiency of McPhail traps and a model for estimating olive fly densities from trap catches using temperature data. *Entomol. Exp. Appl.* 33:20–26

103. Kapatos, E. T., Fletcher, B. S. 1983. Development of a pest management system for *Dacus oleae* in Corfu by utilizing ecological criteria. See Ref. 22, pp. 593–602

104. Kapatos, E. T., Fletcher, B. S. 1984. The phenology of the olive fly, *Dacus oleae* (Gmel.) (Diptera: Tephritidae) in Corfu. *Z. Angew. Entomol.* 97:360–70

105. Kapatos, E. T., Fletcher, B. S. 1986. Mortality factors and life budgets for the immature stage of the olive fly, *Dacus oleae* (Gmel.) (Diptera: Tephritidae) in Corfu. *Z. Angew. Entomol.* In press

106. Kapoor, V. C. 1970. Indian Tephritidae with their recorded hosts. *Orient. Insects* 4:207–51

107. Kapoor, V. C., Agarwal, M. L. 1983. Fruit flies and their natural enemies in India. See Ref. 22, pp. 104–5

108. Katsoyannos, B. I. 1983. Captures of *Ceratitis capitata* and *Dacus oleae* flies (Diptera: Tephritidae) by McPhail and Rebell colour traps suspended on citrus, fig and olive trees on Chios, Greece. See Ref. 22, pp. 451–56

109. Katsoyannos, B. I., Patsouras, G., Vrekoussi, M. 1985. Effect of colour hue and brightness of artificial oviposition substrates on the selection of oviposition site by *Dacus oleae*. *Entomol. Exp. Appl.* 38:205–14

110. Katsoyannos, B. I., Pittara, I. S. 1983. Effect of size of artificial oviposition substrates and presence of natural host fruits on the selection of oviposition site by *Dacus oleae*. *Entomol. Exp. Appl.* 34:326–32

111. Kawai, A., Iwahashi, O., Ito, Y. 1978. Movement of the sterilized melon fly from Kume Is. to adjacent islets. *Appl. Entomol. Zool.* 13:314–15

112. Kawai, A., Yoshikaru, T. 1981. Daily activity rhythms of the melon fly, *Dacus cucurbitae* Coquillett (Diptera: Tephritidae) under laboratory conditions. *Bull. Veg. Ornamental Crops Res. Stn. Jpn. Ser. C* 5:75–78 (In Japanese with English summary)

113. Kawano, Y., Mitchell, W. C., Matsumoto, H. 1968. Identification of the male oriental fruit fly attractant in the golden shower blossom. *J. Econ. Entomol.* 61:986–88

114. Keiser, I., Harris, E. J., Miyashita, D. H., Jacobson, M., Perdue, R. E. 1975. Attraction of ethyl ether extracts of 232 botanicals to oriental fruit flies, melon flies and Mediterranean fruit flies. *Lloydia* 38:141–52

115. Keiser, I., Kobayashi, R. M., Miyashita, D. H., Harris, E. J., Schneider, E. L., Chambers, D. L. 1974. Suppression of Mediterranean fruit flies by oriental fruit flies in mixed infestations in guava. *J. Econ. Entomol.* 67:355–60

116. Keiser, I., Kobayashi, R. M., Miyashita, D. H., Jacobson, M., Harris, E. J., Chambers, D. L. 1967. Trans-6-nonen-1-ol acetate: an ovipositional stimulant of the melon fly. *J. Econ. Entomol.* 60:1355–56

117. Kelly, G. L. 1981. *The effect of temperature, thermal history and maturity on feeding rates of the Queensland fruit fly,* Dacus tryoni. Honours thesis. Univ. Sydney, Australia. 143 pp.

118. Kitching, W., Rilatt, J. A., Fletcher, M. T., Drew, R. A. I., Moore, C. J. 1986. Spiroketals in rectal gland secretions of Australasian fruit fly species. *J. Chem. Soc. Chem. Commun.* In press

119. Kobayashi, R. M., Ohinata, K., Chambers, D. L., Fujimoto, M. S. 1978. Sex pheromones of the oriental fruit fly and the melon fly: mating behavior, bioassay method, and attraction of females by live males and by suspected pheromone glands of males. *Environ. Entomol.* 7:107–12

120. Koyama, J., Chigira, Y., Iwahashi, O., Kakinohana, H., Kuba, H., Teruya, T. 1982. An estimation of the adult population of the melon fly, *Dacus cucurbitae* Coquillett (Diptera: Tephritidae) in Okinawa Island, Japan. *Appl. Entomol. Zool.* 17:550–58

121. Kuba, H., Koyama, J. 1985. Mating behavior of wild melon flies, *Dacus cucurbitae* Coquillet (Diptera: Tephritidae) in a field cage: courtship behavior. *Appl. Entomol. Zool.* 20:365–72

122. Lambrou, P. D., Tzanakakis, M. E. 1978. Inhibition of larval growth of *Dacus oleae* (Diptera: Tephritidae) by streptomycin. II. Effect of treating the parents. *Entomol. Exp. Appl.* 23:163–70

123. Laudého, Y., Canard, M., Liaropoulos, C. 1979. Étude de la phase hypogée de la population de *Dacus oleae* Gmel. (Diptera: Trypetidae) 1. Chute, répartition et devenir des larves migrantes. *Ann. Zool. Ecol. Anim.* 11:19–30

124. Lee, L. W. Y., Chang, T. H., Tsang, C. K. 1983. Sexual selection and mating behaviour of normal and irradiated oriental fruit flies. See Ref. 22, pp. 439–44

125. Lee, W., Tseung, C. K., Chang, T. H. 1985. An ultrastructural study on the ovariole development in the oriental fruit fly. *Bull. Inst. Zool. Acad. Sin.* 24:1–10

126. Leighton, S. M. 1979. *The protein and sugar requirements of the Queensland fruit fly,* Dacus tryoni *(Diptera: Tephritidae).* Honours thesis. Univ. Sydney, Australia. 79 pp.

127. Levinson, H. Z., Levinson, A. R. 1984. Botanical and chemical aspects of the olive tree with regards to fruit acceptance by *Dacus oleae* (Gmelin) and other frugivorous animals. *Z. Angew. Entomol.* 98:136–49

128. Liu, V. C. 1983. Population studies on the oriental fruit fly, *Dacus dorsalis* Hendel, in Central Taiwan. See Ref. 22, pp. 62–65

129. Louskas, C., Liaropoulos, C., Canard, M., Laudého, Y. 1980. Infestation estivale précoce des olives par *Dacus oleae* (Gmel.) (Diptera, Trypetidae) et rôle limitant du parasite *Eupelmus urozonus* Dalm. (Hymenoptera, Eupelmidae) dans une oliveraie grecque. *Z. Angew. Entomol.* 90:473–81

130. MacFarlane, J. R., East, R. W., Drew, R. A. I., Betlinski, G. A. 1986. The dispersal of irradiated Queensland fruit fly *Dacus tryoni* (Froggatt) (Diptera: Tephritidae) in south-eastern Australia. *Aust. J. Zool.* In press

131. Malan, E. M., Giliomee, J. H. 1968. Aspekte van die bionomie van *Dacus ciliatus* Loew (Diptera: Trypetidae). *J. Entomol. Soc. South Afr.* 31:373–89

132. Manoukas, A. G. 1972. Total amino acids in hydrolysates of the olive fly, *Dacus oleae,* grown in an artificial diet and in olive fruits. *J. Insect Physiol.* 18:683–88

133. Manoukas, A. G. 1983. Biological and biochemical parameters of the olive fruit fly with reference to larval nutritional ecology. See Ref. 22, pp. 410–15

134. Matanmi, B. A. 1975. The biology of tephritid fruit flies (Diptera: Tephritidae) attacking cucurbits at Ile-Ite, Nigeria. *Niger. J. Entomol.* 1:153–59

135. Mazomenos, B. E. 1984. Effect of age and mating on pheromone production in the female olive fruit fly, *Dacus oleae* (Gmel.). *J. Insect Physiol.* 30:765–69

136. Mazomenos, B. E., Haniotakis, G. E. 1985. Male olive fruit fly attraction to synthetic sex pheromone components in laboratory and field tests. *J. Chem. Ecol.* 11:397–405

137. Mazomenos, B. E., Pomonis, J. G. 1983. Male fruit fly pheromone: isolation, identification and lab. bioassays. See Ref. 22, pp. 96–103

138. Mazzini, M., Vita, G. 1981. Identificazione submicroscopica del meccanismo di trasmissione del batterio simbionte in *Dacus oleae* (Gmelin) (Diptera: Trypetidae). *Redia* 64:277–301

139. Meats, A. 1976. Seasonal trends in the acclimation to cold by the Queensland fruit fly (*Dacus tryoni,* Diptera) and their prediction by means of a physiological model fed with climatological data. *Oecologia* 27:73–87

140. Meats, A. 1976. Development and long-term acclimation to cold by the Queensland fruit fly (*Dacus tryoni*) at constant and fluctuating temperatures. *J. Insect Physiol.* 22:1013–19

141. Meats, A. 1976. Thresholds for cold torpor and cold survival in the Queensland fruit fly (*Dacus tryoni*) and the predictability of rates of change in survival threshold. *J. Insect Physiol.* 22:1505–9

142. Meats, A. 1981. The bioclimatic potential of the Queensland fruit fly, *Dacus tryoni,* in Australia. *Proc. Ecol. Soc. Aust.* 11:151–63

143. Meats, A. 1983. Critical periods for development acclimation to cold in the Queensland fruit fly, *Dacus tryoni. J. Insect Physiol.* 29:943–46

144. Meats, A. 1983. The response of the Queensland fruit fly, *Dacus tryoni,* to tree models. See Ref. 22, pp. 285–89

145. Meats, A. 1984. Thermal constraints to successful development of the Queensland fruit fly in regimes of constant and fluctuating temperature. *Entomol. Exp. Appl.* 36:55–59

146. Meats, A., Fay, H. A. C. 1976. The effect of acclimation on mating frequency and mating competitiveness in the Queensland fruit fly, *Dacus tryoni,* in optimal and cool mating regimes. *Physiol. Entomol.* 1:207–12

147. Meats, A., Khoo, K. C. 1976. The dynamics of ovarian maturation and oocyte resorption in the Queensland fruit fly, *Dacus tryoni,* in daily rhythmic and constant temperature regimes. *Physiol. Entomol.* 1:213–21

148. Messenger, P. S., Flitters, N. E. 1958. Effect of constant temperature environments on the egg stage of three species of Hawaiian fruit flies. *Ann. Entomol. Soc. Am.* 51:109–19

149. Metcalf, R. L., Metcalf, E. R., Mitchell, W. E., Lee, L. W. Y. 1979. Evolution of olfactory receptor in oriental fruit fly, *Dacus dorsalis. Proc. Natl. Acad. Sci. USA* 76:1561–65

150. Miyahara, Y., Kawai, A. 1979. Movement of sterilized melon fly from Kume Is. to the Amani Islands. *Appl. Entomol. Zool.* 14:496–97

151. Munro, K. H. 1984. A taxonomic treatise on the Dacidae (Tephritoidea, Diptera) of Africa. *Entomol. Mem. Dep. Agric. Repub. South Afr.* 61:1–313

152. Nakamori, H., Soemori, H. 1981. Comparison of dispersal ability and longevity for wild and mass-reared melon flies, *Dacus cucurbitae* Coquillett (Diptera: Tephritidae) under field conditions. *Appl. Entomol. Zool.* 16:321–27

153. Neuenschwander, P., Michelakis, S. 1979. Determination of the lower thermal thresholds and day-degree requirements for eggs and larvae of *Dacus oleae* (Gmel.) (Diptera: Tephritidae) under field conditions in Crete, Greece. *Bull. Soc. Entomol. Suisse* 52:57–74

154. Neuenschwander, P., Michelakis, P., Bigler, F. 1981. Abiotic factors affecting mortality of *Dacus oleae* larvae and pupae in soil. *Entomol. Exp. Appl.* 30:1–9

155. Newell, J. M., Haramoto, F. H. 1968. Biotic factors influencing populations of *Dacus dorsalis* in Hawaii. *Proc. Hawaii. Entomol. Soc.* 20:81–139

156. Ohinata, K., Jacobson, M., Kobayashi, R. M., Chambers, D. L., Fujimoto, M. S., Higa, H. H. 1982. Oriental fruit fly and melon fly: biological and chemical studies of smoke produced by males. *J. Environ. Sci. Health Pt. A* 17:197–208

157. Okumura, M., Ide, T., Takagi, S. 1981. Studies on the effect of temperature on the development of the melon fly, *Dacus cucurbitae* Coquillett. *Res. Bull. Plant Prot. Serv. Jpn.* 17:51–56 (In Japanese with English abstract)

158. O'Loughlin, G. T., East, R. A., Meats, A. 1984. Survival, development rates and generation times of the Queensland fruit fly, *Dacus tryoni,* in a marginally favourable climate: experiments in Victoria. *Aust. J. Zool.* 32:353–61

159. Petri, L. 1910. Untersuchungen über die Darmbakterien der Olivenfliege. *Zentralbl. Bakteriol. Parasitenkd. Infektionskr. Hyg. Abt. 2* 26:357–67

160. Pritchard, G. 1969. The ecology of a natural population of the Queensland fruit fly, *Dacus tryoni.* II. The distribution of eggs and its relationship to behaviour. *Aust. J. Zool.* 60:950–55

161. Pritchard, G. 1970. The ecology of a natural population of the Queensland fruit fly, *Dacus tryoni.* III. The maturation of female flies in relation to temperature. *Aust. J. Zool.* 18:77–89

162. Pritchard, G. 1978. The estimation of natality in a fruit-infesting insect (Diptera: Tephritidae). *Can. J. Zool.* 56:75–79

163. Prokopy, R. J. 1977. Stimuli influencing trophic relations in tephritidae. *Colloq. Int. CNRS* 265:305–36

164. Prokopy, R. J. 1977. Epideictic pheromones that influence spacing patterns of phytophagous insects. In *Semiochemicals: Their Role in Pest Control,* ed. D. A. Nordlund, R. L. Jones, W. J. Lewis 10:181–218. New York: Wiley. 306 pp.

165. Prokopy, R. J., Haniotakis, G. E. 1975. Responses of wild and laboratory cultured *Dacus oleae* to host plant color. *Ann. Entomol. Soc. Am.* 68:73–77

166. Prokopy, R. J., Koyama, J. 1982. Oviposition site partitioning in *Dacus cucurbitae. Entomol. Exp. Appl.* 31:428–32

167. Prokopy, R. J., Owens, E. D. 1983. Visual detection of plants by herbivorous insects. *Ann. Rev. Entomol.* 28:337–64

168. Prokopy, R. J., Roitberg, B. D. 1984. Foraging behavior of true fruit flies. *Am. Sci.* 72:41–49

169. Pucci, C., Forcina, A., Salmistraro, D. 1982. Incidenza della temperatura sulla mortalita degli stadi preimmaginali, sull'impupamento all'interno delle drupe e sull'activita dei parassiti del *Dacus oleae* (Gmel.). *Frustula Entomol.* (NS) 4:143–55

170. Qureshi, Z. A., Ashraf, M., Bughio, A. R., Hussain, S. 1974. Rearing, reproductive behaviour and gamma sterilization of the fruit fly, *Dacus zonatus* (Diptera: Tephritidae). *Entomol. Exp. Appl.* 17:504–10

171. Qureshi, Z. A., Ashraf, M., Bughio, A. R., Siddiqui, Q. H. 1975. Population fluctuation and dispersal studies of the fruit fly, *Dacus zonatus* (Saunders). In *Sterility Principle for Insect Control,* ed. Int. At. Energy Agency, pp. 201–7. Vienna: IAEA

172. Qureshi, Z. A., Bughio, A. R., Siddiqui, Q. H., Najeebullah. 1976. Efficacy of methyl eugenol as a male attractant for *Dacus zonatus* (Saunders) (Diptera: Tephritidae). *Pak. J. Sci. Ind. Res.* 19:22–23

173. Robinson, A., Hooper, G. H., eds. 1987. *Fruit Flies, Their Biology, Natural Enemies and Control.* Amsterdam: Elsevier. In press

174. Saeki, S., Katayama, M., Okumura, M. 1980. Effect of temperature upon the development of the oriental fruit fly and its possible distribution in the mainland

of Japan. *Res. Bull. Plant Prot. Serv. Jpn.* 16:73–76 (In Japanese with English abstract)

174a. Shimada, H., Tanaka, A., Kamiwada, H. 1979. Oviposition behaviour and development of the oriental fruit fly, *Dacus dorsalis* Hendel on *Prunus salicina* Lindl. *Proc. Assoc. Plant Prot. Kyushu* 25:143–46 (In Japanese with English summary)

175. Smith, E. S. C. 1977. Studies on the biology and commodity control of the banana fly, *Dacus musae* (Tryon), in Papua New Guinea. *Papua New Guinea Agric. J.* 28:47–56

176. Smith, P. H. 1979. Genetic manipulation of the circadian clock's timing of sexual behaviour in the Queensland fruit flies, *Dacus tryoni* and *Dacus neohumeralis*. *Physiol. Entomol.* 4:71–78

177. Smith, P. H. 1987. Behavioural partitioning of the day and circadian rhythmicity. See Ref. 173. In press

178. Soultanopoulou-Mantaka, A. 1974. Anatomie des Gehirns von *Dacus*. *Ann. Inst. Phytopathol. Benaki* (NS) 11:59–68

179. Stavraki, H. G. 1974. Mortality of immature stages of *Dacus oleae* (Diptera: Trypetidae) in two areas in Greece in the period 1969–1972. *Z. Angew. Entomol.* 77:210–17

179a. Steffens, R. J. 1983. Ecology and approach to integrated control of *Dacus frontalis* on the Cape Verde Islands. See Ref. 22, pp. 632–38

180. Steiner, L. F., Mitchell, W. C., Baumhover, A. H. 1962. Progress of fruit fly control by irradiation sterilization in Hawaii and the Mariana Islands. *Int. J. Appl. Radiat. Isot.* 13:427–34

181. Suzuki, Y., Koyama, J. 1981. Courtship behavior of the melon fly, *Dacus cucurbitae* Coquillett (Diptera: Tephritidae). *Appl. Entomol. Zool.* 16:164–66

182. Syed, R. A. 1968. Studies on the ecology of some important species of fruit flies and their natural enemies in West Pakistan. *Pak. Commonw. Inst. Biol. Control Stn. Rep., Rawalpindi.* Farnham Royal, Slough, UK: Commonw. Agric. Bur. 20 pp.

183. Syed, R. A. 1969. Studies on the ecology of some important species of fruit flies and their natural enemies in West Pakistan. *Pak. Commonw. Inst. Biol. Control Stn Rep., Rawalpindi.* Farnham Royal, Slough, UK: Commonw. Agric. Bur. 12 pp.

184. Tan, K. H., Lee, S. L. 1982. Species diversity and abundance of *Dacus* (Diptera: Tephritidae) in five ecosystems of Penang, West Malaysia. *Bull. Entomol. Res.* 72:709–16

185. Tsiropoulos, G. J. 1977. Reproduction and survival of the adult *Dacus oleae* (Gmel.) feeding on pollen and honeydew. *Environ. Entomol.* 6:390–92

186. Tsiropoulos, G. J. 1977. Survival and reproduction of *Dacus oleae* (Gmel.) fed on chemically defined diets. *Z. Angew. Entomol.* 84:192–97

187. Tsiropoulos, G. J. 1983. Microflora associated with wild and laboratory reared olive flies (*Dacus oleae* Gmel.). *Z. Angew. Entomol.* 96:337–40

188. Tsitsipis, J. A. 1977. Effect of constant temperatures on the eggs of the olive fly, *Dacus oleae* (Diptera: Tephritidae). *Ann. Zool. Ecol. Anim.* 9:133–39

189. Tsitsipis, J. A. 1980. Effect of constant temperatures on larval and pupal development of olive fruit flies reared on artificial diet. *Environ. Entomol.* 9:764–68

190. Tsitsipis, J. A., Abatzis, C. 1980. Relative humidity effects at 20°C on eggs of the olive fly, *Dacus oleae* (Diptera: Tephritidae) reared on artificial diet. *Entomol. Exp. Appl.* 28:92–99

191. Tychsen, P. H. 1975. Circadian control of sexual drive level in *Dacus tryoni* (Diptera: Tephritidae). *Behaviour* 54:111–41

192. Tychsen, P. H. 1977. Mating behaviour of the Queensland fruit fly, *Dacus tryoni* (Diptera: Tephritidae) in field cages. *J. Aust. Entomol. Soc.* 16:459–65

193. Tychsen, P. H., Fletcher, B. S. 1971. Studies on the rhythm of mating in the Queensland fruit fly, *Dacus tryoni*. *J. Insect Physiol.* 17:2139–56

194. Tzanakakis, M. E. 1974. Tomato as food for larvae of *Dacus oleae* (Diptera: Tephritidae). *Ann. Sch. Agric. For. Univ. Thessalonika* 17:548–52

195. Tzanakakis, M. E., Prophetou, D. A., Vassilou, G. N., Papadopoulos, J. J. 1983. Inhibition of larval growth of *Dacus oleae* by topical application of streptomycin to olives. *Entomol. Hell.* 1:65–70

196. Tzanakakis, M. E., Tsitsipis, J. A., Economopoulos, M. E. 1968. Frequency of mating in females of the olive fruit fly under laboratory conditions. *J. Econ. Entomol.* 61:1309–12

197. Van der Pers, J. N. C., Haniotakis, G. E., King, B. M. 1984. Electroantennogram responses from olfactory receptors in *Dacus oleae*. *Entomol. Hell.* 2:47–53

198. Vargas, R. I., Miyashita, D., Nishida,

T. 1984. Life history and demographic parameters of three laboratory-reared tephritids (Diptera: Tephritidae). *Ann. Entomol. Soc. Am.* 77:651–56

199. Vargas, R. I., Nishida, T. 1985. Life history and demographic parameters of *Dacus latifrons* (Diptera: Tephritidae). *J. Econ. Entomol.* 78:1242–44

200. Vargas, R. I., Nishida, T., Beardsley, J. W. 1983. Distribution and abundance of *Dacus dorsalis* (Diptera: Tephritidae) in native and exotic forest areas on Kauai. *Environ. Entomol.* 12:1185–89

201. Vijaysegaran, S. 1985. Management of fruit flies. In *Integrated Pest Management in Malaysia,* ed. B. S. Lee, W. H. Loke, K. L. Heong, pp. 231–54. Kuala Lumpur: Malays. Plant Prot. Soc. 335 pp.

201a. Vita, G., Barbera, F. 1976. Aspetti biochimici del rapporto piante insetto nel *Dacus oleae* (Gmel.). *Atti Congr. Naz. Ital. Entomol.* 11:151–61

202. Webb, J. C., Sivinski, J., Litzkow, C. 1984. Acoustical behavior and sexual success in the Caribbean fruit fly, *Anastrepha suspensa* (Loew) (Diptera: Tephritidae). *Environ. Entomol.* 13:650–56

203. Wharton, R. A., Gilstrap, F. E. 1983. Key to the status of opiine braconid (Hymenoptera) parasitoids used in biological control of *Ceratitis capitata* and *Dacus s.l.* (Diptera: Tephritidae). *Ann. Entomol. Soc. Am.* 76:721–42

204. Wong, T. T. Y., Mochizuki, N., Nishimoto, J. I. 1984. Seasonal abundance of parasitoids of the Mediterranean and oriental fruit flies (Diptera: Tephritidae) in the Kula area of Maui, Hawaii. *Environ. Entomol.* 13:140–45

205. Yamvrias, C., Panagopoulos, C. G., Psallidas, P. G. 1970. Preliminary study on the internal bacterial flora of the olive fruit fly, (*Dacus oleae* Gmelin). *Ann. Inst. Phytopathol. Benaki* 9:201–6

206. Yao, A., Lee, W. 1978. A population study of the oriental fruit fly, *Dacus dorsalis* Hendel (Diptera: Tephritidae), in guava, citrus fruits and wax apple fruit in Northern Taiwan. *Bull. Inst. Zool. Acad. Sin.* 17:103–8

207. Yao, A., Lee, W. 1979. Abundance and distribution studies of the oriental fruit fly, *Dacus dorsalis* Hendel, and its parasites in Eastern Taiwan. *Natl. Sci. Counc. Mon. ROC* 7:597–601

Ann. Rev. Entomol. 1987. 32:145–62

IMPROVED DETECTION OF INSECTICIDE RESISTANCE THROUGH CONVENTIONAL AND MOLECULAR TECHNIQUES[1,2]

Thomas M. Brown

Department of Entomology, Clemson University, Clemson, South Carolina 29634-0365

William G. Brogdon

Center for Disease Control, 1600 Clifton Road Northeast, Atlanta, Georgia 30333

PERSPECTIVES AND OVERVIEW

The challenge of insecticide resistance for the entomologist is to prevent its development or at least to manage its impact on new, hard-won insecticides (7). Resistance has been documented in 428 species of arthropods, but the documentation has come after failure to control the pest rather than as an early warning (36). Conventional detection of resistance is based on insecticide susceptibility tests which are dosage-mortality experiments usually performed in the laboratory (15).

Resistance management requires more effective techniques for detecting resistance in its early stages of development. Performance of susceptibility tests in the field is one approach to acquiring an earlier diagnosis of the failure of insecticide applications. Another approach is to employ biochemical techniques to determine whether or not individuals possess a mutant resistance allele. These two approaches are complementary in that the former

---

[1]The US Government has the right to retain a nonexclusive, royalty-free license in and to any copyright covering this paper.

[2]Technical contribution No. 2541 of the South Carolina Agricultural Experiment Station.

detects the response to the insecticide and the latter gives information concerning genetic mechanisms of resistance; use of both approaches will be more accurate and mutually confirmatory.

Insecticide resistance is a problem that can be understood best from the perspective of population genetics and ecology (Roush & McKenzie, this volume); however, progress in this area has been slow as compared to that on biochemistry of resistance (14). Improved and more sensitive methods of detecting and monitoring resistance will enhance the study of this phenomenon by providing information about the population dynamics of resistant insects as observed in the field. In order to verify that applied strategies for resistance management are effective, one must be able to detect resistant individuals at frequencies too low for detection by conventional susceptibility tests.

We discuss the progress in development of novel methods for detecting resistance where these methods are more practical than those in common use. In addition, we review recent developments in the related fields of biochemistry and molecular genetics upon which future high-resolution techniques for detecting resistance will be based.

RAPID SUSCEPTIBILITY TESTS IN THE FIELD

Most susceptibility tests in use today were designed for execution in the laboratory, since it is difficult to control conditions in the field. Performance of susceptibility tests in the field would provide the desired data more rapidly, allowing for better informed decisions regarding pest management. A portable environmental chamber facilitates this work; this has been demonstrated for *Lygus hesperus* and *Acyrthosiphon pisum* with self-dosing, contact susceptibility tests rather than the Food and Agriculture Organization (FAO) dip method (9). One very novel approach is the "attracticide" susceptibility test, in which the insecticide is incorporated in the sticky material of a pheromone trap for diagnosis of pyrethroid resistance in *Heliothis virescens* (41). We do not attempt to review the various modifications needed to adapt susceptibility tests to field conditions; however, in most tests insects are placed on insecticide-impregnated papers or insecticide-coated glass surfaces and the behavior of each insect determines the dose it receives.

It must be recognized that many self-dosing methods provide an untreated surface, such as a cotton plug, a screen enclosure, or food, onto which an insect can escape from the insecticide deposit and thereby survive; therefore, these methods are less accurate than topical application, in which insecticide is applied in a known amount directly to the insect. Self-dosing tests cannot discriminate between avoidance behavior or physiological resistance. In fact, *Euxesta notata* adults exposed to treated papers survived by resting on the untreated enclosures of the test container; after 12 generations of selection this strain appeared threefold more resistant through avoidance behavior, but was threefold more susceptible to direct applications of malathion (52). In preliminary susceptibility tests in the field, we have observed that one predatory adult coccinellid, *Hippodamia convergens,* avoided the papers impregnated with methyl parathion, while another, *Coleomegila maculata,* did not (T. M. Brown, unpublished data).

Portable topical application may be a solution to this complication, since hand-held mechanical-syringe repeating dispensers are commercially available (Hamilton Co., Reno, Nevada); however, the self-dosing tests are simpler to perform in the field and are more likely to gain acceptance. The uniformity of test insects and of environmental conditions available in the laboratory cannot be duplicated in most tests in the field; however, these difficulties will be offset by the potential for immediate detection of resistance rather than the wait of several weeks or months required for laboratory diagnoses.

As data accumulate from surveillance programs, it is possible to determine a discriminating dose that kills 99% of the normal, susceptible population. This dose must be based on responses of wild populations and not of laboratory-susceptible strains, since there is often a significant tolerance in field populations prior to introduction of the insecticide; this was documented in *H. virescens* tested prior to the introduction of photostable pyrethroid insecticides (17, 78). The abbreviated susceptibility test using a discriminating dose was successfully used to identify resistant populations of Australian *Heliothis armigera* (38), and this approach is now recommended for surveillance of mosquitoes (16).

Occasionally it is possible to distinguish between heterozygous and homozygous resistant genotypes using a second discriminating dose (13). This requires the prior establishment of a homozygous resistant strain, knowledge of the inheritance of resistance, and the assumption that this same type of inheritance occurs in the field population. These data could be useful to the population geneticist in the study of the development of resistance (Roush & McKenzie, this volume).

DETECTING ENZYMES ASSOCIATED WITH RESISTANCE

Resistance is often based upon increased enzymatic detoxication of an insecticide or reduced sensitivity of a target enzyme to inhibition by the insecticide. These biochemical mechanisms have been reviewed (27, 39, 49, 70, 97, 100). We emphasize recent reports and new techniques that hold promise for detecting these enzymes.

Carboxylester hydrolase (48, 53) conferred resistance in *Myzus persicae* when a specific isozyme, E4, was increased in quantity (29). This enzyme catalyzes the hydrolysis of carboxyl esters and carboxyl amides such as those found in malathion and dimethoate. Furthermore, it was found that resistance to noncarboxyl organophosphorothioate insecticides was conferred as well, owing to the enzyme's rapid phosphorylation and subsequent slow recovery when it reacted with the oxon metabolite of the insecticide. The tremendous quantity of this enzyme in resistant aphids appears to protect acetylcholinesterase by providing an alternative site of phosphorylation. While carboxylester hydrolase activity and turnover of paraoxon did not differ in E4 from resistant and susceptible aphids, this isozyme appeared to recover from phosphorylation more rapidly than mammalian carboxylester hydrolases (18).

This mechanism was also observed in similar investigations in the green rice leafhopper (65). Malaoxon, the metabolite of malathion, can either phosphorylate or be hydrolyzed depending on whether its phosphate ester or carboxyl ester contacts the active site (61); however, the phosphorylation reaction results in inhibition, so that it is by far the less efficient pathway of malaoxon detoxication (65). In some species, e.g. *Tribolium castaneum,* there are some malathion-resistant populations that retain susceptibility to most organophosphorus insecticides, even certain other carboxyl ester–containing compounds (5).

Detection of resistant individuals of *M. persicae, Culex pipiens,* and other species was possible owing to their greater carboxylester hydrolase activity toward 1-naphthyl acetate (45, 75); however, resistance and this activity were not directly correlated in certain other species or strains. In fact, the Indian mealmoth, *Plodia interpunctella,* exhibited an inverse relationship. Homogenates of resistant larvae hydrolyzed malathion 35-fold faster than those of susceptible larvae, while 1-naphthyl acetate hydrolysis measured spectrophotometrically was reduced 65% in the resistant strain. Electropherograms of the resistant strain stained very faintly for this activity (6). In an earlier study malathion-resistant *P. interpunctella* homogenates had half the normal activity against four other acetate esters and methyl butyrate (104). Hemingway (42) observed that 1-naphthyl acetate hydrolysis was similarly uncorrelated with malathion resistance in *Anopheles stephensi* from Pakistan and *Anopheles arabiensis* from Sudan. This recalls the original hypothesis of a "mutant aliesterase," in which organophosphorus-resistant house flies displayed less activity toward methyl butyrate (71). More recently, house flies have been found with positively correlated resistance (79). Malathion-resistant *Drosophila melanogaster* flies showed no increase in malathion hydrolysis, and their malathion carboxylester hydrolase was separated from most 1-naphthyl acetate–hydrolyzing proteins by molecular size exclusion chromatography (50).

While pyrethroid insecticides usually contain carboxyl esters, carboxylester hydrolase activity toward fenvalerate in the midgut of *Trichoplusia ni* was not associated with the isozymes hydrolyzing 1-naphthyl acetate (33). This report demonstrates the importance of studying the specific detoxication reaction of interest rather than relying upon other substrates that indicate only a general type of activity.

Naphthyl acetate is not a definitive substrate for carboxylester hydrolase activity; in fact, it is a general substrate for a variety of other hydrolases including arylester hydrolase, acetylester hydrolase, and cholinesterases. Naphthyl butyrate is more selective for detecting carboxylester hydrolases, since they are more active against butyrates than against acetates (48). Butyrate esters are less soluble than acetates and, in our experience, more difficult to use in gel staining; however, they should be included in research, when possible, to develop more selective methods of detection. Various butyrate thioesters are also available to measure carboxylester hydrolase activity while eliminating much background activity (21); methyl thiobutyrate was employed for this purpose in studies with house flies (55).

It is clear that rapid field assays based on carboxylester hydrolase can be developed for those cases in which resistance is positively correlated with enzyme activity toward a surrogate indicator substrate; e.g. the filter paper assay to detect resistant *C. pipiens* individuals (76). Resistant aphids with increased E4 enzyme were detected using an enzyme-linked immunosorbent assay (ELISA) in which carboxylester hydrolase activity was retained (32). It is equally apparent that this method will not apply to all species or populations and that each case must be investigated thoroughly before conclusions are drawn regarding insecticide detoxication and its relationship to enzyme activity against any surrogate substrate.

Organophosphorus and carbamate resistance due to detoxifying enzymes can be detected when decreased inhibition of a target enzyme by an organophosphate or carbamate may be overcome by an inhibitor of a specific detoxifying mechanism. Herath & Davidson (46) used this sort of procedure to detect hydrolase resistance. They exposed mosquitoes to papers treated with the inhibitor/synergist compounds S,S,S-tributylphosphorothioate (DEF® or TBPT), triphenyl phosphate (TPP), or O,O-dimethyl O-phenyl phosphorothioate (SV-1) prior to routine exposure to organophosphorus-impregnated papers in the World Health Organization (WHO) test kit. The kit test results, showing the effects of exposure to these compounds, provided information on the mechanisms of resistance in insectary strains of *Anopheles albimanus*.

Arylester hydrolase (EC 3.1.1.2), which rapidly hydrolyzes nonionizable phosphate esters such as paraoxon and certain organophosphinates (37), fortunately appears to be absent in insects while it is present in the blood and liver of mammals. Column-chromatographic fractions of resistant house flies contained paraoxon hydrolyzing activity of 2×10^{-6} μmol min^{-1} mg^{-1} protein (55); however, this was 27,000-fold less than the activity of a similar rabbit serum preparation, 5.6×10^{-2} μmol min^{-1} mg^{-1} protein (37). The continuing search for arylester hydrolase in insects could benefit from the use of certain organophosphinate monoesters, which are 10- to 50-fold better substrates than paraoxon for the rabbit enzyme (37).

Glutathione S-transferase (54) and its association with resistance (27) have been reviewed. While spectrophotometric detection of activity toward 3,4-dichloronitrobenzene is straightforward, there has been little effort to adapt laboratory methods to develop a rapid assay based on this enzyme for detecting individual resistant insects. In house flies glutathione S-transferase was apparently altered in a resistant strain since it exhibited a lesser Michaelis constant and was less inducible (73).

Staining of electropherograms for insect glutathione S-transferase has been accomplished by using methyl iodide as substrate and observing the blue color formed from the reaction of the iodine produced with the starch that was incorporated into polyacrylamide gels (23). Mammalian glutathione S-transferases have been detected on starch gels after agarose overlays of substrate (92). ELISA methods are also promising for the detection of this enzyme; a mouse monoclonal antibody for porcine liver enzyme has been successfully employed in detecting DDT-resistant adult *Anopheles albimanus* in Haiti and *Anopheles gambiae* in the laboratory using an ELISA method employing linked peroxidase activity (W. G. Brogdon, unpublished results). DDT-dehydrochlorinase is a glutathione-dependent enzyme that may be a form of glutathione S-transferase (24), and this could account for the results in *Anopheles*.

As in the case of carboxylester hydrolase, there are multiple forms of glutathione S-transferase in each organism (64, 92); therefore, it cannot be assumed that measuring total activity with a nonselective surrogate substrate will provide useful information regarding resistance in which an insecticide is detoxified. While the transferase is a soluble enzyme that can be assayed in homogenates, the assay requires glutathione and there are fewer simple colorimetric assays for its detection than are available for carboxylester hydrolase. Since this glutathione S-transferase can catalyze detoxication of phosphate esters and phosphorothioate esters, it is likely to be more effective than carboxylester hydrolase against the vast majority of organophosphorus insecticides; therefore, additional effort should be expended in devising rapid and accurate means to detect it.

Monooxygenase (26) is a membrane-bound enzyme that catalyzes oxidative detoxication of insecticides in resistant insects (49, 97). Monooxygenase activity is complex, since it is dependent on association with an electron transport system, reduced nicotine adenine dinucleotide phosphate, and molecular oxygen. Activity is inducible (94), and there are multiple forms of the terminal catalytic protein P-450.

Monooxygenase detection on electropherograms has been based on the porphyrin iron contained in the P-450 molecule (63); this technique has not been exploited in insects. There are at least three monooxygenases in the house fly, and the spectrum of activity against various substrates changes upon induction (1). Activity has been assessed by measuring the formation of the oxidized product of aldrin, biphenyl (102), and various other substrates, usually by gas or liquid chromatography. Recently, activity against certain phenoxazones that are oxidized to fluorescent products has been used to categorize the type of monooxygenase induction produced in mammals (20); the increased sensitivity of fluorescence could be exploited in insect assays. Monoclonal antibodies have been produced to recognize human and rabbit P-450 epitopes (4, 87). While detection of individual resistant insects based on monooxygenase activity awaits application of these or other techniques, rapid advances in the molecular genetics of this enzyme in mammals may provide opportunities to develop ELISA detection in insects in the near future.

Acetylcholinesterase in resistant arthropods may be less susceptible to inhibition by carbamate and organophosphorus insecticides. This was demonstrated in spider mites that were heterozygous for resistance; half of their enzyme was resistant and half was susceptible to paraoxon inhibition (90). The less sensitive enzyme has been detected in mosquitoes using the colorimetric assay to determine retention of activity toward acetylthiocholine upon exposure to inhibitor (84, 85). A more sensitive assay has been developed with human tissue, using acetylthiocholine as substrate and the reaction of the hydrolysis product thiocholine with a coumarin derivative to yield blue fluorescence (74). This assay is sensitive enough to make multiple end-point determinations on a single ganglion of *Heliothis virescens* (T. M. Brown, unpublished data).

CURRENT EFFORTS TO ADAPT BIOCHEMICAL DIAGNOSIS TO SITES IN THE FIELD

Up to now, most published biochemical resistance studies, and particularly evaluations of mechanisms, have largely involved laboratory studies of colonized strains. This has been due to the complete lack of methodology for biochemically ascertaining resistance mechanisms in the field. The chemistry of resistance mechanisms has been thoroughly examined (2, 27, 29, 30, 39, 43, 44, 49, 60, 72, 81, 95, 97, 100). Some investigators have extended their work to include studies of genetic and biological factors that are significant for understanding resistance in the field (19, 28, 34, 45, 67, 83, 95).

Unfortunately, the practical field impact of laboratory resistance studies has thus far been limited. For example, there have been few detailed studies of the spatial distribution or level of resistance; however, the lack of practical field information is understandable. For decades, studies on resistance problems in situ have relied on the susceptibility tests for determining resistance. Practical difficulties inherent in these tests, as in the WHO susceptibility test for vectors of disease, are numerous. These are acknowledged (G. Davidson, personal communication) to include:

1. Only one insecticide can be tested per insect.
2. Without a known discriminating dosage, large numbers of insects are needed to produce probit lines. With this method a low level of resistance would not be detected.
3. The use of discriminating dosages establishes the presence or absence of resistance with smaller samples, but does not give an indication of the level of resistance. A discriminating dose previously determined may be grossly inaccurate for the species under consideration, because great variability may exist in different populations of the same species in different ecological settings.
4. Resistance can be indicated falsely when susceptible insects survive because of deterioration of papers, resting on netting, etc, so to confirm resistance detected at a low level it is necessary to rear the progeny of the survivors and test them.

Biochemical tests would have important advantages, especially if a complete field kit for testing biochemical resistance possessed the following attributes:

1. The method should detect resistance and provide information on the likely mechanism(s) involved.
2. It must permit analysis of single insects. Frequently, one must deal with very small sample pools; thus the unit of statistical analysis must be the individual in the small pool.
3. It must permit multiple assays from a single insect. It is difficult to envisage a single biochemical test that would allow simultaneous elucidation of the presence and mechanism(s) of resistance.
4. It must be fast and accurate, since multiple assays on individuals in collections are necessary.
5. It should be usable in undeveloped areas, so the equipment should be simple, inexpensive, and easy to carry.
6. To achieve the most practicality, the kit should be adaptable for most or all species and should provide all needed assessment information in the field.

Methods that would fulfill these rigorous criteria may be biochemical or immunologic, and a number of workers are developing such methods at present (10, 12, 45, 76, 77, 86, 96). The purpose of this methodology is not to elucidate or define the in vivo enzymology (especially in the case of kinetics) of the resistance enzymes of the species studied, but to recognize resistant phenotypes and the mechanisms likely to produce the resistance.

Development of field resistance-detection methods based on the assay of enzyme activities in individual insects is proceeding along two lines. Biochemical microassays are adapted for use of filter paper (75, 76) or microtiter plates (12).

The filter-paper technique was developed for rapid determination of phenotypes with high esterase activity in OP-resistant *Culex quinquefasciatus*. Insects are crushed in 15 μl distilled water on a plexiglass plate using the bottom of a hemolysis tube. The bottom of the test tube is firmly blotted on a piece of Whatman No. 2 filter paper, which has been placed on several layers of tissue paper. After 10–20 mosquito homogenates have been deposited, the filter paper is sequentially immersed in enzyme substrate, stain, and fixing solutions, and is then dried. The particular advantages of this method are that it is simple, it is portable, and a permanent record of results is obtained. Work is under way to adapt other resistance enzyme detection methods for filter paper use.

Brogdon & Dickinson (12) have adapted hydrolase microassays for use in microtiter plates. Each microassay is sufficiently sensitive to allow 30 repetitions from a single mosquito. Microassays for acetylcholinesterase, malathion carboxylester hydrolase (using isomalathion substrate), general hydrolase, and nonspecific esterase have been developed. This approach incorporates some of the advantages of the filter paper method and also provides some unique capabilities. Single mosquitoes are homogenized in 100 μl of the appropriate buffer in ceramic spot plate wells using a teflon or ground glass pestle. Each homogenate is diluted to 3 ml with buffer, providing 30 100-μl aliquots per mosquito. Aliquots are transferred to microtiter plate wells. Aliquots of the appropriate test reagents for each resistance-enzyme assay are added to each well. An incubation period is required at certain stages in some assays. In addition, aliquots of inhibitors may be added at the appropriate time in some assays. Reagents may be added to numerous wells more rapidly using repeating dispensers or pipettes.

In the laboratory, plates are read in an enzyme-immunoassay reader, which reads all wells in a microtiter plate simultaneously and immediately prints the results. There are now battery-operated readers that are highly portable. However, from baselines generated in the laboratory, series of standard reagents or photographic series of standards may be carried to the field for direct comparison and quantitation of field plates without need of instrumentation. Correction for insect size variation is accomplished by protein microassay of each insect (10, 11). Development of microassay methods for glutathione *S*-transferase and monooxygenases is near completion in the laboratory of W. G. Brogdon.

The development of hybridoma technology offers new possibilities for the design of field resistance detection techniques (101). Monoclonal antibodies are being made for resistance enzymes; a monoclonal antibody to glutathione *S*-transferase seems to allow detection of DDT resistance in small fractions of single mosquitoes (W. G. Brogdon, unpublished results). Other laboratories are working on monoclonal antibodies to other inducible resistance enzymes and to receptor sites such as those responsible for altered acetylcholinesterase and DDT resistance due to the *kdr* gene. We foresee a day when application of monoclonal antibody techniques will allow the simultaneous detection of resistance, reproductive state, and presence of malaria parasites in one mosquito.

Recent work in Haiti and Guatemala (W. G. Brogdon, unpublished results) and in Sri Lanka (J. Hemingway, unpublished results) has focused on introducing these biochemical detection methods into the field. It has been possible to identify homozygous resistant and susceptible individual mosquitoes and heterozygotes in the field, and to match elevated hydrolase levels with reduced susceptibility in bioassays. Understanding of the relationship between elevated enzyme levels and efficacy of an insecticide in a particular field situation will be crucial in the development of biochemical field methods.

In Haitian and Guatamalan *Anopheles albimanus* it was possible to correlate levels of 1-naphthyl acetate hydrolysis with WHO-kit susceptibility-test results. Also, the biochemical test showed greater sensitivity for small numbers of resistant phenotypes in the population. In Haiti, where an emerging fenitrothion resistance problem was documented by susceptibility tests, resistant phenotypes were detected six months earlier by biochemical techniques. In Guatemala, it was possible to map insecticide resistance over all regions of the country by testing for 1-naphthyl acetate hydrolysis, malathion carboxylester hydrolase, and propoxur-insensitive acetylcholinesterase in each specimen. Assays were performed with larvae, adults, and blood-fed adults from which abdomena were excised (W. G. Brogdon, unpublished results).

PROSPECTS FOR DETECTING RESISTANCE GENES

With advances in molecular genetics and gene cloning, it may become possible to detect the mutant alleles that confer resistance by using radiolabeled nucleic acid probes. Genes relevant to insecticide resistance are often relevant to human health as well, and certain genes for detoxicative enzymes and for insecticide targets are the present focus of intense research activity. Clones of several genes germane to resistance are already employed in medical and toxicological research.

While direct detection of resistance by nucleic acid probes is an unlikely prospect for the near future, the sensitivity in detection of even single base-pair changes in alleles is increasing. For example, the sickle-cell allele from 1 μg of human DNA can be detected by gene amplification in vitro; a polymerase chain reaction is followed by radiolabeled DNA probe hybridization, and the sequential cleavage of the oligomer by two restriction enzymes results in the release of a characteristic 3-nucleotide fragment of the probe (88). This procedure requires 10 hr, and the authors describe its feasibility for the hospital clinic. Another example is the direct detection of *Plasmodium falciparum* from 50 μl of human blood through a DNA probe for a highly repeated and unique DNA sequence (3).

Once a DNA sequence of the gene of interest has been cloned from one organism, it becomes a useful tool for detecting the analogous gene in another species. For example, the *Ace* gene segment clone of *Drosophila* has been used to identify mRNA of human brain acetylcholinesterase and the human acetylcholinesterase gene from a DNA library (91). Perhaps similar techniques could lead to cloning of resistant acetylcholinesterase genes of insects.

Nucleic acid probes will facilitate the identification and isolation of rare proteins, thereby allowing for the production of ELISA tests for proteins that would otherwise remain obscure. Examples are the noncatalytic receptor proteins and ion channels whose mutant forms would be extremely difficult to isolate by conventional biochemistry.

The acetylcholine receptor, which is the target of insecticides such as nicotine, has been defined through amino acid sequences inferred from the observed nucleotide sequences from cloned DNA of *Torpedo* (22, 69). Similar studies of the molecular genetics of the GABA receptor and the associated chloride channel are in progress (47). This ion channel is the site of action of the cyclodiene insecticides and lindane (62), and it is likely to be involved in cyclodiene resistance, which is rarely due to detoxication.

Pyrethroid insecticides act on sodium channels in the nervous system (59, 62). A recent study has revealed fewer sodium channels of unchanged saxitoxin binding efficiency in fenvalerate-resistant *D. melanogaster* (56). Researchers are investigating the molecular regulation of sodium channel biosynthesis, since resistance is apparently based in a regulatory gene.

Detection of detoxicative resistance will also improve with the knowledge gained through studies of molecular genetics. Several detoxicative enzymes are inducible, so increased activity may be due to alteration in gene regulation, e.g. increased transcription, increased translation, or posttranslational modifications of protein (8). Induction of monooxygenase, glutathione *S*-transferase, and carboxylester hydrolase activities has been demonstrated clearly in the fall armyworm (103).

Monooxygenase P-450 genes have been studied through cDNA clones following phenobarbital, methylcholanthrene, and pregnenolone carbonitrile induction of mRNA in mammalian liver (35, 68, 99). While the P-450p gene from rat was found to be conserved among rabbit, hamster, gerbil, and mouse, response to five inducers was highly variable among these species. This suggests less conservation in some regulatory mechanism (99). In mice, induction of certain monooxygenase activities (66) is regulated at the *Ah* locus by the binding of xenobiotics to a cytosolic receptor protein (40), which controls structural gene translation (68). Mutation of the regulatory gene (which encodes the cytosolic receptor) results in less translation in noninducible D2 mice. This work has included the isolation of candidate cDNA clones for a P-450 structural gene (68).

It is possible, although not established, that certain mechanisms of insecticide resistance are conferred by mutations in regulatory genes. Circumstantial evidence is found in the mapping of detoxicative resistance and the activity of detoxicative enzymes in the house fly; several genes are very closely linked and are possibly due to one locus on chromosome II (25, 80, 89). Testing for a mutant regulatory locus in resistant insects awaits mapping of the structural genes of these detoxicative enzymes, since only their activities have been mapped to date. Genetic regulation of juvenile hormone esterase is under study in *D. virilis* (82). Coordinate regulation of several hydrolases in mice is due to a single gene, *Neu-1* on chromosome 17, which alters activity through control of posttranslational processing of the enzymes (98).

Carboxylester hydrolase–encoding mRNA has been identified from organophosphate-resistant aphids by immunoprecipitation of products translated by a rabbit reticulocyte lysate system (31). This mRNA may allow the production of radiolabeled cDNA probes for the gene that encodes E4 carboxylester hydrolase. While E4 is presumed to be identical to the enzyme found in susceptible aphids, it is produced in far greater quantity in the resistant aphids. The isolation of mRNA will allow further investigation of the regulation of this gene. Radioactive cDNA probes of mammalian protein hydrolases have been developed from a rat pancreatic cDNA library (51).

Glutathion *S*-transferase of the rat has been studied by the cloning of the gene for a transferase subunit following isolation of pentobarbital-induced mRNA from rat liver (57, 93). Nucleotide sequences of these clones have been used to predict the amino acid composition of enzyme subunits.

The examples cited above demonstrate that knowledge of the genes associated with insecticide resistance is rapidly expanding. The new techniques employed in molecular genetics should provide powerful tools for the detection of resistant insects and the basic study of insecticide resistance and its genetic control.

RESEARCH IMPERATIVES

Biochemical techniques offer the capability of detecting the initial stages of resistance in a population and the mechanism of resistance involved. More sensitive tests combined with rapid confirmatory susceptibility tests will enhance the feasibility of managing resistance through operational strategies. Pioneering experiments in the field should continue and should be promoted in both public health and agricultural situations.

Genetic definition of strains of key pests should be increased so that techniques can be developed based on biochemical mechanisms that are known with certainty. Presently linkage relationships for genes relevant to resistance are known only for strains of *Aedes aegypti, Drosophila melanogaster,* and *Musca domestica.* Major agricultural pests must be studied genetically so that the powerful tools of molecular genetics can be applied to such problems as the regulation of resistance genes.

Expansion of ELISA techniques is feasible and should be enhanced by the availability of rare proteins produced from clones of resistance genes and subsequently used as antigen to produce novel antibodies. At present, ELISA offers the best approach for detection of resistance; however, detection of the nucleic acids involved through the use of DNA probes is a prospect to be pursued. It should be noted that gene-detection methods that do not require radioactive labels must be considered, since transport of radioactive materials is difficult, especially when international field research is involved. Perhaps fluorescence or enzyme-linked nucleic acids could provide alternative means of detection, as has been demonstrated with biotin-11-dUTP nick translation probes linked to alkaline phosphatase through streptavidin (58).

Given the urgency of the problem of resistance, the dearth of novel insecticide chemistry and known biochemical targets, and the great potential unleashed by molecular biology, it appears that entomologists will answer the challenge with new and exciting approaches.

ACKNOWLEDGMENTS

We thank R. W. Beeman, C. C. Campbell, W. E. Collins, G. P. Georghiou, and D. G. Heckel for their comments and T. Cothran for preparing the manuscript.

Literature Cited

1. Agosin, M. 1982. Multiple forms of insect cytochrome P-450: role in insecticide resistance. In *Cytochrome P-450, Biochemistry, Biophysics and Environmental Implications,* ed. E. Hietanen, M. Laitinen, O. Hanniner, pp. 661–69. Amsterdam: Elsevier

2. Ayad, H., Georghiou, G. P. 1975. Resistance to organophosphates and carbamates in *Anopheles albimanus* based on reduced sensitivity of acetylcholinesterase. *J. Econ. Entomol.* 68:295–97

3. Barker, R. H., Suebsaeng, L., Rooney, W., Alecrim, G. C., Dourado, H. W., Wirth, D. F. 1986. Specific DNA probe for the diagnosis of *Plasmodium falciparum* malaria. *Science* 231:1434–36

4. Beaune, P., Kremers, P., Letawe-Goujon, F., Gielen, J. E. 1985. Monoclonal antibodies against human liver cytochrome P-450. *Biochem. Pharmacol.* 34:3547–52

5. Beeman, R. W. 1983. Inheritance and linkage of malathion resistance in the red flour beetle. *J. Econ. Entomol.* 76:737–40

6. Beeman, R. W., Schmidt, B. A. 1982. Biochemical and genetic aspects of malathion-specific resistance in the Indianmeal moth (Lepidoptera: Pyralidae). *J. Econ. Entomol.* 75:945–49

7. Brattsten, L. B., Holyoke, C. W. Jr., Leeper, J. R., Raffa, K. F. 1986. Insecticide resistance: challenge to pest management and basic research. *Science* 231:1255–60

8. Bresnick, E. 1980. Induction of enzymes of detoxication. See Ref. 53, 1:69–84

9. Brindley, W. A., Al-Rajhi, D. H., Rose, R. L. 1982. Portable incubator and its use in insecticide bioassays with field populations of lygus bugs, aphids and other insects. *J. Econ. Entomol.* 75:758–60

10. Brogdon, W. G. 1984. Mosquito protein microassay. I. Protein determinations from small portions of single-mosquito homogenates. *Comp. Biochem. Physiol.* 79:457–59

11. Brogdon, W. G. 1984. Mosquito protein microassay. II. Modification for potential field use. *Comp. Biochem. Physiol.* 79b:461–64

12. Brogdon, W. G., Dickinson, C. M. 1983. A microassay system for measuring esterase activity and protein concentration in small samples and in high-pressure liquid chromatography eluate fractions. *Anal. Biochem.* 131:499–503

13. Brown, A. W. A. 1966. Diagnostic dosages to separate genotypes of *Aedes aegypti* for DDT-resistance and dieldrin resistance. *Bull. WHO* 34:311–12

14. Brown, A. W. A. 1976. Chemical foundation of the development of resistance against insecticides. In *Chemie der Pflanzenschutz- und Schädlingsbekämpfungsmittel,* ed. R. Wegler, 3:229–58. Berlin/Heidelberg: Springer-Verlag

15. Brown, A. W. A. 1976. How have entomologists dealt with resistance? *Am. Phytopathol. Soc. Proc.* 3:67–74

16. Brown, A. W. A. 1983. Insecticide resistance as a factor in the integrated control of Culicidae. In *Integrated Mosquito Control Methodologies,* ed. M. Laird, J. W. Miles. 1:161–235. London: Academic

17. Brown, T. M., Bryson, P. K., Payne, G. T. 1982. Pyrethroid susceptibility in methyl parathion–resistant tobacco budworm in South Carolina. *J. Econ. Entomol.* 75:301–3

18. Bryson, P. K., Brown, T. M. 1985. Reactivation of carboxylester hydrolase following inhibition by 4-nitrophenyl organophosphinates. *Biochem. Pharmacol.* 34:1789–94

19. Bunting, S., Van Emden, H. F. 1980. Rapid response to selection for increased esterase activity on small populations of an apomictic clone of *Myzus persicae. Nature* 285:502–3

20. Burke, M. D., Thompson, S., Elcombe, C. R., Halpert, J., Haaparanta, T., Mayer, R. T. 1985. Ethoxy-, pentoxy- and benzyloxyphenoxazones and homologues: a series of substrates to distinguish between different induced cytochromes P-450. *Biochem. Pharmacol.* 34:3337–45

21. Chambers, H. 1973. Thio-esters as substrates for insect carboxylesterases. *Ann. Entomol. Soc. Am.* 66:663–65

22. Changeux, J. P., Devillers-Thiery, A., Chemouilli, P. 1984. Acetylcholine receptor: An allosteric protein. *Science* 225:1335–45

23. Clark, A. G. 1982. A direct method for the visualization of glutathione S-transferase activity in polyacrylamide gels. *Anal. Biochem.* 123:147–50

24. Clark, A. G., Shamaan, N. A. 1984. Evidence that DDT-dehydrochlorinase from the house fly is a glutathione S-transferase. *Pestic. Biochem. Physiol.* 22:249–61

25. Cluck, T., Plapp, F. W. Jr., Johnson, J. S. 1985. Metabolic resistance to in-

secticides: heterozygosity at the chromosome II locus in house flies, *Musca domestica* (Diptera: Muscidae). *J. Econ. Entomol.* 78:1015–19

26. Coon, M. J., Persson, A. V. 1980. Microsomal cytochrome P-450: a central catalyst in detoxication reactions. See Ref. 53, 1:117–35

27. Dauterman, W. C. 1983. Role of hydrolases and glutathione *S*-transferases in insecticide resistance. See Ref. 36, pp. 229–47

28. Denholm, I., Sawicki, R. M., Farnham, A. W. 1985. Factors affecting resistance to insecticides in house-flies, *Musca domestica* L. (Diptera: Muscidae). IV. The population biology of flies on animal farms in south-eastern England and its implications for the management of resistance. *Bull. Entomol. Res.* 75:143–58

29. Devonshire, A. L., Moores, G. D. 1982. A carboxylesterase with broad substrate specificity causes organophosphorus, carbamate and pyrethroid resistance in peach-potato aphids *(Myzus persicae)*. *Pestic. Biochem. Physiol.* 18:235–46

30. Devonshire, A. L., Moores, G. D. 1984. Characterization of insecticide-insensitive acetylcholinesterase: microcomputer-based analysis of enzyme inhibition in homogenates to individual house fly *(Musca domestica)* heads. *Pestic. Biochem. Physiol.* 21:341–48

31. Devonshire, A. L., Moores, G. D., Ffrench-Constant, R. H. 1986. Detection of insecticide resistance by immunological estimation of carboxylesterase activity in *Myzus persicae* (Sulzer) and cross reaction of the antiserum with *Phorodon humuli* (Schrank) (Hemiptera: Aphididae). *Bull. Entomol. Res.* 76:97–107

32. Devonshire, A. L., Searle, L. M., Moores, G. D. 1986. Quantitative and qualitative variation in the mRNA for carboxylesterases in insecticide-susceptible and resistant *Myzus persicae* (Sulzer). *Insect Biochem.* 16:659–65

33. Dowd, P. F., Sparks, T. C. 1986. Characterization of a *trans*-permethrin hydrolyzing enzyme from the midgut of *Pseudoplusia includens* (Walker). *Pestic. Biochem. Physiol.* 25:73–81

34. El-Khatib, Z. I., Georghiou, G. P. 1985. Geographic variation of resistance to organophosphates, propoxur and DDT in the southern house mosquito, *Culex quinquefasciatus*, in California. *J. Am. Mosq. Control Assoc.* 1:279–83

35. Fujii-Kuriyama, Y., Sogawa, K., Mizukami, Y., Suwa, Y., Muramatsu, M., et al. 1985. Molecular multiplicity and gene structure of microsomal cytochrome P-450 in rat liver. *Gann Monogr. Cancer Res.* 30:157–69

36. Georghiou, G. P., Saito, T., eds. 1983. *Pest Resistance to Pesticides.* New York/London: Plenum. 809 pp.

37. Grothusen, J. R., Bryson, P. K., Zimmerman, J. K., Brown, T. M. 1986. Hydrolysis of 4-nitrophenyl organophosphinates by arylester hydrolase from rabbit serum. *J. Agric. Food Chem.* 34:513–15

38. Gunning, R. V., Easton, C. S., Greenup, L. R., Edge, V. E. 1984. Pyrethroid resistance in *Heliothis armiger* (Hübner) (Lepidoptera: Noctuidae) in Australia. *J. Econ. Entomol.* 77:1283–87

39. Hama, H. 1983. Resistance to insecticides due to reduced sensitivity of acetylcholinesterase. See Ref. 36, pp. 299–331

40. Hannah, R. R., Nebert, D. W., Eisen, H. J. 1981. Regulatory gene product of the *Ah* complex. *J. Biol. Chem.* 256:4584–90

41. Haynes, K. F., Miller, T. A., Staten, R. T., Li, W., Baker, T. C. 1986. Pheromone trap for monitoring insecticide resistance in the pink bollworm moth, *Pectinophora gossypiella* (Lepidoptera: Gelechiidae): new tool for resistance management. *Environ. Entomol.* In press

42. Hemingway, J. 1982. The biochemical nature of malathion resistance in *Anopheles stephensi* from Pakistan. *Pestic. Biochem. Physiol.* 17:149–55

43. Hemingway, J. 1983. Biochemical studies on malathion resistance in *Anopheles arabiensis* from Sudan. *Trans. R. Soc. Trop. Med. Hyg.* 77:477–80

44. Hemingway, J., Georghiou, G. P. 1983. Studies on acetylcholinesterase of *Anopheles albimanus* resistant and susceptible to organophosphate and carbamate insecticides. *Pestic. Biochem. Physiol.* 19:167–71

45. Hemingway, J., Georghiou, G. P. 1984. Baseline esterase levels for anopheline and culicine mosquitoes. *Mosq. News* 44:33–35

46. Herath, P. R. J., Davidson, G. 1981. Multiple resistance in *Anopheles albimanus*. *Mosq. News* 41:535–38

47. Heylin, M., ed. 1985. Neuroreceptor study center at Pitt. *Chem. Eng. News* 63:18

48. Heymann, E. 1980. Carboxylesterases and amidases. See Ref. 53, 2:291–324

49. Hodgson, E., Kulkarni, P. 1983. Characterization of cytochrome P-450 in

studies of insecticide resistance. See Ref. 36, pp. 207–28

50. Holwerda, B. C., Morton, R. A. 1983. The *in vitro* degradation of [^{14}C] malathion by enzymatic extracts from resistant and susceptible strains of *Drosophila melanogaster*. *Pestic. Biochem. Physiol.* 20:151–60

51. Honey, N. K., Sakaguishi, A. Y., Lalley, P. A., Quinto, C., MacDonald, R. J., et al. 1984. Chromosomal assignments of genes for trypsin, chymotrypsin B, and elastase in mouse. *Somat. Cell Mol. Genet.* 10:377–83

52. Hooper, G. H. S., Brown, A. W. A. 1965. Development of increased irritability to insecticides due to decreased detoxication. *Entomol. Exp. Appl.* 8: 263–70

53. Jacoby, W. B., ed. 1980. *Enzymatic Basis of Detoxication,* Vols. 1, 2. New York: Academic. 415 pp., 369 pp.

54. Jacoby, W. B., Habig, W. H. 1980. Glutathione transferases. See Ref. 53, 2:63–94

55. Kao, L. R., Motoyama, N., Dauterman, W. C. 1984. Studies on hydrolases in various house fly strains and their role in malathion resistance. *Pestic. Biochem. Physiol.* 22:86–92

56. Kasbekar, D. P., Hall, L. M. 1985. *A Drosophila strain that has reduced sodium channel number is resistant to the pyrethroid fenvalerate.* Presented at Ann. Meet. Genet. Soc. Am., Boston

57. Lai, H.-C. J., Li, N., Weiss, M. J., Reidy, C. C., Tu, C.-P. D. 1984. The nucleotide sequence of a rat liver glutathione *S*-transferase subunit cDNA clone. *J. Biol. Chem.* 259:5536–42

58. Leary, J. J., Brigati, D. J., Ward, D. C. 1983. Rapid and sensitive colorimetric method for visualizing biotin-labeled DNA probes hybridized to DNA or RNA immobilized on nitrocellulose: bio-blots. *Proc. Natl. Acad. Sci. USA* 80:4045–49

59. Lund, A. E., Narahashi, T. 1983. Kinetics of sodium channel modification as the basis for the variation in nerve membrane effects of pyrethroids and DDT analogues. *Pestic. Biochem. Physiol.* 20:203–6

60. MacDonald, R. S., Soloman, K. R., Surgeoner, G. A., Harris, C. R. 1985. Laboratory studies on the mechanisms of resistance to permethrin in a field-selected strain of house flies. *Pestic. Sci.* 16:10–16

61. Main, A. R., Dauterman, W. C. 1967. Kinetics for the inhibition of carboxylesterase by malathion. *Can. J. Biochem.* 45:757–71

62. Matsumura, F. 1985. *Toxicology of In-secticides,* pp. 138–55. New York/London: Plenum. 598 pp. 2nd ed.

63. Moore, R. W., Welton, A. F., Aust, S. D. 1978. Detection of hemoproteins in SDS-polyacrylamide gels. *Methods Enzymol.* 52:324–31

64. Motoyama, N., Hayashi, A., Dauterman, W. C. 1983. The presence of two forms of glutathione *S*-transferases with distinct substrate specificity in OP-resistant and -susceptible housefly strains. *Proc. 5th Int. Int. Union Pure Appl. Chem. (IUPAC) Congr. Pestic. Chem.,* 3:197–202. Oxford: Pergamon

65. Motoyama, N., Kao, L. R., Dauterman, W. C. 1984. Dual role of esterases in insecticide resistance in the green rice leafhopper. *Pestic. Biochem. Physiol.* 21:139–47

66. Nebert, D. W., Eisen, H. J., Hankinson, O. 1984. The *Ah* receptor: binding specificity only for foreign chemicals? *Biochem. Pharmacol.* 33:917–24

67. Needham, P. H., Sawicki, R. M. 1971. Diagnosis of resistance to organophosphorus insecticides in *Myzus persicae* (Sulz.). *Nature* 230:125–26

68. Negishi, M., Swan, D. C., Enquist, L. W., Nebert, D. W. 1981. Isolation and characterization of a cloned DNA sequence associated with the murine *Ah* locus and a 3-methylcholanthrene–induced form of cytochrome P-450. *Proc. Natl. Acad. Sci. USA* 78:800–4

69. Noda, M., Takahashi, H., Tanabe, T., Toyosato, M., Furutani, Y., et al. 1982. Primary structure of α-subunit precursor of *Torpedo californica* acetylcholine receptor deduced from cDNA sequence. *Nature* 299:793–97

70. Oppenoorth, F. J. 1984. Biochemistry of insecticide resistance. *Pestic. Biochem. Physiol.* 22:187–93

71. Oppenoorth, F. J., van Asperen, K. 1961. The detoxication enzymes causing organophosphate resistance in the housefly; properties, inhibition, and the action of inhibitors as synergists. *Entomol. Exp. Appl.* 4:311–33

72. Oppenoorth, F. J., Smissaert, H. R., Welling W., van der Pas, L. J. T. 1977. Insensitive acetylcholinesterase, high glutathione *S*-transferase and hydrolytic activity as resistance factors in a tetrachlorvinphos-resistant strain of house fly. *Pestic. Biochem. Physiol.* 7:34–47

73. Ottea, J. A., Plapp, F. W. Jr. 1984. Glutathione *S*-transferase in the house fly: biochemical and genetic changes associated with induction and insecticide resistance. *Pestic. Biochem. Physiol.* 22:202–8

74. Parvari, R., Pecht, I., Soreq, H. 1983.

A microfluorometric assay for cholinesterases, suitable for multiple kinetic determinations of picomoles of released thiocholine. *Anal. Biochem.* 133:450–56

75. Pasteur, N. 1977. *Recherches de génétique chez* Culex pipiens pipiens *L. polymorphisme enzymatique, autogénèse et résistance aux insecticides organophosphorés.* PhD thesis. Univ. Montpellier, Montpellier, France. 170 pp.

76. Pasteur, N., Georghiou, G. P. 1981. Filter paper test for rapid determination of phenotypes with high esterase activity in organophosphate resistant mosquitoes. *Mosq. News* 41:181–83

77. Pasteur, N., Georghiou, G. P., Iseki, A. 1984. Variation in organophosphate resistance and esterase activity in *Culex quinquefasciatus* Say from California. *Génét. Sél. Evol.* 16:271–84

78. Payne, G. T., Disney, B. J., Brown, T. M. 1985. Field surveillance and laboratory selection for pyrethroid resistance in the tobacco budworm in South Carolina. *J. Agric. Entomol.* 2:85–92

79. Picollo de Villar, M. I., van der Pas, L. J. T., Smissaert, H. R., Oppenoorth, F. J. 1983. An unusual type of malathion-carboxylesterase in a Japanese strain of house fly. *Pestic. Biochem. Physiol.* 19:60–65

80. Plapp, F. W. Jr. 1984. The genetic basis of insecticide resistance in the house fly: evidence that a single locus plays a major role in metabolic resistance to insecticides. *Pestic. Biochem. Physiol.* 22:194–201

81. Prasittisuk, C., Busvine, J. R. 1977. DDT-resistant mosquito strains with cross-resistance to pyrethroids. *Pestic. Sci.* 8:527–33

82. Rausenbach, I. Y., Lukashina, N. S., Korochkin, L. I. 1984. The genetics of esterases in *Drosophila*. VIII. The gene regulating the activity of JH-esterase in *D. virilis*. *Biochem. Genet.* 22:65–80

83. Rawlings, P. 1985. The effects on resistant mosquitoes of interrupted exposure to insecticides. *Pestic. Sci.* 16:186–91

84. Raymond, M., Fournier, D., Berge, J., Cuany, A., Bride, J. M., Pasteur, N. 1985. Single-mosquito test to determine genotypes with an acetylcholinesterase insensitive to inhibition to propoxur insecticide. *J. Am. Mosq. Control Assoc.* 1:425–27

85. Raymond, M., Pasteur, N., Fournier, D., Cuany, A., Bergé, J., Magnin, M. 1985. Le gène d'une acétylcholinestérase insensible au propoxur déterminé la résistance de *Culex pipiens* L. à cet insecticide. *CR Acad. Sci.* 300:509–12

86. Rees, A. T., Field, W. N., Hitchen, J.

M. 1985. A simple method of identifying organophosphate resistance in adults of the yellow fever mosquito, *Aedes aegypti. J. Am. Mosq. Control Assoc.* 1:23–27

87. Reubi, I., Griffin, K. J., Raucy, J. L., Johnson, E. F. 1984. Three monoclonal antibodies to rabbit microsomal cytochrome P-4501 recognize distinct epitopes that are shared to different degrees among other electrophoretic types of cytochrome P-450. *J. Biol. Chem.* 259:5887–92

88. Saiki, R. K., Scharf, S., Faloona, F., Mullis, K. B., Horn, G. T., et al. 1985. Enzymatic amplification of β-globin genomic sequences and restriction site analysis for diagnosis of sickle cell anemia. *Science* 230:1350–54

89. Shono, T. 1983. Linkage group analysis of carboxylesterase in a malathion resistant strain of the housefly, *Musca domestica* L. (Diptera: Musciade). *Appl. Entomol. Zool.* 18:407–15

90. Smissaert, H. B. 1964. Cholinesterase inhibition in spider mites susceptible and resistant to organophosphate. *Science* 143:129–31

91. Soreq, H., Zevin-Sonkin, D., Avni, A., Hall, L. M. C., Spierer, P. 1985. A human acetylcholinesterase gene identified by homology to the *Ace* region of *Drosophila. Proc. Natl. Acad. Sci. USA* 82:1827–31

92. Strange, R. C., Faulder, C. G., Davis, B. A., Hume, R., Brown, J. A. H., et al. 1984. The human glutathione S-transferases: studies on the tissue distribution and genetic variation of the GST1, GST2, and GST3 isozymes. *Ann. Hum. Genet.* 48:11–20

93. Telakowski-Hopkins, C. A., Rodkey, J. A., Bennett, C. D., Lu, A. Y. H., Pickett, C. B. 1985. Rat liver glutathione S-transferases: Construction of a cDNA clone complementary to a Yc mRNA and prediction of the complete amino acid sequence of a Yc subunit. *J. Biol. Chem.* 260:5820–25

94. Terriere, L. C. 1983. Enzyme induction gene amplification and insect resistance to insecticides. See Ref. 36, pp. 265–97

95. Terriere, L. C. 1984. Induction of detoxication enzymes in insects. *Ann. Rev. Entomol.* 29:71–88

96. Villani, F., White, G. B., Curtis, C. F., Miles, S. J. 1983. Inheritance and activity of some esterases associated with organophosphate resistance in mosquitoes of the complex of *Culex pipiens* L. (Diptera: Culicidae). *Bull. Entomol. Res.* 73:153–70

97. Wilkinson, C. F. 1983. Role of mixed-

function oxidases in insecticide resistance. See Ref. 36, pp. 175–205

98. Womack, J. E., Yan, D. L. S., Potier, M. 1981. Gene for neuraminidase activity on mouse chromosome 17 near H-2: pleiotropic effects on multiple hydrolases. *Science* 212:63–65

99. Wrighton, S. A., Schuetz, E. G., Watkins, P. B., Maurel, P., Barwick, J., et al. 1985. Demonstration in multiple species of inducible hepatic cytochromes P-450 and their mRNAs related to the glucocorticoid-inducible cytochrome P-450 of the rat. *Mol. Pharmacol.* 28:312–21

100. Yasutomi, K. 1983. Role of detoxification esterases in insecticide resistance. See Ref. 36, pp. 249–63

101. Yelton, D. E., Scharff, M. D. 1981. Monoclonal antibodies: a powerful new tool in biology and medicine. *Ann. Rev. Biochem.* 50:657–80

102. Yu, S. J. 1983. Induction of detoxifying enzymes by allelochemicals and host plants in the fall armyworm. *Pestic. Biochem. Physiol.* 19:330–36

103. Yu, S. J., Hsu, E. L. 1985. Induction of hydrolases by allelochemicals and host plants in fall armyworm (Lepidoptera: Noctuidae) larvae. *Environ. Entomol.* 14:512–15

104. Zettler, J. L. 1974. Esterases in a malathion-susceptible and a malathion-resistant strain of *Plodia interpunctella* (Lepidoptera: Phycitidae). *J. Ga. Entomol. Soc.* 9:207–13

Ann. Rev. Entomol. 1987. 32:163–79

ARTHROPODS OF ALPINE AEOLIAN ECOSYSTEMS

John S. Edwards

Department of Zoology, University of Washington, Seattle, Washington 98195

PERSPECTIVE AND OVERVIEW

The diversity and ubiquity of insects on the planet Earth is a commonplace premise of every entomology textbook. From the tropics to Antarctica, from deserts and hot pools to high arctic islands, and from open ocean to mountain peaks, insects have found unorthodox niches where other arthropods and microorganisms are commonly their sole companions. Their great capacity for dispersal and their relatively small size, which together at least partially explain insect diversity (71), enable them to find and exploit habitats in which they are protected from macroscale environmental factors. In addition to the benefits of scale, various insects have the physiological capacity to survive ambient temperatures that exclude most Metazoa, salinities that defy most regulatory systems, and/or ability to tolerate deep-freezing and desiccation.

Of all the eccentric habitats colonized by insects, perhaps the most extreme are those where the normal pattern of energy flow from photosynthetic plants is denied, as in areas of extreme aridity, e.g. deserts (64) and lava fields (47); cold, e.g. alpine (65, 66) and polar regions (39); or dark, e.g. caves (20, 49, 69). These environments, like ocean abysses where dwellers depend on organic material filtering down from the photic zone, are all examples of what Hutchinson (53) termed the *allobiosphere,* where resident organisms depend for their subsistence on transported energy and nutrients. Swan (92) drew attention to the widespread distribution of such environments in alpine regions and coined the term *aeolian* for them, thereby explicitly recognizing the role of wind in the transport of nutrients.

The emphasis of this review is on alpine communities, but it is not only in

163

the alpine habitat that wind-transported materials support arthropod communities. For example, arthropod fallout can contribute significantly to the food chain of oligotrophic lakes (74) and streams (73). Gerrids of the genus *Halobates,* which live on the surface of coastal waters (37) and open ocean (15, 16), subsist entirely or in part on arthropod fallout. Sand dunes around the world support arthropod communities that depend on allochthonous organic material (64). In one of the most extreme examples, the Namib desert of southwestern Africa, tenebrionid beetles and others among the 50 dune-inhabiting arthropod species (79) feed predominantly on wind-borne plant detritus that accumulates on the lee side of dunes (56). Arthropods in this arid habitat further depend on wind transport for water, which they take up from nocturnal fog either by "fog basking" (44) or by constructing fog catchment trenches (84). At the other thermal extreme, toward polar regions, the potential exists for aeolian communities. Windblown detritus forms a nutrient resource in the high arctic terrestrial ecosystem of Devon Island (95). An aeolian resource in the Arctic is also implied by the arthropod samples taken in large nets designed to constantly face the wind at Cape Thompson and Point Barrow on the northwestern Alaskan tundra (40). The mites that live furthest south, in alpine Antarctica (39), presumably require allochthonous organic material. Lava fields and pyroclastic debris of volcanoes provide another example of a habitat where spiders and other insect pioneers (7, 36, 47) exploit wind-borne organic material.

THE ALPINE AEOLIAN ZONE

The earlier entomological literature of the alpine zone was predominantly based on anecdotal, casual, or incidental observations, many of which have been compiled by Mani (65, 66). Some notable figures of American entomology, e.g. Van Dyke (97) and Howard (46), contributed early observations, and the European alpine fauna was explored in the early decades of the century (e.g. 9). However, it is appropriate here to begin with a salute to the true founder of alpine ecology, Alexander von Humboldt. In 1802, during his epic climb on Mount Chimborazo in Ecuador, then considered the highest mountain in the world, von Humboldt noted the presence of butterflies and flies on the snowfield at an altitude of 5000 m (52). It was also von Humboldt (51) who first formalized the concept of altitudinal zonation, which lies at the core of alpine ecology.

For the purposes of this review three broad categories of arthropods are distinguished in the alpine zone: (*a*) permanent residents, comprising predators and scavengers that make up the true nival fauna; (*b*) temporary residents, comprising migrants such as coccinellids (42) and hilltopping Lepidoptera, Diptera, and Hymenoptera (3); and (*c*) the diverse allochthonous fauna composed of wind-borne derelicts, the arthropod fallout fauna (most of which is unable to survive for long in the alpine, and which can form the major source of food for the resident nival arthropods) (28), birds (34), reptiles (93), and mammals (89). Although the temporary residents incidentally contribute to the arthropod fallout, and do constitute an unlikely resource for foraging bears (14, 41), they do not feed or reproduce in the alpine zone and are therefore not considered further. The emphasis in this review is on recent studies of the composition of alpine aeolian faunas, the arthropod fallout on which they depend, and aspects of their physiology and behavior. For earlier accounts and Himalayan case studies see Mani (65, 66).

Composition of the Nival Fauna

The general attributes of aeolian alpine arthropods, discussed by Mani & Giddings (67), include tendencies toward melanism, small size, aptery or brachyptery, extended development (86), cold stenothermy (i.e. capability of activity in a low and narrow thermal range), cold hardiness, and predatory or scavenging habits. Mani & Giddings emphasized the diurnal activity patterns of these arthropods and stated that crepuscular and nocturnal species were wholly unknown (67). However, nocturnal behavior is characteristic of widespread alpine faunas (see Activity Patterns of Nival Arthropods). The major representative terrestrial groups in the aeolian alpine arthropod fauna are spiders (especially Lycosidae and Salticidae, but also also Thomisidae, Gnaphosidae, Dassidae, Clubionidae, and Linyphiidae), phalangids, Collembola, and Coleoptera (principally Carabidae and Staphylinidae). Less generally distributed but of particular interest are Thysanura, Dermaptera, and Notoptera (Grylloblattodea).

To start at the top, the reputed highest resident alpine arthropod is the salticid spider *Euophrys omnisuperstes* (98), seen to be active at 6750 m on Mt. Everest. It is possible that these highest of all spiders, along with other arthropods found at extreme altitudes, may feed on fallout and grow in that rigorous environment. It is also possible that they are part of a metapopulation (62), and that they do not renew populations in situ but receive recruits from elsewhere; in this case the spiders may balloon from reproducing populations at lower altitudes. It is perhaps more remarkable that the machilid *Machilanus swani* occurs at 6300 m (91, 94, 99), since these apterygotes are not known to disperse on the wind. They are presumably resident populations, part of the surprisingly complex food net of the Himalayan mountains (65, 91).

An exceptional species that defies the generality that arthropods of the high alpine zone are predators or scavengers has recently been discovered in the high Himalayas. Larvae of the chironomid *Diamesa* sp. grow in meltwater drainage channels in the ice of the Yala glacier at 5100–5600 m in the Langtang region of Nepal, where they subsist on blue-green algae (*Phormidium* sp.) and bacteria (57), which in turn depend on wind-borne nutrients carried to the glacier (58). The brachypterous adults walk on the surface of the glacier (59) and in the subsurface cavities during sunny periods, and descend through snow toward the ablation zone during dark or overcast periods. Adults are reported to be capable of movement at −16°C.

Elsewhere around the world, an increasing number of nival aeolian arthropod communities have been described, e.g. from snowfields in Scotland (8), Lapland (13, 55), and the European Alps (70).

Aeolian nival communities and their food sources are now known in most detail from several studies spanning western America from the subarctic to the tropics. Carabids and staphylinids exploit arthropod fallout on snow patches in montane tundra of central Alaska (28, 34), as in the Scandinavian arctic (13, 55). Along with the scavenging arthropods, birds also exploit the fallout resource. Eight of 11 species breeding at the Alaskan site, for example, foraged for insects on tundra snow patches (34). The nival fauna of Mt. Rainier (68) is typical of other Pacific Northwest volcanoes and the major peaks of the Cascade Range (J. S. Edwards, unpublished). On Mt. Rainier between about 1400 and 2500 m five species of carabid and at least one staphylinid, a grylloblattid, and a phalangid scavenge arthropod fallout. Collembola and an oligochaete ice worm *(Mesenchytraeus solifugus)* forage on pollen and microorganisms in the snowfield fallout. Here also, four bird species that breed in the subalpine zone share the arthropod resource (68).

The bare mineral surface of the blast zone on Mt. Saint Helens has been colonized by 7 of at least 15 species of carabid beetle that have been carried to the site by aerial transport since the 1980 eruption. *Bembidion planatum*, for example, which first appeared in fallout sampling traps during the summer of 1981, one year after the eruption, has now established large local breeding populations that are entirely dependent on arthropod fallout as their food source (33, 35). Similarly, a silphid (*Pteroloma* sp.) established breeding populations in the blast zone four years after the eruption. Thirty-six species of migrant spiders in the blast zone are characteristically western lowland species (18). It is doubtful that during the first four years after the eruption any spiders survived to reproduce, but breeding populations of certain species (e.g. *Pardosa mackenziana*) appear to have been established after five years (35). While the community of the bare mineral surface of the pyroclastic flow at 1000 m is not strictly nival, it is strictly aeolian, and has been so at least during the early posteruption stages. This community is composed principally of carabids (e.g. *Bembidion* spp.) whose niche is the disturbed habitat (96); such habitats also support metapopulations recruited from outside, which do not breed on site (62), as could be common in the nival fauna.

In the central Sierra Nevada of California the snowfield arthropod foragers at 3300 m comprise at least seven species of staphylinids, six carabids, a grylloblattid, a phalangid, and several gnaphosid and lycosid spiders (26, 72). Here again, birds also exploit the arthropod fallout. Farther south in California, on the summit of White Mountain Peak (4342 m), the culmination of a

Great Basin range separated by the Owens Valley from the Sierra Nevada, 9 of the 15 resident arthropods (e.g. a phalangid, an erythraeid mite, a lithobiid centipede, and six species of spider) are generalized feeders or scavengers on insects. They depend on windblown material carried from below. Insectivorous birds and mammals also exploit the fallout on White Mountain Peak (89).

The Mexican volcanoes support a rich nival fauna (93), which includes both arthropods and vertebrates dependent on arthropod fallout. The isolated peaks of the Sierra Nevada de Venezuela at about 4° N form a tropical alpine island with an insular nival fauna; on Pico Espejo (4750 m) at least two species of carabids and several anyphaenid, salticid, and erigonid spiders depend on arthropod fallout (32; J. S. Edwards, unpublished).

Even more isolated than the Venezuelan Andes, the volcanoes of Hawaii are alpine islands in two senses, isolated by ocean and by altitude, and they too have their distinctive aeolian fauna. At least 18 resident species form an aeolian fauna on the summit cinder core of Mauna Kea at 4200 m: six species of predatory mites, a lycosid and three linyphiid spiders, a centipede, three entomobryiid Collembola, two lygaeids, a psocopteran, and a noctuid (50). These include a number of new and undescribed species, among which the lygaeid *Nysius wekiuicola* (6) is of special interest. It is a diurnal scavenger on moribund and dead insects carried from lower elevations, and has the distinctions both of being among the few high-altitude aeolian heteropterans and of being an exceptional scavenger-predator in a predominantly seed-eating family. Larvae of the moth *Hodegia apatela* occupy barren, rocky areas where they feed on windblown leaf fragments (48), further attesting to the uniqueness of the Hawaiian alpine fauna. These aberrant insects share the aeolian habitat with spiders, mites, and Collembola, but the signature family of alpine aeolian communities around the world, the Carabidae, are apparently absent.

Ashmole's recent studies in the Canary Islands (7; M. J. Ashmole, manuscript in preparation) also revealed an insular aeolian arthropod fauna. Bare young lava flows at 1500–2100 m have a diverse resident fauna including carnivores and scavengers from at least 11 orders. Spiders are especially diverse. The food resource is arthropod fallout. Snowfield sampling on Mt. Teide (3712 m) indicated maximum daily fallout rates of about 18 individual insects m^{-2}, mainly aphids and psyllids, with Diptera and Hymenoptera next in numbers. The recent lava communities of the Canary Islands differ significantly from those at similar latitudes and altitudes in Hawaii.

Alpine Arthropod Fallout

The majority of the nival arthropods discussed above depend on arthropod fallout, and to a lesser extent on windblown plant debris. Alpine arthropod fallout is a conspicuous aspect of the more general phenomenon of dispersal;

the constituents are arthropods whose dispersal has carried them to an inappropriate environment from which they cannot escape. In this they share the fate of the enormous biomass of arthropods that is carried out to sea (12, 45) or deposited on barren sites such as volcanic debris (36, 47). Accounts of such aerial dispersal have a distinguished history, including observations on Cook's voyages of discovery, and Darwin's voyage on the Beagle (22). Jonathan Edwards in 1719 reported spiders ballooning out to the Atlantic Ocean (5); von Humboldt noted moribund insects high on Andean snowfields (52); and Charles Elton, a founder of modern animal ecology, recorded migrant insects on the snowfields of Spitzbergen (36). The extensive earlier literature on insect dispersal culminated in Johnson's landmark synthesis (54) and has been treated in a subsequent review and in symposia (21, 24, 88). This literature has built up recognition of the pervasive and profound importance of dispersal and migration in the exploitation of temporally and spatially patchy terrestrial environments by arthropods.

Since Mani (65) and Swan (94) drew attention to the ecological importance of nival arthropod fallout and reviewed earlier accounts, there have been a number of detailed studies that further underline the generality of the phenomenon, e.g. in Alaska (27, 28, 34), Australia (29), Scotland (8), Europe (63, 70), and Western North America (68, 75, 77, 89).

EXAMPLES OF FALLOUT   Aspects of some recent accounts of arthropod fallout are reviewed below.

Cairngorm mountains, Scotland (57°3'N, 1100–1200 m)   The first reports of arthropod fallout on Cairngorm snowfields, dating from 1896 (8), are among the oldest in the literature. More recently, Ashmole et al (8) reported 130 species of insects and spiders in snowfield samples from the Cairngorms; 28 of these species were of lowland origin, 10 were montane, 50 were eurytopic, i.e. of broad altitudinal range, and 42 were of unknown origin. Coleoptera, Hymenoptera, and Diptera were the most diverse groups with 34 species each. At the family level the Diptera were most diverse, with 20 families. Twelve species of spiders were collected. The maximum surface density reported was 11.1 arthropods m^{-2}.

Eagle Summit, Alaska (65°8'N, 1000 m)   At least 83 species of insects and 5 of spiders were taken in snow-patch samples totaling 678 individuals (28, 34). Density ranged from zero to 9.6 m^{-2} on snow patches over 5000 m^2 in area during summer 1968. In a study in the same area in 1971 and 1972, summer insect densities ranged from 1.7 m^{-2} to 10.9 m^{-2}. Diptera dominated numerically (61%), followed by Hymenoptera (28%), Coleoptera (4.4%), and Heteroptera (1.2%). Fallout peaked in early June, and diminished to zero in

September. Alate ants dominated the biomass from mid-June to early July. Aphids also had massive peaks in mid-June. Diptera, with 18 families, and Hymenoptera, with 6 families, were the most diverse orders. The quantity of nutrients carried to the snow patches during summer (June–September) was at least 21.3 mg C m^{-2}, 43 mg N m^{-2}, and 0.25 mg P m^{-2}, estimated according to the assumption that 70% of the fallout came from lower altitudes (34).

Carabids and staphylinids were observed feeding on snowfields during daylight hours in Lapland (55) and Alaska (34). The brevity of the night during summer at high latitudes doubtless accounts for the absence of a nocturnally active fauna comparable to that of sites in Washington and California.

The biomass at the Eagle Summit site is lower than that observed at many comparable sites at lower latitudes, but some incidental observations suggest that considerably larger quantities of arthropod fallout may be deposited at high altitudes in the Alaska Range. Great numbers of syrphid flies (*Syrphus torvus* and *Metasyrphus* sp.) were observed at the head of the Kahiltna Glacier at 5000 m on Mt. McKinley. These insects were derelicts from a massive east–west migration observed in July 1968 (28). Occasional specimens of *Syrphus torvus* were encountered at 5200 m on the West Buttress of Mt. McKinley, where raven footprints along ridges suggested that these high-altitude scavengers include snowbound insects in their diet. Large numbers of birds are said to move in late summer to glaciers of the Alaskan Range, where they feed on stranded insects (D. Sheldon, personal communication). From 22 1-m^2 samples from the Gulkana Glacier (Alaska) Range, taken between 3000 and 1700 m in July 1966 (27), 21 species were collected, including numerous spruce budworm moths (*Choristoneura fumiferana*) and up to 73 aphids m^{-2} (*Euceraphis* prob. *punctipennis,* a species associated with alder and birch). An estimate of total aphid live weight biomass on the 1.5 mi^2 Gulkana neve amounted to 83 kg.

Mount Rainier, Washington State (46°48'N, 4392 m)   This mountain, an alpine island situated 40 km west of the Cascade Range crest, with permanent snowfields down to about 2000 m, is surrounded by forest and agricultural lands. Slopes above the tree line from 2000 m to the summit, with maxima between 2000 and 3000 m, receive arthropod fallout throughout the year, with a peak in mid-June (68). Predominantly southwest winds throughout the summer carry dispersing arthropods from the lowlands. Over 200 insect species representing 105 families from 14 orders have been recognized in the fallout fauna. Twenty-three species of ballooning spiders were found in snowfield samples (19). A typical fallout sample of 40 random 1-m^2 quadrats taken from the Muir snowfield at 2500 m in June 1975 contained 24 insects m^{-2}, representing a biomass of 6.71 mg dry wt m^{-2}. Aphids and various

Heteroptera dominated the samples numerically, but occasionally during late summer massive accumulations of male and female ants (*Formica* sp., *Camponotus* sp.) were encountered. The dominant groups in the fallout are all known to be conspicuous dispersers, e.g. Homoptera such as Aphididae and Cicadellidae; the mirids, however, which are highly dispersive in general, were curiously few, as has been noted on snowfields elsewhere in the Pacific Northwest (83). Among 26 species of Heteroptera from 14 families, some exhibited small-scale dispersal, e.g. *Nysius ericae,* associated with local heather meadows. Others exhibited large-scale dispersal, e.g. *Nabis alternatus,* which was among the five insect species taken on the summit of Mount Rainier, and which is a widespread beneficial insect of agricultural crops such as alfalfa. Numerous Heteroptera have been taken in high-altitude samples on other mountains in the Pacific Northwest (83). Among the 29 families of Diptera, nematocerans were most abundant on the snow. Syrphids are a common component of snowfield fallout in alpine areas (65), but occurred infrequently on Mt. Rainier snows. Of 16 families of Coleoptera, coccinellids and chrysomelids were most frequent. Scolytids occurred occasionally, but were not comparable in number or diversity to those from other sites in Washington and Oregon, where 26 species were taken from summer snowfields between 2000 and 3000 m (38).

Mount Saint Helens, Washington State (46°12'N, 2550 m)   Like Mt. Rainier, Mt. Saint Helens is a volcanic alpine island. Little was known of its ground-dwelling arthropod fauna before the eruption of May 1980, but since that time detailed studies of the colonization of bare pyroclastic surfaces at 1000 m have been in progress (31, 33, 35). It is probable that arthropod fallout recorded since 1980 has not substantially changed from prior to the eruption. At least 70 families of insects from 17 orders have been taken in fallout traps to date. Diptera and Homoptera predominate numerically in the fallout. The major dipteran components in the first two years following the eruption were Chironomidae and Culicidae; this reflects the productivity of organically enriched bodies of fresh water formed as a consequence of the eruption. Otherwise the fallout fauna resembles that described for Mt. Rainier. Within the first two summers 43 species of spiders arrived on the bare mineral surfaces of the blast zone by ballooning from lowland sites to the west. The efficacy of ballooning as a means for dispersal is attested by the presence of three introduced spiders, *Theridion bimaculatum, Euplognatha ovata,* and *Lepthyphantes tenuis,* none of which occurred in Washington State before about 1950 (18). On the basis of data from fallout collectors it is estimated that about 10 mg dry wt biomass m^{-2} day^{-1} arrives at the surface of the blast zone during the approximately hundred days of late spring and summer.

Mount Conness, central Sierra Nevada, California (38°N, 3387 m) Arthropod fallout on persistent snow patches surveyed by Papp (77) from 1972 to 1976 varied in composition in successive years, but each year large numbers of Diptera, Heteroptera, and Homoptera were taken. The density of arthropod fallout was much higher than at the more northern localities discussed above, and the predominance of the homopteran families Aphididae, Psyllidae, and Cicadellidae probably reflects wind transport from fertile agricultural lands of the San Joaquin Valley. Fallout density was positively correlated with wind speed and with local air temperature maxima. Fallout density at Mt. Conness was consistently high in comparison with that from more northern snow surface samples; values between 20 and 40 m^{-2} were found through June and July in years with persistent snow.

White Mountain Peak, California (37°31'N, 4342 m) The White Mountains lie east of and parallel to the Sierra Nevada, from which they are separated by the Owens Valley. The culmination, White Mountain Peak, is a rocky ridge with sparse vegetation (cover 0.2%). A persistent snowfield lies on the eastern slope. Spalding (89) found representatives of 33 families from 8 orders in samples of wind-borne insects from the area; many of these, e.g. the lygaeid *Nysius raphanus* and the coccinellid *Hippodamia convergens,* are derived from agricultural crops such as alfalfa in the lowlands of the Owens Valley. The rocky areas trap arthropod debris, and this resource provides food for the resident small mammals *Peromyscus maniculatus sonoriensis* and *Sorex tenellus.*

Niwot Ridge, Colorado (40°N, 3750 m) The arthropod fallout on a snowfield in the Colorado front range described by Schmoller (82) includes low-altitude species of Hymenoptera, Cicadellidae, Aphididae, Chrysomelidae, Coccinellidae, and diverse Diptera.

GENERAL ASPECTS OF FALLOUT The recent accounts, together with earlier reports, illustrate the taxonomic diversity of arthropod fallout on snowfields. Through a wide range of latitude and altitude a few orders consistently dominate the fallout, notably Diptera, Homoptera, Coleoptera, and Hymenoptera. Whatever the diverse functions ascribed to dispersal and migration (e.g. 21, 24, 43, 88), the end result of aerial transport of arthropods with little control over their broad trajectory is that a significant proportion land in inhospitable environments. Snowfields yield a conspicuous sample of such misdirected insects, which provides an index of the doubtless greater numbers whose end points are less evident (e.g. 4). The constituents range from extremely rare insects, e.g. certain parasitic Hymenoptera (27), to major pests, e.g. spruce budworm, *(Choristoneura)* (27), frit fly *(Oscinella frit)*

(63), and bark beetles (38). Their source may be local (28) or distant (36). Some insects (e.g. birch aphids) arise from permanent vegetation, while others (e.g. alfalfa bugs, *Nysius raphanus*) arise from annual crops (89). While the numbers may vary from year to year according to seasonal variation in productivity and wind patterns, the snowfield fallout is sufficiently reliable to provide a food source for alpine predators and scavengers, both invertebrate and vertebrate, in all localities where it has been investigated.

PHYSIOLOGICAL CONSTRAINTS ON THE RESIDENT NIVAL ARTHROPOD FAUNA

The challenge of the alpine environment for ectotherms is in the alternation of thermal extremes on daily and annual cycles. To be fit in the alpine environment an organism must either accomodate to or evade the extremes. Both strategies are found among alpine insects. Adaptive strategies involve on the one hand physiological mechanisms that enable the animal to survive at low temperatures, a property referred to as cold hardiness (80), or on the other hand avoidance of exposure to periods of extreme cold by various means.

Cold hardiness is achieved either through tolerance of freezing, i.e. the formation of ice in the body fluids (but not within cells), or through supercooling, which inhibits ice formation in a body that may not be able to withstand freezing (11). Supercooling is achieved by the addition to body fluids of one or more antifreeze components, such as glycerol, sorbitol, mannitol, and trehalose. As a result of the presence of these cryoprotectants, arthropods may avoid freezing at temperatures as low as $-40°C$ (85). Two further strategies among cold-hardy insects are the synthesis of thermal-hysteresis proteins (25), which prevent or restrict the growth of ice nuclei, and conversely, the synthesis of ice-nucleating compounds, which induce ice formation in a controlled fashion within extracellular compartments (100). Increases in the proportion of bound water in cells and modulation of the enzymes of intermediary metabolism also contribute to cold hardiness (90). There are parallels yet to be fully explored between responses to freezing (10) and desiccation (17).

An arthropod need not be physiologically cold-hardy to occupy a niche in the alpine environment (32, 86). The seemingly paradoxical occurrence of alpine insects that lack any of the attributes of cold hardiness outlined above is explicable on the basis of alternative strategies, of which three are salient. Firstly, it is commonly overlooked that $0°C$ is not the freezing point of the body fluids of any insect. Temperate and tropical insects that are never challenged by cold supercool to at least $-5°C$. Insects are thus able to live in cold environments without special protection, provided ambient temperatures do not dip below the unspecialized supercooling temperatures. Secondly, the size of insects relative to the surface texture of the alpine environment allows for a degree of protection in subsurface spaces not open to larger animals. Thirdly, the insulation afforded by dry snowpack assures that arthropods living beneath it are protected from the extreme atmospheric temperatures, radiative cooling, and wind chill encountered by larger animals, and can thus remain active throughout the winter (1, 2). Not only are ground-dwelling arthropods thermally protected by snowpack to such an extent that they can continue to be active; but the snow column itself can be occupied by the ground fauna (61). Seasonal changes in cold hardiness are found in animals from these protected environments (85).

The capacity to function at temperatures around 0°C rather than to resist or tolerate freezing is the decisive factor in the lives of many, and perhaps most, arthropods of aeolian habitats, at least in the temperate, subtropical, and tropical zones. The ability to tune the metabolic machinery to function around 0°C is widespread in diverse marine organisms as well as ectotherms of cold terrestrial environments. Thus the challenges for survival of alpine ectotherms lie in metabolic tuning to relatively low temperatures, in the selection of protective microenvironments within the alpine habitat, and in the adaptation of behavioral patterns to evade threatening low temperatures. All three major strategies may be found in insects in one area (87).

The Mt. Rainier grylloblattid (*Grylloblatta* sp. undescribed), a nocturnal forager that appears on snowfields in summer, exemplifies the cold evasion strategy (30, 72). It has limited capacity for supercooling, lacks known cryoprotectants, and dies on freezing, at about −6.5°C. At temperatures above about 14°C it undergoes heat convulsions. Neural conduction and synaptic transmission, as measured by giant interneuron responses to stimulation of cercal mechanoreceptors, are detectable down to −5.8°C, i.e. within 0.4°C of the supercooling point. In effect, the insect is metabolically tuned to function in a thermal window between −6 and 12°C. It forages over the snow surface on summer nights when surface temperatures are generally in the range −1 to 5°C. Winter temperatures are frequently well below the supercooling point, but the insects are not exposed to these lethal conditions within their winter habitat of shattered rock under deep insulating snow. Other arthropods in the Mt. Rainier nocturnal forager guild are more cold-hardy; the phalangid *Leiopilio glaber* appears to be freeze-tolerant at its October supercooling point (approximately −11.5°C), and the carabid *Nebria van-dykei* collected in October tolerates freezing at −10.5°C (32). Both species have activity patterns similar to those of the grylloblattid, but may overwinter in more exposed microhabitats (J. S. Edwards, unpublished).

ACTIVITY PATTERNS OF NIVAL ARTHROPODS

Perhaps because human activity and ecological study in the alpine zone are strongly diurnal, and because the activity of the more conspicuous elements of the diurnal arthropod fauna (Diptera, Lepidoptera, lycosids) is more or less dependent on direct insolation, it has been asserted that the alpine zone lacks a nocturnal fauna (67). It has become evident during the last decade, however, that a nocturnal forager-predator guild is widespread and characteristic of the alpine and subalpine zones. Schmoller (81) first drew attention to the nocturnal fauna of the alpine tundra in the Colorado Range, where 20 species of carabid, three species of phalangids, and 14 gnaphosid spiders were trapped (predominantly by pitfall) during night hours. The transition from light to dark is commonly the limiting factor in determining the onset of activity in carabids (96); with some exceptions this applies to the alpine carabids of middle latitudes (23, 60, 68, 81) and of the tropics (32; J. S. Edwards, unpublished results). Nocturnal foraging guilds composed of diverse arthropods (e.g. carabids, staphylinids, grylloblattids, phalangids) have remarkably similar diel activity patterns. Activity begins after disappearance of twilight, reaches a numerical peak near midnight, and diminishes to zero before dawn. Given the extreme thermal range and rapid fluctuations of the alpine environment, it is to be expected that the timing and duration of surface activity of ectotherms should also be limited by the pattern of intervals of permissive temperatures (66, 78, 91). While members of the nocturnal arthropod foraging guild are mobile at remarkably low temperatures for ectotherms, their surface activity proves to be temperature-limited. Detailed quantitative studies by Ottesen (76) on carabids in south Scandinavian high tundra at Finse (60°36'N, 1200 m) showed that the lower temperature limit for activity is in the range -2.0 to $0.8°C$ (mean $-0.5°C$), confirming observations made elsewhere (68, 77, 81). The aggregate number of nights that meet this criterion during spring and summer may be the decisive factor determining the distribution of alpine nocturnal foragers. Cloudy conditions, which retard cooling, are favorable to activity, while clear-sky radiation, which promotes cooling, can limit activity. Rain, new snow, and extreme wind can also inhibit foraging. Termination of the foraging period is influenced by food satiation; higher densities of arthropod fallout on the snowfield of Mt. Rainier correlated positively with shorter foraging period in a population of the carabid *Nebria paradisi* (68). Reproductive behavior may also determine the length of the active period (68).

The long midsummer day length at higher latitudes restricts nocturnal activity (76). In tropical alpine regions, e.g. Pico Espejo, Venezuela (32), extremely high daytime surface temperatures in the 35–40°C range followed by rapid cooling under clear night conditions at high altitude restrict activity to the period from dusk (7 PM) to about 11 PM. It is perhaps the compromise between these high-latitude and equatorial extremes that has engendered the highest taxonomic diversity in the nocturnal foraging guild in mountains of the temperate zone.

Not all alpine foragers are nocturnal, but diurnal foragers, notably carabids of the genus *Bembidion,* are characteristically much smaller than their nocturnal counterparts. Of the three most probable factors underlying this difference, thermal balance, water balance, and predator pressure, it is agreed (68, 76) that the last is the most significant. As noted above, birds exploit the snowfield arthropod fauna; the snowfield is a dangerous place for a bite-sized arthropod by day, and it is thus not surprising that birds and minute arthropods should exploit the fallout resource by day, while larger nival arthropods forage by night. Larger foragers are occasionally encountered on snow during daylight (55, 68, 77), but these examples (a muscid fly, carabids, and lycosids) seem to be specialists in the short dash, locating prey on snow but carrying it quickly away, or are accidentals, straying from rock or tundra onto snow patches during periods of heavy arthropod fallout.

IMPLICATIONS AND PROSPECTS

The study of aeolian alpine communities and their food source of arthropod fallout, esoteric as it may seem, does have broad implications for arthropod biology in general. The fallout provides a quantifiable index of the immense biomass of wind-borne arthropods, be they dispersers, migrants, or accidentals, that also rain upon agricultural and forest lands. It reflects the continual or periodical recruitment to arthropod populations in habitats where the phenomenon is less easily measured. It raises questions concerning the importance of allochthonous arthropods in agroecological models of predator-prey interactions. It serves to emphasize the extent of gene flow in widely dispersing insects and the seeming difficulty of genetic isolation. Are genes that promote fitness on alpine islands continually swamped by lowland arrivals? The simplicity of the communities, of their food chains, and of the physical habitat in which they live renders the aeolian communities attractive for ecological modelling. The nutrients transported uphill by the arthropod fallout demonstrably provide a resource for foraging arthropods and birds of the alpine and subalpine zones. The value of the fallout for other organisms, e.g. plants and microorganisms, at higher altitudes has been proposed but not yet adequately measured. The various survival strategies of the aeolian nival arthropods in the physically rigorous allobiosphere raise many unanswered questions in environmental physiology and reproductive biology. The evolution of aeolian communities on isolated alpine islands raises a host of unanswered questions. And along with the arthropods there are other groups, beyond the scope of this review: the cryophilic microorganisms, annelids, and other invertebrates. In most cases, nothing more is known about them than the fact that they are present in the aeolian zone. What are they doing and how do they do it?

A final word of caution to the prospective alpine ecologist: Despite the strength of the rationale presented above, expect a skeptical response from colleagues when you are encountered leaving for your study site with skis or ice axe in hand.

ACKNOWLEDGMENTS

Work by the author was supported in part by a National Park Service contract (Mt. Rainier) and National Science Foundation grant DEB 81 07042 (Mt. Saint Helens).

Literature Cited

1. Aitchison, C. W. 1979. Winter subnivean invertebrates in Southern Canada. I. Collembola, II. Coleoptera, III. Acari, IV. Diptera and Hymenoptera. *Pedobiologia* 19:113–20, 121–28, 153–60, 176–82

2. Aitchison, C. W. 1984. The phenology of winter active spiders. *J. Arachnol.* 12:249–71

3. Alcock, J. 1979. *Animal Behavior: An Evolutionary Approach.* Sunderland, Mass: Sinauer. 532 pp.

4. Alexander, G. 1964. Occurrence of grasshoppers as accidentals in the Rocky Mountains of northern Colorado. *Ecology* 44:77–86

5. Anderson, W. E. 1980. *Jonathan Edwards: Scientific and Philosophical Writings.* New Haven, Conn: Yale Univ. Press. 433 pp.

6. Ashlock, P. D., Gagne, W. C. 1983. A remarkable new micropterous *Nysius* species from the aeolian zone of Mauna Kea, Hawaii Island. *Int. J. Entomol.* 25:47–55

7. Ashmole, M. J., Ashmole, N. P. 1986. Arthropod communities supported by biological fallout on recent lava flows in the Canary Islands. *Entomol. Scand.* (Suppl.) In press

8. Ashmole, N. P., Nelson, J. M., Shaw, M. R., Garside, A. 1983. Insects and spiders on snow fields in the Cairngorms, Scotland. *J. Nat. Hist.* 17:599–613

9. Bäbler, E. 1910. Die wirbellose terrestrische Fauna der nivalen Region Genf. *Rev. Suisse Zool.* 18:761–915

10. Baust, J. 1983. Cryoprotective agents: regulation of synthesis. *Cryobiology* 20:357–64

11. Baust, J. G., Rojas, R. R. 1985. Insect cold-hardiness: facts and fancy. *J. Insect Physiol.* 31:755–59

12. Bowden, J., Johnson, C. G. 1976. Migrating and other terrestrial insects at sea. In *Marine Insects,* ed. L. Cheng, pp. 97–117. Amsterdam: Elsevier/North Holland. 581 pp.

13. Brinck, P., Wingstrand, K. G. 1949. The mountain fauna of the Virihaure area in Swedish Lapland. *Acta Univ. Lund. Sect. 2* 45:21–69

14. Chapman, J. A., Romer, J. I., Stark, J. 1955. Ladybird beetles and army cutworm adults as food for grizzly bears in Montana. *Ecology* 36:156–58

15. Cheng, L., Birch, M. C. 1977. Terrestrial insects at sea. *J. Mar. Biol. Assoc. UK* 57:995–97

16. Cheng, L., Birch, M. C. 1978. Insect flotsam: an unstudied marine resource. *Ecol. Entomol.* 3:87–97

17. Clegg, J. S. 1967. Metabolic studies of cryptobiosis in encysted embryos of *Artemia salinas. Comp. Biochem. Physiol.* 20:801–9

18. Crawford, R. L. 1985. Mt. St. Helens and spider biogeography. *Proc. Wash. State Entomol. Soc.* 46:700–2

19. Crawford, R. L., Edwards, J. S. 1986. Ballooning spiders as a component of arthropod fallout on snowfields of Mt. Rainier, Washington. *Arct. Alp. Res.* 18: In press

20. Culver, D. C. 1982. *Cave Life, Evolution and Ecology.* Cambridge, Harvard Univ. Press. 189 pp.

21. Danthanaryana, W., ed. 1986. *Insect Flight, Dispersal and Migration.* Berlin: Springer-Verlag.

22. Darwin, C. 1839. *Journal of Researches into the Geology and Natural History of the Various Countries Visited by H.M.S. Beagle*, p. xxx London: Colburn

23. DeZordo, I. 1979. Ökologische untersuchungen an Wirbellosen des zentralalpinen Hochgebirges (Obergurgl, Tirol). III. Lebenszyklen und Zönotik von Coleopteren. *Alp.-Biol. Stud. Univ. Innsbruck* 11:1–131

24. Dingle, H., ed. 1978. *Evolution of Insect Migration and Diapause.* New York: Springer Verlag. 284 pp.

25. Duman, J., Horwath, K. 1983. The role of hemolymph proteins in the cold tolerance of insects. *Ann. Rev. Physiol.* 45:261–70

26. Durbin, D. V. 1975. *The ecology of insects and other arthropods found on Sierran snowfields.* MS thesis. San Francisco State Univ., Calif. 163 pp.

27. Edwards, J. S. 1970. Insect fallout on the Gulkana Glacier, Alaska Range. *Can. Entomol.* 102:1169–70

28. Edwards, J. S. 1972. Arthropod fallout on Alaskan snow. *Arct. Alp. Res.* 4:167–76

29. Edwards, J. S. 1974. Insect fallout on snow in the Snowy Mountains, New South Wales. *Aust. Entomol. Mag.* 1:57–59

30. Edwards, J. S. 1982. Habitat, behavior and neurobiology of an American grylloblattid. In *Biology of the Notoptera,* ed. H. Ando, pp. 19–28. Nagano, Japan: Kashiyo-Insatsu. 194 pp.

31. Edwards, J. S. 1985. Derelicts of dispersal: arthropod fallout on Pacific Northwest volcanoes. See Ref. 21, pp. 196–203

32. Edwards, J. S. 1986. How small ectotherms thrive in the cold without really trying. *Cryo-Lett.* 6:388–90

33. Edwards, J. S. 1986. Arthropods as pioneers: recolonisation of the blast zone on Mt. St. Helens. *Northwest Environ. J.* 2:63–73

34. Edwards, J. S., Banko, A. M. 1976. Arthropod fallout and nutrient transport on Alaskan snow patches: a quantitative study. *Arct. Alp. Res.* 8:237–45

35. Edwards, J. S., Crawford, R. L., Sugg, P. M., Peterson, M. 1986. Arthropod colonization in the blast zone of Mt. St. Helens: five years of progress. In *Mt. St. Helens; Five Years Later,* ed. S. Keller, E. Wash. Univ. Press. In press

36. Elton, C. S. 1925. Dispersal of insects to Spitzbergen. *Trans. R. Entomol. Soc. London* 1925:289–99

37. Foster, W. A., Treherne, J. E. 1980. Feeding predation and aggregation behaviour in a marine insect, *Halobates robustus* Barber (Hemiptera) in the Galapagos Islands. *Proc. R. Soc. London Ser. B* 209:539–53

38. Furniss, M. M., Furniss, R. L. 1972. Scolytids (Coleoptera) on snowfields above timberline in Oregon and Washington. *Can. Entomol.* 104:1471–78

39. Gressitt, J. L., ed. 1967. *Antarctic Research Series 10.* Washington, DC: Am. Geophys. Union. 395 pp.

40. Gressitt, J. L., Yoshimoto, C. M. 1974. Insect dispersal studies in northern Alaska. *Pac. Insects* 16:11–30

41. Gurney, A. B. 1953. Grasshopper Glacier of Montana and its relation to long distance flight of grasshoppers. *Smithsonian Inst. Publ. 4121,* pp. 305–25. Washington, DC: Smithsonian Inst.

42. Hagen, K. S. 1962. Biology and ecology of predaceous Coccinellidae. *Ann. Rev. Entomol.* 7:289–326

43. Hamilton, W. D., May, R. M. 1977. Dispersal in stable habitats. *Nature* 269:578–81

44. Hamilton, W. J., Seely, M. K. 1976. Fog basking by the Namib desert beetle, *Onymachris unguiculus. Nature* 262:284–85

45. Holzapfel, E. P., Harrell, J. C. 1968. Transoceanic dispersal studies of insects. *Pac. Insects* 10:115–53

46. Howard, L. O. 1918. A note on insects found on snow at high elevations. *Entomol. News* 29:375–77

47. Howarth, F. G. 1979. Neogeoaeolian habitats on new lava flows on Hawaii island: an ecosystem supported by windborne debris. *Pac. Insects* 20:133–44

48. Howarth, F. G. 1979. Notes and exhibitions: *Hodegia apatela* Walsingham. *Proc. Hawaii. Entomol. Soc.* 23:14

49. Howarth, F. G. 1983. Ecology of cave arthropods. *Ann. Rev. Entomol.* 28:365–89

50. Howarth, F. G., Stone, F. D. 1982. An assessment of the arthropod fauna and aeolian system near the summit of Mauna Kea, Hawaii. In *Mauna Kea Science Reserve Complex Development Plan, Mauna Kea, Hawaii,* Vol. 2, Append. H, *Draft Environmental Impact Statement.*

51. Humboldt, A. von. 1805. *Essai sur la Geographie des Plantes.* Paris: Levrault Schoell. 155 pp.

52. Humboldt, A. von. 1808. *Ansichten der Natur mit Wissenschaftlichen Erläuterungen.* Stuttgart/Tübingen: Cotta. 394 pp.

53. Hutchinson, G. E. 1965. *The Ecological*

Theater and the Evolutionary Play. New Haven, Conn: Yale Univ. Press. 139 pp.

54. Johnson, C. G. 1969. *Migration and Dispersal of Insects by Flight.* London: Methuen. 763 pp.

55. Kaisila, J. 1952. Insects from arctic mountain snows. *Ann. Entomol. Fenn. Suom. Hyonteistiet. Aikak.* 18:8–25

56. Koch, C. 1960. The tenebrionid beetles of South West Africa. *Bull. South Afr. Mus. Assoc.* 7:73–85

57. Kohshima, S. 1984. A novel cold-tolerant insect found in a Himalayan glacier. *Nature* 310:225–27

58. Kohshima, S. 1984. Living micro-plants in the dirt layer dust of Yala Glacier, Nepal Himalaya. In *Report of Glacier Boring Project 1981–82 in the Nepal Himalaya,* pp. 91–97. Nagoya, Japan: Nagoya Univ. Water Res. Inst.

59. Kohshima, S. 1985. Migration of the Himalayan wingless glacier midge (Diamesa sp.): slope direction assessment by sun-compassed straight walk. *J. Ethol.* 3:93–104

60. Lang, A. 1975. Koleopterenfauna und faunation in der Alpinen Stufe der Stubaier Alpen (Kühtai). *Alp.-Biol. Stud. Univ. Innsbruck* 1:1–80

61. Leinass, H. P. 1981. Activity of arthropods in snow within a coniferous forest, with special reference to Collembola. *Holarct. Ecol.* 4:127–38

62. Levin, S. A. 1976. Population dynamic models in heterogenous environments. *Ann. Rev. Ecol. Syst.* 7:287–310

63. Liston, A. D., Leslie, A. D. 1982. Insects from high-altitude summer snow in Austria, 1981. *Entomol. Ges. Basel* 32:42–47

64. Louw, G. N., Seely, M. K. 1982. *Ecology of Desert Organisms.* London/New York: Longman. 194 pp.

65. Mani, M. S. 1962. *Introduction to High Altitude Entomology.* London: Methuen. 302 pp.

66. Mani, M. S. 1968. *Ecology and Biogeography of High Altitude Insects.* The Hague: Junk. 530 pp.

67. Mani, M. S., Giddings, L. E. 1980. *Ecology and Highlands.* The Hague: Junk. 249 pp.

68. Mann, D. H., Edwards, J. S., Gara, R. I. 1980. Diel activity patterns in snowfield foraging invertebrates on Mount Rainier, Washington. *Arct. Alp. Res.* 12:359–68

69. Martin, J. L., Oromi, P. 1986. An ecological study of Cueva de los Roques lava tube (Tenerife, Canary Islands). *J. Nat. Hist.* 20:375–88

70. Masutti, L. 1978. Insetti e nevi stagionali. Riflessioni su reperti relativi alle Alpe Carniche e Guilie. *Boll. Ist. Dei Entomol. Univ. Studi Bologna* 34:75–94

71. May, R. M. 1978. The dynamics and diversity of insect faunas. *Symp. R. Entomol. Soc. London* 9:188–204

72. Morrissey, R., Edwards, J. S. 1979. Neural function in an alpine grylloblattid: a comparison with the house cricket *Acheta domesticus. Physiol. Entomol.* 4:241–50

73. Nelson, J. M. 1965. A seasonal study of aerial insects close to a moorland stream. *J. Anim. Ecol.* 34:573–79

74. Norlin, A. 1967. Terrestrial insects on lake surfaces, their availability and importance as fish food. *Rep. Inst. Freshwater Res. Drottningholm* 47:39–55

75. Norvell, J. R. 1985. *Foraging sites of alpine birds in relation to snow accumulation.* MS thesis. Towson State Univ., Maryland. 57 pp.

76. Ottesen, P. S. 1985. Diel activity patterns of South Scandinavian high mountain ground beetles (Coleoptera, Carabidae). *Holarct. Ecol.* 8:191–203

77. Papp, R. P. 1978. A nival aeolian ecosystem in California. *Arct. Alp. Res.* 10:117–31

78. Pearson, O. P. 1954. Habits of the lizard *Liolaemus multiformis* at high altitudes in southern Peru. *Copeia* 1954:111–16

79. Robinson, M. D., Seely, M. K. 1980. Physical and biotic environments of the southern Namib dune ecosystem. *J. Arid Environ.* 3:183–203

80. Salt, R. W. 1961. Principles of insect cold hardiness. *Ann. Rev. Entomol.* 6:55–74

81. Schmoller, R. R. 1971. Nocturnal arthropods in the alpine tundra of Colorado. *Arct. Alp. Res.* 3:345–52

82. Schmoller, R. R. 1971. Populations and production of major arthropods of the Colorado alpine tundra. In *The Structure and Function of the Tundra Ecosystem, US Int. Biol. Program Tundra Biome. Prog. Rep.,* pp. 135–39

83. Scudder, G. C. E. 1963. Heteroptera stranded at high altitudes in the Pacific Northwest. *Proc. Entomol. Soc. BC* 60:41–44

84. Seely, M. K., Hamilton, W. J. 1976. Fog catchment sand trenches constructed by tenebrionid beetles *Lepidochora* from the Namib Desert. *Science* 193:484–86

85. Sømme, L. 1981. Cold tolerance of alpine arctic and antarctic collembola and mites. *Cryobiology* 18:212–20

86. Sømme, L. 1986. Adaptations of alpine insects to the environment. *Cryo-Lett.* 6:384–87

87. Sømme, L., Zachariassen, K. E. 1981. Adaptations to low temperature in high altitude insects from Mount Kenya. *Ecol. Entomol.* 6:199–204

88. Southwood, T. R. E. 1962. Migration of terrestrial arthropods in relation to habitat. *Biol. Rev.* 37:171–214

89. Spalding, J. B. 1979. The aeolian ecology of White Mountain Peak, California: windblown insect fauna. *Arct. Alp. Res.* 11:83–94

90. Storey, K. B. 1986. Biochemical principles of freeze tolerance in insects. *Cryo-Lett.* 6:410–13

91. Swan, L. W. 1961. The ecology of the high Himalayas. *Sci. Am.* 205:68–78

92. Swan, L. W. 1963. Aeolian zone. *Science* 140:77–78

93. Swan, L. W. 1967. Alpine and aeolian regions of the world. In *Arctic and Alpine Environments*, ed. H. E. Wright, W. H. Osburn, pp. 29–54. Bloomington, Ind: Indiana Univ. Press. 308 pp.

94. Swan, L. 1981. The aeolian region of the Himalaya and the Tibetan Plateau. *Proc. Symp. Qinghai-Xizang (Tibet) Plateau, Beijing, China, 1980*, Vol. 2, *Environment and Ecology of Qinghai-Xizang Plateau*, pp. 1971–76. Beijing: Scientific Press/New York: Gordon & Breach

95. Teeri, J. A., Barrett, P. E. 1975. Detritus transport by wind in a high arctic terrestrial system. *Arct. Alp. Res.* 7:387–91

96. Thiele, H.-U. 1977. *Carabid Beetles in Their Environments.* Berlin: Springer-Verlag. 369 pp.

97. Van Dyke, E. C. 1919. A few observations on the tendency of insects to collect on ridges and mountain snowfields. *Entomol. News* 30:241–44

98. Wanless, F. R. 1975. Spiders of the family Salticidae from the upper slopes of Everest and Makalu. *Bull. Br. Arachnol. Soc.* 3:132–36

99. Wygodzynski, P. 1974. Notes and description of Machilidae from the Old World (Microcoryphia, Insecta). *Am. Mus. Novit.* 2555:1–21

100. Zachariassen, K. E., Hammel, H. T. 1976. Nucleating agents in the haemolymph of insects tolerant to freezing. *Nature* 262:285–87

Ann. Rev. Entomol. 1987. 32:181–99

CULTURAL ENTOMOLOGY

Charles L. Hogue

Natural History Museum of Los Angeles County, 900 Exposition Boulevard, Los Angeles, California 90007

DEFINITIONS

Humans spend their intellectual energies in three basic areas of activity: surviving, using practical learning (the application of technology); seeking pure knowledge through inductive mental processes (science); and pursuing enlightenment of taste and pleasure by aesthetic exercises that may be referred to as the "humanities." Entomology has long been concerned with survival (economic or applied entomology) and scientific study (academic entomology), but the branch of investigation that addresses the influence of insects (and other terrestrial Arthropoda, including arachnids, myriapods, etc) in literature, languages, music, the arts, interpretive history, religion, and recreation has only recently been recognized as a distinct field. This is referred to as "cultural entomology" (89).

Because the term "cultural" is narrowly defined, some aspects normally included in studies of human societies are excluded. Thus ethnoentomology, which is concerned with all forms of insect-human interactions in so-called primitive societies, is not synonymous with cultural entomology. For this reason, entomophagy as practiced to complete the regular diet of an Indian tribe is considered applied entomology and is not covered here; however, where entomophagy occurs for recreational or ceremonial reasons, it assumes a place in the subject of this paper (200). Likewise, pharmacological, manufacturing, or other wholly practical uses of insects, even though unusual, such as applications in forensic science, are not part of the subject. The narrative history of the science of entomology is not part of cultural entomology, while the influence of insects on general history is.

Insects have assumed a position of unusually great significance for certain ethnic assemblages or nations (146). To the ancient Egyptians and neighboring cultures, various insects were revered (19, 44, 105, 158); in particular,

181

0066-4170/87/0101-0181$02.00

several species of dung scarabs (Phaeniini, Coprini) rose in religious and symbolic importance early in history. This is witnessed by the prevalence and persistence (approximately 2200 BC to New Kingdom times, circa 1000 BC and later) of scarab imagery in worship and funereal ceremony (129, 212).

The Japanese have a highly developed tradition of aesthetic appreciation for insects, reflected in their literature, art, and recreational pursuits. This has attracted some sensitive commentary by a few authors, especially Hearn (79, 80, 81) and Kevan (116). Much the same could be said of the Chinese (106, 162), who hold crickets and other musical Orthoptera in particularly high esteem (130).

Few authors have treated the subject of cultural entomology in general terms (33, 37, 115, 127, 175, 189, 214). Literature is sparse and is not referenced to this subject in bibliographies. Information is often oriented geographically (119, 131) or is included in extradisciplinary works, especially works on history, iconography, classics, and anthropology. Because cultural aspects often intersect other insect-related topics, examples are sometimes to be found within literature dealing with entomological history (15, 120, 123, 186), the entomological impact on human welfare (25, 34), or the taxonomy of specific insect groups (see below, Species of Special Cultural Significance).

The subject is popular with entomologists from around the world. Almost 70 persons are listed in a recent directory of investigators (C. L. Hogue, unpublished). The first colloquium on cultural entomology took place at the 17th International Congress of Entomology in Hamburg in 1984, and at that time a list was drawn up of the fields of study comprising the subject. Although some overlap occurs, these topics are used as an outline for the following discussion.

LITERATURE AND LANGUAGE

Insects appear frequently in literature (84, 85, 135, 181a, 184). G. J. Umphrey & C. L. Hogue (unpublished) have collected some 100 titles of modern novels and almost as many short stories in English with fictional plots in which insects have a major role.

Insects are useful for establishing a variety of moods or images, both negative (more usual) or favorable. Among the former are many legitimately injurious or dangerous qualities, such as the ability to entrap (*Woman in the Dunes,* K. Abé), poisonous stings (*The Furies,* K. Roberts), rapaciousness (*Bugged,* D. Glut), and swarming instincts (*The Swarm,* A. Hertzog). Thus, they provide foundations for many tales of fantasy (*Leinigen Versus the Ants,* C. Stephenson) and intrigue (*The Gold Bug,* E. A. Poe), but are most abundant in science fiction, either as conjured earthly villains (*Bugs,* T.

Roszak) or space monsters (*Bug Wars,* R. Asprin). Because they are capable of delivering lethal toxins, some species have been employed as murder weapons in detective novels (honeybees in *A Taste for Honey,* G. Heard). Others with intimate microhabitats act as voyeurs and relate erotic tales (*The Fly,* R. Chopping; *Autobiography of a Flea,* Anonymous). Several stories play on the metamorphosis theme, with humans assuming insect characteristics to a limited (*Spider Girl,* P. Lear) or consuming degree (*Metamorphosis,* F. Kafka).

Positive attributes ascribed to insects and spiders, such as patience or industriousness, are the basis for a variety of proverbs and parables; this is true of several among Aesop's Fables (e.g. against arrogance: "A fly sitting on a chariot wheel said, 'What a dust I raise!' "). Some insects with especially likable traits, such as musical talent (Jiminy Cricket, "grigs" [an old term for orthopteroid insects, revived by Kevan (114)]) or high intelligence (archy the cockroach, *the lives and times of archy and mehatibel,* D. Marquis), have even become famous literary figures. A cute, rotund form speaks a message of friendliness and good humor, and little round beetles, bumblebees, woolly caterpillars, and fat spiders are written of as special insect friends (*Charlotte's Web,* E. White).

Parallels between human and insect societies provide a foundation for interplay between the two life forms (*Consider Her Ways,* F. Grove). The size disparity problem is solved either by magically shrinking the human (*Atta,* F. Bellamy) or enlarging the insect (*Empire of the Ants,* H. G. Wells). As teachers, humanized insects are common in children's literature (97), often because they provide an amiable, impartial narrator or actor with which the child can identify (*James and the Giant Peach,* R. Dahl). Hogue (88) referred to such hexapod characters as "bugfolk"; an example is the caterpillar in *Alice in Wonderland* (L. Carroll) (see also 186a). Some bugfolk have become modern day folk heroes (Spiderman) or villains (Mothra).

Bee society formed the basis for simile in a political satire against governmental hypocrisy in 18th century England ("The Fable of the Bees, or Private Vices Made Public Benefits," Bernard Mandeville, 1723). Other examples of political and social satire employing insects, among several cited by Kevan (121), are "The Spider and the Fly," an enormously long English poem published in 1556 by John Heywood (Protestant versus Roman Catholic Church), and "The Locust," written by an anonymous author in 1704 (an attack on the legal profession of the day).

Insect images appear as frequently in poetry (52, 109, 110, 113, 118, 121, 211) as in prose. The ancient Greeks often referred to them symbolically and aesthetically (82, 113), as did the Romans (113, 173). Shakespeare played on many in his works (143, 156), as did Dante in the Divine Comedy (139). Many other poets have been inspired by insects as well; some better known poems with insect titles are, "To a Louse," by Robert Burns; "To-day, this

Insect, and the World I Breath," by Dylan Thomas; "The Beetle," by James Whitcomb Riley; and "To a Butterfly, the Redbreast and Butterfly," by William Wordsworth. Japanese poetry, particularly haiku, commonly incorporates insect allusions (76, 80). One of the shortest poems ever written was about insects: "Ugh-Bugh!" (D. K. McE. Kevan).

Local names and folk taxonomies often reflect cultural beliefs. Several lists may be consulted: Anglo-Saxon or Old English (36), Australian (141), German (107), Tibetan (125, 126), Latin American (42, 63, 75, 198, 219), and Hellenistic (67). Insect forms were converted into hieroglyphs and pictograms in ancient Egyptian (scarab, bee, and grasshopper syllables in alphabet), Mayan (192), and Chinese (190, 203) writing.

In all languages, numerous insects or their names have been enlisted as figures of speech ("social butterfly"), which are extended into oft-used sayings, epigrams, and the like ("Busy as a bee," "Don't bug me," "What is good for the bee is not good for the swarm") (188). All manner of manufactured and commercial objects bear insect names. Many cocktails ("Grasshopper") or other drinks are so named, sometimes to suggest special potency ("Stinger") or distinctive flavor ("Bee's Kiss"). Even English pubs (193) and automobiles have insect epithets.

MUSIC AND THE PERFORMING ARTS

Insects have invaded the world of music to a considerable degree (60), with composers seizing on various attributes to convey a mood or message. The rapid vibrato of "The Flight of the Bumblebee" (Rimsky-Korsakov) imitates the buzz of the bee; the light of the firefly shines as a beacon to love in "Glow-worm;" and butterflies impart airiness, transience, and frivolity in "Poor Butterfly." The inspiration is less obvious in familiar songs such as "La Cucaracha," "The Boll Weevil," and "The Blue-tailed Fly" and unsung ditties like "Grasshopper Rock" and "Stompin the Bug."

As direct emitters of pleasant sounds, stridulating types have long been esteemed by different cultures. Crickets and katydids are still kept in cages to fill the house with cheerful chirps in Oriental countries (78, 130) and were once a passion of many Hamburgers (111).

The insect has been going on stage for more than two millennia. Since Aristophanes produced "Sphēkes," or "The Wasps" in 422 BC, a number of dramas have utilized metaphorical bugs, such as Jean Paul Sartre's "The Fly" and Karel and Josef Čapek's "Ze Života Hmyzu" ("On the Life of Insects," or "Insect Comedy"). Some insects have reached more elegant heights in operas ("Madame Butterfly," Puccini) and ballet ("Le Festin de L'Araignée," Roussel). (Ritual dances inspired by insects are discussed under Religion and Folklore, below.) The cinema and television films are rife with insect villains (army ants in "Naked Jungle," Paramount, 1954) and with a few comedic and heroic stars as well (140).

GRAPHIC AND PLASTIC ARTS

Artists have exploited the insect form in all media (28, 35, 59). Because of their pleasing colors and curious shapes, many types, especially butterflies and metallic beetles, have been used directly for ornamentation (9, 65, 168, 170). They have also served as models for decorative jewelry (56, 64, 166, 222), ceramics (108, 169), textile designs, and a huge variety of other objects (6, 128) from prehistoric, historic, antique, and modern periods. Serving trays, ashtrays, and scenic montages made from the wings of butterflies (especially from the genus *Morpho* in South America) are familiar decorative objects, and insects are on the postage stamps of many countries (136, 191).

Some particularly fine decorative pieces with insect designs are coveted art treasures; examples are the "Cretan Hornets" (Minoan gold pectoral with a pair of wasps) (40) and solid gold fly pendants ("Order of the Golden Fly") (176) found in the funeral cache of Queen Ahotpe, an 18th-Dynasty ancestor of Tutankhamen.

Insects abound in the pictorial arts (178). They provided motifs for Neolithic artists etching on bone (7) and rendering on rock; numerous insects are depicted in prehistoric petroglyphs and pictographs in Europe (179), South Africa (155), and North America (98). One of the enormous figures laid out on the desert plains of southern Peru by the Nazca Culture (300 BC–900 AD) is a spider (164).

Many portrayals of insects come to us from the pens and brushes of Oriental artists from earliest times (112, 162), and insects are still popular themes for decorative and symbolic renderings throughout the Far East (6, 190).

Several insects appear in early European Christian religious art as universal symbols. Among such symbols are bees [mother: "Mary symbols" (178)], bee hives (the church: "Madonna in the Garden," Mathias Grünewald, 1517/1519), the stag beetle (evil: "The Virgin with a multitude of Animals," Albrecht Dürer, 1503), flies [torment: "The Damnation of Lovers," Mathias Grünewald (32a, 134a)], and scorpions [pain: many depictions of Saint Jerome in Penitence (58)]. A special significance is attached to lepidopterans (symbolized by the goddess Psyche) as signatures of the soul (and hence life after death, change, rebirth) and love. For these reasons they sometimes appear in religious scenes (e.g. Albrecht Dürer's "The Virgin of the Irises"). Accordingly, butterfly or moth wings occasionally give powers of flight to some angelic forms (e.g. cupids) and often to fairies and nymphs (many examples in Reference 157). The historical prototype for the biblical cherubs, however, may have been dung beetles (91).

Insect symbols are personal hallmarks of the works of a few famous contemporary artists such as the surrealist Salvador Dali (grasshopper, groupings of ants, and formations of muscoid flies) (38) and Wolfgang Hutter (butterflies) (178).

Because of their inherently provocative forms, odd species provide the principle themes in many paintings by other well-known western artists, such as Graham Sutherland (aquatint series on "The Bees)" (197), and in drawings and engravings of M. C. Escher ("Möbius Band"), James Ensor ("Odd Insects"), Odilon Redon ("The Spider"), and many others of lesser fame. In illuminated medieval manuscripts, border decorations and elaborate initials were often patterned after insects (99, 100).

Images of bugfolk are common. Some of the earliest are fantastic insectoid demons in paintings by Hieronymus Bosch ("The Last Judgement," detail of fallen angel, 1504) and Pieter Brueghel ("Fall of the Angels," 1562); these apparently spawned a style that was followed by a series of later illustrators, among them Martin Disteli (210), Jean I. I. Gerard, "Grandville" ("Adventures d'un papillon" in "Scènes de la vie privée et publique des animaux," 1842), and Alan Aldridge ("Magician Moth" in the 1975 Grossman version of "The Butterfly Ball and the Grasshopper Feast").

Some of their alien characteristics, including antennae, bulbous and facetted eyes, articulated bodies, armored exterior, and biting mouthparts, have made insects favorite prototypes for the design of dream monsters, extraterrestrial creatures, and even spacecraft by fantasy artists. Numerous examples appear on the covers of science fiction novels, on posters, and in cartoons (195).

Sculpture also utilizes insect motifs and symbolism. Best known from history is the frequent appearance of Psyche (represented by lepidopteran figures) on stone carvings of scarabs from classic Egypt and on Roman sarcophagi. Several contemporary artists working in metal, plastics, and other modern materials have specialized in entomological themes (see References 175, 178 for many extraordinary examples).

Insect and arachnid products have even served as art media. Paintings have been made on cobwebs (22, 31). Wax, from both *Apis* and the tropical American meliponine bees, has been used to fashion lone figures (6a) and positive images for the "lost wax" casting technique practiced by Old World (151) and Incan (11) metallurgists. Lacquer made from lac insects has wide application in Oriental art.

For their symbolic value, insects also appear with regularity on seals, coins, and heraldic (54, 104) and other emblems. Napoleon I replaced the fleur-de-lis with the honeybee as the Bourbon family emblem, and its image was displayed on all manner of surfaces in the royal palace and on the Napoleonic coat of arms (54). Twenty of the United States have designated state insects along with state flowers, trees, and birds; most have chosen the honeybee, a sign of industry and sovereignty.

Advertising art frequently makes use of insect images to convey overt or subliminal messages about products by capitalizing on widespread attitudes, either negative (cockroaches as bearers of filth) or favorable (beauty, freshness, and airiness of butterflies).

It is curious that insects depicted in art often bear only two pairs of legs (117).

INTERPRETIVE HISTORY

Insects have greatly influenced human history, principally by forcing shifts in pivotal political events. Battles have been lost, expeditions foiled, and populations decimated through the direct involvement of insects, usually as carriers of disease (34, 167, 187).

Insect products have also helped determine the direction of civilization's march. It could be said that the Chinese Empire was largely founded on the silk trade. Commerce in dyestuffs derived from the bodies of the cochineal insect reached global proportions by the 18th century, and proved so lucrative that the insect and its cactus host were introduced to various parts of the world from their native America. In the adopted countries the plant spread and became a noxious weed that rendered vast tracts of land unusable. Trade in other insect products such as honey and shellac has had similar economic significance. The Israelite band that founded the Jewish nations survived on "manna" during its extended trek through the Sinai Desert. This nutritious substance is thought to have been extruded by scale insects on the tamarisk plant (16, 180).

There are anecdotes of a number of other ways in which insects have crept into our affairs. A moth is supposed to have prevented an accident to a train on which Queen Victoria was riding (213). Several important personages were aided in difficult times and inspired to lofty deeds by insects and spiders (167, Chapter 7). The Chinese inventor of paper, Ts'ai Lun (89–106 AD), according to legend, was shown the process by wasps making their nests by chewing tree bark and mixing it with their saliva.

PHILOSOPHY

The insect is commonly considered a low form of life that deserves only contempt, but it is justifiable to contemplate the rightful relationships between humans and insects. Most of what has been written in this context deals with the direct competition between insects and humans for food and fiber and the human suffering that results from insect-borne diseases (25). Another favorite thesis is the comparison of insect and human societies (68). Our comparatively shaky dominion of nature has also been a theme (e.g. in the motion picture "The Hellstrom Chronicle," David Wolper, 1971), and the insect is pointed to as the most likely life form to inherit the earth after our own presumed demise. A few authors have tried to look at the world through insect eyes (Benjamin Franklin, "Soliloquy of a venerable Ephemera who had lived four hundred and twenty minutes"), and there is some appreciation of insects as friends and teachers (167, Chapter 7). This is a generally neglected area, however.

RELIGION AND FOLKLORE

Animistic religious practices based on insects have been an important part of the culture of many groups. From the ancient world (32) the best known example is the scarab cult of the Egyptians (148, 159, 196, 216). Evidence in the form of scarab amulets dominates the archaeological record of those worshippers (212). Insect gods and goddesses assumed various roles in the religions of the Aztecs [Xochiquetzal, butterfly goddess (10)], Greeks [Artemis was Mylitta, the mother or bee goddess (178, p. 70)], Chinese [Tschun-Wan, insect lord over crop pests (95)], and Babylonians [scorpion men (73, 206)]. The Hopi personify several insect spirits (Butterfly Man, Assassin Fly, etc.) in the form of Kachina dolls (223). In Bushman mythology, the mantis is an important god of creation, Kaggen (177). The insect deities are served with a variety of rites and rituals; for example, youthful initiates are scourged by stinging ants in puberty ceremonies among various Amazonian Indian tribes ["tucandeira" (*Dinoponera* spp.) rituals] (132).

Within the context of Judaism and Christianity, insects have had no small role. Although most of the numerous references to insects in the Bible (24, 113) are historical (12, 45–49), some are allegorical (91) or reflect deep theological meaning (e.g. stinging locusts in Revelation 9:3–11; 113). Of the ten plagues visited upon Egypt preceding the exodus, three were insects and two or three others may have had entomological connections (146). In the Talmudic literature, locusts are included among the disasters for which the sounding of the ram's horn and a public feast were prescribed in the Ta'anit tractate (Section 3:5). The locust plague theme is favored by many religious artists (18).

Curious applications of entomology in the Christian religion were the exorcisms and animal trials performed by the Roman Catholic Church in medieval and even later times (15, Bd. 1, pp. 233 ff). Because animals, including insects, were supposed to possess human qualities, even a soul, they were held accountable for their misdeeds and were subject to divine control and excommunication.

Involvements of insects in other major world religions (Islam, Hinduism, Buddhism) have been relatively unexplored by entomologists (113, 118). The spider sitting in the center of its web is a spinner of illusion and reminds Hindus of maya, the supernatural force behind the creation of the transient world. Hindu holy writings also teach that ants are divine, the first born of the world; ritually the anthill represents the earth (194).

Entomological references in folklore (legends, mythology, beliefs, fairy tales) abound, but they are generally ensconced in the anthropological literature and are therefore not easily located by the entomologist. There are no general reviews or collections of insect-based folktales, although a few limited treatises are available (33, 92, 119, 131, 181a, 202, 218).

Many classic myths, legends, and beliefs are related to insects (8, 215). The Roman goddess Psyche is portrayed with wings and represents rebirth and metamorphosis to a higher state (41c, 181). Butterflies and chrysalids are found in earlier Minoan iconography ["Ring of Nestor" (150)], but the question of the age and origin of the symbolism is unsettled. Lilith, Adam's first wife and begetter of flies and demons, originated in Assyria-Babylon and makes her way into Mohammedan and Jewish books (45, 57). In a variant of the story of the aging of Tithonus, consort of Eos, he is turned into a cicada (69, 113). Early natural historians told about the ant-lion ("myrmicoleon"), a giant ant that resembled a dog with lion's feet and dug for gold (66, 123); it is portrayed in early bestiaries, sometimes in mongrel form with partial human anatomy. Other hybrids are the "scorpion men" (human torso-legs/scorpion abdomen-tail) from second millenium Mesopotamia and neighboring times and places (73).

Other myths originated in European countries and were carried by emigrants to colonies in America and other continents as folktales of almost infinite variety (90). An exemplary and widespread folkloric theme is "telling the bees" when a death occurs in a beekeeper's family. The insects are believed to respond sympathetically by attending the funeral or absconding (226).

One arachnid, the scorpion, comprises the eighth of the normal twelve signs of the Zodiac ("Scorpio"). A second, the spider, is considered by some astrologers to represent a thirteenth sign ("Arachne") that became lost (208).

Folklore and superstitions involving insects are perhaps more prevalent in indigenous or traditional cultures than among industrialized societies. Every group has its repertoire, with common themes running across cultural lines. Many creation myths involve insects: The Hopis explained the origin of the world by the actions of the Spider Grandmother (147); According to the

Yagua Indians of Peru the Amazon River was created by wood-eating insects (51c); and fire came from a mythical campfire ignited by fireflies, according to the Jicarilla Apaches of New Mexico (55).

Involvement of insects in magic and witchcraft is surprisingly infrequent considering the venomous and metamorphic powers of so many types (217). A few species are thought to be deadly poisonous, such that even the slightest contact with them can cause instant or lingering, agonizing death [e.g. *Fulgora* in tropical America (93)]. One such species described by the earliest explorers of the New World remains unidentified (103). A variety of interesting prophylaxes and remedies are employed against these imagined assassins. A few species have supposed or real hallucinogenic or aphrodesiacal powers if ingested, which give them a place in folk ritual (13).

Insects and their products, especially honey from the many species of wild and domestic bees, are employed often in folk healing. The word "medicine" owes its origin to honey; the first syllable has the same root as "mead," an alcoholic beverage made from honeycomb, which was often consumed as an elixir (199). Cockroaches (101), lice (33), bedbugs ("wall lice") (205), beetles, and galls (166) have also been used as medicines. As treatment for scorpion stings, village curanderos in the mountains of western Mexico tie a dead scorpion to the finger that has just been stung (220).

RECREATION AND CURIOSITIES

Insects are the butt of many a joke or cartoon (41). Several kinds are kept as unusual (51b, 72, 133), or educational pets (204, 207). Some are kept for their pleasant sounds (78, 130). Toys are modeled after insects, such as the familiar snapping "cricket" noisemakers and all manner of mechanical bugs. Other playthings may actually incorporate living insects, including Mexican jumping beans and "fly-powered" airplanes. Insects have inspired diversionary pursuits, particularly in the Orient, where kites, bull-roarers (61), and other noisemakers (26) of entomological engineering are common. In the martial arts, the stealth, strength, and speed of praying mantids form the basis of one system of kung fu (77). Cricket (71, 124) and spider (4) fighting are pastimes long practiced in Far Eastern countries. In the West, "flea circuses" (39) were once widely attended; now they are somewhat hard to find.

Several apocryphal tales about insects, better called "humbugs," have cropped up. There are fictitious species, such as winged spiders (51a); Sir Arthur Conan Doyle's tick, "*Ixodes maloni*," which lives in the "Lost World" (94); iron-eating "railroad or cannon worms" (185); and even alleged new species contrived from the imagination, such as Stecker's "*Gibbicellum sude-*

ticum" (174). Real bugs are thought to be behind some "flying saucer" sightings (30). False fossil insects are common, especially in amber but also from fabricated stone (171).

ETHNOENTOMOLOGY

Ethnoentomology, i.e. applications of insect life in so-called primitive (traditional, aboriginal, or nonindustrialized) societies may be regarded as a special branch of cultural entomology. It has taken its place alongside ethnobotany and as part of ethnozoology (142). It is discussed here only cursorily.

Many present-day Amerind groups (9, 86, 160) have adopted insects as totem figures and as a source of animistic explanations in their religions and cosmologies. This is especially true of groups inhabiting tropical areas, probably because of the richness of insects in their surroundings. The ethnoentomologies of the Warao of the Orinoco Delta (221) and the Gorotire Kayapó of Amazonia (161) have been investigated more than others. Other studies have been carried out with indigenous tribes in Zambia (182, 183), Maoris in New Zealand (144), and Kalahari Bushmen (177). Among the North American Indians, the ethnoentomologies of the Navajo (224, 225) and the Hopi (53, 74) have been best documented, although other groups have received some attention (166, 172). The iconography of the Aztecs of Mexico is liberally sprinkled with insects (134). Insect artifacts and remains have been used as topographic and chronologic indicators in other ethnological work as well (83).

SPECIES OF SPECIAL CULTURAL SIGNIFICANCE

Several types of insects have acquired special cultural importance, often for multiple reasons. Orthopteroids ("grigs") (110, 113, 114, 118, 119, 121), including mantids (122), have a wider variety of meanings than any other insects (113). Locusts command special recognition because of the destructive force of their plagues (17). Butterflies and moths (14, 20) have at least 74 symbolic meanings in Western art, according to Gagliardi (62). They were also very important to ancient cultures in Mexico (41a). Bees are nearly culturally ubiquitous, having evoked a considerable number of superstitions and symbolic applications (41e, 163, 209). Others with a particular place in the humanities are dung scarabs (see Religion and Folklore, above) and cicadas (43, 50, 51, 96). Amulets in the form of cicadas were placed on the tongues of the dead in ancient China, to induce resurrection by sympathetic magic (165). Fleas (29), fireflies (149), flies generally (70), myiasis-producing flies (41b), ectoparasites (3, 205), dragonflies (80), spiders (21), and scorpions (201, 206), all carry exceptional meanings in human culture.

Several erroneous beliefs, superstitions, and myths have evolved from the mimicry existing between the drone fly *(Eristalis tenax)* and the honeybee (5, 153). Most curious is the "bugonia" myth, which is the ancient belief that honeybees may arise from animal carcasses, especially dead oxen or cattle (154). The development of these bee-resembling flies on putrifying flesh must be the basis of the story.

CONCLUDING REMARKS

As a conspicuous part of our environment, insects, along with plants, other animals, and geological features, have captured our imaginations and become incorporated into our thinking from earliest times. Almost no aspect of our culture is untouched by these creatures. Their cultural importance relative to that of other life forms is not known, because a comparative study has not yet been conducted. It is clear that culture is another sphere in which their adaptability has compensated for the basically alien arthropod form and comportment. In spite of a hard external skeleton, extra appendages, and robotlike instincts, arthropods still sufficiently parallel humans in structure and behavior to serve as models of friends, enemies, and teachers.

There are various explanations for the significance of insects in human culture. Their meaning most often rests on their symbolic value (87, 181a). Because of some outstanding part of their appearance or behavior, many species are well-established symbols, some with multifarious meanings. These meanings are sometimes contradictory depending on the society in which they appear (e.g. crickets in the house may signify either good luck or impending doom) (41d, 109, 121). The insect itself or its products may also provide a model (decorative art), a device (toy), or a tool (murder weapon in a detective story).

This review has only touched the surface of a vast and complex subject. Unfortunately, under present space limitations it has been possible merely to skip across the more important points, and give only a few examples and primary references. The latter should be consulted for further reading, and the reader is urged to explore the classics, history, poetry and prose, museums of art, archaeology, anthropology, and all around us for more evidence of the insect in our lives.

ACKNOWLEDGMENTS

I am indebted to Jay Bisno, Steven R. Kutcher, and Roy R. Snelling for contributing references and examples to this review and for making valuable suggestions for its betterment. My special thanks go to Dr. D. Keith McE. Kevan for sharing his vast knowledge on the subject and for making many important corrections to my preliminary manuscript.

Literature Cited

1. Deleted in proof
2. Deleted in proof
3. Abalos, J. W., Wygodzinsky, P. 1951. La vinchuca. Folklore y antecedentes históricos, *Cienc. Invest.* 7:472–75
4. Anima, N. 1983. And now, spider fighting. *Am. Arachnol., Newsl. Am. Arachnol. Soc.* 28:17–19
5. Atkins, E. L. 1948. Mimicry between the drone-fly, *Eristalis tenax* (L.), and the honeybee, *Apis mellifera* L. Its significance in ancient mythology and present-day thought. *Ann. Entomol. Soc. Am.* 41:887–92
6. Ball, K. M. 1927. The bat and the butterfly; the dragon-fly. In *Decorative Motifs of Oriental Art*, pp. 257–72. New York: Lane. 286 pp.
6a. *Bee World*. 1978. British Museum beeswax treasures. 59(1):39–40
7. Bégouen, H., Bégouen, L. 1928. Découvertes nouvelles dans la caverne des trois frères a Montesquieu-avantès (Ariège). *Rev. Anthropol.* 38:358–64
8. Bell, R. E. 1982. *Dictionary of Classical Mythology. Symbols, Attributes & Associations*. Santa Barbara, Calif: ABC Clio. 390 pp.
9. Berlin, B., Prance, G. T. 1978. Insect galls and human ornamentation: the ethnobotanical significance of a new species of *Licania* from Amazonas, Peru. *Biotropica* 10:81–86
10. Beutlespacher, C. R. 1976. La diosa Xochiquetzal. *Bol. Inf. Soc. Méx. Lepid.* 2:2–3
11. Bird, J. 1979. Legacy of the stingless bee. *Nat. Hist.* 88(5):48–51
12. Birdsong, R. E. 1934. Insects of the Bible. *Bull. Brooklyn Entomol. Soc.* 29:102–6
13. Blackburn, T. 1976. A query regarding the possible hallucinogenic effects of ant ingestion in south-central California. *J. Calif. Anthropol.* 3(2):78–81
14. Blatchford, C. H. 1891. The butterfly in ancient literature and art. In *Butterflies of New England*, pp. 1257–63. Boston: Bradlee Whidden. 68 pp.
15. Bodenheimer, F. S. 1928, 1929. *Materialien zur Geschichte der Entomologie bis Linne*, Bd. 1, 2. Berlin: Junk. 498, 486 pp.
16. Bodenheimer, F. S. 1947. The manna of Sinai. *Biblical Archaeol.* 10:2–6
17. Böning, K. 1972. Heuschreckendarstellungen aus dem Altertum und ihre Bedeutung für die Geschichte des Pflanzenschutzes. *Anz. Schädlingskd. Pflanzenschutz* 44:21–31

18. Böning, K. 1977. Schädlingsplagen auf Beichtspiegeln und graphischen Blättern des 15. und beginnenden 16. Jahrhunderts. *Anz. Schädlingskd. Pflanz. Umweltschutz* 10:145–50
19. Brentjes, B. 1964. Einige Bemerkungen zur Rolle von Insekten in der altorientalischen Kultur. *Anz. Schädlingskd.* 37:184–89
20. Brewer, J., Sandved, K. B. 1976. Butterflies in art, heraldry, and religion. In *Butterflies*, pp. 41–54. New York: Abrams. 16 pp.
21. Bristowe, W. S. 1928–1929. Facts and fallacies about spiders. *Proc. South London Entomol. Nat. Hist. Soc.* 5:12–23
22. Bristowe, W. S. 1974. Art on a cobweb. *Animals* 16(2):62–63
23. Deleted in proof
24. Bruce, W. G. 1958. Bible references to insects and other arthropods. *Bull. Entomol. Soc. Am.* 4(3):75–78
25. Brues, C. T. 1947. *Insects and Human Welfare*. Cambridge, Mass: Harvard Univ. Press. 154 pp.
26. Brues, C. T. 1950. The Salagubong gong, a Filipino insect toy. *Psyche* 57:26–28
27. Deleted in proof
28. Burgess, N. R. H. 1981. The insect in art. *Antenna* 5(2):52–53
29. Busvine, J. R. 1980. Fleas, fables, folklore and fantasies. In *Fleas*, ed. R. Traub, H. Stark, pp. 209–14. Rotterdam: Balkema. 420 pp.
30. Callahan, P. S., Mankin, R. W. 1978. Insects as unidentified flying objects. *Appl. Opt.* 17:3355–60
31. Cassiver, I. 1956. Paintings on cobweb. *Nat. Hist.* 65:202–7, 219–20
32. Catherine, G. 1929. Les insectes dans les religions anciennes, les legendes et l'histoire. *Bull. Soc. Linn. Seine-Maritime* 15:20–27
32a. Chastel, A. 1986. A fly in the pigment. Iconology of the fly. *FMR* 4(19):61–81 (English ed.)
33. Clausen, L. W. 1954. *Insect Fact and Folklore*. New York: Macmillan. 194 pp.
34. Cloudsley-Thompson, J. L. 1976. *Insects and History*. New York: St. Martin's. 242 pp.
35. Collins, M. S. 1979. The insect in art. *Black Art Int. Q.* 3(3):14–28
36. Cortelyou, J. V. Z. 1906. Die altenglischen Namen der Insekten Spinnen- und Krustentiere. *Angl. Forsch.* 19:1–124
37. Cowan, F. 1865. *Curious Facts in the*

History of Insects. Philadelphia: Lippincott. 396 pp.

38. Cowles, F. 1960. *The Case of Salvador Dali*. Boston: Little, Brown, 334 pp.
39. Dall, W. H. 1877. Educated fleas. *Am. Nat.* 11:7–11
40. Davenport, H., Richards, O. W. 1975. The Cretan "hornet." *Antiquity* 49:212–13
41. Davis, J. J. 1937. *The Entomologists' Joke Book*. Lafayette, Ind: Exterminators Log. 160 pp.
41a. de la Maza R., R. 1976. La mariposa y sus estilizaciones en las culturas Teotihuacana (200 a 750 D.C.) y Azteca (1325 a 1521 D.C.). *Rev. Soc. Méx. Lepid.* 2(1):39–48
41b. de Megalhaes, P. S. 1902. A myiase dos bovideos na poesia patria. *Rev. Med. São Paulo* 5(3):49–50
41c. de Mirimonde, A. P. 1968. Psyche et le papillon. *L'Oeil* 168:2–11
41d. de Vries, A. 1974. *Dictionary of Symbols and Imagery*. Amsterdam: North-Holland. 515 pp.
41e. de Vuyst-Hendrix, L. M. 1978. Les abeilles devant les hommes et devant la loi. *Parc Natl.* 33(4):100–5
42. Dourojeanni, M. J. 1965, Denominaciones vernaculares de insectos y algunos otros invertebrados en la selva del Peru. *Rev. Peruana Entomol.* 8:131–37
43. Dow, R. P. 1915. The sweet singers of Pallas Athene. *Bull. Brooklyn Entomol. Soc.* 10:54–59
44. Dow, R. P. 1916. The testimony of the tombs. *Bull. Brooklyn Entomol. Soc.* 11:25–33
45. Dow, R. P. 1917. Studies in the Old Testament. I. The vengeful brood of Lilith and Samael. *Bull. Brooklyn Entomol. Soc.* 12:1–9
46. Dow, R. P. 1917. Studies in the Old Testament. *Bull. Brooklyn Entomol. Soc.* 12:64–69
47. Dow, R. P. 1918. The grasshopper of the Old Testament. *Bull. Brooklyn Entomol. Soc.* 13:25–30
48. Dow, R. P. 1918. Studies in the Old Testament. *Bull. Brooklyn Entomol. Soc.* 13:90–93
49. Driver, G. R. 1974. Lice in the Old Testament. *Palest. Explor. Q.* 1974:159–60
50. Egan, R. B. 1983. Cercopids and tettiges: the entomology of an Attic myth and cult. *Proc. Linguist. Circ. Manitoba North Dakota* 23:13–14
51. Egan, R. B. 1984. Jerome's cicada metaphor *(Ep.22.18)*. *Scholia* 1984:175–76
51a. *Entomol. News.* 1894. Insect's deadly sting. 5:16
51b. *Entomol. News.* 1969. A beetle boom is building in Japan. 80:66
51c. Explorama Tours. 1976. *The Peruvian Amazon. Legends, Indians, Plants, Birds,* pp. 1–4. Iquitos, Peru: Explorama Tours. 40 pp.
52. Faulkner, P. 1931. Insects in English poetry. *Sci. Mon.* 33:53–73, 148–63
53. Fewkes, J. W. 1910. The butterfly in Hopi myth and ritual. *Am. Anthropol.* (NS) 12:576–94
54. Fox-Davies, A. C. 1969. Insects. In *A Complete Guide To Heraldry*, pp. 195–96. London: Nelson. 513 pp.
55. Frazier, J. G. 1930. *Myths of the Origin of Fire*. London: Macmillan. 238 pp.
56. Free, J. 1978. Bees fashioned by a modern silversmith. *Bee World* 59(4):162–63
57. Freidus, A. S. 1917. A bibliography of Lilith. *Bull. Brooklyn Entomol. Soc.* 12:9–13
58. Friedmann, H. 1980. *A Bestiary for Saint Jerome. Animal Symbolism in European Religious Art*. Washington, DC: Smithsonian Inst. 378 pp.
59. Frost, S. W. 1937. The insect motif in art. *Sci. Mon.* 4:77–83
60. Frost, S. W. 1959. Insects in music. In *Insect Life and Insect Natural History*, pp. 64–65. New York: Dover. 526 pp. 2nd ed.
61. Fujita, M. 1902. Cicada-sound producing toy. *Konchū Sekai [Insect World]* 6:460–61 (In Japanese)
62. Gagliardi, R. A. 1976. *The butterfly and moth as symbols in Western art*. MS thesis. South. Conn. State Coll., New Haven. 199 pp.
63. Garcia, A. R. J. 1976. Nombre de algunos insectos y otros invertebrados en "Quechua." *Rev. Peru. Entomol.* 19:13–16
64. Gardiner, B. O. C. 1978. Decorative art. Butterflies. *Entomol. Rec. J. Var.* 90:249–50
65. Geijskes, D. C. 1975. The dragonfly wing used as a nose plug adornment. *Odonatologica* 4(1):29–30
66. Gerhardt, M. I. 1965. The ant-lion, nature study on the interpretation of a Biblical text, from the Physiologus to Albert the Great. *Vivarium* 3:1–23
67. Gil Fernandez, L. 1959. *Nombres de Insectos en Griego Antiguo*. Madrid: Inst. Antonio de Nebrija. 262 pp.
68. Gosswald, K. 1961. Insektenstaat und Menschenstaat. *Imkerfreund* 5:146–51
69. Grant, M., Hazel, J. 1973. *Gods and Mortals in Classical Mythology*. Springfield, Mass: Merriam. 447 pp.
70. Greenberg, B. 1973. Flies through history. In *Flies and Disease*, Vol. 2, *Bio-*

logy and Disease Transmission, pp. 3–18. Princeton, NJ: Princeton Univ. press 447 pp.

71. Gressitt, J. L. 1946. Entomology in China. *Ann. Entomol. Soc. Am.* 39:153–64

72. Gressitt, J. L. 1969. Oriental caged insects. *Entomol. News* 80:138

73. Grimal, P., ed. 1965. *Larousse World Mythology*. London: Hamlyn. 560 pp.

74. Grinnell, G. B. 1899. The butterfly and the spider among the Blackfeet. *Am. Anthropol.* (NS) 1:194–96

75. Guallart, J. M. 1968. Nomenclatura jibaro-aguaruna de la fauna del Alto Maranon (Invertebrados). *Biota* 7(56):195–209

76. Hackett, J. W. 1968. *Bug Haiku*. Tokyo: Japan Publ. 55 pp.

77. Hallander, J. 1980. The evolution of praying mantis kung fu. *Black Belt* 18(11):36–40

78. Hearn, L. 1898. Insect musicians. In *Exotics and Retrospectives*, pp. 39–80. Boston: Little, Brown 299 pp.

79. Hearn, L. 1900. Sémi. In *Shadowings*, pp. 71–102. Boston: Little, Brown. 268 pp.

80. Hearn, L. 1901. Dragon-flies. In *A Japanese Miscellany*, pp. 75–118. Boston: Little, Brown. 305 pp.

81. Hearn, L. 1921. *Insect Literature*. Transl. M. Ōtani. Tokyo: Hokuseido. 515 pp.

82. Hearn, L. 1926. *Insects and Greek Poetry*. New York: Rudge. 21 pp.

83. Heller, K. M. 1908. Verwendung von Insekten zu ethnographischen Gegenstanden. *Dtsch. Entomol. Z.* 1908:595–99

84. Herfs, A. 1963. Entomologica in Litteris. *Z. Angew. Entomol.* 1963:151–59

85. Herfs, A. 1973. Entomologica in Litteris. II. Imprimus de apibus. *Anz. Schädlingskd.* 46:33–37

86. Hitchcock, S. W. 1962. Insects and Indians of the Americas. *Bull. Entomol. Soc. Am.* 8(4):181–87

87. Hogue, C. L. 1975. The insect in human symbolism. *Terra* 13(3):3–9

88. Hogue, C. L. 1979. The bugfolk. *Terra* 17(4):36–38

89. Hogue, C. L. 1980. Commentaries in cultural entomology. 1. Definition of cultural entomology. *Entomol. News* 91:33–36

90. Hogue, C. L. 1981. Commentaries in cultural entomology. 2. The myth of the louseline. *Entomol. News* 92:53–55

91. Hogue, C. L. 1983. Commentaries in cultural entomology. 3. An entomological explanation of Ezekiel's wheels? *Entomol. News* 94:73–80

92. Hogue, C. L. 1985. Amazonian insect myths. *Terra* 23(6):10–15

93. Hogue, C. L., Lamas, G. 1986. La machaca, insecto mitológico o real? *Geomundo* In press

94. Hoogstraal, H. 1972. *Ixodes maloni* Doyle, 1912 (nomen nudum) (Ixodoidea: Ixodidae) parasitizing humans in Brazil. *Bull. Entomol. Soc. Am.* 18:141

95. Horn, W. 1937. Über einen Insekten-Gott der Chinesen. *Arb. Physiol. Angew. Entomol. Berlin-Dahlem* 4:67

96. Houghton, W. 1870. Classical allusions to cicadas. *Stud. Intellect. Obs.* 4:430–35

97. Hsiung, C.-C. 1984. Some insect books for children (in English). *Notes Lyman Entomol. Mus. Res. Lab.* 12:1–9

98. Hudson, T., Underhay, E. 1978. *Crystals in the Sky: An Intellectual Odyssey Involving Chumash Astronomy, Cosmology and Rock Art*. Socorro, NM: Ballena. 163 pp.

99. Hutchinson, G. E. 1974. Aposematic insects and the master of the Brussels initials. *Am. Sci.* 62:161–71

100. Hutchinson, G. E. 1978. Zoological iconography in the West after A.D. 1200. *Am. Sci.* 66:675–84

101. Ilingsworth, J. F. 1915. Use of cockroaches in medicine. *Proc. Hawaii. Entomol. Soc.* 3:12–13

102. Deleted in proof

103. Kamen-Kaye, D. 1979. A bug and a bonfire. *J. Ethnopharmacol.* 1:103–10

104. Keiser, I. 1966. Insects and related arthropods in heraldry. *Bull. Entomol. Soc. Am.* 12:314–8

105. Keller, O. 1963. *Die Antike Tierwelt*, 2:395–483. Hildesheim, Germany: Olms. 617 pp.

106. Kellogg, C. 1968 [1967]. *Entomological Excerpts from Southeastern China (Fukien Province). Aborigines: Silkworms, Honeybees and other Insects*. Claremont, Calif: Claremont Manor. 88 pp.

107. Kemper, H. 1959. *Die tierischen Schädlinge im Sprachgebrauch*. Berlin: Duncker & Humblot. 401 pp.

108. Kennedy, C. H. 1943. A dragonfly nymph design on Indian pottery. *Ann. Entomol. Soc. Am.* 36:190–91

109. Kevan, D. K. McE. 1974. (1973). Greetings from the entomological society of Canada and from the entomological society of Quebec. *Proc. Entomol. Soc. Ont.* 104:52–59

110. Kevan, D. K. McE. 1974. The land of the grasshoppers: being some verses on grigs. *Mem. Lyman Entomol. Mus. Res. Lab.* 2(Spec. Publ. 8):i–ix, 1–326

111. Kevan, D. K. McE. 1975. The hopper houses of Hamburg. *Insect World Dig.* 2(6):2–9

112. Kevan, D. K. McE. 1976. Ch'ien Hsuan, thirteenth century naturalist—the oldest known portrayal of predation by dragonflies? *Insect World Dig.* 1(6):26–28

113. Kevan, D. K. McE. 1978. The land of the locusts: being some further verses on grigs and cicadas. Part I. (Before 450 A.D.). *Mem. Lyman Entomol. Mus. Res. Lab.* 6:i–x, 1–530

114. Kevan, D. K. McE. 1980. Grigs, graces, graphics and graffiti: an essay on elements of ethnoentomology. *Metaleptea* 2(2):54–72

115. Kevan, D. K. McE. 1981. Remarks on insects and the humanities, or some human sides of entomology. *Bull. Entomol. Soc. Can.* 13:112–17

116. Kevan, D. K. McE. 1981. Utamaro's "Insect Book," 1788. *Notes Lyman Entomol. Mus. Res. Lab.* 9:1–37

117. Kevan, D. K. McE. 1981. Quadruped hexapods. *Antenna* 5:51–52

118. Kevan, D. K. McE. 1983. The land of the locusts: being some further verses on grigs and cicadas. Part II. (Between 450 and 1500 A.D.) *Mem. Lyman Entomol. Mus. Res. Lab.* 10:i–viii, 1–554

119. Kevan, D. K. McE. 1983. The place of grasshoppers and crickets in Amerindian cultures. *Proc. 2nd Trienn. Meet. Pan Am. Acridol. Soc., Bozeman, Montana, 1979,* pp. 8–74c. Bozeman, Mont: Pan Am. Acridol. Soc.

120. Kevan, D. K. McE. 1983. An historical review of knowledge of the orthopteroid insects of Canada and adjacent regions to 1850. *Mem. Lyman Entomol. Mus. Res. Lab.* 13(1):5–45

121. Kevan, D. K. McE. 1985. The land of the locusts: being some further verses on grigs and cicadas. Part III. The sixteenth to eighteenth centuries. *Mem. Lyman Entomol. Mus. Res. Lab.* 16:i–xiv, 1–466

122. Kevan, D. K. McE. 1985. The mantis and the serpent. *Entomol. Mon. Mag.* 121:1–8

123. Kevan, D. K. McE. 1986 (1985). Soil zoology, then and now—mostly then. *Quaest. Entomol.* 21:371.7–472

124. Kevan, D. K. McE., Hsiung, C.-C. 1976. Cricket-fighting in Hong Kong. *Bull. Entomol. Soc. Can.* 8(3):11–12

125. Kianta-Brink, M. A. J. E. 1976. Some Tibetan expressions for dragonfly with special reference to biological features and demonology. *Odonatologica* 5(2): 143–52

126. Kianta-Brink, M. A. J. E. 1977. An interesting comment on the Tibetan dragonfly expression bla.ma.ma.ni. *Odonatologica* 6(4):259–61

127. Knortz, K. 1910. *Die Insekten in Sage, Sitte und Literatur.* Annaberg, Germany: Grafer. 151 pp.

128. Kühn, H. 1935. Die Zikadenfibeln der Völkerwanderungszeit. *Ipek Jahrb. Prähist. Ethnogr. Kunst* 10:85–106

129. Latreille, P. 1819. Des insects peints ou sculptés sur les monuments de l'Egypte. *Mem. Mus. Hist. Nat.* 5:249–70

130. Laufer, B. 1927. Insect-musicians and cricket champions of China. *Anthropol. Leafl. Field Mus. Nat. Hist.* 22:1–27

131. Lenko, K., Papavero, N. 1979. *Insetos no folclore.* São Paulo: Cons. Estadual Artes Cienc. Hum. 518 pp.

132. Liebrecht, F. 1886. Tocandyrafestes. *Z. Ethnol.* 18:350–52

133. Liu, G. 1939. Some extracts from the history of entomology in China. *Psyche* 46:23–28

134. MacGregor, R. 1969. La représentation des insectes dans l'ancien Mexique. *Entomologiste Paris* 25:2–8

134a. Manganelli, G. 1986. A fly in the pigment. The ambiguous fly. *FMR* 4(19):81–84 (English ed.)

135. Marcovitch, S. 1949. The insect in literature. *J. Tenn. Acad. Sci.* 24:135–42

136. Martin, M. W. 1975. Insects on stamps. *Insect World Dig.* 2(1):17–20

137. Deleted in proof

138. Deleted in proof

139. Melis, A. 1958. La posizione sistematica ed allegorica degli insetti nella Divina Commedia (end). *Redia* 43:v–x

140. Mertins, J. W. 1986. Arthropods on the screen. *Bull. Entomol. Soc. Am.* 32:85–90

141. Meyer-Rochow, V. B. 1975. Local taxonomy and terminology for some terrestrial arthropods in five different ethnic groups of Papua New Guinea and Central Australia. *J. R. Soc. West. Aust.* 58:15–30

142. Meyer-Rochow, V. B. 1978/1979. The diverse uses of insects in traditional societies. *Ethnomedicine* 5:287–300

143. Miller, D. 1948. Shakespearean entomology. *Tuatara* 1(2):7–12

144. Miller, D. 1952. The insect people of the Maori. *J. Polynesian Soc.* 61:1–61

145. Deleted in proof

146. Montgomery, B. E. 1959. Arthropods and ancient man. *Bull. Entomol. Soc. Am.* 5:68–70

147. Mullett, G. M. 1980. *Legends of the Hopi Indians. Spider Woman Stories.*

Tucson, Ariz: Univ. Arizona Press. 142 pp.
148. Myer, I. 1894. *The Scarab.* New York: Nutt. 177 pp.
149. Nicholson, I. 1959. *Firefly in the Night.* London: Faber & Faber. 231 pp.
150. Nilsson, M. P. 1971. *The Minoan-Mycenaean Religion and its Survival in Greek Religion.* New York: Biblo & Tannen. 656 pp. 2nd ed.
151. Noble, J. V. 1975. The wax of the lost wax process. *Am. J. Archaeol.* 79:368–69
152. Deleted in proof
153. Osten Sacken, C. R. 1895. *Eristalis tenax* in Chinese and Japanese literature. *Berl. Entomol. Z.* 40:142–47
154. Osten Sacken, C. R. 1898. On the so-called Bugonia of the ancients and its relation to *Eristalis tenax,* a two-winged insect. *Boll. Soc. Entomol. Ital.* 25:186–271
155. Pager, H. 1971. Rock painting in southern Africa showing bees and honey hunting. *Bee World* 54:61–68
156. Patterson, R. 1841. *Letters on the Natural History of the Insects Mentioned in Shakespeare's Plays, with Incidental Notices of the Entomology of Ireland.* London: Orr. 270 pp.
157. Phillpotts, B. 1979. *The Book of Fairies.* New York: Ballantine. 16 pp.
158. Pieper, M. 1925. Die ägyptischen Scarabaen und ihre Nachbildungen in den Mittelmeerlandern. *Z. Ägypt. Sprach. Altertumskd.* 60:45–50
159. Pieper, M. 1930. Bedeutung der Scarabaen für die Palastinenische Altertumskunde. *Z. Dtsch. Palast.-Ver.* 53:185–99
160. Posey, D. A. 1978. Ethnoentomological survey of Amerind groups in lowland Latin America. *Fl. Entomol.* 61:225–28
161. Posey, D. A. 1983. Ethnomethodology as an *emic* guide to cultural systems: The case of the insects and the Kayapó Indians of Amazonia. *Rev. Brasil. Zool.* 1:135–44
162. Priest, A. 1952. Insects: the philosopher and the butterfly. *Bull. Metrop. Mus. Art* 10:172–81
163. Ransome, H. M. 1937. *The Sacred Bee in Ancient Times and Folklore.* Boston: Houghton Mifflin. 308 pp.
164. Reiche, M. 1949. *Mystery on the Desert.* Lima: Reiche. 67 pp.
165. Riegel, G. T. 1981. The cicada in Chinese folklore. *Melsheimer Entomol. Ser.* 30:15–20
166. Riley, W. A. 1919. A use of galls by the Chippewa Indians. *J. Econ. Entomol.* 12:217–18

167. Ritchie, C. I. A. 1979. *Insects, the Creeping Conquerors.* New York: Elsevier/Nelson. 139 pp.
168. Ritchie, J. M. 1977. An African grasshopper used as an ornament. *Entomol. Gaz.* 28:59–60
169. Rodeck, H. G. 1932. Arthropod designs on prehistoric Mimbres pottery. *Ann. Entomol. Soc. Am.* 25:688–94
170. Ross, E. 1937. Ueber Schmuckenkäfer und deren Verwendung bei verschiedenen Völkern. *Entomol. Z.* 50:457–66
171. Rudy, H. 1925. Die Wanderheuschrecke, *Locusta migratoria* L. phasa *migratoria* L. et phasa *danica* L. I. Die Wanderheuschrecke in Bild-dokumenten aus alter und altester Zeit. *Bad. Bl. Schädlingsbekämpf.* 1925:1–34
172. Rutschky, C. W. 1981. Arthropods in the lives and legends of the Pennsylvania Indians. *Melsheimer Entomol. Ser.* 30:39–42
173. Sauvage, A. 1970. Les insects dans la poésie romaire. *Latomus* 29:269–96
174. Savory, T. 1964. *Arachnida.* London: Academic. 291 pp.
175. Schimitschek, E. 1968. Insekten als Nahrung, in Brauchtum, Kult und Kultur. *Handb. Zool.* 4(2)1/10:1–62
176. Schimitschek, E. 1969. Der altägyptische Ordern der goldenen Fliege. *Anz. Schädlingskd. Pflanzenschutz* 42:73
177. Schimitschek, E. 1974. *Mantis* in Kult und Mythe der Buschmänner. *Z. Angew. Entomol.* 76:337–47
178. Schimitschek, E. 1977. Insekten in der bildenden Kunst, im Wandel der Zeiten in psychogenetischer Sicht. *Veröff. Naturhist. Mus. Wien.* 14:1–119
179. Schimitschek, E. 1978. Ein Schmetterlingsidol im Val Camonica aus dem Neolithikum. *Anz. Schädlingskd. Pflanz. Umweltschutz* 51:113–15
180. Schimitschek, E. 1980. Manna. *Anz. Schädlingskd. Pflanz. Umweltschutz* 53:113–21
181. Schlam, C. C. 1976. *Cupid and Psyche: Apuleius and the Monuments.* University Park, Pa: Am. Philos. Soc. 61 pp.
181a. Siganos, A. 1985. *Les Mythologies de l'Insecte, Histoire d'une Fascination.* Paris: Libr. Méridiens. 397 pp.
182. Silow, C. A. 1976. Edible and other insects of mid-western Zambia. Studies in ethnoentomology. II. *Occas. Pap. Inst. Allm. Jämför. Etnogr. Uppsala Univ.* 5:1–233
183. Silow, C. A. 1983. Notes on Ngangela and Nkoya Ethnozoology. Ants and termites. *Götesborgs Ethnogr. Mus. Etnol. Stud.* 36:1–vii, 1–177

184. Slosson, A. T. 1916. Entomology in literature. *Bull. Brooklyn Entomol. Soc.* 11:49–52

185. Smith, J. B. 1888. A wicked worm. *Entomol. Am.* 3(11):196–97

186. Smith, R. F., Mittler, T. E., Smith, C. N., eds. 1973. *History of Entomology.* Palo Alto, Calif: Annual Reviews. 517 pp.

186a. *Smithsonian.* 1977. A suppressed adventure of "Alice" surfaces after 107 years. 8(9):50–57

187. Smulyan, M. T. 1920. An insect and lack of entomological knowledge an immediate cause of the World War. *Psyche* 27:85–86

188. Sones, W. 1979. Our bug-infested language. *Explorer* 21(1):23

189. Southwood, T. R. E. 1977. Entomology and mankind. *Proc. Int. Congr. Entomol.* 15:36–51

190. Sowerby, A. C. 1940. *Nature in Chinese Art.* New York: Day. 203 pp.

191. Stanley, W. F., ed. 1979. *Insects and Other Invertebrates of the World on Stamps.* Milwaukee: Am. Topical Assoc. 148 pp.

192. Stempell, W. 1908. Die Tierbilder der Mayahandschriften. *Z. Ethnol.* 1908: 704–43

193. Stitt, R. 1981. The British insect pub fauna. *Antenna* 5(1):11–12

194. Stutley, M., Stutley, J. 1977. *Harper's Dictionary of Hinduism.* New York: Harper & Row. 372 pp.

195. Summers, I., ed. 1978. *Tomorrow and Beyond.* New York: Workman. 158 pp.

196. Swift, R. H. 1931. The sacred beetles of Egypt. *Bull. South. Calif. Acad. Sci.* 30:1–14

197. Tassi, R. 1978. *Graham Sutherland. Complete Graphic Work.* New York: Rizzoli. 228 pp.

198. Tastevin, C. 1923. Nomes de plantas e animais em Lingua Tupy. *Rev. Mus. Paulista* 13:687–763

199. Taylor, R. L. 1975. *Butterflies in my Stomach.* Santa Barbara, Calif: Woodbridge. 224 pp.

200. Taylor, R. L., Carter, B. J. 1976. *Entertaining with Insects. Or: The Original Guide to Insect Cookery.* Santa Barbara, Calif: Woodbridge. 159 pp.

201. Tod, M. N. 1939. The scorpion in Graeco-Roman Egypt. *J. Egypt. Archaeol.* 25:55–61

202. Tremblay, E. 1970. La lotta agli insetti nella magia e nel folklore. *Scienze* 6:325–31

203. Tsai, J. H. 1982. Entomology in the People's Republic of China. *J. NY Entomol. Soc.* 90:186–212

204. Tweedie, M. W. F. 1968. *Pleasure from Insects.* Newton Abbot, Devon, England: David & Charles. 170 pp.

205. Usinger, R. L. 1966. Introduction. In *Monograph of Cimicidae (Hemiptera—Heteroptera),* pp. 1–9. College Park, Md: Entomol. Soc. Am. 585 pp.

206. van Buren, D. D. 1937. The scorpion in Mesopotamian art and religion. *Arch. Orientforsch.* 12:1–28

207. Villiard, P. 1973. *Insects as Pets.* New York: Doubleday. 143 pp.

208. Vogh, J. 1977. *Arachne Rising. The Thirteenth Sign of the Zodiac.* London: Granada. 202 pp.

209. Deleted in proof

210. Wälchli, G., ed. 1940. *Martin Disteli, 1802–1844, Romantische Tierbilder zu Fabeln und Versen von A. E. Fröhlich, J. W. v. Goethe, A. Hartmann, F. Krutter und C. Rollenhagen.* Zürich: Amstutz & Herdeg. 120 pp.

211. Walton, W. R. 1922. The entomology of English poetry. *Proc. Entomol. Soc. Wash.* 24:159–206

212. Ward, W. A. 1978. *Studies on Scarab Seals,* Vol. 1, *Pre-12th Dynasty Scarab Amulets.* Warminster, England: Aris & Phillips. 116 pp.

213. Watson, A., Whalley, P. E. S. 1975. *The Dictionary of Butterflies and Moths in Color.* New York: McGraw-Hill. 296 pp.

214. Weidner, H. 1952. Insekten im Volkskunde und Kulturgeschichte. *Arbeitsgem. Mus. Schleswig Hollstein Niederschr., Rendsburg, 1950,* pp. 33–45. Kiel: Schleswig Hollstein Mus. (In German)

215. Weiss, H. B. 1912. Some ancient beliefs concerning insects. *Bull. Brooklyn Entomol. Soc.* 8:21–23

216. Weiss, H. B. 1927. The scarabaeus of the ancient Egyptians. *Am. Nat.* 61:353–69

217. Weiss, H. B. 1930. Insects and witchcraft. *J. NY Entomol. Soc.* 38:127–33

218. Weiss, H. B. 1945. Some early entomological ideas and practices in America. *J. NY Entomol. Soc.* 53:301–8

219. Welling, E. C. 1958. Some Mayan names for certain Lepidoptera in the Yucatan Peninsula. *J. Lepid. Soc.* 12:118

220. Werner, D. 1970. Healing in the Sierra Madre. *Nat. Hist.* 79(8):60–67

221. Wilbert, J. 1985. The house of the swallow-tailed kite: Warao myth and the art of thinking in images. In *Animal Myths and Metaphors,* ed. G. Urton, pp. 145–82. Salt Lake City: Univ. Utah Press. 327 pp.

222. Wilkinson, R. W. 1969. Colloquia entomologica. II: a remarkable sale of Victorian entomological jewelry. *Mich. Entomol.* 2:77–81

223. Wright, B. 1977. Insect and reptile Kachina dolls. In *Hopi Kachinas. The Guide to Collecting Kachina Dolls,* pp. 116–21. Flagstaff, Ariz: Northland. 139 pp.

224. Wyman, L. C. 1973. *The Red Antway of the Navaho.* Santa Fe, NM: Mus. Navaho Ceremonial Art. 276 pp.

225. Wyman, L. C., Bailey, F. L. 1964. Navajo Indian ethnoentomology. *Univ. NM Publ. Anthropol.* 12:1–158

226. Wyndham, R. J. 1960. Strange superstitions about bees and honey. *Glean. Bee Cult.* 88:723–27

Ann. Rev. Entomol. 1987. 32:201–24
Copyright © 1987 by Annual Reviews Inc. All rights reserved

BIOLOGY OF *LIRIOMYZA*

Michael P. Parrella

Department of Entomology, University of California, Riverside, California 92521

PERSPECTIVES AND OVERVIEW

The genus *Liriomyza,* erected in 1894 (52), contains more than 300 species. They are distributed widely but are most commonly found in temperate areas; there are relatively few species in the tropics. Within this genus 23 species are economically important, causing damage to agricultural and ornamental plants by their leafmining activity (94). Many of these damaging species are polyphagous, which is uncommon among the Agromyzidae; of 2450 described species in this family only 11 are considered to be truly polyphagous, and 5 of these are in the genus *Liriomyza* (93).

"Serpentine leafminer" was proposed as a common name for any member of this genus because of the wide distribution, polyphagous nature, and morphological similarity of many of the species (98). The Entomological Society of America has adopted this naming policy, with a few exceptions (112). Indeed, many larvae of *Liriomyza* create serpentine mines, which are initially very narrow and gradually enlarge (57), often twisting through the leaf. However, the type of mine produced by *Liriomyza* may be influenced by the developmental stage of the leaf as well as by the host itself (99). Thus, the mines are not always serpentine in all host plants. In addition, mine location in leaves may vary considerably, and either the upper or lower leaf mesophyll may be mined (5, 71). Some species have larval stages that feed in potato tubers (50), bore through stems (34), and feed within seed heads (86).

Most of our knowledge concerning the biology of this genus comes from studies on economically important species. These data have been developed largely since 1900, with an explosion of information since 1975 (76). This reflects the dramatic rise of *Liriomyza* spp. as major pests of numerous ornamental and agricultural crops over the past ten years (72). As a result of

201

0066-4170/87/0101-0201$02.00

this rise in economic importance, the proceedings from three formal conferences (80, 81, 87) and one informal conference (35) have been published. These, together with published reviews and bibliographies (38, 64, 76, 95, 95a), provide a good starting point for those interested in this genus. A comprehensive review of all economically important Agromyzidae (94) is an outstanding contribution to the biology, ecology, taxonomy, and control of these leafminers.

Out of necessity, this article contains general overviews of various biological parameters within this genus. Where specific studies have been done, these are cited. It is intended that this article not be an exhaustive review of the literature, but rather a selective one covering those studies that best address specific biological aspects of the genus.

ADULTS

Emergence

Adults emerge through the dorsal anterior end of the puparium (the retained last larval integument within which pupation occurs) with the aid of the ptilinum (a temporary bladderlike inflatable structure). This process may take from 5 min to more than 1 hr. Some mortality may occur during this process (61). Newly emerged adults exhibit a positive phototactic response and climb up the sides of a cage or up the stalk of a plant, where they remain quiescent for a period of approximately 20 min while expanding their wings and body. The body is fully sclerotized and colored within 20 min–2 hr (16, 61). Adult females are usually larger than males and emerge from larger puparia (61, 65). Puparium size is positively correlated with adult vigor (65). Males appear to emerge prior to females (J. Yost & M. P. Parrella, unpublished); both sexes generally emerge during early morning hours (61). The time of day of peak emergence varies for different species (5). Studies of sex ratios of adults emerging from pupae indicate a 1 : 1 sex ratio (61, 96) or a slight bias in favor of females (2, 5). Intensive laboratory rearing of *L. trifolii* over the past five years has produced approximately a 1 : 1 sex ratio of emerging adults (M. P. Parrella, unpublished).

Premating and Preoviposition

The majority of adults mate soon after emergence, and almost all females have mated within 24 hr (61, 69, 78). The period of time between adult emergence and mating, i.e. the premating interval, appears to be inversely related to temperature (16) and may differ for the sexes (96). The sexes may remain coupled for as little as 10 min (96), but the norm is 30 min–1 hr. Maximum mating time is about 3 hr. Males and females mate more than once, and multiple matings by the female are needed for maximum egg production

(61). Mating can usually be observed at any time of the day, but it generally occurs during early morning hours (16, 61).

It has been suggested that food, temperature, and relative humidity influence the preoviposition period (16), which may extend up to 5 days after adult emergence. Under greenhouse and laboratory conditions, most females begin oviposition within 24–48 hr after emergence (61, 78).

Mating

During copulation, the male assumes a position behind and alongside the female at about a 45° angle above her body. Occasionally, mounting from the front occurs. In the more typical position, the male's forelegs clasp the mesothorax of the female, his middle legs clasp the female's abdomen, and his hind legs spread the female's wings. The wings of the male are held normally over the body. The male brings his abdomen forward and downward to connect to the female genitalia as the male's hind legs move to rest on the substrate. This position is maintained throughout copulation (61; J. A. Bethke & M. P. Parrella, unpublished). No sex pheromone has been reported by researchers working with *Liriomyza;* however, it is possible that a stridulatory organ present in some males of *Liriomyza* spp. may be used in short-distance vocalizations to attract mates (95). Indeed, the rapid bobbing of males of *L. trifolii* when in close association with females may be a physical manifestation of this auditory signal (J. A. Bethke & M. P. Parrella, unpublished).

Aggressive behavior by male *L. trifolii* during mating has been observed in the laboratory under severely crowded conditions (J. A. Bethke & M. P. Parrella, unpublished). Upon the approach of a rival male, the coupled male will continually flex his wings until the intruder leaves.

Feeding and Oviposition Behavior

Excellent descriptions of leaf puncturing, feeding, and oviposition are available (7, 16, 61, 96, 111). The following description is from observations with *L. trifolii* (7), but similar behavior has been noted for *L. sativae* and *L. huidobrensis* (J. A. Bethke & M. P. Parrella, unpublished). When a female initiates a leaf-puncturing sequence, the first event observed, regardless of host plant, is a bending of the abdomen to position the ovipositor perpendicular to the leaf. The ovipositor contacts the leaf through a series of rapid thrusts. Once the ovipositor has penetrated the leaf surface, the thrusts becomes slower and more deliberate. At this point the female damages mesophyll cells in a specific manner, creating one of two different types of leaf punctures. If the abdomen is twisted from side to side, a large fan-shaped leaf puncture is created. A tubular leaf puncture is produced when no abdominal twisting follows the puncture. Eggs are deposited in tubular leaf punctures. The difference between oviposition behavior and the creation of a tubular leaf puncture without an egg is subtle; oviposition entails a pause in slow thrusting followed by a final thrust to deposit an egg. After every leaf puncture the female backs over the wound and feeds from it. The female feeds from all punctures, regardless of whether or not they are used for oviposition. Hence, all leaf punctures can be considered feeding punctures (7); this should clarify considerable confusion in the literature. Males are unable to create their own punctures, but, as many authors indicate (56, 61), they feed from punctures created by females. Leaf puncturing can reduce photosynthesis (71) and may kill young plants (17). Leaf puncture size varies with the size of the adult female (61).

Feeding and oviposition by adults appear to occur primarily during the morning, and the frequency of these activities is positively correlated with temperature (20, 66). Little adult activity is observed after 1800 hr (20, 61, 66, 78). Leaf puncturing may occur with equal frequency on the abaxial and adaxial leaf surfaces (69), but this may depend on the species.

Leaf puncturing and feeding by adult *Liriomyza* undoubtedly serves an important role in host plant assessment. It has been suggested (7) that host feeding is more important in this regard than leaf puncturing. Several researchers have examined the ratio of total punctures to oviposition punctures in an attempt to determine host plant suitability or a general biological characteristic of *Liriomyza* spp. (20, 61, 115). These ratios have ranged from 1:1 to 40:1 and vary with temperature (66), leaf quality, and host plant. Unless the leaf area exposed to flies (126) and the number of flies released onto the plant can be held constant, these ratios are of little value.

Egg-laying capacity varies considerably within the genus *Liriomyza*. Mean egg production per female ranges from less than 100 (25) to greater than 600 (61). Females generally lay the majority of eggs between days 4 and 10 of adult life, depending on temperature (48, 66, 69, 78). Fecundity is strongly related to food source and temperature (16, 61, 66, 78); maximum oviposition occurs between 20–27°C (16, 66) when a constant food source such as honey is provided. Unfertilized females oviposit hundreds of eggs that fail to develop, although ovarian development, egg laying, and other responses appear to be normal (61). Some fertilized females oviposit infertile eggs. Many studies report that individuals of certain *Liriomyza* spp. do not create feeding punctures or lay eggs (5, 61, 78).

Longevity

Most longevity studies have been conducted using caged flies in close association with a host plant or carbohydrate food source (16, 61, 66, 69, 78). Under these conditions, females live 15–20 days and males 10–15 days. Longevity generally decreases at higher temperatures; the presence of honey dramatically increases longevity. Although no studies have examined the longevity of these flies in the field because of the difficulty associated with studies on individual flies in nature, most laboratory studies have probably overestimated the normal longevity (as well as fecundity) of adults. In laboratory rearing studies (K. Heinz and M. P. Parrella, unpublished) we observed that increasing the size of the rearing container tended to shorten the longevity of *L. trifolii*. In addition, most longevity studies have not considered the importance of the times of adult emergence. For example, the deposition of eggs in leaf material during a 3-hr interval (e.g. during laboratory production) resulted in pupation over a 24-hr period and subsequent adult emergence over a 6-day span. When survivorship of adults emerging from these pupae was examined as cohorts based on the day of adult emergence, different survivorship profiles were produced (K. Heinz & M. P. Parrella, unpublished) (Figure 1). It is possible that a short larval development time may be correlated with adult vigor.

Adults are able to withstand freezing temperatures for short periods (62), so adults that emerge during warmer periods of winter in nearctic and palearctic regions may survive. *Liriomyza trifolii*, which is native to the southern part of the Nearctic, was considered incapable of overwintering in more northern areas. However, the survival of adults and pupae at low temperatures suggests that this species may be able to survive in these areas (53, 99). This factor and others may contribute to their colonization in these habitats.

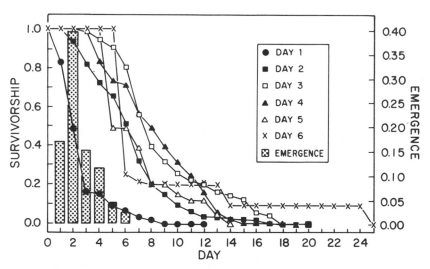

Figure 1 Percent emergence and survivorship of adult *L. trifolii* from 500 pupae. All emerged and pupated within a 24-hr period. Adults emerged over a 6-day period.

EGGS

The whitish, transluscent egg is deposited through the adaxial or abaxial leaf surface. The egg varies in size based on the size of the species: e.g. 0.25 mm × 0.10 mm for *L. congesta* (16) and 0.28 mm × 0.15 for *L. huidobrensis* (2). *Liriomyza* eggs may be confused with thrips eggs oviposited in leaf tissue (54). An egg-staining technique has been developed for detecting eggs of *L. trifolii* in celery, chrysanthemum, and tomato leaves (75). Eggs are laid singly, but often in close proximity to each other. No epideictic (oviposition-deterring) pheromone [present in other Agromizidae (49)] has been discovered in *Liriomyza* (7). Egg nonviability has been estimated to be as high as 20% (5) and is dependent on temperature (16). The eggs increase in size after oviposition, possibly through the imbibition of fluids from plant tissue (16, 102). The period of egg development varies with temperature and ranges from 2–8 days. There appears to be considerable variation in the relationship between temperature and development and in developmental threshold (6.2–13.4°C; Table 1), probably because of differences in species, host plants, and experimental methodology. Sixteen days at 1.1°C is required to cause 100% mortality of *L. trifolii* eggs in celery (42).

As the eggs develop they become opaque, and gradually the brownish cephalopharyngeal skeleton can be differentiated (5). When about to hatch, the larva is oriented with its anterior extremity, which contains the mouth-hook, at the terminus of the egg furthest from the original oviposition puncture made by the female (96). This position results from a 180° rotation of the embryo, whereby the anterior cephalic end moves toward the posterior end of the egg (5). In some species, the larva may eat the eggshell before moving into the leaf mesophyll (5). Pressure exerted by the larva causes the eggshell to become distended longitudinally and eventually to split at its anterior end.

LARVAE

The larva begins feeding immediately after eclosion and feeds incessantly until it is ready to emerge from the leaf (111). Different species of *Liriomyza* feed in different sections of the leaf mesophyll [e.g. *L. trifolii* in the palisade mesophyll, *L. huidobrensis* in the spongy mesophyll (71), and *L. brassicae* in the palisade and spongy mesophyll (94)]. Nonetheless, when larvae are forced to compete for resources because of crowding, they may tunnel into leaf stalks and into the main stem of the plant (96).

The larva is somewhat cylindrical and maggotlike. The anterior end tapers and the posterior end is truncate. Larvae move via peristaltic action of their hydrostatic skeleton. There are four molts and four larval instars. The fourth

Table 1 Temperature-development studies with the egg stage of *Liriomyza* spp.

Species of *Liriomyza*	Host	Development rate regressed on temperature[a]	Estimated threshold temperature for development (°C)	Reference
L. congesta[b]	*Vicia faba*	$y = 0.0355x + 0.529$	14.9	(16)
L. sativae[c]	*Phaseolus* sp.	$y = 0.0214x + 0.133$[d]	6.2	(62)
L. sativae	*Phaseolus* sp.	$y = 0.0166x - 0.116$	7.0	(109)
L. trifolii	*Apium graveolens*	$y = 0.0343x - 0.441$	12.8	(41)
L. trifolii	*Phaseolus* sp.	$y = 0.0266x - 0.266$	10.0	(13)
L. trifolii	*Chrysanthemum morifolium*	$y = 0.0262x - 0.352$[d]	13.4	(13)

[a]Simple linear regression with temperature (x)(°C) = independent variable and development rate (y) (1/day) = dependent variable. The x-intercept method was used to calculate development thresholds.

[b]Reported as *L. trifolii*, but probably *L. congesta* (78).

[c]Reported as *L. pictella*, but probably *L. sativae* (94).

[d]Upper temperatures not used in calculating the regression.

Table 2 Temperature-development studies with the larval stages of *Liriomyza* spp.

Species of *Liriomyza*[a]	Host	Development rate regressed on temperature[a]	Estimated threshold temperature for development (°C)	Reference
L. congesta	Vicia faba	$y = 0.0131x - 0.165$	12.6	(16)
L. sativae	Phaseolus sp.	$y = 0.00657x - 0.0305$[b]	4.6	(109)
L. sativae	Phaseolus sp.	$y = 0.0133x - 0.105$	7.9	(62)
L. trifolii	Apium graveolens	$y = 0.007x - 0.0587$	8.4	(41)
L. trifolii	Phaseolus sp.	$y = 0.0130x - 0.111$	8.5	(13)
L. trifolii	Chrysanthemum morifolium	$y = 0.00509x - 0.0313$	6.1	(9)
L. trifolii	Lycopersicon esculentum	$y = 0.0118x - 0.0926$	7.8	(91)

[a]See footnotes on taxonomy and methodology, Table 1.
[b]All development stages combined for regression analysis.

Table 3 Temperature-development studies with the pupal stage of *Liriomyza* spp.

Species of *Liriomyza*[a]	Host	Development rate regressed on temperature[a]	Estimated threshold temperature for development (°C)	Reference
L. congesta	*Vicia faba*	$y = 0.00662x - 0.0539$	8.1	(16)
L. sativae	*Phaseolus* sp.	$y = 0.00651x - 0.0631$	9.7	(62)
L. trifolii[b]	*Apium graveolens*	$y = 0.00749x - 0.0771$	10.3	(104)
L. trifolii	*Apium graveolens*	$y = 0.00760x - 0.0779$	10.3	(41)
L. trifolii	*Phaseolus* sp.	$y = 0.00662x - 0.0529$	8.0	(13)
L. trifolii	*Chrysanthemum morifolium*	$y = 0.00600x - 0.0539$	9.0	(77)
L. trifolii	*Dendranthema* sp.	$y = 0.00691x - 0.0771$	10.4	(53)

[a]See footnotes on taxonomy and methodology, Table 1.
[b]Reported as *L. sativae*, but probably *L. trifolii*.

Table 4 Temperature-development studies with various combined stages of *Liriomyza* spp.

Species of *Liriomyza*[a]	Life stages	Host	Development rate regressed on temperature[a]	Estimated threshold temperature for development (°C)	Ref.
L. sativae	Egg–adult	*Phaseolus* sp.	$y = 0.00381x - 0.0352$	9.2	(62)
L. sativae	Egg–pupa	*Phaseolus* sp.	$y = 0.00310x - 0.0195$	6.1	(109)
L. trifolii	Egg–adult	*Phaseolus* sp.	$y = 0.00369x - 0.0311$	8.4	(13)
L. trifolii	Egg–adult	*Chrysanthemum morifolium*	$y = 0.00310x - 0.0195^{b}$	6.3	(13)
L. trifolii	Egg–adult	*Apium graveolens*	$y = 0.00281x - 0.0229$	8.1	(41)
L. trifolii	Egg–pupa	*Dendranthema* sp.	$y = 0.00728x - 0.0786$	10.1	(53)

[a]See footnotes on taxonomy and methodology, Table 1.
[b]Upper temperatures not used in calculating the regression.

instar occurs between puparium formation and pupation and is rarely discussed by most authors. Black sclerotized mouthhooks are left within the mine after the molt and can be used to distinguish the duration of the instars (100), because there are distinct mouthhook sizes for each larval instar. Other researchers have used mine width as an indication of instar duration (109). Because interpretation may vary, the method used to determine larval instar duration must be known; unfortunately, many reports do not explain precisely how this was done. Because of the difficulty in separating larval instars, many authors combine all instars and simply refer to larval duration (Table 2). Authors commonly combine egg and larval development to examine the total time of the immature stages spent within the leaf (see Table 4).

Larval development varies with temperature and host plant (Table 2). In addition, larval development time on a single host varies considerably with leaf position and age, but few authors take this into consideration. Many studies can also be criticized for lack of detail in explaining how development times were calculated (e.g. starting and stopping points, time of egg laying, sampling frequency). These points are especially critical for consistent calculations of the total larval duration of many *Liriomyza* spp., considering that it can be as short as 4–6 days at field/greenhouse temperatures (20, 41). Only one attempt to correlate laboratory development with the development of field populations has been reported (53); the development period for eggs and larvae in the laboratory, 147.5 day-degrees above 10.1°C, corresponded well to the observed development in the field.

As the larvae develop, both the diameter of the mine and the rate of mine formation increase (20, 61). For *L. trifolii,* the volume of leaf material consumed by the third stage larva is 643 times greater than that consumed by the first stage larva, and the feeding rate is 50 times greater (20). Mining causes a reduction in leaf photosynthesis, with the amount of reduction varying according to mine location (71). Metabolic rates of larvae double for every 10°C increase in temperature, but the total amount of leaf tissue consumed by larvae apparently remains the same regardless of temperature (9).

Interspecific competition with mites and aphids may inhibit population increases of *Liriomyza* (59) by affecting larval development. Intraspecific competition in both the field and laboratory may reduce survivorship as well as size of larvae and pupae, and may thus reduce adult vigor (59, 65).

When the larva is ready to pupate it cuts a semicircular slit in the leaf surface, usually at or near the end of the mine. This slit may be located on the upper or lower leaf surface, but depends on the mining location of the larva within the mesophyll. The larva emerges with characteristic peristaltic locomotion. When it is three-fourths out of its mine, the anterior portion waves about high above the leaf surface and the larva literally falls out (61).

Movement outside the mine is the same as within and is accompanied by a rolling motion, which usually forces the larva to fall from the leaf to the ground. Larvae occasionally pupate on leaves or at the base of leaves, stems, or stalks, but this is more common on plants with large curled leaves (squash, gerbera, etc). Larvae emerge from leaves during early daylight hours (61), with the majority of emergence occurring before 0800 hr.

PREPUPAE AND PUPAE

The period between larval emergence and puparium formation is generally referred to as the prepupa. The prepupal period is about 2–4 hr (41, 61), but varies considerably with temperature. The prepupa is negatively phototactic and positively thigmotactic. This stage is extended when the prepupa is exposed to constant light in bare containers (62).

The duration of the pupal stage varies inversely with temperature (Table 3), but at least 50% of the total development time of a *Liriomyza* individual is spent in this stage. Total development time of the pupa at greenhouse/field temperatures is about 8–11 days. Estimated temperature thresholds for development are generally consistent for pupae (Table 3), unlike those for eggs and larvae. This is probably due to more standard estimation methods, because the pupae are exposed and easier to observe. Relative humidity between 30 and 70% is optimum for pupation. It has been suggested that the substrate in which pupation occurs influences successful development to the adult stage (62, 63). Pupal weight and development time and percent emergence of adults from pupae appear constant regardless of host plant (62, 128).

Pupae of *L. trifolii* have exhibited a diapause at 16°C in Italy (99). While this diapause has not been observed in Britain, pupae can remain viable outdoors for several months and are able to withstand freezing temperatures (53). Thus, outdoor populations of *Liriomyza* may survive long enough to reinfest subsequent glasshouse crops.

HOST-PLANT INTERACTIONS, MOVEMENT, AND DISPERSION

Adults

Because the larvae of *Liriomyza* are unable to leave one leaf and enter another, the ultimate choice of host selection rests with the ovipositing adult female. This fact has led to numerous studies of adult preference. Unfortunately, only those studies in which an effort was made to standardize the type, age, and size of the leaves offered to adult flies (126) are of value.

Adult females exhibit distinct preferences for host plants, although their feeding and oviposition behavior remains stereotypic regardless of host (7).

The distribution and density of plant trichomes, the phenolic content, and the nutritional value of hosts were found to influence host selection (19, 26, 36). Little quantitative data are available on within-plant preferences of egg-laying adults (3, 30, 88).

The basis for host-plant preference may be genetic. A comparison of laboratory-selected and wild populations of *L. brassicae* suggests an increased tendency to oviposit in the host that this species develops on (101).

Few studies on seasonal abundance and distribution of *Liriomyza* spp. have been performed in wild or agricultural systems where more than one species of plant or one species of *Liriomyza* may be present. Research in this area has involved *L. trifolii* and *L. sativae,* which preferred celery and tomato, respectively (121, 122). There is evidence that *L. trifolii* may be replacing *L. sativae* in tomato (89) and gypsophila (83).

Liriomyza adults have long been known to be attracted to yellow cards, and with the application of adhesive these may be very effective monitoring tools (55) and may be useful in studies on movement and dispersal (31). Studies have confirmed that yellow is more attractive to adults than other colors (105), and that high reflectance through the yellow part of the spectrum increases catch (1, 11). Trap location with respect to the crop influences the number and species of flies captured (12, 123). The variance/mean relationship of *L. trifolii* and *L. sativae* trapped on yellow cards has been shown to be very consistent over time in chrysanthemum and tomato (70, 124).

An understanding of the movement and dispersal of economically important *Liriomyza* has been the objective of numerous studies over the past 35 years. In agricultural fields the within-field spread of leafminers begins slowly at first, generally originating in weed hosts in borders adjacent to field crops (22). Prevailing winds influence the rate and direction of dispersal from the center of origin. Densities are greatest at the point of origin and generally decrease with distance from the source (105, 116). In the greenhouse, where wind was not a factor, the mean distance flown by female flies (21.5 m) was greater than that flown by males (18.0 m) over a 7-day period (31). In addition, it was shown using a generalized distance dispersal model that density decreases more rapidly with distance for males than for females, and that males have slightly more aggregated distributions than females. Data on the sex ratio of adult *Liriomyza* caught on sticky yellow cards vary (12, 31), but most studies show capture of more males than females (11, 123). Based on the fact that females tend to live longer than males, one would expect that more females than males would be caught on yellow traps over a given period of time. Thus the data collected to date suggests differential attraction between the sexes.

The movement and dispersal of adults may be affected by aluminum-foil mulch around tomato and squash plants. This has been shown to reduce infestations of *Liriomyza* (119), presumably by repelling adult flies. Insecticides have also been shown to repel adults, although effectiveness varies with the chemical and method of application (85).

Larvae

In most cases, experimentally transplanted agromyzid larvae develop on plants phylogenetically related to their natural host and die on plants not related to the normal hosts (10). Some *Liriomyza* species, however, were successfully transferred among Compositae (*Eupatorium* sp.), Urticaceae (*Cannabis* sp.), and Labiatae (*Galeopsis* spp.). Larvae may be far less sensitive to repellent or deterrent chemicals than adults. This stage has evolved a completely parasitic mode of life within the plant; thus larvae have poorly developed sense organs, and hence little ability to discriminate among various host species (92). Consequently, if a female oviposits in plants outside its normal range, there is a possibility that the larva could complete development. This may offer one avenue of host-range expansion in *Liriomyza*.

The possibility of sympatric genetic divergence in *L. brassicae* was investigated when genetic variation in survival and development time was demonstrated for three strains collected from different host-plant species (101). Formal quantitative genetic studies of *L. sativae* have been undertaken to determine the amount of genetic variation and to correlate this variation with parameters of host-plant utilization (107, 108). These techniques may be useful in the examination of host races in *Liriomyza* and other genera. A host race is defined as a group of individuals that genetically differs in host plant–related characters from individuals on other hosts. Individuals of the host race do not interbreed with individuals from other hosts because of divergent host preferences (107, 108). Genotype-environment interactions may be examined through correlations of various fitness components (e.g. development time) with environmental variables. These correlations provide estimates of local differentiation and the potential for future evolutionary change (108). Examination of *L. sativae* over a variety of adjacent crops revealed a significant genotype-environment interaction for development time within populations, which suggests that selection on individuals residing on a crop could lead to host-plant specialization at the species level. The absence of host races in *L. sativae* has been attributed to frequent migration among closely spaced crops and to crop rotation (107). This promotes interbreeding and prevents groups from becoming isolated, which is necessary for the formation of races. Thus in the absence of agricultural manipulations or in large agricultural monocultures, host races may develop in *Liriomyza* (93). Monoculture may already have been responsible for the development of a host race in *L. sativae* isolated on melons (misidentified as *L. pictella*) (61, 94).

Most members of the Agromyzidae are homogametic (8). They generally have obscure polytene chromosomes, which may be why so few species have been studied cytologically (8). Only *Liriomyza urophorina* has been investigated (46, 47); it showed chromosomal polymorphism for six paracentric inversions.

In chrysanthemum, celery, and tomato the distribution of larvae and pupae from field samples is generally clumped (3, 30, 88, 124). When the variance/mean relationship is examined for larvae using Taylor's power law, the *a* (a sampling factor) and *b* (a species-specific aggregation constant) values calculated for celery (3) and chrysanthemum (30) are very similar. This suggests that further work on celery may lead to sampling plans similar to those developed on chrysanthemum (30).

Leafminers respond favorably to high nitrogen content in leaves (23, 82, 120). Most of the data stem from studies in which nitrogen content is varied by manipulation of general fertilizer regimes. In many cases it is difficult to tell whether mines were fewer with lower fertilizer concentrations because of poor larval survival or reduced egg-laying by adults. A similar problem is encountered when one reviews the large body of information in which cultivars of one plant species are ranked as to sensitivity or resistance to leafminer damage (33, 90, 110).

ECONOMIC IMPORTANCE

The economic impact of *Liriomyza* leafminers in the United States and throughout the world has been considerable; in California alone it was estimated that the chrysanthemum industry lost approximately 93 million dollars to *L. trifolii* from 1981 through 1985 (57a).

Liriomyza leafminers can impact crops in at least six ways: (*a*) by vectoring disease (127), (*b*) by destroying young seedlings (17), (*c*) by causing reductions in crop yields (39, 117), (*d*) by accelerating leaf drop above developing tomatoes, thus causing "sunburning" of the fruit (26a), (*e*) by reducing the aesthetic value of ornamental plants (67), and (*f*) by causing some plant species to be quarantined (42, 54, 69, 72). While the results of most of these six types of damage are obvious [e.g. heavy mining and stippling in young seedlings or transplants can kill a plant and/or dramatically slow growth (17, 103)], it has been difficult to accurately associate specific levels of mining activity with reductions in crop yield. Reductions in photosynthesis and other physiological parameters have been measured in vegetable crops (29, 103) but have not been correlated to yield loss. Studies have shown that greenhouse-raised tomatoes can tolerate high levels of damage by *L. sativae* without suffering appreciable losses in yield (43). However, in a recent study (39), *L. bryoniae* on tomato caused yield loss, which was greatest when mining occurred in the leaves close to a young, developing fruit. Predetermined damage thresholds have been incorporated into sequential sampling plans for tomato (28, 114), but these plans have not received wide acceptance. An experimental threshold on *L. sativae* in southern coastal pole-tomato fields in California calls for treatment when an average of 10 pupae per sampling tray per day accumulate over a 3–4 day period (26a).

RESPONSE TO INSECTICIDES

Numerous articles have focused on chemical control of *Liriomyza* leafminers (76) because of their potential for causing damage. Control with insecticides is usually complicated by the insect's biology, i.e. fast development time; smallness and high mobility of adults; a relatively long pupal stage occurring in the soil; high reproductive capability; and egg and larval stages within and protected by leaf tissue. In addition, a mine created by the larva remains in the leaf as long as the leaf survives; thus insecticide application may have little use in preserving the aesthetic value of ornamentals or for preventing yield reduction of vegetables. Insecticide applications have commonly been responsible for outbreaks of *Liriomyza* because the insecticides used are often more toxic to the large parasite complex holding these leafminers in check than to the leafminers themselves (60). Thus it is very important to base spray application on accurate damage thresholds. Much more research is needed in this area. Another possible reason for these outbreaks is that leafminers that receive a sublethal insecticide dose may be physiologically stimulated and thus may cause more damage (60). This theory has remained untested, and not all sublethal effects of insecticides are stimulatory (84).

An important part of the biology of *Liriomyza* is the ability to develop resistance to insecticides. Insecticide resistance has been responsible for failure to control these leafminers for many years (21, 72, 118). However, definitive studies documenting insecticide resistance have only been done with *Liriomyza trifolii* (32, 74), and more work with this and other species is needed. Research must be conducted on resistance, cross resistance, genetics of resistance, biological attributes of resistant populations, and duration of resistance in the absence of insecticide selection pressure. The insecticide-resistance capability of *L. trifolii* has been speculated as a mechanism in its gradual replacement of *L. sativae* as the primary leafminer in several crops (83, 89, 122). It has been demonstrated that *L. trifolii* is more tolerant of insecticides than several other Agromyzid species (45, 73), but much more comparison of different species of *Liriomyza* is needed. Haynes et al (23a) have developed a bioassay with insecticide-laced yellow sticky cards that allows rapid assessment of resistance levels in field populations of *L. trifolii*.

RESPONSE TO PARASITES AND PREDATORS

Forty parasite species have been found to use members of the genus *Liriomyza* as hosts (27, 58, 76; M. W. Johnson, unpublished). Only larval and pupal parasites have been found, and under natural conditions parasitism is usually low early in crop development and gradually increases as the crop matures. Under greenhouse conditions, inoculation and augmentation of parasites are needed for effective control. In such situations both larval and pupal parasites have been used (15). Many studies involving these leafminers and their parasites have been concentrated on identifying the parasite complex, estimating its ability to effect control, and determining the impact of insecticides on the parasites. Much more research is needed on the detailed biology (instar preference, fecundity, searching ability, etc) of these parasites, as they use different members of the genus *Liriomyza* as hosts. Few studies are available in this regard (24, 40, 129). For example, it was only recently discovered that *L. trifolii* is capable of encapsulating the eggs of *Dacnusa sibirica* (113). To maximize the potential of parasites in biological control of *Liriomyza* spp., research should focus on biology and proper selection of parasites most likely to succeed in the different cropping systems where these leafminers are a problem. Only limited studies have been done with predators of these leafminers (68).

POSTSCRIPT

The movement of *Liriomyza* spp. within infested plant material has caused a worldwide problem, the magnitude of which is difficult to comprehend. This involves primarily two species, *L. trifolii* (44, 54) and *L. huidobrensis* (69). *Liriomyza trifolii*, which until recently had a relatively limited distribution, can now be considered cosmopolitan and is a major pest on numerous ornamental and vegetable crops almost everywhere it occurs. This polyphagous leafminer has dramatically increased its host range as it has spread into new areas. In 1965 (97), 59 plant hosts were listed for *L. trifolii*. By 1984 this number had increased to 122 (18). In 1986, the number of hosts for *L. trifolii* exceeds 400 (K. A. Spencer, personal communication). This increased host range, coupled with a phenomenal increase in insecticide resistance, has created difficulty in control of *L. trifolii* in a number of crops.

Liriomyza huidobrensis is primarily a problem for importers of cut chrysanthemum flowers into the United States and other areas from South America, where this species is common. It is of concern to the United States Department of Agriculture, Plant Protection and Quarantine, that *L. huidobrensis* is found in the United States, but only in California. Because most cut-flower shipments originating in South America are destined for the eastern United States, the pest has potential for establishment there.

Movement of both species should diminish in the future; those responsible for shipping plant material are fully aware of the potential problem, and considerable research has focused on how to assure that plant material is kept free of *Liriomyza* (41, 54). In addition, several outstanding quarantine guidelines are now available (14, 79).

During the past ten years there has been a dramatic increase in the economic importance of *Liriomyza* leafminers. Researchers have not responded commensurately to this increase. Two factors that have complicated researcher response have been misidentification of species and lack of basic biological information. Regarding the former point, most researchers are aware of potential problems with taxonomy of *Liriomyza* spp. and therefore keep voucher specimens and have experts make the identifications. In addition, more emphasis has been placed on electrophoresis and ovipositor morphology as taxonomic tools (37, 51, 125). Electrophoresis can now be used to separate heretofore indistinguishable immatures of closely related species (51). Other methods of separating larval stages (e.g. larval morphology, mouthhook structure) appear promising (4, 6). With greater emphasis on taxonomy and increased researcher awareness of taxonomic problems, fewer misidentifications should occur in the future (94).

Our knowledge concerning the biology of these leafminers is still undeveloped. Most of the information presented in this review is very general and has come from studies on relatively few species completed over the past 25 years. The next 25 years should yield considerable data on the basic biology of this fascinating group of flies. This data will contribute to the overall understanding of host plant utilization by insects and to the development of more comprehensive and accurate pest management strategies.

Acknowledgments

The critical reviews of Ken Spencer, Karen Robb, Kevin Heinz, Jeff Yost, Roger Youngman, and John Trumble are gratefully appreciated. The use of Marshall Johnson's unpublished data on parasites is also appreciated. Support from the American Florists Endowment and the University of California has made much of my research on *Liriomyza* possible.

Literature Cited

1. Affeldt, H. A., Thimijan, R. W., Smith, F. F., Webb, R. E. 1983. Response of the greenhouse whitefly (Homoptera: Aleyrodidae) and the vegetable leafminer (Diptera: Agromyzidae) to photospectra. *J. Econ. Entomol.* 76(6):1405–9
2. Aguilera, A. P. 1972. Biología de *Liriomyza langei* Frick (Dipt., Agromyzidae) y evaluación de los parásitos que emergen del puparium. *Idesia* 2:71–85
3. Beck, H. W., Musgrave, C. A., Strandberg, J. O., Genung, W. G. 1981. Spatial dispersion patterns of *Liriomyza* spp. on celery. See Ref. 87, pp. 129–40
4. Beri, S. K. 1971. Immature stages of Agromyzidae (Diptera) from India. VII. Taxonomy and biology of fifteen species of *Liriomyza* Mik. *Orient. Insects* 1:85–118 (Suppl.)
5. Beri, S. K. 1974. Biology of a leaf miner *Liriomyza brassicae* (Riley) (Diptera: Agromyzidae). *J. Nat. Hist.* 8:143–51
6. Beri, S. K. 1983. Comparative morphological studies on the cephalopharyngeal skeleton of larval leaf-mining flies (Agromyzidae: Diptera). *Indian J. For.* 6(4):301–8
7. Bethke, J. A., Parrella, M. P. 1985. Leaf puncturing, feeding and oviposition behavior of *Liriomyza trifolii. Entomol. Exp. Appl.* 39:149–54
8. Block, K. 1976. Chromosomal variation in Agromyzidae (Diptera). VI. Comparative chromosome studies. *Hereditas* 84:177–212
9. Bodri, M. S., Oetting, R. D. 1985. Assimilation of radioactive phosphorus by *Liriomyza trifolii* (Burgess) (Diptera: Agromyzidae) from feeding at different temperatures on different chrysanthemum cultivars. *Proc. Entomol. Soc. Wash.* 87(4):770–76
10. Buhr, H. 1937. Parasitenbefall und Pflanzenverwandtschaft. *Bot. Jahrb.* 68:142–98
11. Chandler, L. D. 1981. Evaluation of different shapes and color intensities of yellow traps for use in population monitoring of dipterous leafminers. *Southwest. Entomol.* 6:233–37
12. Chandler, L. D. 1985. Flight activity of *Liriomyza trifolii* (Diptera: Agromyzidae) in relation to placement of yellow traps in bell peppers. *J. Econ. Entomol.* 78(4):825–28
13. Charlton, C. A., Allen, W. W. 1981. The biology of *Liriomyza trifolii* on beans and chrysanthemums. See Ref. 87, pp. 42–49
14. Commonwealth of Australia. 1984. Leafminers of chrysanthemum. *Commonwealth Dep. Health Plant Quarantine Leafl. No. 39.* Canberra: Aust. Gov. Publ. Serv. 4 pp.
15. Cross, J. V., Wardlow, L. R., Hall, R., Saynor, M., Bassett, P. 1983. Integrated control of chrysanthemum pests. *Proc. Work. Group Integrated Control Glasshouses, Darmstadt, West Germany. Bull. Int. Org. Biol. Control/West. Palearct. Reg. Sect.* 6(3):181–85
16. Dimetry, N. Z. 1971. Biological studies on a leaf mining Diptera, *Liriomyza trifolii* (Burgess) attacking beans in Egypt. *Bull. Soc. Entomol. Egypte.* 55:55–69
17. Elmore, J. C., Ranney, C. A. Jr. 1954. Injury to pepper plants by the pea leaf miner. *J. Econ. Entomol.* 47(2):357–58
18. European and Mediteranean Plant Protection Organization. 1984. EPPO Data sheets on quarantine organisms, No. 131. *EPPO Bull.* 14(1):29–37
19. Fagoonee, I., Toory, V. 1983. Preliminary investigations of host selection mechanisms by the leafminer *Liriomyza trifolii. Insect Sci. Appl.* 4(4):337–41
20. Fagoonee, I., Toory, V. 1984. Contribution to the study of the biology and ecology of the leafminer *Liriomyza trifolii* and its control by Neem. *Insect Sci. Appl.* 5(1):23–30
21. Genung, W. G. 1957. Some possible cases of insect resistance to insecticides in Florida. *Proc. Fla. State Hortic. Soc.* 70:148–52
22. Genung, W. G., Janes, M. A. 1975. Host range, wild host significance, and in-field spread of *Liriomyza trifolii* and population buildup and effects of its parasites in relation to fall and winter celery (Diptera: Agromyzidae). *Belle Glade Agric. Res. Ed. Cent. Res. Rep. EV-1975-5,* Univ. Fla. Res. Stn., Belle Glade 18 pp.
23. Harbaugh, B. K., Price, J. F., Stanley, C. D. 1983. Influence of leaf nitrogen on leafminer damage and yield of spray chrysanthemum. *HortScience* 18(6): 880–81
23a. Haynes, K. F., Parrella, M. P., Trumble, J. T., Miller, T. A. 1986. Monitoring insecticide resistance with yellow sticky cards. *Calif. Agric.* In press
24. Hendriske, A., Zucchi, R. 1979. The importance of observing parasite behavior for the development of biological control of the tomato leafminer (*Liriomyza bryoniae* Kalt.). *Meded.*

Fac. Landbouwwet. Rijksuniv. Gent 44(1):107–16

25. Hendriske, A., Zucchi, R., Van Lenteren, J. C., Woets, J. 1980. *Dacnusa sibirica* Telenga and *Opius pallipes* Wesmael (Hym., Braconidae) in the control of the tomato leafminer *Liriomyza bryoniae* Kalt. See Ref. 15, pp. 83–98

26. Ipe, M., Sadaruddin, M. 1984. Infestation and host specificity of *Liriomyza brassicae* Riley and the role of phenolic compounds in host plant resistance. *Entomon* 9(4):265–70

26a. IPM Manual Group. 1985. Integrated pest management for tomatoes. *Univ. Calif. Publ. 3274.* Berkeley: Univ. Calif. 105 pp. 2nd ed.

27. Johnson, M. W., Oatman, E. R., Wyman, J. A. 1980. Natural control of *Liriomyza sativae* (Diptera: Agromyzidae) in pole tomatoes in southern California. *Entomophaga* 25:193–98

28. Johnson, M. W., Oatman, E. R., Wyman, J. A., Van Steenwyk, R. A. 1980. A technique for monitoring *Liriomyza sativae* in fresh-market tomatoes. *J. Econ. Entomol.* 73(4):552–55

29. Johnson, M. W., Welter, S. C., Toscano, N. C., Ting, I. P., Trumble, J. T. 1983. Reduction of tomato leaflet photosynthesis rates by mining activity of *Liriomyza sativae* (Diptera: Agromyzidae). *J. Econ. Entomol.* 76(5):1061–63

30. Jones, V. P., Parrella, M. P. 1986. The development of sampling strategies for larvae of *Liriomyza trifolii* in chrysanthemums. *Environ. Entomol.* 15:268–73

31. Jones, V. P., Parrella, M. P. 1986. The movement and dispersal of *Liriomyza trifolii* (Diptera: Agromyzidae) in a chrysanthemum greenhouse. *Ann. Appl. Biol.* In press

32. Keil, C. B., Parrella, M. P., Morse, J. G. 1985. Method for monitoring and establishing baseline data for resistance to permethrin by *Liriomyza trifolii* (Burgess). *J. Econ. Entomol.* 78(2):419–22

33. Kennedy, G. G., Bohn, G. W., Stoner, A. K., Webb, R. E. 1978. Leafminer resistance in muskmelon. *J. Am. Soc. Hortic. Sci.* 103(5):571–74

34. Kleinschmidt, R. P. 1970. New species of Agromyzidae from Queensland. *Queensl. J. Agric. Sci.* 17:321–37

35. Knodel-Montz, J. J., ed. 1985. *An Informal Conference on Liriomyza Leafminers.* Springfield, Va: USDA, Agric. Res. Serv., Natl. Tech. Inf. Serv. 75 pp.

36. Knodel-Montz, J. J., Lyons, R. E., Poe, S. L. 1985. Photoperiod affects chrysanthemum host plant selection by leafminers (Diptera: Agromyzidae). *HortScience* 20(4):708–10

37. Knodel-Montz, J. J., Poe, S. L. 1982. Ovipositor morphology of three economically important *Liriomyza* species (Diptera: Agromyzidae). See Ref. 80, pp. 186–95

38. Knodel-Montz, J. J., Poe, S. L. 1982. Famulus: an interactive computerized leafminer bibliography. See Ref. 80, pp. 199–205

39. Ledieu, M. S., Heyler, N. L. 1985. Observations on the economic importance of tomato leaf miner *(Liriomyza bryoniae)* (Agromyzidae). *Agric. Ecosyst. Environ.* 13:103–9

40. Lema, K., Poe, S. L. 1979. Age specific mortality of *Liriomyza sativae* due to *Chrysonotomyia formosa* and parasitization by *Opius dimidiatus* and *Chrysonotomyia formosa.* *Environ. Entomol.* 8:935–37

41. Liebee, G. L. 1984. Influence of temperature on development and fecundity of *Liriomyza trifolii* (Burgess) (Diptera: Agromyzidae) on celery. *Environ. Entomol.* 13(2):497–501

42. Liebee, G. L. 1985. Effects of storage at 1.1°C on the mortality of *Liriomyza trifolii* (Burgess) (Diptera: Agromyzidae) life stages in celery. *J. Econ. Entomol.* 78(2):407–11

43. Lindquist, R. K. 1974. Effect of leafminer larvae on yields of greenhouse tomatoes: a preliminary report. *Ohio Agric. Res. Dev. Cent. Res. Summ.* 73:25–29

44. Lindquist, R. K. 1983. New greenhouse pests, with particular reference to the leafminer, *Liriomyza trifolii. Proc. 10th Int. Congr. Plant Prot., Brighton, England,* 3:1087–94. Croydon, England: Br. Crop Prot. Counc.

45. Lindquist, R. K., Casey, M. L., Heyler, N., Scopes, E. A. 1984. Leafminers on greenhouse chrysanthemums: control of *Chromatomyia syngenesiae* and *Liriomyza trifolii. J. Agric. Entomol.* 3:256–63

46. Mainx, F. 1951. Die Verbreitung von Chromosomendislocationen in natürlichen Populationen von *Liriomyza urophorina* Mik. *Chromosoma* 4:521–34

47. Mainx, F., Fiala, Y., Kogever, E. V. 1956. Die geographische Verbreitung der chromosomalen Strukturtypen von *Liriomyza urophorina* Mik. *Chromosoma* 8:18–29

48. McClanahan, R. J. 1980. Biological

control of *Liriomyza sativae* on greenhouse tomatoes. *Proc. Work. Group Integrated Control Glasshouses, Vantaa, Finland. Bull. Int. Org. Biol. Control/West. Palearct. Reg. Sect.* 3(3):135–39

49. McNeil, J. N., Quiring, D. T. 1983. Evidence of an oviposition deterring pheromone in the alfalfa blotch leafminer, *Agromyza frontella* (Rondani) (Diptera: Agromyzidae). *Environ. Entomol.* 12:990–92

50. Mendes, L. O. T. 1940. O minador da batatinha. *Agronomyza braziliensis* Frost (1939) (Diptera: Agromyzidae). *J. Agron. Piracicaba* 3(3):207–20

51. Menken, S. B. J., Ulenberg, S. A. 1983. Diagnosis of the agromyzids *Liriomyza bryoniae* and *L. trifolii* by means of starch gel electrophoresis. *Entomol. Exp. Appl.* 34:205–8

52. Mik, J. 1894. Ueber eine neue Agromyza, deren Larven in den Blüthenknospen von Lilium martagon leben. *Wein. Entomol. Z.* 13:284–90

53. Miller, G. W., Isger, M. B. 1985. Effects of temperature on the development of *Liriomyza trifolii* (Burgess) (Diptera: Agromyzidae). *Bull. Entomol. Res.* 75:321–28

54. Mortimer, E. A., Powell, D. F. 1984. Development of a combined cold storage and methyl bromide fumigation treatment to control the American serpentine leaf miner *Liriomyza trifolii* (Diptera: Agromyzidae) in imported chrysanthemum cuttings. *Ann. Appl. Biol.* 105: 443–54

55. Musgrave, C. A., Poe, S. L., Bennett, D. R. 1975. Leaf miner population estimation in polycultured vegetables. *Proc. Fla. State Hortic. Soc.* 88:156–60

56. Musgrave, C. A., Poe, S. L., Weems, H. V. 1975. The vegetable leafminer, *Liriomyza sativae* Blanchard (Diptera: Agromyzidae) in Florida. *Fla. Dep. Agric. Consum. Serv. Entomol. Circ. No. 162,* Div. Plant Ind., Gainesville, Fla. 4 pp.

57. Needham, J. G. S., Frost, S. W., Tothill, B. H. 1928. *Leafmining Insects,* pp. 231–65. Baltimore, Md: Williams & Wilkins. 351 pp.

57a. Newman, J. P., Parrella, M. P. 1986. A license to kill. *Greenhouse Manage.* 5(3)86–92

58. Oatman, E. R. 1959. Natural control studies of the melon leafminer, *Liriomyza pictella. J. Econ. Entomol.* 52(5): 895–98

59. Oatman, E. R. 1960. Intraspecific competition studies of the melon leafminer,

Liriomyza pictella (Thomson). *Ann. Entomol. Soc. Am.* 53(1):130–31

60. Oatman, E. R., Kennedy, G. G. 1976. Methomyl induced outbreak of *Liriomyza sativae* on tomato. *J. Econ. Entomol.* 69(5):667–68

61. Oatman, E. R., Michelbacher, A. E. 1958. The melon leafminer, *Liriomyza pictella* (Thomson) (Diptera: Agromyzidae). *Ann. Entomol. Soc. Am.* 51(6):557–66

62. Oatman, E. R., Michelbacher, A. E. 1959. The melon leaf miner *Liriomyza pictella* (Thomson) (Diptera: Agromyzidae). II. Ecological studies. *J. Econ. Entomol.* 52(1):83–89

63. Oetting, R. D. 1983. The influence of selected substrates on *Liriomyza trifolii* emergence. *J. Ga. Entomol. Soc.* 18(1):120–24

64. Parrella, M. P. 1982. A review of the history and taxonomy of economically important serpentine leafminers (*Liriomyza* spp.) in California (Diptera: Agromyzidae). *Pan-Pac. Entomol.* 58:302–8

65. Parrella, M. P. 1983. Intraspecific competition among larvae of *Liriomyza trifolii* (Diptera: Agromyzidae): Effects on colony production. *Environ. Entomol.* 12(5):1412–14

66. Parrella, M. P. 1984. Effect of temperature on oviposition, feeding and longevity of *Liriomyza trifolii* (Diptera: Agromyzidae). *Can. Entomol.* 116:85–92

67. Parrella, M. P., Allen, W. W., Morishita, P. 1981. Leafminer species causes California mum growers new problems. *Calif. Agric.* 35(9, 10):28–30

68. Parrella, M. P., Bethke, J. A. 1982. Biological studies with *Cyrtopeltis modestus* (Hemiptera: Miridae): A faculative predator of *Liriomyza* spp. (Diptera: Agromyzidae). See Ref. 80, pp. 180–85

69. Parrella, M. P., Bethke, J. A. 1984. Biological studies of *Liriomyza huidobrensis* (Diptera: Agromyzidae) on chrysanthemum, aster and pea. *J. Econ. Entomol.* 77(2):342–45

70. Parrella, M. P., Jones, V. P. 1985. Yellow traps as monitoring tools for *Liriomyza trifolii* in chrysanthemum greenhouses. *J. Econ. Entomol.* 78(1): 53–56

71. Parrella, M. P., Jones, V. P., Youngman, R. R., Lebeck, L. M. 1985. Effect of leaf mining and leaf stippling of *Liriomyza* spp. on photosynthetic rates of chrysanthemum. *Ann. Entomol. Soc. Am.* 78(1):90–93

72. Parrella, M. P., Keil, C. B. 1984. Insect

pest management: The lesson of *Liriomyza*. *Bull. Entomol. Soc. Am.* 30(2):22–25

73. Parrella, M. P., Keil, C. B. 1985. Toxicity of methamidophos to four species of Agromyzidae. *J. Agric. Entomol.* 2(3):234–37

74. Parrella, M. P., Keil, C. B., Morse, J. G. 1984. Insecticide resistance in *Liriomyza trifolii*. *Calif. Agric.* 38(1,2):22–23

75. Parrella, M. P., Robb, K. L. 1982. A technique for staining the eggs of *Liriomyza trifolii* in chrysanthemum, celery and tomato leaves. *J. Econ. Entomol.* 75(2):383–84

76. Parrella, M. P., Robb, K. L. 1985. Economically important members of the genus *Liriomyza* Mik: A selected bibliography. *Misc. Publ. Entomol. Soc. Am.* 59:1–26

77. Parrella, M. P., Robb, K. L., Bethke, J. A. 1981. Oviposition and pupation of *Liriomyza trifolii* (Burgess). See Ref. 87, pp. 50–55

78. Parrella, M. P., Robb, K. L., Bethke, J. A. 1983. Influence of selected host plants on the biology of *Liriomyza trifolii* (Diptera: Agromyzidae). *Ann. Entomol. Soc. Am.* 76(1):112–15

79. Piedrahita, O. 1985. Two serpentine leafminers attacking vegetables and ornamentals. *Ont. Minist. Agric. Food, Factsheet No. 85–006*, Hortic. Res. Inst. Ont., Vineland Stn. 4 pp.

80. Poe, S. L., ed. 1982. *Proc. 3rd Annu. Ind. Conf. Leafminer, San Diego, Calif.* Alexandria, Va: Soc. Am. Florists. 216 pp.

81. Poe, S. L., ed. 1984. *Proc. 4th Annu. Ind. Conf. Leafminer, Sarasota, Fla.* Alexandria, Va: Soc. Am. Florists. 191 pp.

82. Poe, S. L., Green, J. L., Shih, C. I. 1976. Cultural practices affect damage to chrysanthemum by *Liriomyza sativae* Blanchard. *Proc. Fla. State Hortic. Soc.* 89:229–301

83. Price, J. F., Stanley, C. D. 1982. Gypsophila, leafminer and parasitoid relationships on two farms of different pesticide use patterns. See Ref. 80, pp. 66–78

84. Robb, K. L., Parrella, M. P. 1984. Sublethal effects of two insect growth regulators applied to larvae of *Liriomyza trifolii* (Diptera: Agromyzidae). *J. Econ. Entomol.* 77(5):1288–92

85. Robb, K. L., Parrella, M. P. 1985. Antifeeding and oviposition deterring effects of insecticides on adult *Liriomyza trifolii* (Diptera: Agromyzidae). *J. Econ. Entomol.* 78(3):709–13

86. Rode, H., Vorsatz, E. 1969. Zur Biologie und Bekämpfung der Lilienfliege (*Liriomyza urophorina* Mik). *Nachrichtenbl. Dtsch. Pflanzenschutzdienst Berlin* 20:107–12

87. Schuster, D. J., ed. 1981. *Proc. IFAS-Ind. Conf. Biol. Control Liriomyza Leafminers, Lake Buena Vista, Fla.* Gainesville: Univ. Fla. 325 pp.

88. Schuster, D. J., Beck, H. W. 1981. Sampling and distribution of *Liriomyza* on tomatoes. See Ref. 87, pp. 106–28

89. Schuster, D. J., Everett, P. H. 1982. Laboratory and field evaluations of insecticides for control of *Liriomyza* spp. on tomatoes. See Ref. 80, pp. 20–30

90. Schuster, D. J., Harbaugh, B. K. 1979. Chrysanthemum cultivars differ in foliar leafminer damage. *HortScience* 14:271–72

91. Schuster, D. J., Patel, K. J. 1985. Development of *Liriomyza trifolii* (Diptera: Agromyzidae) larvae on tomato at constant temperatures. *Fla. Entomol.* 68(1):158–61

92. Sehgal, V. K. 1971. Biology and host plant relationships of an oligophagous leafminer *Phytomyza matricariae* Hendel (Diptera: Agromyzidae). *Quaest. Entomol.* 7:255–80

93. Spencer, K. A. 1965. The species-host relationship in the Agromyzidae (Diptera) as an aid to taxonomy. *Proc. 12th Int. Congr. Entomol., London, 1964*, 1:101–2. London: R. Entomol. Soc. London

94. Spencer, K. A. 1973. Agromyzidae (Diptera) of economic importance. *Ser. Entomol.* 9:1–418

95. Spencer, K. A. 1981. A revisionary study of the leaf-mining flies (Agromyzidae) of California. *Univ. Calif. Spec. Publ. No. 3273.* 489 pp. Berkeley: Univ. Calif.

95a. Spencer, K. A., Steyskal, G. C. 1986. *Manual of the Agromyzidae (Diptera) of the United States. USDA Agric. Handb. No. 638.* Washington, DC: USDA. 478 pp.

96. Speyer, E. R., Parr, W. J. 1949. Animal pests. I. Tomato leaf-miner (*Liriomyza solani*, Hering). *Rep. Exp. Res. Stn. Cheshunt* 35:48–56

97. Stegmaier, C. E. Jr. 1966. Host plants and parasites of *Liriomyza trifolii* in Florida (Diptera: Agromyzidae). *Fla. Entomol.* 49:81–86

98. Steyskal, G. C. 1973. The strange fate of the "serpentine leafminer" (*Liriomyza* spp., Agromyzidae, Diptera). *USDA Coop. Econ. Insect Rep.* 23(43):735–36

99. Suss, L., Agosti, G., Costanzi, M. 1984. *Liriomyza trifolii*, note di biologia. *Inf. Fitopatol.* 2:8–12

100. Tauber, M. J., Tauber, C. A. 1968. Biology and leafmining behavior of *Phytomyza lanati* (Diptera: Agromyzidae). *Can. Entomol.* 100:341–49

101. Tavormina, S. J. 1982. Sympatric genetic divergence in the leaf-mining insect *Liriomyza brassicae*. (Diptera: Agromyzidae). *Evolution* 36(3):523–34

102. Tilden, J. W. 1950. Oviposition behavior of *Liriomyza pusilla* (Meigen). *Pan-Pac. Entomol.* 26:119–21

103. Trumble, J. T., Ting, I. P., Bates, L. 1985. Analysis of physiological, growth and yield responses of celery to *Liriomyza trifolii*. *Entomol. Exp. Appl.* 38:15–21

104. Tryon, E. H. Jr., Poe, S. L. 1981. Developmental rates and emergence of vegetable leafminer pupae and their parasites reared from celery foliage. *Fla. Entomol.* 64(4):477–83

105. Tryon, E. H. Jr., Poe, S. L., Cromroy, H. L. 1980. Dispersal of vegetable leafminer onto a transplant production range. *Fla. Entomol.* 63(3):292–96

106. Deleted in proof

107. Via, S. 1984. The quantitative genetics of polyphagy in an insect herbivore. I. Genotype-environment interaction in larval performance on different host plant species. *Evolution* 38(4):881–95

108. Via, S. 1984. The quantitative genetics of polyphagy in an insect herbivore. II. Genetic correlations in larval performance within and among host plants. *Evolution* 38(4):896–905

109. Webb, R. E., Smith, F. F. 1969. Effect of temperature on resistance in lima bean, tomato and chrysanthemum to *Liriomyza munda*. *J. Econ. Entomol.* 62(2):458–62

110. Webb, R. E., Stoner, A. K., Gentile, A. G. 1971. Resistance to leaf miners in *Lycopersicon* accessions. *J. Am. Soc. Hortic. Sci.* 96(1):65–67

111. Webster, F. M., Parks, T. H. 1913. The serpentine leaf-miner. *J. Agric. Res.* 1(1):59–87

112. Werner, F. G. 1982. *Common Names of Insects and Related Organisms*. College Park, Md: Entomol. Soc. Am. 132 pp.

113. Woets, J., Van Der Linden, A. 1985. First experiments on *Chrysocharis parksi* Crawford (Hymenoptera: Eulophidae) as a parasite for leafminer control (*Liriomyza* spp.) (Diptera: Agromyzidae) in European greenhouse tomatoes.

Meded. Fac. Landbouwwet. Rijksuniv. Gent 50(2b):763–68

114. Wolfenbarger, D. A., Wolfenbarger, D. O. 1966. Tomato yields and leaf miner infestations and a sequential sampling plan for determining need for control treatments. *J. Econ. Entomol.* 59(2):279–83

115. Wolfenbarger, D. O. 1947. The serpentine leafminer and its control. *Fla. Agric. Exp. Stn. Press Bull. 639*, Univ. Fla., Gainesville. 6 pp.

116. Wolfenbarger, D. O. 1949. Border effects of serpentine leaf miner abundance in potato fields. *Fla. Entomol.* 31(1):15–20

117. Wolfenbarger, D. O. 1954. Potato yields associated with control of aphids and the serpentine leafminer. *Fla. Entomol.* 37(1):7–12

118. Wolfenbarger, D. O. 1958. Serpentine leafminer: Brief history and a summary of a decade of control measures in south Florida. *J. Econ. Entomol.* 51(3):357–59

119. Wolfenbarger, D. O., Moore, W. D. 1968. Insect abundances on tomatoes and squash mulched with aluminum and plastic sheetings. *J. Econ. Entomol.* 61(1):34–36

120. Woltz, S. S., Kelsheimer, E. G. 1968. Effect of variation in nitrogen nutrition of chrysanthemums on attack by serpentine leafminer. *Proc. Fla. State Hortic. Soc.* 7:404–6

121. Zehnder, G. W., Trumble, J. T. 1984. Host selection of *Liriomyza* species (Diptera: Agromyzidae) and associated parasites in adjacent plantings of tomato and celery. *Environ. Entomol.* 13:492–96

122. Zehnder, G. W., Trumble, J. T. 1984. Intercrop movement of leafminers. *Calif. Agric.* 38:7–8

123. Zehnder, G. W., Trumble, J. T. 1984. Spatial and diel activity of *Liriomyza* species (Diptera: Agromyzidae) in fresh market tomatoes. *Environ. Entomol.* 13(5):1411–16

124. Zehnder, G. W., Trumble, J. T. 1985. Sequential sampling plans with fixed levels of precision for *Liriomyza* spp. (Diptera: Agromyzidae) in fresh market tomatoes. *J. Econ. Entomol.* 78(1):138–42

125. Zehnder, G. W., Trumble, J. T., White, W. R. 1983. Discrimination of *Liriomyza* species (Diptera: Agromyzidae) using electrophoresis and scanning electron microscopy. *Proc. Entomol. Soc. Wash.* 85(3):564–74

126. Zepp, D. B., MacPherson, R., Davenport, R. E. 1982. Cage for experimental

observations of *Liriomyza trifolii* (Diptera: Agromyzidae) adults. *J. Econ. Entomol.* 75(6):1161–63

127. Zitter, T. A., Tsai, J. H. 1977. Transmission of three polyviruses by the leafminer *Liriomyza sativae* (Diptera: Agromyzidae). *Plant Dis. Rep.* 61: 1025–29

128. Zoebisch, T. G., Schuster, D. J., Gilreath, J. P. 1984. *Liriomyza trifolii:* oviposition and development in foliage of tomato and common weed hosts. *Fla. Entomol.* 67(2):250–54

129. Zucchi, R., van Lenteren, J. C. 1978. Biological characteristics of *Opius pallipes* Wesmael (Hymenoptera: Braconidae), parasite of the tomato leafminer, *Liriomyza bryoniae* Kalt. *Meded. Fac. Landbouwwet. Rijksuniv. Gent* 43(2): 455–62

Ann. Rev. Entomol. 1987. 32:225–51

ECOLOGICAL CONSIDERATIONS FOR THE USE OF ENTOMOPATHOGENS IN IPM

J. R. Fuxa

Department of Entomology, Louisiana Agricultural Experiment Station, Louisiana State University Agricultural Center, Baton Rouge, Louisiana 70803

PERSPECTIVES AND OVERVIEW

A sound ecological basis is essential for the use of entomopathogens in IPM. Virtually every definition of IPM, including the 1967 definition of the Food and Agricultural Organization of the United Nations (FAO) (57), states or implies the basic importance of ecology.

Pathogens have their own ecological niches, which must be thoroughly understood for the manipulation of epizootics. Epizootics are characterized by rapid change in prevalence of the disease. This change depends on massive reproduction of the pathogen interacting with or reinforced by host and environmental factors. Something usually triggers an epizootic, such as stress of host insects or an increase in the proportion of susceptible hosts in the population. Efficient transmission is necessary. Initially the host population must be largely susceptible to the microorganism, but later the proportion of susceptible hosts declines and the increased number of pathogen units becomes less important.

Several approaches have evolved for the use of pathogens in IPM: permanent introduction and establishment, inundative augmentation, inoculative augmentation, and environmental manipulation.

Introduction is the establishment of an organism in a pest population where it does not naturally occur, which results in more or less permanent suppression of the pest (modified from 73). Burges & Hussey (30) and Burges (27) cited a total of 41 successful introductions. The most notable are those of

225

Bacillus popilliae for control of *Popillia japonica* (102), and nuclear polyhedrosis virus (NPV) for control of *Gilpinia hercyniae* (41).

Pathogens used for inundative, remedial, or "insecticidal" augmentation are not usually intended to recycle (i.e. produce progeny which in turn infect other insect hosts). Pathogens used in this way include *Bacillus thuringiensis* (Bt) (108), *Heliothis* NPV (88), and *Vairimorpha necatrix* (111). Certain pathogens, e.g. *Beauveria bassiana* (55), are borderline; the agent can recycle, but new generations of the pathogen often must be further augmented within the same season.

In inoculative augmentation, releases are required on some recurring basis yet result in recycling, most often limited to a single season (modified from 129). Perhaps the majority of examples fall in this category, including the fungi *Verticillium lecanii* (71), *Hirsutella thompsonii* (117), and *Nomuraea rileyi* (85); the protozoan *Nosema locustae* (77); the nematode *Romanomermis culicivorax* (134); and the NPVs of *Orgyia pseudotsugata* and *Lymantria dispar* (40).

Environmental (ecosystem) manipulation or conservation involves the enhancement of naturally occurring pest control by means other than direct addition to the pathogen units already present (129). Cultural manipulations can permit the pathogen to reproduce more than usual or can preserve or enhance those already present. This is a logical approach, since pathogens are dependent to various degrees on a favorable environment for their survival or activity. For example, livestock movement has aided transport of virus and thus transmission (95); altered insecticide thresholds and harvest regimes have increased control of *Hypera postica* by *Erynia* sp. (24), and differences in planting date have enhanced control of *Heliothis* spp. by *N. rileyi* (149). Elimination of fungicide treatments has been used to enhance *N. rileyi* epizootics; this is an example of conservation of a naturally occurring pathogen (94). Ignoffo (86) has pointed out that because some species produce large numbers of pathogen units, a small reduction in natural attrition could greatly improve insect control.

Ecological considerations for the use of entomopathogens in IPM include pathogen characteristics, host characteristics, and the ecosystem.

PATHOGEN CHARACTERISTICS

Pathogen Activity

For practical purposes, entomopathogens fall into two categories based upon the speed of their action: "quick-damage" and "slow" pathogens. The quick-damage pathogens mainly include those that produce toxins (e.g. Bt) but also include a few that kill directly or indirectly (e.g. by initiating a bacterial septicemia) when they enter the host [e.g. *Steinernema feltiae* (*Neoaplectana carpocapsae)* (138) and *V. necatrix* (111)]. These pathogens stop feeding by the host within 24 hr, although death may take several days. Unless there are extenuating circumstances, such as high damage thresholds or lack of urgency, only quick damage pathogens, with action thresholds similar to those of chemical insecticides, are accepted for inundative augmentation.

The slow pathogens debilitate their hosts after more than 24 hr by expending at least a portion of their life cycle at the host's expense. These comprise the great majority of entomopathogens. They have been unreliable, ineffective, or too slow in inundative augmentation, but have been more successfully used in other approaches.

Success of Control Approach in Relation to Pathogen Group

VIRUSES   The viruses are clearly slow pathogens; however, through genetic engineering some may become carriers of genes for toxin production. The baculoviruses [granulosis viruses (GVs) and NPVs] have particular potential for microbial control in the near future. These viruses are relatively host-specific and are found in many economically important insects, mainly Lepidoptera and sawflies. They generally are virulent but require ~2–8 days or longer to kill the host. Their persistence in soil is measured in years (158, 168), but on leaf surfaces they are inactivated within days or even hours (137). They are horizontally transmitted by environmental contamination and ingestion, but mechanical horizontal transmission by insect vectors is also known (160), and vertical transovum transmission is common (8). True latency occurs (119), although its extent is uncertain (130).

Introduction and establishment, inoculative augmentation, and environmental manipulation are the methods that take advantage of the strengths of the baculoviruses, particularly their ability to cause impressive natural epizootics (96, 97), the large number of species and strains in the group, and their great persistence in soil. Inoculative augmentation and environmental manipulation can compensate for an important weakness, their often ineffective movement from the soil to the insect's feeding substrate.

Several NPVs and a GV have been introduced and established for control of forest pests (30, 137). Introductions have partially suppressed pest populations in row crops (74); presumably, if the soil becomes contaminated with the virus, the long-lived viral polyhedral inclusion bodies (PIBs) can persist until the pest population returns, at which time the virus population increases rapidly. There also are examples of introductions with no carryover (e.g. 47). Federici (53) hypothesized that successful introductions would be few and would be limited primarily to control of introduced pests. Entwistle (45) supported this view. However, Morris (123) believed that introductions would have the most promise for control of forest pests. Successful control of forest Lepidoptera and sawflies has almost exclusively been achieved either by introduction and establishment or by inoculative augmentation (159).

Inoculative augmentation with baculoviruses has controlled *Oryctes rhinoceros* (45, 74), *Trichoplusia ni, Pieris rapae, Orgyia pseudotsugata,* and *Malacosoma disstria* (86). The rapid generation time of many viruses aids inoculative augmentation. For example, three generations of the NPV of *O. pseudotsugata* can be produced during one generation of the host insect (74).

Research has been directed toward environmental manipulation in two cases involving NPVs of lepidopterous pests in pastures (63, 95) and in one exploiting movement by irrigation water (45).

Novel strategies have been investigated frequently for this group. Auto-dissemination has been used to distribute virus into a host population (66). Vertical transmission is a valuable trait for such releases. Lattice introduction with subsequent spread has been demonstrated (45), and spread of NPV by biotic agents, including birds, mammals, and invertebrate predators, is well known (137). Mortar bombs have even been used to move virus-contaminated soil to insect feeding substrates (137).

Viruses generally have been neither efficacious nor commercially successful when forced into inundative augmentation in row crops or against dense pest populations (137, 168). A major problem has been slow action (131). Other reasons for failure are instability on foliage (137), host behavior (e.g. the insect feeds only a short while in an exposed situation before entering a plant part) (50, 168), and rapid plant growth, which makes untreated foliage available to the insect. Spray adjuvants have partially compensated for some of these weaknesses, particularly by extending the life of the exposed virions and by stimulating host ingestion (137, 168). The specificity of viruses not only hinders commercial development, but also makes inundative augmentation impractical in pest complexes. This deters use of *Laspeyresia pomonella* GV (131) and *Anticarsia gemmatalis* NPV in soybeans. Vertical transmission is not particularly advantageous in these situations, although virulence is.

The cytoplasmic polyhedrosis viruses, entomopoxviruses, and chronic GVs are disadvantageous in inundative augmentation because they have the same weaknesses as NPVs but not their virulence. For some, such as *L. pomonella* GV, inoculative augmentation is more effective than inundative (14). There is little potential for any approach against aquatic insect pests (105).

FUNGI   The fungi are potentially the most versatile entomopathogens. Some have toxins and potential for quick damage (167), but generally they are slow pathogens. Many have wide host ranges, infect different ages and stages of their hosts (54, 153), and are virulent. The variety of species and strains provides screening possibilities. Transmission is almost entirely horizontal, by environmental contamination. Fungi often cause natural epizootics that devastate insect populations (30, 34) inhabiting water (35), soil (69), and aerial plant surfaces (97, 135). Hosts with sucking as well as chewing mouthparts are susceptible (34). However, some authors have stated that fungi rarely keep pest numbers below economic injury levels and that their role as natural regulating agents has not been sufficiently documented (169).

The fungi have some advantages that are almost unique among the entomopathogens. They are able to infect through the host's integument, so ingestion is unnecessary and infection is not limited to chewing insects; certain aquatic fungi have some ability to search for hosts; and they disperse naturally through air movements, although the effective distance and importance to epizootics of such dispersal are poorly understood.

A major disadvantage, though perhaps not as formidable as once thought (87), is the dependence of most terrestrial species of fungi on high relative humidity (RH) at several points in their life cycles (54). Also, they usually take at least one week to kill the host or even to stop its feeding. Certain aquatic fungi require alternate hosts such as an insect and a copepod or ostracod (72). Their persistence in abiotic reservoirs, especially in soil, is known to be poor in at least one important terrestrial species but good in others (5, 72).

The three "ecological" approaches, introduction and establishment, inoculative augmentation, and environmental manipulation, have been more successful than inundative augmentation for fungi (169). Their spreading capability, numerous species and strains, and natural epizootics are attractive features. Fungi are promising for control of mosquitoes (105), sucking insects, soil insects (34, 142), and rangeland grasshoppers (34, 148), and for integrated control with other pathogens (113) or methods (72). Burges & Hussey (30) and Burges (27) cited 16 examples of introductions that resulted in lasting control of insect populations.

Inoculative augmentation is a promising approach because fungi can be raised on artificial media, have wide host ranges, and produce natural epizootics (e.g. 148). Fungi used in this strategy include *Nomuraea rileyi* against lepidopterous soybean defoliators (34, 84), *Hirsutella thompsonii* against mites (117), *Beauveria bassiana* for *Leptinotarsa decemlineata* (148), and *Entomophthora sphaerosperma* in cabbage (86). *B. bassiana* and *Metarhizium anisopliae* are used for *Ostrinia nubilalis* and spittlebugs, and are combined with other treatments against *Oryctes* sp. (148).

The dependence of fungi on environmental factors, particularly RH, makes environmental manipulation a promising approach, but only a few applications have been reported. Cutting time of alfalfa was changed to increase prevalence of *Erynia* sp. in *H. postica;* pesticide usage was reduced to increase *Entomophthora muscae* in *Delia antiqua* (34); and soybean planting time was changed to aid *N. rileyi* (149).

Inundative augmentation has not been particularly successful with fungi (34, 169), probably because of their dependence on environmental conditions, their slow action, and "environmental saturation," i.e. natural presence of fungi such that inundative inoculation does not additionally suppress the insect population (52). Gottwald & Tedders (69) reported possible self-inhibition of *B. bassiana* at a certain pathogen population density.

PROTOZOA   Control with protozoa in the near future is likely to be limited to the microsporidia and neogregarines (22). These are clearly slow pathogens, although quick damage by massive microsporidian spore doses is possible and has been achieved in a field experiment (62). Host ranges are somewhat specific, but there are exceptions. There are apparently many species, but strain selection has not been demonstrated. Natural epizootics of protozoan diseases are common in aquatic insects (105). Transmission is typically horizontal, by environmental contamination and ingestion, but this is the only group besides the viruses with well-demonstrated, mechanical horizontal transmission by insect parasitoids (110) as well as vertical transovum transmission (22, 33). Dispersal is usually ineffective, and persistence in soil has not been determined (22). Recently it has been shown that at least one microsporidium of an aquatic host requires an alternate host during the life cycle (152).

Introduction and inoculative augmentation have been successful with protozoa (86). The difficulty of producing them and the chronic diseases they cause are drawbacks for inoculative augmentation, but their debilitating, sublethal effects on insect hosts (e.g. 9, 68), the large number of species found in every insect order, their frequently high natural prevalence (21, 76), and their array of transmission routes have contributed to their usefulness in these two approaches. Environmental manipulation should prove useful in certain situations, but there are no data that indicate whether this approach is feasible. Limited evidence indicates that protozoa persist outside the host for at most one year (22). Also, they act in an oscillatory fashion rather than keeping host abundance in stable equilibrium (7, 33). Protozoa have been used with insecticides, pheromones, and host-plant resistance in IPM (76).

The protozoa seem suited for introduction or inoculative augmentation only in certain ecosystems. The most important success has been the inoculative augmentation of *Nosema locustae* for grasshopper control on rangeland (76). This approach has also been demonstrated against forest Lepidoptera and sawflies (76). Brooks (22) believes protozoa could be useful for control of insects in stored products. Protozoa have not been particularly successful in control of mosquito larvae (76), but it is too early to disregard them, because only recently have there been breakthroughs in the understanding of their complex life cycles and transmission. Inoculative augmentation has been demonstrated in row crops with two protozoa of *Anthonomus grandis* and with *Nosema pyrausta* (76). *N. pyrausta* was probably introduced accidently along with its host, *O. nubilalis,* and contributes significantly to its natural control (33).

Virtually all authors agree that inundative augmentation is a poor use for the protozoa owing to their initiation of chronic diseases (76). *Vairimorpha necatrix* can kill quickly (62), but this protozoan has no persistence from year to year (J. R. Fuxa, unpublished), and in terms of cost effectiveness it cannot compete with Bt.

BACTERIA   Of the five groups of entomopathogens, the bacteria have the fewest species that are candidates for control; yet, curiously enough, they include the two greatest success stories. Bt and Bt var. *israelensis* (Bti) are clearly quick-damage, inundative agents, whereas the etiological agents of the milky diseases, in particular *B. popilliae,* are slow, permanent introduction agents. The bacteria have the most potential as quick-damage agents, owing to their toxins. The disadvantage of there being few available species is counteracted by the ease of isolating different strains (e.g. 12, 43). Many bacteria are also relatively easy to produce and genetically manipulate. Transmission is horizontal, by environmental contamination and ingestion. Many of these bacteria, though not all, have wide host ranges. Persistence in the soil reservoir is good for *B. popilliae* but poor for other important species such as Bt and *Bacillus sphaericus* (5, 29). The bacteria persist on plant surfaces for a few days at most. Inefficient dispersal or movement between hosts is again a major problem for this group (29, 74); for this reason, along with poor spore and crystal production in dead insects, Bt epizootics are rare (51). However, epizootics of Bt have been observed (156).

Inundative augmentation certainly has been effective. Bt has been or is likely to be accepted as a control agent for certain pests of forest and agriculture, and particularly for leaf feeders near human dwellings or recreational areas (108) and for mosquitoes (65, 105). Some authors (e.g. 97) believe that Bt is the best entomopathogen to develop for forest insect control; Jaques (92), however, proposed that other pathogens have brighter outlooks because they can recycle and because satisfactory spray coverage is difficult in this ecosystem. Bt has not been particularly successful in row crops, usually because of insect feeding habits (91) or perhaps because of problems with persistence (108), coverage, or efficacy of strains.

There are relatively few examples of successful control through the more ecological approaches. *Bacillus popilliae* has been permanently introduced (74), and *B. sphaericus* has been used for long-term control (65). Inoculative augmentation has had only limited success (86). Further introductions or environmental manipulations are unlikely, except perhaps with *B. sphaericus,* unless new naturally occurring or laboratory-produced species or strains are found.

NEMATODES   Nematodes are mainly slow pathogens, but some can be used as quick-damage agents, particularly those that carry a mutualistic bacterium. On the other hand, the life cycle of some nematodes can be up to a year long (132). Many are relatively host-specific, but others have wide host ranges (138). Nematodes include more promising species than bacteria but fewer than the other groups, and strains are not easily separated. Nematode transmission is mainly horizontal, by environmental contamination and infection through the integument or gut. Limited searching ability is common (67). However, the nematodes, like fungi, are susceptible to desiccation, sometimes even in high RH (132). Therefore their use is likely to be restricted to special situations, such as aquatic, soil, or cryptic habitats but not aerial plant surfaces (67, 126). Insects are less likely to develop maturation immunity to nematodes than to other entomopathogens. Also, some nematodes can infect different stages of the host insect. Long-term persistence in soil or other substrates is not yet known to be a particular strength of this group, but high natural prevalence of infection is common (99, 107, 163).

It is difficult to generalize about applications of nematodes, owing to their variety of life cycles and production of both slow and quick damage. This is perhaps the only group in which all four approaches have some promise. Inoculative and inundative augmentation (105, 133, 166) have worked only in habitats with high RH (67). Bark beetles may be a suitable target (98). The slow action of nematodes is beneficial for control of mosquito larvae because the dying, infected larvae compete for resources with uninfected ones (122). Environmental manipulation has not been reported, but since nematodes cause natural epizootics and are sensitive to environmental conditions, this approach cannot be disregarded. Based on Anderson's (5) criteria this would not appear to be a good group for permanent introductions, yet the percentage increase of successful introductions in 1971–1981 over the previous years was greater for nematodes than for any other pathogen group (27, 30). *Deladenus siricidicola* has successfully controlled *Sirex noctilio* (162). Nickle (127) discussed specific nematode groups attractive for introduction or inoculative augmentation.

SUMMARY OF APPROACHES   Modeling studies (5) have indicated that a pathogen introduced to regulate a host population should exhibit moderate pathogenicity (virulence), reduce host reproduction, have good transmission including a vertical component, and produce numerous progeny with a persistent infective stage. Anderson & May (7) hypothesized that virulence and persistence are necessary for lasting though cyclic control. The strategy of introduction and establishment often results in only partial control, i.e. the pathogen is only one component in an IPM system and by itself is not likely to keep insect populations below the economic injury level (EIL). Pathogens are likely to be more successful than parasitoids or predators in row crops owing to their persistence in the absence of hosts, rapid generation time, and lack of density dependence within specific time-space frameworks, i.e. the "lying in wait" strategy of Murdoch et al (125). When a pathogen is too specific to control all pests in a pest complex, introduction and establishment, even if it results in only partial suppression, is the method of choice over augmentation procedures.

Inoculative augmentation and environmental manipulation require pathogens with characteristics similar to those of pathogens used for introductions. These methods deserve more research emphasis because the pathogen is already present and is presumably well adapted to the ecosystem. Inoculative augmentations can include a prophylactic approach with periodic "booster shots" to bolster pathogen population density (83). Partial suppression may be used in conjunction with other types of control. Autodissemination is a promising method for inoculative augmentations (83).

The requirements for pathogens to be used in inundative augmentation are completely different. Virulence and speed are important, as is persistence at the substrate where pathogen and host come into contact (153). Partial suppression or suppression of only one pest in a complex is usually not acceptable, except in the case of key pests. A major problem to be solved is the waste of pathogen; because of current methods of dissemination, a major proportion of the pathogen units does not have an opportunity for host contact (74). Unless genetic engineering has an impact, future success with this approach is likely to be obtained for the most part with better strains of organisms already in use.

Transmission and Dispersal

This section includes discussion of the ecological significance of transmission for control by pathogens. Transmission methods for the various pathogen groups have been thoroughly reviewed elsewhere (8, 79, 99, 131, 155, 157).

Definitions of various types of transmission have been discussed by Andreadis (8) and Fine (56). Smith (146) outlined nine host, pathogen, or environmental factors that affect the frequency of transmission. Fine (56) reviewed four basic types of entomopathogen host-to-host transfer; he concluded that efficient vertical transmission is not necessarily associated with high prevalence and that other variables are also important.

Modeling studies and literature reviews provide some indication that a vertical transmission component allows the pathogen to maintain itself at relatively low host-population densities (24), in hosts with nonoverlapping generations, or when its development and multiplication are limited to certain stages of the host. Pathogens with such a component act in a relatively

density-independent manner (8). Reliance on horizontal transmission makes
the pathogen dependent on host population density for its survival and dispersal; disease prevalence from horizontally transmitted pathogens is proportional to the number of encounters between uninfected host and pathogen
or between infected and uninfected hosts (8). Prevalence of such diseases
often increases dramatically during a season or host generation. Vertical
transmission is the principal means of transfer for some pathogens, but in
others it augments horizontal transmission when host population densities are
low or when hosts at stages susceptible to horizontal transmission are not
present. Vertical transmission also provides foci of disease for later horizontal
routes (8). Horizontal transmission by parasitoid vectors allows the pathogen
to maintain itself at low host population densities (24).

Transmission is interrelated with pathogen virulence, reproductive rate,
and persistence in host population regulation (5). For example, a pathogen
with infrequent vertical transmission can persist in hosts that have low
population density if it also has low virulence and produces chronic disease
(8).

Transmission is relatively unimportant to inundative augmentation. A major emphasis in past control efforts has been on artificial dissemination to
compensate for inefficient transmission or dispersal of a pathogen. This can
work but has major drawbacks, particularly waste and lack of the natural
dispersion needed for most efficient host contact, as dispersion is not a
completely random process in nature (74). Lack of natural dispersion can
perhaps be corrected by autodissemination (74). This technique should be
useful if the early host population growth rate is high relative to transmission
efficiency, although it could adversely affect parasitoid populations if mis-
timed (24).

Transmission is important to the other three approaches. Anderson & May
(6, 7) have presented evidence that a vertical transmission component lowers
the threshold of host population density for successful introduction of an
entomopathogen as well as the equilibrium population density for host population regulation. Similarly, Anderson (5) hypothesized that effective
transmission lowers the minimum introduction rate and is necessary for
introduction and establishment. Either vertical transmission or some form of
persistence is important for successful introduction and establishment in a
temporary habitat (74). Inoculative augmentation and environmental manipulation depend on transmission because they depend on recycling of the
pathogen in the host population.

Virulence

Greater virulence results in better short-term control of insect populations
(157). But modeling studies (5) have indicated that for introduction and

perhaps for inoculative augmentation, moderate virulence ("pathogenicity") is better. For example, moderately virulent strains of *B. popilliae* provide better insect control than highly virulent ones (103).

Pathogen Population Density

Pathogen population density is one of the most important determinants of the course of disease epizootics in insects. This statement is supported by mathematical models (5, 7) and experimental evidence from laboratory determinations of median lethal concentrations, dose-response field research, and studies of natural epizootics (157). All other factors being equal, a greater pathogen population density simply increases the chance of contact between pathogen and uninfected host. Dose-response data are available for bacteria (124), fungi (54, 70), and viruses (96). However, there is practically no information about either the dose actually delivered to insects in the field (136) or thresholds of pathogen numbers necessary for initiating epizootics (157).

The effect of pathogen population density on epizootics is interdependent with other factors such as susceptibility of the host and infectivity and persistence of the pathogen. Unfortunately these interdependencies have been largely ignored, and most research has focused only on introducing pathogen units to increase pathogen population density.

Pathogen Population Distribution or Dispersion

Dispersion or distribution of pathogen units most likely affects their probability of encountering members of the host population (157). Two mathematical descriptions of entomopathogen dispersion (60, 147) both indicated that fungus-infected insects were clumped. Epicenters of disease have been studied more frequently (61), usually with regard to their importance in spread of disease through a host population. Tanada & Fuxa (157) reviewed other distribution patterns, particularly of pathogens in soil.

There is speculation as well as some evidence about the importance of pathogen distribution in control. Harper (74) pointed out that the pattern of a sprayed pathogen is rarely likely to resemble its natural distribution, and that much of it goes to waste. Of course, such waste is common among parasites (i.e. the majority are never able to find a new host), which is a reason for their production of numerous progeny. Burges & Hussey (30) hypothesized that pathogens are naturally unevenly distributed and that this explains how those that must be applied in high doses to produce infection in the laboratory are able to infect hosts in nature. This could account for poor results when many pathogens are applied by spraying. Degree of clumping, as determined by droplet size, has affected control experiments, as has application to different parts of a host plant (157). Additionally, baits increase the efficacy of pathogens (157), though this may be due to change in feeding behavior rather than, or in addition to, clumping of the pathogen.

Host Range

A pathogen that infects several host species in an ecosystem has a better chance of contacting susceptible hosts, which increases pathogen population density and disease prevalence. However, the situation is seldom so straightforward; it can be complicated by factors such as persistence of infectious units or prevalence or number of generations of the hosts. Relatively host-specific pathogens (e.g. *B. popilliae* and various NPVs) as well as generalists (e.g. *N. locustae* and some fungi) have resulted in efficacious introductions and inoculative augmentations (74). Strains of certain pathogens may not have as wide a host range as the species. Ignoffo (83) suggested that the fungus *N. rileyi* should be augmented in a relatively innocuous host species in the field so that it would recycle and increase its population before invasion of the crop by more harmful pest species. However, use of a strain with a narrower host range can defeat such a strategy (J. R. Fuxa, unpublished data).

For inundative augmentation, the relatively narrow host range of even the least host-specific pathogens is one of the greatest disadvantages for control of pest complexes, yet one of the greatest advantages for conservation of other natural control agents such as parasitoids and predators.

Persistence

Persistence is an important factor in pathogen regulation of host populations (7). Persistence gives certain baculoviruses, fungi, and microsporidia good potential as residual agents (5). Models by May & Anderson (116) indicate that persistence in the host population is facilitated by low virulence and long duration of infection. In turn, persistence is one factor that influences frequency of transmission (146). Apparently only persistence, plus an efficient mechanism of movement from reservoir to host substrate, can compensate for lack of efficient transmission. Modeling studies also indicate that even pathogens that induce lifelong immunity or that are very virulent can be endemic if they can survive long in their free-living stage (6). With long-lived transmission stages, the density of the pathogen population as well as that of infected and susceptible hosts becomes a factor (4). The short persistence of pathogens on vegetation is detrimental to inundative augmentation. Introduction and establishment and inoculative augmentation are also greatly affected.

Some of the conclusions from modeling are supported by field research. Anderson & May (7) reviewed field data for their modeling of pathogen regulation of forest pest populations. Persistence of baculoviruses in forest pests has been correlated with the level of control (159). Some pathogens with poor persistence, such as *N. rileyi*, fail to give timely control of row-crop pests, whereas others with good persistence, such as the NPVs of *A. gemmatalis* and *Pseudoplusia includens,* have provided a level of control in soybeans (74).

PEST CHARACTERISTICS

Pest Population Growth Characteristics

One way to conceptualize pest populations is according to the theory of r- and K-selection (39). There is some doubt about whether management strategy should be based on this theory, but as a generalization it could prove useful, and it is beginning to be considered in control with entomopathogens.

Unique characteristics of pathogens have been ignored in past controversies about biological control of r pests, particularly in agroecosystems (44, 139). For example, there may be r-selected entomopathogens that are not readily overwhelmed by r-pest population growth. Andrews & Rouse (10) proposed applying the theory of r- and K-selection to plant pathogens. They listed characteristics of r- and K-selected pathogens, and concluded that r pathogens tend to destroy their habitat (the host) and that K pathogens tend to stress but coexist with the host. Similarly, Mitchell (121) proposed that there are l- and m-selected animal parasites. Whereas l-parasites kill the host before it reaches reproductive age, m-species tend to affect host fecundity.

Anderson & May (6, 7) did not specifically mention r- and K-selection, but their models can be interpreted according to the theory. For example, they concluded (7) that if transmission stages are produced in infected hosts at a sufficiently fast rate, then the disease will regulate the host population if the disease-induced death rate, α, exceeds the host's birth rate minus its natural death rate. Based on their literature review, α was generally highest in bacteria, followed by viruses, fungi, and protozoa. Stated another way (6), for disease to regulate a host population the case mortality rate must be high relative to the intrinsic growth rate of the disease-free population.

Entwistle (45) and Evans & Entwistle (46) considered entomopathogenic viruses in relation to the theory of r- and K-selection. Entwistle (45) listed viruses that infect various r and K pests and made several generalizations. For example, r pests are commonly beset by viral epizootics that result in deposition of numerous persistent virions in the environment for the host's return, and the hosts tend to transport their own viruses and thus latent infections that can be activated by other NPVs. Viruses associated with K hosts tend to produce enzootic rather than epizootic disease.

Hypotheses can be made about the relation of r- and K-selection to control by pathogens. Anderson (5) proposed that the minimum rate of introduction of a pathogen is based partly on the reproductive rate of the host. For example, applications every few years suffice for control of certain forest pests, owing to their low r, but much more frequent application is necessary in food crop pests, which have high r. Harper (74) hypothesized that their short generation times and fast population growth give pathogens more potential than parasites and predators for control of r pests. Service (145) stated that most mosquitoes are r pests or intermediates and that Bti, used inundatively, has good potential for both r and K pests. However, *Romanomermis* may be more of a K parasite, and hence of limited usefulness in mosquito control. Entwistle (45) concluded that the effectiveness of strategies for manipulation of viruses is related to the pest's place in the r-K spectrum.

Generalizations can be made about pathogens that have not been considered in light of r- and K-selection. The r pests are primary targets for the quick-damage pathogens, particularly toxin-producing bacteria and certain nematodes such as *Steinernema feltiae*. The r pests might be amenable to control by r-selected slow entomopathogens used in inoculative augmentation or introduction in certain situations. The NPVs, with their long persistence in soil, explosive population growth, and quick generation times, have been permanently introduced for partial control of r pests in an agroecosystem (74), and have been used in inoculative augmentation to control *A. gemmatalis* in soybeans in Brazil. Certain fungi, because of their inherent spreading capability, may control r pests in agroecosystems. The persistence of entomopathogens is likely to be beneficial because r pests rebound readily from control attempts (39).

Inoculative augmentation and introduction are the best methods for control of K pests. These may be the least amenable to inundative augmentation; even chemical pesticides are often not the treatment of choice except when small pest populations cause high losses (39), a situation in which pathogens usually are not effective. K-strategist pathogens such as most microsporidia, many fungi, and certain nematodes (e.g. mermithids) are possible choices (99). Persistence, whether in the host insect or habitat, and efficient transmission, whether vertical or vectored, are likely to be desirable characteristics.

Against the intermediate pests biological control and selective pesticides are the methods of choice (39), so control with pathogens is a good possibility. These are the pests best suited to biological control by slow pathogens. However, it should be remembered that most intermediate pests in temperate climates tend toward the r end of the spectrum (39). All four control strategies can be successful, and quick-damage, r-selected slow, and K-selected slow pathogens all have potential on a case-by-case basis.

Host Population Composition

Immunity and resistance to pathogens have been reviewed (17, 19, 20, 26, 103, 138, 161). Differences in susceptibility to viruses, fungi, bacteria, protozoa, and a nematode have been demonstrated in field or laboratory populations of single insect species, and populations resistant to viruses, fungi, bacteria, and nematodes have been selected in the laboratory (19). Increased resistance in a host population has not been demonstrated in any field experiment on control by pathogens, but selection of this kind was a probable cause of reduced infection rates in the field in at least one case, involving a GV in *Phthorimaea operculella* (20).

Though field data are few, host-population susceptibility has been a major feature of theoretical discussions of both natural and applied epizootiology. There is little doubt that increased exposure of field populations will result in selection for less susceptible individuals (5) just as in the case of pesticides. It is expected that as selective pressures reduce the average susceptibility of a host population to a pathogen, that pathogen's regulation of that population will decrease (116). However, it should be easier for a pathogen to regulate a host population if there is no individually acquired immunity, which is more nearly the case for invertebrates than vertebrates (115). It has been postulated that changes in host-population susceptibility are more important than changes in pathogen virulence (164). On the other hand, Brown (24) stated that epizootics do not cease owing to a lack of susceptible hosts, but his model was based on oversimplified assumptions and did demonstrate a substantial reduction in the number of susceptible hosts.

The potential of certain recent genetic engineering approaches is questionable because of possible resistance. One approach is to transfer the Bt δ-endotoxin gene to native soil- or leaf-dwelling bacteria (144). Another approach is to cause host plants themselves to produce the toxin (38, 100). Thus the toxin will be produced in the environment and available whenever the pest insect arrives. However, it has been well demonstrated that nonselective, widespread application of insecticides is one of the most efficient ways to induce resistance; resistance by insects to δ-endotoxin has been detected in populations of a stored-grain pest (118).

Host Population Density and Distribution

There is little doubt that entomopathogens are host-density dependent (161). Increased host density leads to increased contact between uninfected and infected hosts and between uninfected hosts and pathogen units. It also can stress insects in the population, predisposing them to disease, and it makes substrate and nutrients (i.e. hosts) more readily available for pathogen growth and reproduction. Density dependence is known in the viruses (96), fungi (70, 72, 143), protozoa (59, 75), bacteria (51), and nematodes (78). Entomopathogens also have been described as imperfectly density dependent (15) and delayed density dependent (96). Entomopathogens do not invariably cause epizootics at high host densities (154). Huffaker et al (82) pointed out that weather does not cause pathogens to act independently of host density as some workers had concluded, but that weather can determine whether a disease will appear in a host population. If the disease does appear, then it can still be density dependent.

Density dependence interacts with other factors. It is more dominant, or at least more evident, in pathogens that are transmitted horizontally than in those transmitted vertically (8). If the transmission rate is low, an obligate pathogen requires a high host population density or a host with a long lifespan in order to persist (103). If a pathogen has a high transmission rate, it can persist with low host density. Some workers believe that latent infections may be important for virus persistence at low host population densities. Long persistence is essential during periods when there are no hosts (96), e.g. during winter in a temperate climate.

Modeling studies have supported the concept of density dependence (e.g. 24). A model by Anderson & May (7) indicated that host density must exceed a certain threshold in order to maintain infections, and that the threshold is higher for virulent pathogens or those with low transmission rates. Also, there are host-density thresholds for disease persistence, establishment of infectious disease (4), and occurrence of epizootics.

Though few workers doubt that entomopathogens are density dependent, it is clear that density dependence must be carefully defined in a space-time framework. Entomopathogens can be density independent for at least two to three seasons owing to the presence of large numbers of pathogen units in the environment (53, 63, 157) or perhaps to transovum transmission or latency (53). Federici (53) believed that viruses can cycle between density-dependent and density-independent phases, but this is probably better described as a very delayed density-dependent phase due to long-lived transmission stages.

Density dependence has an important bearing on entomopathogenic control. The delayed density dependence of viruses, due to persistence, and of fungi, perhaps due to dispersal, makes it possible to suppress insect populations in row crops by introduction and establishment (140). Density dependence is an important consideration when a pathogen is intended to recycle or multiply in a host or to be transmitted to new hosts, and a density-dependent pathogen may be at an advantage when it is intended to regulate a population continually. Density dependence is unimportant to inundative augmentation, which simulates density independence due to environmental contamination. However, augmentation is effective because it speeds up epizootics for insect control by bypassing the lag effect of the pathogen's numerical response to host-population density (74). Inundative augmentation can be relatively ineffective when the target population is too dense (103, 124).

The effect of dispersion or distribution of the host population has not been well studied. Steinhaus (150, 151) hypothesized that tendencies to aggregate or disperse may be important to the severity or extent of disease, and that distribution of disease organisms is likely to be important in control. *Delia antiqua* aggregates where the habitat favors persistence and germination of *Entomophthora muscae* conidia, and this leads to higher disease prevalence (34). Increasing aggregation of *Heliothis zea* in different crops correlates with increasing prevalence of *Nosema heliothidis* infection (110). Aggregation of *Oryctes rhinoceros* females contributes to baculovirus transmission (11).

Host Behavior

Behavior frequently either affects whether a pathogen infects hosts in a population or changes after a host becomes infected. Grooming, cannibalism, aggregation patterns, level of activity, removal of infected by uninfected hosts, and feeding in protected situations are known to affect whether hosts become infected (15, 99, 109, 110, 161). Changes in behavior after infection include changes in flight habits or capability, feeding during daylight rather than at night, reduced activity, failure to burrow deeply into soil, failure to return to the nest, seclusion in debris, and movement to unusually high positions on host plants (61). Some of these behavior changes, such as movement to higher positions, are thought to be important in epizootiology, e.g. in disease transmission.

Host behavior is an important consideration in control, particularly by inundative (49) or inoculative augmentation. These approaches can fail when larvae feed in protected places (109) and application is not designed to deposit the pathogen at feeding sites (91). Because most entomopathogens are passive there is little doubt that active pests are generally more amenable than sedentary ones to control by any approach. For example, introduction of NPV was more successful against a pest whose females deposit eggs singly and thus start many foci of infection than with a similar pest that lays eggs in clumps (16). Behavior of *Trogoderma* spp. has been modified with a pheromone to aid neogregarine dissemination (32), and other autodissemination techniques similarly take advantage of host behavior. In some cases a pathogen may be particularly suited for controlling an insect in a protected habitat. For example, certain nematodes can search to a limited extent for protected insects (67).

THE ECOSYSTEM

The use of pathogens or any other control agent in IPM necessitates knowledge of the ecosystem. The ecosystem components that affect control, other than the pathogen and the pest, include the pest and mortality-agent complex, environmental factors, type of ecosystem, and community structure. Additionally, the pest's interactions with its host and with humans are important.

Economic or Action Threshold

The economic or action threshold so important to IPM is critical to entomopathogenic control. The concept that only a certain portion of a pest population need be eliminated is important, because pathogens are even less likely than insecticides to eliminate near 100% of an insect population. In medical entomology, eradication is now seldom deemed possible; even the most efficient control measures are not likely to do more than reduce disease transmission (145). It is generally recognized that pests with low economic injury levels (EILs) require a virulent pathogen applied inundatively or repeatedly, since virulent pathogens tend not to be persistent. Pests with high EILs might be suppressed adequately with a less virulent, chronic pathogen (157).

One problem in control with entomopathogens is that whereas the EIL is the same regardless of control measure, the economic or action threshold is not likely in many cases to be the same for pathogens as it is for insecticides. Action thresholds are dynamic and depend on a wide range of factors, including the activity and speed of the control agent. The quick-damage pathogens could be used in inundative augmentation when action thresholds are expected to be similar to those for chemicals. The slow pathogens, however, are not likely to be successful with these thresholds. For example, the action threshold for control of *A. gemmatalis* by NPV in soybean is different from that for control by insecticides (140). Modeling efforts like those of Anderson (5) could help in selection of method. If the EIL is lower than Anderson's pest population critical density (H_T) for an epizootic, then only augmentation is likely to be practical. If the EIL is greater than H_T, then introduction could be considered. The disease threshold is dynamic and could perhaps be lowered by environmental manipulation.

However, EILs are not easily determined, and action thresholds for pathogens are even less so. Both thresholds are dynamic and can change in response to factors such as consumer attitudes. Also, maturation immunity, a common phenomenon with infectious disease in insects, can cause confusion. Is improved control with earlier application a function of exposing the more easily infected, younger insects, or one of allowing the pathogen more time to work before the pest population reaches the EIL? In the case of the *A. gemmatalis* NPV in soybean, improved control could be attributed to the increased time for action, partly because this host has relatively little maturation immunity to NPV (18).

Pest and Mortality Agent Complexes

The specificity of entomopathogens was once thought to be their major advantage in IPM. However, it has turned out to be a disadvantage as well, partly because it contributes to their ineffectiveness in pest complexes. Regardless of the advantages, growers are reluctant to apply more than one specific agent when a single broad-action agent will suffice. Thus if a pathogen does not control all the pests in a complex, then neither inundative augmentation, inoculative augmentation, nor environmental manipulation is likely to be useful. Control attempts with pathogens in such situations should be limited to introduction and establishment. One exception is when the target is a key pest in a pest complex; in this case inundative and perhaps inoculative augmentation can be acceptable, particularly when the key pest population reaches action thresholds before occasional pests, as with *Heliothis* spp. in cotton in the southeastern United States. One of the most valuable contributions of genetic engineering to control with pathogens is likely to be in the manipulation of pathogen host ranges.

The mortality agent complex is important. Pathogens sometimes infect parasitoids as well as the common host (e.g. 23), but more commonly antagonism that occurs results from competition for the same host. This is usually less severe than with chemical insecticides (e.g. 140). Thus a major advantage of entomopathogens is that by affecting other mortality agents less severely, they seldom induce outbreaks of secondary pests. Another consideration is that parasitoids (e.g. 40), predatory arthropods (e.g. 1), or mammals (e.g. 40) can aid the dispersal and thus the efficacy of a pathogen. A pathogen can be one component of a mortality agent complex that suppresses a pest population, as in control of *Gilpinia hercyniae* (40).

Environmental Factors

Environmental factors have generally been recognized as one of the three or four major factors that affect epizootiology of entomopathogens and, hence, their use in control of insects (64, 153). For example, pathogen persistence is a critical factor in epizootiology and introductions (5), and environmental factors are important to persistence. Abiotic components are more important than biotic ones for inundative augmentation, but all components influence the other three approaches and interact in a complex way (58). Environmental factors are probably the only component of the ecosystem besides the pathogen population that can be manipulated for control, although it is possible that the usefulness of this approach will be limited.

The effects of environmental factors have recently been reviewed thoroughly (8, 15, 157, 161) and are thus only summarized here. Abiotic factors (e.g. temperature, humidity, rain or other moisture, wind, sunlight, soil, substrate pH, pesticides, and antimicrobial substances) are known to affect host susceptibility and pathogen persistence, transmission, dispersal, development, germination, and production. Biotic factors (e.g. host population, insect host plant, predatory and parasitic insects, alternate hosts, small mammals, grazing mammals, birds, and other microorganisms) can affect the life cycle, dispersal, transmission, and persistence of entomopathogens. Benz (15) stressed that interactions of environmental factors affect epizootics and that testing of the effects of only one factor at a time may not always yield valid results.

In most attempts to manipulate the environment for control, formulation additives have been used in conjunction with augmentation. The most common additives have been UV-sunlight protectants (93). Other substances have been added to neutralize and buffer the foliage substrate (170) or to retard evaporation (89). Another method has been to apply pathogens between dusk and dawn to avoid sunlight (86).

Environmental manipulations without augmentation have had negative as well as positive effects on epizootics. Deep tilling of soil can make virus unavailable, but it can also disturb the upper layers of soil, allowing virus to be dispersed (79). Movement of cattle or farm machinery might move virus from the soil (63, 95). Applications of sublethal doses of insecticides have induced epizootics by stressing insects, whereas normal doses can selectively remove diseased, more susceptible insects, thereby reducing the pathogen reservoir (101). Fungal epizootics have been encouraged by irrigation or sprinkling water (141). Soybeans have been planted early in narrow rows at high seeding rates to accelerate canopy closure and thus raise humidity to enhance *N. rileyi* epizootics (149). Early cutting of alfalfa (24) and cultural manipulations of field borders (34) have concentrated host insects in areas with high RH, which resulted in enhanced epizootics and insect control. Finally, disregard of pathogens in selection for resistance in host plants has proved disadvantageous; a corn cultivar resistant to *Pyrausta nubilalis* lost its advantage because the insects on this cultivar were relatively resistant to the microsporidium *Nosema (Perezia) pyrausta* (37).

Types of Ecosystems

The type of ecosystem affects the use of entomopathogens in IPM (27, 58). In relatively stable habitats, such as forests and grasslands, introduction and establishment, low-frequency inoculative augmentation, and occasionally inundative augmentation and environmental manipulation (11, 27, 58, 74, 80, 117) have been effective. In unstable ecosystems, particularly agricultural crops, only the quick-damage pathogens are likely to be able to compete with insecticides when applied in inundative augmentation. Many experimental inundative augmentations have been successful, particularly with baculoviruses, but have not been put into practical use (168), probably because they cannot compete with chemicals in speed, cost, and reliability. Inoculative augmentation is the best approach for the slow pathogens, although environmental manipulation and partial control by introduction and establishment are possible (74). In aquatic ecosystems, inundative augmentation and introduction and establishment have been effective, and inoculative augmentation has potential (27, 105), particularly for pathogens that are vertically transmitted.

Community Structure

Community structure can be affected by entomopathogens and should be considered in the selection of control agents. Entomogenous fungi are known to cause collapse of parasitoid populations, allowing pests of cotton to reach high population levels. After two months, the fungi become inactive, the parasite populations recover, and the pest problem diminishes (48). Miller (120) pointed out that patterns in parasite community composition may provide a framework for selecting biocontrol agents. Weiser (165) discussed patterns of entomopathogens in certain geographical pest analogs. He found similar types of pathogens in similar pest insects from different parts of the world.

CONCLUSION

There is little doubt that epizootiology and ecology are central to the successful use of entomopathogens in IPM. Future development should be based more on the ecosystem and less on one pest or pathogen. Successful inundative augmentations in the future are likely to be limited owing to the limited number of quick-damage pathogens and the unsuitability of slow pathogens, unless genetic manipulations increase the number of the former. The greatest potential for IPM lies in the inoculative augmentation and introduction and establishment approaches, including introduction of new strains of pathogen species already present. These approaches have been effective, yet have received only a fraction of the research devoted to similar approaches with parasitoids and predators. These approaches minimize certain entomopathogen weaknesses such as slow debilitation of pest individuals and populations, and take advantage of ecological strengths such as recycling, persistence, and rapid generation times. Success is most likely if pathogens are used as relatively inexpensive partial suppression agents in conjunction with treatment thresholds and other control methods. Unfortunately, the research required to evaluate their potential for use in this manner is unglamorous and relatively unfunded. Environmental manipulation is a promising but little-studied approach that is likely to be used only to a limited extent.

Ecological research on entomopathogens is complex and time-consuming, but not so overwhelming that great strides cannot be made. Such research is necessary if data and suggestions are to be available when IPM generalists must evaluate components to include in interactive research or IPM programs.

ACKNOWLEDGMENTS

I thank W. M. Brooks, L. D. Newsom, R. N. Story, and Y. Tanada for critically reading the manuscript and for providing helpful suggestions.

Literature Cited

1. Abbas, M. S. T., Boucias, D. G. 1984. Interaction between nuclear polyhedrosis virus-infected *Anticarsia gemmatalis* (Lepidoptera: Noctuidae) larvae and predator *Podisus maculiventris* (Say) (Hemiptera: Pentatomidae). *Environ. Entomol.* 13:599–602
2. Allen, G. E., Ignoffo, C. M., Jaques, R. P., eds. 1978. *Microbial Control of Insect Pests: Future Strategies in Pest Management Systems.* Gainesville, Fla: NSF-USDA-Univ. Fla. 290 pp.
3. Anderson, J. F., Kaya, H. K. 1976. *Perspectives in Forest Entomology.* London/New York: Academic. 428 pp.
4. Anderson, R. M. 1981. Population ecology of infectious disease agents. See Ref. 114, pp. 318–55
5. Anderson, R. M. 1982. Theoretical basis for the use of pathogens as biological control agents of pest species. *Parasitology* 84:3–33
6. Anderson, R. M., May, R. M. 1979. Population biology of infectious diseases: Part I. *Nature* 280:361–67
7. Anderson, R. M., May, R. M. 1980. Infectious diseases and population cycles of forest insects. *Science* 210:658–61
8. Andreadis, T. G. 1987. Transmission. See Ref. 64, In press
9. Andreadis, T. G., Hall, D. W. 1979. Significance of transovarial infections of *Amblyospora* sp. (Microspora: Thelohaniidae) in relation to parasite maintenance in the mosquito *Culex salinarius* Coquillet. *J. Invertebr. Pathol.* 34:152–57
10. Andrews, J. H., Rouse, D. I. 1982. Plant pathogens and the theory of r- and K-selection. *Am. Nat.* 120:283–96
11. Bedford, G. O. 1981. Control of the rhinoceros beetle by baculovirus. See Ref. 28, pp. 409–26
12. Beegle, C. C. 1979. Use of entomogenous bacteria in agroecosystems. In *Developments in Industrial Microbiology,* Vol. 20, ed. L. A. Underkofler, pp. 97–104. Arlington, Va: Soc. Ind. Microbiol. 793 pp.
13. Beltsville Agricultural Research Center Symposium V Committee. 1981. *Biological Control in Crop Production.* London/Toronto/Sydney: Allanheld, Osmun. 444 pp.
14. Benz, G. 1981. Use of viruses for insect suppression. See Ref. 13, pp. 259–72
15. Benz, G. 1987. Environment. See Ref. 64, In press
16. Bird, F. T. 1961. Transmission of some insect viruses with particular reference to ovarial transmission and its importance in the development of epizootics. *J. Insect Pathol.* 3:352–80
17. Boman, H. G. 1981. Insect responses to microbial infections. See Ref. 28, pp. 769–84
18. Boucias, D. G., Johnson, D. W., Allen, G. E. 1980. Effects of host age, virus dosage, and temperature on the infectivity of a nucleopolyhedrosis virus against velvetbean caterpillar, *Anticarsia gemmatalis,* larvae. *Environ. Entomol.* 9:59–61
19. Briese, D. T. 1981. Resistance of insect species to microbial pathogens. See Ref. 42, pp. 511–45
20. Briese, D. T., Podgwaite, J. D. 1985. Development of viral resistance in insect populations. See Ref. 112, pp. 361–98
21. Brooks, W. M. 1978. Induced epizootics: protozoa. See Ref. 2, pp. 37–42
22. Brooks, W. M. 1980. Production and efficacy of protozoa. *Biotechnol. Bioeng.* 22:1415–40
23. Brooks, W. M., Cranford, J. D. 1978. Host-pathogen relationships of *Nosema heliothidis* Lutz and Splendore. *Misc. Publ. Entomol. Soc. Am.* 11:51–63
24. Brown, G. C. 1987. Modeling. See Ref. 64, In press
25. Bulla, L. A. Jr., ed. 1973. *Regulation of Insect Populations by Microorganisms,* Ann. NY Acad. Sci., Vol. 217. New York: NY Acad. Sci. 243 pp.
26. Burges, H. D. 1971. Possibilities of pest resistance to microbial control agents. See Ref. 31, pp. 445–57
27. Burges, H. D. 1981. Strategy for microbial control of pests in 1980 and beyond. See Ref. 28, pp. 797–836
28. Burges, H. D. 1981. *Microbial Control of Pests and Plant Diseases 1970–1980.* London/New York: Academic. 949 pp.
29. Burges, H. D. 1982. Control of insects by bacteria. *Parasitology* 84:79–117
30. Burges, H. D., Hussey, N. W. 1971. Past achievements and future prospects. See Ref. 31, pp. 687–709
31. Burges, H. D., Hussey, N. W., eds. 1971. *Microbial Control of Insects and Mites.* London/New York: Academic. 861 pp.
32. Burkholder, W. E. 1980. Linking insect-behavior-modifying chemicals and pathogens. See Ref. 90, p. 50
33. Canning, E. U. 1982. An evaluation of protozoal characteristics in relation to biological control of pests. *Parasitology* 84:119–49
34. Carruthers, R. I., Soper, R. S. 1987. Fungal diseases. See Ref. 64, In press

35. Chapman, H. C., Petersen, J. J., Fukuda, T. 1972. Predators and pathogens for mosquito control. *Am. J. Trop. Med. Hyg.* 21:777–81
36. Cheng, T. C. 1984. *Comparative Pathobiology*, Vol. 7, *Pathogens of Invertebrates. Application in Biological Control and Transmission Mechanisms.* New York/London: Plenum. 278 pp.
37. Chiang, H. C., Holdaway, F. G. 1960. Relative effectiveness of resistance of field corn to the European corn borer, *Pyrausta nubilalis,* in crop protection and in population control. *J. Econ. Entomol.* 53:918–24
38. Comai, L., Stalker, D. M. 1984. Impact of genetic engineering on crop protection. *Crop Prot.* 3:399–408
39. Conway, G. 1981. Man versus pests. See Ref. 114, pp. 356–86
40. Cunningham, J. C. 1982. Field trials with baculoviruses: control of forest insect pests. See Ref. 104, pp. 335–86
41. Cunningham, J. C., Entwistle, P. F. 1981. Control of sawflies by baculovirus. See Ref. 28, pp. 379–407
42. Davidson, E. W., ed. 1981. *Pathogenesis of Invertebrate Microbial Diseases.* Totowa, NJ: Allanheld, Osmun. 562 pp.
43. Dulmage, H. T. and Cooperators. 1981. Insecticidal activity of isolates of *Bacillus thuringiensis* and their potential for pest control. See Ref. 28, pp. 193–222
44. Ehler, L. E., Miller, J. C. 1978. Biological control in temporary agroecosystems. *Entomophaga* 23:207–12
45. Entwistle, P. F. 1986. Epizootiology and strategies of microbial control. In *Biological Plant and Health Protection,* ed. J. M. Franz, M. Lindaur. Stuttgart: Fischer. In press
46. Evans, H. F., Entwistle, P. F. 1987. Viral diseases. See Ref. 64, In press
47. Falcon, L. A. 1971. Microbial control as a tool in integrated control programs. See Ref. 81, pp. 346–66
48. Falcon, L. A. 1973. Biological factors that affect the success of microbial insecticides: development of integrated control. See Ref. 25, pp. 173–86
49. Falcon, L. A. 1976. Problems associated with the use of arthropod viruses in pest control. *Ann. Rev. Entomol.* 21:305–24
50. Falcon, L. A., Kane, W. R., Bethell, R. S. 1968. Preliminary evaluation of a granulosis virus for control of the codling moth. *J. Econ. Entomol.* 61:1208–13
51. Faust, R. M., Bulla, L. A. Jr. 1982. Bacteria and their toxins as insecticides. See Ref. 104, pp. 75–208
52. Fawcett, H. S. 1944. Fungus and bacterial diseases of insects as factors in biological control. *Bot. Rev.* 10:327–48
53. Federici, B. A. 1980. Disease prevalence and epizootics in insect populations. In *Characterization, Production and Utilization of Entomopathogenic Viruses,* ed. C. M. Ignoffo, M. E. Martignoni, J. L. Vaughn, pp. 17–30. Washington, DC: Am. Soc. Microbiol. 216 pp.
54. Ferron, P. 1978. Biological control of insect pests by entomogenous fungi. *Ann. Rev. Entomol.* 23:409–42
55. Ferron, P. 1981. Pest control by the fungi *Beauveria* and *Metarhizium.* See Ref. 28, pp. 465–82
56. Fine, P. E. M. 1984. Vertical transmission of pathogens of invertebrates. See Ref. 36, pp. 205–41
57. Flint, M. L., van den Bosch, R., eds. 1981. *Introduction to Integrated Pest Management.* New York/London: Plenum. 240 pp.
58. Franz, J. M. 1971. Influence of environment and modern trends in crop management on microbial control. See Ref. 31, pp. 407–44
59. Franz, J. M., Huger, A. M. 1971. Microsporidia causing the collapse of an outbreak of the green tortrix (*Tortrix viridana* L.) in Germany. *Proc. Int. Colloq. Insect Pathol., 4th, College Park, Md.,* pp. 48–53. College Park, Md: Soc. Invertebr. Pathol.
60. Fuxa, J. R. 1984. Dispersion and spread of the entomopathogenic fungus *Nomuraea rileyi* (Moniliales: Moniliaceae) in a soybean field. *Environ. Entomol.* 13:252–58
61. Fuxa, J. R. 1987. Ecological methods. See Ref. 64, In press
62. Fuxa, J. R., Brooks, W. M. 1979. Effects of *Vairimorpha necatrix* in sprays and corn meal on *Heliothis* species on tobacco, soybeans, and sorghum. *J. Econ. Entomol.* 72:462–67
63. Fuxa, J. R., Geaghan, J. P. 1983. Multiple-regression analysis of factors affecting prevalence of nuclear polyhedrosis virus in *Spodoptera frugiperda* (Lepidoptera: Noctuidae) populations. *Environ. Entomol.* 12:311–16
64. Fuxa, J. R., Tanada, Y., eds. 1987. *Epizootiology of Insect Diseases.* New York/Chichester: Wiley. In press
65. Garcia, R. 1983. Mosquito management: ecological approaches. *Environ. Manage.* 7:73–78
66. Gard, I. E., Falcon, L. A. 1978. Autodissemination of entomopathogens: virus. See Ref. 2, pp. 46–54
67. Gaugler, R. 1981. Biological control

potential of neoaplectanid nematodes. *J. Nematol.* 13:241–49

68. Gaugler, R. R., Brooks, W. M. 1975. Sublethal effects of infection by *Nosema heliothidis* in the corn earworm, *Heliothis zea. J. Invertebr. Pathol.* 26:57–63

69. Gottwald, T. R., Tedders, W. L. 1984. Suppression of pecan weevil populations with entomopathogenic fungi. *Environ. Entomol.* 12:471–74

70. Hagen, K. S., van den Bosch, R. 1968. Impact of pathogens, parasites, and predators on aphids. *Ann. Rev. Entomol.* 13:325–84

71. Hall, R. A. 1981. The fungus *Verticillium lecanii* as a microbial insecticide against aphids and scales. See Ref. 28, pp. 483–98

72. Hall, R. A., Papierok, B. 1982. Fungi as biological control agents of arthopods of agricultural and medical importance. *Parasitology* 84:205–40

73. Hamm, J. J. 1984. Invertebrate pathology and biological control. *J. Ga. Entomol. Soc.* 19:6–13

74. Harper, J. D. 1987. Applied epizootiology: microbial control of insects. See Ref. 64, In press

75. Henry, J. E. 1972. Epizootiology of infections by *Nosema locustae* Canning in grasshoppers. *Acrida* 1:111–20

76. Henry, J. E. 1981. Natural and applied control of insects by protozoa. *Ann. Rev. Entomol.* 26:49–73

77. Henry, J. E., Oma, E. A. 1981. Pest control by *Nosema locustae,* a pathogen of grasshoppers and crickets. See Ref. 28, pp. 573–86

78. Hominick, W. M., Tingley, G. A. 1984. Mermithid nematodes and the control of insect vectors of human disease. *Biocontrol News Inf.* 5:7

79. Hostetter, D. L., Bell, M. R. 1985. Natural dispersal of baculoviruses in the environment. See Ref. 112, pp. 249–84

80. Huber, J., Dickler, E. 1977. Codling moth granulosis virus: its efficiency in the field in comparison with organophosphorus insecticides. *J. Econ. Entomol.* 70:557–61

81. Huffaker, C. B., ed. 1971. *Biological Control.* New York/London: Plenum/Rosetta. 511 pp.

82. Huffaker, C. B., Messenger, P. S., De-Bach, P. 1971. The natural enemy component in natural control and the theory of biological control. See Ref. 81, pp. 16–67

83. Ignoffo, C. M. 1978. Strategies to increase the use of entomopathogens. *J. Invertebr. Pathol.* 31:1–3

84. Ignoffo, C. M. 1980. Soybeans. See Ref. 90, pp. 27–28

85. Ignoffo, C. M. 1981. The fungus *Nomuraea rileyi* as a microbial insecticide. See Ref. 28, pp. 513–38

86. Ignoffo, C. M. 1985. Manipulating enzootic-epizootic diseases of arthropods. In *Biological Control in Agricultural Integrated Pest Management Systems,* ed. M. A. Hoy, D. C. Herzog, pp. 243–62. New York: Academic. 589 pp.

87. Ignoffo, C. M., Anderson, R. F. 1979. Bioinsecticides. In *Microbial Technology,* ed. H. J. Pepller, D. Perlman, Vol. 1, pp. 1–28. London/New York: Academic. 594 pp. 2nd ed.

88. Ignoffo, C. M., Couch, T. L. 1981. The nucleopolyhedrosis virus of *Heliothis* species as a microbial insecticide. See Ref. 28, pp. 329–62

89. Ignoffo, C. M., Hostetter, D. L., Sutter, D. B. 1976. Gustatory stimulant, sunlight protectant, evaporation retardant: three characteristics of a microbial insecticidal adjuvant. *J. Econ. Entomol.* 69:207–10

90. Insect Pathology Resource Center. 1980. *Proceedings of Workshop on Insect Pest Management with Microbial Agents: Recent Achievements, Deficiencies, and Innovations.* Ithaca, NY: Boyce Thompson Inst.-USDA-Cornell Univ. Dep. Entomol. 71 pp.

91. Jaques, R. P. 1973. Biological factors that affect the success of microbial insecticides. See Ref. 25, pp. 109–19

92. Jaques, R. P. 1976. Use of microbial agents for insect control in Canada. *Proc. 1st Int. Colloq. Invertebr. Pathol., 9th Ann. Meet. Soc. Invertebr. Pathol., Kingston, Canada,* pp. 64–68. Kingston, Canada: Queens's Univ. Print. Dep.

93. Jaques, R. P. 1978. Manipulation of the environment to increase effectiveness of microbial agents. See Ref. 2, pp. 72–84

94. Johnson, D. W., Kish, L. P., Allen, G. E. 1976. Field evaluation of selected pesticides on the natural development of the entomopathogen, *Nomuraea rileyi,* on the velvetbean caterpillar in soybean. *Environ. Entomol.* 5:964–66

95. Kalmakoff, J., Crawford, A. M. 1982. Enzootic virus control of *Wiseana* spp. in the pasture environment. See Ref. 104, pp. 435–48

96. Kaupp, W. J., Sohi, S. S. 1985. The role of viruses in the ecosystem. See Ref. 112, pp. 441–65

97. Kaya, H. K. 1976. Insect pathogens in natural and microbial control of forest defoliators. See Ref. 3, pp. 251–63

98. Kaya, H. K. 1984. Nematode parasites of bark beetles. See Ref. 128, pp. 727–54
99. Kaya, H. K. 1987. Diseases caused by nematodes. See Ref. 64, In press
100. Kirschbaum, J. B. 1985. Potential implication of genetic engineering and other biotechnologies to insect control. *Ann. Rev. Entomol.* 30:51–70
101. Klassen, W. 1975. *Impressions of Applied Insect Pathology in the U.S.S.R.* Washington, DC: USDA Agric. Res. Serv. 47 pp.
102. Klein, M. G. 1981. Advances in the use of *Bacillus popilliae* for pest control. See Ref. 28, pp. 183–92
103. Krieg, A. 1987. Diseases caused by bacteria and other prokaryotes. See Ref. 64, In press
104. Kurstak, E. 1982. *Microbial and Viral Pesticides.* New York/Basel: Dekker. 720 pp.
105. Laird, M. 1981. Biocontrol of biting flies: recent history, status, and priorities. See Ref. 106, pp. 196–228
106. Laird, M., ed. 1981. *Biocontrol of Medical and Veterinary Pests.* New York: Praeger. 235 pp.
107. Legner, E. F., Sjorgren, R. D., Hall, I. M. 1974. The biological control of medically important arthropods. *CRC Crit. Rev. Environ. Control* 4(1):85–113
108. Luthy, P., Cordier, J.-L., Fischer, H.-M. 1982. *Bacillus thuringiensis* as a bacterial insecticide: basic considerations and application. See Ref. 104, pp. 35–74
109. MacBain Cameron, J. W. 1963. Factors affecting the use of microbial pathogens in insect control. *Ann. Rev. Entomol.* 8:265–86
110. Maddox, J. V. 1987. Protozoan diseases. See Ref. 64, In press
111. Maddox, J. V., Brooks, W. M., Fuxa, J. R. 1981. *Vairimorpha necatrix,* a pathogen of agricultural pests: potential for pest control. See Ref. 28, pp. 587–94
112. Maramorosch, K., Sherman, K. E., eds. 1985. *Viral Insecticides for Biological Control.* Orlando, Fla: Academic. 809 pp.
113. Marschall, K. J. 1980. Coconuts. See Ref. 90, p. 31
114. May, R. M. 1981. *Theoretical Ecology. Principles and Applications.* Sunderland, Mass: Sinauer 489 pp.
115. May, R. M. 1983. Parasitic infections as regulators of animal populations. *Am. Sci.* 71:36–45
116. May, R. M., Anderson, R. M. 1979. Population biology of infectious diseases: Part II. *Nature* 280:455–61
117. McCoy, C. W. 1981. Pest control by the fungus *Hirsutella thompsonii.* See Ref. 28, pp. 499–512
118. McGaughey, W. H. 1985. Insect resistance to the biological insecticide *Bacillus thuringiensis. Science* 229:193–95
119. McKinley, D. J., Brown, D. A., Payne, C. C., Harrap, K. A. 1981. Cross-infectivity and activation studies with four baculoviruses. *Entomophaga* 26:79–90
120. Miller, J. C. 1983. Ecological relationships among parasites and the practice of biological control. *Environ. Entomol.* 12:620–24
121. Mitchell, R. 1975. Models for parasite populations. In *Evolutionary Strategies of Parasitic Insects and Mites,* ed. P. W. Price, pp. 49–65. New York/London: Plenum. 224 pp.
122. Mogi, M. 1981. Population dynamics and methodology for biocontrol of mosquitoes. See Ref. 106, pp. 140–72
123. Morris, O. N. 1980. Entomopathogenic viruses: strategies for use in forest insect pest management. *Can. Entomol.* 112:573–84
124. Morris, O. N. 1982. Bacteria as pesticides: forest applications. See Ref. 104, pp. 239–87
125. Murdoch, W. W., Chesson, J., Chesson, P. L. 1985. Biological control in theory and practice. *Am. Nat.* 125:344–66
126. Nickle, W. R. 1981. Mermithid parasites of agricultural insect pests. *J. Nematol.* 13:262–66
127. Nickle, W. R. 1981. Nematodes with potential for biological control of insects and weeds. See Ref. 13, pp. 181–99
128. Nickle, W. R., ed. 1984. *Plant and Insect Nematodes.* New York/Basel: Dekker. 925 pp.
129. Nordlund, D. A. 1984. Biological control with entomophagous insects. *J. Ga. Entomol. Soc.* 19:14–27
130. Payne, C. C. 1981. Cytoplasmic polyhedrosis viruses. See Ref. 42, pp. 61–100
131. Payne, C. C. 1982. Insect viruses as control agents. *Parasitology* 84:35–77
132. Petersen, J. J. 1982. Current status of nematodes for the biological control of insects. *Parasitology* 84:177–204
133. Petersen, J. J. 1984. Nematode parasites of mosquitoes. See Ref. 128, pp. 797–820
134. Petersen, J. J., Cupello, J. M. 1981.

Commercial development and future prospects for entomogenous nematodes. *J. Nematol.* 13:280–84

135. Pickford, R., Reigert, P. W. 1964. The fungous disease caused by *Entomophthora grylli* Fres., and its effect on grasshopper populations in Saskatchewan in 1963. *Can. Entomol.* 96:1158–66

136. Pinnock, D. E. 1975. Pest populations and virus dosage in relation to crop productivity. In *Baculoviruses for Insect Pest Control: Safety Considerations*, ed. M. Summers, R. Engler, L. A. Falcon, P. V. Vail, pp. 145–54. Washington, DC: Am. Soc. Microbiol. 186 pp.

137. Podgwaite, J. D. 1985. Strategies for field use of baculoviruses. See Ref. 112, pp. 775–97

138. Poinar, G. O. Jr. 1979. *Nematodes for Biological Control of Insects.* Boca Raton, Fla: CRC. 277 pp.

139. Price, P. W. 1981. Relevance of ecological concepts to practical biological control. See Ref. 13, pp. 3–19

140. Richter, A. R., Fuxa, J. R. 1984. Timing, formulation, and persistence of a nuclear polyhedrosis virus and a microsporidium for control of the velvetbean caterpillar (Lepidoptera: Noctuidae) in soybeans. *J. Econ. Entomol.* 77:1299–306

141. Robert, Y., Rabasse, J.-M., Scheltes, P. 1973. Facteurs de limitation des populations d'*Aphis fabae* Scop. dans l'Ouest de la France. I. Epizootiologie des maladies à Entomophthorales sur fèverole de printemps. *Entomophaga* 18:61–75

142. Roberts, D. W. 1973. Means for insect regulation: fungi. See Ref. 25, pp. 76–84

143. Roberts, D. W., Yendol, W. G. 1971. Use of fungi for microbial control of insects. See Ref. 31, pp. 125–49

144. Schneiderman, H. A. 1984. What entomology has in store for biotechnology. *Bull. Entomol. Soc. Am.* 30:55–61

145. Service, M. W. 1981. Ecological considerations in biocontrol strategies against mosquitoes. See Ref. 106, pp. 173–95

146. Smith, C. E. G. 1971. The spread and maintenance of infections in vertebrates and arthropods. *J. Invertebr. Pathol.* 18:i–xi

147. Soper, R. S., MacLeod, D. M. 1981. Descriptive epizootiology of an aphid mycosis. *US Dep. Agric. Tech. Bull.* 1634:1–17

148. Soper, R. S., Ward, M. G. 1981. Production, formulation and application of fungi for insect control. See Ref. 13, pp. 161–80

149. Sprenkel, R. K., Brooks, W. M., Van Duyn, J. W., Dietz, L. L. 1979. The effects of three cultural variables on the incidence of *Nomuraea rileyi*, phytophagous lepidoptera, and their predators on soybeans. *Environ. Entomol.* 8:334–39

150. Steinhaus, E. A. 1949. *Principles of Insect Pathology.* New York/Toronto/London: McGraw-Hill. 757 pp.

151. Steinhaus, E. A. 1954. The effects of disease on insect populations. *Hilgardia* 23:197–261

152. Sweeney, A. W., Hazard, E. I., Graham, M. F. 1985. Intermediate host for an *Amblyospora* sp. (Microspora) infecting the mosquito, *Culex annulirostris*. *J. Invertebr. Pathol.* 46:98–102

153. Tanada, Y. 1963. Epizootiology of infectious diseases. In *Insect Pathology, An Advanced Treatise*, Vol. 2, ed. E. A. Steinhaus, pp. 423–75. New York/London: Academic. 689 pp.

154. Tanada, Y. 1976. Ecology of insect viruses. See Ref. 3, pp. 265–83

155. Tanada, Y. 1976. Epizootiology and microbial control. In *Comparative Pathobiology*, Vol. 1, *Biology of the Microsporidia*, ed. L. A. Bulla, Jr., T. C. Cheng, pp. 247–79. New York/London: Plenum. 371 pp.

156. Tanada, Y. 1984. *Bacillus thuringiensis:* integrated control—past, present, and future. See Ref. 36, pp. 59–90

157. Tanada, Y., Fuxa, J. R. 1987. The pathogen population. See Ref. 64, In press

158. Thompson, C. G., Scott, D. W., Wickman, B. E. 1981. Long-term persistence of the nuclear polyhedrosis virus of the Douglas-fir tussock moth, *Orgyia pseudotsugata* (Lepidoptera: Lymantriidae), in forest soil. *Environ. Entomol.* 10:254–55

159. Tinsley, T. W., Entwistle, P. F., Robertson, J. S. 1980. Advances in insect pest control. I. Insect pathogenic viruses. *Commonw. For. Rev.* 59:327–36

160. Vinson, S. B., Iwantsch, G. F. 1980. Host suitability for insect parasitoids. *Ann. Rev. Entomol.* 25:397–419

161. Watanabe, H. 1987. The host population. See Ref. 64, In press

162. Webster, J. M. 1980. Biocontrol: the potential of entomophilic nematodes in insect management. *J. Nematol.* 12:270–78

163. Webster, J. M., Thong, C. H. S. 1984. Nematode parasites of orthopterans. See Ref. 128, pp. 697–726

164. Webster, L. T. 1946. Experimental epidemiology. *Medicine* 25:77–109
165. Weiser, J. 1987. Patterns over place and time. See Ref. 64, In press
166. Wouts, W. H. 1984. Nematode parasites of lepidopterans. See Ref. 128, pp. 655–96
167. Wright, V. F., Vesonder, R. F., Ciegler, A. 1982. Mycotoxins and other fungal metabolites as insecticides. See Ref. 104, pp. 559–83
168. Yearian, W. C., Young, S. Y. 1982. Control of insect pests of agricultural importance by viral insecticides. See Ref. 104, pp. 387–423
169. Yendol, W. G., Roberts, D. W. 1971. Is microbial control with entomogenous fungi possible? *J. Ser. Pa. State Univ. Agric. Exp. Stn.* 3972:28–42
170. Young, S. Y. 1978. Manipulation of the environment: viruses. See Ref. 2, pp. 95–99

Ann. Rev. Entomol. 1987. 32:253–73

BIOLOGY OF RIFFLE BEETLES

H. P. Brown

Department of Zoology, University of Oklahoma, Norman, Oklahoma 73019

INTRODUCTION

What are riffle beetles? This question is not new to me, nor is it unexpected, even when it comes from an entomologist. No formal designation has been proposed for the use of this common name, but many of us have unofficially adopted the term "riffle beetles" for the aquatic dryopoid beetles that typically occur in flowing streams, especially in the shallow riffles or rapids. Unlike the familiar and conspicuous water beetles such as dytiscids, gyrinids, and hydrophilids, riffle beetles do not swim and do not come to the surface for air. Most of them are slow in their movements and cling tenaciously to the substrate. Most are also quite small, about the size of household ants, although giants among them exceed the dimensions of a housefly. Some aquatic biologists restrict the term "riffle beetles" to members of the family Elmidae, or even to the subfamily Elminae, but in this review I also include the riffle-dwelling members of the family Dryopidae, the genus *Lutrochus* (Lutrochidae, formerly in the family Limnichidae), and the water penny beetles (family Psephenidae *sensu lato*). Most attention is devoted to the elmids, which are represented by the greatest numbers of both species and individuals.

All elmids are aquatic as larvae. Adults of the subfamily Elminae and of some Larinae live and feed underwater, often alongside their larvae. Among the Dryopidae, the known larvae are mostly terrestrial or semiaquatic, as are the adults of many species (4, 10, 14, 16). Only the adults of *Helichus* would qualify as familiar riffle beetles, but those of *Elmoparnus* deserve mention as interesting neotropical rarities (20, 95). All Psephenidae are aquatic as larvae; the adults are usually riparian or terrestrial.

The small, dermestidlike limnichids are associated with water and are frequently abundant at or near the water's edge, but I am aware of none that

253

inhabits streams except *Lutrochus* and close neotropical relatives, which Crowson (36) puts in a separate family, the Lutrochidae. I heartily agree that they merit separation. Paulus (79) has described the terrestrial larvae of two European genera of limnichids, and I have examined those of two American genera; they bear little resemblance to the larvae of *Lutrochus* (22, 49, 112).

Of the remaining families placed by Crowson (36) in the superfamily Dryopoidea, I omit the Eulichadidae because I have never found them in riffles and am unfamiliar with the group. Among the ptilodactylids, the larvae of *Anchytarsus* inhabit riffles, but I don't think of them as riffle beetles; I shall leave coverage of these to LeSage & Harper (65) and Spangler (93) (ecology and life history), and to Stribling (100) (systematics). Spangler (90) also spares us the trouble of considering the Chelonariidae as "almost" riffle beetles. Both larvae and adults of Heteroceridae are closely associated with water (62, 78), but are not aquatic, and are definitely not riffle dwellers.

I outline the life histories of the groups covered and discuss their means of remaining permanently underwater, some aspects of their ecology, predators and parasites, and something of their past history. In no case is my treatment exhaustive. Perhaps it is most extensive in least-known subject areas.

LIFE HISTORY

Elmidae

ELMINAE   Members of this subfamily are the most completely aquatic of all beetles. Little is known of courtship behavior, but it seems minimal. Although there is nothing unusual about mating in most elmids, copulating pairs of *Zaitzevia* are so firmly attached they commonly remain linked when preserved or dried and pinned.

The female genitalia include sensory styli used in selecting oviposition sites. Few observations of elmid oviposition have been reported (67, 111), but it is probable that most elmids glue their eggs singly or in small clusters to the undersides of submerged rocks, wood, or plant stems, depending upon habitat preference of the species.

Incubation time is rather short (5–15 days), varying with the temperature. Duration of larval stage (6–36 mo) and number of instars (5–8) also vary with temperature, as well as with body size and available food. Development is faster at higher temperatures; northern populations of a given species may require an additional year in the larval stage. Other factors being equal, small beetles need less time to develop and exhibit fewer larval instars than large beetles (7, 8, 59, 67, 74, 87, 111).

Mature larvae pupate in protected sites above the water line. They use various strategies to get from larval habitat to pupation site; direct crawling, if feasible, is the simplest method. Streams in which the water level fluctuates provide another opportunity, as demonstrated by White (111) and confirmed by Seagle (84); mature larvae simply wait in the shallow water at the stream edge until the water level drops, allowing them to pupate in situ. One result of this pattern is relatively synchronized emergence of adults.

Drifting is a widely used means of transport among stream insects (12, 30, 60, 77, 82, 107, 109). Larvae can drift from midstream to a snag or boulder upon which to crawl out. Perez (80) was the first to report larval drift of elmids and to suggest its mechanism, although he noted it simply as a device for escaping poor environmental conditions such as water with low dissolved oxygen. Mature larvae develop tracheal air sacs that can be compressed or expanded by contraction or relaxation of tergo-sternal muscles, respectively; these provide a cartesian diver system for controlling the specific gravity of the body (41, 96, 102). The fact that air sacs do not occur in earlier instars suggests that their primary adaptive function is probably for transport toward pupation sites rather than simply for the search for greener pastures or less foul water.

Having found a satisfactory site in moist sand, humus, moss, or the like, the prepupal larva proceeds to form an ovoid cell or chamber by appropriate body movements. Xylophagous larvae such as *Macronychus* usually pupate in rotten wood or beneath loose bark (8, 66, 67). In temperate zones pupation commonly occurs from late spring through summer and requires 1–2 wk. In tropical regions where there is no pronounced dry season, pupation occurs throughout the year.

Following eclosion, commonly at dusk of the first warm and humid night, recently emerged alate elmids take flight. This is their only dispersal flight. At this stage they are capable of surviving for several days out of water (84, 111). Many species are positively phototactic, to the delight of coleopterists equipped with light traps. In fact, some species and even genera are known only from light-trapped specimens, and are only presumed to deserve the appellation "riffle beetles" (38, 91). Collectors are understandably pleased with light-trapped material, not only because great numbers may be taken with minimal effort, but also because the specimens are cleaner than those collected from streams. (In some mineral-laden streams, elderly elmids may become so encrusted that their stony armor far outweighs the beetles themselves.) Some species are presumably always flightless, e.g. those that inhabit warm springs, such as *Microcylloepus thermarum* or *Zaitzevia thermae* (22). Others are dimorphic with respect to flight wings (37). *Macronychus glabratus,* for example, though often abundant in streams of the eastern United States, is rarely taken in light traps. This is not because it is not attracted to lights, but because only a small percentage of the population possesses functional wings; most individuals are virtually apterous as described by Segal (86). Although Segal depicted *Helichus fastigiatus* and *Promoresia*

("Limnius") *elegans* with fully developed wings, he made the surprising statement that no species of Dryopidae (including elmids) was known to take flight. Furthermore, having examined over 100 specimens of *Stenelmis crenata* taken at different seasons and sites, and having found all to have vestigial wings, he ascribed this condition to the genus. Actually, *Stenelmis* is our best-represented genus in light-trap collections. Many genera are absent from light-trap collections because the beetles fly during the day rather than at night. It is my impression that most montane elmids are in this category. W. D. Shepard (personal communication) suggests, quite reasonably, that the nocturnal temperatures near most western streams are too low for flight by small insects.

Once they have entered the water, most elmids never again emerge under natural conditions. In fact, they may then be unable to survive out of water for more than a few hours or to successfully reenter the water once they have dried out (111). As yet, very little experimental evidence is available on this point; it may be that some elmids can leave the water and perhaps fly during the first few days or weeks of adult life, and become permanently bound to water as their flight muscles atrophy to provide nutrients for egg production, etc. At any rate, some flying individuals appear to have spent a bit of time underwater. Adults are long-lived, but there are no precise data on specimens under natural conditions. In the laboratory, some individuals survive for several years at room temperature, even under adverse conditions (24, 34).

LARINAE   Most members of this subfamily deserve designation as riffle beetles only during the larval stage, when they are completely aquatic and indistinguishable as a group from larval Elminae. Adults of North American genera are riparian, have a relatively short life cycle, and probably enter the water only to oviposit. When disturbed, they fly readily and quickly, sometimes swarming in the spray of waterfalls or cascades. However, adults of some tropical genera normally occur beneath the water surface (22, 25, 101).

Steedman (96) is the only person who has made a detailed study of the life history of a larine elmid. Of *Lara avara,* which inhabits streams of the Coast Range in Oregon, he wrote: "Adults live approximately three weeks, and occur from May to August. Females lay 100–150 eggs on submerged wood. Larvae grow through seven instars, taking about one year for instars 1–3, and from three to six years for instars 4–7. Last-instar larvae leave the water in the spring, and burrow into moss at the edge of the stream. Pupation occurs when the moss dries in early summer. The pupal stage lasts at least two weeks." The larvae feed on decaying wood.

Dryopidae

Adults of *Helichus* occur widely in riffles and are often among the most abundant insects in rivers of the southwestern United States. Of special interest is the fact that *Helichus* larvae are terrestrial rather than aquatic (4, 10, 21, 112), whereas virtually all other aquatic insects are aquatic either only as immatures or as both immatures and adults. *Elmoparnus* inhabits riffles in Central and South America, and may share this unusual developmental feature, but its larva is as yet unknown; adults are less well adapted to aquatic life than are those of *Helichus,* but deserve mention for their peculiar use of antennae in replenishing a ventral air bubble (95). Like the elmids, many dryopids make dispersal flights, and great numbers of some species are taken in light traps. In the eastern United States, *Helichus lithophilus* is the dryopid most commonly collected in light traps.

No complete life history of *Helichus* has yet been published, but G. Ulrich (personal communication) is describing the last-instar larvae and pupae of *H. suturalis* and *H. productus.* He has reported evidence that larval development requires 2–3 yr for *H. suturalis* and 4–5 yr for the larger *H. productus.* Larvae of both species were found in slightly moist sand about 5 m from the stream edge and 2 m above the water level, where they presumably feed upon roots and/or decaying vegetation.

Lutrochidae (Formerly Part of Limnichidae)

Adults of *Lutrochus laticeps* in the eastern United States, *L. arizonicus* in Arizona, and some neotropical species are typically riparian and are capable of flight when disturbed (22, 28). Adults of *L. luteus* in Texas and Oklahoma are more inclined to be aquatic, commonly remaining submerged long enough to acquire heavy calcareous coats. They graze upon travertine-forming algae, mate briefly, and oviposit by inserting their eggs in the travertine, into which the larvae burrow (H. P. Brown, unpublished observations). Pupation occurs in cells above the water line, often within the travertine or rotten logs. Recently eclosed adults are attracted to light while on dispersal flights. In Oklahoma, adults are present from late May through September.

Psephenidae

PSEPHENINAE   In the eastern United States, adult males of *Psephenus herricki* are usually noticed as black, soft-bodied beetles approximately the size of houseflies, nervously scurrying over stones that project above the water in stream rapids. When approached, they fly quickly to other rocks and resume their busy search for females. Copulation is brief and has seldom been observed. Mated females crawl down into the water, where they spend the rest of their lives (probably no more than a week or so) seeking suitable sites and ovipositing on the lower surfaces of stones in the riffle. Adults eat little or nothing (1, 75, 110).

The bright yellow eggs, clustered in a mass of up to 600 in a single layer, are familiar to most stream biologists. In Ohio and New York, hatching occurs in 10–15 days and eggs can be found throughout the summer (75, 110). In the tropics, psephenid adults and eggs are present at all times of year, but the adults are rather short-lived and may die before their eggs hatch.

Psephenid larvae (water pennies) are reminiscent of trilobites in form, broadly oval, conspicuously flattened, and often colored so much like the rocks to which they cling that casual observers fail to detect most of them. Because they are present in streams year-round and are frequently abundant, water pennies are probably recognized by more biologists and laymen than any other dryopoids. West (110) observed that they graze during the night and on very overcast days upon the crop of diatoms and other algae covering stones in riffles, and retire to the undersides of the stones during sunny hours. Larvae are inactive during the winter; they do not feed at temperatures below 13°C (75). There are about six larval instars. Mature larvae leave the water during spring or summer of their second year. The larvae do not construct pupal cases in soft, moist soil as stated by White et al (112), nor do they pupate within the last larval skin. Rather, they pupate beneath the carapace of the larval cuticle, which is plastered to the substrate all around by its marginal ciliary fringe. The rest of the larval exuviae forms a wad, also beneath the carapace. Pupation requires about 10–11 days, and occurs on a solid substrate in a moist site beneath or on the side of a rock or log anywhere from a few centimeters to several meters from the water's edge.

In the case of the small tropical and subtropical psephenids, it is possible that the entire life cycle may require less than a year.

EUBRIANACINAE   Although it has not been studied in detail, the life history of *Eubrianax* appears to be quite similar to that of *Psephenus,* except that the pupa is not completely beneath the last larval carapace. Instead, the last several abdominal segments of the larval skin are replaced by rather similar ones that belong to the pupa; thus, these tergites of the pupa are exposed, whereas no part of a *Psephenus* pupa is exposed (11, 53). Despite Essig's statement (40), pupation does not occur under water. Clark & Ralston (33) kept 40 larvae in an aerated aquarium; about half of them pupated, in every case on rocks above the water line. I have collected many pupae and eclosing adults, all out of the water and in sites like those where *Psephenus* pupae are found.

PSEPHENOIDINAE (OR PSEPHENOIDIDAE) I have not observed living specimens of the Asian/African genus *Psephenoides,* and no life history study has
been published. From Champion (31) we learn that adults and larvae exhibit
behavior much like that of *Psephenus:* "[Adults were] Found flying over
shallow running water and settling on the partially submerged stones, the
females sitting on the undersides of the stones like an Elmid, twenty males
seen on one stone, these flying off when approached. . . . Larvae and pupae
. . . [are] found in abundance on the submerged stones." Unlike those of
Psephenus, however, the pupae are aquatic (31, 42, 52, 53, 55), and the
adults, at least those of *P. volatilis,* are attracted to light (32). Immatures of an
African species, *P. marlieri,* occur on the undersides of submerged rocks on
the shore of Lake Tanganyika, much as larvae of the American *Psephenus
herricki* live on rocky or gravelly wave-washed shoals in Lake Erie, where I
first saw water pennies.

Morphologically, the larvae of *Psephenoides* resemble those of eubriines
more closely than they do those of *Psephenus* or *Eubrianax* (13, 53).

EUBRIINAE (OR EUBRIIDAE) Adults of *Eubria* were known in Europe for
over a century before the larva was described, and even then it was described
as that of a dryopid (64). In fact, most of the eubriine larvae now identifiable
to genus were initially considered dryopids because Kellicott (63) described
and illustrated a larva he supposed to be *Helichus,* but which was actually
Ectopria. It was not until 1939 that anyone correctly identified eubriine
larvae, when Bertrand reared *Eubria* (9).

Eubriine larvae ("false water pennies") superficially resemble the *Psephenus* larva, and some share its habitat and food. Overwintering occurs in
the larval stage. The mature larvae leave the water for pupation, but do not
make use of the last larval skin as a tent or shield beneath which to
pupate.

Adults are even less like *Psephenus* in behavior. Those of *Ectopria* are
nocturnal and attracted to light. They exhibit little interest in water. Nothing is
known of mating behavior or oviposition except that both must be different
from those of *Psephenus.* In suitable streams of eastern Oklahoma, larvae of
the two genera are common and may occupy the same rock; they may also
pupate side by side. Then they part company. *Psephenus* males spend the
sunny hours on exposed rocks in riffles while females oviposit within the
riffles. In contrast, *Ectopria* adults spend their days resting on the lower sides
of leaves of trees and shrubs near the stream. I have never seen them on rocks
or in the water, nor have I seen any sign of their eggs. At night they become
active, but their activities, aside from movement toward light, have not been
described. It is surprising that a species as widely distributed and locally

abundant as *Ectopria thoracica* in the eastern United States should be so poorly known.

RESPIRATION

Perhaps the most interesting feature of riffle beetles for the average biologist is the underwater respiratory device used by adult elmids and the dryopid *Helichus*. When the beetles first enter the water, they take with them a film of air covering an appreciable portion of their body surface. This thin sheet of air, the plastron, functions as a gill, and is in contact with the air space beneath the elytra, into which the spiracles open. As oxygen is removed from this reservoir by absorption for respiration, it is replaced through diffusion of dissolved oxygen from the surrounding water (2, 15, 44, 45, 104, 105). Existence and maintenance of the plastron is made possible by a dense coating of specialized hydrofuge setae. Scanning electron microscopy reveals remarkable diversity in the microstructure of these setae among elmid genera (56–58). In some genera, such as *Pagelmis* and *Stenhelmoides*, other structures provide a plastron (91). [Some authors have found it convenient to use the term "plastron" to refer to the area on the body occupied by these hydrofuge hairs, scales, etc, as well as to the gas film that is present only when the beetle is immersed (70, 91)]. Some elmids (e.g. *Riolus*) supplement their plastron from oxygen bubbles released by photosynthesizing aquatic plants; others (e.g. *Elmis*) remain permanently submerged without such supplementation as long as the dissolved oxygen level in the surrounding water is adequate (15, 49, 104, 105). Messner & Langer (70) have provided evidence that gaseous exchange between enclosed air spaces of the elytra and the surrounding water may occur directly through thin regions of elytral cuticle.

As one might expect, reliance upon plastron respiration restricts most elmids to waters nearly saturated with dissolved oxygen, hence to typically shallow, fast-flowing, cool, or cold streams. A few enterprising species have succeeded in colonizing other habitats. Probably the easiest transition has been to wave-washed lake shores, as evidenced by *Dubiraphia brunnescens* in California's Clear Lake (22). In shallow midwestern lakes *Stenelmis quadrimaculata*, which does not require wave action, thrives in marl concretions (22, 102). In isolated western localities endemic species of *Microcylloepus, Stenelmis,* and *Zaitzevia* are known from thermal pools and springs. But most remarkable by far is *Austrelmis consors*, which occurs in Lake Titicaca at depths as great as 11 m and in hot springs at temperatures up to 29°C (27, 50). Even in cool, clear, well-aerated streams, other riffle beetles are rarely taken at depths greater than 1–2 m.

Spangler (92) recently described the eyeless genus *Anommatelmis* from wells in Haiti and suggested that its extensive plastron might represent an adaptation to low concentrations of dissolved oxygen. Other genera with comparably extensive plastrons *(Pagelmis, Stenhelmoides)* have been taken only in light traps; perhaps they, too, live in some sort of stagnant habitat that has not yet been sampled (91).

Although adult elmids of the subfamily Larinae lack plastrons and are typically riparian rather than aquatic, the behavior and structure of adults of the African genus *Potamodytes* enable them to remain indefinitely beneath the water surface. I know of no mystery novel more spellbinding than Stride's article (101), in which Bernouilli's principle is shown to play a pivotal role in the formation and maintenance of a large respiratory bubble. Such neotropical genera as *Hispaniolara* and *Potamophilops* apparently use the same strategy (25).

Pollution by soaps and detergents is understandably disastrous to adult elmids because such wetting agents make it impossible for the beetles to maintain either a plastron or a bubble (17). This is a widespread problem in less developed regions where streams are used for bathing and clothes washing. In arid regions the problem is most acute, for such pollution is concentrated at the source of the only available water.

All elmid larvae have basically the same type of respiratory apparatus (39, 49, 71, 80, 102). Tripartite tufts of tracheal gills are extruded from the anus, but can be quickly withdrawn and covered by a hinged operculum. Larvae of *Lutrochus*, Eubriinae, and Psephenoidinae use almost identical mechanisms (3, 49, 68). Larvae of Psepheninae and Eubrianacinae have instead a series of paired non-retractile tracheal gills that arise from the ventral sides of four or five abdominal segments (49, 53).

The pupae of *Psephenoides* are remarkable in that they are aquatic and have tufts of spiracular gills, which arise from the lateral margins of the abdominal segments (13, 49, 52, 53, 55, 68). Some specimens of *Psephenus murvoshi* successfully pupate under water in certain warm springs in Arizona and Chihuahua (76). However, they show no perceptible respiratory adaptations, so the feat seems all the more astonishing; most *Psephenus* pupae and prepupal larvae whose gills have withered simply drown if the rocks to which they are attached are submerged. The respiratory mechanism has not yet been explained. Perhaps in these springs the prepupal larva somehow manages to trap and retain a bubble beneath the carapace that functions like the plastron of adult elmids.

ECOLOGY

Riffle beetles exhibit appropriate adaptations to their particular benthic habitats. Adult elmids that inhabit shifting sand and gravel are commonly plump, compact, and heavily sclerotized (e.g. *Austrolimnius, Elmis, Optioservus,*

Oulimnius, and *Xenelmis*). If they are dislodged by turbulent water or subjected to abrasion by tumbling sand and pebbles, as must frequently happen, they further minimize injury by retracting vulnerable appendages. Larvae of such genera are also relatively compact and are capable of curling up like armadillos or pillbugs. In contrast, beetles that occupy submerged vegetation or dead wood or that live in slower water tend to have elongated bodies, long legs, and prominent claws (e.g. *Ancyronyx, Dubiraphia, Hintonelmis,* and *Macronychus*). Collecting techniques such as kick-sampling or Surber sampling are very effective for the former group, but useless for the latter.

Because of their depressed onisciform structure, which makes boundary layer control possible (89), water pennies and their ilk require a solid and relatively level substrate to which they can cling, preferably rocks.

Larvae of *Lutrochus* and the elmid *Hexacylloepus* tunnel in the porous travertine of southwestern streams, presumably deriving their nourishment from the embedded algae. Most elmid larvae ingest algae and debris with less crunchy roughage. Xylophagous larvae such as *Lara* and *Macronychus* chew their own grooves in sunken wood, probably gaining most nutrition from the fungi and bacteria responsible for the decay of the wood (41, 96). The great quantities of frass they generate indicate that only a small proportion of their ingested food is digestible. It is conceivable that the caeca at the anterior end of the midgut of most adult elmids (1, 51) harbor symbiotic microorganisms that aid in digestion; however, no evidence has been adduced for this, and the caeca of xylophagous beetles are no more prominent than those of beetles that feed on algae and detritus.

Feeding ecology varies among species, and even more among genera (67, 85). Judging chiefly from gut contents, it appears that most riffle beetles, both as larvae and as adults, subsist on microorganisms and debris scraped from the substrate. The siliceous tests of diatoms are often the only recognizable items in the gut, but they are gratifyingly identifiable (43, 67, 85).

Steffan (97) showed in central Europe that the largest species within any genus inhabits the zone of least temperature fluctuation, and that cold stenothermic species are always bigger than eurythermic species. No comparable study has been made in the New World, but the principles seem to apply only in selected genera, e.g. *Heterelmis*.

Turcotte & Harper (106) found that in a small Andean stream the density of elmids was largely regulated by spates. This may be the case in many nonseasonal streams subject to spates.

Several authors have mentioned apparent gregariousness. The suggestion is understandable. I have counted 29 *Ectopria* larvae on a fist-sized rock, and both larvae and pupae of *Psephenus* sometimes seem crowded on a stone. Dozens of male *Psephenus* may be seen on a projecting rock in stream rapids, and scores of *Lutrochus* adults may be squeezed into streamside crevices.

Nevertheless, such aggregations are probably attributable to the microenvironment rather than to true gregariousness; a combination of factors makes the favored spot the most logical and suitable site for any of the individuals to find and settle upon. The popular spot never seems an inappropriate one, and usually can be seen to be the most favorable. In other words, I think that the probability that each individual would have found that rock or crevice is independent of the presence, absence, or proximity of other individuals. This hypothesis should be amenable to experimental investigation.

During my early years as a riffle beetle enthusiast (23) I assumed that elmids and perhaps other riffle beetles required permanent streams, and that no obviously temporary stream was worth checking. How many interesting specimens I missed for this reason I shall never know, but in arid and semiarid regions the loss was greatest. Then I noted that water pennies clinging to boulders could crawl down the rock surface into the sand to remain beneath the level of the water as it receded into the sandy/pebbly substrate when the stream bed appeared to have dried up completely. By digging into damp sand, especially sand in contact with large stones, I found that the last part of a riffle to dry up could be a profitable collecting site. A dried-up creek near Waxahatchee, Texas yielded a total of 285 adult and 173 larval *Stenelmis,* 2 adult *Heterelmis,* and 1 *Helichus.* Bell (5) found *Narpus* and *Zaitzevia* adults in moving water 40–50 cm beneath the gravelly surface of a dry stream bed in California. Some of my best collecting in Baja California has been in such stream beds, especially where an occasional trickle has suggested recent surface flow.

A few pragmatic biologists have capitalized upon the fact that many riffle beetles (and other creatures) are finicky about their choice of habitats, and have devised a means of evaluating the water quality of streams by sampling and scoring the dryopoid fauna (47, 88, 108). At times I find it handy to be able to cite such potentially utilitarian aspects of the work I do.

RELATIONS WITH OTHER ORGANISMS

Predators

As far as we know, predators have but a minor role in the affairs of riffle beetles. I know of only one report that contains pertinent data, that of Stewart et al (99) concerning food habits of hellgrammites, larvae of the megalopteran *Corydalus cornutus.* Three elmid genera were among the prey: *Dubiraphia, Heterelmis,* and *Stenelmis.* However, it appears that hellgrammites did not find the elmids exactly delectable, for elmids represented only 0–1.6% of the organisms found in the predator guts, whereas these elmids represented 3–12% of the organisms available as potential food.

The coloration of many adult riffle beetles is quite conspicuous. Some, such as *Macronychus,* are jet-black and shiny, with contrasting silvery-white areas covered by the air film of the plastron. Others, e.g. *Optioservus, Promoresia, Gonielmis,* and *Dubiraphia,* are black with colorful spots or streaks of yellow, orange, or red on their elytra. In *Ampumixis, Heterlimnius,* and *Narpus* the bright colors form broad transverse bands, and in *Ancyronyx* they produce eye-catching loops. It seems reasonable to surmise that the possessors of such presumably aposematic pigments and patterns would probably be unpalatable to predators. D. S. White (personal communication) has tested this hypothesis. He found that none of the tested elmids were acceptable as food to any of various predators that share the habitats of local elmids (e.g. crayfish, turtles, sunfish, and darters), although the predators readily consumed other prey organisms of comparable dimensions. When the predators accidentally ingested elmids with other food, they typically spit them out. The beetles apparently possess some sort of chemical defense that repels diverse types of potential predators. This line of investigation should be interesting to pursue.

Homo sapiens is the only species I know of that has developed a taste for riffle beetles. Over a century ago Philippi (81) described a new species of elmid from Peru, which he named *Elmis condimentarius.* The generic name was shifted to *Cylloepus,* then to *Macrelmis,* and finally to *Austrelmis* (27), but it is the specific epithet, *condimentarius,* that suggests the species' distinction. I take the liberty of presenting my free translation of the first few sentences of Philippi's intriguing paper:

> A short time ago I received from Dr. Barranca in Lima some insects in a paper package, and in a bit of paper a clump of small beetles with the following note: "Insects that people form into lumps of paste known here under the term chiche; they serve as seasoning for food which they call chupe de chiche. They are found in the quiet waters of brooks and streams of the mountains, and their commercial value is not inconsiderable."
>
> Concerning the taste of the beetles I cannot say anything, for I promptly threw them into alcohol before I noticed the note. At least 90% of the mass consisted of little beetles like *Elmis* that were well preserved, so that I could extract a couple of hundred specimens and am in the position to provide almost all fellow coleopterists with specimens for their collections, if not for their palates . . .

Had I read Philippi's paper before my 1971 trip to South America, I might have been able to say something about the taste of chiche and whether it is still an item of commerce in Lima. Before I read his paper, all the specimens of *Austrelmis* that I had collected in the vicinity of Lima and elsewhere had long been in alcohol. Perhaps they were once piquant and flavorful, but I can attest that they now do little to titillate the taste buds.

I suspect that riffle beetle larvae are appreciably more palatable to predators than are the adults, but are less readily detected. Larvae dislodged by disturbance of the substrate are devoured by fishes. In streams of the western United States, Mexico, Central America, and the Andean portions of South

America, there are almost always abundant predatory benthic bugs (chiefly naucorids, and perhaps helotrephids in the Andes). These probably include elmid larvae among their prey. Triclad planarian worms, which are often very abundant, are also likely to take their toll on larval elmids and psephenids.

Predation upon elmid eggs would be quite difficult to study under natural conditions, but psephenids provide an ideal setup for both predator and investigator. The gluttonous worms *Dugesia* spp. snuggle right beside ovipositing *Psephenus* spp. with their proboscides extending beneath or alongside the beetles, and ingest the colorful contents of the eggs. The scent of ruptured eggs probably entices other predators and scavengers. I have even seen *Psephenus* larvae and *Stenelmis* adults eat such planarian-damaged eggs, but this could hardly be considered cannibalism.

My report of cannibalism by a larva of *Microcylloepus* (29), which attacked first an adult of its own species and then a fellow larva, probably described aberrant behavior; however, nutritionally deprived individuals perhaps exhibit such tendencies more frequently than we suspect.

Parasites

HYMENOPTERA   In 1863, Perez described the parasitic behavior of the tiny hymenopteran *Pteromalus macronychivorus* on pupae of the elmid *Macronychus quadrituberculatus* in France (80). Although he did not make a complete life-history study of the parasite, he succeeded in rearing larvae to pupae and one pupa to an adult female. It might seem surprising that no one else has discovered other such parasites on elmids during the dozen decades since Perez's publication, but very few elmid pupae have been observed under natural conditions. Enterprising investigators are likely to observe additional hymenopteran parasites or parasitoids.

Psephenid pupae are much larger and easier to find than elmid pupae, and are often present in considerable numbers. Two very different species of wasps parasitize pupae and prepupal larvae of *Psephenus* (18, 19). *Psephenivorus,* a eulophid chalcidoid, is so tiny that as many as 39 wasps may develop within a single water penny. Only one individual of the larger wasp *Trichopria,* a diapriid proctotrupoid, develops within a water penny. But parasitism by either wasp is invariably fatal to the host, and the incidence of parasitism may exceed 90% of the water penny population. Thus far, these wasps have not been reported north of the Rio Grande valley.

ACARINA   I know of no published studies concerning water mites on riffle beetles, and I presume that they are of little importance. I have found a few attached to adult elmids in Costa Rica, Venezuela, the Dominican Republic, and Brazil. I have also encountered what I took to be eggs of Hydracarina attached to adults of *Microcylloepus* in Texas and *Oulimnius* in Tennessee.

MISCELLANEOUS ARTHROPODS   I have noted hemipteran eggs (Corixidae?) on the dorsum of a *Psephenus* larva in Mexico, and dipterous larvae (Chironomidae) on six larvae of the elmid *Potamophilops* in Brazil.

FUNGI   Members of the primitive ascomycete order Laboulbeniales are obligate but benign ectoparasites of arthropods. Of 116 genera, 90 are known only from beetles and 9 others parasitize both beetles and other arthropods (6). Although 14 genera occur on other water beetles, only two are known from dryopoids: *Helodiomyces* on *Dryops* and *Cantharomyces* on the dryopids *Dryops, Helichus,* and *Pelonomus,* on *Lutrochus,* and on the larine elmid *Phanocerus* (6, 103; R. K. Benjamin, personal communication). Of these beetles, all but *Helichus* are riparian.

PROTOZOA   The organisms most commonly noticed on riffle beetles are attached peritrich ciliates, which are often abundant. Since the ciliate fauna of hydrophilids is quite diverse (69), that of riffle beetles merits investigation. Possibly more exciting are the sessile suctorians that find elmids prime substrates for attachment (98).

A group of internal parasites that deserves mention is the gregarine sporozoans, whose opaque whitish trophozoites and gametocysts are often big enough to be seen through the body wall of dryopoid larvae under low magnification or even with the naked eye. They probably represent undescribed species and perhaps new genera. These sporozoans occur in both adults and larvae (1, 43), but dissection or sectioning is required to observe those in adults. They seem to do little harm to their hosts.

OTHER PROTISTA   Diatoms are frequently attached to the cuticle of adult elmids, and filamentous bacteria may render the beetles hairy-looking.

ANCESTRY

The paleontological record provides as yet but little bearing on the ancestry of modern riffle beetles. Material from glacial and postglacial deposits represents a fauna very like that of today (72, 73). A fossil of *"Psephenus"* from the Miocene (83) is not a psephenid or even a member of the superfamily Dryopoidea. In Crowson's opinion (35) the earliest fossil dryopoid is from the Upper Jurassic. He dates the origin of the superfamily from the Lower Jurassic (36) or, if we accept his earlier estimate (35), the late Triassic—about 190 million years ago. His phylogeny is based upon Hennig's principles (46).

The most convincing evidence for the antiquity of riffle beetles lies in the geographic distribution of extant forms, interpreted according to present understanding of plate tectonics and continental drift. Long before the concept of continental drift was accepted, Hinton (48) described a new species of the flightless dryopid *Protoparnus* from the West Indies, pointed out that all previously known members of the genus inhabited New Zealand, and reasoned that there must have once existed land connections between the two regions. The presence of the elmid *Hydora* in South America as well as in Australia and New Zealand supports this interpretation (94). The elmid genus *Austrolimnius,* however, provides the most striking evidence. Of the 78 species now known, 60 occur in Australia and New Guinea, and 18 in South and Central America (26, 54, 56, 57, 61). Some of those in Australia are much more closely related to certain South American species than to any of their fellow Australian species. Thus, if it is true that these continents have been separated since the Lower Cretaceous, such genera as *Austrolimnius* must have been well differentiated at least that long ago.

Presumably the elmid genera *Oulimnius, Macronychus,* and *Stenelmis* similarly antedate the separation of Europe from North America, since they are found in Europe and eastern North America, but not in far western North America, Central America, or South America (26).

The dryopids *Dryops* and *Helichus* are prominent on all continents except Australia and Antarctica (26). Perhaps these genera, like placental mammals, are absent in Australia because they did not evolve until after the separation of Australia from South America, whereas the dryopid genus *Protoparnus,* as mentioned above, is enough older to have been around before those continents drifted apart.

Only three genera of aquatic dryopoids have been reported from both Africa and South America: *Dryops, Helichus,* and the psephenid *Eubrianax* (26). Whether any of these provide strong evidence for migration across the land bridge connecting these continents during much of the Cretaceous is debatable, since all three also occur in North America and Eurasia. Interestingly, no elmids are included in this list. Investigation should surely reveal at least sister groups of elmids in Africa and South America.

SUMMARY

For purposes of this article, "riffle beetles" include the following groups of nonswimming dryopoid beetles that typically occur in shallow, fast-flowing water: all Elmidae and Psephenidae *(sensu lato),* the dryopids *Helichus* and *Elmoparnus,* and the genus *Lutrochus* (Lutrochidae or Limnichidae).

Members of the elmid subfamily Elminae are aquatic as both larvae and adults, and feed chiefly upon algae and detritus scraped from the substrate. Larvae have tracheal gills that are retractable into the rectum. Both larvae and adults are benthic in or on various substrates, but in many species both are capable of controlling their specific gravity and may be taken in drift samples.

In mature larvae of many species, tracheal air sacs provide buoyancy, and drifting assists the larvae in reaching pupation sites. Pupation occurs in a pupal cell in a protected site outside the water. Recently eclosed adults commonly make dispersal flights (often at night) before they enter the water. Some species of elmids lack functional flight wings; others are dimorphic, with alate forms comprising only a fraction of the population. After they enter the water their flight muscles probably atrophy, precluding subsequent flights, and the beetles remain underwater for the rest of their lives, which may be more than a year. They respire by means of a plastron, a film of air that functions as a gill. Mating and oviposition take place underwater.

Members of the elmid subfamily Larinae differ from Elminae in that adults are less long-lived, lack a plastron, and are commonly riparian. Adults of certain tropical genera are aquatic, but have evolved a different respiratory technique employing a large bubble. Larvae of the two subfamilies are essentially alike, but adult ecology and behavior are very different. Adult larines are quite agile; they fly readily and rapidly, in contrast with the slow and water-bound Elminae.

Psephenid larvae are flattened and aquatic, and possess tracheal gills. Members of the subfamilies Psepheninae and Eubrianacinae are quite similar. Their larvae, the true water pennies, exhibit a series of paired, ventral, abdominal, nonretractile gills. By day they cling to the undersides of stones, moving at night to the upper surfaces of the stones to graze upon the algae. They pupate beneath the carapace of the last larval skin, usually in protected sites not far from the water. Adults are short-lived; the males are diurnal, riparian, and fast-flying. After mating, females crawl down into the water and spend the rest of their lives at the task of oviposition.

Larvae of the Eubriinae and Psephenoidinae have retractile anal gills like those of elmids. Eubriines pupate out of the water, but not beneath a larval carapace; adults are nocturnal and commonly spend the day on foliage. *Psephenoides* pupates underwater on the stones inhabited by the larvae; the pupa respires by means of filamentous spiracular gills; adult behavior is apparently similar to that of *Psephenus*.

Adults of *Helichus* behave much like those of elmines except that females presumably leave the water to deposit their eggs in the larval habitat, e.g. sand or soil. No known dryopid larvae live in stream riffles; most are terrestrial. *Lutrochus* larvae are aquatic and resemble those of elmids. The adults are riparian or in some cases aquatic. No complete life history study has been published for either *Helichus* or *Lutrochus*.

Predators seem to pose little hazard to adult elmids, but eggs and larvae are more vulnerable. Lethal parasitic wasps include a pteromalid that attacks pupae of the elmid *Macronychus* in France and a eulophid and a diapriid that attack *Psephenus* pupae and prepupal larvae in Mexico and southern Texas.

Nonlethal parasites include Hydracarina, fungi (Laboulbeniales), and gregarine sporozoans; other protists commonly attached to riffle beetles are peritrich ciliates, suctorians, and diatoms.

Modern concepts of plate tectonics and continental drift not only explain how closely related species happen to occur in regions as widely-separated as Australia and South America, but also emphasize the antiquity of the group.

Literature Cited

1. Arrington, R. Jr. 1966. *Comparative morphology of some dryopoid beetles.* PhD dissertation. Univ. Okla., Norman. 198 pp.
2. Beier, M. 1948. Zur Kenntnis von Körperbau und Lebensweise der Helminen (Coleoptera, Dryopidae). *Eos* 24(2): 123–211
3. Beier, M. 1950. Zur Kenntnis der Larve von *Eubria palustris* L. (Col., Dascillidae). *Eos* Tomo extraordinario:59–85
4. Beling, T. 1882. Beitrag zur Biologie einiger Käfer aus den Familien Dascillidae und Parnidae. *Verh. Zool.-Bot. Ges. Wien* 32:435–42
5. Bell, L. N. 1972. Notes on dry-season survival in two species of Elmidae (Coleoptera). *Pan-Pac. Entomol.* 48(3): 218–19
6. Benjamin, R. K. 1973. Laboulbeniomycetidae. In *The Fungi,* Vol. IV A, ed. G. C. Ainsworth, F. K. Sparrow, A. S. Sussman, pp. 223–46. New York: Academic. 621 pp.
7. Berthélemy, C., Ductor, M. 1965. Taxonomie larvaire et cycle biologique de six espèces d'*Esolus* et d'*Oulimnius* Europeens (Coleoptera Dryopoidea). *Ann. Limnol.* 1(2):257–76
8. Berthélemy, C., Olmi, M. 1978. Dryopoidea. In *Limnofauna Europaea,* ed. J. Illies, pp. 315–18. Stuttgart: Fischer. 532 pp. 2nd ed.
9. Bertrand, H. 1939. Les premiers états des *Eubria* Latr. *Bull. Mus. Natl. Hist. Nat.* 1–2(2e Ser.):129–36, 291–99
10. Bertrand, H. P. I. 1972. *Larves et Nymphes des Coléoptères Aquatiques du Globe.* Paris: Paillart. 804 pp.
11. Blackwelder, R. E. 1930. The larva of *Eubrianax edwardsi* (Lec.) (Coleoptera Psephenidae). *Pan-Pac. Entomol.* 6(3):139–41
12. Bournaud, M., Thibault, M. 1973. Le derive des organismes dans les faux courantes. *Ann. Hydrobiol.* 4:11–49
13. Böving, A. G. 1926. The immature stages of *Psephenoides gahani* Champ. *Trans. R. Entomol. Soc. London* 74(2):381–88
14. Brigham, A. R., Brigham, W. U., Gnilka, A., eds. 1982. *Aquatic Insects and Oligochaetes of North and South Carolina.* Mahomet, Ill: Midwest Aquatic Enterp. 837 pp.
15. Brocher, F. 1912. Recherches sur la respiration des insectes aquatiques adultes—Les Elmides. *Ann. Biol. Lacustre* 5:136–79
16. Brocher, F. 1913. *L'Aquarium de Chambre.* Paris: Librairie Pagot & Cie. 451 pp.
17. Brown, H. P. 1966. Effects of soap pollution upon stream invertebrates. *Trans. Am. Microsc. Soc.* 85(1):167 (Abstr.)
18. Brown, H. P. 1967. Psephenids (Coleoptera: Dryopoidea) parasitized by eulophid and darapriid [sic] wasps (Hymenoptera: Chalcidoidea and Proctotrupoidea). *Am. Zool.* 7(2):193 (Abstr.)
19. Brown, H. P. 1968. *Psephenus* (Coleoptera: Psephenidae) parasitized by a new chalcidoid (Hymenoptera: Eulophidae). II. Biology of the parasite. *Ann. Entomol. Soc. Am.* 16(2):452–56
20. Brown, H. P. 1970. Neotropical dryopoids. II. *Elmoparnus mexicanus* sp. n. (Coleoptera, Dryopidae). *Coleopt. Bull.* 24(4):124–27
21. Brown, H. P. 1970. A key to the dryopid genera of the New World (Coleoptera, Dryopoidea). *Entomol. News* 81(7):171–75
22. Brown, H. P. 1972. *Aquatic Dryopoid Beetles (Coleoptera) of the United States. Biota of Freshwater Ecosystems Identification Manual No. 6.* Washington, DC: US Environ. Prot. Agency. 82 pp.
23. Brown, H. P. 1972. Trials and tribulations of a riffle beetle buff (or Why didn't I stick with the protozoa?). *Bios Madison NJ* 43(2):51–60
24. Brown, H. P. 1974. Survival records for elmid beetles, with notes on laboratory rearing of various dryopoids (Coleoptera). *Entomol. News* 84(9):278–84
25. Brown, H. P. 1981. Key to the world

genera of Larinae (Coleoptera, Dryopoidea, Elmidae), with descriptions of new genera from Hispaniola, Colombia, Australia, and New Guinea. *Pan-Pac. Entomol.* 57(1):76–104

26. Brown, H. P. 1981. A distributional survey of the world genera of aquatic dryopoid beetles (Coleoptera: Dryopidae, Elmidae, and Psephenidae sens. lat.). *Pan-Pac. Entomol.* 57(1):133–48

27. Brown, H. P. 1984. Neotropical dryopoids III. Major nomenclatural changes affecting *Elsianus* Sharp and *Macrelmis* Motschulsky, with checklists of species (Coleoptera: Elmidae: Elminae). *Coleopt. Bull.* 38(2):121–29

28. Brown, H. P., Murvosh, C. M. 1970. *Lutrochus arizonicus*, new species, with notes on ecology and behavior (Coleoptera, Dryopoidea, Limnichidae). *Ann. Entomol. Soc. Am.* 63(4):1030–35

29. Brown, H. P., Shoemake, C. M. 1969. Cannibalism by a "herbivore," *Microcylloepus pusillus* (Coleoptera: Elmidae). *Proc. Okla. Acad. Sci.* 48:15

30. Brusven, M. A. 1970. Drift periodicity of some riffle beetles (Coleoptera: Elmidae). *J. Kans. Entomol. Soc.* 43(4):364–71

31. Champion, G. C. 1920. Some Indian Coleoptera (3). *Entomol. Mon. Mag.* 56:194–96

32. Champion, G. C. 1924. Some Indian Coleoptera. *Annu. Mag. Nat. Hist. London* (9)13:249–64

33. Clark, W. H., Ralston, G. L. 1974. *Eubrianax edwardsi* in Nevada with notes on larval pupation and emergence of adults (Coleoptera: Psephenidae). *Coleopt. Bull.* 28(4):217–18

34. Cole, L. C. 1957. A surprising case of survival. *Ecology* 38(2):357

35. Crowson, R. A. 1978. Problems of phylogenetic relationships in Dryopoidea (Coleoptera). *Entomol. Ger.* 4(3/4):250–57

36. Crowson, R. A. 1981. *The Biology of the Coleoptera.* New York: Academic. 802 pp.

37. Delève, J. 1945. Contribution à l'étude des Dryopidae. III. Le genre *Pseudomacronychus* Grouvelle et le dimorphisme alaire de ses espèces. *Bull. Mus. R. Hist. Nat. Belg.* 21(9):1–12

38. Delève, J. 1967. Contribution à la faune du Congo (Brazzaville). Mission A. Villiers et A. Descarpentries. XLIV. Coléoptères Elminthidae. *Bull. Inst. Fondam. Afr. Noire Ser. A* 29(1):318–43

39. Dufour, L. 1862. Etudes sur la larve de *Potamophilus. Ann. Sci. Nat. Zool. Biol. Anim.* 17(4):162–73

40. Essig, E. O. 1926. *Insects of Western North America.* New York: Macmillan. 1035 pp.

41. Gage, K. L. 1983. *Wood utilization by Macronychus glabratus (Say) (Coleoptera: Elmidae).* MS thesis. Univ. Okla., Norman. 57 pp.

42. Gahan, C. J. 1914. A new genus of Coleoptera of the family Psephenidae. *Entomologist* 47:188–89

43. Green, E. A. 1972. *Comparative larval morphology of representative New World dryopoid beetles.* PhD dissertation. Univ. Okla., Norman. 401 pp.

44. Harpster, H. T. 1941. An investigation of the gaseous plastron as a respiratory mechanism in *Helichus striatus* LeConte. *Trans. Am. Microsc. Soc.* 60(3):329–58

45. Harpster, H. T. 1944. The gaseous plastron as a respiratory mechanism in *Stenelmis quadrimaculata* Horn (Dryopoidae). *Trans. Am. Microsc. Soc.* 63(1):1–26

46. Hennig, W. 1966. *Phylogenetic Systematics.* Urbana, Ill: Univ. Ill. Press. 263 pp.

47. Hilsenhoff, W. L. 1982. Using a biotic index to evaluate water quality in streams. *Tech. Bull. 132.* Madison, Wis: Dep. Nat. Resourc. 22 pp.

48. Hinton, H. E. 1937. *Protoparnus pusillus,* new species of Dryopidae from St. Vincent. *Col. Rev. Entomol.* 7(2–3):302–6

49. Hinton, H. E. 1939. An inquiry into the natural classification of the Dryopoidea, based partly on a study of their internal anatomy (Col.). *Trans. R. Entomol. Soc. London* 89(7):133–84

50. Hinton, H. E. 1940. The Percy Sladen Trust Expedition to Lake Titicaca in 1937 under the leadership of Mr. H. Cary Gilson. VII. The Peruvian and Bolivian species of *Macrelmis* Motsch. (Coleoptera, Elmidae). *Trans. Linn. Soc. London Ser. 3* 1:117–47

51. Hinton, H. E. 1940. A monographic revision of the Mexican water beetles of the family Elmidae. *Novit. Zool.* 42(2):217–396

52. Hinton, H. E. 1947. The gills of some aquatic beetle pupae (Coleoptera. Psephenidae). *Proc. R. Entomol. Soc. London Ser. A,* 22:52–60

53. Hinton, H. E. 1955. On the respiratory adaptations, biology, and taxonomy of the Psephenidae, with notes on some related families (Coleoptera). *Proc. Zool. Soc. London* 125(3/4):543–68

54. Hinton, H. E. 1965. A revision of the Australian species of *Austrolimnius* (Col. Elmidae). *Aust. J. Zool.* 13(1):97–172

55. Hinton, H. E. 1966. Respiratory adaptations of the pupae of beetles of the family Psephenidae. *Phil. Trans. R. Soc. London Ser. B Biol. Sci.* 251(771):211–45

56. Hinton, H. E. 1968. The subgenera of *Austrolimnius* (Coleoptera: Elminthidae). *Proc. R. Entomol. Soc. London Ser. B* 37(7/8):98–102

57. Hinton, H. E. 1971. Some American *Austrolimnius* (Coleoptera: Elmidae). *J. Entomol. Ser. B Taxon.* 40(2):93–99

58. Hinton, H. E. 1976. Plastron respiration in bugs and beetles. *J. Insect Physiol.* 22:1529–50

59. Holland, D. G. 1972. A key to the larvae, pupae and adults of the British species of Elminthidae. *Freshwater Biol. Assoc. Sci. Publ. No. 26.* Ambleside, Westmorland, England: 58 pp.

60. Hynes, H. B. N. 1972. *The Ecology of Running Waters.* Toronto: Univ. Toronto Press. 555 pp.

61. Jäch, M. A. 1985. Beitrag zur Kenntnis der Elmidae und Dryopidae New Guineas (Coleoptera). *Rev. Suisse Zool.* 92(1):229–54

62. Kaufmann, T., Stansly, P. 1979. Bionomics of *Neoheterocerus pallidus* Say (Coleoptera: Heteroceridae) in Oklahoma. *J. Kans. Entomol. Soc.* 52(3):565–77

63. Kellicott, D. S. 1883. *Psephenus lecontei.* On the external anatomy of the larva. *Can. Entomol.* 15:191–96

64. Lauterborn, D. 1921. Faunistische Beobachtungen aus der Oberrheins Bodensee. *Mitt. Bad. Landesanst. Naturgesch. Freiburg* 1921:197

65. LeSage, L., Harper, P. P. 1976. Notes on the life history of the toed-winged beetle *Anchytarsus bicolor* (Melsheimer) (Coleoptera: Ptilodactylidae). *Coleopt. Bull.* 30(3):233–38

66. LeSage, L., Harper, P. P. 1976. Description de nymphes d'Elmidae néarctiques (Coléoptères). *Can. J. Zool.* 54(1):65–73

67. LeSage, L., Harper, P. P. 1977. Life cycles of Elmidae (Coleoptera) inhabiting streams of the Laurentian Highlands, Quebec. *Ann. Limnol.* 12(2):139–74

68. Marlier, G. 1960. La morphologie et la biologie de la larve de l'*Afropsephenoides* (Coléoptère, Psephenoide). *Rev. Zool. Bot. Afr.* 61:1–14

69. Matthes, D., Guhl, W. 1975. Systematik, Anpassungen und Raumparasitis-

mus auf Hydrophiliden lebender operculariformer Epistyliden. *Arch. Protistenkd.* 117:110–86

70. Messner, B., Langer, C. 1984. Atmungsfähige Deckflügel als Anpassung an die submerse Lebensweise bei Käfern. *Zool. Jahrb. Abt. Anat. Ontog. Tiere* 111:469–84

71. Metzky, J. 1950. Morphologie und Physiologie des Branchialorgans der Helminenlarven. *Osterreich Zool. Z.* 2(5/6):585–604

72. Miller, R. F., Morgan, A. V. 1982. A postglacial coleopterous assemblage from Lockport Gulf, New York, U.S.A. *Quat. Res. NY* 17(2):258–74

73. Morgan, A. V., Morgan, A. 1980. Faunal assemblages and distributional shifts of Coleoptera during the late Pleistocene in Canada and the northern United States. *Can. Entomol.* 112:1105–28

74. Moubayed, Z. 1983. The larval development of *Grouvellinus coyei* (Coleoptera: Elmidae) collected from the Bekaa plain, Lebanon. *Ann. Limnol.* 19(2):115–20

75. Murvosh, C. M. 1971. Ecology of the water penny beetle *Psephenus herricki* DeKay. *Ecol. Monogr.* 41:79–96

76. Murvosh, C. M., Brown, H. P. 1977. Underwater pupation by a psephenid beetle. *Ann. Entomol. Soc. Am.* 70(6):975–76

77. Newman, D. L., Funk, R. C. 1984. Drift of riffle beetles (Coleoptera: Elmidae) in a small Illinois stream. *Great Lakes Entomol.* 17(4):211–14

78. Pacheco, F. 1964. *Sistematica, Filogenia y Distribucion de los Heteroceridos de America (Coleoptera: Heteroceridae).* Chapingo, Mexico: Monogr. Col. Postgrad. 115 pp.

79. Paulus, H. F. 1970. Zur Morphologie und Biologie der Larven von *Pelochares* Mulsant & Rey (1869) und *Limnichus* Latreille (1829) (Coleoptera: Dryopoidea: Limnichidae). *Senckenbergiana Biol.* 51(1/2):77–87

80. Perez, M. 1863. Histoire des metamorphoses du *Macronychus quadrituberculatus* et de son parasite. *Ann. Soc. Entomol. France* (4e Ser.)3:621–36

81. Philippi, R. A. 1864. Ein Käferchen, das als Gewurz dient. *Stettiner Entomol. Ztg.* 25:93–96

82. Reisen, W. K. 1977. The ecology of Honey Creek, Oklahoma: downstream drift of three species of aquatic dryopoid beetles (Coleoptera: Dryopoidea). *Entomol. News* 88(7/8):185–91

83. Scudder, S. H. 1900. Adephagous and

clavicorn Coleoptera from the Tertiary deposits at Florissant, Colorado with descriptions of a few other forms and a systematic list of the non-rhynchophorous Tertiary Coleoptera of North America. *US Geol. Surv. Monogr. 40.* Washington, DC: USGS. 117 pp.

84. Seagle, H. H. 1980. Flight periodicity and emergence patterns in the Elmidae (Coleoptera: Dryopoidea). *Ann. Entomol. Soc. Am.* 73(3):300–7

85. Seagle, H. H. 1982. Comparison of the food habits of three species of riffle beetles, *Stenelmis crenata, Stenelmis mera,* and *Optioservus trivittatus* (Coleoptera: Dryopoidea: Elmidae). *Freshwater Invertebr. Biol.* 1(2):33–38

86. Segal, B. 1933. The hind wings of some Dryopidae in relation to habitat (Coleop.). *Entomol. News* 44:85–88

87. Shepard, W. D. 1980. *A morphological study of larval forms of North American species of* Stenelmis *(Coleoptera: Elmidae).* PhD dissertation. Univ. Okla., Norman. 127 pp.

88. Sinclair, R. M. 1964. *Water Quality Requirements of the Family Elmidae (Coleoptera), with Keys to the Larvae and Adults of the Eastern Genera.* Nashville, Tenn: Stream Pollut. Control Board, Tenn. Dep. Public Health. 14 pp.

89. Smith, J. A., Dartnell, A. J. 1980. Boundary layer control by water pennies (Coleoptera: Psephenidae). *Aquat. Insects* 2:65–72

90. Spangler, P. J. 1980. Chelonariid larvae, aquatic or not? (Coleoptera: Chelonariidae). *Coleopt. Bull.* 34(1):105–14

91. Spangler, P. J. 1981. *Pagelmis amazonica,* a new genus and species of water beetle from Ecuador (Coleoptera: Elmidae). *Pan-Pac. Entomol.* 57(1):286–94

92. Spangler, P. J. 1981. Amsterdam expeditions to the West Indian Islands, Report 15. Two new genera of phreatic elmid beetles from Haiti, one eyeless and one with reduced eyes (Coleoptera: Elmidae). *Bijdr. Dierkd.* 51(2):375–87

93. Spangler, P. J. 1983. Immature stages and biology of *Tetraglossa palpalis* Champion (Coleoptera: Ptilodactylidae). *Entomol. News* 94(5):161–75

94. Spangler, P. J., Brown, H. P. 1981. The discovery of *Hydora,* a hitherto Australian-New Zealand genus of riffle beetles, in austral South America (Coleoptera: Elmidae). *Proc. Entomol. Soc. Wash.* 83(4):596–606

95. Spangler, P. J., Perkins, P. D. 1977. Three new species of the neotropical water beetle genus *Elmoparnus* (Cole-

optera: Dryopidae). *Proc. Biol. Soc. Wash.* 89(63):743–60

96. Steedman, R. J. 1983. *Life history and feeding role of the xylophagous aquatic beetle,* Lara avara LeConte *(Dryopoidea: Elmidae).* MS thesis. Oregon State Univ., Corvallis. 110 pp.

97. Steffan, A. W. 1963. Beziehungen zwischen Lebensraum und Körpergrösse bei mitteleuropäischen Elminthidae (Coleoptera: Dryopoidea). *Z. Morphol. Ökol. Tiere* 53:1–21

98. Steffan, A. W. 1967. Über die biozönotische und geographische Verbreitung der europäischen Vertreter der Hydrophiloidea: Georissidae; Dascilloidea: Clambidae, Elodidae, Dascillidae; Dryopoidea: Psephenidae, Heteroceridae, Dryopidae, Limnichidae, Elminthidae; Sphaeroidea: Hydroscaphidae, Sphaeriidae (Coleoptera). In *Limnofauna Europaea,* ed. J. Illies, pp. 269–75. Stuttgart: Fischer. 1st ed.

99. Stewart, K. W., Friday, G. P., Rhame, R. E. 1973. Food habits of hellgrammite larvae, *Corydalus cornutus* (Megaloptera: Corydalidae), in the Brazos River, Texas. *Ann. Entomol. Soc. Am.* 66(5):959–63

100. Stribling, J. B. 1986. Revision of *Anchytarsus* (Coleoptera: Dryopoidea) and a key to the New World genera of Ptilodactylidae. *Ann. Entomol. Soc. Am.* 79(1):219–34

101. Stride, G. O. 1955. On the respiration of an aquatic African beetle, *Potamodytes tuberosus* Hinton. *Ann. Entomol. Soc. Am.* 48(5):344–51

102. Susskind, M. E. C. 1936. A morphological study of the respiratory system in various larval instars of *Stenelmis sulcatus* Blatchley. *Pap. Mich. Acad. Sci. Arts Lett.* 21:697–714

103. Tavares, I. I. 1979. The Laboulbeniales and their arthropod hosts. In *Insect-Fungus Symbiosis,* ed. L. R. Batra, pp. 229–58. Montclair, NJ: Allanheld, Osmun. 276 pp.

104. Thorpe, W. H. 1950. Plastron respiration in aquatic insects. *Biol. Rev.* 25:344–90

105. Thorpe, W. H., Crisp, D. J. 1949. Studies on plastron respiration. IV. Plastron respiration in the Coleoptera. *J. Exp. Biol.* 26(3):219–60

106. Turcotte, P., Harper, P. P. 1982. The macroinvertebrate fauna of a small andean stream, South America. *Freshwater Biol.* 12(5):411–20

107. Waters, T. F. 1972. The drift of stream insects. *Ann. Rev. Entomol.* 17:253–72

108. Weber, C. I. 1973. *Biological Field and*

Laboratory Methods for Measuring the Quality of Surface Waters and Effluents. Cincinnati, Ohio: Off. Res. Dev., US Environ. Prot. Agency. 176 pp.

109. West, L. S. 1929. The behavior of *Macronychus glabratus* Say (Coleopt. Helmidae). *Entomol. News* 40(6):171–73

110. West, L. S. 1929. Life history notes on *Psephenus lecontei* Lec. (Coleoptera: Dryopoidea; Psephenidae). *Battle Creek Coll. Bull.* 3(1):3–20

111. White, D. S. 1978. Life cycle of the riffle beetle, *Stenelmis sexlineata* (Elmidae). *Ann. Entomol. Soc. Am.* 71(1):121–25

112. White, D. S., Brigham, W. U., Doyen, J. T. 1984. Aquatic Coleoptera. In *An Introduction to the Aquatic Insects of North America,* ed. R. W. Merritt, K. W. Cummins, pp. 361–437. Dubuque, Iowa: Kendall/Hunt. 722 pp.

Ann. Rev. Entomol. 1987. 32:275–95
Copyright © 1986 by Annual Reviews Inc. All rights reserved

SCORPION BIONOMICS

S. C. Williams

Department of Biology, San Francisco State University, San Francisco, California 94132

INTRODUCTION

Scorpions comprise a diverse and highly successful order of Arachnids. They have a distinctive morphology by which living and extinct members are readily recognized. The scorpion body is conspicuously metameric with a prosoma covered by an unsegmented carapace, a broad mesosoma consisting of seven segments, a narrow taillike metasoma consisting of five scleroma, and a telson modified into a stinging apparatus. The body is protected by a sclerotized exoskeleton that efficiently retards water loss, and is set with a variety of sensory setae and other sophisticated sensillae. Scorpions utilize an interesting engorgement type of predatory feeding, and are adapted for surviving long periods between feedings. They use water thriftily; retention is facilitated by excretion of guanine, uric acid, or xanthine as primary nitrogen excretion products. Females are viviparous and, unlike most arthropods, show maternal association with the young. Perhaps scorpions are best known because of their venomous sting, which in some species is lethal to humans. The antiquity of scorpions gives them a special significance. They are the oldest known terrestrial metazoans, occurring in silurian fossils. With all the evolutionary changes in the terrestrial environment it is remarkable that scorpions have changed little in basic external morphology over geologic time.

Over the past two decades the scientific community has begun to discover the potential of scorpions as subjects and tools of biological research, and as a result much literature has been published. This review is limited to a summary of recent findings in the areas of life histories, reproduction, feeding biology, habitats, cyclic behavior, and predators.

0066-4170/87/0101-0275$02.00

SYSTEMATICS AND DIVERSITY

Order Scorpiones is composed of eight living families: Bothriuridae, Buthidae, Chactidae, Chaerilidae, Diplocentridae, Iuridae, Scorpionidae, and Vaejovidae. In addition, 14 extinct families are recognized. Living members of the order show less diversity among families than insects, mites, and spiders. Stahnke (89) estimated the numbers of described families, genera, and species, and Williams (104) modified this estimate accounting for subsequent taxonomic work. Francke (42) reviewed scorpion nomenclature. Today it is estimated that there are about 1500 described living species in 112 recognized genera.

There are many apparently hospitable habitats in which scorpions are not permanently established, and also extensive habitats occupied by single species. Single-species communities have been reported for *Paruroctonus boreus* (3, 92), *Vaejovis gertschi* (91), and *Uroctonus mordax* (8). Such limited diversity tends to occur in extreme environments, e.g. at higher latitudes and higher elevations. In more favorable environments it is sometimes possible to find as many as eight sympatric species. Eight species each were found in Sonoran Desert habitats in Arizona (100) and in an intermontane basin in Coahuila, Mexico (101). Gertsch & Allred (46) reported nine species from the Nevada Test Site in southern Nevada, but this count included data from a number of adjacent habitats, and it is unlikely that all nine species coexisted sympatrically. Six species were reported in a California dune community (67), five in a coastal dune habitat in Sonora, Mexico (107), and four in a seasonally wet tropical forest in Jalisco, Mexico (105). Within favorable regions three to five species are commonly encountered. Regional species diversity of scorpions is low compared to that of spiders and mites. This low diversity is probably due to similar habitat requirements, behavior, and feeding biology among scorpions.

LIFE HISTORY

Life histories and population structures are hard to evaluate because of problems in rearing litters through the life cycle and difficulties in sampling all stadia in natural populations. The problems in determining the number of instars in a life cycle have been analyzed (44). Life history information has been compiled from field, laboratory, and theoretical studies. Life tables have been constructed for *Paruroctonus mesaensis* (82), *Urodacus abruptus* (86), and *Urodacus yaschenkoi* (82). All three species have high early-instar and adult mortality, whereas intermediate instars exhibit relatively low mortality.

Life Stages

Following fertilization, young develop viviparously within the ovarian tube of the female (41). Young are born in an early stage of development called the larva (first instar of life cycle). The larval stage differs from the subsequent stages in that larvae lack ungues, a fully formed telson, and a fully sclerotized exoskeleton; they also differ in their response to the environment, in part because of their smaller surface-to-volume ratio. They do not feed, are not capable of significant independent locomotion, and cannot sting. They normally assume a position on the mother's dorsum and molt to the second instar stage (first nymphal instar) in about 1–2 wk. Scorpions of most species disperse from the mother as second instars; in some species this stage may be the primary dispersal stage. Nymphs molt several times before reaching sexual maturity. Scorpions of the few taxa studied have 5–7 stadia. The number of instars in the life cycle (including larva and imago) has been estimated at seven in *Uroctonus mordax* (37), *Paruroctonus mesaensis* (70), and *Centruroides gracilis* (43) and six in *Buthotus minax* (90), *Urodacus abruptus* (86), and *Urodacus yaschenkoi* (83). Some species may mature in more than one stadium. Some males of *Tityus trivittatus fasciolatus* mature in the fifth instar, while others mature in the sixth (57, cited in 44). Females, however, appeared to all mature as sixth instars. A similar pattern of males maturing in more than one instar has been reported for several species (44). It is believed that scorpions do not molt after reaching sexual maturity, but they may live for several years in the adult stadium.

Growth Rate

Many species reach maturity during the first year of life. This has been determined for *C. gracilis* (41). However some, such as *P. mesaensis*, have been estimated to mature in 19–24 mo (72), and *U. abruptus* requires 3 yr (86). In a tabulation of the time required for 10 species to reach sexual maturity (43), *Buthotus minax* was the fastest, reaching sexual maturity in 164–307 days, while *Tityus trivittatus* was the slowest, requiring 426–810 days.

Most scorpions experience the first molt within 2 wk, and the following five or six molts occur at various intervals during the next 9–12 mo. Growth, as a result of molting, is in linear dimensions only, because mass decreases slightly as a result of ecdysis. Comparison of postmolt to premolt lengths has revealed the following increases: *Uroctonus mordax*, 20–45% (37); *Isometrus maculatus*, 30% (74); *Buthus occitanus*, 26% (7); and *Tityus bahiensis*, 22% (62, cited in 44).

Longevity

Scorpions vary considerably in longevity. In a number of taxa females tend to live longer than males; mature males may have a higher mortality rate because of increased exposure to predation and suboptimum habitat conditions during the mate-seeking season. Estimates of maximum longevity for some species are as follows: *Paruroctonus mesaensis,* 5 yr (68); *Urodacus abruptus,* 10 yr (86); *Centruroides gracilis,* 4 yr (43); *Centruroides sculpturatus,* longer than 5 yr (88); *U. mordax,* longer than 579 days (37); *Tityus bahiensis,* 1417 days (62, cited in 44); and *Tityus serrulatus,* 1365–1565 days (64).

REPRODUCTIVE BIOLOGY

Courtship and Mating

Mating is preceded by an elaborate courtship, referred to as the "promenade a deux," which has similar behavioral components throughout the order. Courtship and insemination normally occur during the night on exposed habitat surfaces. Fertilization is internal. Males, however, lack a penis and depend on a spermatophore for sperm transfer. The courtship is a curious behavioral sequence during which mates locate and orient to each other. They grip each other by the pedipalp chelae and sometimes also by the chelicerae, and walk forward and backward in coordinated movements that resemble a dance until a suitable substrate for spermatophore deposition is found. The male emits an elongated spermatophore, which is usually glued to a firm surface; he then pulls the female over the spermatophore, and insemination takes place.

Courtship has been described numerous times in a variety of taxa (9, 10, 84, 87, 88, 102). The use of spermatophores was not detected until 1955 (1, 2, 4–6). It now appears that male scorpions universally inseminate females indirectly in this way. The uptake of sperm by courted females may not always be successful (2).

Polis & Farley (71) described an episode of courtship and insemination in *Paruroctonus mesaensis.* The initial contact phase of courtship lasted about 22 min and involved physical contact, movement, and a distinct vibration of the body called "juddering." Following initial contact the pair went into the courtship dance, which lasted 9.3 min and ended in spermatophore deposition and insemination. During the promenade the pair traveled approximately 8 m. Insemination took 3–5 sec.

Parthenogenesis is known in scorpions but is not common. It has been reported in *Tityus serrulatus* (61, 63, 64, 79) and in *Liocheles australasiae* (59, 60). In *L. australasiae* thelytokous parthenogenesis was observed in the laboratory, and field observations suggest that some, if not most, populations of this species reproduce only in this fashion.

Gestation

Many temperate taxa, such as members of *Vaejovis, Uroctonus, Hadrurus, Serradigitus,* and *Paruroctonus,* appear to normally have one litter per year, with births more or less synchronized within a given population. Some more tropical taxa, such as members of *Centruroides,* may have one to several litters per year, and data suggest that gestation intervals are 2–4 mo. The gestation period for a number of species has been estimated as follows: *Centruroides insulanus,* 4.5–7.5 mo (10); *Centruroides vittatus,* 8 mo (10); *Urodacus abruptus,* 16 mo (86); and *Paruroctonus mesaensis,* 1 yr (72). Probable causes for variation in gestation among species include size, number of young born, and general climatic conditions experienced by the population (102).

Scorpions are categorized as apoikogenic and katoikogenic based on the nature of their internal development (41). In apoikogenic forms the embryos are each covered by a birth membrane formed by the fusion of the embryonic serosa and amnion. In contrast, katoikogenic embryos are not covered by extraembryonic membranes; they develop in large diverticulae opening into the reproductive tube. Katoikogenic forms are consistently born tail first, while apoikogenic forms are born either head or tail first.

Parturition

The birth process has been described a number of times and for a variety of taxa (40, 41, 52, 85, 102). The mother stilts high above the substrate on her hind two pairs of walking legs; her genital opercula open and the young emerge, one by one, through the genital aperture. In most families, each young is covered at birth by a birth membrane. In contrast, members of the Diplocentridae and Scorpionidae are not covered by such membranes (41). As the young emerge, the mother usually catches them in a "birth basket" formed by the first two pairs of walking legs. In the birth basket, young that are enclosed in birth membranes shed them. Then they climb the mother's anterior walking legs to her dorsum. As larvae ascend to the mother's dorsum they may assume either an ordered orientation *(Vaejovis)* or, more commonly, a random orientation *(Centruroides, Uroctonus, Syntropis,* and *Megacormus)* (37, 38, 51, 52, 102). The significance of the highly ordered orientation in *Vaejovis* is still not known.

The length of parturition varies among individuals of the same species and among members of different species. In general, the time spent in parturition depends on the number of young delivered, the size of young, and the occurrence of complications. One *Hadrurus arizonensis* female delivered 10 young in 1 hr, while one *Vaejovis spinigerus* delivered 69 young in 7.5 hr (102). Apoikogenic forms tend to have more rapid parturition, ranging from

55 min to 37 hr, while parturition for katoikogenic forms ranged from 36 to 74 hr (41).

Parturition is a critical point in the life cycle; if the young are unable to shed their birth membranes or ascend to the mother's dorsum their probability of survival is low. In *U. abruptus* the mortality during birth is about 27% (86).

Maternal Care

Once on the mother's dorsum larvae assume a characteristic posture and remain there throughout the first stadium. Within 2 wk the larvae synchronously molt to the second instar stage (the first nymphal instar) and remain on the mother's dorsum for 1–2 wk, at which time they begin dispersing. In some genera, such as *Vaejovis, Uroctonus, Paruroctonus,* and *Hadrurus,* mothers remain inactive within their protective shelters while with young. In others, such as *Centruroides,* the females may be seen during nocturnal hours in exposed areas with young on their dorsum.

During the first stadium there is no obvious nutrient exchange between mother and offspring. Second instars have been observed feeding on remains of their mother's prey, but they generally begin dispersal in this stadium, soon after their cuticle hardens. Thus, maternal feeding of young is probably insignificant. In a contrasting observation, however, a *Palamnaeus longimanus* mother provided food for her litter, which was reported to disperse in the third stadium (81). Vannini and coworkers (95) recently investigated the question of nutrient exchange between mother and larvae. In experiments with *Euscorpius flavicaudis,* mothers given tritiated water transferred tritium to larvae on their dorsum. Tritium may have been incorporated in epicuticular lipids which were then transferred to the larvae by contact; or it may have been absorbed by the larvae in the form of liquid water or vapor from the mother. This study suggests that a reexamination of the adaptive scope of mother-larval association is in order. The mother-larval association was recently investigated in three species of *Euscorpius* (96–98). Larvae were observed to mount substitute mothers, and in the absence of a mother larvae mounted artificial models and dead scorpions, but they did not settle on these permanently. Larvae settled permanently on the mother's dorsum owing to substratum recognition. Chemical recognition was not species-specific, nor did it carry information on sex or reproductive phase. Three adaptive benefits of mother-larval association have been hypothesized: antipredator defense; continuous maternal selection of the optimal microclimate for larvae; and trophic exchange between mother and larvae (97).

Fecundity

The number of young in a litter varies. Scorpions have been reported to have 6–105 young per litter. Litter sizes have been reported as follows: *Centrur-*

oides insulanus, 20–105 (10); *Centruroides sculpturatus,* 12–36 (88); *Centruroides vittatus,* 20–47 (10); *Diplocentrus bigbendensis,* 37 (40); *Diplocentrus spitzeri,* 9 (39); *Hadrurus arizonensis,* 10 (102); *Megacormus gertschi,* 19–75 (38); *Paruroctonus boreus,* average of 34 (92); *Paruroctonus mesaensis,* 18–21 (68); *Uroctonus mordax,* 28–34 (37); *Urodacus abruptus,* 9–17 (86); and *Vaejovis spinigerus,* 66–69 (88, 102). Litter size and number of litters produced per female are perhaps related to species group, population density, and availability of prey; however, the influence of these factors has not been established.

FEEDING BIOLOGY

Scorpions are essentially obligate predators. They seldom ingest dead prey. However, adults occasionally accept freshly killed prey in the laboratory.

Prey Encounter

The several strategies that scorpions have evolved to encounter prey can be put in three primary classes: waiting on exposed substrates, active stalking, and waiting in their burrows. The most common strategy is to leave the diurnal shelter in early evening, settle in an exposed location on the ground surface or other exposed substrate, and wait for prey. Prey is captured only if it encounters the scorpion or comes close enough to stimulate the sensory receptors of the quiescent scorpion. Most species of *Vaejovis, Paruroctonus, Vejovoidus, Uroctonus, Serradigitus,* and *Hadrurus* use this strategy. Members of *Centruroides,* in contrast, sometimes actively traverse exposed substrates, apparently searching for prey. Members of *Anuroctonus phaiodactylus* use their burrows to trap prey. The scorpion remains quiescent just inside the entrance to its burrow and captures insects that enter. Examination of the remains of prey in the burrow tumuli has suggested that these scorpions may be diurnal predators. *Superstitionia donensis* has been observed living in a shallow cell under cattle dung, where it preys on dung-feeding insects. Studies have revealed that *Paruroctonus mesaensis* has a sophisticated ability to locate and orient toward prey using prey-generated low-frequency compressional vibrations in sand (12–15). It uses basitarsal slit sensillae to detect these vibrations, and can effectively locate moving prey at a distance of up to 50 cm. In *P. mesaensis* this appears to be the primary means of locating prey. Even prey burrowing under the sand surface are detected and caught by a thrust with the pedipalps. A similar pattern of orientation has been observed in *Hadrurus arizonensis,* but the range and accuracy of this scorpion's response to prey was less than that observed for *P. mesaensis* (15). Scorpions have other sense organs (mechanoreceptive setae, chemoreceptive setae, and chemoreceptive tarsal organs) that may aid in predation (36).

Prey Capture

Scorpions firmly grasp encountered prey with the chelae of the pedipalps. Some scorpions, such as *Centruroides exilicauda,* then sting the prey before ingestion. Others, such as *Hadrurus arizonensis* and *Anuroctonus phaiodactylus,* often do not attempt to sting the prey although they are capable of doing so. When the sting is not used or is not lethal, the scorpion firmly immobilizes its prey with the pedipalp chelae and ingests it alive. Some species have venoms that show no effect on prey. Scorpions probably kill their prey more often by eating them alive than by injecting venom.

Scorpions engorge when prey are available, and may remain for a long time in a quiescent state between feedings. Feeding scorpions ingest a high proportion of their prey's biomass, may increase their own biomass substantially through engorgement, and utilize a preoral exudate that appears to initiate digestion in the carcass of the prey before ingestion. *Vaejovis gertschi* increased body weight by 9.6–16.5% during a single feeding (91). Weight increases of 9–33% in *A. phaiodactylus* and 10–36% in *U. mordax* have been measured (35).

Ingestion

Following immobilization of prey, the scorpion uses its chelicerae to tear it open, and pumps a preoral exudate from the scorpion's buccal region into the carcass. A rhythmic flow of this fluid from the scorpion to the carcass and back is established shortly and lasts throughout ingestion. *Anuroctonus phaiodactylus* fed on crickets ingested 8–97% of the prey biomass, and *Uroctonus mordax* showed similar ingestion efficiency of 5–87% (35). The efficiency of ingestion is striking, and is probably due to digestive enzymes in the preoral exudate, a setal mechanism in the cheliceral region that minimizes fluid loss, and an efficient gnathobase press that separates fluids from particulate matter. During the evening, under natural conditions, scorpions ingest prey on their habitat surface or at burrow entrances. Most prey of observed *Paruroctonus mesaensis* were eaten at the capture site, but about one quarter were carried to nearby vegetation for ingestion (67).

Predation Frequency

Observations suggest that scorpions normally capture and ingest prey infrequently, and that they are well adapted for intermittent feeding. *Vaejovis gertschi* feeding on termites lost weight in a linear fashion after engorgement (91). On the average these scorpions returned to their prefeeding weights in 13 days; this suggests that individuals need to feed about once every two weeks to maintain their biomass (at 22°C and near 100% RH) (91). In a similar study, *Anuroctonus phaiodactylus* individuals feeding on crickets returned to their prefeeding weights in 10–23 days, while *Uroctonus mordax* individuals returned to their prefeeding weights in 5–68 days (35); individuals of these species apparently need to feed once every one to ten weeks to maintain normal biomass. Under natural environmental conditions individual scorpions need to feed only periodically; the periodicity is probably determined by size of prey, temperature, relative humidity, and physical activity of the scorpion.

Prey

Prey selectivity by scorpions has been reviewed (8–10, 35, 65, 67, 87, 88, 91, 92). Scorpions will capture almost any prey encountered that they can physically immobilize and ingest. They feed primarily on insects, arachnids, and other arthropods. Predation on mollusks has also been reported (55) but is not common. Some larger species also prey on small vertebrates. In one study *Vaejovis gertschi* accepted geometrid caterpillars, ichneumonids, silverfish, machilids, tipulids, termites, and ground-dwelling spiders as prey (91), but rejected crickets, small roaches, mirids, carabids, tenebrionids, and millipedes. *Vaejovis confusus, Vaejovis spinigerus,* and *Hadrurus arizonensis* prey on a wide variety of arthropods (100). *Hadrurus arizonensis* is distinctive in that it readily eats lizards and mice. *Paruroctonus mesaensis* also preys on a variety of arthropods, and moreover shows a substantial tendency toward cannibalism (50, 67). In a California dune community 9% of the diet of *P. mesaensis* was estimated to be supplied by cannibalism (70). This impressive level of cannibalism was not detected in a different study on this species in another area (50). Polis (67) recorded 95 prey species accepted by *P. mesaensis* in a sand dune habitat; 81 of these comprised only 1% of the scorpion's diet. He reported that 10% of the prey were fossorial, 80% were cursorial, and 10% were aerial. The major prey composition was 42% tenebrionids, 17% orthopterans, 16% other scorpions, and 12% hymenopterans. Worm snakes *(Leptotyphlops)* were also eaten. *Urodacus abruptus* has been reported to accept a variety of prey, mainly insects and spiders; they exhibited no sign of cannibalism or aggressive intraspecific behavior when they were well fed (87).

Scorpion predation on vertebrates has been summarized (65). Vertebrate prey has included 2 species of mammals, 11 species of lizards, and 2 species of snakes. Ten species of scorpions have been reported to prey on vertebrates: *Pandinus pallidus, Opisthophthalmus carinatus, Hadogenes* sp., *Parabuthus villosus, Vaejovis* sp., *Paruroctonus mesaensis, Hadrurus arizonensis, Hadrurus spadix,* and *Hadrurus hirsutus.*

HABITATS

Scorpions occupy a greater range of habitats than is generally recognized. They are most abundant and diverse in arid environments of lower temperate latitudes. The most diverse assemblage of scorpions known is in Baja California, Mexico, where 61 species, 11 genera, and 5 families have been reported in the narrow peninsula (104). In this region scorpions are found in stabilized and unstabilized sand dunes, littoral wrack and cobblestones, basaltic cliffs, rock slides, burrows in soil, rock crevices, mud cracks, vegetation, and skeletons of dead cacti; and under stones, bark of trees, surface debris, and cattle dung. Some species, particularly members of the genus *Centruroides,* are noted for adapting to human habitation, where they may be numerous in roof thatching, between bricks and stones in walls, in firewood, and in yard debris. In moist tropical forests of Mexico scorpions are found in the forest canopy, in bromeliads, and on tree trunks. Some of these scorpions, such as *Centruroides thorelli,* complete their life cycle in the forest canopy and perhaps never descend to the ground. Some scorpions are also adapted for cave existence (58, 66).

Scorpions are generally thought of as inhabitants of xeric environments; indeed some forms, such as *Serradigitus deserticola,* are found in habitats that may only receive rainfall every few years. At the other extreme, *Uroctonus mordax* may be found under damp mosses in moist habitats of northern California.

Scorpions have generally been considered dwellers of low elevations, but populations have recently been detected at elevations of 2133 m in the Sierra Nevada of California (106). The range of scorpions extends down to sea level, where species like *Vaejovis littoralis* live in the littoral wrack zone along the shore; *Centruroides exilicauda* and *Vaejovis janssi* are active predators along beaches (104). Some scorpions have established populations that extend to high latitudes. In North America, *Paruroctonus boreus* is found to about 52° N latitude in British Columbia, Alberta, and Saskatchewan (16, 47).

Even though some species are widely distributed, there is evidence of habitat preference and a habitat-related limitation of distribution. Allred (3) reported habitat preferences among 283 *P. boreus* individuals collected in 8 of 12 different plant communities in Idaho. The ideal habitats were those vegetated with 85% ground cover composed of at least 60% broadleaf shrubs and 5% grass. In a similar study in southern Nevada, *P. boreus* was most abundant in pinyon pine and juniper communities (46). This species and most individuals of eight others were broadly distributed among several habitat types, but also showed preferences. *Vaejovis gertschi* has shown preference for three natural habitats in central California: open rock surface, rock crevice, and vegetation-covered soil (91). The habitat preference changed seasonally. The main population migrated to elevated rock outcrops during the winter (rainy season). In spring they migrated back to the adjacent grassland, where they used fissures in the drying soils as diurnal shelters. *Vaejovis confusus* and *Vaejovis spinigerus* both occurred in the same region of the Sonoran desert, and although their distributions overlapped, their different microhabitat preferences allowed them to avoid competitive interactions (103). Habitat preferences have also been reported in the Australian species *Urodacus abruptus* (86).

Burrowing Biology

Burrowing is a major adaptation of many scorpion species for survival in extreme environments (48, 99). Burrows may be as simple as a cell excavated under rocks, logs, or other surface cover, or they may be elaborate and deep. *Paruroctonus mesaensis* and *Vejovoidus longiunguis* dig relatively shallow, temporary burrows, while other scorpions such as *Anuroctonus phaiodactylus* construct permanent burrows, sometimes to depths of up to 42 cm (99).

All 19 species of the African genus *Opisthophthalmus* burrow in the ground for shelter (56). Lamoral (56) found that habitat selection, distribution, and range of each of these species were correlated with particular soil types. The most important soil factor was hardness and texture. Differential interspecific soil-hardness requirements among burrowing species of *Opisthophthalmus* insured that different species in the same area were ecologically allopatric, and thus prevented competition for burrowing sites.

The structure of burrows has been related to their functions. Koch (54) found significant structural differences in burrows of *Urodacus hoplurus* and *Urodacus yaschenkoi*. Both of these Australian species resemble the North American *Anuroctonus phaiodactylus* in that they construct burrows with a distinct entrance followed by a tunnel (tortuous and spiral in *Urodacus,* usually simpler in *Anuroctonus*) and a terminal chamber. *Urodacus yaschenkoi* excavated normal burrows within 8–10 hr in the field (83). The burrows were maintained throughout the year; the most obvious maintenance was clearing of debris from the entrance. Shorthouse & Marples (83) concluded that this burrow structure is adaptive because vertically spiralling tunnels are less likely to intersect adjacent burrows than are oblique straight tunnels; the entrance chamber facilitates predation; and the deep terminal chamber may facilitate avoidance of environmental extremes. Burrowing behavior results in avoidance of high daytime temperatures and low humidities, minimization of water loss, and probably lowering of metabolic needs (82).

The subterranean environment of burrowing scorpions differs from the surface environment. Crawford & Riddle (24) recognized that the environment of the burrowing *Diplocentrus spitzeri* is essentially mesic, even though the ground surface is xeric. This species frequently constructs its burrow entrance under a surface rock, which gives additional protection to the burrow. These authors suggested that *D. spitzeri* could survive in the absence of free water if sufficient prey were available; thus drought should not normally cause significant mortality for individuals in established burrows. They also concluded that high temperature does not usually cause significant mortality because of the tempering effect of the burrow, but that exposure to extreme winter cold could occasionally result in mortality.

Home Range

A few species, such as *Anuroctonus phaiodactylus* and *Didymocentrus comondae*, appear to be restricted to their burrows throughout their lives. These species have home ranges limited to the burrow and a modest surface area surrounding the burrow entrance. *Paruroctonus mesaensis* usually remains within 0.5 m of its burrow entrance at night (50). Tourtlotte (92), in a mark-recapture study of *Paruroctonus boreus,* noted individual recaptures in approximately the same locations in two successive years.

Dispersal

Scorpions probably disperse less than is generally believed. The suggestion that scorpions are transient by nature has come from frequent observations of scorpions entering human habitation. Most of those observed have been mature males that have become more ambulatory during courtship season. Williams (100) found that most juvenile stadia of *Vaejovis spinigerus, Vaejovis confusus,* and *Hadrurus arizonensis* were not commonly encountered on the habitat surface in the Sonoran desert. Dispersal was primarily accomplished by nymphs in the first postlarval instar and adults, especially males. Some island scorpions disperse by rafting on logs and other floating debris. In the Gulf of California, rafting species such as *Centruroides exilicauda* are widely distributed among the numerous islands, while nonrafting species, such as members of *Vaejovis,* have more restricted distributions and more extensive allopatric speciation (104).

Some scorpions are dispersed by humans. The large tropical American species *Centruroides gracilis* is commonly found alive in imported produce and forest products at ports of entry throughout the world. Over the past several decades campers from the Colorado River region have repeatedly introduced *Centruroides exilicauda* into southwestern California, and there is now evidence that it has become established there (45, 78).

CYCLES

Scorpions exhibit conspicuous activity cycles. In captivity, and probably under natural conditions as well, they spend most of their time in an immobile state. During these periods of suppressed activity their sensory awareness is also depressed. Periodically, however, scorpions become physically active and engage in predation, burrowing, and courtship. In the inactive state, scorpions are usually located in their protective shelters. During the active phase of their cycle they may come to the surface. Perhaps their extensive inactivity is related to their dependence on relatively inefficient booklungs for their oxygen supply. The oxygen consumption of scorpions suggests that their respiratory rates are among the lowest known for terrestrial animals (24–27, 49, 76–77, 93). Their minimal physical activity, coupled with low energy requirements, is an adaptive strategy for survival in unfavorable or highly cyclic environments. In the discussion that follows, the term "surface activity" is used to indicate movement from the shelter to the exposed surface of the environment, even if this does not entail physical activity on the surface.

On any given night only a small subset of a scorpion population surfaces, and the proportion of the population that surfaces is influenced by environmental parameters such as temperature, rainfall, relative humidity, moonlight, and season. In New Mexico, about 5% of a population of *Paruroctonus utahensis* came to the surface on any night during a favorable season (11). About 15% of a population of *Uroctonus mordax* studied in central California between May and September was active on the surface at any time (including those in burrow entrances) (8). About half of a *P. mesaensis* population was on the surface of a southern California sand dune during a given evening (69).

Seasonal Cycles

Activity cycles of scorpions are correlated with seasonal changes. In Idaho, *Paruroctonus boreus* was observed on the ground surface only from March through October, with highest surface densities in July and August (92). Similar seasonal patterns in surface activity have been reported in *Anuroctonus phaiodactylus* (46, 99), *Paruroctonus becki* (46), *Paruroctonus boreus* (3, 46), *Paruroctonus mesaensis* (69), *Paruroctonus utahensis* (11, 77), *Uroctonus mordax* (8), *Vaejovis confusus* (46, 100), *Vaejovis gertschi* (91), *Vaejovis hirsuticauda* (46), *Vaejovis spinigerus* (100), *Vaejovis wupatkiensis* (46), and *Hadrurus arizonensis* (46). In the western United States scorpions can generally be observed on the surface from March through October, with highest surface densities from May through September.

Seasonal surface-activity patterns may be influenced by temperature, rainfall, courtship season, and moonlight. Positive correlations between temperature and surface activity have been reported in *Paruroctonus boreus* (92), *Uroctonus mordax* (8), and *Vaejovis gertschi* (91). Critical temperatures below which scorpions do not ordinarily surface have been identified for a few species: *P. boreus*, 10°C (92); *P. mesaensis*, adults 20°C, juveniles 9°C (69); *U. mordax*, 4°C (8); and *V. gertschi*, 4°C (91).

Precipitation also generally affects surfacing. During and immediately after precipitation, surface densities of *P. mesaensis* were reduced (69). *P. boreus* exhibited no surface activity during precipitation, and the surface densities remained reduced for 3–5 days following substantial precipitation (92). The surfacing of *V. gertschi* was negatively correlated with relative humidity (91).

Illumination influences the surfacing of some species. The surface densities of *P. mesaensis* (50), *U. mordax* (8), *V. confusus* (50, 100), *V. gertschi* (91), and *V. spinigerus* (100) appear to be negatively correlated with intensity of moonlight. By contrast, surfacing of *C. sculpturatus* (50) and *P. mesaensis* (69) was reported to be uncorrelated with moonlight. *P. mesaensis* was inhibited by moonlight in a coastal dune community but not in an inland dune community; this suggests a difference in adaptation to predation pressures. Ninety-five percent of observed matings by *P. mesaensis* occurred on moonless nights (69). During the full moon only 1.6% of surfaced individuals of this species were feeding, in contrast to 3.8% during the new moon.

Surface activity usually increases significantly during courtship season, particularly among mature males. Moreover, males appear to abandon their normal burrows or other shelters and become more nomadic, and thus show increased surface locomotion. This results in observed sex ratios that strongly favor males. The influence of courtship on surface activities is evident for *Anuroctonus phaiodactylus* (99), *Uroctonus mordax* (8), *Paruroctonus boreus* (3, 46, 92), *P. mesaensis* (69), *Vaejovis confusus* (100), *V. gertschi* (91), and *V. spinigerus* (100).

Scorpions generally remain relatively inactive following a feeding to satiation. Bradley (11) found that when *Paruroctonus utahensis* had fed, they did not return to the ground surface for an average of 16–20 days. He concluded that they remained in their protective burrows to minimize exposure to predation.

Circadian Cycles

Some of the most obvious activity cycles exhibited by scorpions are circadian. Most species exhibit such cycles. Most are considered nocturnal, in that they reside in their shelters during the day and come to the surface after sunset. In contrast, a few species display primarily diurnal activity. *Vaejovis littoralis,* a coastal littoral dweller, has activity cycles correlated with tidal movement, and is active during the incoming tide, even during midday (104). Circadian surface activity has been reported in the forest-dwelling *Pandinus imperator* (diurnal) and in *Buthus hottentotta* (nocturnal) (94). *Centruroides sculpturatus* and *Diplocentrus spitzeri* both have nocturnal activity under a normal photoperiod, but when kept in constant darkness *C. sculpturatus* displayed an endogenous circadian rhythm, while *D. spitzeri* did not (23). The contrasting circadian rhythms of these two species may create a temporal separation that ultimately facilitates sympatric coexistence (23). Endogenous circadian rhythms in physical activity have also been reported in a number of other species (17–22).

Nocturnal species do not come to the surface every night. A periodic pattern of surfacing has been reported in several species. Those species studied characteristically surfaced shortly after sunset and remained on the surface for varying periods. Surface density is usually highest during the first few hours following sunset. Surfaced scorpions gradually return to their shelters throughout the evening, and by midnight a large proportion have done so. By sunrise few if any normally remain on the surface. This pattern has been observed in a number of species (23, 50, 70, 82, 91, 94).

Enzymatic and neurophysiological functions also have circadian rhythms and are correlated with the activity cycle of scorpions (48). Muscle dehydrogenase activity in *Heterometrus fulvipes* is highest at times of highest locomotion and oxygen consumption (75). Circadian rhythms are found in the phosphorylase activity and glycogen content of the heart muscle of *H. fulvipes* (53). The neurophysiological basis of circadian rhythms has been studied in the North African species *Androctonus australis* and four other species (29–31, 33, 34). These studies reveal that light sensitivity of the ocelli is controlled by a circadian clock located in the central nervous system. The median eyes were found to be 1000 times more sensitive during the subjective night than during the subjective day. Efferent neurosecretory fibers mediate the circadian signal from the central nervous system. The sensitivity cycle of median and lateral ocelli is synchronized, but the median ocelli are far more responsive to the sensitivity cycle. This circadian rhythm is synchronized to varying environmental conditions by an external zeitgeber (28, 31, 33). Both median and lateral ocelli are capable of mediating zeitgeber stimuli, but lateral ocelli are more sensitive in this function (31, 32). The lateral ocelli sacrifice image information in favor of increased sensitivity, and show small fluctuations in sensitivity (80). The combination of both types of ocelli is believed to enable the central nervous system to calculate the time of dusk and dawn more precisely than it could with input from either type alone (28, 29). Electroretinograms from five species of Buthidae and Scorpionidae suggest that use of ocelli and zeitgeber stimuli may be a general mechanism for regulating and synchronizing circadian rhythms in scorpions. Extraocular light reception through the exoskeleton of the metasoma has been observed in a species of *Urodacus* (108); these scorpions were able to orient to dark and light habitats exclusively by this means (109).

PREDATORS OF SCORPIONS

Scorpions are potentially attractive prey. Factors contributing to their desirability as prey include large body size, rich nutrient content, abundant populations, wide distribution, relative lack of defense, and predictable surface behavior. At least 124 vertebrates and 26 invertebrates prey on scorpions (73). These data suggest that vertebrates are more important than invertebrates as scorpion predators.

North American vertebrates that may utilize scorpions as a major part of their diet include elf owls *(Micranthene)*, burrowing owls *(Speotyto)*, barn owls *(Tyto)*, grasshopper mice *(Onychomys)*, desert shrews *(Notiosorex)*, and a variety of lizards. Some of these protect themselves from scorpion sting by removing the telson at the time of capture. This is a curious behavior, since the venom of most scorpions appears to have little if any effect on vertebrate predators.

The importance of invertebrates as scorpion predators has been underestimated because (a) they have been recorded as predators only when actually observed preying on scorpions; (b) invertebrate stomach contents have not been evaluated; (c) invertebrate predators do not usually leave conspicuous remains similar to owl pellets, scats, or unconsumed remains; and (d) after scorpions have fed to satiation and have returned to a catatonic state in their shelter, they may fall prey to small insects that otherwise might not prey on them. In the laboratory, inactive scorpions are commonly killed by meal worms, crickets, and other small insects supplied as prey. Black widow spiders *(Latrodectus)*, Jerusalem crickets *(Stenopelmatus)*, darkling beetles *(Eleodes)*, centipedes *(Scolopendra)*, and solpugids are known to feed on scorpions.

In a California dune community, 11 species of scorpion predators were identified (67). These included loggerhead shrikes, great horned owls, kit foxes, grasshopper mice, harvester ants, black widow spiders, and five species of scorpion (70). Extensive predation on scorpions by other scorpions was observed.

CONCLUSIONS

There has been much activity in scorpion research, yet studies have been based on relatively few species. Of the estimated 1500 species, only three to four dozen have been studied extensively. The results of recently introduced ultraviolet sampling suggest that our knowledge of the composition and distribution of the scorpion fauna is far from complete. Recent field studies indicate that many new taxa remain to be discovered. It is hoped that discovery of new taxa will stimulate a much-needed reinvestigation of the systematic and biogeographic relationships among scorpions. Scorpions should continue to be an important tool in the study of circadian rhythms and their neurological bases. The importance of scorpions as predators and stabilizers within terrestrial communities is still not fully appreciated. The impressive adaptations of scorpions to rigorous environments continue to attract the attention of physiologists, behaviorists, and ecologists, and our understanding of these adaptations is certain to be significantly enhanced. Scorpions will most likely become even more popular in research as their taxonomy becomes better known.

ACKNOWLEDGMENTS

Much appreciation is due my colleagues Paul H. Arnaud, Jr., Jack T. Tomlinson, and John E. Hafernik for kindly reading and criticizing this manuscript. Thanks also to Paul H. Arnaud, Jr., Mont A. Cazier, Joel F. Gustafson, John S. Hensill, George E. Lindsay, and James R. Sweeney for years of encouragement of my studies on scorpions.

Literature Cited

1. Alexander, A. J. 1956. Mating in scorpions. *Nature* 178:867–68
2. Alexander, A. J. 1957. The courtship and mating of the scorpion, *Opisthophthalmus latimanus*. *Proc. Zool. Soc. London* 128:529–44
3. Allred, D. M. 1973. Scorpions of the National Reactor Testing Station, Idaho. *Great Basin Nat.* 33:251–54
4. Angermann, H. 1955. Indirekte Spermatophorenübertragung bie *Euscorpius italicus* (Hbst.) (Scorpiones, Chactidae). *Naturwissenschaften* 42(10):303–5
5. Angermann, H., Schaller, F. 1955. Die spermatophore von *Euscorpius italicus* und ihre übertragung. *Verh. Dsch. Zool. Ges.* 1955:459–62
6. Angermann, H., Schaller, F. 1956. Spermatophorenbau und -bildung bei arthropoden mit indirekter spermatophoren-übertragung. *Ber. Hundertjahrfeier Dtsch. Entomol. Ges., Berlin,* pp. 228–37. Berlin: Akademie-Verlag
7. Auber, M. 1963. Reproduction et croissance de *Buthus occitanus* Amx. *Ann. Sci. Nat. Zool. Biol. Anim.* 5:273–86
8. Bacon, A. D. 1972. *Ecological studies on a population of* Uroctonus mordax *Thorell (Scorpionida: Vaejovidae).* MA thesis. San Francisco State Univ., San Francisco. 56 pp.
9. Baerg, W. J. 1954. Regarding the biology of the common Jamaican scorpion. *Ann. Entomol. Soc. Am.* 47:272–76
10. Baerg, W. J. 1961. Scorpions: Biology and effect of their venom. *Agric. Exp. Stn. Bull. 647,* Univ. Arkansas, Fayetteville. 34 pp.
11. Bradley, R. 1982. Digestion time and reemergence in the desert grassland scorpion *Paruroctonus utahensis* (Williams) (Scorpionida, Vaejovidae). *Oecologia* 55:316–18
12. Brownell, P. N. 1977. Compressional and surface waves in sand: Used by desert scorpions to locate prey. *Science* 197:479–82
13. Brownell, P. H. 1984. Prey detection by the sand scorpion. *Sci. Am.* 251:86–97
14. Brownell, P., Farley, R. D. 1979. Detection of vibrations in sand by tarsal sense organs of the nocturnal scorpion *Paruroctonus mesaensis. J. Comp. Physiol.* 131:23–30
15. Brownell, P., Farley, R. D. 1979. Prey-localizing behaviour of the nocturnal desert scorpion, *Paruroctonus mesaensis:* Orientation to substrate vibrations. *Anim. Behav.* 27:185–93
16. Buckle, D. J. 1972. Scorpions in Saskatchewan. *Blue Jay* 30:109
17. Cloudsley-Thompson, J. L. 1962. Some aspects of the physiology of *Buthotus minax* (Scorpiones: Buthidae) with remarks on other African scorpions. *Entomol. Mon. Mag.* 23:243–46
18. Cloudsley-Thompson, J. L. 1973. Entrainment of the "circadian clock" in *Buthotus minax* (Scorpiones: Buthidae). *J. Interdiscipl. Cycle Res.* 4:119–23
19. Cloudsley-Thompson, J. L. 1975. Entrainment of the "circadian clock" in *Babycurus centrurimorphus* (Scorpiones: Buthidae). *J. Interdiscip. Cycle Res.* 6:185–88
20. Cloudsley-Thompson, J. L., Constantinou, C. 1985. The circadian rhythm of locomotory activity in a Neotropical forest scorpion, *Opisthacanthus* sp. (Scorpionidae). *Int. J. Biometeorol.* 29:87–89
21. Constantinou, C. 1980. Entrainment of the circadian rhythm of activity in desert and forest inhabiting scorpions. *J. Arid Environ.* 3:133–39
22. Constantinou, C., Cloudsley-Thompson, J. L. 1980. Circadian rhythms in scorpions. *C. R. 8th Congr. Int. Arachnol, Vienna,* pp. 53–55. Vienna: Egermann
23. Crawford, C. S., Krehoff, R. C. 1975. Diel activity in sympatric populations of the scorpions *Centruroides sculpturatus* (Buthidae) and *Diplocentrus spitzeri* (Diplocentridae). *J. Arachnol.* 2:195–204
24. Crawford, C. S., Riddle, W. A. 1975. Overwintering physiology of the scorpion *Diplocentrus spitzeri. Physiol. Zool.* 48:84–92

25. Dresco-Derouet, L. 1961. Le métabolisme respiratoire des scorpions. I. Existence d'un rythme nycthéméral de la consommation d'oxygène. *Bull. Mus. Natl. Hist Nat. Ser.* 2 32:553–57

26. Dresco-Derouet, L. 1964. Le métabolisme respiratoire des scorpions. II. Mesures de l'intensité respiratoire chez quelques espèces a différentes températures. *Bull. Mus. Natl. Hist. Nat. Ser.* 2 36:97–99

27. Dresco-Derouet, L. 1967. Le métabolisme respiratoire des scorpions. III. Influences des variations de température sur l'intensité respiratoire de deux espèces d'*Euscorpius. Bull. Mus. Natl. Hist. Nat. Ser.* 2 39:308–12

28. Fleissner, G. 1975. A new biological function of the scorpion's lateral eyes as receptors of zeitgeber stimuli. *Proc. Int. Arachnol. Congr., 6th, Amsterdam,* pp. 176–82. Amsterdam: Ned. Entomol. Ver.

29. Fleissner, G. 1977. Differences in the physiological properties of the median and lateral eyes and their possible meaning for the entrainment of the scorpion's circadian rhythm. *J. Interdiscip. Cycle Res.* 8:15–26

30. Fleissner, G. 1977. Entrainment of the scorpion's circadian rhythm via the median eyes. *J. Comp. Physiol.* 118:93–99

31. Fleissner, G. 1977. Scorpion lateral eyes: Extremely sensitive receptors of zeitgeber stimuli. *J. Comp. Physiol.* 118:101–8

32. Fleissner, G. 1977. The absolute sensitivity of the median and lateral eyes of the scorpion, *Androctonus australis* (Buthidae, Scorpiones). *J. Comp. Physiol.* 118:109–20

33. Fleissner, G., Fleissner, G. 1985. Neurobiology of a circadian clock in the visual system of scorpions. In *Neurobiology of Arachnids,* ed. F. G. Barth, pp. 351–75. Berlin: Springer

34. Fleissner, G., Heinrichs, S. 1982. Neurosecretory cells in the circadian-clock system of the scorpion, *Androctonus australis. Cell Tissue Res.* 224:233–38

35. Floyd, L. E. 1977. Feeding biology of two California scorpions *Uroctonus mordax* Thorell and *Anuroctonus phaiodactylus* (Wood) (Scorpionida: Vaejovidae). MA thesis. San Francisco State Univ., San Francisco. 84 pp.

36. Foelix, R. F., Schabronath, J. 1983. The fine structure of scorpion sensory organs. I. Tarsal sensilla. *Bull. Br. Arachnol. Soc.* 6:53–67

37. Francke, O. F. 1976. Observations on the life history of *Uroctonus mordax* Thorell (Scorpionida, Vaejovidae). *Bull. Br. Arachnol. Soc.* 3:254–60

38. Francke, O. F. 1979. Observations on the reproductive biology and life history of *Megacormus gertschi* Diaz (Scorpiones: Chactidae; Megacorminae). *J. Arachnol.* 7:223–30

39. Francke, O. F. 1981. Birth behavior and life history of *Diplocentrus spitzeri* Stahnke (Scorpiones: Diplocentridae). *Southwest. Nat.* 25:517–23

40. Francke, O. F. 1982. Birth behavior in *Diplocentrus bigbendensis* Stahnke (Scorpiones, Diplocentridae). *J. Arachnol.* 10:157–64

41. Francke, O. F. 1982. Parturition in scorpions (Arachnida, Scorpiones): A review of the ideas. *Rev. Arachnol.* 4:27–37

42. Francke, O. F. 1985. Conspectus genericus scorpionorum 1758–1982 (Arachnida: Scorpiones). *Occas. Pap. Mus. Tex. Tech Univ.* 98:1–32

43. Francke, O. F., Jones, S. K. 1982. The life history of *Centruroides gracilis* (Scorpiones, Buthidae). *J. Arachnol.* 10:223–39

44. Francke, O. F., Sissom, W. D. 1984. Comparative review of the methods used to determine the number of molts to maturity in scorpions (Arachnida), with analysis of the post-birth development of *Vaejovis coahuilae* Williams (Vaejovidae). *J. Arachnol.* 12:1–20

45. Geck, R. 1980. Introduction of scorpions to Orange County. *Proc. Pap. Annu. Conf. Calif. Mosq. Vector Control. Assoc.* 48:136

46. Gertsch, W. J., Allred, D. M. 1965. Scorpions of the Nevada Test Site. *Brigham Young Univ. Sci. Bull. Biol. Ser.* 6(4):1–15

47. Gertsch, W. J., Soleglad, M. E. 1966. The scorpions of the *Vejovis boreus* group (subgenus *Paruroctonus*) in North America (Scorpionida: Vejovidae). *Am. Mus. Novit.* 2278:1–54

48. Hadley, N. F. 1974. Adaptational biology of desert scorpions. *J. Arachnol.* 2:11–24

49. Hadley, N. F., Hill, R. D. 1969. Oxygen consumption of the scorpion *Centruroides sculpturatus. Comp. Biochem. Physiol.* 29:217–26

50. Hadley, N. F., Williams, S. C. 1968. Surface activities of some North American scorpions in relation to feeding. *Ecology* 49:726–34

51. Haradon, R. M. 1972. Birth behavior of the scorpion *Uroctonus mordax* Thorell

(Vaejovidae). *Entomol. News* 83:218–21

52. Hjelle, J. T. 1974. Observations on the birth and post-birth behavior of *Syntropis macrura* Krapelin (Scorpionida: Vaejovidae). *J. Arachnol.* 1:221–27

53. Jayaram, V., Chandra Sekara Reddy, D., Padmanabha Naidu, B. 1978. Circadian rhythmicity in phosphorylase activity and glycogen content in the heart muscle of the scorpion *Heterometrus fulvipes* C. L. Koch. *Experientia* 34:1184–85

54. Koch, L. E. 1978. A comparative study of the structure, function and adaptation to different habitats of burrows in the scorpion genus *Urodacus* (Scorpionida, Scorpionidae). *Rec. West. Aust. Mus.* 6:119–46

55. Lamoral, B. H. 1971. Predation on terrestrial molluscs by scorpions in the Kalahari desert. *Ann. Natal Mus.* 21:17–20

56. Lamoral, B. H. 1978. Soil hardness, an important and limiting factor in burrowing scorpions of the genus *Opisthophthalmus* C. L. Koch, 1837 (Scorpionidae, Scorpionida). *Symp. Zool. Soc. London* 42:171–81

57. Lourenco, W. R. 1978. *Étude sur les scorpions appartenant au "complexe"* Tityus trivittatus *Kraeplein, 1898 et, en particulier, de la sous-espèce* Tityus trivittatus fasciolatus *Pessoa 1935*. 3rd cycle thesis. Univ. Pierre Marie Curie, Paris. 128 pp.

58. Lourenco, W. R. 1981. Scorpions cavernicoles de l'équateur: *Tityus demangei* n.sp. et *Ananteris ashmolei* n.sp. (Buthidae); *Troglotayosicus vachoni* n.gen., n.sp. (Chactidae), Scorpion troglobie. *Bull. Mus. Natl. Hist. Nat. Ser. 4* 2:635–62

59. Makioka, T., Koike, K. 1984. Parthenogenesis in the viviparous scorpion, *Liocheles australasiae. Proc. Jpn. Acad. Ser. B* 60:374–76

60. Makioka, T., Koike, K. 1985. Reproductive biology of the viviparous scorpion, *Liocheles australasiae* (Fabricius) (Arachnida, Scorpiones, Scorpionidae). I. Absence of males in two natural populations. *Int. J. Invertebr. Reprod. Dev.* 8:317–23

61. Matthiesen, F. A. 1962. Parthenogenesis in scorpions. *Evolution* 16:255–56

62. Matthiesen, F. A. 1969. Le développement post-embryonnaire du scorpion Buthidae: *Tityus bahiensis* (Perty, 1834). *Bull. Mus. Natl. Hist. Nat. Ser. 2* 41:1367–70

63. Matthiesen, F. A. 1971. The breeding of *Tityus serrulatus* Lutz & Mello, 1927, in captivity (Scorpions, Buthidae). *Rev. Bras. Pesqui. Med. Biol.* 4:299–300

64. Matthiesen, F. A. 1971. Observations on four species of Brazilian scorpions in captivity. *Rev. Bras. Pesqui. Med. Biol.* 4:301–2

65. McCormick, S., Polis, G. A. 1982. Arthropods that prey on vertebrates. *Biol. Rev.* 57:29–58

66. Mitchell, R. W. 1971. *Typhlochactas elliotti*, a new eyeless cave scorpion from Mexico (Scorpionida, Chactidae) (1). *Ann. Spéléol.* 26:135–48

67. Polis, G. A. 1979. Prey and feeding phenology of the desert sand scorpion *Paruroctonus mesaensis* (Scorpionida: Vaejovidae). *J. Zool.* 188:333–46

68. Polis, G. A. 1980. The significance of cannibalism on the demography and activity in a natural population of desert scorpions. *Behav. Ecol. Sociobiol.* 7:25–35

69. Polis, G. A. 1980. Seasonal patterns and age-specific variation in the surface activity of a population of desert scorpions in relation to environmental factors. *J. Anim. Ecol.* 49:1–18

70. Polis, G. A., Farley, R. D. 1979. Characteristics and environmental determinants of natality, growth and maturity in a natural population of the desert scorpion, *Paruroctonus mesaensis* (Scorpionida: Vaejovidae). *J. Zool.* 187:517–42

71. Polis, G. A., Farley, R. D. 1979. Behavior and ecology of mating in the cannibalistic scorpion, *Paruroctonus mesaensis* Stahnke (Scorpionida: Vaejovidae). *J. Arachnol.* 7:33–46

72. Polis, G. A., Farley, R. D. 1980. Population biology of a desert scorpion: Survivorship, microhabitat and the evolution of life history strategy. *Ecology* 61:620–29

73. Polis, G. A., Sissom, W. D., McCormick, S. J. 1981. Predators of scorpions: Field data and a review. *J. Arid Environ.* 4:309–27

74. Probst, P. 1972. Zur fortpflanzungsbiologie und zur entwicklung der /giftdrüsen beim skorpion *Isometrus maculatus* (De Geer, 1778) (Scorpiones: Buthidae). *Acta Trop.* 29:1–87

75. Rao, K. P., Govindappa, S. 1967. Dehydrogenase activity and its diurnal variations in different muscles of the scorpion, *Heterometrus fulvipes. Proc. Indian Acad. Sci. Sect. B* 66:243–49

76. Riddle, W. A. 1978. Respiratory physiology of the desert grassland scor-

pion *Paruroctonus utahensis*. *J. Arid Environ.* 1:243–51

77. Riddle, W. A. 1979. Metabolic compensation for temperature change in the scorpion *Paruroctonus utahensis*. *J. Therm. Biol.* 4:125–28

78. Russell, F. E., Madon, M. B. 1984. Introduction of the scorpion *Centruroides exilicauda* into California and its public health significance. *Toxicon* 22:658–64

79. San Martin, P., de Gambardella, L. A. 1966. Nueva comprobacion de la partenogenesis en *Tityus serrulatus* Lutz y Mello-Campos 1922. *Rev. Soc. Entomol. Argent.* 28:79–84

80. Schliwa, M., Fleissner, G. 1980. The lateral eyes of the scorpion, *Androctonus australis*. *Cell Tissue Res.* 206:95–114

81. Schultze, W. 1927. Biology of the large Philippine scorpion. *Philipp. J. Sci.* 32:375–89

82. Shorthouse, D. J. 1971. *Studies on the biology and energetics of the scorpion* Urodacus yaschenkoi *(Birula 1904)*. PhD thesis. Aust. Natl. Univ., Canberra

83. Shorthouse, D. J., Marples, T. G. 1980. Observations on the burrow and associated behavior of the arid-zone scorpion *Urodacus yaschenkoi* (Birula). *Aust. J. Zool.* 28:581–90

84. Shulov, A., Amitai, P. 1958. On mating habits of three scorpions: *Leiurus quinquestriatus* H et F., *Butothus judaicus* E. Sim. and *Nebo hierichonticus* E. Sim. *Arch. Inst. Pasteur Alger.* 36:351–69

85. Shulov, A., Rosin, R., Amitai, P. 1960. Parturition in scorpions. *Bull. Res. Counc. Isr. Sect. B* 9:65–69

86. Smith, G. T. 1966. Observations on the life history of the scorpion *Urodacus abruptus* Pocock (Scorpionidae), and an analysis of its home sites. *Aust. J. Zool.* 14:383–98

87. Southcott, R. V. 1955. Some observations on the biology, including mating and other behaviour, of the Australian scorpion *Urodacus abruptus* Pocock. *Trans. R. Soc. South Aust.* 78:145–54

88. Stahnke, H. L. 1966. Some aspects of scorpion behavior. *Bull. South. Calif. Acad. Sci.* 65:65–80

89. Stahnke, H. L. 1974. An estimate of the number of taxa in the order Scorpionida. *BioScience* 24:339

90. Stockmann, R. 1979. Développement postembryonnaire et cycle d'intermue chez un scorpion Buthidae: *Buthotus minax occidentalis* (Vachon et Stockmann). *Bull. Mus. Natl. Hist. Nat. Ser. 4* 1:405–20

91. Toren, T. J. 1973. Biology of the Cali-fornia coast range scorpion *Vaejovis gertschi striatus* Hjelle (Scorpionida: Vaejovidae). MA thesis. San Francisco State Univ., San Francisco. 78 pp.

92. Tourtlotte, G. I. 1974. Studies on the biology and ecology of the northern scorpion *Paruroctonus boreus* (Girard). *Great Basin Nat.* 34:167–79

93. Towle, B. A. 1984. *The effect of starvation on the oxygen consumption of the scorpion* Uroctonus mordax *Thorell*. MA thesis. San Francisco State Univ., San Francisco. 71 pp.

94. Toye, S. A. 1970. Some aspects of the biology of two common species of Nigerian scorpions. *J. Zool.* 162:1–9

95. Vannini, M., Balzi, M., Becciolini, A., Carmignani, I., Ugolini, A. 1985. Water exchange between mother and larvae in scorpions. *Experientia* 41:1620–21

96. Vannini, M., Ugolini, A. 1980. Permanence of *Euscorpius carpathicus* (L.) larvae on the mother's back (Scorpiones, Chactidae). *Behav. Ecol. Sociobiol.* 7:45–47

97. Vannini, M., Ugolini, A., Carmignani, I. 1985. Mother-young relationships in *Euscorpius* (Scorpiones): Trophic exchange between mother and larvae. *Monit. Zool. Ital.* 19:172

98. Vannini, M., Ugolini, A., Marucelli, C. 1978. Notes on the mother-young relationship in some *Euscorpius* (Scorpiones Chactidae). *Monit. Zool. Ital.* 12:143–54

99. Williams, S. C. 1966. Burrowing activities of the scorpion *Anuroctonus phaeodactylus* (Wood) (Scorpionida: Vejovidae). *Proc. Calif. Acad. Sci.* 34:419–28

100. Williams, S. C. 1968. *Habitat preferences and surface activities of the scorpions* Hadrurus arizonensis, Vejovis confusus *and* Vejovis spinigerus. PhD thesis. Arizona State Univ., Tempe. 101 pp.

101. Williams, S. C. 1968. Scorpions from northern Mexico: Five new species of *Vejovis* from Coahuila, Mexico. *Occas. Pap. Calif. Acad. Sci.* 68:1–24

102. Williams, S. C. 1969. Birth activities of some North American scorpions. *Proc. Calif. Acad. Sci.* 37:1–24

103. Williams, S. C. 1970. Coexistence of desert scorpions by differential habitat preference. *Pan-Pac. Entomol.* 46:254–67

104. Williams, S. C. 1980. Scorpions of Baja California, Mexico, and adjacent islands. *Occas. Pap. Calif. Acad. Sci.* 135:1–127

105. Williams, S. C. 1986. A new species of *Vaejovis* from Jalisco, Mexico (Scorpiones: Vaejovidae). *Pan-Pac. Entomol.* 62(4): In press

106. Williams, S. C. 1986. A new species of *Uroctonus* from the Sierra Nevada of California (Scorpiones: Vaejovidae). *Pan-Pac. Entomol.* 62(4): In press

107. Williams, S. C., Hadley, N. F. 1967. Scorpions of the Puerto Penasco area (Cholla Bay), Sonora, Mexico, with description of *Vejovis baergi*, new species. *Proc. Calif. Acad. Sci.* 35:103–16

108. Zwicky, K. T. 1968. A light response in the tail of *Urodacus*, a scorpion. *Life Sci.* 7:257–62

109. Zwicky, K. T. 1970. Behavioral aspects of the extraocular light sense of a *Urodacus* scorpion. *Experientia* 26:747–48

Ann. Rev. Entomol. 1987. 32:297–316
Copyright © 1987 by Annual Reviews Inc. All rights reserved

VISUAL ECOLOGY OF BITING FLIES

Sandra A. Allan

Department of Biological Sciences, Old Dominion University, Norfolk, Virginia 23508

Jonathan F. Day

Florida Medical Entomology Laboratory, IFAS—University of Florida, Vero Beach, Florida 32962

John D. Edman

Department of Entomology, University of Massachusetts, Amherst, Massachusetts 01003

PERSPECTIVES AND OVERVIEW

Vision has a complex role in the location of resources and mates by insects, and is often integrated with other senses, especially olfaction. Prokopy & Owens (102) reviewed the literature on visual detection of plants by herbivorous insects. A recent increase of interest in the importance of vision in host and resource location by biting flies prompted this review.

Visual ecology may be defined as the description and analysis of an animal's natural optical environment in terms of the animal's visual system and the relationship between environment, visual perception, and behavior (89, 102). This approach is used in our discussion of the visual systems of biting flies. Description of the optical environment entails delineation of parameters such as shape, size, color, contrast, light intensity, and texture. These parameters can be described quantitatively (89, 102) and can be controlled in experiments to discern their importance to insects in their natural environment. Many studies of vision of biting flies have been hampered by insufficient information on the natural habitat or the substrate being tested, or by failure to control the specific orientation behavior being tested; conclusions can often only be inferred.

297

0066-4170/87/0101-0297$02.00

The photon detection system of Diptera consists of ocelli and compound eyes. Since vision is defined as the ability to detect and utilize patterns of light, our discussion deals primarily with the compound eye, which possesses specialized receptors to detect patterns, movement, contrast, and color. Ocelli primarily detect light versus shaded areas and possibly polarized light; they may be important for entrainment of circadian rhythms (72).

Biting flies display a diversity of activity patterns and are either diurnal, nocturnal, or crepuscular. Species that are crepuscular and nocturnal are unlikely to have well-developed color vision; however, their abilities to detect intensity contrast may be well-developed, enabling them to locate open areas and to discern form and movement. Diurnal species may have well-developed color sensitivity. The importance of vision varies considerably among different biting flies. In a few species, resource location appears to be primarily visual. In many species, visual location of a resource may be only one part of a series of integrated steps. In other species, vision may be only of supplemental importance.

In this review we discuss the present state of knowledge of vision in biting fly behavior and indicate areas that require further research. The review is restricted to the true biting flies; synanthropic flies and immature stages will not be included.

NEMATOCERA

Culicidae

Visual perception of the environment is an important aspect of mosquito behavior in all life stages. Vision, including simple responses to illumination levels as well as the perception of objects, is important in all of the major activities including mating, dispersal, appetitive flight, and the location of sugar sources, hosts, and resting, oviposition, and overwintering sites. Since female mosquitoes feed on blood and then must oviposit, their behavior patterns are more diverse than those of males. The visual ecology of females will be emphasized in this section.

Although controversy surrounds the function of male swarms in mating behavior, there is little question about the importance of vision to swarm formation and maintenance. Swarms commonly form over conspicuous visual markers such as corners of buildings and human observers. Cohesion and horizontal width of the swarm appear to depend on visual characteristics of the swarm marker (49, 97). Males within a swarm apparently respond to each other visually, and females entering a swarm respond visually both to males within the swarm and to the swarm marker (41). Swarming has been observed among nocturnally active, drably colored mosquitoes in Africa (60) and South America (112). Not all mosquito species form swarms. In diurnal forest species from genera such as *Haemagogus* and *Toxorhynchites* and from the subgenus *Stegomyia* of *Aedes,* mating is associated with hovering behavior. The bright metallic coloration and unique scale patterns of these species may be used in location of conspecifics (60). *Aedes triseriatus* and *Aedes aegypti* occasionally mate in flight in small swarms that form close to host animals, or mate on the hosts themselves (88, 94). Males orient to and grasp conspecific and heterospecific females and other males as they fly; they appear to be responding to moving visual targets (88).

Adult mosquitoes feed on nectar (17, 90), cane sugar (40), extrafloral nectaries, and honeydew (61) in the field. How mosquitoes locate natural sources of sugar remains unknown. Some flowers have ultraviolet-reflecting nectar guides, and many insects utilize these to locate nectar (114). Many Diptera are known to have UV receptors that are sensitive in the 340–360 nm range (120). It is possible that diurnally active mosquitoes use visual patterns, and UV reflectance in particular, to locate flowers in the field. Crepuscular species are less likely to use such cues, owing to the low light levels at dawn and dusk.

Migration, dispersal, and appetitive flight occur at specific times daily and throughout the life of adult mosquitoes. Both visual and nonvisual factors influence these flights. In general, response to light intensity levels initiates flight activity, whereas orientation is controlled by visual perception of the terrain once the mosquito is in flight.

Circadian rhythms have been reported for most mosquito species studied. Crepuscular activity peaks clearly have a physiological basis in monitoring of daily light changes by the mosquito eye (77). Moonlight appears to influence the level of activity of nocturnal species (103). For example, the number of *Aedes taeniorhynchus* collected in Florida was 546% higher during the full moon than on moonless nights (14). Activity of females of all other species sampled increased 122% during the full-moon periods. Light intensity on nights of a full moon is equal to that at twilight, which is when maximum flight activity occurs for many species; the increased light levels from the full moon presumably resulted in increased flight (14).

Extensive field work and laboratory studies (16, 18–20) have shown mosquitoes in flight depend on visual input for orientation. Flight paths of mosquitoes in the field are also affected by vegetative patterns (16). On the basis of their flight preferences, Bidlingmayer (16) grouped mosquitoes as field species, edge species (occurring in ecotones between forests and fields), and woodland species. These groups responded differently to artificial and natural barriers. Field species responded sooner to the barriers, presumably because they were more adapted than woodland species to flying with a horizon in view. These results were further supported by a study in which (*a*) distant visible objects were attractive to mosquitoes, (*b*) a change in flight direction occurred when mosquitoes were in close proximity to visible objects, and (c) woodland species came closer to visible objects than field species (18).

Bidlingmayer & Hem (19) used an elaborate grid system of suction traps to investigate the distance at which various mosquito species respond to visual cues in the field (19). Most species (*Aedes vexans, Psorophora columbiae, Culex nigripalpus,* and *Culiseta melanura*) responded to suction traps from 15.5–19.0 m. *Uranotaenia sapphirina* and *Culex quinquefasciatus* were found to have visual ranges less than 7.5 m. In another study, woodland species that flew into open areas appeared to maintain visual contact with tall silhouettes, particularly the woodland edge (20). Other authors (27, 52, 53) have also observed the reliance of mosquitoes on visual targets in the field. Giglioli (50) found that in searching flights *Anopheles gambiae* are channelled along visible edges such as the junctions between woodlands and grasslands.

Among the important activities undertaken by adult female mosquitoes are host location and blood-feeding. The detection of movement is important for host location by some species (117). Gillett (51) recounted that field-workers who walked into areas of high mosquito density (particularly *Aedes*) during the daylight hours suspected that their movement was responsible for the initial attraction of host-seeking mosquitoes.

Laboratory studies with artificial targets and anaesthetized hosts (117, 150) and field studies using visual targets in combination with odor (27, 53) have shown the importance of vision in host location. Host-seeking females are generally more attracted to low-intensity colors such as blue, black, and red than to high-intensity colors such as white and yellow (27). Females reportedly exhibit the strongest phototactic responses to blue-green (27) and UV (39). The shape of visual targets also affects attraction (27), and complex patterns appear to attract diurnal mosquitoes (117). Visual cues such as solid dark objects appear to be important to host-seeking nocturnal species (53). The combination of visual stimuli with other cues such as odor, heat, and moisture, however, increases attraction and stimulates landing and initiation of blood-feeding behavior (71).

Mosquitoes must select suitable resting sites each day and must also find sites in which to endure periods of adverse weather and successfully overwinter. The preference of dark surfaces for landing appears to be related to selection of daily resting sites (69). Mosquito species respond to visual characteristics of the terrain according to species-inherent recognition of patterns and illumination levels. Visual images of resting sites within habitats appear to guide the adults toward potential sites (15). Other factors important in selection of sites are temperature and humidity.

Visual factors involved in the selection of suitable hibernation sites are probably similar to those used in the selection of daily resting sites. Overwintering adults generally choose dark, humid, sheltered areas such as caves, animal burrows, and man-made shelters. The physiological factors leading to hibernation are complex and include response to changing photoperiods (108), which are detected visually.

A gravid female's search for an oviposition site is similar to the host-searching flight in that a variety of cues, including visual ones, are used. The importance of color and of container size and shape has been shown in a number of field and laboratory studies (12, 45, 70, 82, 118). Reflected light from water surface was observed to influence oviposition site location in the field (12) and in the laboratory. Kennedy (80) observed that gravid mosquitoes prepared for oviposition while flying over a mirror. Some gravid mosquitoes respond to the sight of water (95) and movement of larvae (93) at the oviposition site.

The responsiveness of many night-flying insects, including mosquitoes, to artificial lights is considered an aberrant phenomenon. Several theories have been proposed to explain the physiological mechanisms involved (reviewed in 73). Regardless of the mechanisms involved, responsiveness to artificial light (as well as to other visual cues, including motion and contrasting colors) has been exploited in the design of extremely efficient traps for mosquito surveillance (9, 68, 69, 111). Traps have been designed to sample emerging, dispersing, appetitively flying, host-seeking, and ovipositing mosquitoes and have proved invaluable in sampling of pest and vector populations.

Simuliidae

Black flies are for the most part diurnally active (35, 110), with limited activity observed at night (76, 110, 141). Only females feed on blood, and males are not found near hosts; immatures develop in streams and rivers. Most information on the ecology of black flies is derived from field observations and anecdotes.

After emergence during daylight hours, adults move directly to the edge of the river in a flight that is probably visually oriented. Unfed flies show an aversion to highly reflective surfaces such as water, and this aversion may direct the newly emerged flies towards shore (33).

Mating occurs near the emergence site. Two distinct mating strategies have been described (37), and both involve visual orientation. Most commonly, columnar swarms of male flies (144) form over specific visual markers such as shrubs and tree limbs (41). Females are attracted to the swarms, either by visual markers or by the sound of swarming males. Initial contact between the sexes occurs in flight. Orientation of males to females is visual, as shown by experiments in which female mimics elicited mating attempts by males from the swarms (81, 145). A less common strategy entails mating activity on the substrate with little discrimination until body contact is made (116). Males that mate in swarms have more facets in the eye and larger facets in the upper part of the eye than males that mate on substrate. These morphological adaptations allow swarming males to detect small objects such as females against a bright background of skylight (38, 81).

Black flies use plants as nectar sources, and nectar-feeding flies are usually observed on inconspicuous flowers. It is thought that location of nectar sources is primarily by olfaction (121).

Black flies can be divided into forest species that live near small canopied streams, species that are found in large rivers in open prairies or savannas, and species that are associated with mid-sized rivers and forest clearings and are frequently found at the forest edge (124). It is generally assumed that visual cues for habitat orientation differ among the groups, although there have been few studies. Reliance on visual orientation seems to be greatest among species associated with open terrain and least among forest dwellers (124). Resting at greater vertical heights may result in increased visual detection of movement for forest and edge-dwelling species (41, 124). Long-range migration and short-range dispersal have been documented for many species; however, the role of visual cues in these flights is unknown.

In a comprehensive review, Sutcliffe (121) discussed host location processes in the context of appetitive flight and directed orientation. He discussed long-, middle-, and close-range flight orientation in terms of visual and olfactory cues. Although it is commonly assumed that most black flies embark on appetitive flights actively seeking host cues, there is little evidence to support this. Most research on the role of vision in black flies has been based on close-range orientation. However, vision is thought to be more important in long- and middle-range orientation than current literature indicates. Host location strategies also differ in that many species living in open terrain actively seek hosts, while most forest-dwelling species "passively" locate hosts, often first detecting them from a resting position. The types of visual cues used are assumed to differ between the two strategies.

Information concerning the presence of color receptors is generally inferred from experimental data. Black flies appear to be able to detect UV (115), blue (26), and green (37, 100). Host-seeking flies in the field are most attracted to low-reflectance colors such as blue, black, and red, while white, yellow, and UV-reflective colors are least attractive (23, 26, 36). Strong contrast of traps or hosts against the background is an important factor in the attraction of host-seeking females (23, 26). Thus, low intensity traps placed against highly reflective background vegetation are very attractive (34, 36, 37, 100, 124). Field experiments combining visual traps with odor indicated that flies discriminated traps on the basis of color, independent of the amount of CO_2 (24). Black flies do not exhibit preferences for particular shapes (23, 100). However, increased size of objects increases attraction (7). The importance of host movement to orientation has been examined, but results are equivocal (124, 132).

Many black fly species exhibit specific preferences for feeding sites on hosts and landing sites on traps that may be located by visual cues (13, 141). Some species preferentially land on projecting edges of trap silhouettes (13, 23, 115); this behavior may be related to feeding-site preferences. The very host-specific species *Simulium euryadminiculum* orients toward uropygial gland extracts of loons. However, landing depends on visual stimulus provided by the silhouette of the extended head and neck (13). Simmons (115) observed that flies preferred to land on cow mimics with legs rather than on those without, and he observed species-specific landing preferences that corresponded to behavior on hosts. Preference for the underbelly region is well documented (141) and may be related to the tendency of flies to land on shaded parts of the animal (6, 36). Species that normally feed on cattle (55, 115) sometimes swarm around the heads of humans, but seldom land or bite. When humans assume the quadriped position flies orient to the chest region, but few land (115). General visual and olfactory cues may lead flies to abnormal hosts; however, in the absence of specific landing cues blood-feeding may not occur.

Attraction of host-seeking females to visual stimuli is greatly enhanced when olfactory cues such as CO_2 are added (24, 37, 121, 124). Few flies land on odorless host mimics (100, 124), and those that do are usually campestral species (124). Species such as avian feeders within forests could be collected only in traps baited with CO_2 (43). However, to what extent the traps presented visual cues for orientation is not known.

Black flies can have two types of oviposition behavior; some species drop eggs singly while flying over water, and others lay egg masses on either cobble or trailing vegetation in water (75). Gravid females, in contrast to host-seeking females (36, 115), are attracted to UV-reflecting surfaces (11), which mimic water. Unlike host-seeking females, gravid females of species that oviposit on trailing vegetation are reported to be highly attracted to substrates of high-intensity colors (11) such as yellow and green (37, 100). Choice of oviposition substrate appears to involve more than just visual cues (75).

BRACHYCERA

Tabanidae

Horse flies and deer flies, which are mainly diurnal, belong to the family Tabanidae. Females of this family are hematophagous, although both sexes feed on nectar. Upon emergence or after autogenous gonotrophic cycles, females may make a dispersal flight prior to obtaining blood meals (113). The role of vision in this flight is unknown.

Mating in tabanids generally occurs after the initial in-flight contact of males and females. In species such as *Tabanus nigrovittatus* (8) and *Chrysops atlanticus* (5), males hover, occasionally darting after other flying insects. If a receptive conspecific is encountered, mating occurs. In other species such as *Chrysops fuliginosus,* males rest on vegetation and dart after passing females (29). Initial detection of females by males is visual, as determined by the fact that males pursued small objects such as stones thrown near them. The importance of vision to this behavior in tabanids is not as well studied as in male bibionids, which behave similarly and have complex visual specializations and flight behavior (151).

Both sexes of tabanids feed on a wide variety of nectar and pollen sources, ranging from grasses to oak trees (91, 148). Nectar is considered a major source of energy for tabanids, yet little is known about their mechanisms for locating nectar sources.

On the basis of electrophysiological and behavioral techniques, blue and green photoreceptors have been detected in *Tabanus bromius* (480 nm and 515 nm, respectively) and *Hybomitra schineri* (487 nm and 528 nm, respectively) (92). Spectral sensitivity curves for *T. nigrovittatus* indicate the presence of photopigments maximally sensitive to UV (350 nm), blue (477 nm), and green (520 nm) (2). Spectral sensitivity curves of males have a higher proportion of blue photoreceptors than those of females, and also differ in having decreasing sensitivity with age. Parous and older females are considerably more sensitive than younger nulliparous females (2). Hanec & Bracken (63) reported that adult tabanids were photopositive to light between 380 and 430 nm and between 500 and 550 nm.

There are numerous reports that tabanids are more attracted to dark-colored than to light-colored animals (125) and objects (57, 104). Response of host-seeking tabanids to color, as determined from field experiments with traps, is generally greatest to blue, red, and black, and for some species, white (21, 26, 57). Green, silver, yellow, and generally white are much less attractive (21, 57). Attraction or lack of attraction to certain hues is independent of intensity of background (3).

Host-seeking tabanids are strongly attracted to solid dark objects, and they have been used in a wide variety of visual traps (21, 22, 126, 128). Information about the factors important in visual orientation of flies to hosts has been largely derived from experimental modifications of visual traps and decoys. Attraction to traps increases as trap size increases (22, 127, 128), possibly owing to the greater visibility of larger objects from a distance. Three-dimensional decoys were more attractive than two-dimensional decoys (21, 129) as a result of their visibility from a greater number of directions and a greater distance. Some host-seeking tabanids are attracted to specific shapes; for example, they may be more attracted to spheres than to vertical or horizontal cylinders or cubes (21, 129). Others are not attracted to shapes (26). This distinction reflects species-specific differences in host location strategies. Visual attraction decreased when trap surfaces were interrupted by patterning such as stripes (22, 26, 64), which decreased the distance from which traps could be detected. Simple, strongly contrasting patterns were attractive and could be detected from a distance (2). Reports on the importance of movement in attraction of host-seeking tabanids are contradictory (21, 22, 26); more research is needed to clarify this.

High-intensity contrast of objects or hosts against the background enhances their attractiveness to host-seeking female tabanids. This is true whether the background or object is light or dark (3, 21, 104). In a field study on *T. nigrovittatus,* two-dimensional panels with the greatest contrast against the background, regardless of whether object or background was darker, collected the most flies, possibly because they were detected from a greater distance (4). A hue capable of eliciting attraction (i.e. blue) could do so independently of intensity (4). Color contrast is also an important factor in visual attraction. According to Hailman (62), the maximum contrast against a background of grass would be provided by a dark, saturated blue. Host-seeking *T. nigrovittatus,* which are visually sensitive to both blue and green, are strongly attracted to blue traps against a highly reflective background of salt marsh grass (4).

Attraction of host-seeking tabanids to visual traps increases if CO_2 is used as an odor source (113, 126). The level of attraction to visual traps without a CO_2 source varies among species; usually a combination of visual (dark-light contrast) and olfactory (CO_2) stimuli are necessary. *Chrysops* are readily collected in traps with animal baits or CO_2 (149), but are not collected in unbaited stationary traps (64). Attraction to stationary traps is considered the result of active searching behavior, and *Chrysops* spp., which spend much of their time resting, are generally more attracted to traps that emit odor, which elicits flight activity and then visual orientation.

Some tabanid species exhibit preferences for specific feeding sites on hosts (96, 125). Orientation to CO_2 emissions may explain feeding preferences near the head (96); however, the location of other feeding sites may involve visual cues. For example, *T. nigrovittatus* feeds on the undersides of cattle and is commonly trapped in box traps (Manning traps) (2, 64) that provide only visual stimuli and require flies to fly under the trap and then upward through a screen funnel.

Tabanids are reported to oviposit on grass (56), vegetation overhanging water, rocks, or other objects in aquatic environments (130). Visual factors found to be important for attraction of *Chrysops* spp. are the proximity of plants to water (130) and color of oviposition substrate (105). In general, however, little is known about the importance of vision in location of oviposition sites or substrates.

CYCLORRHAPHA

Muscidae

STOMOXYS SPP.   The stable fly, *Stomoxys calcitrans*, is an economically important diurnal species. Both sexes feed on blood. Flies require blood meals about once every 24 hr and are found in close proximity to host aggregations.

Spectral sensitivity curves for adult flies indicate a peak of sensitivity at 360 nm (UV), a broader peak at 450–550 nm (blue-green), and a plateau at 625 nm (orange-red) (1). Both sexes exhibit strong phototactic responses to light in the UV and blue portions of the spectrum (123, 140), which indicates that the flies not only perceive but are also attracted to these wavelengths. Other portions of the spectrum were not found to attract flies.

Both sexes assemble and rest on light-colored objects and surfaces in the vicinity of hosts. Males occasionally dart out from their resting sites at passing flies in attempts to locate a mate (28). Adhesive-coated fiberglass panels are highly attractive to flies in the field (147). The most attractive fiberglass, Alsynite®, displayed substantial reflectance (5–20%) of UV in the range of 300–420 nm. Attractiveness of panels decreased when UV reflectance was decreased by either weathering of the panel or application of sticky adhesives (1).

Assembly of stable flies on surfaces with UV-reflecting properties was related to mating behavior and thermoregulation behavior in cool weather. Flies that rested in the sun on white (UV-reflecting) surfaces at low ambient temperatures (~6°C) increased their internal body temperature to 22–28°C; this permitted them to remain active at relatively low temperatures. When fly body temperatures reached 31–34°C flies were found primarily in shaded resting areas (28). Further support for the role of UV-reflecting surfaces in fly thermoregulation was provided by the observation that attraction to Alsynite® panels increased during cool periods (1).

CO_2 and other host emissions are important cues for host location (48). These cues, in combination with hunger, increase random flight activity, thus increasing the probability that flies will encounter visual cues that they can use to locate hosts (142, 143). High contrast against a background is important in the detection and subsequent orientation of flies to objects or hosts (84, 101). Both sexes of host-seeking flies are more attracted to low-intensity than to high-intensity colors (48, 84, 101). When low-intensity colors are used in traps, collections are increased by the addition of olfactory cues (142, 143).

HAEMATOBIA SPP.   The horn fly, *Haematobia irritans irritans,* is an economically important livestock pest. Both sexes blood-feed and are continually associated with their hosts. Horn flies are photopositive to light in the near UV and blue portions of the spectrum (1), and are attracted to UV light traps (65). Spectral sensitivity studies indicate a peak of sensitivity at 360 nm (UV) and a broad peak at 460 nm (blue) (1). There have been a few reports of higher numbers of flies on dark cattle and dark areas of cattle (99). Otherwise, little is known about the role of vision in host orientation; it may be of only minor importance in this diurnal, host-associated species.

Glossinidae

Vision is critical in the biology of tsetse. Spontaneous activity of tsetse, although low, generally follows a diurnal trend, increasing after sunrise, decreasing during mid-day owing to high temperatures, and then increasing again until sunset (25). Starvation increases activity levels (25). The visual responsiveness of females parallels the level of spontaneous activity and is minimal just prior to and during larviposition (79).

Tsetse tend to have a strong photopositive response, which switches to a photonegative response when ambient temperatures exceed 30°C. Prior to larviposition, females become strongly photonegative and tend to rest in dark sites. Immediately after larviposition, they become strongly photopositive regardless of temperature (106).

Tsetse visually choose well-defined resting sites such as woody parts of vegetation (trunks, branches, logs, twigs) and undergrowth, and they appear to spend all but 15–30 min of the day resting (30). The preference of unmated females (107) and males (131) in the laboratory for vertical surfaces and borders of visual contrast is believed to be related to innate alighting and resting behavior. Pregnant females, however, showed strong preferences in the laboratory for landing on horizontal surfaces, simple and unbroken areas, and subterranean cavities (106, 107). This is congruent with larviposition observed in the field, which occurred in shaded areas. Pregnant females show no preference for substrates of particular color in the laboratory (106).

Spectral efficiency curves of *Glossina morsitans morsitans* indicate a peak of sensitivity in the UV (350–365 nm) and a broad peak at 450–550 nm (blue-green). An additional shoulder is also observed at 600–625 nm (orange-red) (59), but this does not necessarily indicate the presence of a related photopigment (146). Major differences in sensitivity of flies of different sexes or ages were not observed. As is typical with diurnal, fast-flying insects, *G. m. morsitans* has a high flicker fusion (131), which confers an ability to detect image movement. Tsetse exhibited strong positive responses to UV light and moderate responses to blue. No photopositive response was seen to green light (59). Green (58) concluded that *G. m. morsitans* has red-green dis-cimination ability.

The color most attractive to host-seeking tsetse is blue (31, 78, 122), followed by black (10, 42, 137) and white (42). Yellow and green are consistently unattractive. The relative attractiveness of black, blue, and white depend on the species, the type of trap, and the placement of color on the trap (30, 42). Some trap parameters result in landing, whereas others result in attraction but no landing (67, 134, 136).

UV reflectance does not appear to be important in attracting tsetse. Jordon & Green (78) reported that tsetse were most attracted to fluorescent blue and a nonfluorescent white, whereas other fluorescent colors were not as attractive. Of interest, however, are findings that the brightness of white cloth used for traps correlated with the number of insects collected. As the white cloth yellowed, trap collections decreased 90%; when the cloth was painted white, the traps' attractiveness returned (44).

Tsetse are more attracted to uniform black targets than to targets with numerous or complex edges such as stripes or checkerboards (106, 107, 131). Increasing the size of moving or stationary targets increased the alightment response (67, 134). Many tsetse traps mimic the solid dark shape of a host animal (137). The similarity in shape may be of little importance, however; a very effective visual trap, the biconical trap, bears little resemblance to any animal shape. The attractiveness of this trap is the result of its color or intensity contrast against the background (31).

Strong contrast of an object against the background is an important factor in attraction of host-seeking tsetse (32); *G. morsitans* is thought to be more attracted to brightness of contrast than to color (85, 131). Lambrecht (85) and Barrass (10) obtained conflicting results concerning the attractiveness of black and white. However, both authors suggested that the response of the *G. morsitans* flies reflects their sensitivity to contrast, as it is related to their habitat preference for edge vegetation where contrasts are promi-nent. Tsetse tend to land on black edges of black and white stripes in labor-atory studies; this is believed to be related to resting-site preference (131). Although traps may have certain innate attractive attributes (i.e. size, color, shape), contrast against the background is also important. Hosts are gen-erally seen as dark objects in maximum contrast to vegetation (light back-ground).

High sensitivity to contrast (131) permits tsetse to detect distant objects. Detection is heightened as the brightness contrast between an object and the background increases. The range at which a biconical trap could be detected was estimated to be 10–15 m for *Glossina brevipalpis* and 15–20 m for *Glossina pallidipes* (42). *G. pallidipes* and *G. morsitans* detected an ox from less than 30 m (135). Waage (139) has proposed that the striping pattern of zebras is a defense against biting flies, since it decreases detection of zebras from a distance. He pointed out that zebra species that are not sympatric with tsetse do not have well-defined striping patterns.

Tsetse take flight during the day in response to stimuli such as odor or movement (136), but may also take flight in the absence of such stimuli (25). Flight occurring in the latter case aids in orientation to stationary hosts (136). Snow (119) presented two models for orientation behavior based on the work of Vale (134, 135). The first model describes flying tsetse that encounter an odor, move upwind toward the host, and enter a limited range of visual attraction. The second model describes resting tsetse that are stimulated by host odors to fly upwind and enter the limited range of visual attraction. Stimuli such as size, contrast with background, and movement are important visual components in attraction of flies to moving hosts. Moving hosts are probably located by a combination of visual and olfactory cues, as location by olfaction alone is considered unlikely (133).

The formation of loose swarms of male and female tsetse behind large slow-moving animals or objects is related to location of hosts for both blood-feeding and mating (25, 134). These swarms consist of both food-seeking and non-food-seeking flies; the latter may be present only for mating purposes. Tsetse evidently express the same behavior in mating as in blood-meal searching (119). The visual attraction of males to large dark moving objects is well documented, as is the attraction of females to objects displaying interrupted movement (47). The formation of swarms by tsetse is thought to increase encounters between the sexes (54), which is important since the flies have low population densities. Males locate females visually (74, 86), dart after them, and attempt to grasp them and copulate. Short-range recognition of females also involves sex pheromones (87). The visual responsiveness of males and females to moving stimuli increases with starvation, except in unresponsive pregnant females. Orientation of tsetse to movement is enhanced by host odor (25). Field studies have shown that moving black and white models are less attractive for landing than stationary models (10). Owaga & Challier (98) concluded that flies attracted to traps with rotating screens were not hungry but were seeking mates. Those attracted to stationary traps were mostly hungry and were seeking a blood meal.

Owing to the economic importance of tsetse, considerable effort has been put into the development of convenient, efficient, and cost-effective traps. Because of different trap designs (98, 137), species-specific responses to traps (66, 133), and varying use of odors in conjunction with trap color or type (122), different traps appear to attract different species and different components of the tsetse population. Many of the traps currently in use are stationary and include an odor source (CO_2, acetone) for long-range attraction combined with visual stimuli for short-range attraction (134).

Other Flies

The visual ecology of biting flies belonging to Ceratopogonidae, Psychodidae, Rhagionidae, Athericidae, Hippoboscidae, Streblidae, and Nycteribiidae has not been well studied. Although vision does not appear to be very important for host location, it may be important for other behaviors (7, 46, 83, 109).

SUMMARY

Many of the similarities in visual ecology between the Nematocera and Brachycera and within the Cyclorrhapha may reflect the evolution of blood-feeding in these groups. In Nematocera and Brachycera, blood-feeding is thought to have evolved from predatory or nectar-feeding behavior (138). Only females feed on hosts, and association with hosts generally occurs when hosts are close to the aquatic or semiaquatic habitats of the immatures. Flies feed on nectar, make appetitive flights, disperse, or migrate prior to blood-feeding, and then oviposit in water. Many species are nocturnal or crepuscular. In Cyclorrhapha, flies are closely associated with hosts. They may have arisen from compost-feeding flies that developed a larval dependence on vertebrate-produced microhabitats. Both sexes blood-feed, and mating occurs on or near hosts. Flies generally emerge in the proximity of hosts and maintain close contact with them. These species are diurnal, and their visual systems are well developed. Comparisons between closely related blood-feeding and non-blood-feeding species may provide insight into the visual ecology of blood-feeding species.

Despite the different origins of hematophagy, there appears to be a convergence of morphology and behavior that is related to ecology rather than to phylogenetic relationships. This is clearly seen in host-location strategies by tsetse and tabanids. Even within groups such as mosquitoes, species that are active at the same time of day and in the same habitat have more in common than closely related species in different habitats. For this reason, an ecological review would be more cohesive than this phylogenetic discussion. However, because of the disproportionate amount of literature on a small number of groups, the phylogenetic approach is the most practical for this subject. However, this review does point out the great need for research on the less well-studied groups and behaviors.

ACKNOWLEDGMENTS

We thank W. L. Bidlingmayer, G. F. O'Meara, E. D. Owens, and C. J. Geden for helpful suggestions on the manuscript. We thank J. F. Sutcliffe for sharing his unpublished manuscript. This review was supported in part by NIH Grant No. AI-20983.

Literature Cited

1. Agee, H. R., Patterson, R. S. 1983. Spectral sensitivity of stable, face, and horn flies and behavioral responses of stable flies to visual traps (Diptera: Muscidae). *Environ. Entomol.* 12:1823–28
2. Allan, S. A. 1984. *Studies on vision and visual attraction of the salt marsh horse fly,* Tabanus nigrovittatus *Macquart.* PhD thesis. Univ. Mass., Amherst. 178 pp.
3. Allan, S. A., Stoffolano, J. G. 1986. The effects of hue and intensity on visual attraction of adult *Tabanus nigrovittatus* (Diptera: Tabanidae). *J. Med. Entomol.* 23:83–91
4. Allan, S. A., Stoffolano, J. G. 1986. Effects of background contrast on visual attraction and orientation of *Tabanus nigrovittatus* (Diptera: Tabanidae). *Environ. Entomol.* 15:689–94
5. Anderson, J. F. 1971. Autogeny and mating and their relationship to biting in the saltmarsh deer fly, *Chrysops atlanticus* (Diptera: Tabanidae). *Ann. Entomol. Soc. Am.* 64:1421–24
6. Anderson, J. R., DeFoliart, G. R. 1961. Feeding behavior and host preferences of some black flies (Diptera: Simuliidae) in Wisconsin. *Ann. Entomol. Soc. Am.* 54:716–29
7. Anderson, J. R., Hoy, J. B. 1972. Relationship between host attack rates and CO_2 baited insect flight trap catches of certain *Symphoromyia* sp. *J. Med. Entomol.* 9:373–92
8. Bailey, N. S. 1948. The hovering and mating of Tabanidae: A review of the literature with some original observations. *Ann. Entomol. Soc. Am.* 41:403–12
9. Barr, A. R., Smith, T. A., Boreham, M., White, K. E. 1963. Evaluation of some factors affecting the efficiency of light traps on collecting mosquitoes. *J. Econ. Entomol.* 56:123–27
10. Barrass, R. 1960. The settling of tsetse flies *Glossina morsitans* Westwood (Diptera: Muscidae) on cloth screens. *Entomol. Exp. Appl.* 3:59–67
11. Bellec, C. 1976. Captures d'adultes de *Simulium damnosum* Theobald, 1903 (Diptera: Simuliidae) a l'aide de plaques d'aluminium, en Afrique de l'ouest. *Cah. ORSTOM Sér. Entomol. Méd. Parasitol.* 14:209–17
12. Belton, P. 1967. Effect of illumination and pool brightness on oviposition by *Culex resturans* (Theo.) in the field. *Mosq. News* 27:66–68
13. Bennett, G. F., Fallis, A. M., Campbell, A. G. 1972. The response of *Simulium (Eusimulium) euryadminiculum* Davies (Diptera: Simuliidae) to some olfactory and visual stimuli. *Can. J. Zool.* 50:793–800
14. Bidlingmayer, W. L. 1964. The effect of moonlight on the flight activity of mosquitoes. *Ecology* 45:87–94
15. Bidlingmayer, W. L. 1971. Mosquito flight paths in relation to the environment. 1. Illumination levels, orientation, and resting areas. *Ann. Entomol. Soc. Am.* 64:1121–31
16. Bidlingmayer, W. L. 1975. Mosquito flight paths in relation to the environment. Effect of vertical and horizontal visual barriers. *Ann. Entomol. Soc. Am.* 68:51–57
17. Bidlingmayer, W. L., Hem, D. G. 1973. Sugar feeding by Florida mosquitoes. *Mosq. News* 33:535–38
18. Bidlingmayer, W. L., Hem, D. G. 1979. Mosquito (Diptera: Culicidae) flight behaviour near conspicuous objects. *Bull. Entomol. Res.* 69:691–700
19. Bidlingmayer, W. L., Hem, D. G. 1980. The range of visual attraction and the effect of competitive visual attractants upon mosquito (Diptera: Culicidae) flight. *Bull. Entomol. Res.* 70:321–42
20. Bidlingmayer, W. L., Hem, D. G. 1981. Mosquito flight paths in relation to the environment. Effect of the forest edge upon trap catches in the field. *Mosq. News* 41:55–59
21. Bracken, G. K., Hanec, W., Thorsteinson, A. J. 1962. The orientation of horseflies and deerflies (Tabanidae: Diptera). II. The role of some visual factors in the attractiveness of decoy silhouettes. *Can. J. Zool.* 40:685–95
22. Bracken, G. K., Thorsteinson, A. J. 1965. The orientation behavior of horseflies and deerflies (Tabanidae: Diptera). IV. The influence of some physical modification of visual decoys on orientation of horseflies. *Entomol. Exp. Appl.* 8:314–18
23. Bradbury, W. C., Bennett, G. F. 1973. Behavior of adult Simuliidae (Diptera). I. Response to color and shape. *Can. J. Zool.* 52:251–59
24. Bradbury, W. C., Bennett, G. F. 1974. Behavior of adult Simuliidae (Diptera). II. Vision and olfaction in near-orientation and landing. *Can. J. Zool.* 52:1355–64
25. Brady, J. 1972. The visual responsiveness of the tsetse fly *Glossina morsitans* Westw. (Glossinidae) to moving objects:

the effects of hunger, sex, host odour and stimulus characteristics. *Bull. Entomol. Res.* 62:257–79

26. Browne, S. M., Bennett, G. F. 1980. Color and shape as mediators of host-seeking responses of simuliids and tabanids (Diptera) in the Tantramar marshes, New Brunswick, Canada. *J. Med. Entomol.* 17:58–62

27. Browne, S. M., Bennett, G. F. 1981. Response of mosquitoes (Diptera: Culicidae) to visual stimuli. *J. Med. Entomol.* 18:505–21

28. Bushman, L. L., Patterson, R. S. 1981. Assembly, mating and thermoregulatory behavior of stable flies under field conditions. *Environ. Entomol.* 10:16–21

29. Catts, E. P., Olkowski, W. 1972. Biology of Tabanidae (Diptera): Mating and feeding behavior of *Chrysops fuliginosus*. *Environ. Entomol.* 1:448–53

30. Challier, A. 1982. The ecology of tsetse (*Glossina* spp.) (Diptera, Glossinidae): A review (1970–1981). *Insect Sci. Appl.* 3:97–143

31. Challier, A., Eyraud, M., Lafaye, A., Laveissière, C. 1977. Amélioration du rendement du piège biconique pour glossines (Diptera, Glossinidae) par l'emploi d'un cône inférieur bleu. *Cah. ORSTOM Sér. Entomol. Méd. Parasitol.* 15:283–86

32. Chapman, R. F. 1961. Some experiments to determine the methods used in host finding by tsetse flies, *Glossina medicorum*. *Bull. Entomol. Res.* 52:83–97

33. Colbo, M. 1977. Diurnal emergence patterns of two species of Simuliidae (Diptera) near Brisbane, Australia. *J. Med. Entomol.* 13:514–15

34. Das, S. C., Bhuyan, M., Das, N. G. 1984. Attraction of Simuliidae to different colors on humans—field trial. *Mosq. News* 44:79–80

35. Davies, D. M. 1952. The population and activity of adult female black flies in the vicinity of a stream in Algonquin Park, Ontario. *Can. J. Zool.* 30:287–321

36. Davies, D. M. 1972. The landing of blood-seeking female black-flies (Simuliidae: Diptera) on coloured materials. *Proc. Entomol. Soc. Ont.* 102:124–55

37. Davies, D. M. 1978. Ecology and behavior of adult black flies (Simuliidae): a review. *Quaest. Entomol.* 14:3–12

38. Davies, D. M., Peterson, B. V. 1956. Observations on the mating, feeding, ovarian development and oviposition of adult black flies (Simuliidae: Diptera). *Can. J. Zool.* 34:615–55

39. Delong, D. M. 1954. Fundamental studies on the behavior of larval and adult mosquitoes and evaluation of mosquito repellents. *Ohio State Univ. Eng. Exp. Stn. News* 26:51–55

40. DeMeillon, B., Sebastian, A., Khan, Z. H. 1967. Cane-sugar feeding in *Culex pipiens fatigans*. *Bull. WHO* 36:53–65

41. Downes, J. A. 1969. The swarming and mating flight of Diptera. *Ann. Rev. Entomol.* 14:271–98

42. Dransfield, R. D., Brightwell, R., Onah, J., Okolo, C. J. 1982. Population dynamics of *Glossina morsitans submorsitans* Newstead and *G. tachinoides* Westwood (Diptera: Glossinidae) in sub-Sudan savanna in northern Nigeria. 1. Sampling methodology for adults and seasonal changes in numbers caught in different vegetation types. *Bull. Entomol. Res.* 72:175–92

43. Fallis, A. M. 1964. Feeding and related behaviour of female Simuliidae (Diptera). *Exp. Parasitol.* 15:439–70

44. Flint, S. 1985. A comparison of various traps for *Glossina* spp. (Glossinidae) and other Diptera. *Bull. Entomol. Res.* 75:529–34

45. Focks, D. A., Sackett, S. R., Dame, D. A., Bailey, D. L. 1983. Ability of *Toxorhynchites amboinensis* (Doleschall) (Diptera: Culicidae) to locate and oviposit in artificial containers in an urban environment. *Environ. Entomol.* 12:1073–77

46. Fritz, G. N. 1983. Biology and ecology of bat flies (Diptera: Streblidae) on bats in the genus *Carollia*. *J. Med. Entomol.* 20:1–10

47. Gatehouse, A. G. 1972. Some responses of tsetse flies to visual and olfactory stimuli. *Nat. New Biol.* 236:63–64

48. Gatehouse, A. G., Lewis, C. T. 1973. Host location behaviour of *Stomoxys calcitrans*. *Entomol. Exp. Appl.* 16:275–90

49. Gibson, G. 1985. Swarming behavior of the mosquito *Culex pipiens quinquefasciatus*: a quantitative analysis. *Physiol. Entomol.* 10:283–96

50. Giglioli, M. E. C. 1965. The influence of irregularities in the bush perimeter of the cleared agricultural belt around a Gambian village on the flight range and direction of approach of a population of *Anopheles gambiae melas*. *Proc. Int. Congr. Entomol.* 12:757–58

51. Gillett, J. D. 1972. *The Mosquito: Its Life, Activities and Impact on Human Affairs*. New York: Doubleday. 109 pp.

52. Gillies, M. T., Wilkes, T. J. 1974. The range of attraction of birds as baits for some West African mosquitoes (Diptera, Culicidae). *Bull. Entomol. Res.* 63:573–81

53. Gillies, M. T., Wilkes, T. J. 1982. Responses of host-seeking *Mansonia* and *Anopheles* mosquitoes (Diptera: Culicidae) in West Africa to visual features of a target. *J. Med. Entomol.* 19:68–71

54. Glasgow, J. P. 1963. *The Distribution and Abundance of Tsetse.* Oxford: Pergamon. 241 pp.

55. Golini, V., Davies, D. M. 1971. Upwind orientation of female *Simulium venustum* Say (Diptera) in Algonquin Park, Ontario. *Proc. Entomol. Soc. Ont.* 101:49–54

56. Graham, N. L., Stoffolano, J. G. 1983. Oviposition behavior of the salt marsh greenhead, *Tabanus simulans* (Diptera: Tabanidae). *Ann. Entomol. Soc. Am.* 76:703–6

57. Granger, C. A. 1970. Trap design and color as factors in trapping the salt marsh greenhead fly. *J. Econ. Entomol.* 63:1670–72

58. Green, C. H. 1984. A comparison of phototactic responses to red and green light in *Glossina morsitans morsitans* and *Musca domestica. Physiol. Entomol.* 16:165–72

59. Green, C. H., Cosens, D. 1983. Spectral responses of the tsetse fly, *Glossina morsitans morsitans. J. Insect Physiol.* 29:795–800

60. Haddow, A. J., Corbet, P. S. 1961. Entomological studies from a high tower in Mpanga forest, Uganda. V. Swarming activity above the forest. *Trans. R. Entomol. Soc. London* 133:284–300

61. Haeger, J. S. 1955. The non–blood feeding habits of *Aedes taeniorhynchus* (Diptera, Culicidae) on Sanibel Island, Florida. *Mosq. News* 15:21–26

62. Hailman, J. P. 1979. Environmental light and conspicuous colors. In *The Behavioral Significance of Color,* ed. E. H. Burtt, Jr., pp. 288–616. New York: Garland. 450 pp.

63. Hanec, W., Bracken, G. K. 1962. Responses of female horse flies (Tabanidae: Diptera) to light. *Ann. Entomol. Soc. Am.* 55:720–21

64. Hansens, E. J., Bosler, E. M., Robinson, J. W. 1971. Use of traps for study and control of salt marsh flies. *J. Econ. Entomol.* 64:1481–86

65. Hargett, R. T., Goulding, R. L. 1962. Studies on the behavior of the horn fly, *Haematobia irritans* (Linn.). *Oreg. State Univ. Agric. Stn. Tech. Bull.* 67:1–61

66. Hargrove, J. W. 1976. The effect of human presence on the behaviour of tsetse (*Glossina* spp.) (Diptera, Glossinidae) near a stationary ox. *Bull. Entomol. Res.* 66:173–78

67. Hargrove, J. W. 1980. Improved estimates of the efficiency of traps for *Glossina morsitans morsitans* Westwood and G. *pallidipes* Austen (Diptera: Glossinidae) with a note on the effect of the concentration of accompanying host odour on efficiency. *Bull. Entomol. Res.* 70:579–87

68. Haufe, W. O. 1964. Visual attraction as a principle in design of mosquito traps. *Can. Entomol.* 96:118

69. Hecht, O., Hernandez-Corzo, J. 1963. On the visual orientation of mosquitoes in their search of resting places. *Entomol. Exp. Appl.* 6:63–74

70. Hilburn, L. R., Willis, N. L., Seawright, J. A. 1983. An analysis of preference in the color of oviposition sites exhibited by female *Toxorhynchites r. rutilus* in the laboratory. *Mosq. News* 43:302–6

71. Hocking, B. 1971. Blood-sucking behavior of terrestrial arthropods. *Ann. Rev. Entomol.* 16:1–26

72. Horridge, G. A. 1975. *The Compound Eye and Vision in Insects.* Oxford: Clarendon. 595 pp.

73. Hsiao, H. S. 1972. *Attraction of Moths to Light and to Infrared Radiation.* San Francisco: San Francisco Press. 89 pp.

74. Huyton, P. M., Langley, P. A., Carlson, D. A., Coates, T. W. 1980. The role of sex pheromones in initiation of copulatory behaviour by male tsetse flies, *Glossina morsitans morsitans. Physiol. Entomol.* 5:243–52

75. Imhoff, J. E., Smith, S. M. 1979. Oviposition behaviour, egg-masses, and hatching responses of eggs of five Nearctic species of *Simulium* (Diptera: Simuliidae). *Bull. Entomol. Res.* 69: 405–25

76. Johnson, C. G., Crosskey, R. W., Davies, J. B. 1982. Species composition and cyclical changes in numbers of savanna blackflies (Diptera: Simuliidae) caught by suction traps in the Onchocerciasis Control Programme area of West Africa. *Bull. Entomol. Res.* 72:39–63

77. Jones, M. D. R. 1982. Coupled oscillators controlling circadian flight activity in the mosquito, *Culex pipiens quinquefasciatus. Physiol. Entomol.* 7:281–89

78. Jordan, A. M., Green, C. H. 1984. Visual responses of tsetse to stationary traps. *Insect Sci. Appl.* 5:331–34

79. Karim, E. I. A., Brady, J. 1984. Changing visual responsiveness in pregnant and larvipositing tsetse flies, *Glossina morsitans. Physiol. Entomol.* 9:125–31

80. Kennedy, J. S. 1941. On water finding

and oviposition by captive mosquitoes. *Bull. Entomol. Res.* 32:279–301

81. Kirschfeld, K., Wenk, P. 1976. The dorsal compound eye of simuliid flies: an eye specialized for detection of small, rapidly moving objects. *Z. Naturforsch.* 31:764–65

82. Kloter, K. O., Bowman, D. D., Carrol, M. K. 1983. Evaluation of some ovitrap materials used for *Aedes aegypti* surveillance. *Mosq. News* 43:438–41

83. Kunz, T. H. 1976. Observations on the winter ecology of the bat fly *Trichobius corynorhini* Cockerall (Diptera: Streblidae). *J. Med. Entomol.* 12:631–36

84. LaBreque, G. C., Meifert, D. W., Rye, J. 1972. Experimental control of the stable fly, *Stomoxys calcitrans* (Diptera: Muscidae) by releases of chemosterilized adults. *Can. Entomol.* 104:885–87

85. Lambrecht, F. L. 1973. Colour attraction of *Glossina morsitans* in N'Gamiland, Botswana. *J. Trop. Med. Hyg.* 76:94–96

86. Langley, P. A., Huyton, P. M., Carlson, D. A., Schwarz, M. 1981. Effects of *Glossina morsitans morsitans* Westwood (Diptera: Glossinidae) sex pheromone on behaviour of males in field and laboratory. *Bull. Entomol. Res.* 71:57–63

87. Langley, P. A., Pimley, R. W., Carlson, D. A. 1975. Sex recognition pheromone in tsetse fly, *Glossina morsitans. Nature* 254:51–52

88. Loor, K. A., DeFoliart, G. R. 1970. Field observations on the biology of *Aedes triseriatus. Mosq. News* 30:60–64

89. Lythgoe, J. N. 1979. *The Ecology of Vision.* Oxford: Clarendon. 244 pp.

90. Magnarelli, L. A. 1983. Nectar sugars and caloric reserves in natural populations of *Aedes canadensis* and *Aedes stimulans* (Diptera: Culicidae). *Environ. Entomol.* 12:1482–86

91. Magnarelli, L. A., Anderson, J. F., Thorne, J. H. 1979. Diurnal nectar-feeding of salt marsh Tabanidae (Diptera). *Environ. Entomol.* 8:544–48

92. Mazokhin-Porshnyakov, G. A., Cherkosov, A. D., Burakova, A., Vischnevskaya, T. M. 1975. On the color vision of Tabanidae (Diptera). *Zool. Zh.* 54:574–76

93. McCrae, A. W. R. 1984. Oviposition by African malaria vector mosquitoes. II. Effects of site tone, water type and conspecific immatures on target selection by freshwater *Anopheles gambiae* Giles, sensu lato. *Ann. Trop. Med. Parasitol.* 78:307–18

94. McIver, S. B. 1980. Sensory aspects of mate-finding behavior in male mosquitoes (Diptera: Culicidae). *J. Med. Entomol.* 17:54–57

95. Muirhead-Thompson, R. C. 1940. Studies on the behaviour of *Anopheles minimus*. Part II. The influence of water movement on the selection of the breeding place. *J. Malar. Inst. India* 3:295–322

96. Mullens, B. A., Gerhardt, R. R. 1979. Feeding behavior of some Tennessee Tabanidae. *Environ. Entomol.* 8:1047–51

97. Neilson, E. T., Haeger, J. S. 1960. Swarming and mating in mosquitoes. *Misc. Publ. Entomol. Soc. Am.* 1:71–95

98. Owaga, M. L., Challier, A. 1985. Catch composition of the tsetse *Glossina pallidipes* Austen in revolving and stationary traps with respect to age, sex ratio and hunger stage. *Insect Sci. Appl.* 6:711–18

99. Parr, H. C. M. 1962. Studies on *Stomoxys calcitrans* (L.) in Uganda, East Africa. II. Notes on life-history and behaviour. *Bull. Entomol. Res.* 53:437–43

100. Peschken, D. P., Thorsteinson, A. J. 1965. Visual orientation of black flies (Simuliidae: Diptera) to colour, shape and movement of targets. *Entomol. Exp. Appl.* 8:282–88

101. Pospisil, J., Zdarek, J. 1965. On the visual orientation of the stable fly (*Stomoxys calcitrans* L.) to colours. *Acta Entomol. Bohemoslov.* 62:85–91

102. Prokopy, R. J., Owens, E. D. 1983. Visual detection of plants by herbivorous insects. *Ann. Rev. Entomol.* 28:337–64

103. Provost, M. W. 1959. The influence of moonlight on light-trap catches of mosquitoes. *Ann. Entomol. Soc. Am.* 52:261–71

104. Roberts, R. H. 1970. Color of malaise trap and collection of Tabanidae. *Mosq. News* 29:236–38

105. Roth, A. R., Lindquist, A. W. 1948. Ecological notes on the deer flies at Summer Lake, Oregon. *J. Econ. Entomol.* 41:473–76

106. Rowcliffe, C., Finlayson, L. H. 1981. Factors influencing the selection of larviposition sites in the laboratory by *Glossina morsitans morsitans* Westwood (Diptera: Glossinidae). *Bull. Entomol. Res.* 71:81–96

107. Rowcliffe, C., Finlayson, L. H. 1982. Active and resting behavior of virgin and pregnant females of *Glossina morsitans morsitans* Westwood (Diptera: Glossinidae) in the laboratory. *Bull. Entomol. Res.* 72:271–88

108. Saunders, D. S. 1981. Insect photoperiodism—the clock and the counter: a review. *Physiol. Entomol.* 6:99–116

109. Schmidtman, E. T., Jones, C. J., Gollands, B. 1980. Comparative host-seeking activity of *Culicoides* (Diptera: Ceratopogonidae) attracted to pastured livestock in Central New York State, USA. *J. Med. Entomol.* 17:221–31

110. Schreck, C. E., Smith, N., Posey, K., Smith, D. 1980. Observations on the biting behavior of *Prosimulium* and *Simulium venustum. Mosq. News* 40:113–15

111. Service, M. W. 1976. *Mosquito Ecology: Field Sampling Methods,* New York: Wiley. 583 pp.

112. Shannon, R. C. 1931. On the classification of Brazilian Culicidae with special reference to those capable of harboring the yellow fever virus. *Proc. Entomol. Soc. Wash.* 33:125–57

113. Sheppard, C., Wilson, B. H. 1976. Flight range of Tabanidae in a Louisiana bottomland hardwood forest. *Environ. Entomol.* 4:752–54

114. Silberglied, R. E. 1979. Communication in the ultraviolet. *Ann. Rev. Ecol. Syst.* 10:373–98

115. Simmons, K. R. 1984. Reproductive ecology and blood feeding behavior of the black fly, *Simulium venustum* (L.) (Diptera: Simuliidae). PhD thesis. Univ. Mass., Amherst. 204 pp.

116. Simmons, K. R., Edman, J. D. 1981. Sustained colonization of the black fly *Simulium decorum* Walker (Diptera: Simuliidae). *Can. J. Zool.* 59:1–7

117. Sippel, W. L., Brown, A. W. A. 1953. Studies on the responses of the female *Aedes* mosquito. Part B. The role of visual factors. *Bull. Entomol. Res.* 43:567–74

118. Snow, W. F. 1971. The spectral sensitivity of *Aedes aegypti* (L.) at oviposition. *Bull. Entomol. Res.* 60:683–96

119. Snow, W. F. 1980. Host location and feeding patterns in tsetse. *Insect Sci. Appl.* 1:23–30

120. Stark, W. S., Tan, K. E. W. P. 1982. Ultraviolet light: photosensitivity and other effects on the visual system. *Photochem. Photobiol.* 36:371–80

121. Sutcliffe, J. F. 1986. Black fly host-location: a review. *Can. J. Zool.* 64:1041–53

122. Takken, W. 1984. Studies on the biconical traps as a sampling device for tsetse (Diptera: Glossinidae) in Mozambique. *Insect Sci. Appl.* 5:357–61

123. Thimijan, R. W., Pickens, L. G., Morgan, N. O. 1973. Responses of the house fly, stable fly, and face fly to electromagnetic radiant energy. *J. Econ. Entomol.* 66:1269–70

124. Thompson, B. H. 1976. Studies on the attraction of *Simulium damnosum* s.l. (Diptera: Simuliidae) to its hosts. I. The relative importance of sight, exhaled breath, and smell. *Tropenmed. Parasitol.* 27:455–73

125. Thompson, P. H., Pechuman, L. L. 1970. Sampling populations of *Tabanus quinquefasciatus* about horses in New Jersey, with notes on the identity and ecology. *J. Econ. Entomol.* 63:1515

126. Thornhill, A. R., Hays, K. L. 1972. Dispersal and flight activities of some species of *Tabanus* (Diptera: Tabanidae). *Environ. Entomol.* 1:602–6

127. Thorsteinson, A. J., Bracken, G. K. 1965. The orientation behavior of horse flies and deer flies (Tabanidae: Diptera). III. The use of traps in the study of orientation of tabanids in the field. *Entomol. Exp. Appl.* 8:189–92

128. Thorsteinson, A. J., Bracken, G. K., Hanec, W. 1964. The Manitoba horse fly trap. *Can. Entomol.* 96:166

129. Thorsteinson, A. J., Bracken, G. K., Tostawaryk, W. 1966. The orientation behavior of horse flies and deer flies (Tabanidae: Diptera). VI. The influence of the number of reflecting surfaces on attractiveness to tabanids of glossy black polyhedra. *Can. J. Zool.* 44:275–79

130. Tidwell, M. A., Hays, K. L. 1971. Oviposition preferences of some Tabanidae (Diptera). *Ann. Entomol. Soc. Am.* 64:547–49

131. Turner, D. A., Invest, J. F. 1973. Laboratory analyses of vision in tsetse flies (Dipt., Glossinidae). *Bull. Entomol. Res.* 62:343–57

132. Underhill, G. W. 1940. Some factors influencing feeding activity of simuliids in the field. *J. Econ. Entomol.* 33:915–17

133. Vale, G. A. 1974. New field methods for studying the responses of tsetse flies (Diptera, Glossinidae) to hosts. *Bull. Entomol. Res.* 64:199–208

134. Vale, G. A. 1974. The responses of tsetse flies (Diptera, Glossinidae) to mobile and stationary baits. *Bull. Entomol. Res.* 64:545–88

135. Vale, G. A. 1977. The flight of tsetse flies (Diptera: Glossinidae) to and from a stationary ox. *Bull. Entomol. Res.* 67:297–303

136. Vale, G. A. 1980. Field studies of the responses of tsetse flies (Glossinidae) and other Diptera to carbon dioxide, acetone and other chemicals. *Bull. Entomol. Res.* 70:563–70

137. Vale, G. A. 1982. The trap-oriented behaviour of tsetse flies (Glossinidae)

and other Diptera. *Bull. Entomol. Res.* 72:71–93

138. Waage, J. K. 1979. The evolution of insect/vertebrate associations. *Biol. J. Linn. Soc.* 12:187–224

139. Waage, J. K. 1981. How the zebra got its stripes—biting flies as selective agents in the evolution of zebra coloration. *J. Entomol. Soc. South. Afr.* 44:351–58

140. Waldbilling, R. C. 1968. Color vision of the female stable fly, *Stomoxys calcitrans*. *Ann. Entomol. Soc. Am.* 61:789–91

141. Walsh, J. F. 1978. Light trap studies on *Simulium damnosum* s.l. in northern Ghana. *Tropenmed. Parasitol.* 29:492–96

142. Warnes, M. L., Finlayson, L. H. 1985. Responses of the stable fly, *Stomoxys calcitrans* (L.) (Diptera: Muscidae) to carbon dioxide and host odours. I. Activation. *Bull. Entomol. Res.* 75:519–27

143. Warnes, M. L., Finlayson, L. H. 1985. Responses of the stable fly, *Stomoxys calcitrans* (L.) (Diptera: Muscidae), to carbon dioxide and host odours. II. Orientation. *Bull. Entomol. Res.* 75:717–27

144. Wenk, P. 1981. Bionomics of adult black flies. In *Blackflies: The Future for Biological Methods in Integrated Control,* ed. M. Laird, pp. 259–84. London: Academic

145. Wenk, P. 1986. Mating behavior of black flies. *Proc. Int. Workshop Ecol. Manage. Black Flies.* University Park, Pa: Univ. Pa. Press. In press

146. White, R. H. 1978. Insect visual pigments. *Adv. Insect Physiol.* 13:35–67

147. Williams, D. F. 1973. Sticky traps for sampling populations of *Stomoxys calcitrans*. *J. Econ. Entomol.* 66:1279–80

148. Wilson, B. H., Lieux, M. 1972. Pollen grains in the guts of field collected tabanids in Louisiana. *Ann. Entomol. Soc. Am.* 63:1264–66

149. Wilson, B. H., Tugwell, N. P., Burns, E. C. 1966. Attraction of tabanids to traps baited with dry ice under field conditions in Louisiana. *J. Med. Entomol.* 3:148–49

150. Wood, P. W., Wright, R. H. 1968. Some responses of flying *Aedes aegypti* to visual stimuli. *Can. Entomol.* 100:504–13

151. Zeil, 1983. Sexual dimorphism in the visual system of flies: The free flight behaviour of male Bibionidae (Diptera: *J. Comp. Physiol.* 150:395–412

Ann. Rev. Entomol. 1987. 32:317–40

FACTORS AFFECTING INSECT POPULATION DYNAMICS: Differences Between Outbreak and Non-Outbreak Species[1]

W. E. Wallner

USDA Forest Service, Center for Biological Control of Northeastern Forest Insects and Diseases, 51 Mill Pond Road, Hamden, Connecticut 06514

INTRODUCTION

Even though outbreaks of insects are temporally rare events, they have aroused the interest of theoretical and applied entomologists alike. The population dynamics of spatially rare species have been the subject of long-term studies concerned with the extirpation of these species, the unusual habitats they occupy, or the unique adaptations they possess (5, 107, 112). A comparison of the features of insects possessing different population processes may lead to a more thorough understanding of insect population dynamics. Past *Annual Review of Entomology* articles have addressed a number of population dynamics topics including density dependence (100), life tables (54), dynamics (160), stability and diversity (153), and pest management (159). However, no review has specifically addressed insect pest outbreaks or attempted to compare population dynamics of outbreak and nonoutbreak species. Richards (118) pointed out that this comparison is essentially an issue of population balance versus population fluctuation.

Populations of organisms are never truly stable, but rise from some low density and then fall to approximately their original size. They may exhibit stable equilibrium points, stable cyclic oscillations between two population points, stable cycles, or a regime of aperiodicity (87). Insect pests usually

---

exhibit gradient, cyclic, or irruptive patterns (14, 44, 114); on the other hand, endemic or rare species approximate stable equilibrium. The change in numbers and the extent of damage have been the principal concerns of applied entomologists (4). Insects are classified as pests based not solely on numbers but on socioeconomic sensitivity and on biological tolerance of the host to pest attack (102, 132).

In this review emphasis is placed upon phytophagous insects of perennial plant habitats, since extended temporal observations are essential for describing endemic populations and the processes by which populations change from latent to epidemic. Comparisons between agroecosystems and natural systems are difficult to make, and will be avoided since factors such as weather, plant genetics, natural enemies, and vegetational simplification may have quite different roles (1, 119). "Endemic" is sometimes used to denote the low-density phase of insects that frequently or infrequently irrupt; the term "latent" best describes this period. This is consistent with terms used in Europe, where an outbreak is called "culmination" and the period of low population is called "latent" (64). In this review "endemic" refers to insects that are either rare or uncommonly abundant and therefore seldom, if ever, occur at densities sufficient for them to be considered pests. This consistency in terminology makes comparisons of worldwide literature possible so that nonoutbreak, endemic insect dynamics can be contrasted with those of outbreak, epidemic species. Expectations should be tempered by Morris' (95) pronouncement that "outbreak species constitute a minor proportion of all phytophagous insects and may represent the exception rather than the rule, and that broad principles should not be drawn from this behavior alone." The voluminous literature on insects from varied habitats dictates that this review be focused on (a) how ecological factors such as natural enemies, weather, host, and site of perennial plant habitats (mainly forests) influence endemic and epidemic insect dynamics and (b) how heritable features of individual populations are important in differentiating between endemic insects and epidemic pests.

POPULATION STABILITY AND CHANGE

Natural multi-species assemblages of plants and animals, including insect pests, may possess several equilibrium points (88, 93, 133, 134). Although equilibrium points are sometimes discontinuously stable, they are not necessarily static temporally or spatially; population stress may cause sudden and unexpected shifts in numbers. Turnock (152) studied 13 native phytophagous pests of Canadian prairie ecosystems from presettlement times to the 1970s and found three population patterns: (a) opportunistic populations with two equilibrium states; (b) populations with a single, unstable state, apparently lacking effective natural enemies; and (c) populations with a single, stable equilibrium and effective natural enemies. The terms "resilience" and "stability" have been used to describe certain ecological systems. Pest systems, while highly variable in time and space, are affected by dispersal and therefore possess high resilience (60).

Studies on tropical insects are scarce, but Wolda (174) and Bigger (16) have shown that tropical insect populations are cyclical, with amplitudes comparable to those of temperate insects. Because of the highly specialized and coevolved feeding relationships between herbivorous insects and host plants in the tropics and the apparently stable insect populations that fluctuate seasonally there, the frequency of insect outbreaks with any one host plant species is expected to be low (175). However, there are indications that tropical communities have a much greater number of species and contain more unique species than similar temperate communities (68). Similarities are apparent in comparative stability indexes for temperate and tropical invertebrates: populations of common species have lower densities than populations of less common species. Among the more variable insects were agricultural pests, migrating moths in Great Britain, and Japanese rice pests; among the most stable were German forest insects, Canadian spiders, and Austrian caddisflies (174).

The spectrum of change in stability is associated with increasing temporal constancy of the habitat. The type and stability of an insect habitat governs its population regulation strategy and the environmental conditions that favor large increases vary among species and are more restricted for rare than for outbreak species (53). MacArthur & Wilson (82) used the term "r and K selection" to describe different strategies for survival in crowded and uncrowded environments. These terms are also used to distinguish animals in temporary and permanent habitats. Insects that are r-strategists exist in unpredictable environments, have high capacities for population increase, and are poor competitors. Many agricultural pests of annual crops fit this description. K-strategists inhabit fairly constant environments, have low capacities for population increase, and are good competitors. The periodical cicada is an example of a K-strategist. Interestingly, some of the most effective, numerous, endemic insect parasites are K-selected, but these species are often neglected in biological control programs because establishment of colonies in host populations is difficult (42). A synoptic model incorporating the r-K continuum with population growth, population density, and habitat stability provides a framework for comparing population dynamics of different species (134).

WEATHER

On a global scale, seasonal temperatures and rainfall patterns constitute the major factors that determine the distribution of organisms in space (17). Insects and plants become adapted to combinations of these factors through natural selection (8), yet insects with periodic outbreaks occur especially in areas that are physically severe. A short growing season and a cycle peak and outbreak are preceded by unfavorable environmental conditions (9, 41, 58, 101, 140). A propensity for outbreaks in other cases has been attributed to climatic warming, which enables herbivores to invade the cooler ranges of their environment. In these cooler regions, plants that were previously protected from these herbivore invasions became susceptible because they had evolved weak rapid-defense responses and strong delayed responses and were therefore highly susceptible to herbivore attack (117). Such effects are generally not expected in the tropics, but highly synchronized population cycles in a seasonal environment and long-term reduction in host plant diversity are expected to lead to outbreaks of herbivorous insects in the tropics (175). Tropical insects on the average have the same annual variability as insects from temperate zones, but insect populations from dry areas, temperate or tropical, tend to fluctuate more than those from wet areas (172).

Outbreaks of both temperate and tropical insects have followed periods of drought (7, 49, 64, 76, 108, 114, 129, 146, 147, 165, 169), high sunspot activity (11), decreased storminess and increased atmospheric circulation (11, 165), or combinations of drought and excessive moisture (6, 43, 51, 83, 86, 92, 150, 168). Severe fluctuations in precipitation can be extremely detrimental to phytophagous insects; drought caused the rapid decline of populations of common and rare British butterflies by reducing the number of nectar plants (107). Temperature can have an array of direct or indirect effects on populations, such as asynchrony between insect and host (31) or predator and insect prey (65), mortality to overwintering stages due to density-dependent developmental lag (91), selective mortality that shifts population genetics (20), appearance of refuge areas for insect survival (135), or synchronization of host flowering and increased larval survival (95). In isolated, fragile ecosystems, unusual and severe temperature and precipitation regimes can lead to extinction, as was the case with a subalpine lycaenid butterfly (40).

Weather influences are often not density-dependent and therefore should not affect endemic and epidemic insect species differently. In fact, simultaneous irruptions of different species have been reported, e.g. eastern spruce budworm and forest tent caterpillar in Canada (165) and pine caterpillar and *Selenephera lunigera* and *Dasychira abietis* and geometrid species in Siberia (66). However, seasonality and abundance may be more strictly controlled by food availability in some insect groups than in others (173). Abundance can be influenced by small aberrations in rainfall, as happened with two species of grasshoppers in North and South American desert ecosystems (104). *Ligurotettix coquilletti*, a K-selected species, lives on the resilient, perennial creosote bush and maintains constant low populations. Conversely, an r-selected ground-inhabiting grasshopper, *Trimerotropis pallidipennis*, feeds on annual plants associated with the creosote bush, but its population fluctuates dramatically. Populations of *L. coquilletti* are regulated by male territorial defense, whereas populations of the nonterritorial *T. pallidipennis* reach very high densities by responding to luxuriant annual plant growth created by precipitation.

The effects of environmental stresses such as weather on insect dynamics cannot be explained as easily in most cases. While environmental stresses such as drought and temperature fluctuations have been recorded preceding insect outbreaks, the precise "mode of action" of these stresses is unknown. The bulk of evidence suggests that such environmental stresses affect insects indirectly by reducing phenotypic variability in the host-plant population (81). For example, a psyllid population irrupted when drought caused elevated amino acid levels in eucalyptus (167); populations of two Scandinavian geometrids irrupted as a result of lowered levels of antiherbivory compounds in birch foliage caused by cold summer temperatures (101); and excessive rainfall during the wet season produced new flushes of vegetative growth on widely scattered *Bunchosia* trees, promoting outbreaks of species of the Costa Rican chrysomelid *Uroplata* (175). Thus, weather stresses may trigger simultaneous metabolic and physiological responses among individual trees in a variety of climatic zones, and may thereby reduce phenotypic variability. This increased homogeneity in host traits results in rapid selection of phenotypes that can reproduce successfully on these plants; when this occurs, an outbreak ensues (81). The concept of climate release is therefore inextricably linked with an insect population's genetic composition and spatial distribution (164). Species capable of surviving periods of extreme climatic fluctuations are highly resilient but unstable (60, 161), and are usually considered pests. The lack of reports on dramatic changes in densities of nonoutbreak species associated with weather variation suggests that weather is not as critical a factor in their dynamics as it is in those of outbreak species.

HABITAT

Site

Insect outbreaks in forests commonly occur in stands that have passed peak efficiency in biomass production (86). Conversely, insect outbreaks on arable crops occur on highly productive plants in early successional communities (132). Resource availability and variation in the carrying capacity of the habitat are often the most important determinants of year-to-year fluctuations in the populations of insects (33). Area for area, trees have more herbivore species than bushes, which in turn have more than herbs (137). The distribution of a plant is determined indirectly by the prevailing physical conditions; more importantly, the changing physical environment of the growing site influences a plant's ability to resist being displaced by other plant species. Site conditions determine the number of different host species present and the type and abundance of pests found there (15). Environmentally severe or deficient sites are believed to reduce host genetic variability; as a result, environmental stress could create a monoculture (81). Typically, an outbreak occurs when the physiological state of the plant permits a herbivore phenotype with a high reproductive capacity to become dominant. Thus, cyclic and irregular fluctuations in herbivore populations result from adaptation and selection of both the host plant and the herbivore feeding on it (58). Agricultural and forest monocultures consisting of extensive plantings of hosts with narrow genetic variability are havens for pest outbreaks (47).

European scientists have used the term "primary foci" to denote optimal sites for survival of certain insects under adverse weather, habitat, and host conditions. Primary foci are associated with forests growing under unfavorable conditions (25, 64). These sites are characterized by soils that are unsuitable for a given species, sharply changing humidity conditions, or vulnerability to drought or human disturbance (121). The term "foci" has had limited use in North American, Asian, and Australian literature, but it is analogous to "epicenters" (12, 55), "refuges" (169), "reserves" (144), and "susceptible forests" (62). Outbreaks of forest defoliators are usually first noted in open forests that are subject to drought, and are associated with poor growing sites such as ridgetops, upper slopes, or deep sands (62, 64, 71, 135). Disturbances such as fire and clear-cutting have increased the extent, frequency, and severity of outbreaks associated with such sites (18).

The progressive redistribution of populations and the resulting defoliation are described as a spreading from primary foci to secondary and tertiary foci (64, 114). Primary foci may contain dense, stable populations all year; secondary foci do not always contain larvae in the spring but do in the fall; in tertiary foci it is difficult to find insects during latency (122). Differences were noted in numbers and ratio of the dark phenotype of *Dendrolimus superans sibericus* in primary, secondary, and tertiary foci. The percentage of dark-colored third instars in primary foci is an important predictor of future population increases and an important indicator of population migration and spread. The outbreak spread characteristics for Canadian forest defoliating insects that are based upon habitat restriction (125) are comparable to the characteristics used in the European concept.

The level of migratory activity in different populations of a single species differs with population conditions such as insect density and food availability. Butterflies that have spatially segregated larval and adult (i.e. temporary) habitats are more vagile than those with unsegregated adult and larval habitats (i.e. permanent) (126, 130). Berryman (12) has demonstrated that pests with low dispersal properties have short, intense, restricted outbreaks, whereas those with high vagility have long, extended outbreaks. Species in the same genus may have quite different population dynamics (endemic and epidemic) depending on habitat, host selection, and heritable adaptations. For example, Nakamura & Ohgushi (97) compared two phytophagous congeneric lady beetles, *Henosepilachna pustulosa* (K-selected) on thistle and *H. vigintiocto-punctata* (r-selected) on potato. The thistle-infesting lady beetle maintained a constant population below the food limit; its fecundity and reproductive rate were low, and adults were long-lived and had low vagility. The potato-inhabiting lady beetle displayed violent fluctuations and high fecundity; adults were short-lived and very mobile.

For annual crop pests and most economically important pests, migration is the most important process in determining whether an outbreak will occur (130). A case in point is the nymphalid butterfly, *Chlosyne* [-*Melitaea* auctt.] *harrisii*, which has low adult vagility and lays its eggs in masses on individual aster plants, and whose larvae feed gregariously for the first two instars and then disperse and feed solitarily. The insect seldom consumes more than a fraction of its host resource; limited larval dispersal in respect to food plant density is the factor that keeps populations low. Populations do not remain at a constant density, but fluctuate according to the availability of food, which is in turn regulated by succession (36).

In perennial crops (forests, orchards, plantations) many of the insects have low vagility, and migration seldom initiates outbreaks. It may, however, contribute to outbreaks. For example, dispersal by adult spruce budworms can significantly augment local population densities via deposition of egg masses; the resulting larvae then overwhelm natural enemy response (120). In perennial host systems, variation in natality and mortality are most important, since the permanence of the host habitat allows many chances for invasion by different species including parasites and predators as well as other pests (130). However, in very temporary habitats the level of migratory activity is high and the number of migrants is a major factor in population irruptions. Agricultural practices have expanded temporarily favorable habitats into areas where overwintering conditions are unsuitable for certain multivoltine, highly vagile species. The "pied piper" effect of these cropping systems has undermined the survival strategies of these insects (113).

There is extensive evidence that certain pests capitalize on conditions that are less favorable to the host (13, 15). Examples include gypsy moth (25, 50, 62), pine caterpillar (122), woodwasp (84), Swaine sawfly (93), European pine sawfly (73, 76), various bark beetles (7, 48, 115), locusts (149), Douglas fir tussock moth (171), lodgepole needle miner (135), eucalyptus psyllids (169), eastern (55) and western (71) spruce budworms, and false spruce webworm (123). Interestingly, many of these insects are very responsive to variations in weather and are classified as irruptive. Specific, spatially well-defined habitats may provide highly selective survival conditions for insects during inhospitable periods in tropic (78, 129) or temperate (144) ecosystems. The spatial extent of an epicenter, focus, or reserve varies depending upon the requirements of the insect involved; it can be limited to part or all of a tree or shrub (7, 48, 84, 115, 169), or can encompass several hectares or more (62, 76, 93, 122, 143). While such select habitats provide escape from enemies (114, 128), there is evidence that they are also repositories of highly nutritious food. White (169) reviewed the numerous factors that stress plants, depress their resistance, and enrich their tissues with nitrogen, conditions advantageous to the survival and growth of young insects. Certain epidemic pests persist during latent periods in epicenters of good nutritional resources in otherwise inhospitable environments. It is unlikely that rare insects survive similarly in such epicenters because of their inability to overcome host chemical defenses. However, economic entomologists or forest managers should not overlook potential foci, since they may be the source of future pest epidemics (12, 64). It was proposed that the western tent caterpillar (143, 144) and the Swaine sawfly (93) could be controlled more effectively over the long term by combatting insects in these refugia during a population decline rather than by intervening less selectively at the culmination of an outbreak.

Hosts

The geographic range, size, growth form, and variety of resources offered by a host influences the number and species of insects that feed upon it. Rafes (114) synthesized Eastern European literature and described three general types of insect population dynamics based upon insect-host relationships: (*a*) background or endemic insects react weakly to environmental conditions, but are limited by host defenses; (*b*) cyclic or gradient insects have abundance directly proportional to fluctuations in food resources; and (*c*) flareups or epidemics recur at irregular intervals and are mediated through the host by meteorological or environmental stress conditions. A comparable synthesis was suggested by Turnock (152) for insect pests of prairie ecosystems. Rhoades (117) described stable populations as "stealthy" and those with variable populations as "opportunistic." Stealthy insects are controlled by rapid decline in plant nutritional quality, while opportunistic insects preferentially attack stressed plants or naive plants that have not been immunized by recent attack. Interestingly, these classifications are synonymous with K-selected and r-selected species (82, 134).

Insects are recognized as being mono-, oligo-, or polyphagous depending upon host-plant range; polyphagous species are favored in unpredictable habitats and monophagous species are favored in predictable ones (170). Polyphagous or generalist species tend to have more variable populations than specialists (161), and are more likely to become widespread introduced pests than are species with a restricted host range (24, 127). The actual abundance and diversity of insects on forest trees is not related to breadth of diet. Rather, differences in the properties of trees cause variations in their insect fauna (46, 109). Host-plant selection is believed to be a behavioral process, and the emergence of specific insect–host plant relationships probably results from evolutionary change in the insect's chemosensory system (69). Dramatic changes occur in insect physiology and development when populations fluctuate from latent to epidemic levels (9, 20, 149). Herbivores prone to irrupt to high levels exhibit such phase changes when they alternate between strategies of stealth and opportunism (117).

Responses of trees to herbivore pest attack have been demonstrated to have adverse effects on several lepidopteran defoliators (57, 116, 129, 151, 156) and aphids (38, 72, 106, 162). Insects feeding on previously browsed foliage exhibited decreased growth and increased mortality, which was intensified after two years of defoliation. It is not known if plants have evolved a defensive response to insect browsing, if browsing causes coincidental wound repair by the plants, or if browsing indirectly affects the nutritional quality of the foliage (39). Not all defoliators and their hosts react similarly. Intense feeding by western tent caterpillar on alder did not cause subsequent population reductions (96). Heavy damage by larvae of the stem boring moth *Archanara geminipuncta* causes its host, *Phragmites australis,* to develop thinner shoots that cannot support any infestation, and local extinction of the moth occurs every three years (94). Effects of prior feeding may not affect other insects feeding on the same host similarly. Over a 30-year period, 74 outbreaks of the major oak pests gypsy moth, brown-tail moth, lackey moth, and green leafroller were documented in the USSR. Gypsy moth defoliation did not inhibit outbreaks of the other three pests, but heavy feeding by each species prevented its further outbreak (114). The extent to which wound-induced responses affect other phytophagous species depends on the persistence of damaged tissue, the effect upon undamaged tissues, and the nature of the chemicals involved. Immobile insects such as leafminers and gall-formers are more likely to suffer from interactions with defoliators (166). This could explain why insect guilds exhibit patterns of high proportions of defoliators and lower proportions of sap-sucking phytophages or vice versa.

Timing of prior herbivory can have an effect on insect responses. Birches previously defoliated by autumnal moths exhibit defense against succeeding populations (57, 58). However, birches browsed by moose during the winter harbor larger populations of ants, psyllids, leaf galls, leafminers, and leaf-eating insects than do unbrowsed birches during the following summer. After browsing by moose, subsequent foliage had lower C:N ratio, lower protein-precipitating capacity, and higher dry-matter digestibility (30).

Yearly variation, site variation, and the carryover effect upon subsequent insect populations suggest that the effects of defoliation on plants are more a response to stress than an active defense response. Nutrient stress from defoliation is believed to rearrange the plant carbon/nutrient balance, and on nutrient poor soils such stress can result in increased production of carbon-based allelochemicals. Plants with compensatory nutrient uptake slowed by defoliation may have a relaxation period of several years (151). Plant species with slow growth rates and high levels of defense chemicals are selected for low-resource environments; plants with faster growth rates and weaker defense are favored in sites with higher availability of resources (19, 27). Tropical, fast-growing trees are eaten six times faster than slow-growing species in the same microhabitat. This pattern is compatible with observations made in agroecosystems and certain tropical forests, but is not entirely consistent with observations of temperate forest insects. On stressed sites, oak trees grow more slowly and are defoliated more often by gypsy moth than those on more productive sites, yet they suffer less mortality (62).

Rhoades (117) suggested that attack strategies of herbivores may be an overriding factor in determining population densities. Innocuous or endemic species with low, invariant populations should exhibit highly efficient conversion of ingested matter and food digestion and low feeding rates. Variable populations such as pests should exhibit less efficient conversion of ingested matter and food digestion and variable feeding rates. In general, rare or endemic insects are characterized by an inability to respond to change in host constituency; *Pardia tripunctana* and *Notocelia roborana* on rose (10), *Fiorinia externa* on hemlock (90), *Saissetia oleae* on olive (105), and *Melitaea harrisii* on aster (36) are examples. There is evidence that outbreak species and generalist feeders are less affected by variable food quality than are rare species feeding on the same host (53).

Epidemic pest populations increase rapidly and commonly exhaust their hosts, dramatically altering their environments, whereas endemic species rarely alter their host environments significantly. Numbers of gradient pests are regulated by food availability and may increase dramatically; such insects rarely alter their habitat or significantly damage their hosts. Gradient pests include seed and cone insects such as coneworms (*Dioryctria* spp.), seed-worms (*Laspeyresia* spp.), seed bugs (*Leptoglosus* spp.) and cone beetles (*Conophthorus* spp.) (85), fruit insects such as codling moth (114), fir engraver beetle (14), and tropical bruchids on acacia (67). Most of these pests reduce host reproduction and do not directly damage the host, but may also alter the structure of the vegetation and the characteristics of individual plants. Some may actually increase host-plant growth by killing flowers, conelets, and cones that would otherwise use considerable amounts of nutrients, photosynthates, and energy. These resources thus become available for cambia, shoots, roots, and leaves; tree growth is improved and annual flower production may be less variable (85).

NATURAL ENEMIES

Andrewartha & Birch (4) attribute insect irruptions to an unusual increase in food quality, reduced natural enemy pressure, or an extended period of favorable weather. Resource concentration can best explain outbreaks in simple, annual cropping systems, but natural enemies are a more important factor in complex, perennial systems (119). However, there are several examples in which attack of conifers by bark beetles is food dependent (52, 115). The equilibrium population size and dynamic behavior of many phytophagous insects are largely determined by their predators and host foods. Natural enemies are key factors in keeping many phytophagous insects rare (77, 110). Huffaker et al (63) reasoned that the natural enemy model for r- and K-strategies (134) should be extended to include even the most stable end of the habitat stability dimension. This habitat stability dimension may also be caused by host defenses, as in the case of bark beetles (12).

The influence of the majority of the natural enemies of the winter moth is not density-dependent, but this influence differs from place to place. For this species and most phytophagous insects, enemies take many forms, from microsporidia to mammals and polyhedrosis viruses to parasitoids (137). Other evidence suggests that natural enemies may have a minor role in population dynamics of rare, endemic, and gradient insects (114), but a major role in those of cyclic and irruptive pests (12, 64, 73). Examples of the interactions of parasites, predators, and diseases with insect pests abound in the literature; the scope of this review permits only a few illustrative examples.

Predators

Vertebrate and invertebrate predators have been implicated in the control of numerous temperate and tropical insects (12, 22, 26, 29, 45, 52, 53, 59, 61, 64, 73, 79, 128, 139). Because their rate of reproduction is limited, predators usually cannot compensate quickly enough to suppress the dramatic insect irruptions characteristic of r-selected pests. For example, the gypsy moth population increases so rapidly that it saturates the feeding response of its vertebrate predators (22). Mountain pine beetle, an irruptive pest, has a complex of predators and parasites dependent on beetle density (2). However, a clerid beetle preyed upon a higher proportion of mountain pine beetles in latent than in epidemic infestations; this suggests that it may be important in maintaining sparse populations. In the case of K-selected endemic species such as two tortricid moths on rose and two diprionid sawflies on pine, polyphagous pupal and cocoon predators maintained populations at equilibrium levels (10, 53).

The diversity and abundance of birds and small mammals that prey on gypsy moth can be determined by physiographic differences in forest sites

(128). Sites classified as susceptible to defoliation by gypsy moth contain abundant above-ground resting locations that permit the insects to avoid predators (62). Highly susceptible forest sites on ridgetops or deep sands have historically been defoliated more often than adjacent mesic forests (155). Evidence presented by Hanski & Otronen (53) for rare and common pine sawflies further supports the contention that outbreaks are unlikely to develop in every habitat; outbreaks were most common in rocky sites or bogs, which had fewer small mammal predators.

Avian predators have been more thoroughly studied and used in pest control in Europe (139) and Asia (45) than in North America. They are important regulators of sparse insect populations. Of the 283 insectivorous bird species in North America, 70% are migrants from the Neotropics (139). Many of these species make annual pilgrimages between the same winter and summer forest sites. The alarming deforestation of wintering sites in tropical highland forests (141) can be expected to result in major reductions in the numbers and variety of predatory bird species in North American woodlands. This is an example of the complexity of ecological systems and demonstrates how insect dynamics can be influenced by human activities hundreds or thousands of miles away.

Parasites

There is a paucity of literature dealing with parasites of endemic insect populations, since entomologists tend to study species that have outbreaks rather than those that maintain a steady level (131). Parasites have greater influence on high populations of *Diprion pini* than on latent populations (111), and have a minor role in western spruce budworm gradations (21, 148). Parasite populations lagged behind larch bud moth populations (32), followed leaf miner cycles on cocoa (16), and exhibited a delayed response to fall webworm populations (95). Less vigorous western tent caterpillar larvae were more heavily parasitized than more active larvae (163). Parasites have aggregated in high-density patches of winter moth larvae (56). They have been identified as important contributors to mortality in low to moderate Douglas fir tussock moth populations (29), and they maintained low densities of hemlock scale (91) and European spruce sawfly for many years (99).

Polyphagous parasites can prevent successful establishment of migrants in areas where one pest precedes or accompanies another (53, 64). The success of parasite guilds of the larch casebearer is determined by the seral stage of forest succession (136). Native parasites that are well adapted to plant communities have a major role in regulating introduced herbivores. It is also clear that the interaction of these natural enemies and the host plant has an important effect on the population dynamics of phytophagous insects (77, 110, 116). It is very likely that this trophic relationship is more important for epidemic pests, which are better adapted to host-plant variability, than for endemic and gradient insects.

Disease

The action of insect diseases is most obvious at high population densities where more contact occurs among individuals and the unfavorable effects of overcrowding are evident (154). Horizontal transmission of viral, bacterial, protozoan, and fungal infections usually requires high host densities or host aggregating behavior (89). Endemic insects lack these population characteristics, and there is little published evidence that their numbers are regulated by pathogens. Striking exceptions are the K-selected 13- and 17-year periodical cicadas, whose densities are determined by infection with the fungus *Massospora cicadina* (80).

Anderson & May (3) analyzed host-pathogen mortalities for 28 insects and concluded that pathogens contribute wholly or in part to the regulation of their host populations. In fact, the authors presented evidence that 5–12–year population cycles of temperate-forest insect pests were attributable to either a nucleopolyhedrosis virus (either alone or in combination with a microsporidian protozoan) or a granulosis virus. Zelinskaya (176) reported that four microsporidian agents are major regulators of gypsy moth populations; they increased larval mortality and egg sterility and lowered fecundity. The influence of the type and condition of the plant host can significantly affect disease expression. Gypsy moth susceptibility to nucleopolyhedrosis virus increased with the duration of host defoliation (155). The cyclic behavior of the larch bud moth has been attributed to foliage quality, with the period of outbreak set by host-virus dynamics and the amplitude of the outbreak cycle set by the insect-plant relationship (89). Studies on viral and microsporidian infections of the larch bud moth also suggest that the wide-amplitude, stable-limit cycles are probably driven by long delays built into the infection and transmission process (3). The host-pathogen system is complicated by long stages during which the pathogen exists in the external environment, protected from ultraviolet degradation by the soil of temperate forests. Finally, evidence supports the insect-pathogen relationships proposed by the r-K continuum and Southwood's synoptic model (134). That is, populations of endemic, rare, or innocuous insects are regulated more by heritable traits and profound host plant influences and to a lesser extent by natural enemies. It is unclear how pervasive diseases such as microsporidian infections affect endemic insect dynamics. Gradient, cyclic, and irruptive pests respond quickly to host plant availability and susceptibility and are "tracked" by natural enemies. Predators are most important in regulating or maintaining low insect densities, whereas diseases tend to be most effective in reducing dense populations.

HERITABLE TRAITS

The prevailing environment determines immediate factors such as food, parasites, predators, and disease, and these in turn affect the degree of competition for required resources. Changes in the environment thus change the rate of increase of the population and its genetic properties (63). Natural populations of plants and animals contain relatively high levels of genetic polymorphism and heterogeneity, and there is considerable evidence that genetic variation is maintained through natural selection resulting from adaptation of organisms to particular habitats (136). Most entomologists consider genes to be stable in a population or to undergo minor unimportant shifts (98). However, genetic response to environmental variables may be a key factor in endemic and epidemic insect population dynamics.

Changes in the population size and structure of six endangered butterflies were correlated with subtle as well as drastic habitat changes (5). Some introduced insects require a period of genetic adaptation before population spread and increase can occur (127). Both of these examples indicate that phenotypic properties related to survival and reproduction change in response to environmental conditions. Those genetic traits that produce the proper response to environmental factors ensure persistence of the species (57). Insects with low fertility require low environmental resistance, while pests exhibit high fertility, fluctuate widely in abundance, and must be controlled by high environmental resistance, both temporally and spatially (64). Insect herbivores must adjust to leaf persistence, leaf chemistry, reproductive strategy, longevity, and successional status of plants that resist the insects feeding on them (103). Since habitats of natural insect populations are heterogeneous, chances of survival differ on different sites, and the probability of wide population fluctuations is unequal for a number of local groups within the population (34). One carabid species in which subpopulations fluctuated unequally survived ten times better than another inhabiting the same heath area whose subpopulations fluctuated in parallel (35). Since no population is composed of individuals with identical genotypes, various individuals contribute unequally to the genetic makeup of the next generation (75).

Polymorphic expression and behavioral change are common in a number of epidemic insects and are coincidental with fluctuations in insect populations (8, 20, 25, 149, 163). In *Choristoneura* spp. and *Malacosoma disstria* such qualitative changes are the direct result of shifts in allelic frequency of a major sex-linked gene (20). Natural enemies and reduction of previously abundant high-quality food dampen expanding populations; however, when these controls fail (88, 133, 134) enormous increases in genetic variability occur as the population reaches its culmination (23). This increase in genetic variability is evident as polymorphism in some species (9, 25, 149, 165), and reflects the change described by Rhoades (116) from stealthiness to opportunism. Such changes in genetic variability are seldom if ever observed in endemic insects, but are common in insects with "flush-crash" cycles (outbreaks) that possess substantial genetic diversity. In some cases, populations irrupt through the multiplication of highly fit individuals (23). Outbreeding is considered prerequisite to outbreaks, and "luxuriant heterosis" is important for population increase.

During latency, outbreak species such as *Neodiprion sertifer* are less abundant than rare *Gilpinia verticolis* and *G. socia* on Scots pine. Certain characteristics of *N. sertifer* outbreaks can select for further outbreaks (53). Western tent caterpillar larvae that hatch from large egg masses are larger, more active, and better oriented than larvae from small egg masses. These active larvae transform into more active adults that can fly much farther than those that develop from sluggish larvae (163). Egg-mass size, synchrony of eclosion, and activity patterns of larvae were the most important predictors of gypsy moth population dynamics (74). Active larvae aggregated more rapidly and withstood stressful environmental conditions more effectively than did less active larvae, which had depressed protease activity. In simulated population-control studies with the western tent caterpillar, the harvesting of sluggish larvae resulted in dramatic increases in subsequent populations, whereas the harvesting of active colonies caused the population to collapse (145).

Insects that are adapted to widely different environments, especially those whose populations are density-dependent, exhibit genetic variation that affects population dynamics. Rare, endemic species with small numbers of offspring per generation may differ from adjacent populations even under uniform selection pressure (98). In cyclic, epidemic insects, adaptations are not in constant equilibrium, hence the genetics and behavior of the populations change. Furthermore, it is not clear if epidemics are started by subpopulations or by a species reacting to uniform environmental conditions. In the case of irruptive pests, which are rarely epidemic, subpopulations may initially have similar gene frequencies, but each may evolve with its own unique dynamics. Most will remain latent or become extinct, but occasionally a subpopulation may irrupt to epidemic proportions from a unique genetic epicenter.

Characteristics of insects reflect their genetic selection and are useful for comparing rare, gradient, cyclic, and irruptive insects (Table 1). Adult vagility permits out-migration and colonization of a spectrum of habitats, including those that serve as refuges during unfavorable periods (164). Multibrooded insects have greater propensity to disperse than single-brooded species, which are adapted to specific habitats (124). Butterfly species, which have a high threshold of flight activity, persist in high-density local populations with short flights. Insects with short flights are likely to be clustered into wholly or partially isolated populations, which may become extinct (124). With highly vagile species, however, dispersion and colonization leads directly to dispersion of gene frequencies (23). The most vagile species are likely to be those from disturbed sites with spatially segregated larval and adult habitats (126). Spruce budworm and forest and western tent caterpillars are polymorphic for size; small adults colonize new habitats, whereas large, inactive adults exploit local favorable conditions. Thus polymorphism, which is believed to be transmitted by the X chromosome, permits these species to cope with fluctuating environments (37). The ability to maintain physiological plasticity ensures the dispersal capacity of some individuals, which contributes to species persistence in refuges during unfavorable periods and to colonization of marginal habitats under favorable conditions (142).

Species that cluster their eggs in batches have high realized fecundity (28, 74). Egg-clustering is advantageous because food-plant searching by larvae and adults is reduced. As a result, preproductive adults have lower mortality and can devote more of their energy to egg production (28). Mortality of both early and late instar larvae is higher for defoliators that lay their eggs singly on surfaces of their host plants than for those that lay eggs in clusters (157), and is useful in characterizing r- and K-strategists (138). While increased mortality may be due to natural enemies or to host plant defenses, aggregations (whether due to direct egg-clumping, limited larval dispersion, or thigmotrophic response) can affect the manner and rate of gene flow in successive

Table 1 Relative importance of biological characteristics of different insect population types[a]

Biological trait	K-selected Rare	←————————→ Gradient	Cyclic	r-selected Irruptive
Adult life span	+++	+++	+	+
Adult feeding	+++	+++	+	+
Habitat restriction	+++	+++	+	+
Response to plant defense	+++	+	+	++
Degree of host specificity	+++	+++	+	++
Flush-crash cycles	+	+++	+++	+++
Adult vagility	+	++	++	+++
Fecundity	+	++	+++	+++
Alteration of environment	+	+	++	+++
Degree of egg clumping	+	+	+++	+++
Degree of larval aggregation	+	+	+++	++
Response to weather	+	++	+++	+++
Importance of biological control	+	++	+++	+++
Utilization of foci or refuges	+	+	+++	+++
Incidence of polymorphism	+	+	+++	+++

[a]References: Rare: 5, 10, 35, 36, 90, 97, 104; Gradient: 14, 68, 85, 115; Cyclic: 9, 25, 76, 140, 171; Irruptive: 52, 114, 120, 122, 146, 175.

generations (158). Behavioral adaptations such as aggregation of conspecifics clearly benefit many herbivores during feeding (57, 70, 117), and ensure maintenance of genetic demes in select habitats (23, 145).

Fecundity and survival are equal, interchangeable components of fitness and persistence (57, 82). High fecundity among certain gypsy moth populations, due to either physiological or genetic changes, was observed in certain North American populations characterized by successive periods of proliferation and collapse (21). Directional selection for two fitness types, which is triggered by epidemic-induced changes in food quality, is believed to cause the epidemic cycles in larch bud moth; density antibiosis selects for a slow-growing heterozygous insect ecotype for three to four years after the outbreak, which permits the host to recover; then the larch bud moth produces a dark-colored fast-developing ecotype that capitalizes on the improved food resource. Thus, epidemics of this insect are tied closely to strong directional selection for the heterozygous ecotype, long-range adult dispersal, and persistence of the bud-moth ecotype under nonoutbreak conditions (8).

It is evident that a number of heritable traits are interrelated in contributing to pest outbreaks. Egg clumping, larval aggregation, phenotypic plasticity, adult vagility, and utilization of select habitats during periods of latency are common to cyclic and irruptive pests. The normal balance of climate, host condition, and natural enemies provides a complex series of interactions so that an epidemic can be expected to occur periodically or rarely. The common latent condition of pests between cycles, irruptions, or gradient increases may be a result of the failure of the genetic system to generate appropriate genotypes at a suitable time to initiate an outbreak (98). It also may be that insufficient phenotypic variability in rare or endemic insects is largely responsible for their scarcity.

CONCLUSIONS

Comparisons of the biological attributes of endemic and epidemic insects demonstrate that these population types are significantly different in many ways. Direct comparisons are unlikely to yield solutions for pest problems or to rescue rare and endangered species. This review underscores that each population type has evolved a set of physical and behavioral traits for species survival. It does not necessarily hold that any one characteristic has overwhelming survival value. Rather, combinations of features and relative capacity for adaptation determine the success of a given insect species.

ACKNOWLEDGMENTS

I wish to thank Katherine McManus for assistance in preparing various versions of this article, Phyllis Grinberg and Daniel Starr for literature review and retrieval, Alice Vandel for translating numerous articles, and Drs. Alan Berryman, Joseph Elkinton, Clive Jones, Michael Montgomery, and Harry Valentine for critically reviewing earlier versions.

Literature Cited

1. Altieri, M. A., Letourneau, D. K. 1984. Vegetation diversity and insect pest outbreaks. In *Critical Reviews in Plant Science*, ed. B. V. Konger, pp. 131–69. Boca Raton, Fla: CRC
2. Amman, G. D. 1984. Mountain pine beetle (Coleoptera: Scolytidae) mortality in three types of infestations. *Environ. Entomol.* 13:184–91
3. Anderson, R. M., May, R. M. 1980. Infectious diseases and population cycles of forest insects. *Science* 210:658–61
4. Andrewartha, H. G., Birch, L. C. 1982. Theory of the distribution and abundance of animals. In *The Ecological Web: More on the Distribution and Abundance of Animals*, pp. 185–211. Chicago: Univ. Chicago Press. 495 pp.
5. Arnold, R. A. 1983. *Ecological Studies of Six Endangered Butterflies (Lepidoptera: Lycaenidae): Island Biogeography, Patch Dynamics, and the Design of Habitat Preserves.* Berkeley, Calif: Univ. Calif. Press. 161 pp.
6. Bailey, W. J., McCrae, A. W. R. 1978. The general biology and phenology of swarming in the East African tettigoniid, *Ruspolia differens* (Serville) (Orthoptera). *J. Nat. Hist.* 12:259–88
7. Bakke, A. 1983. Host tree and bark beetle interaction during a mass outbreak of *Ips typographus* in Norway. *Z. Angew. Entomol.* 96:118–25
8. Baltensweiler, W. 1976. Natural control factors operating in some European forest insect populations. *Proc. 15th Int. Congr. Entomol., Washington, DC,* pp. 617–21. Washington, DC: US For. Serv.
9. Baltensweiler, W., Benz, G., Bovey, P., Delucchi, V. 1977. Dynamics of larch bud moth populations. *Ann. Rev. Entomol.* 22:79–100
10. Bauer, G. 1985. Population ecology of *Pardia tripunctana* Schiff. and *Notocelia roborana* Den. and Schiff. (Lepidoptera: Tortricidae)—an example of "equilibrium species." *Oecologia* 65:437–41
11. Benkevich, V. I. 1972. Gypsy moth outbreaks in the European part of the USSR as related to solar activity fluctuations, atmospheric circulation, climatic and weather conditions. *Proc. 13th Int. Congr. Entomol., Moscow, 1968,* 3: 14–15. Leningrad: Nauka
12. Berryman, A. A. 1978. Towards a theory of insect epidemiology. *Res. Popul. Ecol.* 19:181–96
13. Berryman, A. A. 1985. Site characteristics and population dynamics: A theoretical perspective. See Ref. 15, pp. 1–7
14. Berryman, A. A., Stark, R. W. 1985. Assessing the risk of forest insect outbreaks. *Z. Angew. Entomol.* 99:199–208
15. Bevan, D., Stoakley, J. T., eds. 1985. *Site Characteristics and Population Dynamics of Lepidopteran and Hymenopteran Forest Pests.* Sherborne, Dorset, UK: Sawtells. 139 pp.
16. Bigger, M. 1976. Oscillations of tropical insect populations. *Nature* 259:207–9
17. Birch, L. C. 1957. The role of weather in determining the distribution and abundance of animals. *Cold Spring Harbor Symp. Quant. Biol.* 22:203–18
18. Blais, J. R. 1983. Trends in the frequency, extent and severity of spruce budworm outbreaks in eastern Canada (*Choristoneura fumiferana*). *Can. J. For. Res.* 13:539–47
19. Bombosch, S. 1985. Investigations on the food quality of spruce needles for larvae of *Gilpinia hercyniae*. See Ref. 15, pp. 77–84
20. Campbell, I. M. 1964. Genetic variation related to survival in lepidopteran species. In *Breeding Pest-Resistant Trees,* ed. H. D. Gerhold, R. E. McDermott, E. J. Schreiner, J. A. Winieski, pp. 129–35. Oxford: Pergamon
21. Campbell, R. W. 1981. Evidence for high fecundity among certain North American gypsy moth populations. *Environ. Entomol.* 10:663–67
22. Campbell, R. W., Sloan, R. J. 1977. Release of gypsy moth populations from innocuous levels. *Environ. Entomol.* 6:323–30
23. Carson, H. L. 1968. The population flush and its genetic consequences. In *Population Biology and Evolution,* ed. R. C. Lewontin, pp. 123–37. Syracuse, NY: Syracuse Univ. Press
24. Cherrett, J. M., Sagar, G. R. 1977. *Origins of Pest, Parasite, Disease, and Weed Problems.* Oxford: Blackwell

25. Chugunin, Ya. 1949. Focal periodicity of gypsy moth outbreaks. *Zool. Zh.* 28:431–38. Transl. L. W. Murphy, 1976, *Gypsy Moth Tech. Inf. Proj. Transl. No. 7144T*
26. Clark, L. R. 1964. The population dynamics of *Cardiaspina albitextura* (Psyllidae). *Aust. J. Zool.* 12:362–80
27. Coley, P. D., Bryant, J. P., Chapin, F. S. 1985. Resource availability and plant antiherbivore defense. *Science* 250:895–99
28. Courtney, S. P. 1984. The evolution of egg clustering by butterflies and other insects. *Am. Nat.* 123:276–81
29. Dahlsten, D. L., Luck, R. F., Schlinger, E. I., Wenz, J. M., Copper, W. A. 1977. Parasitoids and predators of the Douglas-fir tussock moth, *Orgyia pseudotsugata* (Lepidoptera: Lymantriidae), in low to moderate populations in central California. *Can. Entomol.* 109:727–46
30. Danell, K., Muss-Danell, K. 1985. Feeding by insects and hares on birches earlier affected by moose browsing. *Oikos* 44:75–81
31. Day, K. 1984. The growth and decline of a population of the spruce aphid, *Elatobium abietinum,* during a three year study, and the changing pattern of fecundity, recruitment and alary polymorphism in a Northern Ireland forest. *Oecologia* 64:118–24
32. DeLucchi, V. 1982. Parasitoids and hyperparasitoids of *Zeiraphera diniana* (Lep.: Tortricidae) and their role in population control in outbreak areas. *Entomophaga* 27:77–92
33. Dempster, J. P., Pollard, E. 1981. Fluctuations in resource availability and insect populations. *Oecologia* 50:412–16
34. den Boer, P. J. 1970. Stabilization of animal numbers and the heterogeneity of the environment: The problem of the persistence of sparse populations. *Proc. Adv. Study Inst. Dynamics Numbers Popul., Oosterbeek, Commun. Biol. Stn. Wijster* 158:77–97
35. den Boer, P. J. 1981. On the survival of populations in a heterogeneous and variable environment. *Oecologia* 50:39–53
36. Dethier, V. G. 1959. Food-plant distributions and density and larval dispersal as factors affecting insect populations. *Can. Entomol.* 91:581–96
37. Dingle, H. 1972. Migration strategies of insects. *Science* 175:1327–35
38. Dixon, A. F. G. 1977. Aphid ecology: Life cycles, polymorphism, and population regulation. *Ann. Rev. Ecol. Syst.* 8:329–53
39. Edwards, P. J., Wratten, S. D. 1985. Induced plant defences against insect grazing: fact or artefact? *Oikos* 44:70–74
40. Ehrlich, P. R., Breedlove, D. E., Brussard, P. F., Sharp, M. A. 1972. Weather and the regulation of subalpine populations. *Ecology* 53:243–47
41. Elton, C. S. 1975. Conservation and the low population density of invertebrates inside Neotropical rain forest. *Biol. Conserv.* 7:3–15
42. Force, D. C. 1975. Succession of r and K strategists in parasitoids. See Ref. 109, pp. 112–29
43. Fuentes, E. R., Campusano, C. 1985. Pest outbreaks and rainfall in the semi-arid region of Chile. *J. Arid Environ.* 8:67–72
44. Furuta, K. 1976. Studies on the dynamics of the low density populations of gypsy moth and Todo-fir aphid: Analysis of the environmental resistance factors by artificial host increase method. *Bull. Gov. For. Exp. Stn. No. 279,* Tokyo. 85 pp.
45. Furuta, K. 1982. Natural control of *Lymantria dispar* L. (Lepidoptera: Lymantriidae) populations at low density levels in Hokkaido (Japan). *Z. Angew. Entomol.* 93:513–22
46. Futuyma, D. J., Gould, F. 1979. Associations of plants and insects in a deciduous forest. *Ecol. Monogr.* 49:33–50
47. Gibson, I. A. S., Jones, T. 1977. Monoculture as the origin of major forest pests and diseases. See Ref. 24, pp. 139–61
48. Gray, B. 1975. Distribution of *Hylurdrectonus araucariae* Schedl (Coleoptera: Scolytidae) and progress of outbreak in major hoop pine plantations in Papua, New Guinea. *Pac. Insects* 16:383–94
49. Greenbank, D. O. 1956. The role of climate and dispersal in the initiation of outbreaks of the spruce budworm in New Brunswick. *Can. Entomol.* 34:453–76
50. Grimalskii, V. I. 1974. Resistance of tree stands against needle- and leaf-eating pests with respect to the trophic theory of dynamics in insect numbers. *Zool. Zh.* 53:189–98 (In Russian)
51. Hain, F. P., ed. 1979. *Population Dynamics of Forest Insects at Low Levels.* *Work Conf.* Raleigh, NC: North Carolina State Univ. 32 pp.
52. Hain, F. P., McClelland, W. T. 1979. Studies of declining and low level populations of the southern pine beetle in North Carolina. See Ref. 51, pp. 9–26
53. Hanski, I., Otronen, M. 1985. Food quality induced variance in larval per-

formance: Comparison between rare and common pine-feeding sawflies (Diprionidae). *Oikos* 44:165–74

54. Harcourt, D. G. 1969. The development and use of life tables in the study of natural insect populations. *Ann. Rev. Entomol.* 14:175–96

55. Hardy, Y. V., Lafond, A., Hamel, L. 1983. The epidemiology of the current spruce budworm outbreak in Quebec. *For. Sci.* 29:715–25

56. Hassell, M. P. 1982. Patterns of parasitism by insect parasitoids in patchy environments. *Ecol. Entomol.* 7:365–77

57. Haukioja, E. 1980. On the role of plant defences in the fluctuation of herbivore populations. *Oikos* 35:202–13

58. Haukioja, E., Hakala, T. 1975. Herbivore cycles and periodic outbreaks. Formulation of a general hypothesis. *Rep. Kevo Subarct. Res. Stn.* 12:1–9

59. Holling, C. S. 1959. The components of predation as revealed by a study of small-mammal predation of the European pine sawfly. *Can. Entomol.* 91:293–320

60. Holling, C. S. 1973. Resilience and stability of ecological systems. *Ann. Rev. Ecol. Syst.* 4:1–23

61. Holmes, R. T., Schultz, J. C., Nothnagle, P. 1979. Bird predation on forest insects: An exclosure experiment. *Science* 206:462–63

62. Houston, D. R., Valentine, H. T. 1977. Comparing and predicting forest stand susceptibility to gypsy moth. *Can. J. For. Res.* 7:447–61

63. Huffaker, C. B., Dahlsten, D. L., Janzen, D. H., Kennedy, G. G. 1984. In *Ecological Entomology*, ed. C. B. Huffaker, R. L. Rabb, pp. 679–91. New York: Wiley Interscience. 844 pp.

64. Il'inskii, A. I., Evlakhova, M. I., Shvetsova, S., Shvetsova, O. I., Andreeva, G. I., et al. 1975. *Inspection, Calculation, and Prognosis of Forest Insects in the USSR*. Transl. USDA For. Serv., 1980. 397 pp. (From Russian)

65. Inozmetsev, A. A. 1976. The dynamics of the trophic links of forest insectivorous birds and their significance in the invertebrate population control. *Zh. Obshch. Biol.* 37:192–204 (In Russian)

66. Isaev, A. S., Khlebopros, R. G., Kondakov, Iu. P. 1974. Regularities of the insect numbers of dynamics. *Lesovedenie* 3:27–42 (In Russian)

67. Janzen, D. H. 1975. See Ref. 109, pp. 154–86

68. Janzen, D. H., Schoener, T. W. 1968. Differences in insect abundance and diversity between wetter and drier sites during a tropical dry season. *Ecology* 49:96–110

69. Jermy, T. 1984. Evolution of insect/host plant relationships. *Am. Nat.* 124:609–30

70. Kalin, M., Knerer, G. 1977. Group and mass effects in diprionid sawflies. *Nature* 267:427–29

71. Kemp, W. P., Moody, U. L. 1984. Relationships between regional soils and foliage characteristics and the western spruce budworm (Lepidoptera: Tortricidae) outbreak frequency. *Environ. Entomol.* 13:1291–97

72. Kidd, N. A. 1985. The role of the host plant in the population dynamics of the large pine aphid, *Cinara pinea*. *Oikos* 44:114–22

73. Kolomiets, N. G., Stadnitskii, G. V., Vorontzov, A. I. 1972. *The European Pine Sawfly. Distribution, Biology, Economic Importance, Natural Enemies and Control.* Transl. Amerind, New Delhi, 1979. 138 pp. (From Russian)

74. Konikov, A. S., Chernysheva, L. V. 1981. Hereditary mechanisms of the regulation of behavior and the population of gypsy moth *(Porthetria dispar)*. In *Fauna i Ekologia Chlenistonogikh: Materialy v Soveshchaniia Entomologov Sibiri*, pp. 57–60. Novosibirsk, USSR: Nauka (In Russian)

75. Labeyrie, V. 1969. The variability of the physiological and ethological activities and the population growth of insects. *Int. Symp. Stat. Ecol., New Haven, Conn.*, Vol. 2, *Sampling and Modeling*, pp. 331–36

76. Larsson, S., Tenow, O. 1984. Areal distribution of a *Neodiprion sertifer* (Hym.: Diprionidae) outbreak on Scots pine as related to stand condition. *Holarct. Ecol.* 7:81–90

77. Lawton, J. H., McNeil, S. 1979. Between the devil and the deep blue sea: on the problem of being a herbivore. *Br. Ecol. Soc. Symp.* 20:223–44

78. Leigh, E. G. Jr., Rand, A. S., Windsor, D. M., eds. 1982. *The Ecology of a Tropical Forest: Seasonal Rhythms and Long-Term Changes*. Washington, DC: Smithsonian Inst.

79. Leigh, E. G. Jr., Smythe, N. 1982. Leaf production, leaf consumption and the regulation of folivory on Barro Colorado Island. See Ref. 78, pp. 33–50

80. Lloyd, M., Dybas, H. S. 1966. The periodical cicada problem: I. Population ecology. *Evolution* 20:133–49

81. Lorimer, N. 1980. Pest outbreaks as a function of variability in pests and plants. In *Resistance to Diseases and Pests in Forest Trees. Proc. 3rd Int.*

Workshop Genet. Host-Parasite Interactions For., Wageningen, pp. 287–94

82. MacArthur, R. H., Wilson, E. O. 1967. *The Theory of Island Biogeography.* Princeton, NJ: Princeton Univ. Press. 203 pp.

83. Maceljski, M., Balarin, I. 1974. Factors influencing the population density of the looper *Autographa gamma* L. in Yugoslavia. *Acta Entomol. Jugoslov.* 10:63–76 (In Serbo-Croatian)

84. Madden, J. L. 1975. An analysis of an outbreak of the woodwasp, *Sirex noctilio* F. (Hymenoptera: Siricidae), in *Pinus radiata. Bull. Entomol. Res.* 65:491–500

85. Mattson, W. J. 1978. The role of insects in the dynamics of cone production of red pine. *Oecologia* 33:327–49

86. Mattson, W. J., Addy, N. D. 1975. Phytophagous insects as regulators of forest primary production. *Science* 190:515–22

87. May, R. M. 1974. Biological populations with nonoverlapping generations: Stable points, stable cycles, and chaos. *Science* 186:645–47

88. May, R. M. 1977. Thresholds and breakpoints in ecosystems with a multiplicity of stable states. *Nature* 269:471–77

89. May, R. M. 1983. Parasitic infections as regulators of animal populations. *Am. Sci.* 71:36–45

90. McClure, M. S. 1979. Self-regulation in populations of the elongate hemlock scale, *Fiorinia externa* (Homoptera: Diaspididae). *Oecologia* 39:25–36

91. McClure, M. S. 1983. Population dynamics of a pernicious parasite: Density-dependent vitality of red pine scale. *Ecology* 64:710–18

92. McCrae, A. W. R., Visser, S. A. 1975. *Paederus* in Uganda. I. Outbreaks, clinical effects, extraction and bioassay of the vesicating toxin. *Ann. Trop. Med. Parasitol.* 69:109–20

93. McLeod, J. M. 1979. Discontinuous stability in a sawfly life system and its relevance to pest management strategies. *Proc. 15th Int. Congr. Entomol., Washington, DC: US For. Serv. Gen. Tech. Rep.* 8:61–81. Washington, DC: Sup. Doc. Print. Off.

94. Mook, J. H., van der Toorn, J. 1985. Delayed response of common reed, *Phragmites australis* to herbivory as a cause of cyclic fluctuations in the density of the moth, *Achranara geminipuncta. Oikos* 44:142–48

95. Morris, R. F. 1964. The value of historical data in population research, with particular reference to *Hyphantria cunea* Drury. *Can. Entomol.* 96:356–68

96. Myers, J. A., Williams, K. S. 1984. Does tent caterpillar attack reduce the food quality of red alder foliage? *Oecologia* 62:74–79

97. Nakamura, K., Ohgushi, T. 1981. Studies on the population dynamics of a thistle-feeding lady beetle, *Heno. ?pilachna pustulosa* (Kono), in a cool temperate climax forest. II. Life table s, key factor analysis, and detection of re,'-ulatory mechanisms. *Res. Popul. Ecoι* 23:210–31

98. Namkoong, G. 1979. The dynamics of population genetics in forest insects. See Ref. 51, pp. 6–8

99. Neilson, M. M., Morris, R. F. 1964. The regulation of European spruce sawfly numbers in the Maritime Provinces of Canada from 1933 to 1963. *Can. Entomol.* 96:773–84

100. Nicholson, A. J. 1958. Dynamics of insect populations. *Ann. Rev. Entomol.* 3:107–36

101. Niemela, P. 1980. Dependence of *Oporinia autumnata* (Lep.: Geometridae) outbreaks on summer temperature. *Rep. Kevo Subarct. Res. Stn.* 16:27–30

102. Norton, G. A., Conway, G. R. 1977. The economic and social context of pest disease and weed problems. See Ref. 24, pp. 205–26

103. Opler, P. A. 1978. Interaction of plant life history components as related to arboreal herbivory. In *Arboreal Folivores,* ed. G. G. Montgomery, pp. 23–31. Washington, DC: Smithsonian Inst.

104. Otte, D., Joern, A. 1975. Insect territoriality and its evolution: population studies of desert grasshoppers on creosote bushes. *J. Anim. Ecol.* 44:29–54

105. Paraskakis, M., Neuenschwander, P., Michelakis, S. 1980. *Saissetia oleae* (Oliv.) (Hom.: Coccidae) and its parasites on olive trees in Crete, Greece. *Z. Angew. Entomol.* 90:450–64

106. Perrin, A. M. 1976. The population dynamics of the stinging nettle aphid, *Microlophium carnosum* (Bukt.). *Ecol. Entomol.* 1:31–40

107. Pollard, E. 1984. Fluctuations in the abundance of butterflies, 1976–82. *Ecol. Entomol.* 9:179–88

108. Polyakov, I. Ya., Khomyakova, V. O., Kub'yas, L. M. 1977. Causes of mass outbreaks of the meadow moth. *Zashch. Rast.* 2:40–41 (In Russian)

109. Price, P. W. 1975. *Evolutionary Strategies of Parasitic Insects and Mites.* New York: Plenum

110. Price, P. W., Bouton, C. E., Gross, P.,

McPheron, B. A., Thompson, J. N., et al. 1980. Interactions among three trophic levels: Influence of plants on interactions between insect herbivores and natural enemies. *Ann. Rev. Ecol. Syst.* 11:41–65

111. Pschorn-Walcher, H. 1980. Population fluctuations and parasitization of the birch-alder casebearer (*Coleophora serratella* L.) in relation to habitat diversity. *Z. Angew. Entomol.* 89:63–81

112. Pyle, R., Bentzien, M., Opler, P. 1981. Insect conservation. *Ann. Rev. Entomol.* 26:233–58

113. Rabb, R. L., Stinner, R. E. 1979. The role of insect dispersal and migration in population processes. In *Radar, Insect Population Ecology, and Pest Management,* ed. C. R. Vaughn, W. Wolf, W. Klassen, pp. 3–16. Wallops Island, Va: NASA

114. Rafes, R. M. 1978. Biogeocoenological theory of the population dynamics of phytophagous forest insects. Transl. P. Rubtzoff, in *Mathematical Modeling in Ecology,* ed. A. M. Molchanov, pp. 34–51. Moscow: Nauka (From Russian)

115. Raffa, K. F., Berryman, A. A. 1983. The role of host plant resistance in the colonization behavior and ecology of the bark beetles (Coleoptera: Scolytidae). *Ecol. Monogr.* 53:27–49

116. Rhoades, D. F. 1983. Herbivore population dynamics and plant chemistry. In *Variable Plants and Herbivores in Natural and Managed Systems,* pp. 155–220. New York: Academic

117. Rhoades, D. F. 1985. Offensive-defensive interactions between herbivores and plants: Their relevance in herbivore population dynamics and ecological theory. *Am. Nat.* 125:205–38

118. Richards, O. W. 1961. The theoretical and practical study of natural insect populations. *Ann. Rev. Entomol.* 6:147–62

119. Risch, S. J. 1987. Agricultural ecology and insect outbreaks. In *Insect Outbreaks,* ed. P. Barbosa, J. C. Schultz. New York: Academic. In press

120. Royama, T. 1984. Population dynamics of the spruce budworm, *Choristoneura fumiferana. Ecol. Monogr.* 54:429–62

121. Rudnev, D. F. 1972. Causes of mass reproduction of forest pests and the nature of resistance in plantings. See Ref. 11, 3:86–87

122. Ryapolov, V. Ya., Ryapolova, L. M. 1983. Analysis of change in the population of defoliating insects in residual foci of *Dendrolimus sibiricus. Lesovedenie* 4:49–55 (In Russian)

123. Schmutzenhofer, H. 1985. Site characteristics and mass outbreaks of *Cephalcia abietis* in Austria. See Ref. 15, pp. 27–35

124. Scott, J. A. 1975. Flight patterns among eleven species of diurnal lepidoptera. *Ecology* 56:1367–77

125. Shepherd, R. F. 1977. A classification of western Canadian defoliating forest insects by outbreak spread characteristics and habitat restriction. *Tech. Bull. Agric. Exp. Stn. Univ. Minn.* 310:80–88

126. Shreeve, T. G. 1981. Flight patterns of butterfly species in woodlands. *Oecologia* 51:289–93

127. Simmonds, F. J., Greathead, O. J. 1977. Introductions and pest and weed problems. See Ref. 24, pp. 109–24

128. Smith, H. R. 1985. Wildlife and the gypsy moth. *Wildl. Soc. Bull.* 13:166–74

129. Smith, N. G. 1982. Population irruptions and periodic migrations in the day-flying moth *Urania fulgens.* See Ref. 79, pp. 331–44

130. Southwood, T. R. E. 1972. The role and measurement of migration in the population system of an insect pest. *Trop. Sci.* 13:275–78

131. Southwood, T. R. E. 1975. The dynamics of insect populations. In *Insects, Science, and Society,* ed. D. Pimentel, pp. 151–99. New York: Academic

132. Southwood, T. R. E. 1977. The relevance of population dynamic theory to pest status. See Ref. 24, pp. 35–54

133. Southwood, T. R. E. 1981. Stability in field populations of insects. In *The Mathematical Theory of the Dynamics of Biological Populations II,* ed. R. W. Hiorns, D. Cooke, pp. 31–45. London/New York: Academic

134. Southwood, T. R. E., Comins, H. N. 1976. A synoptic population model. *J. Anim. Ecol.* 45:949–65

135. Stark, R. W. 1959. Climate in relation to winter mortality of the lodgepole needle miner, *Recurvaria starki* Free., in Canadian Rocky Mountain parks. *Can. J. Zool.* 37:753–61

136. Stark, R. W., Burnell, D. G., Neuenschwander, L. F., Stock, M. W., Nathanson, R. A. 1985. Dynamics of parasite guilds and insect herbivores in forest successional stages. See Ref. 15, pp. 110–23

137. Strong, D. R., Lawton, J. H., Southwood, R., eds. 1984. *Insects on Plants: Community Patterns and Mechanisms.* Cambridge, Mass: Harvard Univ. Press. 313 pp.

138. Stubbs, M. 1977. Density dependence in the life-cycles of animals and its im-

portance in K- and r-strategies. *J. Anim. Ecol.* 46:677–88

139. Takekawa, J. Y., Garton, E. O., Langelier, L. A. 1982. Biological control of forest insect outbreaks: the use of avian predators. *Trans. North Am. Wildl. Nat. Resour. Conf.* 47:393–409

140. Tenow, O. 1972. The outbreaks of *Oporinia autumnata* Bkh. and *Operophthera* spp. (Lep.: Geometridae) in the Scandinavian mountain chain and northern Finland 1862–1968. In *Zoologiska Bidrag, Suppl. 2.* Uppsala: Almqvist & Wiksells. 107 pp.

141. Terborgh, J. W. 1980. The conservation status of neotropical migrants: Present and future. In *Migrant Birds in the Neotropics: Ecology, Behavior, Distribution, and Conservation,* ed. A. Keast, E. S. Morton, pp. 21–30. Washington, DC: Smithsonian Inst.

142. Thompson, W. A., Cameron, P. J., Wellington, W. G., Vertinsky, I. B. 1976. Degrees of heterogeneity and the survival of an insect population. *Res. Popul. Ecol.* 18:1–13

143. Thompson, W. A., Vertinsky, I. B., Wellington, W. G. 1979. The dynamics of outbreaks: Further simulation experiments with the western tent caterpillar. *Res. Popul. Ecol.* 20:188–200

144. Thompson, W. A., Vertinsky, I. B., Wellington, W. G. 1981. Intervening in pest outbreaks: Simulation studies with the western tent caterpillar. *Res. Popul. Ecol.* 23:27–38

145. Thompson, W. A., Wellington, W. G., Vertinsky, I. B., Matsumura, E. M. 1977. Harvesting strategies, control styles and information levels: A study of planned disturbances to a population. *Res. Popul. Ecol.* 18:160–76

146. Thomson, A. J., Shrimpton, D. M. 1984. Weather associated with the start of mountain pine beetle outbreaks. *Can. J. For. Res.* 14:255–58

147. Thomson, A. J., Shepherd, R. F., Harris, W. E., Silversides, R. H. 1984. Relating weather to outbreaks of western spruce budworm, *Choristoneura occidentalis* (Lepidoptera: Tortricidae), in British Columbia. *Can. Entomol.* 116:375–81

148. Torgersen, T. R., Campbell, R. W. 1982. Some effects of avian predators on the western spruce budworm in North Central Washington. *Environ. Entomol.* 11:429–31

149. Tsyplenkov, E. P. 1978. *Harmful Acridoidea of the USSR.* Transl. Natl. Tech. Inf. Serv.

150. Tucker, M. R., Pedgley, D. E. 1983.

Rainfall and outbreaks of the African armyworm, *Spodoptera exempta* (Walker) (Lepidoptera: Noctuidae). *Bull. Entomol. Res.* 73:195–99

151. Tuomi, J., Niemela, P., Haukioja, E., Siren, S., Neuvonen, S. 1984. Nutrient stress: an explanation for plant antiherbivore responses to defoliation. *Oecologia* 61:208–10

152. Turnock, W. J. 1977. Adaptability and stability of insect pest populations in prairie agricultural ecosystems. *Tech. Bull. Agric. Exp. Stn. Univ. Minn.* 310:89–101

153. van Emden, V. H. F., Williams, G. C. 1974. Insect stability and diversity in agro-ecosystems. *Ann. Rev. Entomol.* 19:445–75

154. Viktorov, G. A. 1971. Some general principles of insect population density regulation. See Ref. 11, 1:573–74

155. Wallner, W. E. 1983. Gypsy moth host interactions: A concept of room and board. In *Forest Defoliator-Host Interactions: A Comparison Between Gypsy Moth and Spruce Budworms, USDA For. Serv. Gen. Tech. Rep. NE-85* ed. R. L. Talerico, M. E. Montgomery, pp. 5–8. Broomall, Pa: Northeast. For. Exp. Stn.

156. Wallner, W. E., Walton, G. S. 1979. Host defoliation: A possible determinant of gypsy moth population quality. *Ann. Entomol. Soc. Am.* 72:62–67

157. Watanabe, M. 1976. A preliminary study on population dynamics of the swallowtail butterfly, *Papilio xuthus* L., in a deforested area. *Res. Popul. Ecol.* 17:200–10

158. Waters, W. E. 1962. The ecological significance of aggregation in forest insects. *Proc. 11th Int. Kongr. Entomol., Vienna, 1960,* pp. 205–10. Herausgeber, Austria: Organ. Kom. XI Int. Kongr. Entomol.

159. Waters, W. E., Stark, R. W. 1979. Forest pest management: Concept and reality. *Ann. Rev. Entomol.* 25:479–509

160. Watt, K. E. F. 1962. Use of mathematics in population ecology. *Ann. Rev. Entomol.* 7:243–60

161. Watt, K. E. F. 1964. Comments on fluctuations of animal populations and measures of community stability. *Can. Entomol.* 96:1434–42

162. Wellings, P. W., Chambers, R. J., Dixon, A. F. G., Aikman, D. P. 1985. Sycamore aphid numbers and population density. I. Some patterns. *J. Anim. Ecol.* 54:411–24

163. Wellington, W. G. 1957. Individual dif-

ferences as a factor in population dynamics: The development of a problem. *Can. J. Zool.* 35:293–323

164. Wellington, W. G. 1980. Dispersal and population change. In *Dispersal of Forest Insects: Evaluation, Theory, and Management Implications. Proc. Int. Union For. Res. Org. Sandpoint, Idaho, 1979,* ed. A. A. Berryman, L. Safranyik, pp. 11–24. Pullman, Wash: Washington State Univ. Coop. Ext. Serv. 278 pp.

165. Wellington, W. G., Fettes, J. J., Turner, K. B., Belyea, R. M. 1950. Physical and biological indicators of the development of outbreaks of the spruce budworm, *Choristoneura fumiferana* (Clem.) (Lepidoptera: Tortricidae). *Can. J. Res.* 28:308–31

166. West, C. 1985. Factors underlying the late seasonal appearance of the lepidopterous leaf-mining guild on oak. *Ecol. Entomol.* 10:111–20

167. White, T. C. R. 1969. An index to measure weather-induced stress of trees associated with psyllids in Australia. *Ecology* 50:905–9

168. White, T. C. R. 1976. Weather, food and plagues of locusts. *Oecologia* 22:119–34

169. White, T. C. R. 1984. The abundance of invertebrate herbivores in relation to the availability of nitrogen in stressed food plants. *Oecologia* 63:90–105

170. Wiklund, C. 1974. The concept of oligophagy and the natural habitats and host plants of *Papilio machaon* L. in Fennoscandia. *Entomol. Scand.* 5:151–60

171. Williams, C. B. Jr., Wenz, J. M., Dahlsten, D. L., Norick, N. X. 1979. Relation of forest site and stand characteristics to Douglas-fir tussock moth outbreaks in California. *Bull. Soc. Entomol. Suisse* 52:297–307

172. Wolda, H. 1978. Fluctuations in abundance of tropical insects. *Am. Nat.* 112:1017–45

173. Wolda, H. 1978. Seasonal fluctuations in rainfall, food and abundance of tropical insects. *J. Anim. Ecol.* 47:369–81

174. Wolda, H. 1983. "Long term" stability of tropical insect populations. *Res. Popul. Ecol. Suppl.* 3:112–26

175. Young, A. M. 1979. Notes on a population outbreak of the beetle, *Uroplata* sp. (Coleoptera: Chrysomelidae) on the tree *Bunchosia pilosa* (Malpighiaceae) in Costa Rica. *J. NY Entomol. Soc.* 87:289–98

176. Zelinskaya, L. M. 1980. The role of microsporidia in the population dynamics of the gypsy moth (*Porthetria dispar* L.) in forest plantations in the lower Cis-Dneiper region. *Vestn. Zool.* 1:57–62 (In Russian)

Ann. Rev. Entomol. 1987. 32:341–60

INSECT PESTS OF SUGAR BEET

W. Harry Lange

Department of Entomology, University of California, Davis, California 95616

PERSPECTIVES AND OVERVIEW

In more than 40 countries, the sugar beet, *Beta vulgaris,* is an important agricultural asset in that it produces a pure food source, sucrose. World sugar production in 1984–1985, including sugar cane, was 97.5 million metric tons (MMT). In 1983–1984 the highest production of beet sugar was attained by the European community, with 11.36 MMT, followed by the USSR with 8.50 MMT and North America with 2.58 MMT. In 1983 the United States harvested 9,047.4 × 10³ acres of sugar beets and 771.2 × 10³ acres of sugar cane (132). World sugar beet production is shown in Table 1. The center of origin of the genus *Beta* is thought to be eastern Asia, and some of the wild relatives of the sugar beet, such as *Beta maritima,* have ranged along the Atlantic and North Sea coastlines (9). Wild beets were very resistant to drought, tolerant of rather high salt concentrations, and adaptable to many varied ecological situations. Some of these characteristics are present in the modern-day sugar beet. The sugar beet can grow in any state in the United States, but commercial production requirements limit it to about 13 states. In 1983 Minnesota harvested 257.0 × 10³ acres, California 171.0 × 10³ acres, North Dakota 144.0 × 10³ acres, and Idaho 143.0 × 10³ acres (132).

Research has been and continues to be the main reason that the sugar beet remains an important agricultural crop and can compete successfully with other sources of sugar. Sugar beet researchers had to produce a cultivar with not only satisfactory production qualities such as high sucrose content and nonbolting characteristics, but also resistance or tolerance to a continual succession of disease-producing organisms. In addition to seedling diseases, root and crown rots, leaf spots, virus diseases such as curly top, the yellowing virus complex, and beet mosaic, and about 150 or more insect and mite pests,

341

0066-4170/87/0101-0341$02.00

Table 1 World sugar beet production, 1979–1980[a]

Locality	Area planted (10^3 ha[b])	Yield (tons/ha)	Production (10^3 metric tons)
North America	479	43.1	20,629
South America	34	26.3	893
Western Europe	2,078	42.9	89,209
Eastern Europe	1,538	30.9	47,355
USSR	3,650	20.8	76,000
Asia	369	31.2	11,524

[a]Reference 133, p. 19.
[b]1 ha = 2.471 acres.

the list of pests has recently come to include powdery mildew, Erwinia root rot, lettuce infectious yellows, and rhizomania.

The most helpful review of insects that affect sugar beet is a volume with color plates published in several languages by the Delephanque Company in Paris, France, in collaboration with the Institut Technique de la Betterave Industrielle (ITB). The English text is distributed by Broom's Barn Experimental Station in England (36). The bulletin of Jones & Dunning (70) on sugar beets is practical and very complete. A series of important contributions covering insect pests and/or pertinent literature was initiated in 1903 by Chittenden (16) and includes work by Maxson (89), Lange (75, 76), and Meyerdick et al (93), as well as other important contributions of a more general nature (7, 19, 59). Several helpful publications dealing with sugar beet pests outside the United States include those of Lüdecke & Winner (87), Gram & Bovien (49), and Kheyri (73). The book edited by Nault & Rodriguez (96) covers much of the literature on leafhoppers. Russell's review of plant breeding for pest and disease resistance (113) not only covers most of the original references, but offers his ideas on the prospects of breeding for resistance.

This paper is not intended to cover the multitudinous contributions to the subject of sugar beet pests, but takes an encyclopedic approach that would assist nonexperts in the field to delve deeper into specific aspects of it. This review should also raise some new ideas on how to better and more economically solve some persistent sugar beet pest problems. The emphasis is primarily on California problems. With the average cost of production of sugar beets in California at $900/acre, it takes an average of 24 tons/acre and 15% sucrose to break even (48). In 1985, 204,000 acres were planted in California and 192,000 acres were harvested, with an average of 24.3 tons/acre. A two-ton decrease in 1985 from the harvest of the previous year has been attributed to an increase in virus yellows, a severe infestation by the bean aphid, *Aphis fabae* (115), high populations of *Empoasca* spp., and the presence of rhizomania.

SUGAR BEET ECOSYSTEM

A knowledge of the sugar beet ecosystem is essential to the development of pest control components necessary for an integrated pest management (IPM) program. In terms of many field crops beets could be considered a long-term cropping system, taking 6.5–15 months from time of planting to harvest. Spring plantings can be harvested in the fall of the same year or can be overwintered and harvested the next year. Fall plantings are harvested the following year in late summer or fall. The overwintering of sugar beets in some areas of California not only allows many insects and mites to overwinter, but creates a source of virus inoculum for new plantings. To correct this situation nine beet-free districts were developed in California to regulate planting and harvesting and thereby prevent proximity of old and new plantings.

Weeds are important in sugar beet production, as weeds not only are a source of viruses, but also harbor many of the same species of insects that infest sugar beets. This is particularly true for weeds in the family Chenopodiaceae, such as *Chenopodium* and *Amaranthus* spp., which harbor identical species of aphids, thrips, cutworms, leafhoppers and spider mites. The sugar beet ecosystem also allows for the presence and interaction of a large complex of beneficial insects. A knowledge of the species present and their roles in the ecosystem can be essential for deciding whether or not to use pesticides.

The production of sugar beets depends primarily on temperature and water. In the United States sugar beets grow best in a zone with summer isotherms between 67° and 72°F (9). The climatic adaptability of the sugar beet in the United States and other parts of the world means that the crop is exposed to many different complexes of insect pests. Even with the same or related insect pests it is usually necessary to tailor pest-management strategies to suit specific geographic areas.

INSECT AND MITE PESTS

Sugar beets are subject to attack by more than 150 species of insects and mites, and 40–50 of these species can cause economic damage. The species of most concern have been recorded by several investigators (7, 36, 70, 76, 81, 102). Like many agroecosystems, that of sugar beets has a few key or primary pests that may actually limit production under certain conditions. A few of these have worldwide distribution. In addition to the primary pests there are numerous species that cause periodic losses to sugar beets, and a few species that may occur with such low population levels that no serious damage occurs. In California at least 28 species of Lepidoptera feed on sugar beet, and a few cause serious damage. Open blade species include the beet armyworm, *Spodoptera exigua;* the clover cutworm, *Scotogramma trifolii;* the salt marsh caterpillar, *Estigmene acrea;* the western yellow-striped armyworm, *Spodop-tera praefica;* the beet webworm, *Loxostege sticticalis;* the spotted cutworm, *Amathes c-nigrum;* and the variegated cutworm, *Peridroma saucia.* Leaf rollers and folders include *L. sticticalis;* the omnivorous leafroller, *Platynota stultana;* the celery leaftier, *Udea rubigalis;* and the false celery leaftier, *Udea profundalis.* Crown and root feeders include *P. stultana; S. exigua;* the black cutworm, *Agrotis ipsilon;* the granulate cutworm, *Feltia subterranea;* and the rough skinned cutworm, *Proxenus mindara.* The actual complex of species and the numbers vary from locality to locality and year to year. A constant surveillance system is necessary in order to monitor fields for species and numbers present.

Insects attack the sugar beet crop from the time the seed is planted until harvest of roots or seed. At the time of planting, wireworms, springtails, seed-corn maggots, symphylans, root worms, soil mites, cutworms, army-worms, and others are often already present and feed on the germinating seeds or young seedlings (3, 70, 76, 106). Many of the pests of sugar beet in the western part of the United States are listed in Table 2.

Table 2 Pests of sugar beet in the western United States

Key pests	Pests of older plants
Green peach aphid, *Myzus persicae* (vector of beet yellows, beet western yellows, and beet mosaic)	Clover cutworm, *Scotogramma trifolii*
	Other lepidopterous larvae
Beet leafhopper, *Circulifer tenellus* (vector of curly top)	Leaf-mining flies, *Liriomyza* spp.
	Blotch miners, *Pegomya* spp.
Beet armyworm, *Spodoptera exigua*	Mites, *Tetranychus* spp.
	Bean and other thrips
	Lygus and stink bugs
Seedling pests	Ephydrid leaf miner, *Psilopa leucostoma*
	Leafhoppers, *Empoasca* spp.
Wireworms (*Limonius* spp., *Melanotus* spp., *Anchastus* sp.)	Aphids
	Blister beetles, *Epicauta* spp., *Lytta* spp.
Seed-corn maggot, *Hylemya platura*	Petiole borer, *Cosmobaris americana*
Springtails, *Onychiurus* spp.	Grasshoppers
Spinach crown mite, *Tyrophagus similis*	
Garden symphylan (garden centipede), *Scutigerella immaculata*	**Pests of beet seed production**
Root aphid, *Pemphigus populivenae*	Lygus bugs, *Lygus hesperus, L. elisus*
Rootworms, *Diabrotica* spp.	Beet armyworm
Cutworms and armyworms, *Agrotis* spp., *Proxenus* spp.	False chinch bug, *Nysius ericae*
	Omnivorous leaf tier, *Platynota stultana*
Beet crown borer, *Hulstia undulatella*	Other lepidopterous larvae
Webworms, *Loxostege* spp.	Leafhoppers, *Empoasca* spp.
Root maggot, *Tetanops myopaeformis*	Green peach aphid, *Myzus persicae*
Darkling ground beetles, *Blapstinus* spp.	Bean aphid, *Aphis fabae*
Bean aphid, *Aphis fabae* complex	Beet leafhopper
Other aphids	

APHID-BORNE VIRUSES

Beet yellows disease is often a limiting factor in the economic production of a sugar beet crop in many parts of the United States and Europe. In the United States it is produced by two aphid-borne viruses, beet yellows virus (BYV) and beet western yellows virus (BWYV) (64). In Europe it is produced by BWYV and a relative, beet mild yellows virus (BMYV), in addition to BYV (28, 29). Beet mosaic virus (BMV) is also aphid-borne. The chief vector of these aphid-borne viruses is the green peach aphid, *Myzus persicae,* although 14 other species of aphids can play a part in their transmission. These aphid-borne viruses have strains or isolates that range in virulence from mild to severe. These strains may be host-oriented, may need "helper" viruses to produce successful infections, or may be transmitted only by certain aphid species or aphid biotypes (119). Such strain differences are commonly found in the luteovirus group to which BWYV and BMYV belong (108).

Virus Vectors

Myzus persicae has worldwide distribution and is a vector of over 100 viruses. It is the most efficient vector of the yellowing viruses and beet mosaic, with a host range of several hundred plant species. The bean aphid, in the *Aphis fabae* complex, is a vector of about 29 viruses (72) including BYV, beet yellow net (BYN), and BMYV in Europe (130). *A. fabae* has 200 host plants in 25 plant families. It is generally not as good a vector of sugar beet viruses as *M. persicae,* but it injects a toxin into sugar beet foliage, causing stunting, curling, yellowing of the leaves, and even death of the plants.

The symptoms of BYV are partial to complete yellowing of the leaves followed by development of necrotic tissue in some cases and a thickening of the leaves. The leaves feel leathery and are brittle to the touch. With resistant beets there may be light green vein banding and a slight lessening of the normal green color. Mild BYV, BMYV, and WBYV are often difficult to separate in the field. A typical mosaic pattern of light and dark areas is typical of BMV.

Biology and Ecology

M. persicae has both holocyclic and anholocyclic life histories (101, 127). In northern California, as in Washington state, eggs overwinter on peach trees (nonhosts of the beet viruses). Resulting spring alates fly to weed hosts, pick up the virus (BWYV), and carry it to sugar beet fields (101, 123). Because weeds play an essential part as a source of virus, weed control is of prime importance in suppressing BWYV (135). In areas of less severe winter temperatures *M. persicae* has no sexual cycle, and continuous generations live on weed hosts and in commercially grown crops such as sugar beets. Some biotypes or strains of *M. persicae* are apparently unable to produce sexual forms or may produce only a few males in the fall. In central California some strains lay eggs in the fall on peach trees, but there is no indication of survival. *A. fabae* also has a sexual cycle in California, and lays eggs on tree hosts such as *Euonymus*. However, it can survive most winters asexually on its many weed hosts or in overwintering sugar beet fields or other agriculturally grown crops. The biology, ecology, and host relationships of *M. persicae* have been presented in depth in a series of publications by Tamaki and his associates in Washington state (124–128).

Intraspecific Variation

One form of variation in aphids is the occurrence of clones or strains that seem to differ genetically (119). These clones can be maintained in laboratories for years without losing their disease-transmitting capabilities, ability to colonize certain host plants, behavior, or other characteristics. Lowe (83) collected 58 *M. persicae* clones, adapted them to Chinese cabbage, *Brassica pekinensis,* and used them to test for resistance in breeding sugar beet aphid-resistant stock. W. H. Lange (unpublished) collected 26 clones of *M. persicae* from weeds and sugar beets in widely separated areas of California, adapted them for two years on *B. pekinensis,* and determined their ability to transmit BYV by exposing them to indicator plants. Nine of the clones adapted readily to *B. pekinensis*. Two of these clones, one a good vector of BYV and the other a poor vector, were maintained for another three years. The vectoring ability of the clones seemed stable as long as they were maintained (Table 3).

A common species on sugar beet is the potato aphid, *Macrosiphum euphorbiae*. Two strains occur on sugar beets, pink and green. The pink form readily selects sugar beet and many weeds in the family Chenopodiaceae, and

Table 3 Beet yellows virus transmission by two selected clones of green peach aphid in a greenhouse test

Clone	Test plant	Total number of aphids in test	Number of aphids transmitting virus	% Transmission
19[a]	*Chenopodium capitatum*	100	55	
	Sugar beet	$\frac{100}{200}$	$\frac{30}{85}$	42.5
11[b]	*Chenopodium capitatum*	100	16	
	Sugar beet	$\frac{100}{200}$	$\frac{2}{18}$	9.0

[a]Originally from *Brassica* sp., Eureka, California; then 5 yr on *B. pekinensis* prior to this test.
[b]Originally from *Malva parviflora*, Santa Maria, California; then 5 yr on *B. pekinensis* prior to this test.

it is a pest of tomato, but it does not occur on potato. It increases during the warmest times of the year and aggregates more than the green form, which has a different preferential host range and is most common at cooler times of year.

It appears that the differences in ability of *M. persicae* strains to transmit BYV are behavioristic, in that certain strains appear to feed in the parenchymous tissues instead of in the phloem (15, 51, 52). Lowe & Russell (86) found no indication that inhibition of probing causes varietal resistance to aphids in sugar beets.

Integrated Pest Management

Yellowing disease of sugar beet was identified in California in 1951, although records indicate that it was probably present at an earlier date. Serious losses in 1957–1958 instigated research that culminated in a pest management program for the yellowing viruses that began in 1968 and is still in operation. In 1984–1985 an outbreak of the yellows disease together with an unprecedented increase of *A. fabae,* universal beet mosaic, and high numbers of *Empoasca* leafhoppers resulted in decreases in root yields of 4–8 tons/acre and a drop of several percentage points in sucrose content (48).

In 1968–1969 ten beet-free districts (26) were set up in California with the cooperation of the California Beet Growers Association, processors, and growers, to separate older beets (virus sources) from new plantings. In districts where a conflict exists between overwintered and spring-planted beets, an isolation of 5 mi is enforced. The value of the beet-free program was shown in a 12-year survey taken in northern California (Table 4). The 1985 outbreak of yellowing disease was attributed to increased flights of *M. persicae* and *A. fabae* bean aphid and to the planting of new beets within 5 mi of virus sources (old beets).

It now seems that because of epidemiological differences between BYV, BWYV, BMYV, and BMV, they will have to be managed in many instances as separate entities. With the suppression of BYV there could be a higher incidence or more virulent strains of BWYV in the United States and BMYV in Europe (34, 110). Usually BMV can be managed with measures similar to those used to suppress BYV. In England BMV incidence declined as control for BYV became effective (55). The yellowing disease of sugar beets stimulated international cooperation in the management of pests and diseases; in 1931 the I.I.R.B., the International Institute for Sugar Beet Research, was formed in Belgium (33). In 1972 Dunning (32), under the auspices of I.I.R.B., summarized information from 20 European countries, including excellent statistics on the pests, use of pesticides, amount of crop damage, and current problems. The virus yellows problem and the green peach and

Table 4 Summary of 12-yr survey (1968–1981) taken in Glenn, Butte, and Tehama counties (northern California) on the role of overwintering in sugar beets and the extent of viruses and root yields[a]

Year	Overwintering within 5 mi	% Yellows	% Beet mosaic	Tons/acre (fall harvest)
1968	N	5	0	24.3
1969	Y	76	54	16.1
1970	N	5	1	20.8
1971	N	33	4	19.7
1972	N	16	1	25.9
1973	Y[b]	15	8	21.2
1974	Y[c]	6	1	21.0
1975	N	11	2	23.3
1976	N	28	3	28.0
1977	N	31	5	22.9
1978		No survey made		
1979		No survey made		
1980	Y[d]	52	36	22.3
1981	N[d]	24	7	22.7
12-yr average				22.4

[a]Surveys by R. G. Sailsbery, Cooperative Extension Service.
[b]Large areas treated with insecticide.
[c]Only two partial fields overwintered.
[d]Glenn and Butte Counties; Tehama county now in overwintered area.

bean aphids were listed among the most damaging pests. IPM procedures have been developed in many countries to combat yellows disease of sugar beets. Successful programs have been reported from England, France, Germany, and the Netherlands (32, 34, 35, 37, 50, 54, 55, 57, 68, 69, 84, 110). In the United States IPM systems have been established in California, Oregon, Washington, and the Central Plains States (26, 64, 79, 80, 107, 126, 140).

To develop an IPM approach one must not only understand the ecology of the aphid vectors of aphid-borne viruses, but one must also be able to predict population increases, monitor populations, and institute warning systems so growers can modify their planting dates, utilize pesticides, or take other action rather than sustain losses. Weather conditions are one obstacle to prediction-making, as mild winters, adverse temperatures, and moisture all influence aphid buildup (40, 101, 134).

In California the IPM program that was successful in controlling virus yellows for about 17 years included many recommendations similar to those of European programs: (*a*) Plant resistant seed. (*b*) Avoid virus sources and follow the "beet-free" recommended planting and harvesting dates for each district. (*c*) Avoid peaks in aphid flights. (*d*) Practice good cultural methods such as crop cleanup following harvest, weed control, and proper irrigation, fertilization, and spacing. (*e*) Use pesticides judiciously. (*f*) Protect natural enemies when possible by using systemic granular insecticides. (*g*) Watch for resurgences of minor pests, and monitor insect populations on beets during the season and particularly during the early developmental period.

Duffus (24, 25, 27) has followed the losses and recoveries in production of sugar beets related to aphid-borne viruses. It is difficult to estimate total losses, but until the IPM program was initiated losses ran into millions of dollars yearly. A combined research effort is credited for the recovery from the 1958–1959 epidemic. The development of resistant or tolerant cultivars, the application of pesticides, the timing of planting to avoid peak flights of aphids, and better knowledge of virus sources all played a part in making sugar beets a profitable crop in California. USDA plant breeders are constantly working on the development of resistant varieties for better production and disease tolerance (91, 92). Russell (109) has published extensively on breeding for resistance. Resistance to aphid feeding, deterrence of colonization habits, and reduced attractiveness to winged forms are all possible characteristics for tolerance in sugar beet breeding lines. These characteristics also include resistance to probing of leaf tissues (86) and genetic variation that could affect host selection (83, 85).

The fact that we not only have strains of aphids involved in virus transmission, but also strains of the viruses, makes field identification difficult. Several methods of detection have been used: indicator plants, a study of the aphid gut, serological techniques, and enzyme-linked immunosorbent assays (17, 39). Smith & Hinckes (118) determined how to follow certain viruses in the field to identify the speed of movement of a virus strain carried by particular aphids from a known virus source. Thielemann & Nagi (131) found that there were no differences in symptoms of BMYV transmitted by different strains of *A. fabae,* but symptoms occurred in 10–14 days with *M. persicae* and in 10–14 wk with *A. fabae.*

Warning systems to alert growers of impending flights of aphids are highly organized in Europe, and are particularly helpful for timing planting to avoid flights or for properly timing the application of pesticides (122, 129, 136, 137). Water traps were used in California to keep growers in one locality informed daily of *M. persicae* flights. They were also used in a 10-yr study of the flight patterns of about 25 species of aphids in more than 10 localities in California. The mean flight patterns for each area were used in determining planting and harvesting dates in the formation of the "beet-free district" strategy. Although peak flights varied from year to year, characteristic flight patterns became evident. Flight patterns for four areas in California are shown in Figure 1. In Europe, potato plants are used as traps to determine flight activity of winged aphids in sugar beet fields (56).

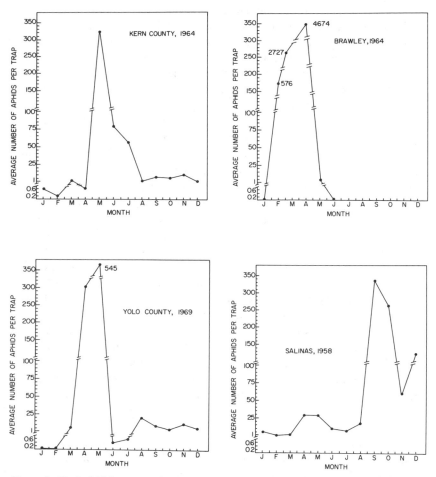

Figure 1 Typical flight patterns for the green peach aphid, *Myzus persicae,* captured in yellow water pan traps in four California localities (after 79).

Insecticides

Properly applied insecticides have proven successful for controlling the aphids vectoring the yellows viruses (11, 38, 78). Early work in California (1961–1971) indicated a root-yield increase of 3–9.5 tons/acre when oxy-demeton-methyl (Metasystox-R®) was sprayed at 7–10 day intervals, with 0.5 apterous aphids/plant used as the action point. The sprays were also timed to bracket a decreased flight period of the green peach aphid. Thus in northern California planting was delayed until about May 1. The usual high temperatures in June, ranging from 85°F to over 100°F, caused an almost immediate decrease in green peach aphid populations owing to a marked decrease in multiplication of the aphid (see Figure 1, Yolo County) (60–63, 77, 79). In 1969 (W. H. Lange, unpublished) the application of insecticides largely prevented spread of viruses by aphids within the fields. In many years a low percentage of aphids brought in the viruses from outside sources. We collected 223 aphids in a beet field and tested them individually on indicator plants for BWYV and BYV. BWYV tests were positive for 15 of the 101 *M. persicae* (14.9%) and 1 of the 6 *Aphis craccivora* (16.7%), but none of the 13 other species represented contained BWYV. A similar group of 173 aphids (86 *M. persicae*) was tested for BYV; none was recovered from the 14 species represented.

In Europe insecticides have been used for the suppression of viruses and direct damage by the bean aphid (32). In 1953 Hull (66–68) reported insecticidal control of *Aphis fabae* for about 340,000 acres. Systemic granules of phorate and aldicarb, applied in the seed rows or side-dressed into the beds, have been used successfully in many parts of the world (8). In California tests in 1971 (79) aldicarb and phorate gave an increase in root yield of 3–4 tons/acre with single time-of-planting applications. A factorial experiment in 1968 (60) identified improvement in root yield as due 13% to insecticide, 14% to resistant variety, and 39% to resistant variety plus insecticide. Several investigators have found that the use of mineral oil alone and with an insecticide can interfere with the transmission of viruses by aphids (44, 111). Broadbent (10) published an early review of the use of insecticides in the spread of viruses.

Pyrethroids are now being tested for aphid control. Elliott & Janes (41) discussed the difficulties of tailoring a specific pyrethroid to fit a specific pest problem. Other materials that are being investigated for possible use on sugar beets, or that have been used, include thiofanox, benzamidazole, ethiofencarb, and terbufos (2, 112, 114, 117). Owing largely to resistance problems, the potential for alternative chemical methods of preventing virus transmission remains of interest to many investigators (97, 105). Pirimicarb is one answer to the problem, as it is effective against some resistant strains of aphids. The aphid pheromone, (E)-β-farnesene, and carboxylic acids are being investigated (20, 21, 99).

A LEAFHOPPER-BORNE VIRUS: CURLY TOP

For many years curly top disease has caused severe and widespread losses of sugar beets, tomatoes, beans, cucurbits, and several other crops in the arid regions of California and other western states. While the host range of the virus includes some 300 species of wild and cultivated plants in 44 families, the only known natural vector in the United States is the beet leafhopper, *Circulifer tenellus*. In Turkey *Circulifer opacipennis* is a vector. These insects are necessary for dissemination and probably perpetuation of the disease, since the virus is neither seed-transmitted nor readily transmitted mechanically (4, 5, 23, 120). The 1979 report by Oldfield & Kaloostian (98) of *C. tenellus* as a vector of citrus stubborn disease, *Spiroplasma tenella*, has focused even more attention on this insect. Hundreds of papers have been written on the beet leafhopper, but I mention a few outstanding contributions. The 1983 bibliography by Meyerdick et al on *C. tenellus* and *S. tenella* is excellent (93). Bennett's classical monograph on the curly top disease includes 392 references (6). Carter's article on ecological studies of the beet leafhopper is a model of thorough research (14). Severin's 37 contributions to the ecology and transmission of curly top covered the period from 1919 to 1947 (cited in 6).

Many other species of leafhoppers occur on sugar beets. Small green leafhoppers in the genus *Empoasca* include several species on sugar beet: *E. abrupta*, *E. arida*, *E. mexara*, *E. fabae*, and *E. solana*. The latter two are the most important economically. Damage from *Empoasca* often resembles virus symptoms and can cause decreases in yield of sugar beets, cotton, and other crops (74, 94, 100, 103, 104). The leafhopper *Aceratogallia calcaris* is a vector of yellow vein disease of sugar beet (121).

Economic Importance

Since the advent of the California State Department of Agriculture beet leafhopper control program in the 1940s, a program of spraying and weed host management in an area on the west side of the San Joaquin Valley, losses to agricultural crops have been kept to a minimum (22, 71, 138). The United States Department of Agriculture estimated an annual loss of 1% due to curly top in susceptible crops for the period 1951–1960. Without an effective control program it is estimated that in certain years 10–40% of susceptible crops could be lost (138). New strains of curly top with increased virulence can account for increased losses in the future (5, 30, 31, 45, 46). An enzyme-linked immunosorbent assay can now be used to identify beet leafhopper populations carrying beet curly top (95).

Biology and Ecology

The biology and ecology of *C. tenellus* have been investigated by many researchers (e.g. 18, 88, 116, 139). Harries & Douglas (53) determined that the egg stage ranges from 5.5 days at 37.8°C to 43.8 days at 15.6°C. The nymphal stage consists of five instars that last from 13.0 days at 35°C to 75.4 days at 18.3°C. The authors established that the threshold temperature for development was 14.4°C and that 111.1 day-degrees were required for the egg stage and 250.0 day-degrees for nymphal development. Yokomi (138) investigated the thermal requirements for development of the beet leafhopper and found that 13.0°C was minimal; 414.6 day-degrees were required for male development and 425.7 day-degrees for female development from emergence to oviposition. A total life history time from oviposition to oviposition was 505.6 day-degrees. Cook (18) reported three broods of leafhoppers a year and possibly five in some southern localities. Symptoms and host plants are given in Bennett's monograph (6).

In the fall *C. tenellus* migrates from the floor of the San Joaquin Valley to Russian thistle, *Salsola* spp., along the western perimeter of the valley. As the thistle senesces in the fall and early winter, the leafhoppers move to perennial holdover plants such as *Atriplex polycarpa, Lepidospartum squamatum,* or *Gutierrezia* spp. After rainfall *C. tenellus* can move to the annual weed hosts on the slopes, such as *Erodium* spp., *Lepidium* spp., and *Plantago* spp. This makes the leafhoppers amenable to sampling and to control by insecticide application (1). Lawson et al (82) and Yokomi (138) studied the movement of leafhoppers from the annual host plants into cultivated areas. In a 3-yr study (138), Yokomi found that the triggering mechanism for dissemination was drying of host plants in the foothill breeding areas. Knowledge of the time of dispersion of leafhoppers to the valley and into cultivated crops is very important, as control measures must be applied prior to the insects' arrival in the crops.

Control

Economic losses arise in a multitude of ways. Weakening or stunting of established plants significantly reduces yields. A loss of seedlings may necessitate use of additional seed, production of the crop with missing plants, or the costly mechanical replanting of entire fields. Options for crop selection may be reduced based on the dissemination patterns of viruliferous leafhoppers. Fields located near breeding grounds are poor risks if planted with susceptible crops. Insecticides add to direct costs and to environmental contamination, and do not provide effective control of the disease when applied as sprays. Systemic materials such as phorate and aldicarb have been used in sugar beets for leafhopper control; experimental work at the Westside Field Station at Five Points, California, has shown as high as a 26–28% average increase in yield over a 6-yr period (Table 5) (12, 107).

As already mentioned, the state of California's beet leafhopper control program involves all aspects of control, including insecticide application if populations reach economic threshold levels. Work on host resistance still continues as more potent strains of the virus appear (13, 42, 47, 90). Natural enemies have long been sought for control of the beet leafhopper. In Idaho, Henderson (58) has attempted to use two egg parasites, *Aphelinoidea plutella* and *Abbella subflava*. California workers (43, 65) have attempted to colonize five species of introduced trichogrammatid and mymarid egg parasites and have made surveys of natural egg parasites for possible releases.

Table 5 Results of chemical treatments with resistant (R) and susceptible (S) varieties at Westside Field Station, Five Points, California[a,b]

		% Increase in root yield over untreated sugar beets					
		Phorate[c]		Aldicarb[d]		Oxydeme-tonmethyl[e]	
Year	% Curly top	R	S	R	S	R	S
1972	93	31	55	17	22	—	—
1973	50	51	50	30	32	45	45
1974	43	19	25	17	22	22	22
1975	38	6	0	6	2	6	8
1976	25	17	5	17	11	8	0
1977	86	34	37	25	26	10	13
Mean	56	26	28	19	19	18	18

[a]R = US10B, S = S101.
[b]W. H. Lange, unpublished.
[c]1.5 lb/acre, active as granule, 1×.
[d]2.0 lb/acre, active as granule, 1×.
[e]0.5 lb/acre, active as spray, 3×.

CONCLUSIONS

Modern sugar beet production is the product of research. Private industry and federal and state governments have all cooperated to solve many and almost constant production problems. Insect and disease problems have often threatened the existence of the industry, but in each case cooperative research has come to the rescue. Worldwide conferences and organizations have assisted by making research contributions available to almost all of the countries that produce sugar beets. Costs of production continue to increase, and as this trend continues the producer will need more economical, safe, and effective means to protect these plants from insects and plant diseases.

In some countries older types of pesticides cannot be used owing to contamination problems and increased knowledge of the dangers involved in use of the chemicals. We will need alternative approaches to the wide use of pesticides. Development of resistant sugar beets would curtail the need for excessive amounts of pesticides. New and safer chemicals for control of pests, e.g. insect growth regulators or repellents such as aphid alarm pheromones, may replace conventional types of chemicals.

More concerted effort is needed to determine the natural complex of beneficial insects in the sugar beet ecosystem and the role of introduced predators and parasites. Changes in the virus complexes in sugar beets in many areas of the world strengthen the need for proper identification of the strains of viruses and their sources. The use of serological techniques and enzyme-linked immunosorbent assay methods to determine the presence and movement of specific viruses may enhance our knowledge of transmission. More work on the genetics of the insects is needed to elucidate differences in virus transmission among aphid clones and biotypes. Cultural aspects of pest and disease management, such as the replacement of hosts in some areas with nonhosts or less desirable hosts, may assist in solving some of the pest problems.

ACKNOWLEDGMENTS

The author thanks fellow researchers J. E. Duffus and F. J. Hills, who assisted with literature and comments during the preparation of this paper. Ben Goodwin and James Scharf of the California Beet Growers Association furnished statistical information. Andrew Dunning of Brooms Barn Experiment Station in England assisted with literature and unpublished comments. Mike Brewer and Hans Rocke assisted in the review of literature. Louise Hope provided invaluable help in the preparation and editing of the manuscript.

Literature Cited

1. Armitage, H. M. 1952. Controlling curly top virus in agricultural crops by reducing populations of overwintering beet leafhoppers. *J. Econ. Entomol.* 45:432–35
2. Aston, J. L., Sisto, A. M., Martin, R., Tayler, P. 1975. Recent experiences with terbufos against sugar beet pests in Europe. *Proc. 8th Br. Insectic. Fungic. Conf., Brighton,* 2:477–88. Nottington, UK: Boots
3. Baker, A. N., Dunning, R. A. 1975. Association of populations of onychiurid Collembola with damage to sugar-beet seedlings. *Plant Pathol.* 24:150–54
4. Ball, E. D. 1909. The leafhoppers of the sugar beet. *USDA Bull.* 66:1–52
5. Bennett, C. W. 1967. Apparent absence of cross-protection between strains of the curly top virus in the beet leafhopper, *Circulifer tenellus. Phytopathology* 57:207–9
6. Bennett, C. W. 1971. The curly top disease of sugarbeet and other plants. *Am. Phytopathol. Soc. Monogr.* 7:1–81
7. Blickenstaff, C. C. 1976. Sugarbeet insects: How to control them. *USDA Farmer's Bull.* 2219:1–20
8. Blickenstaff, C. C. 1982. The effect of aldicarb on sugarbeet insects and yield. *Agric. Res. Results, No. ARR-W-23.*

23 pp. Washington, DC: USDA Agric. Res. Serv.
9. Brandes, E. W., Coons, G. H. 1941. Climatic relations of sugarcane and sugar beet. In *USDA Yearbook of Agriculture, Climate and Man,* pp. 421–28. Washington, DC: USDA
10. Broadbent, L. 1957. Insecticidal control of the spread of plant viruses. *Ann. Rev. Entomol.* 2:339–54
11. Bryan, K. M. G. 1979. Control of soil pests of sugar beet with a granular formulation of bendiocarb. *Proc. Br. Crop. Prot. Conf.* 1979:231–37
12. Burtch, L. 1968. The use of insecticides for curly top control. *Spreckels Sugar Beet Bull.* 32:14–15, 24
13. Carsner, E. 1933. Curly top resistance in sugarbeets and tests of the resistant variety USI. *USDA Tech. Bull.* 360:1–68
14. Carter, W. 1930. Ecological studies of the beet leafhopper. *USDA Tech. Bull.* 206:1–116
15. Chang, V. C. 1968. *Intraspecific variation in the ability of the green peach aphid to transmit sugar beet virus yellows.* PhD thesis. Univ. Calif., Davis
16. Chittenden, F. H. 1903. Principal insect enemies of the sugar beet. *USDA Div. Entomol. Bull.* 43:1–71
17. Clark, M. F., Adams, A. N. 1977.

Characteristics of the microplate method of enzyme-linked immunosorbent assay for the detection of plant viruses. *J. Gen. Virol.* 34:475–83

18. Cook, W. C. 1967. Life history, host plants, and migrations of the beet leafhopper in the western United States. *USDA Tech. Bull.* 1365:1–122

19. Coons, G. H. 1953. Some problems in growing sugar beets. In *Plant Diseases/ The Yearbook of Agriculture 1953*, pp. 509–24. Washington, DC: USDA

20. Dawson, G. W., Griffiths, D. C., Pickett, J. A., Smith, M. C., Woodcock, C. M. 1984. Natural inhibition of the aphid alarm pheromone. *Entomol. Exp. Appl.* 36:197–99

21. Devonshire, A. L. 1977. Properties of a carboxylesterase from the peach-potato aphid *Myzus persicae* (Sulz.) and its role in conferring insect resistance. *Biochem. J.* 167:675–83

22. Dilley, D. R., Morrison, A., Hise, J. 1971. Curly top virus control program. *Calif. Sugar Beet* 1971:32–33

23. Douglas, J. R., Cook, W. C. 1954. The beet leafhopper. *USDA Circ.* 942:1–21

24. Duffus, J. E. 1960. Radish yellows, a disease of radish, sugar beet, and other crops. *Phytopathology* 50:389–94

25. Duffus, J. E. 1961. Economic significance of beet western yellows (Radish yellows) on sugar beet. *Phytopathology* 51:605–7

26. Duffus, J. E. 1977. Beet free periods— the key to higher sugarbeet yields. *Calif. Agric.* 31:18–19

27. Duffus, J. E. 1978. The impact of yellows control on California sugarbeets. *J. Am. Soc. Sugar Beet Technol.* 20:1–5

28. Duffus, J. E., Russell, G. E. 1970. Serological and host range evidence for the occurrence of beet western yellows virus in Europe. *Phytopathology* 60:1199–202

29. Duffus, J. E., Russell, G. E. 1975. Serological relationship between beet western yellows and beet mild yellowing viruses. *Phytopathology* 65:811–15

30. Duffus, J. E., Skoyen, I. O. 1976. Beet curly top damage. Losses from new strains. *Calif. Sugar Beet* 1976:42–44

31. Duffus, J. E., Skoyen, I. O. 1977. Relationships of age of plants and resistance to a severe isolate of the beet curly top. *Phytopathology* 67:151–54

32. Dunning, R. A. 1972. Sugar beet pest and disease incidence and damage, and pesticide usage. *J. Inst. Rech. Better.* 6:19–34

33. Dunning, R. A. 1975. International cooperation in the development of control of pests and diseases of sugar beet. See Ref. 2, pp. 1013–19

34. Dunning, R. A. 1985. Integrated control of sugar-beet virus yellows: current practices and future prospects. *Inst. Int. Rech. Better. Congr. Hiver C. R.* 48:173–90

35. Dunning, R. A., Byford, W. J. 1979. Weed, disease and pest control: costs, profitability and possible improvements for certain programmes. Part II: Disease and pest control. *Inst. Int. Rech. Better. Congr. Hiver C. R.* 42:85–103

36. Dunning, R. A., Byford, W. J. 1982. *Pests, Diseases and Disorders of the Sugar Beet.* Paris: Deleplanque & Cie. 167 pp.

37. Dunning, R. A., Heijbroek, W. 1981. Improved plant establishment through better control of pest and disease damage. *Inst. Int. Rech. Better., Congr. Hiver C. R.* 44:37–58

38. Dunning, R. A., Winder, G. H. 1976. Seed-furrow application of granular pesticides and their biological efficiency on sugar beet. *Br. Crop Prot. Counc. Monogr.* 18:37–45

39. Edwards, J. S. 1965. On the use of gut characters to determine the origin of migrating aphids. *Ann. Appl. Biol.* 55:485–94

40. Elliott, W. M. 1973. A method of predicting short term population trends of the green peach aphid, *Myzus persicae* (Homoptera: Aphididae), on potatoes. *Can. Entomol.* 105:11–20

41. Elliott, W. M., Janes, N. F. 1983. Evolution of pyrethroid insecticides for crop protection. *10th Int. Congr. Plant Prot., Brighton, England,* 1:216–23. Croydon, England: Br. Crop Prot. Counc.

42. Finkner, R. E. 1976. Cultivar blends for buffering against curly top and leafspot disease of sugar beet. *J. Am. Soc. Sugar Beet Technol.* 19:74–82

43. Flock, R. A., Doutt, R. C., Dickson, R. C., Laird, E. F. 1962. A survey of beet leafhopper egg parasites in the Imperial Valley, California. *J. Econ. Entomol.* 55:277–81

44. Fritzsche, R., Proeseler, G., Thiele, S., Geissler, K., Dubnik, H., Kramer, W. 1979. Insektizid-Mineralölkombination zur Senkung von Virusinfektionen durch Blattläuse im Zuckerrübenund Pflanzkartoffelanbau. *Nachrichtenbl. Pflanzenschutzdienst DDR* 33:177–79

45. Giddings, N. J. 1950. Combinations and separation of curly-top virus strains. *Am. Soc. Sugar Beet Technol. Proc.* 6:502–7

46. Giddings, N. J. 1950. Some inter-

relationships of virus strains in sugar-beet curly top. *Phytopathology* 40:377–88

47. Giddings, N. J. 1954. Relative curly-top resistance of sugar beet varieties in the seedling stage. *Am. Soc. Sugar Beet Technol. Proc.* 8(Part 1):197–200

48. Goodwin, B. 1985. Directors report. *Calif. Sugar Beet* 1985:9

49. Gram, E., Bovien, P. 1944. *Rodfrugternes Sygdomme og Skadedyr.* Copenhagen: Danske Forlag. 125 pp.

50. Gut, J. 1980. Aphids in sugar beet. In *Integrated Control of Insect Pests in the Netherlands,* ed. A. K. Minks, P. Gruys, pp. 71–74. Wageningen: Cent. Agric. Publ. Doc.

51. Haniotakis, G. E. 1971. *The nature of beet yellows virus resistance in the sugar beet.* PhD thesis. Univ. Calif., Davis. 146 pp.

52. Haniotakis, G. E., Lange, W. H. 1974. Beet yellows virus resistance in sugar beets: mechanism of resistance. *J. Econ. Entomol.* 67:25–28

53. Harries, F. H., Douglas, J. R. 1948. Bionomic studies on the beet leafhopper. *Ecol. Monogr.* 18:45–79

54. Heathcote, G. D. 1972. Influence of cultural factors on incidence of aphids and yellows in beet. *J. Inst. Int. Rech. Better.* 6:6–14

55. Heathcote, G. D. 1973. Beet mosaic—a declining disease in England. *Plant Pathol.* 22:42–45

56. Heathcote, G. D., Engsbro, B., Häni, A., Månsson, B. 1982. Potato plants used as traps to determine when winged aphids which can carry yellowing viruses invaded some European sugar-beet crops, 1975–1977. *Inst. Int. Rech. Better. Congr. Hiver C. R.* 45:239–48

57. Heijbroek, W., van der Bund, C. F., Maas, P. W. T., Maenhout, C. A. A. A., Simons, W. R., Tichelaar, G. M. 1980. Approaches to integrated control of soil arthropods in sugar-beet. See Ref. 50, pp. 83–85

58. Henderson, C. F. 1955. Overwintering, spring emergence and host synchronization of two egg parasites of the beet leafhopper in southern Idaho. *USDA Circ.* 967:1–16

59. Hills, F. J., Johnson, S. S. 1973. The sugar beet industry in California. *Univ. Calif. Agric. Exp. Stn. Circ. 562,* Div. Agric. Sci., Berkeley. 14 pp.

60. Hills, F. J., Lange, W. H., Kishiyama, J. 1969. Varietal resistance to yellows, vector control, and planting date as factors in the suppression of yellows and

mosaic of sugar beet. *Phytopathology* 59:1728–31

61. Hills, F. J., Lange, W. H., Loomis, R. S., Reed, J. L., Hall, D. H. 1964. Sugar beets damaged by yellows viruses improved by aphid control. *Calif. Agric.* 18:11

62. Hills, F. J., Lange, W. H., Reed, J. L., Hall, D. H., Loomis, R. S. 1962. Response of sugar beet to date of planting and infection by yellows viruses in northern California. *J. Am. Soc. Sugar Beet Technol.* 12:210–15

63. Hills, F. J., Lange, W. H., Reed, J. L., Loomis, R. S. 1966. Aphid control and planting date for the control of yellows of sugar beet. *J. Am. Soc. Sugar Beet Technol.* 14:118–26

64. Hills, F. J., Lange, W. H., Shepherd, R. J., McFarlane, J. S. 1982. Sugarbeet pest management: Aphid-borne viruses. *Univ. Calif. Spec. Publ. 3277,* Div. Agric. Sci., Berkeley. 12 pp.

65. Huffaker, C. B., Holloway, J. K., Doutt, R. G., Finney, G. L. 1954. Introduction of egg parasites of the beet leafhopper. *J. Econ. Entomol.* 47:785–89

66. Hull, R. 1969. Brooms Barn Experimental Station. *Rothamsted Exp. Stn. Rep. 1969 Pt. I,* p. 308. Harpenden, UK: Rothamsted Exp. Stn.

67. Hull, R. 1971. Conclusions on the symposium on integrated control of pests and diseases of sugar beet. *J. Inst. Int. Rech. Better.* 5:191–98

68. Hull, R. 1974. Integrated control of pests and diseases of sugar beet. *13th Symp. Br. Ecol. Soc., Oxford, England, 1972,* pp. 269–76. New York/Oxford: Wiley/Blackwell

69. Jadot, R. 1976. Aspects des epidemies de jaunisse et de mosaique de la betterave. I. Symptomes, modes de transmission, methodes d'identification, vecteurs. II. Sources d'infection, allure des epidemies, lutte. *Rev. Agric. Brussels* 29:555–82, 843–76

70. Jones, F. G. W., Dunning, R. A. 1972. Sugar beet pests. *Minist. Agric., Fish. Food Bull. 162,* London. 133 pp.

71. Jones, M. 1956. Regulatory control projects. *Bull. Calif. State Dep. Agric.* 45:157–59

72. Kennedy, J. S., Day, M. F., Eastop, V. F. 1962. *A Conspectus of Aphids as Vectors of Plant Viruses.* London: Commonw. Inst. Entomol. 114 pp.

73. Kheyri, M. 1966. *The Important Pests of Sugarbeet in Iran.* Tehran: Sugarbeet Seed Inst., Sugarbeet Pest Res. Cent. Karadj. 76 pp.

74. Lamp, W. O., Morris, M. J., Armbrust, E. J. 1984. Suitability of common weed species as host plants for the potato leafhopper, *Empoasca fabae*. *Entomol. Exp. Appl.* 36:125–31

75. Lange, W. H. 1947. Sugarbeet insects and nematodes and their control. *Spreckels Sugar Beet Bull.* 11:1–8

76. Lange, W. H. 1971. Insects and mites of sugar beet and their control. In *Advances in Sugarbeet Production: Principles and Practices,* ed. R. T. Johnson, J. T. Alexander, G. E. Rush, G. R. Hawkes, pp. 287–333. Ames, Iowa: Iowa State Univ. Press

77. Lange, W. H., Hills, F. J. 1966. Virus suppression on sugar beets by means of properly timed insecticides. *Calif. Sugar Beet* 1966:49–51

78. Lange, W. H., Hills, F. J., Loomis, R. S., Kishiyama, J. 1967. Early aphid control increases beet production. *Calif. Agric.* 21:14–15

79. Lange, W. H., Hills, F. J., Shepherd, R. J. 1971. Yellows control. *Calif. Sugar Beet* 1971:26–29

80. Lange, W. H., Kishiyama, J. S., Hills, F. J. 1978. Integrated pest management of sugar beet insects. *Calif. Agric.* 32:21–22

81. Lange, W. H., Suh, J. B. 1980. Leaf-feeding and other caterpillars of sugar beet. *Calif. Sugar Beet* 1980:35–36

82. Lawson, F. R., Chamberlin, J. C., York, G. T. 1951. Dissemination of the beet leafhopper in California. *USDA Tech. Bull.* 1030:1–59

83. Lowe, H. J. B. 1974. Intraspecific variation of *Myzus persicae* on sugar beet *(Beta vulgaris)*. *Ann. Appl. Biol.* 78:15–26

84. Lowe, H. J. B. 1975. Crop resistance to pests as a component of integrated control systems. See Ref. 2, 1:87–92

85. Lowe, H. J. B., Russell, G. E. 1969. Inherited resistance of sugar beet to aphid colonization. *Ann. Appl. Biol.* 63:337

86. Lowe, H. J. B., Russell, G. E. 1974. Probing by aphids in the leaf tissues of resistant and susceptible sugar beet. *Entomol. Exp. Appl.* 17:468–76

87. Lüdecke, H., Winner, C. 1959. *Farbtafelatlas der Krankheiten und Schaedigungen der Zuckerruebe,* pp. 55–80. Frankfurt: DLG

88. Magyarosy, A. C., Duffus, J. E. 1976. Feeding preference and reproduction of the beet leafhopper on two Russian thistle plant species. *J. Am. Soc. Sugar Beet Technol.* 19:16–18

89. Maxson, A. C. 1920. *Principal Insect Enemies of the Sugar Beet.* Denver: Great Western Sugar Beet Co. 157 pp.

90. McFarlane, J. S. 1969. Breeding for resistance to curly top. *J. Int. Inst. Sugar Beet Res.* 4:73–83

91. McFarlane, J. S., Skoyen, I. O. 1968. New sugar beet varieties reduce losses from virus yellows. *Calif. Agric.* 22:14–15

92. McFarlane, J. S., Skoyen, I. O., Lewellen, R. T. 1969. Development of sugarbeet breeding lines and varieties resistant to yellows. *J. Am. Soc. Sugar Beet Technol.* 15:347–60

93. Meyerdick, D. E., Oldfield, G. N., Hessein, N. A. 1983. Bibliography of the beet leafhopper, *Circulifer tenellus* (Baker), and two of its transmitted plant pathogens, curly top virus and *Spiroplasma citri* Saglio et al. *Bibliogr. Entomol. Soc. Am.* 2:17–55

94. Moffitt, H. R., Reynolds, H. T. 1972. Bionomics of *Empoasca solana* DeLong on cotton in southern California. *Hilgardia* 41:247–97

95. Mumford, D. L. 1982. Using enzyme-linked immunosorbent assay to identify beet leafhopper populations carrying beet curly top virus. *Plant Dis.* 66(10):940–41

96. Nault, L. R., Rodriguez, J. G., eds. 1985. *The Leafhoppers and Planthoppers.* New York: Wiley. 500 pp.

97. Needham, P. H., Sawicki, R. M. 1971. Diagnosis of resistance to organophosphorus insecticides in *Myzus persicae* (Sulz.). *Nature* 230:125–27

98. Oldfield, G. N., Kaloostian, G. H. 1979. Vectors and host range of the citrus stubborn disease pathogen, *Spiroplasma citri. Proc. ROC-US Coop. Sci. Semin. Mycoplasma Dis. Plants,* ed. H.-J. Su, R. H. McCoy, pp. 119–25. Washington, DC: Natl. Sci. Counc.

99. Phelan, P. L., Miller, J. R. 1982. Postlanding behavior of alate *Myzus persicae* as altered by (E)-β-farnesene and three carboxylic acids. *Entomol. Exp. Appl.* 32:46–53

100. Poos, F. W., Wheeler, N. H. 1943. Studies on host plants of the leafhoppers of the genus *Empoasca. USDA Tech. Bull.* 850:1–51

101. Reed, J. L. 1964. *The influence of certain physical factors on the relative abundance of the green peach aphid, Myzus persicae (Sulzer) (Homoptera: Aphididae).* PhD thesis. Univ. Calif., Davis. 151 pp.

102. Reed, J. L. 1965. The mite problem in northern California sugar beet fields. *Spreckels Sugar Beet Bull.* 29:20–21, 24

103. Reynolds, H. T., Deal, A. S. 1956. Control of the southern garden leafhopper, a new pest of cotton in southern California. *J. Econ. Entomol.* 49:356–58

104. Reynolds, H. T., Dickson, R. C., Hannibal, R. M., Laird, E. F. 1967. Effect of the green peach aphid, southern garden leafhopper and *Tetranychus cinnabarinus* Bvd. populations upon yield of sugar beets in Imperial Valley, California. *J. Econ. Entomol* 60:1–7

105. Rice, A. D., Devonshire, A. L., Gibson, R. W., Gooding, A. R., Moores, G. D., Stribley, M. F. 1985. The problem of aphid resistance to aphicides, and alternative chemical methods of preventing virus transmission. *Inst. Int. Rech. Better. Congr. Hiver C. R.* 48:209–28

106. Rimsa, V. 1979. Protection of emerging sugar beet in Czechoslovakia. See Ref. 11, pp. 245–50

107. Ritenour, G., Hills, F. J., Lange, W. H. 1970. Effect of planting date and vector control on the suppression of early top and yellows in sugarbeet. *J. Am. Soc. Sugar Beet Technol.* 16:78–84

108. Rochow, W. F., Duffus, J. E. 1981. Luteovirus and yellows diseases. In *Handbook of Plant Virus Infections and Comparative Diagnosis,* ed. E. Kurstak, pp. 147–70. Amsterdam: Elsevier/North Holland Biomedical

109. Russell, G. E. 1966. Breeding for resistance to infection with yellowing viruses in sugar beet. *Ann. Appl. Biol.* 57:313–20

110. Russell, G. E. 1968. The distribution of sugar beet yellowing viruses in East Anglia from 1965–1968. *Br. Sugar Beet Rev.* 37:77–84

111. Russell, G. E. 1970. Effects of mineral oil on *Myzus persicae* (Sulz.) and its transmission of beet yellows virus. *Bull. Entomol. Res.* 59:691–94

112. Russell, G. E. 1977. Some effects of benzimidazole compounds on the transmission of beet yellows virus by *Myzus persicae. Proc. 9th Br. Insectic. Fungic. Conf., Brighton, England,* pp. 831–34. London: Br. Crop Prot. Counc.

113. Russell, G. E. 1978. *Plant Breeding for Pest and Disease Resistance,* pp. 231–90, 325–400. London/Boston: Butterworth. 485 pp.

114. Schaner, R. L. 1975. Field trials in sugarbeet made with thiofanox insecticide in the United Kingdom in 1975. See Ref. 2, 2:473–77

115. Schardt, J. 1985. The bean aphid. *Calif. Sugar Beet* 1985:28, 34

116. Severin, H. H. P. 1930. Life history of the beet leafhopper, *Eutettix tenellus* Baker, in California. *Univ. Calif. Berkeley Publ. Entomol.* 5:37–38

117. Smailes, A., Rose, P. W., Rollett, A. C. 1977. Ethiofencarb, a new carbamate aphicide for use on sugar beet and potatoes. See Ref. 112, pp. 499–504

118. Smith, H. G., Hinckes, J. A. 1985. The two viruses: the effects on their different epidemiologies on control. *Inst. Int. Rech. Better. Congr. Hiver C. R.* 48:191–98

119. Smith, K. M. 1967. *Insect Virology,* pp. 202–215. New York/London: Academic. 256 pp.

120. Stahl, C. F., Carsner, E. 1923. A discussion of *Eutettix tenella* Baker as a carrier of curly top of sugar beets. *J. Econ. Entomol.* 16:476–79

121. Staples, R., Jansen, W. P., Andersen, L. W. 1970. Biology and relationship of the leafhopper *Aceratagallia calcaris* to yellow vein disease of sugar beet. *J. Econ. Entomol.* 63:460–63

122. Steudel, W. 1971. Entwicklung und Notwendigkeit von Schadvoraussagen und Spritzwarnsystemen bei der Zuckerrübe, speciell bei Viruskrankheiten. *Zucker* 24:465–70

123. Tamaki, G. 1973. Spring populations of the green peach aphid on peach trees and the role of natural enemies in their control. *Environ. Entomol.* 2:186–91

124. Tamaki, G., Fox, L., Butt, B. A. 1979. Ecology of the green peach aphid as a vector of beet western yellows virus of sugarbeets. *USDA Tech. Bull.* 1599:1–16

125. Tamaki, G., Fox, L., Butt, B. A., Richards, A. W. 1978. Relationships among aphids, virus yellows, and sugar beet fields in the Pacific Northwest. *J. Econ. Entomol.* 71:654–56

126. Tamaki, G., Fox, L., Featherston, P. E. 1980. Evaluation of green peach aphid activity and the occurrence of beet western yellows virus in sugar beet fields. *J. Am. Soc. Sugar Beet Technol.* 20:578–82

127. Tamaki, G., Nawrocka, B., Fox, L., Annis, B., Gupta, R. K. 1982. Comparison of yellow holocyclic and green anholocyclic strains of *Myzus persicae* (Sulzer): transmission of beet western yellows virus. *Environ. Entomol.* 11(1):234–38

128. Tamaki, G., Weiss, M. A., Long, G. E. 1982. Effective growth units in population dynamics of the green peach aphid (Homoptera: Aphididae). *Environ. Entomol.* 11(6):1134–36

129. Tatchell, G. M., Woiwod, I. P. 1983.

The interpretation and dissemination of aphid monitoring data. *Proc. 10th Int. Congr. Plant Prot.* 1:169

130. Thielemann, R., Nagi, A. 1977. Further contributions to the incidence of 2 yellow viruses on sugar beets in West Germany with reference to their vectors and yield losses. *Z. Pflanzenkr. Pflanzenschutz* 84:257–69

131. Thielemann, R., Nagi, A. 1979. Significance of the strains of *Aphis fabae* in the transmission of beet mild yellowing virus. *Z. Planzenkr. Pflanzenschutz* 86:161–68

132. United States Department of Agriculture. 1983. *Sugar and Sweeteners Outlook and Situation Report, December*, p. 16. Washington, DC: USDA Econ. Res. Serv.

133. United States Department of Agriculture. 1984. Sugar, molasses and honey. *Foreign Agric. Circ. FS3-84*, pp. 6, 17–22. Washington, DC: USDA

134. Van Emden, H. F., Eastop, V., Hughes, R. D., Way, M. J. 1969. The ecology of *Myzus persicae*. *Ann. Rev. Entomol.* 14:197–270

135. Wallis, R. L., Turner, J. E. 1969. Burning weeds in drainage ditches to suppress populations of green peach aphids and the incidence of beet western yellows disease in sugarbeets. *J. Econ. Entomol.* 62:307–10

136. Watson, M. A., Heathcote, G. D., Lauckner, F. B., Sowray, P. A. 1975. The use of weather data and counts of aphids in the field to predict the incidence of yellowing viruses of sugarbeet crops in England in relation to the use of insecticides. *Ann. Appl. Biol.* 81:181–98

137. Wiktelins, S. 1977. The importance of southerly winds and other weather data on the incidence of sugar beet yellowing viruses in southern Sweden. *Swed. J. Agric. Res.* 7:89–96

138. Yokomi, R. K. 1979. *Phenological studies of* Circulifer tenellus *(Baker) (Homoptera: Cicadellidae) in the San Joaquin Valley of California*. PhD thesis. Univ. Calif., Davis. 92 pp.

139. Young, D. A., Frazier, N. W. 1954. A study of the leafhopper genus *Circulifer* Zakhvatkin (Homoptera: Cicadellidae). *Hilgardia* 23:25–52

140. Yun, Y. M., Sullivan, E. F. 1980. Pest management systems for sugarbeets in the North American central Great Plains region. *J. Am. Soc. Sugar Beet Technol.* 20:455–76

Ann. Rev. Entomol. 1987. 32:361–80

ECOLOGICAL GENETICS OF INSECTICIDE AND ACARICIDE RESISTANCE

Richard T. Roush

Department of Entomology, Mississippi State University, Mississippi State, Mississippi 39762

John A. McKenzie

Department of Genetics, University of Melbourne, Parkville 3052, Victoria, Australia

PERSPECTIVES AND OVERVIEW

Resistance to one or more insecticides had been reported in at least 447 species of insects and mites by 1984 (34, 69). As a result of cross and multiple resistance (65), many insect and mite pests are able to tolerate virtually all pesticides available for their control. An extreme example is the Colorado potato beetle, *Leptinotarsa decemlineata,* on Long Island, New York, which has developed resistance to all major classes of modern synthetic insecticides (28). By 1983, the only insecticide effective against this pest was synergized rotenone (69).

The monetary and human costs of resistance are difficult to assess, but loss of pesticide effectiveness almost invariably entails increased application frequencies and dosages and, finally, more expensive replacement compounds (69), as new pesticides become increasingly more difficult to discover, develop, register, and manufacture (65). Therefore, it is essential to develop strategies to delay or minimize the probability of resistance evolution. This will be possible only if the genetics of resistance and population ecology of pest species are defined in a manner that is relevant to actual pesticide use. Although this has been recognized since the 1950s (21), relevant data have not generally been available. The only widely applied

361

0066-4170/87/0101-0361$02.0(

method for avoiding resistance has been reduction of pesticide use, which has been a primary motivation for the development of integrated pest management systems (97).

Although resistance may also have a behavioral basis (31), the importance of behavioral resistance is a matter of debate (14, 37, 51). We confine ourselves to physiological resistance in this review.

By the mid-1970s the basic inheritance of insecticide resistance was well understood from laboratory studies. Specific biochemical and physiological mechanisms of resistance were often assigned to individual genes (73). In some species, the genes responsible for physiological resistance were mapped to specific chromosomes (e.g. 96). In contrast, studies of the genetic expression of resistance under field conditions were barely underway in the 1970s (104). Factors promoting resistance development had been categorized (35) and some resistance frequencies had been estimated (104), but few key population parameters, such as the fitness of resistant genotypes, had been measured.

The emphases in genetic studies reflected the broad trends of the times. The intellectual communities of genetics (including population genetics) and population ecology advanced almost independently (8). In general, population ecologists were concerned with the dynamics and classification of demographic units; population geneticists were concerned with population structure, genetic variability, and the mechanisms by which variability was maintained and passed to subsequent generations. These two approaches are integrated in the experimental study of natural selection, the discipline of ecological genetics (27).

The integration of ecology and genetics soon had an impact on resistance studies. Between the mid-1970s and mid-1980s, resistance researchers gradually came to agree that population biology had much to contribute to resistance management (69), and that resistance management could be considered an aspect of applied ecological genetics (104). In the future, the integration of genetics and ecology will probably be more reciprocal; not only can resistance management benefit from ecological research, the study of resistance can provide paradigms for ecological genetics.

MODE OF INHERITANCE OF RESISTANCE

When Crow wrote on resistance inheritance in the mid-1950s (18), no general conclusion could be drawn concerning the number of loci responsible for resistance. Resistance seemed to be due to many loci in some cases, but to only one major locus in others. During the 1960s, however, as a result of an increase of examples, improvements in assays and data analysis, and use of genetic markers, it became widely recognized that economically important insecticide and acaricide resistance was usually due to allelic variants at just one or two loci (12, 14, 30, 39, 66). A similar conclusion has since been reached in genetic studies of toxicant resistance in vertebrates and plants (9, 38, 52, 54, 107).

In hindsight, much of the confusion over the number of genes involved in resistance appears to have resulted from laboratory selection studies, which usually produce polygenic resistance (66, 71, 101). This may be an outcome of selective channeling of variation. Selection within an initial phenotypic distribution tends to use existing, common variation, thereby producing a polygenic response; selection outside the distribution draws on novel variation, i.e. a rare allele at a single locus (58). Almost all laboratory regimes select within existing phenotypic distributions, often at 80–90% mortality, in order to provide survivors for the next generation.

Even when rare variants are specifically screened for, most laboratory populations lack the necessary alleles to respond. The alleles that confer high levels of resistance are initially very rare, as discussed later in this review. To reliably sample alleles at these low frequencies from field populations, it would be necessary to collect thousands to billions of individuals (82). As genetic bottlenecks occur during laboratory colonization, it is unlikely that any laboratory strain would carry major resistance alleles for a pesticide to which its ancestors had not been exposed. Thus, a laboratory strain may fail to develop monogenic resistance simply because no major alleles are present.

The dichotomy between laboratory and field selection in the evolution of resistance shows that the genetic basis of character variation is not intrinsic to the character, but rather to the selection regime applied (58, 101). Selection implies the differential survival and reproduction of genotypes (19). It is helpful to think of selection for pesticide resistance as discrimination among the dose response regressions of different genotypes, as illustrated in Figure 1. At very high or very low doses, there is no discrimination among genotypes in favor of resistance, so no selection can occur. Selection can only occur at those doses that permit discrimination between susceptible and resistant genotypes.

The selecting dose in laboratory programs can be very closely controlled and precise, permitting discrimination even among very similar genotypes (Figure 2A) within the normal physiological distribution of phenotypes (58). Because of uneven coverage, field doses are unlikely to be nearly so precise, but may still discriminate between susceptible and rare resistant genotypes when major alleles are involved (Figure 2B). Laboratory selection programs are designed to allow at least some survivors in each generation. Thus, they are well suited for the accumulation of many genes, each of small effect, over several generations, which results in polygenically inherited resistance.

In contrast, field applications are usually designed to kill every individual

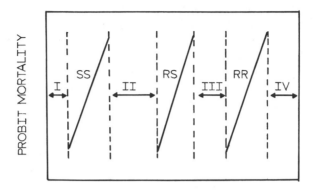

Figure 1 Four ranges of pesticide concentrations: *I*, no mortality of any genotype (no selection); *II*, mortality of SS only (resistance dominant); *III*, mortality of RS and SS (resistance recessive); *IV*, all genotypes killed (no selection).

with which contact is made. Selection occurs outside the normal range of the physiological distribution. Modern synthetic pesticides are very specific in their modes of action, such that subtle changes in target site or pesticide metabolism can confer high levels of resistance. Therefore, it is not surprising that there are many resistance mechanisms that confer high levels of resistance (often on the order of 10–100 fold) that are under the control of a single locus (see 73). When a single locus can produce such a high level of resistance, other loci that contribute only very little resistance may be superfluous. Polygenic resistance could still develop in the field, but it would be expected to develop very slowly owing to poor genotypic discrimination. Field treatment tends to select for major genes that can confer large increases in resistance immediately.

This model is supported by an experiment with *Lucilia cuprina,* the sheep blow fly. Four populations of *L. cuprina,* already nearly fixed for a major allele after more than 20 years of diazinon selection in the field, were brought into the laboratory for eight further generations of selection. The level of resistance approximately doubled in each strain. The response was polygenic, involving at least four chromosomes (59). Eight generations of selection in the laboratory produced a polygenic resistance, whereas more than 100 generations of field selection did not.

Another factor contributing to the rarity of polygenic resistance in the field is that monogenic resistance is more likely to spread than polygenic resistance, particularly when pesticide exposure varies across time or space. If a polygenically resistant individual emigrates to a susceptible population, the resistance alleles are likely to be diluted by hybridization if the population is

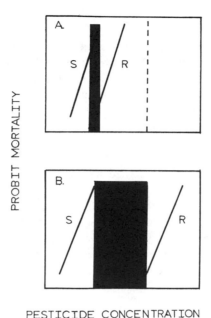

Figure 2   Selection for minor and major resistance alleles. Shaded area in *A* represents the tightly controlled doses characteristic of laboratory selection; in *B*, shaded area represents variable doses characteristic of field selection.

not treated for a few generations. Thus, there may not be any phenotypes with sufficient resistance to survive subsequent treatment. An individual with monogenic (especially nonrecessive) resistance (95) that successfully emigrates to a susceptible population is more likely to leave some resistant descendants, even after several untreated generations.

The hypothesis that monogenic resistance spreads more rapidly is supported by experiments with insecticide-resistant predatory mites. When Hoy et al (42) released a polygenic pyrethroid-resistant strain of *Metaseiulus occidentalis* in an orchard where native susceptible mites were abundant, resistance was lost before the first pyrethroid application was made. When the strain was released in an orchard after a pyrethroid application had eliminated the susceptible natives, the resulting population was resistant for several generations. In contrast, a carbaryl-resistant strain in which resistance is monogenically inherited (83) has been established in several orchards without prior treatments and appears to disperse rapidly (41, 81).

There have been cases in which apparently polygenic resistance has evolved in the field, but these have usually involved intensely sprayed or relatively isolated populations in which some of the resistance alleles had been selected by previously used insecticides (e.g. 45, 50, 104).

Implications for Resistance Management

The contrasts between resistance selection under laboratory and field conditions imply that experimental studies of resistance-management strategies should be based on resistances that are genetically similar to those that commonly develop in the field. Such studies should involve strains that include major resistance genes isolated from field populations, rather than strains selected only in the laboratory (69). At the very least, field strains will provide the most rigorous tests of the management approaches considered.

A common practice for evaluating the likelihood that new insecticides and acaricides will effect resistance is to try to induce resistance by selecting strains in the laboratory for many generations. This is probably of limited value for pesticides of truly novel chemistry. Laboratory selection occasionally produces monogenic resistance, but usually only when genetic material is incorporated into the laboratory strain from field populations that have already had extensive exposure to the specific pesticide or to one that shows cross-resistance to it (e.g. 83). Failure of a strain to develop resistance certainly cannot be taken as evidence that resistance will not develop in the field, as historical examples show (11, 70).

In addition, if laboratory selection tends to produce strains that are genetically different from field strains, it seems likely that the physiological resistance mechanisms selected will also differ. Laboratory selection programs have often produced levels and mechanisms (e.g. increased metabolism) of resistance qualitatively similar to those of field strains. However, general mechanism information is usually obvious from mode of action and metabolism studies, since significant resistance almost always involves changes in the target or enhancement of usual metabolic capabilities. The resistance manager needs more specific and realistic information.

The work of Kikkawa (46) suggests a possible solution to this problem. He used X-rays to induce a parathion resistance gene in *Drosophila melanogaster* that was apparently allelic with a parathion resistance gene of a resistant field strain. If further studies show that X-rays or other mutagens can reliably generate resistance genes similar to those found in the field, this method might accelerate the isolation of useful strains.

INFLUENCES ON THE RATES OF RESISTANCE EVOLUTION

The rate of change in allele frequency at any given locus in a closed population is a function of allele frequency, dominance, and the relative fitness of the various genotypes. Populations are rarely completely closed in nature. Therefore, a fourth factor, population structure, i.e. the subdivision of populations into smaller interbreeding units, is important because it affects gene flow ("migration" in the genetic sense) from area to area (29). Both fitness and structure depend on the biotic and abiotic environment, as discussed later in this review.

Dominance refers to the phenotypic expression of a character in heterozygotes relative to expression in the homozygous parents. High doses of insecticides can make the toxicological phenotype of resistance effectively (20) or functionally (95) recessive by killing the heterozygotes (e.g. dose range III in Figure 1). A lower dose (range II in Figure 1) can make resistance effectively dominant. However, dominance is a property of phenotypic characters, not of alleles (27). The dominance of resistance alleles for other characters, such as fitness in the absence of pesticide exposure, does not have to be the same as dominance for toxicological phenotype (84).

Since the four factors listed above have a profound influence on the evolution of resistance, we need to know as much about them as possible in order to manage resistance effectively; this is a major goal of ecological genetics research in pesticide resistance. Much of our knowledge of the mechanics of the population genetics of resistance is based on mathematical models. These models may be helpful in a general, descriptive way but depend, often critically, on the assumptions used in their construction (91, 92, 94). Resistance is usually assumed, with reason, to be inherited as a single gene (as described above). The common assumption that genotypes occur in Hardy-Weinberg proportions before selection also seems to be generally supported by data (57, 99, 104). However, it has been necessary to make rather broad assumptions about fitness, dominance, migration rates, and the proportion of the population that escapes treatment. The idiosyncracies of particular systems with respect to these factors may seriously limit the utility of general models in comparison of management strategies for specific cases. For example, in studies of clines in melanic moths, data on the major source of fitness differences (bird predation) were sufficient to develop a general but only partial picture of the impact of evolutionary forces (7).

Initial Allele Frequency

With one possible exception, as discussed below, no one has yet measured frequencies of resistance alleles prior to selection in any field population. Measurement of initial frequencies is difficult. The resistance phenotype, and therefore an efficient method to detect it, cannot be determined until resistance develops, which generally occurs only after the pesticide has been in widespread use (80, 82).

In the absence of data, researchers have assumed that resistance allele frequencies may range from 10^{-2} (35) to 10^{-13} (101) based on mutation-selection equilibrium theory (19). This theory is that the frequency of any allele prior to selection in its favor is maintained by an equilibrium between the generation of new alleles by mutation and selection against the heterozygous genotypes. (Resistant homozygotes will be so rare that they can be ignored.)

However, there may be cases where mutation-selection equilibria calculations are inappropriate. Allele frequencies were measured for dieldrin resistance in populations of *Anopheles gambiae* that were presumed to have been unexposed to dieldrin or related synthetic insecticides. The frequencies observed seem unreasonably high, ranging from 3% to 72%, indicating prior natural selection by unknown compounds (104). A candidate compound has emerged from recent work on dieldrin-resistant *Blattella germanica* cockroaches and on mosquitofish, *Gambusia affinis,* resistant to endrin (another cyclodiene insecticide). These strains, and presumably most other cyclodiene-resistant species, are cross-resistant to picrotoxinin, a plant-derived nerve excitant (44, 107). Picrotoxinin is derived from *Anamirta* sp. (Menispermaceae). Perhaps this or a related plant is found in the African areas where *A. gambiae* exhibited unusually high resistance-allele frequencies, and perhaps it occasionally decays in the water where the mosquitoes breed. Based on this example, we can suppose that other resistance alleles may be maintained in nature at much higher frequencies than ordinarily assumed.

Accurate estimates of initial resistance-allele frequencies may be important for evaluating management tactics that depend on low resistance frequencies. Improved resistance-detection techniques that facilitate direct measurement of allele frequencies will provide better estimates (80).

Dominance of Resistance Phenotypes

Dominance of the toxicological aspects of resistance under field conditions depends on the pesticide dose applied (20, 95), as illustrated in Figure 1. However, it may be difficult to estimate dominance under field conditions on the basis of laboratory studies alone.

In most species, all three genotypes (RR, RS, and SS) resistant or susceptible to lindane and cyclodiene insecticides, including dieldrin, can usually be clearly discriminated (12). In fact, Figure 1 probably represents the usual dose-responses more closely for cyclodienes than for any other insecticide group. Thus, one would expect that some doses (or concentrations) in field applications should kill all susceptible and heterozygous insects but not all resistant homozygotes (doses in range III of Figure 1). Rawlings et al (78) demonstrated that this occurred with *Anopheles culicifacies*. Adults representing RR, RS, and SS genotypes were released into village huts in Pakistan that had been sprayed with lindane at a range of typical field rates. The higher concentrations initially killed all the genotypes (corresponding to dose IV of Figure 1), but gradually decayed to allow some RR individuals to survive (dose III of Figure 1).

In similar experiments, eggs of RR, RS, and SS sheep blow flies *(L. cuprina)* were implanted into wounds that simulated those resulting from myiasis in sheep sprayed with standard field concentrations of dieldrin or lindane. Developed larvae were collected from the wounds, reared to adulthood, and identified to genotype with a discriminatory dose. In contrast to the mosquitoes in the experiment described above, RS blow flies survived throughout the course of the experiment; RS survivorship values were roughly intermediate between those of the RR and SS larvae (62, 63), much as one might expect from laboratory bioassays (100).

The sheep blow fly experiment was repeated for diazinon resistance, with very different results (62, 63). Diazinon resistance is incompletely dominant in laboratory assays of blow fly larvae (3, 60, 87). Therefore, one might expect that RR and RS genotypes would have similar relative viabilities in the field, i.e. that resistance would be dominant at some concentrations. Experiments showed that RS genotypes actually had relative viabilities very similar to those of the SS genotypes for a significant time after treatment (i.e. resistance was effectively recessive under field conditions), even as the diazinon residues decayed and began to allow considerable survival of the SS homozygotes.

IMPLICATIONS FOR RESISTANCE MANAGEMENT   In theory, resistance development can be delayed by using doses of insecticides that are at least high enough to kill heterozygotes (92). The sheep blow fly studies seem consistent with this idea. Resistance to dieldrin, which was incompletely dominant in the field, developed in two years, whereas diazinon resistance, recessive in the field, took ten years to develop (62). However, selection for dieldrin resistance may have occurred prior to the use of dieldrin for blow fly control, through the use of lindane for louse control (63).

There are several practical problems in the use of high doses. One of these is the difficulty of estimating the dose necessary to kill heterozygotes (92). The studies cited here show that one cannot simply extrapolate from laboratory data to estimate this dose. Field studies of effective dominance, particularly as pesticide residues decay, are needed for efficient implementation of a high-dose strategy.

Relative Fitness

There are very few quantitative estimates of the relative fitness of resistant genotypes under pesticide exposure. Based on data of increases in the frequencies of DDT- and dieldrin-resistant phenotypes in field populations of *Anopheles* mosquitoes, fitness of resistant phenotypes was estimated to be 1.3–6.1 times that of susceptible phenotypes (20, 105). These estimates are a function of genotypic survival, reproductive potential, and emigration from untreated habitats.

In many circumstances, the relative fitness of resistant genotypes is probably much higher than 6.1. For example, when pyrethroid-treated ear tags are used for horn fly control on cattle, survival of susceptible flies is nearly zero for several horn fly generations (89). Furthermore, as shown in studies of sheep blow flies and mosquitoes, fitness varies with decay of pesticide residues (62, 63, 78) and possibly with resistance frequency (20, 60). Importantly, when there is density-dependent competition, selection for resistance may continue even after there is no net increase in pesticide-induced mortality (e.g. 62, 63). This is an example of "soft" selection (see Reference 29 for an explanation of the terminology).

RELATIVE FITNESS IN ABSENCE OF PESTICIDES   One resistance-management tactic is to alternate different pesticides. This tactic is based on the assumption that frequencies of resistance to each compound will decline fairly rapidly in absence of the compound (33), either because of dilution of resistance by immigration of susceptible individuals, or because of natural selection against carriers of R alleles, or both. However, unless migration is unidirectional (i.e. the susceptible pool escapes contamination), which is unlikely, permanent avoidance of resistance would have to depend on natural selection against RR or RS genotypes. Resistant genotypes must be at some fitness disadvantage in the absence of pesticides or resistance alleles would be much more common prior to selection (18), but the difference may be small. Thus, it is important to determine whether such selective disadvantages are sufficiently large to be useful in practical situations.

Resistant strains are often reported to show disadvantages in life-history characteristics, but strains may differ in fitness for reasons independent of resistance (2, 10, 53, 80, 84). Even when RR and SS genotypes actually differ, a more critical issue is whether differences exist between RS and SS genotypes, because the RS heterozygotes will be the most common carriers of resistance during the early stages of a resistance episode (84).

Two general methods have been used to assess the fitness disadvantages that result from resistance. One is to measure the components of fitness, such as fecundity, development time, fertility, and mating competitiveness, for each genotype. The other method, which can be referred to as the "population cage," is to follow changes in genotypic frequencies in replicate populations held for a number of discrete or overlapping generations. There are a number of problems in estimation of genotypic fitness using either method (74), but the population cage method is probably preferable, especially when conducted across a range of environmental conditions (80). We disregard changes in LC_{50}s or resistance ratios, which do not correlate well with allele frequencies (82) and are therefore qualitative rather than quantitative measures. We also generally ignore studies in which neither genotype nor homozygosity of strains for a major resistance gene could be established.

One generality emerging from fitness studies is that large disadvantages in resistant arthropods, unlike those in some other biological taxa (53, 72, 75), seem to be the exception rather than the rule.

From field data on declines in resistance frequencies after the cessation of spraying, DDT-resistant phenotypes of *Anopheles stephensi* were estimated to have 91–93% of the fitness of susceptible phenotypes. In support of the validity of laboratory studies, laboratory population-cage data on a field-collected strain of the same species yielded an estimate of 96–98% (20). Fitness of DDT- and dieldrin-resistant phenotypes of *Anopheles culicifacies* was estimated to be somewhat lower, 62–97% for DDT and 44–79% for dieldrin (20, 79), but the difference may be due to relatively greater immigration of *A. culicifacies* (20, 104). Using similar methods, Muggleton (68) estimated relative fitness of about 63–76% for malathion-resistant phenotypes of the stored products pest *Oryzaephilus surinamensis* in the laboratory at 25°C, but the resistant phenotypes might have had an advantage at 30°C (68). In the cattle tick *Boophilus microplus,* the frequency of DDT-resistant homozygotes remained at about 55% for at least 10 generations in a heterogeneous laboratory population, which suggested similar fitness for each genotype; however, the frequency of dieldrin-resistant phenotypes declined (90).

In the *Anopheles* and *Oryzaephilus* studies, resistant individuals could not be separated into RR and RS genotypes, so it was not possible to assign fitness to specific genotypes. However, this has been done in seven separate fitness-component studies. The intrinsic growth rate, r, of RR, RS, and SS strains was measured for temephos-resistant *Culex quinquefasciatus* collected in California. The r of the RR strain was 79% that of the SS strain, but in RS heterozygotes r was 95% that of the SS strain. Since the RR and SS strains had different origins, heterosis may have inflated the fitness of the heterozygotes (24, 26). The selective disadvantages of the temephos resistance allele were confirmed in a population-cage study (40). However, in a fitness-component study of chlorpyrifos-resistant *C. quinquefasciatus* from Tanzania, differences between isogenic RR and SS homozygotes were less than 10% (2).

In a study of house flies, *Musca domestica,* diazinon-resistant (RR) strains had about 57–89% of the reproductive potential of an SS strain, but no fitness disadvantages could be found in heterozygotes. This suggests that the reproductive effects of resistance were recessive (84). A similar phenomenon was observed in a population-cage study of diazinon resistance at a secondary resistance locus (Rop-2) (3) in sheep blow flies; RR genotypes had a fitness of 0.61 relative to that of RS and SS flies (98).

Other fitness-component studies on DDT or dieldrin resistance in *Anopheles gambiae* (25), DDT resistance in *Musca domestica* (10), malathion resistance in *Anopheles arabiensis* (49), and carbaryl resistance in the beneficial predatory mite *Metaseiulus occidentalis* (81, 83) showed few or no component disadvantages in any genotype.

COADAPTATION   In the context of resistance, coadaptation refers to the selection and integration of resistance genes with other loci that ameliorate the deleterious effects of resistance (1, 31, 56). Coadaptation was postulated to account for observations that resistance tended to revert in the early stages of selection, but appeared to become more stable after further cycles of selection and relaxation of selection. Accordingly, the estimates cited above might underestimate the disadvantages associated with an R allele when it is initially rare. However, recent experimental studies suggest that coadaptation may not be common in field-evolved resistance.

The best way to isolate a major gene for resistance is through repeated backcrossing and selection (see Reference 30 for an excellent description of this process). This procedure can "cull out" modifiers of any type, including fitness modifiers when a strain without prior pesticide exposure is used as the backcross stock; thus a uniform genetic background can be established for fitness comparisons. Of five published reports on repeated backcrossing followed by observations on fitness (2, 5, 39, 64, 99), only one (64) showed the presence of modifiers that improved fitness of the R allele. However, the importance of genetic background to accurate fitness measurement was apparent in three of the studies (2, 39, 64). This emphasizes the need for repeated backcrossing in such work.

The first such report was on the Leverkusen strain of the two-spotted spider mite, *Tetranychus urticae* (39). The organophosphorus susceptible Leverkusen strain was selected for more than 30 generations to produce an RR strain. This strain proved to be inferior to the SS strain in several fitness components, and resistance reverted after relaxation of selection. However, after repeated backcrossing and selection, and after intercrossing to produce RR homozygotes, no differences could be detected in fitness of the new RR strain and the SS strain (39, 88). Similarly, a chlorpyrifos-resistant strain of *Culex quinquefasciatus* had reduced fitness compared to a standard susceptible strain. After five generations of repeated backcrossing to the SS strain, the fitness of resistant homozygotes improved (2).

No fitness deficits could be associated with resistance in either parental strains or backcrossed strains of diazinon-resistant house flies (99) or malathion-resistant *Tribolium castaneum* (5); this suggests that resistance was not deleterious and that fitness modifiers were not present. However, a backcrossing study on the sheep blow fly has demonstrated that resistance coadaptation can occur in the field. Diazinon resistance was not deleterious in population cages established from F_1 and BC_3 RS flies (backcrossed for three generations to the SS strain). However, it was significantly deleterious in cages established from BC_6 and BC_9 RS flies; by this stage the R allele was in a genetic background similar to that in which it first occurred (64). In this background the relative fitness of the RR and RS genotypes may have been as low as 50% and 75%, respectively (80). The resistance modifier(s) were on a different chromosome than the major resistance locus (61). Thus, the coadaptation of fitness and resistance did not occur through the evolution of linked complexes (supergenes), as in some ecological traits (27).

Resistance evolution in sheep blow flies in Australia differed significantly from most resistance episodes. Fitness modifiers can offer an advantage only in the presence of the resistance allele. Thus, selection for fitness modification must be fairly weak until the resistance allele reaches high frequency (see Reference 16 for a similar argument). In most cases a new insecticide is substituted when resistance appears. However, diazinon continued to be useful against the sheep blow fly in spite of resistance, so the resistance allele was maintained in very high frequency by continuous diazinon use for more than 10 years after first detection of resistance (64). Thus, it is reasonable to presume that modification occurred in sheep blow fly but would be unlikely in other resistant pests.

This conclusion conflicts with the traditional view of resistance evolution, but there are alternative interpretations for the studies on which that view is based. Reports that resistance appears to revert briefly before a higher level of resistance develops are usually anecdotal. Such observations were seldom part of a planned experiment, so unselected controls were often not maintained. The importance of maintaining a control is illustrated by selection for resistance in *Heliothis virescens;* an unselected control strain showed even greater variation in susceptibility between generations than the selected line (102).

A second problem is the genetics of laboratory strains. The classical studies on which coadaptation theory is based (47) used polygenic traits. With these traits allelic substitution is at many loci and may require reorganization of the genome. Major resistance genes may have a smaller impact on the total genotype of the organism because allelic substitution is at only one or a few loci. Where resistance is polygenic, coadaptation might be important, but it is irrelevant in the field. Laboratory strains selected for resistance seem to have lower fitness than their susceptible counterparts, but this is not generally true for strains whose resistance developed in the field (80).

The report by Abedi & Brown (1) is an important example for reanalysis. Resistance was isolated from field mosquitoes and controlled at a single locus (13), but data on control strains were not reported. Although there were trends between viability and resistance (Reference 1, Tables III and IV), correlational analysis shows no statistically significant relationships. Furthermore, the sources of reduction in the numbers of mature larvae varied (poor hatch rate in the F_2 generation, low oviposition in the F_4 generation; Reference 1, Table VII), which seems difficult to explain genetically. Another commonly cited study (56) used strains in which resistance was probably polygenic (39). Field observations on coadaptation have been confounded by dilution of resistance by susceptible migrants (103).

FITNESS IN APHIDS   The concept of relative genotypic fitness loses some of its meaning for parthenogenic organisms. In aphids, resistance remains stable in most clones for long periods, but when reversion starts (often after 10–30 generations of stable resistance), complete susceptibility usually results within four generations (4, 6). Resistance in at least some of these clones is apparently due to gene duplications that increase the quantity of a carboxylesterase. In highly resistant clones of *Myzus persicae,* this esterase may amount to 3% of the total body protein (23). However, in such clones a few individuals with lower resistance appear spontaneously each generation, presumably owing to mutational loss of the gene duplications that produce the high enzyme levels. This may account for abrupt losses of resistance (85). A similar phenomenon may occur in *Phorodon humuli* (48).

HYPOTHESES ON THE RELATIONSHIP BETWEEN RESISTANCE MECHANISM AND FITNESS IN ABSENCE OF PESTICIDES   The limited data currently available suggest some interesting relationships between fitness disadvantages and specific resistance mechanisms. The most serious and consistent disadvantages often seem to be associated with resistance due to generalized esterases (2, 23, 24, 43). This may be because esterases comprise a relatively large portion of body protein, as in the green peach aphid (71), in strains with this mechanism of resistance (67a). In contrast, little or no reproductive disadvantages have been associated with altered acetylcholinesterase (39, 88), increased oxidative detoxication (81, 83, 84), malathion-specific carboxylesterases (5, 49) or cases where *kdr*-type resistance is probable (10, 86, 90).

Dispersal of Resistant and Susceptible Individuals

The conclusion that resistance does not generally confer large fitness disadvantages implies that resistance management must rely heavily on pools of susceptibility to delay resistance. Various models but few experimental studies have illustrated the importance of immigration of susceptible individuals on resistance development (35, 91, 92, 94, 95, 106). However, the significance of immigration is also inferred from numerous examples of one species or population developing resistance when similar species or populations failed to do so. For example, lindane resistance was widespread in *Sitophilus oryzae* in Queensland, Australia by 1963. In contrast, lindane resistance was not detected in contemporaneous surveys of *Sitophilus zeamais* (15) in Australia through 1980, even though it had been reported in populations of this species from five other countries (32). In Queensland, *S. oryzae* infestations occur primarily in grain stored on farms. In contrast, *S. zeamais* attacks maize heavily both in the field and in storage, so *S. zeamais* populations exposed to lindane in storage are significantly diluted by susceptible field populations (15). Such examples emphasize the need for more detailed studies of population structure and migration rates.

In general, migration rates for susceptibility and resistance are variable and somewhat difficult to quantify (9, 76, 77, 80), particularly since one needs to know not only whether individuals disperse but also what genetic impact they have when they join a population (80). One way to quantify this impact is to use visible genetic markers (22) or the expression of the pesticide resistance itself (67). Although the results of dispersal studies can be expected to vary with species and habitat, two such studies have shown that overwintering sites can provide foci for resistance to develop and spread (22, 67).

CONCLUSIONS

Problems in the direct investigation of selection in the field continue to plague ecological genetics. It is necessary to identify mechanistic associations between selective agents and phenotypic differences and to be able to manipulate a system to test predictions (17, 36). These needs demand an interface between ecology and genetics (8). In practice, an unambiguous definition of the selective agent is the most difficult requirement (17).

Like heavy-metal tolerance in plants and warfarin resistance in rodents, insecticide resistance provides a rare opportunity to study selection that is direct, usually easily measured, and clearly fitness-related. Some studies have already quantitatively examined the impact of specific details of population biology on resistance development (22, 55, 67, 93). As emphasized in this review, many important aspects of resistance evolution, such as migration rates, are not easy to measure. Thus the study of the ecological genetics of resistance is no trivial problem, particularly in light of the importance of resistance to food production and human health. Nonetheless, because the selective agent can be readily identified, resistance should be somewhat easier to study than many other traits, and therefore should provide fertile ground for developing approaches to tackle more difficult genetic problems.

We propose that studies of pesticide resistance based on the philosophy of ecological genetics will not only have significant benefits for the derivation of pest-management strategies, but will also make fundamental contributions to an understanding of evolutionary change.

ACKNOWLEDGMENTS

We thank M. J. Whitten, F. W. Plapp, C. F. Curtis, B. E. Tabashnik, B. A. Croft, T. M. Brown, R. G. Luttrell, G. P. Georghiou, and S. B. Ramaswamy for comments on an earlier draft.

Literature Cited

1. Abedi, Z. H., Brown, A. W. A. 1960. Development and reversion of DDT-resistance in *Aedes aegypti*. *Can. J. Genet. Cytol.* 2:252–61
2. Amin, A. M., White, G. B. 1984. Relative fitness of organophosphate-resistant and susceptible strains of *Culex quinquefasciatus* Say (Diptera: Culicidae). *Bull. Entomol. Res.* 74:591–98
3. Arnold, J. T. A., Whitten, M. J. 1976. The genetic basis for organophosphorus resistance in the Australian sheep blowfly, *Lucilia cuprina* (Wiedemann) (Diptera, Calliphoridae). *Bull. Entomol. Res.* 66:561–68
4. Bauernfeind, R. J., Chapman, R. K. 1985. Nonstable parathion and endosulfan resistance in green peach aphids (Homoptera: Aphididae). *J. Econ. Entomol.* 78:516–22
5. Beeman, R. W., Nanis, S. M. 1986. Malathion resistance alleles and their fitness in the red flour beetle (Coleoptera: Tenebrionidae) *J. Econ. Entomol.* 79: 580–87
6. Beranek, A. P. 1974. Stable and nonstable resistance to dimethoate in the peach-potato aphid *(Myzus persicae)*. *Entomol. Exp. Appl.* 17:381–90
7. Bishop, J. A. 1972. An experimental study of the cline of industrial melanism in *Biston betularia* (L.) (Lepidoptera) between urban Liverpool and rural North Wales. *J. Anim. Ecol.* 41:209–43
8. Bishop, J. A. 1973. The proper study of populations. In *Insects: Studies In Population Management*, ed. P. W. Geier, L. R. Clark, D. J. Anderson, H. A. Nix, pp. 52–68. Canberra, Australia: Ecol. Soc. Aust. 294 pp.
9. Bishop, J. A., Hartley, D. J. 1976. The size and age structure of rural populations of *Rattus norvegicus* containing individuals resistant to the anticoagulant poison warfarin. *J. Anim. Ecol.* 45:623–46
10. Bøggild, O., Keiding, J. 1958. Competition in house fly larvae: experiments involving a DDT-resistant and a susceptible strain. *Oikos* 9:1–25
11. Brown, A. W. A. 1967. Insecticide resistance—genetic implications and applications. *World Rev. Pest Control* 6: 104–14
12. Brown, A. W. A. 1967. Genetics of insecticide resistance in insect vectors. In *Genetics of Insect Vectors of Disease*, ed. J. W. Wright, R. Pal, pp. 505–52. New York: Elsevier. 794 pp.
13. Brown, A. W. A., Abedi, Z. H. 1962. Genetics of DDT-resistance in several strains of *Aedes aegypti*. *Can. J. Genet. Cytol.* 4:319–32
14. Busvine, J. R. 1963. The present status of insecticide resistance. *Bull. WHO* 29:31–40 (Suppl.)
15. Champ, B. R., Cribb, J. N. 1965. Lindane resistance in *Sitophilus oryzae* (L.) and *Sitophilus zeamais* Motsch. (Coleoptera, Curculionidae) in Queensland. *J. Stored Prod. Res.* 1:9–24
16. Charlesworth, B. 1979. Evidence against Fisher's theory of dominance. *Nature* 278:848–49
17. Clarke, B. 1975. The contribution of ecological genetics to evolutionary theory: detecting the direct effects of natural selection on particular polymorphic loci. *Genetics* 79:101–13 (Suppl.)
18. Crow, J. F. 1957. Genetics of insect resistance to chemicals. *Ann. Rev. Entomol.* 2:227–46
19. Crow, J. F., Kimura, M. 1970. *An Introduction to Population Genetics Theory*. New York: Harper & Row. 591 pp.
20. Curtis, C. F., Cook, L. M., Wood, R. J. 1978. Selection for and against insecticide resistance and possible methods of inhibiting the evolution of resistance in mosquitoes. *Ecol. Entomol.* 3:273–87
21. Davidson, G., Pollard, D. G. 1958. Effects of simulated field deposits of gamma-BHC and dieldrin on susceptible, hybrid and resistant strains of *Anopheles gambiae* Giles. *Nature* 182: 739–40
22. Denholm, I., Sawicki, R. M., Farnham, A. W. 1985. Factors affecting resistance to insecticides in house-flies, *Musca*

domestica L. (Diptera: Muscidae). IV.
The population biology of flies on animal farms in south-eastern England and
its implications for the management of
resistance. *Bull. Entomol. Res.* 75:143–
58
23. Devonshire, A. L., Moores, G. D.
1982. A carboxylesterase with broad
substrate specificity causes organophosphorus, carbamate and pyrethroid resistance in peach-potato aphids *(Myzus persicae)*. *Pestic. Biochem. Physiol.* 18:
235–46
24. El-Khatib, Z. I., Georghiou, G. P.
1985. Comparative fitness of temephos-
resistant, susceptible, and hybrid phenotypes of the southern house mosquito
(Diptera: Culicidae). *J. Econ. Entomol.*
78:1023–29
25. Emeka-Ejiofor, S. A. I., Curtis, C. F.,
Davidson, G. 1983. Tests for effects of
insecticide resistance genes in *Anopheles
gambiae* on fitness in the absence of
insecticides. *Entomol. Exp. Appl.* 34:
163–68
26. Ferrari, J. A., Georghiou, G. P. 1981.
Effects of insecticidal selection and
treatment on reproductive potential of
resistant, susceptible, and heterozygous
strains of the southern house mosquito.
J. Econ. Entomol. 74:323–27
27. Ford, E. B. 1975. *Ecological Genetics*.
London: Chapman & Hall. 442 pp. 4th
ed.
28. Forgash, A. J. 1984. History, evolution,
and consequences of insecticide resistance. *Pestic. Biochem. Physiol.* 22:
178–86
29. Futuyma, D. J. 1979. *Evolutionary Biology*. Sunderland, Mass: Sinauer. 565
pp.
30. Georghiou, G. P. 1969. Genetics of resistance to insecticides in houseflies and
mosquitoes. *Exp. Parasitol.* 26:224–55
31. Georghiou, G. P. 1972. The evolution
of resistance to pesticides. *Ann. Rev.
Ecol. Syst.* 3:133–68
32. Georghiou, G. P. 1981. *The Occurrence
of Resistance to Pesticides in Arthropods: An Index of Cases Reported
Through 1980*. Rome: FAO. 172 pp.
33. Georghiou, G. P. 1983. Management of
resistance in arthropods. In *Pest Resistance to Pesticides*, ed. G. P. Georghiou, T. Saito, pp. 769–92. New York:
Plenum. 809 pp.
34. Georghiou, G. P., Mellon, R. 1983.
Pesticide resistance in time and space.
See Ref. 33, pp. 1–46
35. Georghiou, G. P., Taylor, C. E. 1977.
Operational influences in the evolution
of insecticide resistance. *J. Econ. Entomol.* 70:653–58

36. Deleted in proof
37. Gould, F. 1984. Role of behavior in the
evolution of insect adaptation to insecticides and resistant host plants. *Bull.
Entomol. Soc. Am.* 30(4):34–40
38. Greaves, J. H., Redfern, R., Ayres, P.
B., Gill, J. E. 1977. Warfarin resistance: a balanced polymorphism in the
Norway rat. *Genet. Res.* 30:257–63
39. Helle, W. 1965. Resistance in the acarina: mites. *Adv. Acarol.* 2:71–93
40. Hemingway, J., Georghiou, G. P. 1984.
Differential suppression of organophosphorus resistance in *Culex quinquefasciatus* by the synergists IBP, DEF, and
TPP. *Pestic. Biochem. Physiol.* 21:1–9
41. Hoy, M. A. 1982. Aerial dispersal and
field efficacy of a genetically improved
strain of the spider mite predator
*Metaseiulus occidentalis. Entomol. Exp.
Appl.* 32:205–12
42. Hoy, M. A., Knop, N. F., Joos, J. L.
1980. Pyrethroid resistance persists in
spider mite predator. *Calif. Agric.* 34
(11–12):11–12
43. Hughes, P. B., Raftos, D. A. 1985. Genetics of an esterase associated with resistance to organophosphorus insecticides in the sheep blowfly, *Lucilia
cuprina* (Weidemann) (Diptera: Calliphoridae). *Bull. Entomol. Res.* 75:
535–44
44. Kadous, A. A., Ghiasuddin, S. M.,
Matsumura, F., Scott, J. G., Tanaka, K.
1983. Difference in the picrotoxinin receptor between the cyclodiene-resistant
and susceptible strains of the German
cockroach. *Pestic. Biochem. Physiol.*
19:157–66
45. Keiding, J. 1977. Resistance in the
house fly in Denmark and elsewhere. In
*Pesticide Management and Insecticide
Resistance*, ed. D. L. Watson, A. W. A.
Brown, pp. 261–302. New York: Academic. 638 pp.
46. Kikkawa, H. 1964. Genetical studies on
the resistance to parathion in *Drosophila
melanogaster*. II. Induction of a resistance gene from its susceptible allele.
Botyu-Kagaku 29:37–42
47. Lerner, I. M. 1958. *The Genetic Basis of
Selection*. New York: Wiley. 284 pp.
48. Lewis, G. A., Madge, D. S. 1984. Esterase activity and associated insecticide
resistance in the damson-hop aphid,
Phorodon humuli (Schrank) (Hemiptera:
Aphididae). *Bull. Entomol. Res.* 74:
227–38
49. Lines, J. D., Ahmed, M. A. E., Curtis,
C. F. 1984. Genetic studies of malathion
resistance in *Anopheles arabiensis* Pat-
ton (Diptera: Culicidae). *Bull. Entomol.
Res.* 74:317–25

50. Liu, M.-Y., Tzeng, Y.-J., Sun, C.-N. 1982. Insecticide resistance in the diamondback moth. *J. Econ. Entomol.* 75:153–55

51. Lockwood, J. A., Sparks, T. C., Story, R. N. 1984. Evolution of insect resistance to insecticides: a reevaluation of the roles of physiology and behavior. *Bull. Entomol. Soc. Am.* 30(4):41–51

52. Macnair, M. R. 1983. The genetic control of copper tolerance in the yellow monkey flower, *Mimulus guttatus.* *Heredity* 50:283–93

53. Macnair, M. R., Watkins, A. D. 1983. The fitness of the copper tolerance gene of *Mimulus guttatus* in uncontaminated soil. *New Phytol.* 95:133–37

54. MacSwiney, F. J., Wallace, M. E. 1978. Genetics of warfarin-resistance in house mice of three separate localities. *J. Hyg.* 80:69–75

55. May, R. M., Dobson, A. P. 1986. Population dynamics and the rate of evolution of pesticide resistance. See Ref. 69, pp. 170–93

56. McEnroe, W. D., Naegele, J. A. 1968. The coadaptive process in an organophosphorus-resistant strain of the two-spotted spider mite, *Tetranychus urticae.* *Ann. Entomol. Soc. Am.* 61:1055–59

57. McKenzie, J. A. 1984. Dieldrin and diazinon resistance in populations of the Australian sheep blowfly, *Lucilia cuprina,* from sheep-grazing areas and rubbish tips. *Aust. J. Biol. Sci.* 37:367–74

58. McKenzie, J. A. 1985. Genetics of resistance to chemotherapeutic agents. In *Resistance in Nematodes to Anthelmintic Drugs,* ed. N. Anderson, P. J. Waller, pp. 89–95. Sydney, Australia: CSIRO Div. Anim. Health, Aust. Wool Corp. 189 pp.

59. McKenzie, J. A., Dearn, J. M., Whitten, M. J. 1980. Genetic basis of resistance to diazinon in Victorian populations of the Australian sheep blowfly, *Lucilia cuprina.* *Aust. J. Biol. Sci.* 33:85–95

60. McKenzie, J. A., Fegent, J. C., Weller, G. 1986. Frequency-dependent selection at the diazinon resistance locus of the Australian sheep blowfly, *Lucilia cuprina.* *Heredity* 56:373–80

61. McKenzie, J. A., Purvis, A. 1984. Chromosomal localisation of fitness modifiers of diazinon resistance genotypes of *Lucilia cuprina.* *Heredity* 53: 625–34

62. McKenzie, J. A., Whitten, M. J. 1982. Selection for insecticide resistance in the Australian sheep blowfly, *Lucilia cuprina.* *Experientia* 38:84–85

63. McKenzie, J. A., Whitten, M. J. 1984. Estimation of the relative viabilities of insecticide resistance genotypes of the Australian sheep blowfly, *Lucilia cuprina.* *Aust. J. Biol. Sci.* 37:45–52

64. McKenzie, J. A., Whitten, M. J., Adena, M. A. 1982. The effect of genetic background on the fitness of diazinon resistance genotypes of the Australian sheep blowfly, *Lucilia cuprina.* *Heredity* 49:1–9

65. Metcalf, R. L. 1980. Changing role of insecticides in crop protection. *Ann. Rev. Entomol.* 25:219–56

66. Milani, R. 1960. Genetic studies on insecticide-resistant insects. *Misc. Publ. Entomol. Soc. Am.* 2:75–83

67. Miller, R. W., Croft, B. A., Nelson, R. D. 1985. Effects of early season immigration on cyhexatin and formetanate resistance of *Tetranychus urticae* (Acari: Tetranychidae) on strawberry in central California. *J. Econ. Entomol.* 78:1379–88

67a. Mouchès, C., Fournier, D., Raymond, M., Magnin, M., Bergé, J.-B., et al. 1985. Association entre l'amplification de séquences d'ADN, l'augmentation quantitative d'estérases et la résistance à des insecticides organophosphorés chez des Moustiques du complexe *Culex pipiens,* avec une note sur une amplification similaire chez *Musca domestica* L. *C. R. Acad. Sci. Sér. III* 301(16):695–700

68. Muggleton, J. 1983. Relative fitness of malathion-resistant phenotypes of *Oryzaephilus surinamensis* L. (Coleoptera: Silvanidae). *J. Appl. Ecol.* 20:245–54

69. National Research Council. 1986. *Pesticide Resistance: Strategies and Tactics for Management.* Washington, DC: Natl. Acad. Press. 471 pp.

70. Nolan, J., Roulston, W. J. 1979. Acaricide resistance as a factor in the management of acari of medical and veterinary importance. In *Recent Advances in Acarology,* Vol. 2, ed. J. G. Rodriguez, pp. 3–13. New York: Academic. 569 pp.

71. Oppenoorth, F. J. 1984. Biochemistry of insecticide resistance. *Pestic. Biochem. Physiol.* 22:187–93

72. Partridge, G. G. 1979. Relative fitness of genotypes in a population of *Rattus norvegicus* polymorphic for warfarin resistance. *Heredity* 43:239–46

73. Plapp, F. W. Jr. 1976. Biochemical genetics of insecticide resistance. *Ann. Rev. Entomol.* 21:179–97

74. Prout, T. 1971. The relation between fitness components and population prediction in *Drosophila.* I. The estimation

of fitness components. *Genetics* 68:127–49

75. Radosevich, S. R., Holt, J. S. 1982. Physiological responses and fitness of susceptible and resistant weed biotypes to triazine herbicides. In *Herbicide Resistance in Plants*, ed. H. M. LeBaron, J. Gressel, pp. 163–83. New York: Wiley. 401 pp.

76. Rawlings, P., Curtis, C. F., Wickramasinghe, M. B., Lines, J. 1981. The influence of age and season on dispersal and recapture of *Anopheles culicifacies* in Sri Lanka. *Ecol. Entomol.* 6:307–19

77. Rawlings, P., Davidson, G. 1982. The dispersal and survival of *Anopheles culicifacies* Giles (Diptera: Culicidae) in a Sri Lankan village under malathion spraying. *Bull. Entomol. Res.* 72:139–44

78. Rawlings, P., Davidson, G., Sakai, R. K., Rathor, H. R., Aslamkhan, K. M., Curtis, C. F. 1981. Field measurement of the effective dominance of an insecticide resistance in anopheline mosquitos. *Bull. WHO* 59:631–40

79. Rawlings, P., Herath, P. R. J., Kelly, S. 1985. *Anopheles culicifacies* (Diptera: Culicidae): DDT resistance in Sri Lanka prior to and after cessation of DDT spraying. *J. Med. Entomol.* 22:361–65

80. Roush, R. T., Croft, B. A. 1986. Experimental population genetics and ecological studies of pesticide resistance in insects and mites. See Ref. 69, pp. 257–70

81. Roush, R. T., Hoy, M. A. 1981. Laboratory, glasshouse, and field studies of artificially selected carbaryl resistance in *Metaseiulus occidentalis. J. Econ. Entomol.* 74:142–47

82. Roush, R. T., Miller, G. L. 1986. Considerations for design of insecticide resistance monitoring programs. *J. Econ. Entomol.* 79:293–98

83. Roush, R. T., Plapp, F. W. Jr. 1982. Biochemical genetics of resistance to aryl carbamate insecticides in the predaceous mite, *Metaseiulus occidentalis. J. Econ. Entomol.* 75:304–7

84. Roush, R. T., Plapp, F. W. Jr. 1982. Effects of insecticide resistance on biotic potential of the house fly (Diptera: Muscidae). *J. Econ. Entomol.* 75:708–13

85. Sawicki, R. M., Devonshire, A. L., Payne, R. W., Petzing, S. M. 1980. Stability of insecticide resistance in the peach-potato aphid, *Myzus persicae* (Sulzer). *Pestic. Sci.* 11:33–42

86. Schnitzerling, H. J., Roulston, W. J., Schunter, C. A. 1970. The absorption and metabolism of [^{14}C]DDT in DDT-resistant and susceptible strains of the cattle tick, *Boophilus microplus. Aust. J. Biol. Sci.* 23:219–30

87. Shanahan, G. J. 1979. Genetics of diazinon resistance in larvae of *Lucilia cuprina* (Wiedemann) (Diptera: Calliphoridae). *Bull. Entomol. Res.* 69:225–28

88. Smissaert, H. R. 1964. Cholinesterase inhibition in spider mites susceptible and resistant to organophosphate. *Science* 143:129–31

89. Sparks, T. C., Quisenberry, S. S., Lockwood, J. A., Byford, R. L., Roush, R. T. 1985. Insecticide resistance in the horn fly, *Haematobia irritans* (L.). *J. Agric. Entomol.* 2:217–33

90. Stone, B. F. 1962. The inheritance of DDT-resistance in the cattle tick, *Boophilus microplus. Aust. J. Agric. Res.* 13:984–1007

91. Tabashnik, B. E. 1986. Computer simulation as a tool for pesticide resistance management. See Ref. 69, pp. 194–206

92. Tabashnik, B. E., Croft, B. A. 1982. Managing pesticide resistance crop-arthropod complexes: interactions between biological and operational factors. *Environ. Entomol.* 11:1137–44

93. Tabashnik, B. E., Croft, B. A. 1985. Evolution of pesticide resistance in apple pests and their natural enemies. *Entomophaga* 30:37–49

94. Taylor, C. E. 1983. Evolution of resistance to insecticides: the role of mathematical models and computer simulations. See Ref. 33, pp. 163–73

95. Taylor, C. E., Georghiou, G. P. 1979. Suppression of insecticide resistance by alteration of gene dominance and migration. *J. Econ. Entomol.* 72:105–9

96. Tsukamoto, M. 1983. Methods of genetic analysis of insecticide resistance. See Ref. 33, pp. 71–98

97. Van den Bosch, R., Stern, V. M. 1962. The integration of chemical and biological control of arthropod pests. *Ann. Rev. Entomol.* 7:367–86

98. White, R. J., White, R. M. 1981. Some numerical methods for the study of genetic changes. In *Genetic Consequences of Man Made Change*, ed. J. A. Bishop, L. M. Cook, pp. 295–342. New York: Academic. 409 pp.

99. Whitehead, J. R., Roush, R. T., Norment, B. R. 1985. Resistance stability and coadaptation in diazinon-resistant house flies (Diptera: Muscidae). *J. Econ. Entomol.* 78:25–29

100. Whitten, M. J., Dearn, J. M., McKenzie, J. A. 1980. Field studies on insecticide resistance in the Australian sheep blowfly, *Lucilia cuprina. Aust. J. Biol. Sci.* 33:725–35

101. Whitten, M. J., McKenzie, J. A. 1982. The genetic basis for pesticide resistance. *Proc. 3rd Aust. Conf. Grassland Invertebr. Ecol.*, pp. 1–16. Adelaide, Australia: South Aust. Gov. Print. 402 pp.

102. Wolfenbarger, D. A., Raulston, J. R., Bartlett, A. C., Donaldson, G. E., Lopez, P. P. 1982. Tobacco budworm: selection for resistance to methyl parathion from a field-collected strain. *J. Econ. Entomol.* 75:211–15

103. Wood, R. J. 1981. Strategies for conserving susceptibility to insecticides. *Parasitology* 82:69–80

104. Wood, R. J., Bishop, J. A. 1981. Insecticide resistance: populations and evolution. See Ref. 98, pp. 97–127

105. Wood, R. J., Cook, L. M. 1983. A note on estimating selection pressures on insecticide resistance genes. *Bull. WHO* 61:129–34

106. Wool, D., Noiman, S. 1983. Integrated control of insecticide resistance by combined genetic and chemical treatments: a warehouse model with flour beetles (*Tribolium;* Tenebrionidae, Coleoptera). *Z. Angew. Entomol.* 95:22–30

107. Yarbrough, J. D., Roush, R. T., Bonner, J. C., Wise, D. A. 1986. Monogenic inheritance of cyclodiene resistance in mosquitofish, *Gambusia affinis. Experientia.* 42:851–53

Ann. Rev. Entomol. 1987. 32:381–413
Copyright © 1987 by Annual Reviews Inc. All rights reserved

BIOSYNTHESIS OF ARTHROPOD EXOCRINE COMPOUNDS

Murray S. Blum

Laboratory of Chemical Ecology, Department of Entomology, University of Georgia, Athens, Georgia 30602

INTRODUCTION

The remarkable success of the arthropods is highly correlated with their ability to produce a variety of natural products that regulate diverse inter- and intraspecific interactions. Numerous distinctive pheromones (21) and allomones (18) have already been identified as exocrine products of these invertebrates, and it seems certain that these animals will continue to be an outstanding source of natural products. Furthermore, the structural eclecticism of these compounds emphasizes the great biogenetic potential of this animal phylum.

Research, particularly in the last decade, has probed the biosynthesis of some of the novel acetogenins, alkaloids, and steroids identified in the glandular secretions of arthropods. While investigations into the metabolic derivations of both pheromones and allomones are still in the embryonic stage, they indicate that this is an especially fruitful topic in insect biochemistry and physiology. Except in a 1981 review on the biogenesis of arthropod defensive allomones (18), however, the biochemistry of these exocrine compounds has not been examined in great detail. In the present review, the biosynthetic elegance of these wondrous animals, as well as their great potential as subjects for studies on the metabolic derivations of natural products, becomes abundantly manifest.

BIOSYNTHESIS OF EXOCRINE COMPOUNDS

Researchers have studied the biogenesis of a large variety of arthropod natural products, particularly compounds produced by species in the orders Poly-

381

0066-4170/87/0101-0381$02.00

desmida, Dictyoptera, Coleoptera, Lepidoptera, and Hymenoptera. For the most part, these investigations have determined whether candidate precursors could be converted to an exocrine compound, and metabolic intermediates are largely unknown. Although putative precursors of certain exocrine compounds have sometimes been identified in secretions along with the compounds (18), these are generally not discussed in this review, which is limited to demonstrated biosynthetic conversions. Exocrine compounds are grouped into major chemical classes whenever possible, but some exceptions are made to emphasize selected biosynthetic interrelationships.

Nonterpenoid Hydrocarbons

The incorporation of labeled acetate into hydrocarbons was established some time ago (67), and considerable evidence now supports that these compounds are derived by elongation-decarboxylation of fatty acids (26). For the most part, studies on the biosynthesis of hydrocarbons classified as exocrine compounds have been limited to dipterous pheromones that are cuticular constituents (54).

Muscalure, Z-9-tricosene, the primary sex pheromone produced by female *Musca domestica*, can be biosynthesized from [1-^{14}C]acetate, [1-^{14}C]stearate, [1-^{14}C]oleate, and [9,10-^{3}H]oleate (32). Acetate is converted to stearic acid, which is desaturated to oleic acid and then elongated to Z-10-tetracosenoic acid. Decarboxylation yields Z-9-tricosene. Two other components of the sex pheromone, Z-9,10-epoxytricosane and Z-14-tricosen-10-one, are also synthesized from muscalure (17). Females produce both compounds from [9,19-^{3}H]Z-9-tricosene, the epoxide constituting more than 80% of the oxygenated compounds. Biosynthesis of muscalure and its oxygenated metabolites is induced by 20-hydroxyecdysone (1, 16) (Figure 1).

Another dipterous sex pheromone, 15,19,23-trimethylheptatriacontane, produced by female *Glossina morsitans*, is readily synthesized from [2,3-^{14}C]succinate (65).

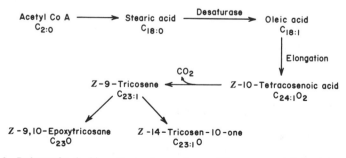

Figure 1 Pathway for the biosynthesis of muscalure and its oxygenated derivatives, modified from Dillwith et al (32).

n-Undecane, a Dufour's gland constituent of the ant *Lasius fuliginosus,* is highly labeled from ingested [2-^{14}C]mevalonate (111). Similarly, the pentatomid *Nezara viridula* converts [1-^{14}C]acetate into *n*-tridecane and *n*-dodecane (44).

Carboxylic Acids

The short-chain fatty acids studied in ants, beetles, and lepidopterous larvae are all derived from amino acids.

The pygidial gland secretion of the beetle *Carabus taedatus* is dominated by methacrylic acid, which is in admixture with ethacrylic acid. Adults of this carabid efficiently convert DL-[4-^{14}C]valine into methacrylic acid (7) (Structure 1), in a manner similar to that in which mammals produce other short-chain unsaturated acids from isoleucine (89), the probable precursor of ethacrylic acid. Half of the radioactivity from the terminal methyl groups of valine was isolated in the methylene carbon of methacrylic acid (7).

$$\underset{\text{Valine}}{\text{(NH}_2\text{)-COOH}} \longrightarrow \underset{}{\text{(O)-COOH}} \longrightarrow \underset{\text{Methacrylic acid}}{\text{=COOH}} \qquad 1.$$

The alcoholic moiety of isopentyl acetate, an alarm pheromone of *Apis mellifera,* is reported to be derived from leucine (99), presumably after reduction of the derived isovaleric acid.

Both [*U*-^{14}C]leucine and [2-^{14}C]acetate are converted into β-hydroxybutyric acid in the osmeteria of early-instar larvae of the papilionid *Papilio aegus* (99). On the other hand, last-instar larvae produce only isobutyric and α-methylbutyric acids in their osmeteria, and both of these compounds are derived from branched-chain amino acids (100). Isobutyric acid was synthesized either in vivo or in vitro from DL-[^{3}H]valine or [*U*-^{14}C]valine (Structure 2), whereas α-methylbutyric acid was derived from [*U*-^{14}C]isoleucine (Structure 3). Strong branched-chain amino acid transaminase activity, which is only expressed in the osmeteria of last-instar larvae, results in the ready transamination of α-ketoglutarate when either leucine, isoleucine, or valine are administered as substrates. Although valine was transaminated less efficiently by osmeterial enzymes than either leucine or isoleucine, the reverse was true when pyruvic acid was incubated instead of α-ketoglutarate (100).

$$\underset{\text{Valine}}{\text{(NH}_2\text{)-COOH}} \longrightarrow \underset{}{\text{(O)-COOH}} \longrightarrow \underset{\text{Isobutyric acid}}{\text{-COOH}} \qquad 2.$$

Isoleucine   α-Methylbutyric acid   3.

Both the biogenesis and mechanism of accumulation of formic acid, the hallmark of formicine ants, have been elucidated. This strongly cytotoxic acid is produced in very high concentrations in formicine venoms, and its biosynthetic regulation appears to present a special problem. However, the combination of an active C-1 metabolism in the poison gland cells with the accumulation of metabolic end product in an isolated and insulated reservoir constitutes an operational system for biosynthesis and storage of formic acid (50).

In both *Camponotus* and *Formica* species, formic acid can be derived from the α or β carbons of serine or the α carbon of glycine (51). Serine can be easily converted to glycine following a β-elimination reaction, and the latter can be totally degraded to yield its α carbon for formate synthesis. In addition, the C-2 of the imidazole ring of histidine is incorporated into formic acid, probably via *N*-formylglutamate and formimino tetrahydrofolate (103). Hefetz & Blum (51) concluded that any compound capable of contributing a C-1 fragment to tetrahydrofolate (H_4-folate) can be regarded as a potential precursor of formic acid.

Studies with a cell-free system established an absolute requirement for H_4-folate. Incubation with [3-^{14}C]serine resulted in the synthesis of labeled [5,10-^{14}C]methenyl formate, which can only arise from 5,10-methylene folate (51). Since [^{14}C]formic acid is the end product when poison glands are incubated with [3-^{14}C]serine, 5,10-methenyl H_4-folate must be converted to 10-formyl H_4-folate prior to the generation of the formic acid moiety (Figure 2). Furthermore, the demonstration that the four enzymes transferring the β-carbon of serine to formic acid (serine hydroxymethyltransferase, 5,10-methylene H_4-folate dehydrogenase, 5,10-methylene H_4-folate cyclohydrolase, and 10-formyl H_4-folate synthetase) were all present in high concentrations in the poison gland provides further strong support for the H_4-folate pathway. Indeed, 10-formyl H_4-folate synthetase, which directly catalyzes the formation of formic acid, is the most active enzyme present in the formicine poison gland (51).

Compartmentalization of formic acid effectively isolates this powerful cytotoxin. The last step in the conversion of H_4-folate intermediates to formic acid in the poison gland cells also generates ATP, which is probably utilized in the transport of formic acid through the cell membrane to the gland lumen. The carrier system would be saturated after the gland lumen was filled because of the back pressure created by the accumulated formic acid (51). Production of the acid would then cease, since the equilibrium between formic acid and 10-formyl H_4-folate is 1 : 20, favoring the latter. This equilib-

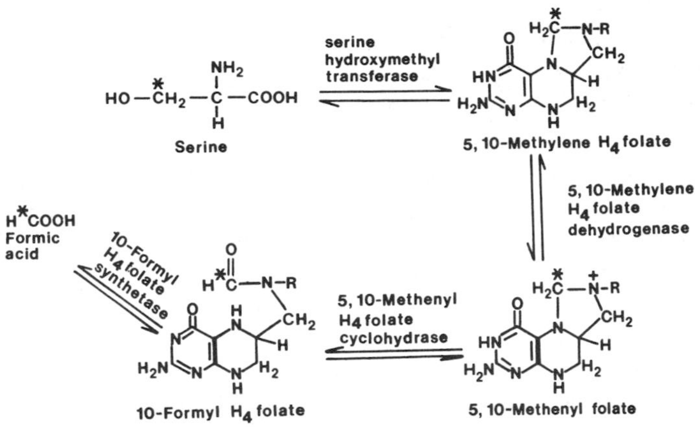

Figure 2   Pathway for the biosynthesis of formic acid in formicine ants, after Hefetz & Blum (51).

rium guarantees that the concentration of formic acid in the gland cells will always be low; this cytotoxin can only accumulate by being transferred, in this case to the impermeable poison-gland reservoir. Reactivation of this system will only occur after discharge of formic acid from the poison-gland reservoir.

Acetogenins

A multitude of compounds derived from the acetate pool have been studied recently, particularly the sex pheromones of moths. These acetogenins generally possess unbranched carbon skeletons with an oxygen-containing functional group and one or two double bonds. Most of them are presented together so it is possible to focus on their biogenetic relationships as components of pheromonal blends. The biosyntheses of a few acetogenic, novel allomones are also described.

PEDERIN   A polyketide origin has been demonstrated for the complex secondary amide pederin (Structure 4), a product of the beetle *Paederus fuscipes*. In beetles fed [1-^{14}C]acetate, [2-^{14}C]acetate, [1,2-^{14}C]glycine, and [2-^{14}C]propionate, pederin was labeled in all cases. Specific degradation studies with pederin derived from beetles fed either [1-^{14}C]acetate or [2-^{14}C]acetate demonstrated that carbon atoms at positions 2 and 2a were more efficiently labeled than those at 5, 6, 7, 8, 10, and 11 (23). The C-2 atom of acetate was utilized almost exclusively for C-2 of pederin. The acetates also provided extensive labeling of carbon atoms 10–18. The labeling pattern after administration of ^{14}C-acetates is consistent with a polyketide origin for pederin.

ALIPHATIC ESTERS, ALCOHOLS, ALDEHYDES, AND ACIDS   The C_{12}–C_{16} sex pheromones (mono- and diunsaturated esters, alcohols, and aldehydes) of many moth species have been the subject of intensive research in the last several years (90). Most of these investigations have been limited to in vivo studies, and with few exceptions (77, 120) the relevant enzymes have not been characterized. Analyses of candidate compounds as fatty acyl precursors, combined with the utilization of labeled precursors usually applied to the sex pheromone gland, have made it possible to identify some major metabolic pathways for the biogenesis of these chemical releasers of sexual behavior.

The major components of the sex pheromone of *Argyrotaenia velutinana, Z*- and *E*-11-tetradecenyl acetates (91:9), and other acetates (9) have been used as models for biosynthetic studies. The analogous acids, *Z*- and *E*-11-tetradecenoic acids (2:3), are present only in the sex pheromone gland, primarily in the triglycerides (15). However, neither labeled *E*- nor *Z*-11-tetradecenoate were incorporated into pheromonal components when glands were incubated with synthetic triacylglycerols (14). Whereas application of [1-^{14}C]acetate to the gland resulted in incorporation of label into these acids, very little label was found in the C_{16} and C_{18} unsaturated acids (10). A similar pattern of incorporation occurred with [*U*-^{14}C]hexadecanoic acid, whereas less label from [1-^{14}C]hexadecanoic acid was incorporated into the 14-carbon acyl moieties, indicating that the latter arose from chain-shortening of hexadecanoyl moieties. Decrease in proportions of both labeled hexadecanoate and tetradecanoate with time, and increase in both *Z*- and *E*-11-tetradecenoate were considered to be consistent with chain-shortening of the C_{16} fatty acyl moiety and desaturation of the C_{14} moiety (12).

Studies with cell-free preparations of glands of *Argyrotaenia citrana* supported the idea that *Z*-11-tetradecenyl acetate is synthesized via the chain-shortening pathway (119). Since *E*-11-[1-^{14}C]tetradecenoic acid is preferentially converted into *E*-11-tetradecenyl acetate, most of the labeled acetate must have arisen from reduction and acetylation of the labeled fatty acyl moiety. There was no evidence for the presence of *Z/E*-11-isomerases in the gland. Evidence was obtained for the presence of *Z*- and *E*-11-desaturases specific for the tetradecanoyl chain (10) (Figure 3).

Subsequent studies with *Argyrotaenia velutinana* in which either mass-labeled deuterated or C_{13} pheromone precursors were utilized have demonstrated, however, that triacylglycerols are not the source of *Z*- and *E*-11-

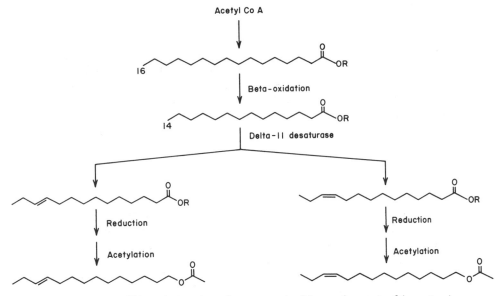

Figure 3   Proposed biosynthetic pathway for components of the sex pheromone of *Argyrotaenia velutinana*, after Roelofs & Bjostad (90).

tetradecenoate (14). Previously observed increases in the relative proportion of ^{14}C-labeled *E*-11-tetradecenoate with time after incubation were believed to be due to a 10:1 difference in the respective rates of conversion of the triacylglycerol *E*- and *Z*-11-tetradecenoates to their corresponding pheromonal components (12). It is now suggested that Δ-11-desaturation of tetradecanoate may produce twice as much *Z*- as *E*-11-tetradecenoate (14). Some of the *Z*-isomer would be converted to *Z*-11-tetradecenoate, and the remainder would be converted to triacylglycerols, which would constitute a metabolic dead end. Only a trace of the *E*-isomer would be converted to the Δ-11-unsaturated fatty acid; the remainder would be incorporated into triacylglycerols. According to this new scheme the triacylglycerols are regarded as a sink for the C_{14} precursors of the pheromonal components (13). The scheme remains to be verified.

The biosynthesis of *Z*-7-dodecenyl acetate, the sex pheromone of *Trichoplusia ni*, has also been shown to occur via chain-shortening of *Z*-11-tetradecenoate generated from *Z*-11-hexadecenoate (11). The major labeled unsaturated fatty acid in the gland after treatment with [1-^{14}C]acetate was *Z*-11-hexadecenoate, accompanied by *Z*-7-dodecenoate, the pheromone precursor. The *Z*-11-desaturase, present only in the sex pheromone gland, has been characterized (116); Bjostad & Roelofs (11) have suggested that this enzyme is the key desaturase for generating tortricid pheromones. On the other hand, the presence of an unusual Z-10-desaturase has been postulated for moths producing Z-8-tetradecenyl acetate in the sex pheromone gland (66).

Similar results were obtained in studies on the biosynthesis of bombykol, the *E,Z*- and *E,E*-isomers of 10,12-hexadecadienol, the sex pheromone of *Bombyx mori*. Hexadecanoate has been demonstrated to be a precursor of bombykol (56), and the fatty acyl precursors of the pheromones, *E,Z*-10,12-hexadecadienoate and *E,E*-10,12-hexadecadienoate, are present in the sex-pheromone gland, primarily in the triacylglycerols (13). In addition, Z-11-hexadecenol has been detected in the gland (123, 125), along with Z-11-hexadecenoic acid (122). The conversion of Z-$[11,12d_2]$-11-hexadecenoic acid to the *E,Z*- and E,E-isomers of bombykol, Z-11-hexadecenol, and bombyk acid (*E,Z*-10,12-hexadecadienoic acid) (125) demonstrates that the Z-11-16:acyl moiety is a major precursor of the pheromone components (Figure 4). This pathway would require isomerization about the Z-11 double bond in order to produce the characteristic 10,12 double-bond system of the pheromonal components. The presence of protein-bound *E,Z*-bombykol linolenate in the hemolymph two days before emergence of the adult female (124) may indicate that this ester is a storage form of the pheromone.

The major sex-pheromone components of *Choristoneura fumiferana*, *E*- and Z-11-tetradecenal (96:4), are synthesized from $[1-^{14}C]$acetate via *E*- and Z-11-tetradecenyl acetates (76). However, no evidence of chain-shortening was found, and $[1-^{14}C]$hexadecanoate was incorporated as efficiently as tetradecanoate or dodecanoate, presumably after β-oxidation to acetyl CoA and resynthesis into 11-tetradecenyl acetate. In vitro luminescence assays have further demonstrated that the sex-pheromone gland contains a specific esterase that hydrolyzes the *E*- and Z-isomers of 11-tetradecenyl acetate, an oxidase that converts both isomers to 11-tetradecenol in the presence of oxygen, and an NAD-dependent dehydrogenase that generates the isomers of

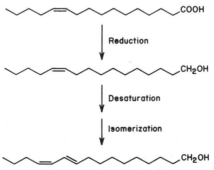

Figure 4 Proposed pathway for the biosynthesis of *E,Z*-10,12-hexadecadienol, a sex pheromone of *Bombyx mori* (125).

11-tetradecenal (77). In vivo studies with tritiated tetradecanyl acetate demonstrated the presence of all three enzymes in the gland.

In vitro and in vivo studies with both [³H]acetyl-CoA and [2-¹⁴C]malonyl-CoA established that the synthesized fatty acids are the same as those in the triglycerides (78). Incorporation into the acetate ester was the same with myristate as it was with the other labeled fatty acids; this again points to β-oxidation acid resynthesis. In *C. fumiferana* free exogenous fatty acids cannot be used efficiently to generate the pheromones.

The presence of an aldehyde reductase, but not a fatty acid reductase, has been demonstrated in the gland of *C. fumiferana*. It has been suggested that this reductase functions in concert with the desaturase so that the former modifies the isomeric ratio produced by the latter (78). The presence in the gland of a specific enzyme for long-chain alcohol acceptors with acetyl CoA as the acyl donor indicates that this enzyme is exclusively involved in pheromone biosynthesis (78). In addition, a specific acetate esterase, which generates the long-chain alcohol, and an alcohol oxidase, which produces the aldehyde, are present in the gland. Since the aldehydic pheromone is only generated when the female "calls," it is likely that the decrease in the acetate ester precursor in the "calling" female reflects production of the pheromone from the ester storage form (Figure 5).

Alcoholic precursors of aldehydic pheromones applied to glands are rapidly converted into pheromone constituents. Two of the major constituents of the sex pheromones of *Heliothis virescens* and *H. zea*, Z-11-hexadecenal and Z-9-tetradecenal, are biosynthesized in large quantities if the glands of these

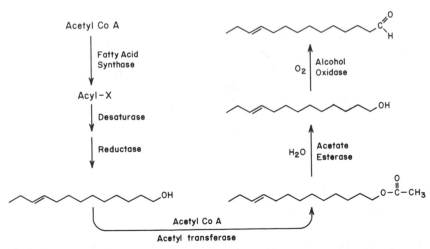

Figure 5 Proposed pathway for the biosynthesis of *E*-11-tetradecenal in *Choristoneura fumiferana*, after Morse & Meighen (78).

moths are treated with Z-11-hexadecenol and Z-9-hexadecenol, respectively (104). Furthermore, the gland of *H. virescens* will produce bombykal (*E,Z*-10,12-hexadecadienal), a component of the pheromone of *B. mori*, after treatment with the alcoholic precursor; this indicates that the alcohol oxidase is not highly specific. These results suggest that a cuticle-bound alcohol oxidase converts the alkenol to the enal as the former diffuses through the cuticle (104).

The sex pheromone of *Chilo suppressalis*, Z-11-hexadecenal, is rapidly produced from deuterated hexadecanoate (mass-labeled) and hexadecanal (3). The absence of detectable 1-hexadecanol in the gland and the failure of deuterated hexadecanoic acid to be reduced to the alcohol are consistent with a biosynthetic pathway in which hexadecanoic acid is reduced directly to the aldehyde and is then desaturated.

The saturated aldehydic sex pheromones produced by males of *Galleria mellonella*, *n*-nonanal and *n*-undecanal, are both produced from ^{14}C-labeled acetate, propionate, and oleate (96). *n*-Nonanal is labeled much more extensively from administered [10-^{14}C]oleate than from [1-^{14}C]oleate, probably because the latter is degraded by β-oxidation. The C_{11} aldehyde was poorly labeled compared to *n*-nonanal.

Allomonal aldehydes have also been demonstrated to have acetogenic origins. *E*-2-Hexenal, *E*-2-decenal, and 4-oxo-*E*-2-hexenal, defensive compounds produced in the metathoracic glands of adults of *Nezara viridula*, are labeled after the pentatomids are injected with [1-^{14}C]acetate (44). Another hemipteran, *Leptoglossus phyllopus*, produces a defensive secretion that is dominated by hexanal, hexyl acetate, and 1-hexanol (2). The concentration of the ester decreases with age, whereas that of the aldehyde increases. The secretion from the glandular reservoir is biphasic, containing both an organic and an aqueous phase. Significantly, the aqueous phase contains an esterase that hydrolyzes the acetate and two dehydrogenases that oxidize 1-hexanol. Thus, the reactive aldehyde is generated in the storage reservoir, away from sensitive glandular cells, by the actions of the esterase and dehydrogenases.

Another ester, Z-vaccenyl acetate, is converted by an enzyme, esterase 6, into an antiaphrodisiacal pheromone, Z-vaccenyl alcohol, in the female reproductive tract of *Drosophila melanogaster* (68). Both the substrate and enzyme are male accessory-gland products introduced into the female vagina as part of the seminal ejaculate.

Another ester, decyl acetate, a Dufour's gland product of the ant *Formica schaufussi*, was labeled after workers were fed [1-^{14}C]acetate (45). Only one sixth of the label was found in the decyl alcohol moiety.

EPOXIDES   Some moth species use alkenes or their epoxides as sex pheromones, and it is likely that they are derived from decarboxylated fatty acids.

The sex pheromone of *Lymantria dispar,* Z-7,8-epoxy-2-methyloctadecane, was produced when adult females were treated with Z-[7,8-^{3}H]-2-methyl-7-octadecene, a precursor that is also found in the gland (60).

MACROLIDE LACTONES   Males of the beetle *Cryptolestes ferrugineus* converted labeled palmitic, oleic, and linoleic acids into one of their macrolide aggregation pheromones (D. Vanderwel, H. D. Pierce, Jr., A. C. Oehlschlager, unpublished data). This metabolic pathway must involve β-oxidation, hydroxylation, desaturation, and ultimately cyclization (Figure 6).

Quinones

The metabolic derivations of several of the 1,4-quinones produced by insects and millipedes have been determined. In an early investigation of the tracheal glands of *Diploptera punctata,* Roth & Stay (91) demonstrated the presence of a phenoloxidase and β-glucosidase capable of generating 1,4-quinones. *D. punctata,* like the tenebrionid beetle *Tribolium castaneum,* produces three quinones, *p*-benzoquinone, methyl-*p*-benzoquinone, and ethyl-*p*-benzoquinone. Happ (47) identified glucosides and *p*-diphenols in the secretory cells of the abdominal defensive glands of tenebrionid beetles, and demonstrated that hydrolysis and oxidation of the quinols generates 1,4-quinones (47). Compartmentalization of the glucosidase and phenoloxidase resulted in the synthesis of quinones in the gland reservoir, spatially removed from the sensitive secretory cells.

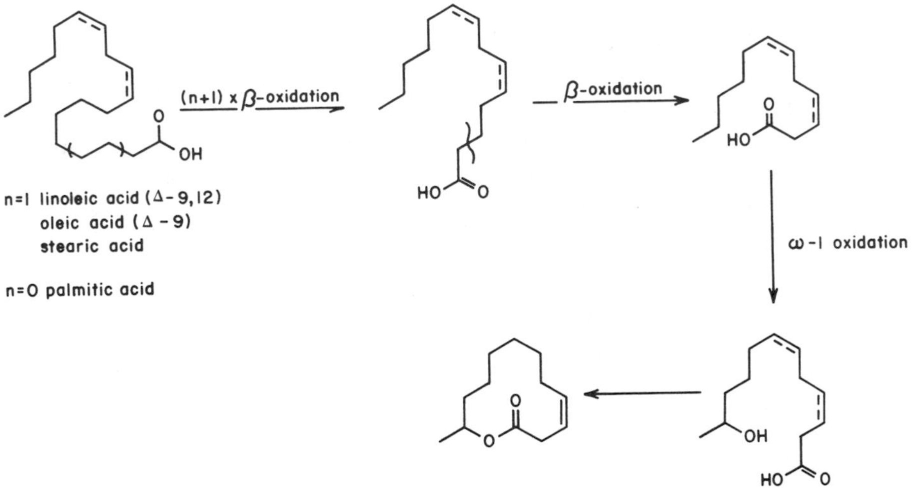

Figure 6 Proposed pathway for the biosynthesis of an aggregation pheromone of *Cryptolestes ferrugineus,* after Vanderwel, Pierce & Oehlschlager (unpublished data).

The enzymatic derivation of 1,4-quinones by bombardier beetles in the genus *Brachinus* has been studied in considerable detail (93). Quinones are explosively discharged after biosynthesis in a bicompartmentalized secretory apparatus. A 25% solution of hydrogen peroxide and a 10% solution of hydroquinones are transferred to the explosion chamber, which contains both catalases and polyphenoloxidases. The peroxide is decomposed into oxygen and water by four catalases (H_2O_2 = oxidoreductase), whereas the quinols are oxidized to *p*-benzoquinone and methyl-*p*-benzoquinone by three polyphenol-oxidases (polyphenol = H_2O_2: oxidoreductase) (Figure 7). The enzymes are very stable in the presence of concentrated H_2O_2 and have remarkably high temperature optima (70–80°C). The reaction rates of the catalases are accelerated by the high reaction temperature, and these enzymes exhibit an unlimited increase in activity with increasing substrate concentration. Hydroquinone was oxidized more effectively than any quinol evaluated; only resorcinol had a similar oxidative rate (93).

Meinwald et al (71) studied the biosynthesis of *p*-benzoquinone, methyl-*p*-benzoquinone, and ethyl-*p*-benzoquinone in *Eleodes longicollis* by administering [2-^{14}C]tyrosine, [ring-^{14}C]phenylalanine, [1-^{14}C]tyrosine, [2-^{14}C]acetate, [2-^{14}C]propionate, and [2-^{14}C]malonate (71). Only the alkylated quinones were synthesized from the aliphatic acids, whereas *p*-benzoquinone was generated from the preformed rings of the aromatic amino acids. Similar results were obtained with another tenebrionid, *Zophobas rugipes* (S. S. Duffey, M. S. Blum, unpublished data).

Degradation of methyl-*p*-benzoquinone established that 30% of the label appeared at C-2, and less than 0.1% of the activity was present in the methyl group. The other 70% of the activity was present in the other ring carbons. On the other hand, although [1-^{14}C]propionate was poorly incorporated into methyl-*p*-benzoquinone, ethyl-*p*-benzoquinone was effectively labeled; 95% of the total activity was localized in C-2 of the ring (71). Since propionate is not incorporated into the methyl homolog of *p*-benzoquinone, this precursor appears to be only slightly incorporated into C-2 units. Condensation of malonate units with propionyl CoA is compatible with the observed labeling (Figure 8).

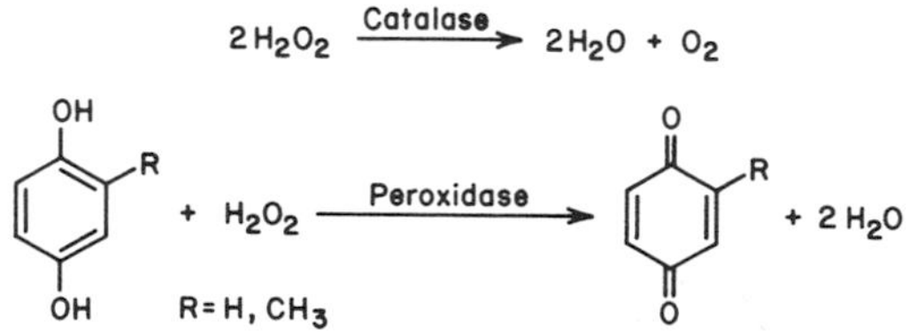

Figure 7 Proposed pathway for the biosynthesis of *p*-benzoquinones in *Brachinus* spp., after Schildknecht et al (93).

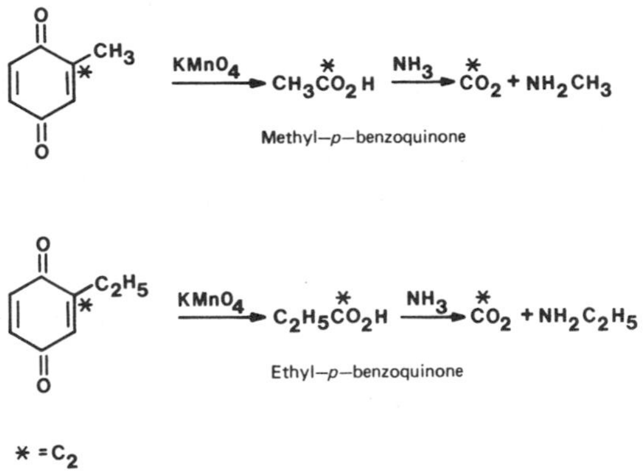

Figure 8 Degradation scheme for labeled *p*-benzoquinones, after Meinwald et al (71).

The methoxy group of 2-methoxy-3-methyl-1,4-benzoquinone, a product of the millipede *Narceus gordanus,* is reported to be derived from methionine; most of the label is found in the methyl group of the alkoxy group (112). In addition, both this quinone and methyl-*p*-benzoquinone were highly labeled after administration of $[^{14}C]$6-methylsalicylic acid; this indicates that they arise via the acetate pathway. On the other hand, the millipedes *Narceus annularis, Cambala annulata,* and *Rhinocricus holomelanus* required the preformed aromatic ring of tyrosine in order to synthesize these 1,4-quinones; no labeled quinones were produced after administration of ^{14}C-labeled acetate or malonate (S. S. Duffey, M. S. Blum, unpublished data).

Terpenes

Arthropods produce a large variety of terpenoid constituents including mono-, sesqui-, di-, and even sesterterpenes (18). In these invertebrates it appears that the biosynthesis of polyisoprenoids generally occurs via the classical pathway of acetate condensation and utilizes mevalonic acid as an intermediate. In some cases it has been demonstrated that insects, most commonly scolytid bark beetles, can metabolize host-derived monoterpenes into pheromones (114). Although it has also been demonstrated that fungi and bacteria associated with insects may be important in the production of terpenoid pheromones (20), their exact role remains to be determined.

MONOTERPENES   Citral and citronellal, two cephalic alarm pheromones of *Acanthomyops claviger,* were extensively labeled after the ants were fed $[1-^{14}C]$acetate, $[2-^{14}C]$acetate, and $[2-^{14}C]$mevalonate (48). Similarly, injection of these compounds into *Anisomorpha buprestoides* resulted in extensive labeling of anisomorphal, a terpenoid defensive compound produced by these stick insects (70). The same precursors were incorporated into a defensive allomone, terpinolene, a cephalic product of soldiers of *Nasutitermes octopilis* (88). However, in at least some cases an insect may synthesize the same monoterpenes via the mevalonate pathway and/or from exogenously derived terpenoid precursors.

All pheromonal components of grandlure, a quaternary blend produced by males of *Anthonomus grandis,* can be derived from labeled acetate, mevalonate, or glucose (73). However, since insects developing on cotton produce more pheromone than those reared on an artificial diet, a possible biosynthetic role has been suggested for exogenous terpenoid precursors. Administration of a mixture of [^{3}H]geraniol-nerol to male weevils resulted in far more extensive labeling of all four monoterpenes (105) than when the pheromones were derived from either acetate or mevalonate (73).

It has been suggested that males of *A. grandis* may also produce pheromonal constituents from respired monoterpenes, since both myrcene and limonene were allylically oxidized to alcohols after weevils were exposed to vapors of either of these terpenoid hydrocarbons (49). Oxidation of volatile monoterpenes has been well established for bark beetles as well. For example, ipsenol and ipsdienol, two of the aggregation pheromones produced by the bark beetle *Ips paraconfusus,* were synthesized in the hindgut of males exposed to vapors of myrcene (55). The concentration of pheromones was directly related to the myrcene concentration (22). The conversion of myrcene to ipsenol and ipsdienol by males of *I. paraconfusus* was confirmed with deuterated myrcene (52), and it was also demonstrated that ipsenol was stereoselectively produced by reduction of (−)- but not (+)-ipsdienol (43). Furthermore, ipsdienone, the ketone analog of ipsdienol, was suggested to be in equilibrium with the latter. It was subsequently determined that males convert exogenous ipsdienone to a 36:64 ratio of (+)- and (−)-ipsdienol and a 93:7 ratio of (+)- and (−)-ipsenol. Thus, (−)-ipsenol can be produced indirectly by oxidation of ipsdienol to ipsdienone, followed by enantioselective reduction of ipsdienone to ipsdienol, which is reduced to ipsenol (42) (Figure 9). A number of additional host-derived monoterpene hydrocarbons are reported to be oxidized to allylic alcohols (D. Vanderwel, H. D. Pierce, Jr., A. C. Oehlschlager, unpublished data), probably by microsomal mixed-function oxidases (114). Some bark beetles have evidently exploited this normal detoxicative system to generate pheromones.

Among insect-derived monoterpenes, cantharidin, a product of meloid beetles, has been studied in most detail. Males of *Lytta vesicatoria* produce this terpenoid anhydride from [1-^{14}C]acetate and [2-^{14}C]mevalonate, but adult females are unable to synthesize it (72, 95). Females receive their cantharidin as a copulatory bonus; more than 90% of the labeled terpene was transferred from the seminal plasma of males injected with an appropriate precursor 24–30 hr prior to a copulation of 24 hr duration (101). More than

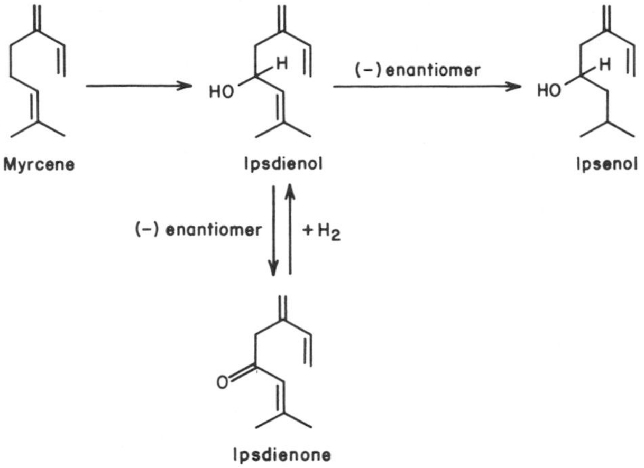

Myrcene   Ipsdienol   Ipsenol

Ipsdienone

Figure 9 Proposed biosynthetic pathways for ipsenol and ipsdienol in males of *Ips paraconfusus*, after Fish et al (42).

10% of labeled cantharidin was transferred to females even when the males were injected during a relatively short (4 hr) copulatory period. Cantharidin appears to be produced in the male accessory glands (101).

When cantharidin is biosynthesized from [1-^{14}C]acetate two thirds of the labeling occurs in carbon atoms 2 and 3, whereas when it is biosynthesized from mevalonate one sixth of the activity is at atoms 8 and 9 and one sixth is at atoms 10 and 11 (72, 85). These results demonstrate that cantharidin is not synthesized by the head-to-head linkage of two isoprene units. Following these findings, it was demonstrated that E,E-11-[^{14}C;^{3}H]-farnesol was efficiently converted to cantharidin (101); this established that this sesquiterpene alcohol was a key precursor. On the other hand, labeled geraniols were not converted to cantharidin via farnesol, even when the C_{10} and C_{15} alcohols were injected together (84). Injection of beetles with doubly labeled farnesol demonstrated that cantharidin was derived from farnesol after cleavage between C-1 and C-2, C-4 and C-5, and C-7 and C-8. Thus elimination of carbon atoms C-1, C-5, C-6, C-7, and C-7' results in the transformation of farnesol to cantharidin (84).

Injection of doubly labeled (^{14}C, ^{3}H) farnesol further demonstrated that E,E-farnesol was the actual precursor of cantharidin (118) (Structure 5). Furthermore, simultaneous injection of E,E-[11,12-^{3}H]farnesol and [2-^{14}C]mevalonolactone yielded doubly labeled cantharidin. The remainder of the farnesol, isolated after incubation, had incorporated ^{14}C activity according to the isoprene rule (83). An intramolecular conversion of farnesol to cantharidin is indicated both by the retention in cantharidin of the [2-^{14}C]/[11'-^{14}C] ratio present in the injected precursor, E,E-[2-^{14}C; ^{14}C,11'-^{3}H]farnesol, and by the randomization of the $11'$-^{14}C label at C-9 and C-11 after injection of E,E-[2,4-^{14}C; $11'$,12-^{3}H$_2$]farnesol (117).

Farnesol   Cantharidin   5.

SESQUITERPENES   A variety of sesquiterpene exocrine compounds are derived via mevalonic acid. Workers of *Lasius fuliginosus* rapidly biosynthesize the novel furanoterpene dendrolasin (Structure 6) from [2-^{14}C]mevalonate (24) as well as from [^{14}C]acetate precursors (111). Mevalonate was incorporated into carbon atoms 3, 4, 5, 6, 7, 7', 8, 9, 10, 10', 11, 11', and 12. Although the labeling pattern of dendrolasin was consistent with the usual terpene biosynthetic pathway, the poor utilization of both acetate and mevalonate for dendrolasin biosynthesis indicates that *L. fuliginosus* may use another precursor specifically for the production of this furanoterpene.

6.

Sesquiterpene hydrocarbons, products of the osmeteria of larval *Papilio*, are deuterated when these organs are treated with d_4-acetic acid (53). E-β-farnesene (produced by *P. helenus*) and β-caryophyllene, germacrene-A, and germacrene-B (produced by *P. protenor*) all incorporated deuterium after osmeterial treatment with labeled acetate.

An unusual aggregation pheromone produced by males of *Cryptolestes ferrugineus*, E,E-4,8-dimethyl-4,8-decadien-10-olide, was biosynthesized from labeled acetate and mevalonate (D. Vanderwel, H. D. Pierce, Jr., A. C. Oehlschlager, unpublished data). E,E-farnesol may be a terpenoid precursor of this macrolide lactone, since this sesquiterpene can be converted into the pheromone by oxidative cleavage of the terminal double bond and subsequent cyclization (Figure 10).

DITERPENES   Soldiers of the termite *Nasutitermes octopilis* incorporated injected ^{14}C-labeled acetates and mevalonates into their distinctive tri- and tetracyclic diterpene allomones (88).

TRITERPENES   Although insects lack a cholesterogenic pathway, some species produce a large variety of triterpene allomones. These are derived from

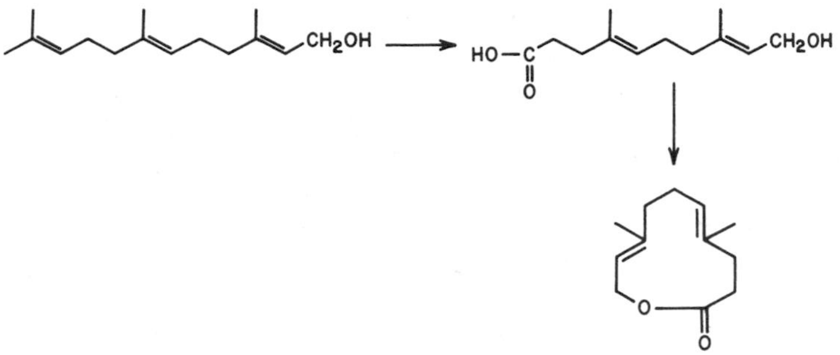

Figure 10 Proposed biosynthetic pathway for a terpenoid aggregation pheromone of *Cryptolestes ferrugineus*, after Vanderwel, Pierce & Oehlschlager (unpublished data).

exogenous sterols (Structure 7). The prothoracic glands of dytiscid beetles are a particularly rich source of these steroids, and such compounds as cortexone, 6,7-dehydrocortexone, cybisterone, and 6,7-dihydrocybisterone have been identified in their secretions. Injection of [4-^{14}C]cholesterol or [4-^{14}C]progesterone into adults of *Acilius sulcatus* resulted in labeling of all of these steroids (92). Although [4-^{14}C]cholesterol served equally well as a precursor for both enones and dienones, [4-^{14}C]progesterone was incorporated well only into the enones. The biosynthetic pathways for the enones and dienones presumably diverge before the biosynthesis of progesterone, which is apparently the precursor of dihydrocybisterone and cortexone. This conclusion is supported by the greater incorporation of [4,6-^{14}C]cholestadien-3-one into dienones than into enones (92). Injection of [4-^{14}C]cholesterol and [^{3}H]cholesterol into *A. sulcatus* revealed that the Δ^4 and Δ^6 bonds are formed by elimination of the 4β and 7β hydrogens, respectively (25).

Aromatic Compounds

Several arthropod pheromones and allomones are biosynthesized from aromatic precursors. 2-Phenylethyl alcohol, the sex pheromone of males of *Mamestra configurata*, was derived from [^{14}C]phenylalanine both in vivo and in vitro (113). [3-^{14}C]Cinnamic acid was also efficiently converted to phenethanol, and the in vitro diminution of the acid was correlated with an increase in phenethanol. Thus, it appears that this aromatic pheromone, whose immediate precursor is 2-phenethyl alcohol glucoside, is derived by the conversion of phenylalanine to 2-phenyllactic acid via cinnamic acid. Glucosylation of the hydroxyl group of phenyllactate followed by reductive decarboxylation would yield the glucosidic precursor of 2-phenylethyl alcohol (113) (Figure 11).

Cortexone

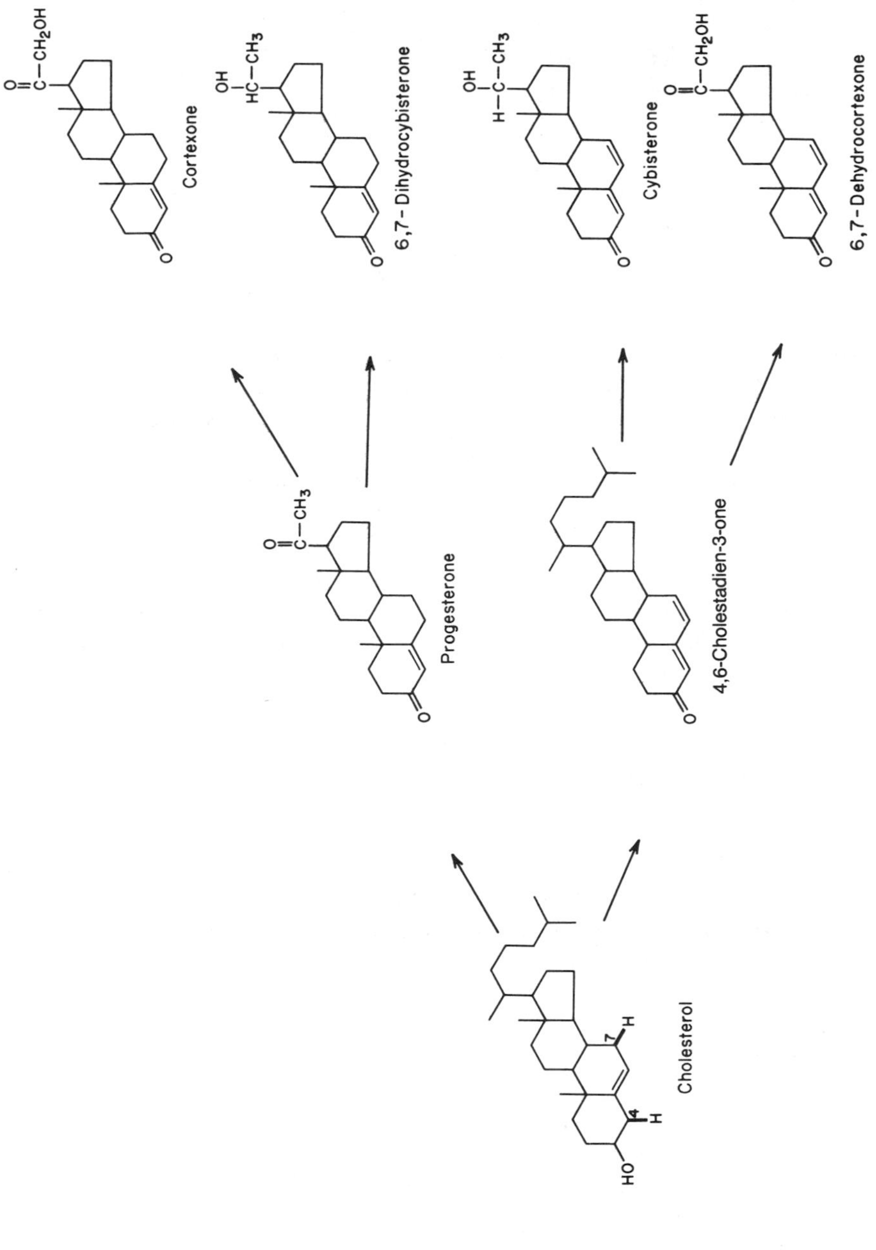

6,7 – Dihydrocybisterone

Cybisterone

6,7 – Dehydrocortexone

Progesterone

4,6-Cholestadien-3-one

Cholesterol

7.

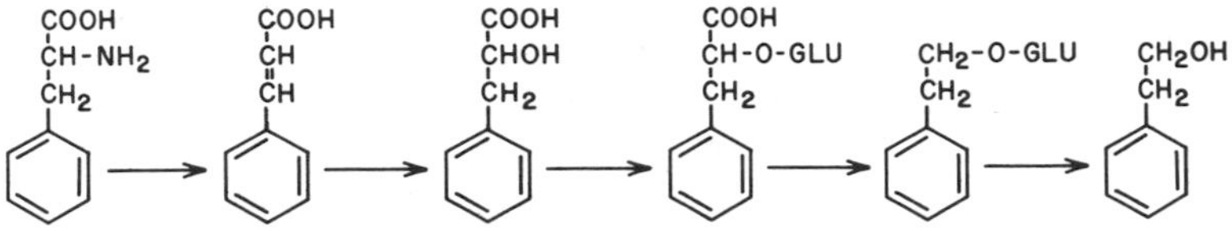

Figure 11   Proposed biosynthetic pathway for 2-phenylethyl alcohol in males of *Mamestra configurata*, after Weatherston & Percy (113).

Guaiacol and phenol, defensive allomones of the millipedes *Oxidus gracilis, Euryurus maculatus,* and *Pseudopolydesmus erasus,* are synthesized from [U-^{14}C]tyrosine (Structure 8) but not from phenylalanine (35). Both in vivo and in vitro studies have demonstrated that the allomones are synthesized in the cyanogenic gland. Duffey et al (33) have isolated the enzyme tyrosine-phenol lyase from the glandular area of *O. gracilis,* and have shown that it reversibly catalyzes the conversion of phenol to tyrosine and vice versa. The same enzyme catalyzes biosynthesis of phenol and guaiacol in the subgenital glands of males of the coreid *Leptoglossus phyllopus* (33).

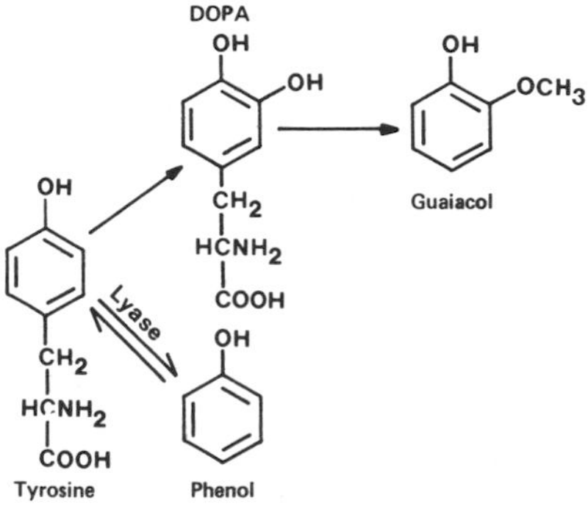

8.

Larvae of two chrysomelid species, *Phratora vitellinae* and *Chrysomela tremulae,* convert salicin, a phenylglucoside, into their defensive allomone, salicylaldehyde (82) (Structure 9). Salicin, a constituent of these insects' host plants, *Salix* or *Populus* spp., is metabolized to an aglycone by the action of a β-glucosidase concentrated in the defensive glands.

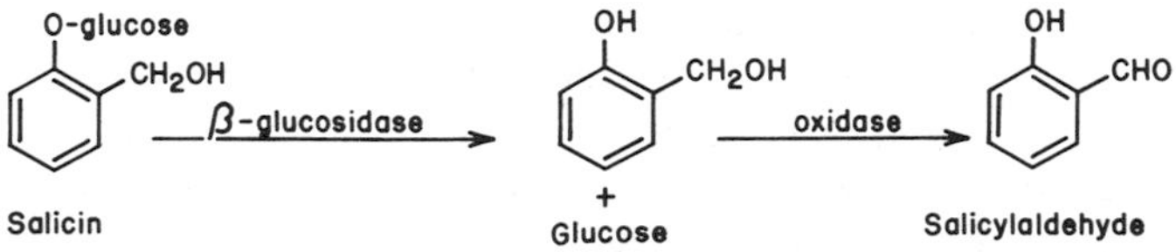

9.

The sex pheromone of the ticks *Amblyomma americanum* and *A. maculatum*, 2,6-dichlorophenol, was labeled in adult females that had been injected with $Na^{36}Cl$ as engorged nymphs (8). This unusual chlorinated pheromone may be derived from chlorination of a phenolic precursor such as tyrosine.

Cyanogens

Several cyanogens produced by arthropods have been studied recently, particularly in millipedes and lepidopterans. The most common cyanogen generated in arthropod defensive glands, HCN (18), frequently appears to be derived from common aromatic precursors in both millipedes and insects.

The defensive secretions of polydesmoid millipedes generally contain a mixture of HCN and benzaldehyde; in some species other cyanogens may be present as well (18). Mandelonitrile (27, 36, 41), benzoyl cyanide (27, 36), and mandelonitrile benzoate (36) have all been identified as defensive allomones of millipedes, and all three compounds can serve as precursors of HCN. The first two compounds have also been detected in exudates of geophilomorph centipedes (58). However, the metabolic derivation has been established unambiguously only for mandelonitrile.

In the millipede *Oxidus gracilis,* labeled HCN was produced from ingested DL-[2-^{14}C]phenylalanine but not from ingested DL-[2-^{14}C]tyrosine (106). Since DL-[3-^{14}C]phenylalanine was converted to labeled benzaldehyde, this amino acid clearly serves as the precursor for both of the major defensive compounds produced by this millipede. Many of the metabolic steps leading from phenylalanine to HCN and benzaldehyde from mandelonitrile have been elucidated for another polydesmid species, *Harpaphe haydeniana* (37).

N-Hydroxyphenylalanine is an efficient precursor of HCN in *H. haydeniana,* whereas phenylpyruvic acid oxime is not. Thus it is unlikely that the hydroxy acid is metabolized to mandelonitrile through its conversion into the free amino acid and oxime. Although phenylpyruvic acid is not utilized efficiently as a precursor of benzaldehyde, it is a better precursor than phenylalanine. In addition, the phenylpyruvic acid oxime reduces the utilization of *N*-hydroxyphenylalanine when these two compounds are fed simultaneously. These facts may indicate that the α-oximino acid is a functional intermediate in the biosynthesis of mandelonitrile (37). Since phenylacetaldoxime does not suppress the incorporation of the oxime into mandelonitrile, it is possible that the oxime enters the metabolic pathway by direct conversion to phenylacetonitrile.

The efficient incorporation of phenylacetaldoxime, phenylacetonitrile, and 2-hydroxyphenylacetaldoxime into cyanogen makes these compounds good candidates for precursors of mandelonitrile. Since labeled phenylacetaldoxime and phenylacetonitrile have been isolated from *H. haydeniana*, derivation of mandelonitrile via hydroxylation of the nitrile (37) seems likely.

Both mandelonitrile lyase and a β-glycosidase are present in the defensive glands of *H. haydeniana* (37). The high activity of the lyase indicates that it is capable of generating the rapid bursts of HCN production observed in this millipede. On the other hand, β-glycosidase is not very active, and is probably involved in the slow resynthesis of glycosidic cyanogen. However, mandelonitrile is clearly the major stored cyanogen (Figure 12).

HCN and benzaldehyde are also probably derived from mandelonitrile in carabids of the genus *Megacephala*. All three compounds have been identified in their defensive secretions (19). The larval defensive secretions of paropsine chrysomelid beetles may be derived from a glucoside of mandelonitrile, as glucose has been identified in their exudates along with HCN and benzaldehyde (75).

Adults of zygaenid moths liberate HCN when injured. It has been demonstrated that they derive this compound from linamarin and lotaustralin, two cyanogenic glucosides that are also produced by plant species (30). Although zygaenids often develop on cyanogenic plants, they also produce these glucosides if they develop on acyanogenic plants. The glucosides are found in eggs, larvae, and pupae of many species (31). In larvae they are abundant in the cuticular defensive secretion and are also found in the hemolymph (115). β-Glucosidase is much more active in the hemolymph than in the secretion, which suggests that the metabolism of the cyanoglucosides occurs primarily in the blood.

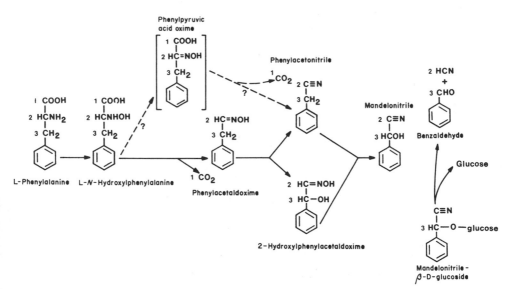

Figure 12 Pathway for the production of benzaldehyde and HCN in *Harpaphe haydeniana*, after Duffey et al (37).

Cyanogenesis has also been detected in adult butterflies of two nymphalid genera, *Acraea* and *Heliconius* (79). Both linamarin and lotaustralin have been identified in a 4:1 ratio in several species in both genera. Recent studies have demonstrated that these two cyanoglucosides are widespread in the tribe Heliconiini; in some genera the linamarin:lotaustralin ratio may exceed 10:1 (80).

The biosynthesis of linamarin and lotaustralin was recently studied in both *Zygaena* and *Heliconius* spp. Larvae fed [U-^{13}C]valine and [U-^{13}C]isoleucine converted these two amino acids into linamarin and lotaustralin, respectively (121). C-13 NMR data indicated that the intact units of the amino acids, minus the carboxyl groups, were incorporated during biogenesis of the glucosides. Results were similar when several heliconiine species were fed [U-^{14}C]valine and [U-^{14}C]isoleucine (81) (Figure 13). Cyanogenesis in adult *Heliconius* occurred but was reduced when these two amino acids were absent from the diet (81).

β-Cyanoalanine, a detoxification product of HCN that is produced by millipedes (35, 37), has also been identified in the larval defensive secretion of *Zygaena trifolii* (115). This known neurotoxin may enhance the defensive efficacy of the secretion. Whether millipedes derive any defensive benefit from detoxifying HCN by producing β-cyanoalanine in their defensive glands (37) remains to be determined.

Alkaloids

Preliminary investigations have been carried out on the biogenesis of several arthropod alkaloids. The quinazilinones produced by the millipede *Glomeris*

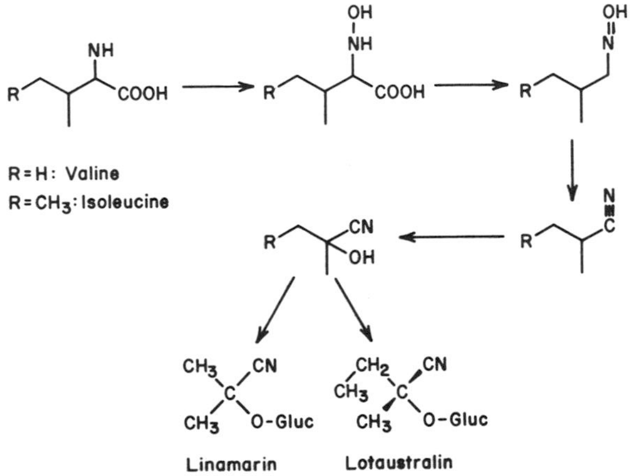

Figure 13 Proposed pathway for the production of linamarin and lotaustralin in *Zygaena* and *Heliconius* spp., after Wray et al (121).

marginata, 1-methyl-2-ethyl-4(3H)-quinazilinone and 1,2-dimethyl-4(3H)-quinazilinone, are derived from anthranilic acid, as are plant-derived quinazilinones. Administration of [^{14}COOH]anthranilic acid to millipedes resulted in extensive labeling of both quinazilinones (94). Alkaline hydrolysis of 1-methyl-2-ethyl-4(3H)-quinazilinone yielded *N*-methyl anthranilic acid (Structure 10), which possessed the same level of specific activity as the intact alkaloid. Thus, the biosynthesized quinazilinone was labeled entirely from the carboxylic acid group of anthranilic acid.

10.

The coccinellines, tricyclic alkaloids produced by coccinellid beetles, have a polyketide origin. Adults of *Coccinella septempunctata* to which [1-^{14}C]acetate or [2-^{14}C]acetate had been administered produced labeled coccinelline, which, after derivatization and oxidation to acetic acid, accounted for about 16% of the original activity (107). Since administered C-2–labeled [^{14}C]acetate yielded labeled acetic acid (corresponding to carbon atom C-10) after oxidation, the methyl group of the alkaloids cannot come from the one-carbon pool. Coccinelline thus appears to be formed by linear condensation of seven acetate units via decarboxylation of the original condensation product, through a precoccinelline intermediate (Figure 14).

Males of certain species of butterflies and moths ingest pyrrolizidine alkaloids produced by plants in the families Boraginaceae, Leguminosae, and Compositae and convert them into sex pheromones that are often referred to as aphrodisiacs (86). These alkaloids are produced in abdominal hair pencils of danaids and are transferred to the females' antennae. Males of *Danaus gilippus berenice* denied access to pyrrolizidine alkaloid–containing plants lacked reproductive competence and were also deficient in 2,3-dihydro-7-methyl-1H-pyrrolizin-1-one, a hair-pencil constituent (87). It was later established that this ketone was present on the hair pencils of males of *D. chrysippus petilea* that had fed on an alkaloid-rich boraginaceous plant, but

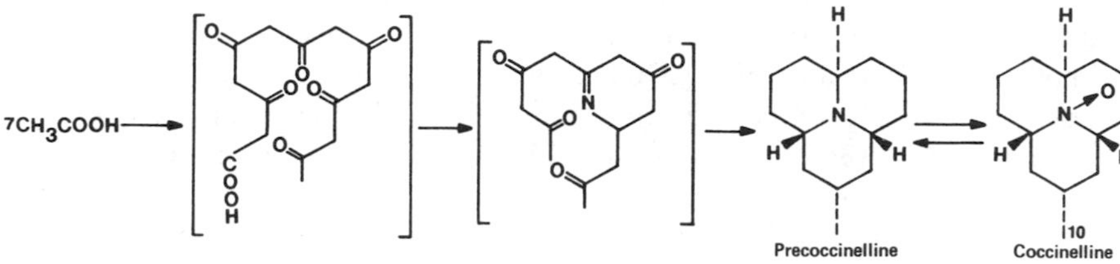

Figure 14 Proposed biosynthetic pathway for coccinelline in *Coccinella septempunctata*, after Tursch et al (107).

was lacking in males that had not fed on these plants (40). Similar results have been obtained with *D. chrysippus dorippus;* lycopsamine, a plant-derived alkaloid, appears to be the precursor of the ketone (97) (Structure 11). In some danaid species, lycopsamine itself is used by the males as a sex pheromone (38). It also appears that male ithomiines derive a unique pyrrolizidine lactone from ingested alkaloids (39).

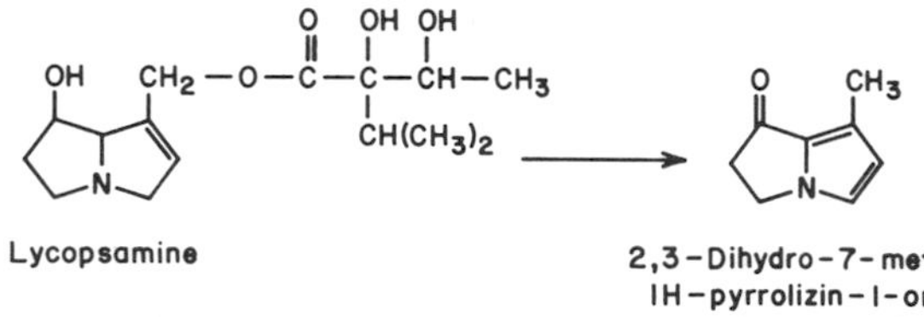

11.

Males of arctiid moths in the genus *Creatonotus* require pyrrolizidine alkaloids in order to maximally develop their scent organs, the coremata. Larvae reared on food plants lacking these alkaloids possessed underdeveloped coremata and contained only traces of 7-hydroxy-6,7-dihydro-5H-pyrrolizine-1-carboxyaldehyde (hydroxydanaidal) (98). Each male reared on alkaloid-rich plants produced nearly 0.5 mg hydroxydanaidal. It has been demonstrated that this compound is derived from heliotrine with conversion of configuration at the remaining asymmetric center (C-7) (5) (Structure 12). Larvae readily converted dietary heliotrine into (*R*)-(−)-hydroxydanaidal; thus precursors with an "incorrect" configuration could readily be metabolized to an active pheromone (6).

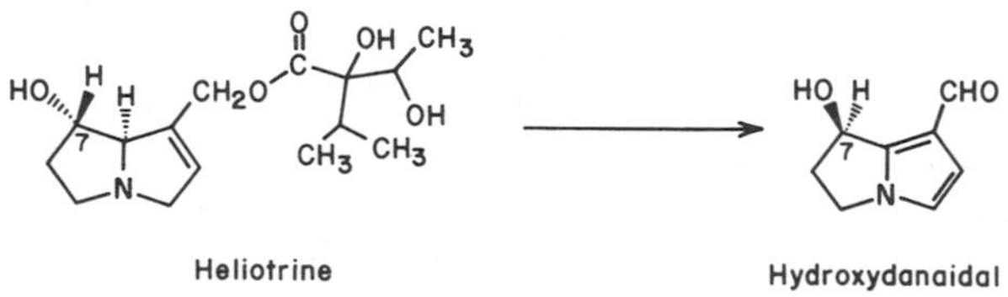

12.

Alkyl Sulfides

Ants in a few ponerine genera are known to produce alkyl and aryl sulfides in their mandibular glands. Workers of *Paltothyreus tarsatus* biosynthesized both dimethyldisulfide and dimethyltrisulfide from methionine (28, 29) (Structure 13). Both L-[^{35}S]methionine and L-[*methyl*-^{14}C]methionine were incorporated into both compounds, probably as the intact methanethiol group. The origin of the third sulfur atom in dimethyltrisulfide is unknown.

Dimethyldisulfide

$CH_3 - S - S - CH_3$

$CH_3 - S - S - S - CH_3$

Methionine

Dimethyltrisulfide

13.

Peptides and Proteins

Studies on the metabolic derivation of two peptides and an enzyme in honey bee venom essentially constitute the only analyses of the biosynthesis of proteinaceous exocrine compounds in arthropods. The defensive secretion of honey bees is one of the few proteinaceous venoms produced by invertebrates for which the structures of most of the components have been established.

The phospholipase of worker venom, unlike the pancreatic enzyme, is not reported to have an enzymatically inactive precursor (57). Phospholipase A_2, which constitutes about 10% of worker bee venom, is virtually absent from the venom of queen bees (69); this fact emphasizes that the venoms of queens and workers of a species have major biosynthetic and quantitative differences.

The biosynthesis of melittin, the main constituent in honey bee venom, has been worked out in considerable detail. This system has become a model for understanding how peptides are produced via larger precursors. Melittin, a lytic peptide comprising 26 amino-acid residues, is generated from pro-melittin, its immediate precursor (4). Queens and workers fed [4,5-^{3}H]leu-cine, one of the main amino acids in melittin, incorporated it into both melittin and promelittin in their venom. In workers the nontoxic promelittin reached peak production in about 10 days, at which time its conversion to melittin was also maximal. On the other hand, in queens the rate of conver-sion of promelittin to melittin was maximal at emergence, so that these reproductives possessed a functional, toxic venom at eclosion (4). Thus queens and workers have very different control systems for the synthesis and conversion of venom peptides. Since queens commonly have fatal encounters immediately after eclosion, the adaptive significance of maximal production of the toxic venom peptide at that time is evident.

Promelittin is generated from prepromelittin, a precursor that contains 70 amino acid residues (102). This precursor has been produced in vitro by translation of melittin messenger RNA from queen bee venom glands in a cell-free system (wheat germ). Prepromelittin contains four different discernible parts. At the amino end a peptide of 21 residues (mostly hydrophobic) constitutes the pre- or signal peptide. This is followed by an acidic preregion of 22 amino acids with either proline or alanine at the even-numbered positions. The 26 amino acids of melittin follow, and the peptide is terminated by an extra glycine residue at the carboxyl end (107) (Figure 15). Cloned mRNA has confirmed the structure of prepromelittin; the 70 amino acids represent the total coding capacity of the mRNA (110). Prepromelittin constitutes the smallest presecretory peptidic precursor known.

The liberation of melittin from prepromelittin involves a stepwise process involving three types of enzymatic reactions. The first step requires the endoproteolytic cleavage of the signal peptide, prepromelittin (63). This reaction is catalyzed by a nonspecific signal peptidase, as demonstrated by the production of promelittin by frog oocytes injected with melittin mRNA (64).

Like other presecretory peptides, prepromelittin can be converted by a truncating enzyme in the presence of microsomes and a detergent (59). The signal peptidase is a cryptic membrane enzyme, and membrane-perturbing agents are required to unmask its activity. After treatment with detergent the membrane-bound endoprotease in rat-liver microsomes cleaved prepromelittin only at the pre-pro peptide bond (74). Although the mechanism of the amidation reaction of prepromelittin has not been determined, its sequence indicates that the extra glycine has a role in the formation of the carboxy-terminal amide of melittin (46).

Promelittin, in which every other residue in the propart is either proline or alanine, is activated by a poison gland enzyme, a dipeptidyl aminopeptidase (61). Kreil et al (61) used an acidic fragment prepared from prepromelittin, containing the entire pro region and the amino-terminal hexapeptide of melittin (residues 22–49 of prepromelittin), to obtain evidence that the signal peptide is activated by stepwise cleavage of dipeptides. This type of exoproteolytic activation, regulated by the structure of the second amino acid,

NH$_2$ Met-Lys-Phe-Leu-Val-Asn-Val-Ala-Leu-Val-
Phe-Met-Val-Val-Tyr-Ile-Ser-Tyr-Ile-Tyr-Ala-
ALA-Pro-Glu-Pro-Glu-Pro-Ala-Pro-Glu-pro-Glu-
Ala-Glu-Ala-Asp-Ala-Glu-Ala-Asp-Pro-Glu-Ala-
GLY-Ile-Gly-Ala-Val-Leu-Lys-Val-Leu-Thr-Thr-
Gly-Leu-Pro-Ala-Leu-Ile-Ser-Trp-Ile-Lys-Arg-
Lys-Arg-Gln-Gln-Gly COOH

Figure 15   Amino acid sequence of honeybee prepromelittin (102, 110). The first residues of the propeptide and of melittin are underlined.

exhibits great specificity even in the presence of diverse polypeptides (61). Thus, the premature release of the final lytic product, melittin, is very unlikely; it is released only after it is delivered into the chitin-lined venom reservoir, spatially removed from sensitive cell membranes.

Preprosecapin, a precursor of secapin, a minor peptide in worker honey bee venom, has been obtained as cDNA clones from venom glands of queen bees (109). Secapin is apparently more concentrated in queen than in worker venom. It appears to be produced from prosecapin, which, unlike promelittin, does not contain even-numbered residues of proline or alanine. Activation of prosecapin is clearly different from that of promelittin; this indicates that these peptides are produced by different mechanisms in the poison gland.

ARTHROPOD NATURAL PRODUCTS: A BIOSYNTHETIC TREASURE TROVE

The pheromones and allomones biosynthesized by insects and their relatives constitute a great potpourri of compounds. The biosynthesis of these products represents an area of exocrinology that is as distinct to the arthropods as any topic in physiology or biochemistry. As more and more facets of arthropod behavior are demonstrated to have exocrinological bases, the need to understand how the biosynthesis of pheromones and allomones is regulated has become manifest. Until we know how these compounds are metabolically derived, we will be unable to appreciate the factors involved in regulating their production.

The lack of a cholesterogenic system in insects notwithstanding, dytiscids, lampyrids, and chrysomelids generate a remarkable diversity of steroids. Even if they cannot synthesize the triterpene skeleton de novo, arthropods produce many unique triterpenoid constituents (18). Beetles rival fungi by producing naphthoquinones and anthroquinones, and the biosynthetic prowess of higher plants is challenged by hymenopterans that synthesize a multitude of methyl and ethyl ketones as well as abundant pyrazines. If such biosynthetic virtuosity provides fertile grounds for the natural-products chemists, it presents the molecular biologist with a veritable gold mine.

As Kreil and his colleagues have demonstrated (62), the compounds in insect venoms are eminently suitable paradigms for the study of the stepwise generation of toxic peptides. Enough scorpion, spider, and insect venoms have been qualitatively characterized to make these secretions excellent subjects for investigation of the biosynthesis of a variety of toxic proteinaceous compounds. These studies of invertebrate toxins further emphasize that the biosynthesis of arthropod natural products is a subject whose time has come.

ACKNOWLEDGMENTS

I thank J. R. Aldrich, L. B. Bjostad, A. Nahrstedt, A. C. Oehlschlager, P. E. A. Teal, and J. H. Tumlinson for providing me with preprints of their work. Special thanks are due to H. G. Cutler and N. A. Blum for critical review of the manuscript.

Literature Cited

1. Adams, T. S., Dillwith, J. W., Blomquist, G. J. 1984. The role of 20-hydroxyecdysone in housefly sex pheromone biosynthesis. *J. Insect Physiol.* 30:287–94
2. Aldrich, J. R., Blum, M. S., Hefetz, A., Fales, H. M., Lloyd, H. A., Roller, P. 1978. Proteins in a nonvenomous defensive secretion: biosynthetic significance. *Science* 201:452–54
3. Arai, K., Ando, T., Tatsuki, S., Usui, K., Ohguchi, Y., et al. 1984. The biosynthetic pathway of (Z)-11-hexadecenal, the sex pheromone component of the rice stem borer, *Chilo suppressalis* Walker (Lepidoptera: Pyralidae). *Agric. Biol. Chem.* 48:3165–68
4. Bachmeyer, H., Kreil, G., Suchanek, G. 1972. Synthesis of promelittin and melittin in the venom gland of queen and worker bees: patterns observed during maturation. *J. Insect Physiol.* 18:1515–21
5. Bell, T. W., Boppré, M., Schneider, D., Meinwald, J. 1984. Stereochemical course of pheromone biosynthesis in the arctiid moth, *Creatonotus transiens*. *Experientia* 40:713–14
6. Bell, T. W., Meinwald, J. 1986. Pheromones of two arctiid moths (*Creatonotus transiens* and *C. gangis*): Chiral components from both sexes and achiral female components. *J. Chem. Ecol.* 12:385–409
7. Benn, M. H., Lencucha, A., Maxie, S., Telang, S. A. 1973. The pygidial defensive secretion of *Carabus taedatus*. *J. Insect Physiol.* 19:2173–76
8. Berger, R. S. 1974. Incorporation of chloride from Na^{36}Cl into 2,6-dichlorophenol in lone star and gulf coast ticks. *Ann. Entomol. Soc. Am.* 67:961–63
9. Bjostad, L., Linn, C., Roelofs, W., Du, J.-W. 1985. Identification of new sex pheromone components in *Trichoplusia ni* and *Argyrotaenia velutinana*, predicted from biosynthetic precursors. In *Semiochemistry: Flavors and Pheromones*, ed. T. E. Acree, D. M. Soderlund, pp. 223–37. Berlin: de Gruyter
10. Bjostad, L. B., Roelofs, W. L. 1981. Sex pheromone biosynthesis from radiolabeled fatty acids in the redbanded leafroller moth. *J. Biol. Chem.* 256:7936–40
11. Bjostad, L. B., Roelofs, W. L. 1983. Sex pheromone biosynthesis in *Trichoplusia ni*: Key steps involve delta-11 desaturation and chain-shortening. *Science* 220:1387–89
12. Bjostad, L. B., Roelofs, W. L. 1984. Biosynthesis of sex pheromone components and glycerolipid precursors from sodium [1-^{14}C] acetate in redbanded leafroller moth. *J. Chem. Ecol.* 10:681–91
13. Bjostad, L. B., Roelofs, W. L. 1984. Sex pheromone biosynthetic precursors in *Bombyx mori*. *Insect Biochem.* 14:275–78
14. Bjostad, L. B., Roelofs, W. L. 1986. Sex pheromone biosynthesis in the redbanded leafroller moth studied by mass-labeling with stable isotopes and analysis with mass spectrometry. *J. Chem. Ecol.* 12:431–50
15. Bjostad, L. B., Wolf, W. A., Roelofs, W. L. 1981. Total lipid analysis of the sex pheromone gland of the red-banded leafroller moth, *Argyrotaenia velutinana*, with reference to pheromone biosynthesis. *Insect Biochem.* 11:73–79
16. Blomquist, G. J., Adams, T. S., Dillwith, J. W. 1984. Induction of female sex pheromone production in male houseflies by ovary implants or 20-hydroxyecdysone. *J. Insect Physiol.* 30:295–302
17. Blomquist, G. J., Dillwith, J. W., Pomonis, J. G. 1984. Sex pheromone of the housefly. Metabolism of (Z)-9-tricosene to (Z)-9,10-epoxytricosane and (Z)-14-tricosen-10-one. *Insect Biochem.* 14:279–84
18. Blum, M. S. 1981. *Chemical Defenses of Arthropods*. New York: Academic. 562 pp.

19. Blum, M. S., Jones, T. H., House, G. J., Tschinkel, W. R. 1981. Defensive secretions of tiger beetles: Cyanogenetic basis. *Comp. Biochem. Physiol.* 69B:903–4

20. Brand, J. M., Bracke, J. W., Britton, L. N., Markovetz, A. J., Barras, S. J. 1976. Bark beetle pheromones: production of verbenone by a mycangial fungus of *Dendroctonus frontalis*. *J. Chem. Ecol.* 2:195–99

21. Brand, J. M., Young, J. C., Silverstein, R. M. 1979. Insect pheromones: A critical review of recent advances in their chemistry, biology and application. *Fortschr. Chem. Org. Naturst.* 37:1–190

22. Byers, J. A., Wood, D. L., Browne, L. E., Fish, R. H., Piatek, B., Hendry, L. B. 1979. Relationship between a host plant compound, myrcene and pheromone production in the bark beetle *Ips paraconfusus*. *J. Insect Physiol.* 25:477–82

23. Cardani, C., Fuganti, C., Ghiringhelli, D., Grasselli, P., Pavan, M., Valcurone, M. D. 1973. The biosynthesis of pederin. *Tetrahedron Lett.* 1973:2815–18

24. Castellani, A. A., Pavan, M. 1966. Prime ricerche sulla biogenesi della dendrolasina. *Boll. Soc. Ital. Biol. Sper.* 42:221

25. Chapman, J. C., Lockley, W. J. S., Rees, H. H., Goodwin, T. W. 1977. Stereochemistry of olefinic bond formation in defensive steroids of *Acilius sulcatus* (Dytiscidae). *Eur. J. Biochem.* 81:293–98

26. Chu, A. J., Blomquist, G. J. 1980. Decarboxylation of tetracosanoic acid to *n*-tricosane in the termite *Zootermopsis angusticollis*. *Comp. Biochem. Physiol.* 66B:313–17

27. Conner, W. E., Jones, T. H., Eisner, T., Meinwald, J. 1977. Benzoyl cyanide in the defensive secretion of polydesmoid millipedes. *Experientia* 33:206–7

28. Crewe, R. M., Ross, F. P. 1975. Biosynthesis of alkyl sulphides by an ant. *Nature* 254:448–49

29. Crewe, R. M., Ross, F. P. 1975. Pheromone biosynthesis: The formation of sulphides by the ant *Paltothyreus tarsatus*. *Insect Biochem.* 5:839–43

30. Davis, R. H., Nahrstedt, A. 1979. Linamarin and lotaustralin as the source of cyanide in *Zygaena filipendulae* L. (Lepidoptera). *Comp. Biochem. Physiol.* 64B:395–97

31. Davis, R. H., Nahrstedt, A. 1982.

Occurrence and variation of the cyanogenic glucosides linamarin and lotaustralin in species of the Zygaenidae (Insecta: Lepidoptera). *Comp. Biochem. Physiol.* 71B:329–32

32. Dillwith, J. W., Blomquist, G. J., Nelson, D. R. 1981. Biosynthesis of the hydrocarbon components of the sex pheromone of the housefly, *Musca domestica*. *Insect Biochem.* 11:247–53

33. Duffey, S. S., Aldrich, J. R., Blum, M. S. 1977. Biosynthesis of phenol and guaiacol by the hemipteran *Leptoglossus phyllopus*. *Comp. Biochem. Physiol.* 56B:101–2

34. Deleted in proof

35. Duffey, S. S., Blum, M. S. 1977. Phenol and guaiacol: Biosynthesis, detoxication, and function in a polydesmid millipede, *Oxidus gracilis*. *Insect Biochem.* 7:57–65

36. Duffey, S. S., Blum, M. S., Fales, H. M., Evans, S. L., Roncadori, R. W., et al. 1977. Benzoyl cyanide and mandelonitrile benzoate in the defensive secretions of millipedes. *J. Chem. Ecol.* 3:101–13

37. Duffey, S. S., Underhill, E. W., Towers, G. H. N. 1974. Intermediates in the biosynthesis of HCN and benzaldehyde by a polydesmid millipede, *Harpaphe haydeniana* (Wood). *Comp. Biochem. Physiol.* 47B:753–66

38. Edgar, J. A., Culvenor, C. C. J. 1974. Pyrrolizidine ester alkaloid in danaid butterflies. *Nature* 248:614–16

39. Edgar, J. A., Culvenor, C. C. J., Pliske, T. E. 1976. Isolation of a lactone, structurally related to the esterifying acids of pyrrolizidine alkaloids, from the coastal fringes of male Ithomiinae. *J. Chem. Ecol.* 2:263–70

40. Edgar, J. A., Culvenor, C. C. J., Robinson, G. S. 1973. Hairpencil dihydropyrrolizines of Danainae from the New Hebrides. *J. Aust. Entomol. Soc.* 12:144–50

41. Eisner, T., Eisner, H. E., Hurst, J. J., Kafotos, F. C., Meinwald, J. 1963. Cyanogenic glandular apparatus of a millipede. *Science* 139:1218–20

42. Fish, R. H., Browne, L. E., Bergot, B. J. 1984. Pheromone biosynthetic pathways: conversion of ipsdienone to (−)-ipsdienol, a mechanism for enantioselective reduction in the male bark beetle, *Ips paraconfusus*. *J. Chem. Ecol.* 10:1057–64

43. Fish, R. H., Browne, L. E., Wood, D. L., Hendry, L. B. 1979. Pheromone biosynthetic pathways: Conversions of deuterium labelled ipsdienol with sexual

and enantioselectivity in *Ips paracon-fusus* Lanier. *Tetrahedron Lett.* 1979: 1465–68

44. Gordon, H. T., Waterhouse, D. F., Gilby, A. R. 1963. Incorporation of ^{14}C-acetate into scent constituents by the green vegetable bug. *Nature* 197:818

45. Graham, R. A., Brand, J. M., Markovetz, A. J. 1979. Decyl acetate synthesis in the ant, *Formica schaufussi* (Hymenoptera: Formicidae). *Insect Biochem.* 9:331–33

46. Habermann, E. 1972. Bee and wasp venoms. *Science* 177:314–22

47. Happ, G. M. 1968. Quinone and hydrocarbon production in the defensive glands of *Eleodes longicollis* and *Tribolium castaneum* (Coleoptera, Tenebrionidae). *J. Insect Physiol.* 14:1821–37

48. Happ, G. M., Meinwald, J. 1965. Biosynthesis of arthropod secretions. I. Monoterpene synthesis in an ant *(Acanthomyops claviger)*. *J. Am. Chem. Soc.* 87:2507–8

49. Hedin, P. A. 1977. A study of factors that control biosynthesis of the compounds which comprise the boll weevil pheromone. *J. Chem. Ecol.* 3:279–89

50. Hefetz, A., Blum, M. S. 1978. Biosynthesis and accumulation of formic acid in the poison gland of the carpenter ant *Camponotus pennsylvanicus*. *Science* 201:454–55

51. Hefetz, A., Blum, M. S. 1978. Biosynthesis of formic acid by the poison glands of formicine ants. *Biochim. Biophys. Acta* 543:484–96

52. Hendry, L. B., Piatek, B., Browne, L. E., Wood, D. L., Byers, J. A., et al. 1980. *In vivo* conversion of a labelled host plant chemical to pheromones of the bark beetle *Ips paraconfusus*. *Nature* 284:485

53. Honda, K. 1983. Evidence for *de novo* biosynthesis of osmeterial secretions in young larvae of the swallowtail butterflies *(Papilio):* deuterium incorporation *in vivo* into sesquiterpene hydrocarbons as revealed by mass spectrometry. *Insect Sci. Appl.* 4:255–61

54. Howard, R. W., Blomquist, G. J. 1982. Chemical ecology and biochemistry of insect hydrocarbons. *Ann. Rev. Entomol.* 27:149–72

55. Hughes, P. R. 1974. Myrcene: A precursor of pheromones in *Ips* beetles. *J. Insect Physiol.* 20:1271–75

56. Inoue, S., Hamamura, Y. 1972. The biosynthesis of "bombykol", sex pheromone of *Bombyx mori*. *Proc. Jpn. Acad.* 48:323–26

57. Jentsch, J., Dielenberg, D. 1972. Min-

destens zwei Phospholipasen A im Bienengift. *Ann. Chem. Justus Liebig* 757:187–92

58. Jones, T. H., Conner, W. E., Meinwald, J., Eisner, H. E., Eisner, T. 1976. Benzoyl cyanide and mandelonitrile in the cyanogenetic secretion of a centipede. *J. Chem. Ecol.* 2:421–29

59. Kasnitz, R., Kreil, G. 1978. Processing of prepromelittin by subcellular fractions from rat liver. *Biochem. Biophys. Res. Commun.* 83:901–7

60. Kassang, G., Schneider, D. 1974. Biosynthesis of the sex pheromone disparlure by olefin-epoxide conversion. *Naturwissenschaften* 61:130–31

61. Kreil, G., Haiml, L., Suchanek, G. 1980. Stepwise cleavage of the pro part of promelittin by dipeptidylpeptidase IV. *Eur. J. Biochem.* 111:49–58

62. Kreil, G., Hoffman, W., Hutticher, A., Malec, I., Mollay, C., et al. 1983. Biosynthesis of peptides in honeybee venom glands and frog skin: Structure and multi-step activation of precursors. In *Protein Synthesis,* ed. A. K. Abraham, T. S. Eikhom, J. F. Pryme, pp. 101–15. Clifton, NJ: Humana

63. Kreil, G., Lollay, C., Kashnitz, R., Haiml, L., Vilas, V. 1980. Prepromelittin: Specific cleavage of the pre- and propeptide *in vitro*. *Ann. NY Acad. Sci.* 343:338–46

64. Lane, C. D., Champion, J., Haiml, L., Kreil, G. 1981. The sequestration, processing and retention of honeybee promelittin made in amphibian oocytes. *Eur. J. Biochem.* 113:273–81

65. Langley, P. A., Carlson, D. A. 1983. Biosynthesis of contact sex pheromone in the female tsetse fly, *Glossina morsitans morsitans* Westwood. *J. Insect Physiol.* 29:825–31

66. Löfstedt, C., Roelofs, W. L. 1985. Sex pheromone precursors in two primitive New Zealand tortricid moth species. *Insect Biochem.* 15:729–34

67. Louloudes, S. J., Kaplanis, J. N., Robbins, W. E., Monroe, R. E. 1961. Lipogenesis from C[14] acetate by the American cockroach. *Ann. Entomol. Soc. Am.* 54:99–103

68. Mane, S. D., Tompkins, L., Richmond, R. C. 1983. Male esterase 6 catalyzes the synthesis of a sex pheromone in *Drosophila melanogaster* females. *Science* 222:419–21

69. Marz, R., Mollay, C., Kreil, G. 1981. Queen bee venom contains much less phospholipase than worker bee venom. *Insect Biochem.* 11:685–90

70. Meinwald, J., Happ, G. M., Labows,

J., Eisner, T. 1966. Cyclopentanoid terpene biosynthesis in a phasmid insect
and in catnip. *Science* 151:79–80
71. Meinwald, J., Koch, K. F., Rogers, J.
E. Jr., Eisner, T. 1966. Biosynthesis of
arthropod secretions. III. Synthesis of
simple *p*-benzoquinones in a beetle
*(Eleodes longicollis). J. Am. Chem.
Soc.* 88:1590–92
72. Meyer, D., Schlatter, Ch., Schlatter-
Lanz, I., Schmid, H., Bovey, P. 1968.
Die Zucht von *Lytta vesicatoria* im
Laboratorium und Nachweis der Canth-
aridinsynthese in Larven. *Experientia*
24:995–98
73. Mitlin, N., Hedin, P. A. 1974.
Biosynthesis of grandlure, the pheromone of the boll weevil, *Anthonomus
grandis* from acetate, mevalonate, and
glucose. *J. Insect Physiol.* 20:1825–
31
74. Mollay, C., Vilas, V., Kreil, G. 1982.
Cleavage of honeybee prepromellitin by
an endoprotease from rat liver microsomes: Identification of intact signal
peptide. *Proc. Natl. Acad. Sci. USA*
79:2260–63
75. Moore, B. P. 1967. Hydrogen cyanide
in the defensive secretions of larval
Paropsini (Coleoptera: Chrysomelidae).
J. Aust. Entomol. Soc. 6:36–38
76. Morse, D., Meighen, E. 1984. Aldehyde pheromones in Lepidoptera: Evidence for an acetate ester precursor in
Choristoneura fumiferana. Science
226:1434–36
77. Morse, D., Meighen, E. 1984. Detection of pheromone biosynthetic and deg-
radative enzymes *in vitro. J. Biol.
Chem.* 259:475–80
78. Morse, D., Meighen, E. 1986. Pheromone biosynthesis and role of functional groups in pheromone specificity. *J.
Chem. Ecol.* 12:335–51
79. Nahrstedt, A., Davis, R. H. 1981. The
occurrence of the cyanoglucosides, linamarin and lotaustralin, in *Acraea* and
Heliconius butterflies. *Comp. Biochem.
Physiol.* 68B:575–77
80. Nahrstedt, A., Davis, R. H. 1983.
Occurrence, variation and biosynthesis
of the cyanogenic glucosides linamarin
and lotaustralin in species of the Heliconiini (Insecta: Lepidoptera). *Comp.
Biochem. Physiol.* 75B:65–73
81. Nahrstedt, A., Davis, R. H. 1985.
Biosynthesis and quantitative relationships of the cyanogenic glucosides,
linamarin and lotaustralin, in genera of
the Heliconiini (Insecta: Lepidoptera).
Comp. Biochem. Physiol. 82B:745–49
82. Pasteels, J. M., Rowell-Rahier, M.,

Braekman, J. C., Dupont, A. 1983.
Salicin from host plant as precursor of
salicylaldehyde in defensive secretion of
chrysomeline larvae. *Physiol. Entomol.*
8:307–14
83. Peter, M. G., Waespe, H. R., Woggon,
W. D., Schmid, H. 1977. Einbauver-
suche mit (^{3}H und ^{14}C)-doppelmarkier-
tem Farnesol in Cantharidin. *Helv.
Chim. Acta* 60:1262–72
84. Peter, M. G., Woggon, W. D., Schlatter, C., Schmid, H. 1977. Einbauver-
suche mit Geraniol und Farnesol in
Cantharidin. *Helv. Chim. Acta* 60:844–
66
85. Peter, M. G., Woggon, W. D., Schmid,
H. 1977. Identifizierung von Farnesol
als Zwischenstufe in der Biosynthese des
Cantharidins aus Mevalonsäurelacton.
Helv. Chim. Acta 60:2756–62
86. Pliske, T. E., Edgar, J. A., Culvenor,
C. C. J. 1976. The chemical basis of
attraction of ithomiine butterflies to
plants containing pyrrolizidine alkaloids. *J. Chem. Ecol.* 2:255–62
87. Pliske, T. E., Eisner, T. 1969. Sex pheromone of the queen butterfly: Biology.
Science 164:1170–72
88. Prestwich, G. D., Jones, R. W., Collins, M. S. 1981. Terpene biosynthesis
by nasute termite soldiers (Isoptera:
Nasutitermitinae). *Insect Biochem.*
11:331–36
89. Robinson, W. G., Bachhaus, B. K.,
Coon, M. J. 1956. Tiglyl coenzyme A
and α-methylacetoacetyl coenzyme A,
intermediates in the enzymatic degradation of isoleucine. *J. Biol. Chem.*
218:391–400
90. Roelofs, W., Bjostad, L. 1984.
Biosynthesis of lepidopteran pheromones. *Bioorg. Chem.* 12:279–98
91. Roth, L. M., Stay, B. 1958. The occurrence of *para*-quinones in some arthropods, with emphasis on the quinone-
secreting tracheal glands of *Diploptera
punctata* (Blattaria). *J. Insect Physiol.*
1:305–18
92. Schildknecht, H. 1970. The defensive
chemistry of land and water beetles. *Angew. Chem. Int. Ed. Engl.* 9:1–9
93. Schildknecht, H., Maschwitz, E., Maschwitz, U. 1970. Die Explosionschemie
der Bombardierkäfer: Struktur und
Eigenschaften der Brennkammeren-
zyme. *J. Insect Physiol.* 16:749–89
94. Schildknecht, H., Wenneis, W. F.
1967. Über arthropoden-Abwehrstoffe
XXV. (1). Anthranilsäure als Precursor
der arthropoden-Alkaloide Glomerin
und Homoglomerin. *Tetrahedron Lett.*
1967:1815–18

95. Schlatter, C., Waldner, E. E., Schmid, H. 1968. Zur Biosynthese des Cantharadins. I. *Experientia* 24:994–95

96. Schmidt, S. P., Monroe, R. E. 1976. Biosynthesis of the waxmoth sex attractants. *Insect Biochem.* 6:377–80

97. Schneider, D., Boppré, M., Schneider, H., Thompson, W. R., Boriack, C. J., et al. 1975. A pheromone precursor and its uptake in male *Danaus* butterflies. *J. Comp. Physiol.* 97:245–56

98. Schneider, D., Boppré, M., Zweig, J., Horsley, S. B., Bell, T. W., et al. 1982. Scent organ development in *Creatonotos* moths: Regulation by pyrrolizidine alkaloids. *Science* 215:1264–65

99. Seligman, I. M., Doy, F. A. 1972. β-Hydroxy-*n*-butyric acid in the defensive secretion of *Papilio aegus*. *Comp. Biochem. Physiol.* 42B:341–42

100. Seligman, I. M., Doy, F. A. 1973. Biosynthesis of defensive secretions in *Papilio aegus*. *Insect Biochem.* 3:205–15

101. Sierra, J. R., Woggon, W. D., Schmid, H. 1976. Transfer of cantharidin (1) during copulation from the adult male to the female *Lytta vesicatoria* ("Spanish flies"). *Experientia* 32:142–44

102. Suchanek, G., Kreil, G., Hermodson, M. A. 1978. Amino acid sequence of honeybee prepromellitin synthesized *in vitro*. *Proc. Natl. Acad. Sci. USA* 75:701–4

103. Tabor, H., Mehler, A. H., Hayaishi, O., White, J. 1952. Urocanic acid as an intermediate in the enzymatic conversion of histidine into glutamic and formic acids. *J. Biol. Chem.* 196:121–28

104. Teal, P. E. A., Tumlinson, J. H. 1986. Terminal steps in pheromone biosynthesis by *Heliothis virescens* and *H. zea*. *J. Chem. Ecol.* 12:353–66

105. Thompson, A. C., Mitlin, N. 1979. Biosynthesis of the sex pheromone of the male boll weevil from monoterpene precursors. *Insect Biochem.* 9:293–94

106. Towers, G. H. N., Duffey, S. S., Siegel, S. M. 1972. Defensive secretion: biosynthesis of hydrogen cyanide and benzaldehyde from phenylalanine by a millipede. *Can. J. Zool.* 50:1047–50

107. Tursch, B., Daloze, D., Braekman, J. C., Hootele, C., Pasteels, J. M. 1975. Chemical ecology of arthropods—X. The structure of myrrhine and the biosynthesis of coccinelline. *Tetrahedron* 31:1541–43

108. Deleted in proof

109. Vlasak, R., Kreil, G. 1984. Nucleotide sequence of cloned cDNAs coding for preprosecapin, a major product of

queen-bee venom glands. *Eur. J. Biochem.* 145:279–82

110. Vlasak, R., Unger-Ullman, C., Kreil, G., Frischauf, A. M. 1983. Nucleotide sequence of cloned cDNA coding for honeybee prepromelittin. *Eur. J. Biochem.* 135:123–26

111. Waldner, E. E., Schlatter, Ch., Schmid, H. 1969. Zur Biosynthesis des Dendrolasins, eines Inhaltsstoffes der Ameise *Lasius fuliginosus* Latr. *Helv. Chim. Acta* 52:15–24

112. Weatherston, J., Percy, J. E. 1970. Arthropod defensive secretions. In *Chemicals Controlling Insect Behavior*, ed. M. Beroza, pp. 95–144. New York: Academic

113. Weatherston, J., Percy, J. E. 1976. The biosynthesis of phenethyl alcohol in the male bertha armyworm *Mamestra configurata*. *Insect Biochem.* 6:413–17

114. White, R. A. Jr., Agosin, M., Franklin, R. T., Webb, J. W. 1980. Bark beetle pheromones: evidence for physiological synthesis mechanisms and their ecological implications. *Z. Angew. Entomol.* 90:255–74

115. Witthohn, K., Naumann, C. M. 1984. Qualitative and quantitative studies on the compounds of the larval defensive secretion of *Zygaena trifolii* (Esper, 1783) (Insecta, Lepiodoptera, Zygaenidae). *Comp. Biochem. Physiol.* 79C:103–6

116. Witthohn, K., Naumann, C. M. 1984. Die Verbreitung des β-Cyan-L-alanin bei cyanogenen Lepidopteren. *Z. Naturforsch.* 39c:837–40

117. Woggon, W. D., Hauffe, S. A., Schmid, H. 1983. Biosynthesis of cantharidin: Evidence for the specific incorporation of C-4 and C-11' of farnesol. *J. Chem. Soc. Chem. Commun.* 1983:272–74

118. Woggon, W. D., Peter, M. G., Schmid, H. 1977. Experimente zum Kompetitiven Einbau der Stereoisomeren des Farnesols in Cantharidin. *Helv. Chim. Acta* 60:2288–94

119. Wolf, W. A., Roelofs, W. L. 1983. A chain-shortening reaction in orange tortrix moth sex pheromone biosynthesis. *Insect Biochem.* 13:375–79

120. Wolf, W. A., Roelofs, W. L. 1986. Properties of the 11-desaturase enzyme used in cabbage looper moth sex pheromone biosynthesis. *Arch. Insect Biochem. Physiol.* 3:45–52

121. Wray, V., Davis, R. H., Nahrstedt, A. 1983. Biosynthesis of cyanogenic glucosides in butterflies and moths: Incorporation of valine and isoleucine into lina-

marin and lotaustralin by *Zygaena* and *Heliconius* species (Lepidoptera). *Z. Naturforsch.* 38c:583–88

122. Yamaoka, R., Hayashiya, K. 1982. Daily changes in the characteristic fatty acid (Z)-11-hexadecenoic acid of the pheromone gland of the silkworm pupa and moth, *Bombyx mori* L. (Lepidoptera: Bombycidae) *Jpn. J. Appl. Entomol. Zool.* 26:125–30

123. Yamaoka, R., Honzawa, S., Hayashiya, K. 1983. Further investigations on the pheromone gland components of the

silkworm moth using capillary GC-MS (Occurrence of (Z)-11-hexadecenol). *Jpn. J. Appl. Entomol. Zool.* 27:77–83

124. Yamaoka, R., Nakayama, Y., Hayashiya, K. 1985. The identification of bombykol linolenate in the haemolymph of the female silkworm pupa, *Bombyx mori*. *Insect Biochem.* 15:73–76

125. Yamaoka, R., Taniguchi, Y., Hayashiya, K. 1984. Bombykol biosynthesis from deuterium-labeled (Z)-11-hexadecenoic acid. *Experientia* 40:80–81

Ann. Rev. Entomol. 1987. 32:415–37

COMPUTER-ASSISTED DECISION-MAKING AS APPLIED TO ENTOMOLOGY

Robert N. Coulson

Department of Entomology, Texas A & M University, College Station, Texas 77843

Michael C. Saunders

Department of Entomology, The Pennsylvania State University, North East, Pennsylvania 16428

INTRODUCTION

In the next decade, use of computers to represent and structure knowledge and to solve problems will be a common practice. The need for computer-assisted decision-making in entomology is a consequence of the accelerated research and development programs in integrated pest management (IPM) of the past decade. As principles and concepts of IPM have become widely accepted, the information base on various aspects of agricultural and natural resource management has dramatically increased. The fundamental issue is how best to use existing information for decision-making within a problem domain. The issue is certainly not new, as humankind has sought ways to support and enhance decision-making throughout recorded history. The focus is now on uses of computers to aid in the process.

Several different types of computer software systems have been developed to assist decision-making. Each system has certain appropriate uses, as well as limitations. Accordingly, our objectives are to (*a*) review the history of decision support, (*b*) describe the attributes of effective computer-based decision aids, and (*c*) define the types and characteristics of computer-based decision aids and their implications for entomology.

General articles that deal with topics examined in this review provide overviews and approaches to problem solving and representation of knowledge (3, 4, 11, 12, 25, 31, 55), applications in entomology (35, 46, 57), and applications in agriculture (23, 30, 47). Contemporary texts and proceedings from symposia include the following: decision support systems and data base management systems (2, 5, 18, 26, 49, 52) and expert systems (6, 21, 22, 33, 58, 59).

HISTORY OF DECISION-MAKING

As civilization has evolved, so too have the art and science of decision-making. Today we have a variety of computer-based systems that are capable of augmenting human decision-making in ways our ancestors could not envision. Before dealing with some of these modern advances in computer-aided decision-making, however, we consider the ancient techniques of decision support.

Passive Decision-Making

Perhaps the most widespread of the ancient decision-making tools was the simple recording of sequences of unusual or important events. These "omens" and "omen texts" comprise much of the extant literature from the second and first millennium BC (24). Saggs (38, cited in 24) translated numerous Babylonian omen texts. Although often irrational in content, these omen texts foreshadow the basic construct of today's expert systems in that they consist of precisely worded "if-then" (antecedent-consequent) statements (e.g. "If a town is set on a hill, then it will not be good for the dweller within that town.").

Structured Problems

By the third millennium BC, the Assyrians had developed simple techniques for solving structured problems. A structured problem is one for which all solutions are prespecifiable (26). For an ancient Assyrian, a structured problem might have been, "Which of my sheep is most suitable for the temple sacrifice?" The dominant techniques for supporting decisions about structured problems fall under the heading of "sortilege," which is divination by lots (24). Since the concept of chance (probability) was unknown to the Assyrians, they believed that the outcome of the cast lots represented the intent of the gods, whose wishes were being divined.

Unstructured Problems

As societies and their problems became more elaborate, people developed techniques for dealing with the most complex class of problems, the unstructured problem. An unstructured problem is one for which the full set of possible solutions is not known (26) (e.g. strategies of warfare, the timing and content of religious festivals). By the second millennium BC, decision-making in response to unstructured problems was supported by augury (24). Augury, like sortilege, was a form of divination; it consisted of elaborate rituals designed to elicit signs and portents that were interpretable according to established rules. In contrast to sortilege, augury was designed to divine a great deal more information from the unspeaking gods. Jaynes (24) suggested that the difference between sortilege and augury is functionally equivalent to the difference between digital and analog computers. In Mesopotamia during the first millennium BC, augury had achieved a cultic status. The most elaborate form of augury was "extispicy," the derivation of decision support by examination of the entrails of sacrificed animals. Typically, a bullock was slaughtered and its organs were examined in detail for any abnormalities, the arrangement of intestinal coils, and the locations of pockets of undigested food in the alimentary canal. The "baru," a priest, interpreted these signs in regard to an unresolved unstructured problem. The entrails were often cast in bronze and sent to the ruler for verification; this signifies the advent of hard copy output.

Authorization

A recommendation for any decision has value only if it comes from an accepted authority and conforms to the decision-maker's perception of reality. The ancient Assyrians and Mesopotamians perceived reality as ordered but mysterious, subject to the wishes and whims of gods. To know if a course of action was appropriate, one only needed to know how the gods felt about it. In the modern context of decision-making it is still important that the decision aid address the decision-maker's view of reality. Saaty (37) referred to the work of LeShan and Margenau, *Einstein's Space and Van Gogh's Sky; Physical Reality and Beyond* (28), to point out that there are four models of reality: (*a*) sensory (the model scientists prefer), (*b*) clairvoyant, (*c*) transpsychic, and (*d*) mythic. In the first model, individuals construct reality as if they are detached observers of a larger whole, as in building an analytical model; in the second, as if they are an extension of the whole without a sense of separation, as in dancing, music, and meditation; in the third, as if they are reciprocals to the whole, as in prayer; and in the fourth, as if they are identical to the whole, as in dreaming and play (32). Because of these differing perceptions of reality, a method of problem-solving that is suitable for a scientist may not be suitable for an artist. If we wish to communicate with others about decision-making in, for example, agroecosystems, we must be willing to translate our sensorially obtained data so it will be understood by the decision-maker. This task has become the primary challenge of computer-based decision support. Namely, a decision-making tool must be made capable of sharing the decision-maker's assumptions about a problem domain.

Computer-Aided Decision-Making

With the advent of computers, it became possible to store and gain access to vast amounts of information in the form of facts and rules. Like the active decision-making techniques of early civilizations, those involving computers could initially deal only with structured problems; they gradually evolved to deal with the more complex domain of unstructured problem-solving. Initially, computers had limited capabilities, having been built for fixed purposes. Accordingly, new advances in hardware had to accompany refined techniques for handling information. Within disciplines not directly associated with the computing sciences (i.e. entomology), the development and implementation of computer software has lagged behind that of hardware. As a result, we can find examples of each level of computer-aided decision-making in the evolution of entomological software without paying much attention to the hardware developments that preceded them.

The earliest entomological software was written by scientists for other scientists. Like the Assyrians, entomologists share a particular world view; they study insects. Thus they share access to the same bodies of information. Accordingly, the computer was first applied in entomology to process data (file management systems, data base management systems, and simulation models) and information (management information systems). Some computer functions include record-keeping, development and use of life tables, and delivery of mathematical models (35). The information stored and/or generated was generally interpretable by other entomologists. These uses of computers in entomology are analogous to the Assyrian use of sortilege. The techniques were developed for fixed purposes to deal with prespecified questions and for use by individuals with experience in the same knowledge domain.

As entomologists began to use computers further for decision-making, higher-level programs were needed. Systems were needed to accomodate increasing interdisciplinary research and to disseminate new information for agricultural extension. The problem of building these systems has centered around the distinction between information and knowledge. Whereas information consists of facts and rules, knowledge implies understanding. The entomologist/programmer must develop software with explanatory power, which will be used by individuals who may not share the same knowledge and who lack the capability to organize information to solve specific problems.

"Decision support systems" represent the first computer-based systems for organizing information to deal with unstructured problem-solving. The key feature that distinguishes decision support systems from systems with an information management orientation is the degree to which the information processing task is prespecified (26). The information contained within a decision support system (e.g. data, narrative information, simulation models) may be identical to that in an information management system. The fundamental difference lies in the ability of a decision support system to help the user organize the information in a manner that is relevant to a specific problem and consistent with his or her view of how to express that problem (35).

Although decision support systems represent a significant advancement in computer use for problem-solving, the output of these systems still requires interpretation by the user. The effectiveness of the interpretation depends on the user's experience and knowledge. Because the expertise needed to effectively assemble and interpret information within a problem domain is not widespread, this dependence on user interpretation is the most obvious shortcoming of information-based systems. Fortunately, software systems have been developed that mimic the deductive and inductive reasoning of a human expert. These knowledge-based systems are called "expert systems" and will surely rank among the most significant technologies to emerge in the twentieth century (33).

Expert systems capture the knowledge of a domain expert by encoding the heuristic reasoning process used by that expert for solving problems. A common type of expert system, known as a "rule-based system," reasons by processing rules. Typically, these rules are expressed as "if-then" (condition-consequent) statements from which the expert system makes inferences and draws conclusions. In a sense, we have come full circle from the omens used by Babylonian experts to rule-based expert systems that mimic the reasoning process of modern specialists. Perhaps most significantly, an expert system can explain the logical sequence of its reasoning process. As a result, a decision-maker can review the reasoning steps of a human expert rather than rely simply on outputs consisting of numeric values or rote recommendations.

Because inexactitude is a necessary fact of life, expert systems have inherent weaknesses. This circumstance has led to the concept of an "expert support system," which is a computer system that represents a blend of attributes from decision support systems and expert systems. The need for this type of system is evident when we consider that much of the information base in a particular problem domain may be wrong, imprecise, or incomplete.

ATTRIBUTES OF EFFECTIVE COMPUTER-BASED DECISION AIDS

The landmark publication *Concepts of Pest Management* (34) ushered in the era of integrated pest management. Since then, the multidisciplinary approach and the IPM paradigm have been used in several successful accelerated research and development programs in both agriculture and forestry. The principal methodology used in early IPM research was simulation modeling.

This approach provided a convenient and efficient way of abstracting complex systems in biology and economics. The modeling activity was supported through basic studies dealing with components of the systems. Numerous scientific and technical reports describing this work were produced. Therefore, the information base for a particular subject domain often consists of simulation models and technical reports that supplement existing, although generally undocumented, opinions of experts.

IPM has certainly served as a useful paradigm for organization of research and demonstration of the utility of systems science in solving complex problems. Agriculture and forestry are complex enterprises that require managerial efficiency, broad-based knowledge, and use of advanced technologies for success. The IPM concept has provided focus to the significant level of organization for management decision-making, the agroecosystem (or forest ecosystem), which encompasses components associated with the production system per se (including pest impacts) as well as the policy and marketing systems for each commodity.

In the aftermath of the accelerated research and development programs on IPM we are left with an expanded information base that consists of a variety of simulation models, technical information, and expert opinion. In order to use this information base more effectively in decision-making, we need computer-based systems that (*a*) provide access to the information by serving as a cache, (*b*) integrate and synchronize the information for use, (*c*) interpret the information, and (*d*) serve as implementation coach (32).

Cache for Information

Of the three components of the information base, technical information is the most accessible. The content can be either qualitative or quantitative. Traditional search procedures can be used to identify pertinent citations, and hard copy reports can be obtained from primary and secondary literature sources. It is possible to store, retrieve, edit, sort, and merge technical information electronically using conventional computing systems, i.e. file management, data base management, and management information systems (57).

The application of principles of systems science in entomology was one of the significant accomplishments of the accelerated IPM research and development programs. The impact of simulation modeling on agriculture and forestry extension has been examined in detail (19, 32, 57), and there is little question of the usefulness of the methodology as a reliable shorthand for dynamic information handling. However, the complex problems associated with application of simulation models in IPM decision-making were not anticipated by either the architects of the IPM paradigm or the extension specialists who would be responsible for delivering the technology to the practitioner. Although numerous simulation models are potentially available on a variety of subjects, they are not generally accessible. Few of the models are actually being used by farmers or foresters.

Much of the functional knowledge in IPM, as in other subject domains, is qualitative and retained by experts whose judgments have traditionally guided decision-making at operational levels. In the past there has been little effort devoted to archiving the wisdom of experts in a formal sense. Occasionally an expert records the experiences, observations, and work of a lifetime [e.g. Stoddard's 1932 treatise on the bobwhite quail (45)], but this type of record is rare. The knowledge is eventually lost, perhaps to be partially rediscovered by the next specialist. The magnitude of this loss is compounded by the fact that within any subject domain the knowledge of the multiple experts is rarely collected and assimilated. Even during the life of the expert, the benefits of his or her knowledge are localized and not generally available to a community of potential users. One challenge for the future is to develop means to capture and broadcast the wisdom of experts.

Synchronizer of Information

Given that it is possible to obtain access to the information base in a subject domain, the next issue is integration of the pertinent elements needed and available to solve a particular problem. An individual who is more experienced with a subject domain can more easily identify the types of information needed to solve a problem. Consider the example of a farmer making a decision regarding application of an insecticide to suppress a pest insect. The types of information needed for this decision could include: (*a*) an estimate of projected loss caused by the insect, (*b*) estimates of the projected crop yield with and without the loss, (*c*) an estimate of the projected value of the crop at harvest, (*d*) types and costs of insecticides available for the pest, (*e*) and an estimate of the cost of application. With this information it is possible (*f*) to conduct a cost/benefit analysis and make an informed judgment regarding the problem. As stated, the problem does not appear to be complicated. However, reaching an optimal solution is obviously not a simple task and involves use of simulation models (*a, b,* and *c*), technical information (*d*), and evaluation functions (*e* and *f*). Past experiences (the expert opinion of the farmer) will also probably enter into the selection of a particular insecticide. One of the important functions of a computer-based decision aid is to identify the information needed and to organize it into a coherent framework for solution of a problem.

Interpreter of Information

Once the pertinent components of the information base have been identified for a particular problem, the next issue a user faces is how to interpret their significance and/or utility. This issue is particularly important because the technical information may have questionable reliability, the simulation models may not have been extensively validated, and experts may disagree. It is also quite likely that many of the components comprising the information base were not developed for solving problems. For example, simulation models might be built for exploration, explanation, projection, or prediction (9). The opportunity for error is compounded when the different components of an extensive information base are combined. Obviously, some measure of confidence is needed for decisions supported by diverse information sources.

Implementation Coach

The information content and goals of a decision aid depend on specific attributes of the problem domain. The specificity of the recommendations rendered by a decision aid is largely dependent on its ability to identify the precise nature of the user-supplied problem. The method whereby this task is accomplished can range from a menu, to a problem analysis tree, to natural language processing. Once the nature of the decision goal is identified, the system can select pertinent information from its own archives and request additional information from the user for use in problem-solving. The dialogue that is generated by the decision aid should be as understandable and interactive as possible and should guide or coach the user in problem-solving.

TYPES AND CHARACTERISTICS OF COMPUTER-BASED DECISION AIDS

Although computer-assisted decision-making has been used in industry and engineering since the 1950s, applications in entomology closely followed the development of IPM and therefore are of recent origin. There are several types of systems, each with unique characteristics and functions. The different systems have resulted from developments in computer science, management science, and cognitive science. Emphasis in computing science has centered on advancements in design and application of hardware and software systems. Management science considers matters of optimization, efficiency, and productivity. Cognitive science is concerned with theories of intelligence, and includes issues associated with problem-solving and decision-making by humans.

Computer-assisted decision-making has evolved through four levels of complexity: (a) data processing, (b) information processing, (c) knowledge processing, and (d) decision-making (Figure 1). The nomenclature used to describe the specific types of systems is from the jargon of computing science. The systems examined in this review include file management systems, data base management systems, management information systems, decision support systems, expert systems, and expert support systems. Simulation models and "stand-alone systems," which are often used as components of other types of decision aids, are treated separately, as they have been developed and used independently of the other types of systems. The kinship among the systems, relative to the evolution of computer-assisted problem-solving and decision-making, is illustrated in Figure 1. There are fundamental differences in the tasks performed, structure, function, and appropriate use(s) of the various systems. Major emphasis is placed on decision support systems and expert systems because of their utility in integrating, interpreting, and delivering the information base available for managing agroecosystems and natural resources. Although all of the systems were originally developed on mainframe computers, in most cases the systems can now be developed and implemented on micro- and mini-computers.

File Management Systems

File management systems (FMS) are the forerunner of the more complex data base management systems (DBMS) and management information systems (MIS). In FMS, data are organized into files, which can be viewed as a

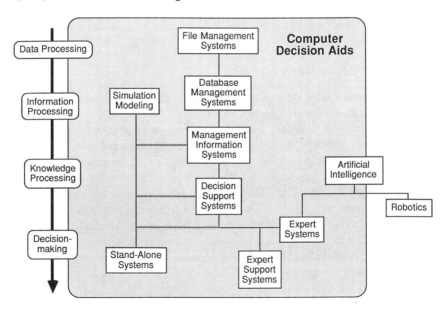

Figure 1 Trends in computer applications (data processing, information processing, knowledge processing, and decision-making) and the types of systems used to address each level of complexity. File management systems, data base management systems, management information systems, and decision support systems were developed from conventional computer science applications. Expert systems were developed from basic research in artificial intelligence. (Robotics is included to indicate that there are pending applications of this technology in agriculture.) Simulation models, used as decision aids, were developed in parallel to the other types of systems and are a product of IPM research. (Knowledge Engineering Laboratory, Departments of Entomology and Agricultural Economics, Texas A & M University)

sequence of similar or related parcels of data. For example, an extension specialist might have several parcels of data on important pest insects. Each parcel is composed of data about a different insect (e.g. common name, scientific name, primary and secondary hosts, etc). A sequence of these data parcels constitutes a file for a specific insect. File management is concerned with organization of data in this file, i.e. how it is arranged and sequenced, what data about the insects are included, etc. Furthermore, file management is concerned with ways of manipulating data in the file, e.g. how a particular parcel of data can be extracted from the file to produce a report or create a new file, how data in one file can be merged with data in another, how the file can be updated, etc. These tasks are accomplished by interfacing the files with specific application programs (Figure 2). In general, FMS consist of several independent files and application programs. Typically, files are developed for a specific purpose. In the example above, the extension specialist might have a completely separate file on ways and means of suppression of pest insects. An extension specialist in plant pathology, perhaps working in the same office as the insect pest specialist, would have an entirely different set of files. In file management, emphasis is on fragmented treatment of data, involving many separate files (5).

Data Base Management Systems

Data base management systems (DBMS) were developed in response to the need to integrate treatment of data. Emphasis in DBMS is on consolidation of data into a common data base with an accompanying reduction of redundancy. A data base consists of many files developed for different uses and users. Interface between the data base and specific application programs is accomplished with a control program, usually referred to as a data base processing system (Figure 2). In effect, DBMS add a layer of software between the application programs and the file system. Continuing the example from above, using a DBMS the extension specialist could access, merge, edit, and file information on pest species of insects and procedures for suppression. Furthermore, the data base would also contain the files of the plant pathologist as well as those of other specialists (52).

Management Information Systems

Management information systems (MIS) represent the next logical progression from DBMS (Figure 1). The components of an MIS include a data base, a model base, a data base management system (DBMS), a model base management system (MBMS), a control program, and a user interface or dialogue generation and management system (DGMS) (Figure 2). The data and applications program bases are consolidated and structured. Access to these subsystems is provided by the DBMS and MBMS. Operation of the MIS

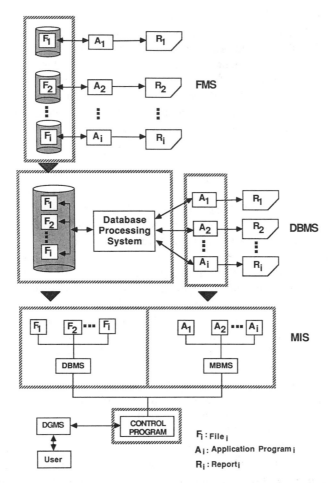

Figure 2   The structure of a file management system (FMS), data base management system (DBMS), and management information system (MIS), and the functional relationships of the three types of systems: F_i = file, A_i = application program (or model), R_i = report. (Knowledge Engineering Laboratory, Departments of Entomology and Agricultural Economics, Texas A & M University)

is directed by a master control program, which links the user to the data and model bases and directs the functioning of the system.

An MIS is designed to address structured problems, i.e. the solution to the problem can be specified in advance. In an MIS, task-processing is predefined and input/output (I/O) procedure is also preformatted. This type of system is often "report-oriented" in the sense that data are extracted and summarized in predefined formats, usually on a periodic basis. Output can be presented in narrative, numeric, and graphic formats.

To continue the example from above, an extension agency might be responsible for conducting surveys for a pest insect at prescribed times during the year. The types of information collected in the survey would likely be standardized and could include variables such as relative density, degree of damage caused, and distribution (e.g. by region of the state, by county, or by farm ownership). An MIS could be used to summarize data on these variables efficiently in predefined formats, and these summaries could be provided to various clients. The MIS might also include a geographic information system (GIS), and might thereby provide graphic representation of the variables of concern.

Decision Support Systems

ATTRIBUTES   Decision support systems (DSS) represent the next logical progression from MIS. DSS concepts have been developed rapidly in recent years in an attempt to focus disparate information on the management decision process (1, 5, 18, 26, 53). A DSS is an interactive computer-based system designed to help decision-makers utilize data, narrative information, and models to solve unstructured problems. The key feature that distinguishes a DSS from the other systems described thus far is the degree to which the information-processing task is specified beforehand by the user, i.e. the flexibility built into the system for using data and selecting, sequencing, and executing models to develop novel solution methods. In the other systems information processing is specified in advance, while in DSS it is not.

A DSS subsumes portions of the other systems, especially the data-summarization and extraction capabilities. However, the procedures used in DSS are more loosely defined, and the user can therefore select the frequency and content of reports on an ad hoc basis. DSS have a more generalized organization because problems are often unanticipated or could have so many variations that it is not possible to specify in advance an appropriate solution, i.e. the problems are unstructured. Therefore, it is essential to maintain as much organizational flexibility within the system as possible to meet the individual needs of the user.

A DSS extends the applicability of the concepts embodied in FMS, DBMS, and MIS by freeing the user from the structures imposed by the systems, thereby allowing the user to create novel problem-solving structures. The system is designed to define, retrieve, process, and organize information in a way that is meaningful for the user's particular problem (35).

STRUCTURE   The general structure of a DSS is illustrated in Figure 3. There are seven basic components: (a) model base, (b) data base, (c) model base management system (MBMS), (d) data base management system (DBMS), (e) problem-analysis program, (f) control program, and (g) dialogue generation and management system (DGMS). The model base can consist of various types of simulation models or application programs (e.g. evaluation functions). The data base can consist of numerical data, technical information, and narrative information (e.g. reports or opinions of experts). The MBMS and DBMS control access to the model base and data base. The problem-analysis routine is an important innovation not found in the previously described systems. It defines the exact nature of the problem presented to the system; i.e. a series of prompts and menus helps the user frame questions (51). The problem definition allows the system, through its control program, to identify pertinent elements in the model and data bases. Since the focus of DSS is on problem-solving, the problem-analysis program is an extremely important conceptual as well as operational component. The user communicates with the system through a dialogue generation and management system (DGMS). The operations of the DSS subsystems are directed by a control program. All subsystems are designed to be transparent to the user, who merely responds to system prompts and then evaluates the reports that are generated.

A DSS does not make decisions, but instead provides the manager with information pertinent to a specific problem. The manager interprets the information and makes the actual decision. The rationale for the DSS

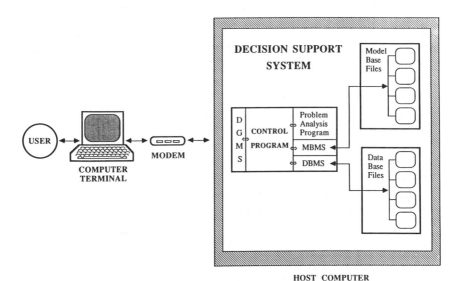

Figure 3 General structure of a decision support system. Key elements include a model base, data base, model base management system, data base management system, user interface (dialogue generation and management system), and control program. The system is housed on a mainframe computer and addressed from a terminal through an acoustic coupler. (Modified from 35)

approach is that managers are capable of making decisions. The quality of a decision depends, in part, on how well-informed the manager is. Therefore, one of the prominent purposes of a DSS is to provide information to managers effectively and efficiently (26). It is also worthwhile to note that psychologists who study the physiology of human behavior define "stress" as "decision-making" (24). A DSS should also lessen the trauma associated with the decision-making process.

An example of the application of DSS concepts to an entomological problem domain is the Southern Pine Beetle Decision Support System (SPBDSS) (8, 39). This system, which is a FORTRAN program housed in a mini-computer, was designed to aid forest managers and pest management specialists in decision-making about IPM of the southern pine beetle, *Dendroctonus frontalis*. The structure of SPBDSS is illustrated in Figure 3. This system provides access to much of the modeling technology and abstracted technical information derived from accelerated research and development programs on IPM of *D. frontalis*.

Expert Systems

An expert system (EXS) is a computer program designed to simulate the problem-solving behavior of a human who is expert in a narrow domain (12). These systems draw and store inferences from information. EXS are also frequently called "knowledge-based systems." EXS were developed via a different path than FMS, DBMS, MIS, and DSS. Concepts came from artificial intelligence (AI) and required a radical departure from conventional computing practices and programming techniques. In the following sections we review the background necessary to understand the operation of EXS, examine the structure and function of an EXS, and describe actual and potential applications in entomology (10, 15, 20, 21, 31, 42).

BACKGROUND   John McCarthy coined the term "artificial intelligence" in 1957 to denote the field dealing with the design and development of machines that exhibit traits and capabilities associated with human intelligence. The study of artificial intelligence (AI) has taken two directions; researchers seek to discover how machines can be made to perform intelligently, or use machines to simulate, model, and characterize human intelligence. Pursuit of the first activity necessitates consideration of the second. AI has both theoretical and practical applications. Currently emphasis is on duplication of certain human abilities (speech/natural-language processing, vision/pattern recognition, and physical dexterity/robotics) and automation of human expertise (expert systems).

Contemporary computer applications of AI principles began with the development of a computer language that used symbolic computations to model simple human problem-solving capabilities. Symbolic computation is the processing of "symbols" and "relationships" instead of the digits, characters, and strings of characters that typify traditional data processing (16). The original symbolic computer language, introduced in 1956, was termed IPL (Information Processing Language), and it served as the basis for the LISP (List Processing) language, which followed in 1957. LISP was designed to facilitate the symbolic representation and processing of arbitrary objects and the relationships among them, and was suited for the exploration and development of symbolic processing technology and artificial intelligence. In EXS the processing of lists of symbols approximates the process of associative memory in humans. Therefore, LISP can be used to examine and simulate different human cognitive faculties. It is the language most commonly used for symbolic computation and artificial intelligence applications, although there are others, e.g. PROLOG.

The next step in the evolution of AI involved the design and development of computers that optimize symbolic computation. The basic attributes required of a computer for symbolic processing include a flexible memory that allows for the storage of symbols, properties, and relationships, and a closely integrated processing unit that can manipulate these symbols as effortlessly as traditional computers can handle data stored as numbers. The first commercial products were introduced in 1980 by Symbolics Corporation, and there are now several vendors that offer symbolic processing computers (e.g. Texas Instruments Corporation and Xerox Corporation) (27, 40, 48, 56).

STRUCTURE AND FUNCTION   There are several different types of EXS. We emphasize "rule-based" systems because of their applicability to a variety of problems in entomology. Denning (12) described the structure and function of an EXS using the analogy of a system of logic. In mathematical logic, a system consists of finite sets of axioms and rules of inference. A proof is a sequence of strings of symbols, e.g. $S_1, \ldots, S_p$, such that each symbol S_i either is an axiom or is derivable from some subset of $S_1, \ldots, S_{p-1}$ by rule of inference. An EXS attempts to compute a sequence of strings of symbols representing the steps in the solution of a problem. The sequence serves as a proof of a solution. The rules of inference are of the simple "if-then" form of problem-solving (6, 12).

The architecture of an EXS can be envisioned to consist of the following components: (a) user interface, (b) knowledge base, (c) inference engine, and (d) memory. In some cases the system may also contain subroutines, which are often referred to as demons (Figure 4).

The user interface, which can include interactive graphics, provides a means for the user to ask questions and provide information to the system.

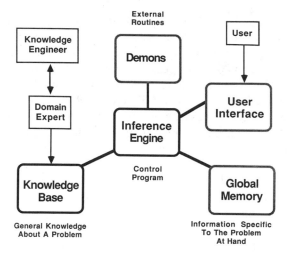

Figure 4 General structure of a "rule based" expert system. Key elements are the inference engine, knowledge base (rule base), global memory, and user interface. External subroutines (demons) may be present to enhance power and versatility of the system. (Courtesy N. D. Stone, Knowledge Engineering Laboratory, Department of Entomology, Texas A & M University)

The unique components of EXS are the knowledge base and the inference engine.

The knowledge base is a dependency network of facts, generalities, opinions, and heuristic knowledge about a subject encoded (using a knowledge-representation language) so that the necessary relationships among all the elements are defined. [See Figure 2 in Stone et al (46) for an example of a dependency network.] The knowledge base contains all of the system's task-specific information. A dependency network for the knowledge base is generally developed through an interview process involving a "knowledge engineer" and domain experts. The content is usually substantiated by assimilation of information from sources such as technical reports, simulation trials, and opinions of experts. The knowledge base is often represented by rules, in which case it is referred to as a "rule base." Rules are "antecedent-consequent" clauses or "if-then" statements (e.g. *if* there is evidence that A and B are true, *then* conclude there is evidence that C is true). The left-hand side of the rule contains conditional statements, while the right-hand side contains instructions to be carried out (consequents) (13). There are other ways to structure the knowledge base (e.g. frames, semantic nets, predicate calculus, and procedures), but the most familiar technology uses rules (11).

The inference engine is the supervisor of the system's reasoning process. In effect it is a control program. Two principal control strategies are used in expert systems to chain a set of rules together for solution of a problem: forward chaining and backward chaining. The forward-chaining (data-driven) strategy uses rules to reason forward from observations (data) to conclusions; the backward-chaining (goal-driven) strategy uses rules to reason backward from a hypothesis or goal to observations (data) that might support or refute the hypothesis (Figure 5) (7). Both strategies can be employed in an EXS. The "working memory" contains information specific to the particular problem being addressed. It provides a continuous record of the "state of the world" as understood by the system, and of the chain of reasoning that led to this state.

In some cases subroutines or demons (i.e. simulation models, a data base manager, communications, etc) from outside the EXS provide information needed by the system that the user cannot supply directly. The EXS calls for subroutines through execution of a rule that might be stated as follows: "*If* the user does not know the value of variable *X*, *Then* execute simulation model *XYZ*." The value of the variable would be provided and the problem-solving episode could continue.

Some EXS can reach a level of performance comparable to that of a human expert in a specialized problem area. The formalized knowledge base and associated control mechanisms of EXS are in essence a model of the expertise of the best practitioners of the relevant problem area. Like books, EXS can

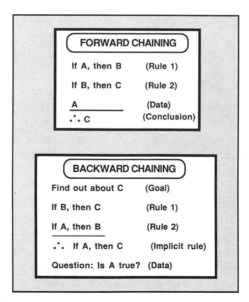

Figure 5   Illustration of the two principal control strategies used to chain a set of rules together for the purpose of problem-solving in a rule-based expert system: forward chaining (data-driven) and backward chaining (goal-driven). (Modified from 7)

make the knowledge of few available to many. EXS are intended for consulting purposes and generally do not provide final conclusions that would displace human decision-making (11, 12, 21, 50, 60).

EXS have a number of important advantages over conventional approaches. First, the knowledge base (rule base) can consist of a combination of objective information (i.e. simulation models and technical information) and expert opinion, which is often subjective. Second, the inference engine that governs the operation of an EXS uses pattern matching, which is a nonprocedural and flexible interactive mechanism, instead of the more standard encoded branching method of conventional systems. Third, the inference engine (control program) and knowledge base (rule base) are separate; this greatly simplifies maintenance and updating of the EXS. Fourth, EXS have the internal data structures necessary to deal with imprecise and incomplete information, e.g. through the assignment of confidence values to inputs and conclusions. Conventional systems generally deal with incomplete information by assigning default values; however, they do not consider the consequences of including defaults instead of measured or known data. For instance, in an EXS the internal reliability rating of output from a simulation model run using default values will be lower than that of output from the same model run using observed data. Consequently, recommendations based on the former model's output would be less reliable. Fifth, because an EXS "remembers" its chain of reasoning, it can explain the steps taken in solving a problem. The user may ask the system to explain why it wants information or why it gave a particular response. Sixth, no special technical expertise is required to use an EXS. The system performs all information analysis and provides the user with a set of recommended options ranked by confidence factors. With conventional systems, the user must interpret the significance of the results presented. [In this comparison we have emphasized negative aspects of conventional approaches; however, it is important to recognize that in many instances these approaches are quite valuable for computer-aided decision-making, and have also just begun to prove useful in entomology (32).]

APPLICATIONS   Expert systems have been developed for use in industry (29), medicine (44), geology (14), chemistry (17), and engineering (2). Entomologists and agriculturalists are just beginning to explore applications (3, 23, 30, 32, 36, 46, 47). The prospect for rapid advancement in EXS applications is great, because a large number of expert-system development tools (shells) designed for application on conventional computers are commercially available. These shells eliminate the requirement for a symbolic processing computer for delivery of EXS. Furthermore, the expense of symbolic processing computers may well be ameliorated by competition among vendors.

In entomology, EXS for taxonomic identification, pesticide recommendations, and delivery of simulation models have been, or are being, developed (30, 46, 47). These are classic applications in the sense that they treat discrete problems and demonstrate how scarce expert knowledge can be encoded into rule-based systems for use by the nonexpert. Stone et al (47) have illustrated how a complex problem like farm management (whose components include production, marketing, and policy decisions) can be addressed using EXS techniques if the subject domain is broken into modules. A rule base can be developed for each module and the various modules can then be integrated. This approach is being applied to cotton and rice farming, viticulture, and IPM of *Dendroctonus frontalis*.

It is likely that in the future there will be many more diagnostic and decision-aid applications for EXS. However, the prospect of using AI techniques to investigate cognitive mechanisms of animals other than humans seems most intriguing. For example, H. Saarenmaa (personal communication), working in the Knowledge Engineering Laboratory in the Department of Entomology at Texas A & M University, has developed a prototype EXS (using LISP and a Symbolics® computer) that mimics the behavior of moose *(Alces alces)* in forests of Finland. Moose are not only significant game animals, but are also important pests that have an impact on forest regeneration and growth. For these reasons the behavior of moose has been studied in detail, and the acquired information was used to build the system. Of perhaps more interest to entomologists is the possibility of understanding the various forms of intelligence displayed by insects. How do insects find their way about, select mates, identify hosts, build nests, etc? All these activities clearly entail information processing. The vast quantities of information available on insect behavior (e.g. for certain species of ants and for the honey bee, *Apis mellifera*) should greatly expedite applications of AI techniques to the study of insect intelligence (3).

Expert Support Systems

The integration of concepts of EXS and DSS has given rise to "expert support systems" (EXSS). In some instances, decision-making can be enhanced by incorporating EXS within the structure of DSS (54). Scott Morton (41) has suggested that EXSS represent a combination of the features of expert consultative systems and those of traditional DSS (Figure 1). This type of system is also referred to as an "executive decision support system" (EDSS). The important attributes of an EXSS are (*a*) capability of supporting decision-making involving several alternatives, (*b*) capability of addressing several (often competing) criteria or factors that influence decision-making, (*c*) incorporation of subjective as well as objective factors, (*d*) accommodation of expert judgments about the relative importance of these factors, (*e*) incorporation of the decision-maker's expertise by allowing of expression of insight or judgment, and (*f*) synthesis or combination of judgments about the main facets of a decision.

Like DSS, EXSS enable the user to make the actual decision regarding a particular problem or question. Moreover, like EXS, EXSS interpret the significance of information presented. EXSS are particularly well suited for problems associated with delivery of information on IPM.

Simulation Models and Stand-Alone Systems

The use of simulation modeling as the principal methodology in IPM research contributed greatly to the need for computer assistance in entomological decision-making. The MIS, DSS, EXS, and EXSS approaches all use simulation models in their problem-solving formats. Some of the simulation models have direct uses in decision-making, although the original intent in building a model may have had little to do with applied aspects of problem-solving. For example, a model of poikilotherm development (43) was constructed to elucidate basic issues of insect development. However, the concepts embodied in the model can be used to predict rates of development of pest insects under field conditions, and this information can be useful for timing applications of insecticides (55). Many of the simulation models, as well as various types of evaluation functions (e.g. a procedure for conducting a cost/benefit analysis), have been modified to include a "user-friendly" interface, which enhances their applicability in problem-solving and decision-making.

In Figure 1, we have indicated that the development and application of simulation models and evaluation functions for the purpose of problem-solving and decision-making paralleled those of the other types of systems described. There is little question that these "stand-alone systems" are valuable tools. However, access to them is often inconvenient, changes and modifications are not easily distributed to all users, and data input requirements often preempt use of the models by practitioners. Furthermore, there is little standardization among the types of computers used in development of the models and in delivery, although these problems are certainly being addressed.

SUMMARY

From the earliest periods of recorded history, humankind has sought ways to enhance decision-making. As we do, our predecessors had to deal with both unstructured and structured problems, and for credibility the solutions required endorsement from recognized authorities. Techniques such as reference to omens, sortilege, and augury were adequate for a time, but as societies became more complex so did the problems of decision-making. This increased complexity was accompanied by an increase in the quantity and quality of information available for decision-making. The advent of computers provided a new tool for the decision-making process. Computers could be used to archive information in the three principal representations available to decision-makers (technical reports, simulation models and evaluation functions, and expert opinion). Furthermore, various types of computer-based systems were developed to assist in synchronization of information, interpretation of the information, and identification of information pertinent to a specific problem. Uses of computers for problem-solving proceeded through four levels of complexity: data processing, information processing, knowledge processing, and decision-making. Conventional computer science techniques provided file management systems, data base management systems, management information systems, and decision support systems. Research representing a blend of contributions from computer, management, and cognitive sciences produced a decision-making tool known as an expert system, which can use technical information, knowledge of experts, and simulation models to mimic the deductive and inductive reasoning processes of humans. The expert system represents one of the profound technologies of this century. In addition to providing a means for problem-solving, expert systems can be used to explore and model fundamental issues of animal behavior. The inexact nature of much of the information used in problem-solving led to the development of another type of system, known as an expert support system, which combines attributes of both expert systems and decision support systems. Simulation models and "stand-alone" systems (e.g. evaluaticn functions) were developed as part of the research conducted on IPM in various forest and agricultural commodities. In many cases these models and functions have been adapted for use in decision-making. Each of the different types of computer decision aids has been applied in entomology, but in few cases are we using the full measure of information available for problem-solving and decision-making.

ACKNOWLEDGMENTS

We acknowledge and thank Drs. D. K. Loh, R. E. Frisbie, J. W. Richardson, and N. D. Stone (Knowledge Engineering Laboratory, Departments of Entomology and Agricultural Economics, Texas A & M University); Dr. H. Saarenmaa (Finnish Forest Research Institute); and Drs. T. L. Payne, P. E. Pulley, P. J. H. Sharpe, L. Hu, E. J. Rykiel, and R. H. Turnbow, developers (with D. K. Loh) of the Southern Pine Beetle Decision Support System, for their contributions to the content and organization of this review. We are grateful to Dr. Molly Stock for her candid review of the manuscript. Special thanks are extended to Ms. A. M. Bunting, Ms. L. Gattis, and Ms. M. R. Bunting for technical assistance in the preparation of the manuscript. This paper is TA No. 21546 of the Texas Agricultural Experiment Station and paper No. 7431 in the Journal Series of the Pennsylvania Agricultural Experiment Station.

Literature Cited

1. Bennett, J. L., ed. 1983. *Building Decision Support Systems.* Reading, Mass: Addison-Wesley
2. Bennett, J. S., Engelmore, R. S. 1979. SACON: A knowledge-based consultant for structural analysis. *Proc. Int. Jt. Conf. Artif. Intell., 6th, Tokyo, 1979,* 1:47–49. Stanford, Calif: Stanford Univ. Comput. Sci. Dep.
3. Bobrow, D. G., Hayes, P. J., eds. 1985. Artificial intelligence—where are we? *Artif. Intell.* 25:375–415
4. Bobrow, D. G., Stefik, M. J. 1986. Perspectives on artificial intelligence programming. *Science* 231:951–57
5. Bonczek, R. H., Holsapple, C. W., Whinston, A. B. 1981. *Foundations of Decision Support Systems.* New York: Academic
6. Buchanan, B. G., Duda, R. O. 1983. Principles of rule-based systems. In *Advances in Computers,* ed. M. C. Yovits, 22:163–216. New York: Academic
7. Buchanan, B. G., Shortliffe, E. H., eds. 1985. *Rule-Based Expert Systems.* Reading, Mass: Addison-Wesley
8. Coulson, R. N., Saunders, M. C., Loh, D. K., Rykiel, E. J., Payne, T. L., et al. 1985. A decision support system for the southern pine beetle. In *Insects and Diseases of Southern Forests,* ed. R. A. Goyer, J. P. Jones, pp. 35–46. Baton Rouge, La: La. Agric. Exp. Stn.
9. Coulson, R. N., Witter, J. A. 1984. *Forest Entomology.* New York: Wiley
10. D'Ambrosio, B. 1985. Expert systems—myth or reality? *Byte* 10:275–82
11. Davis, R. 1986. Knowledge-based systems. *Science* 231:957–63
12. Denning, P. J. 1986. The science of computing: expert systems. *Am. Sci.* 71:18–20
13. Duda, R. O., Gaschnig, J. G. 1981. Knowledge-based expert systems come of age. *Byte* 6:238–78
14. Duda, R. O., Gaschnig, J. G., Hart, P. E. 1979. Model design in the PROSPECTOR consultant system for mineral exploration. In *Expert Systems in the Micro-Electronic Age,* ed. D. Michie, pp. 153–67. Edinburgh: Edinburgh Univ. Press
15. Duda, R. O., Shortliffe, E. H. 1983. Expert systems research. *Science* 220: 261–68
16. Epstein, D. A. 1985. Symbolic structure. *Digital Rev.* 2(8):50–58
17. Feigenbaum, E. A., Buchanan, B. G., Lederberg, J. 1971. On generality and problem solving: a case study using the DENDRAL program. *Mach. Intell.* 6: 165–90
18. Fick, G., Sprague, R. H., eds. 1980. *Decision Support Systems: Issues and Challenges.* Oxford: Pergamon
19. Getz, W. M., Gutierrez, A. P. 1982. A perspective on systems analysis in crop production and insect pest management. *Ann. Rev. Entomol.* 27:447–66
20. Gevarter, W. B. 1983. Expert systems: limited but powerful. *IEEE Spectrum* 20(8):39–45
21. Hayes-Roth, F. 1984. Knowledge-based expert systems. *Computer* 17:263–73
22. Hayes-Roth, F., Waterman, D. A., Lenat, D. B., eds. 1983. *Building Expert Systems.* Reading, Mass: Addison-Wesley
23. Holt, D. A. 1985. Computers in production agriculture. *Science* 228:422–27
24. Jaynes, J. 1976. *The Origin of Consciousness in the Breakdown of the Bicameral Mind.* Boston, Mass: Houghton Mifflin
25. Jennings, D. M., Landweber, L. H., Fuchs, I. H., Farber, D. J., Adrion, W. R. 1986. Computer networking for scientists. *Science* 231:943–50
26. Keen, P. G. W., Scott Morton, M. S. 1978. *Decision Support Systems: An Organizational Perspective.* Reading, Mass: Addison-Wesley
27. Lenat, D. B. 1983. Theory formation by heuristic search, the nature of heuristics. II. Background and examples. *Artif. Intell.* 21:31–59
28. LeShan, L. L., Margenau, H. 1982. *Einstein's Space and Van Gogh's Sky; Physical Reality and Beyond.* New York: MacMillan
29. McDermott, J. 1981. XSEL: A computer salesperson's assistant. *Mach. Intell.* 1981:325–37
30. McKinion, J. M., Lemmon, H. E. 1985. Expert systems for agriculture. *Comput. Electron. Agric.* 1:31–40.
31. Michaelsen, R. H., Michie, D., Boulanger, A. 1985. The technology of expert systems. *Byte* 10:303–7
32. Naegele, J. A., Coulson, R. N., Stone,

N. D., Frisbie, R. E. 1986. The use of expert systems to integrate and deliver IPM technology. In *Integrated Pest Management in Major Agricultural Systems*, ed. R. E. Frisbie, P. L. Adkisson, pp. 692–711. College Station, Tex: Tex. Agric. Exp. Stn.

33. Negoita, C. V. 1985. *Expert Systems and Fuzzy Systems*. Menlo Park, Calif: Benjamin/Cummings

34. Rabb, R. L., Guthrie, F. E., eds. 1970. *Concepts of Pest Management*. Raleigh, NC: NC State Univ. Press

35. Rykiel, E. J., Saunders, M. C., Wagner, T. L., Loh, D. K., Turnbow, R. H., et al. 1984. Computer-aided decision making and information accessing in pest management systems, with emphasis on the southern pine beetle (Coleoptera: Scolytidae). *J. Econ. Entomol.* 77:1073–82

36. Saarenmaa, H. 1985. Within-tree population dynamics models for integrated management of *Tomicus piniperda* (Coleoptera: Scolytidae). *Commun. Inst. For. Fenn.* 128:1–56

37. Saaty, T. L. 1985. *The role of microcomputers in OR/MS: analysis, imagination, and morphic causation*. Presented at Natl. Conf. Decision Support Expert Syst., Washington, DC

38. Saggs, H. W. F. 1962. *The Greatness that was Babylon; A Sketch of the Ancient Civilization of the Tigris-Euphrates Valley*. New York: Hawthorn

39. Saunders, M. C., Loh, D. K., Coulson, R. N., Rykiel, E. J., Payne, T. L., et al. 1985. Development and implementation of the southern pine beetle decision support system. In *Proc. Integrated Pest Manage. Res. Symp., Asheville, NC*, ed. S. J. Branham, R. C. Thatcher, pp. 335–63. New Orleans: USDA For. Serv., South. For. Exp. Stn.

40. Schank, R., Hunter, L. 1985. The quest to understand thinking. *Byte* 10:143–55

41. Scott Morton, M. S. 1985. *Expert support systems*. Presented at Natl. Conf. Decision Support Expert Syst., Washington, DC

42. Shannon, R. E., Mayer, R., Adelsberger, H. H. 1985. Expert systems and simulation. *Simulation* 44:275–84

43. Sharpe, P. J. H., DeMichele, D. W. 1977. Reaction kinetics of poikilotherm development. *J. Theor. Biol.* 64:649–70

44. Shortliffe, E. H. 1976. *Computer-Based Medical Consultations: MYCIN*. New York: American Elsevier

45. Stoddard, H. L. 1932. *The Bobwhite Quail, Its Habits, Preservation and Increase*. New York: Scribner

46. Stone, N. D., Coulson, R. N., Frisbie, R. E., Loh, D. K. 1986. Expert systems in entomology: three approaches to problem solving. *Bull. Entomol. Soc. Am.* In press

47. Stone, N. D., Frisbie, R. E., Richardson, J. W., Coulson, R. N. 1986. Integrated expert system applications for agriculture. In *Proc. Int. Conf. Comput. Agric. Ext. Programs, 2nd, Lake Buena Vista, Fla., 1986*. Gainesville, Fla: Univ. Fla. In press

48. Symbolics, Inc. n.d. *A Short History of Artificial Intelligence*. Cambridge, Mass: Symbolics

49. Thierauf, R. J. 1982. *Decision Support Systems for Effective Planning and Control*. Englewood Cliffs, NJ: Prentice-Hall

50. Thompson, B. A., Thompson, W. A. 1985. Inside an expert system. *Byte* 10:315–18

51. Turnbow, R. H., Hu, L. C., Rykiel, E. J., Coulson, R. N., Loh, D. 1983. *Procedural Guide for FERRET, the Question Analysis Routine of the Decision Support System for Southern Pine Beetle Management*. College Station, Tex: Tex. Agric. Exp. Stn. 21 pp.

52. Vasta, J. A. 1985. *Understanding Data Base Management Systems*. Belmont, Calif: Wadsworth

53. Vazsonyi, A. 1978. Information systems in management science. *Interfaces* 9:72–77

54. Vedder, R., Nestmen, C. H. 1985. Understanding expert systems: companion to DSS and MIS. *Ind. Manage. NY* 27:1–8

55. Wagner, T. L., Wu, H. I., Sharpe, P. J. H., Schoolfield, R. M., Coulson, R. N. 1984. Modeling insect development rates: a literature review and application of a biophysical model. *Ann. Entomol. Soc. Am.* 77:208–25

56. Waldrop, M. M. 1984. Artificial intelligence in parallel. *Science* 225:608–10

57. Welch, S. M. 1984. Developments in computer-based IPM extension delivery systems. *Ann. Rev. Entomol.* 29:359–81

58. Winston, P. H. 1984. *Artificial Intelligence*. Reading, Mass: Addison-Wesley

59. Winston, P. H., Prendergast, K. A. 1985. *The AI Business*. Cambridge, Mass: MIT Press

60. Yaghmai, N. S., Maxin, J. A. 1984. Expert systems: a tutorial. *J. Am. Soc. Inf. Sci.* 35:297–305

Ann. Rev. Entomol. 1987. 32:439–62

PHYSIOLOGY OF OSMOREGULATION IN MOSQUITOES

T. J. Bradley

Department of Developmental and Cell Biology, University of California, Irvine, California 92717

PERSPECTIVES AND OVERVIEW

Mosquito larvae are found in fresh, brackish, and saline waters. Beadle, in 1939 (4), was the first to demonstrate that the larvae found in saline waters possess specific osmoregulatory mechanisms not found in freshwater forms. In the intervening time it has been shown that these distinct mechanisms are a result of morphological and physiological specializations (17, 52). Freshwater larvae produce a dilute urine to rid the body of water and replace lost salts by active ion uptake through the cuticle (55). Saline-water larvae drink the external medium to maintain body volume and produce a hyperosmotic urine to eliminate ingested ions (52). These two distinct strategies led physiologists to suggest that mosquito species can be divided into two groups: those possessing freshwater larvae and those possessing saline-water larvae (8, 52).

This conclusion was based on studies of rather limited taxonomic breadth. The only freshwater species examined were *Culex pipiens* and *Aedes aegypti,* while the saline-water species were largely limited to the genus *Aedes.* More recently, physiological studies have been extended to saline-tolerant mosquito larvae in the genera *Culex* and *Culiseta* (25, 26). In the species examined, these larvae osmoconform to the external medium by maintaining high levels of organic solutes in the hemolymph. This represents an osmoregulatory strategy previously undescribed in mosquito larvae; the strategy is presumably the result of an evolutionary pathway distinct from that of saline-water larvae.

Larval osmoregulation has only been examined in a few mosquito species in sufficient detail that the pattern of osmoregulation can be identified. Only

439

0066-4170/87/0101-0439$02.00

Aedes aegypti and *Culex pipiens* have been rigorously shown to be freshwater forms. The species known to have saline-water larvae are *Aedes campestris*, *A. detritus*, *A. dorsalis*, *A. taeniorhynchus*, *A. togoi*, and *Opifex fuscus*. The brackish-water species identified to date are *Culiseta inornata* and *Culex tarsalis*.

Within the genus *Aedes*, saline-water forms may have evolved more than once from freshwater ancestors. The monotypic saline-water species in the genus *Opifex* may also have evolved from a species of *Aedes*. Within the genera *Culex* and *Culiseta*, different evolutionary pathways led to saline-tolerant forms. To date, osmoregulatory mechanisms have been examined in the larvae of mosquitoes from the genera *Aedes*, *Opifex*, *Culex*, and *Culiseta*. Saline-tolerant forms are known to occur in at least five other genera.

Although the osmoregulatory mechanisms of larval mosquitoes have undergone numerous radiations, the adults of these same species are surprisingly uniform with regard to the morphology and physiology of their osmoregulatory organs. Thus while the larval stages have evolved into numerous osmoregulatory niches, the adults are still, like their ancestors, terrestrial, flying insects that imbibe liquid meals. Adaptation to this life-style involves a set of physiological mechanisms radically different from those found in the larvae. Adult adaptations include the capacity to conserve water in a terrestrial environment, as well as the ability to rapidly void water following blood-feeding.

This review is intended to provide a current account of our knowledge of osmoregulatory mechanisms in mosquitoes. These mechanisms have evolved in the context of the developmental complexity common to the Diptera and the diverse aquatic and terrestrial habitats available for exploitation. Current research in the field is directed at characterizing in increasing detail the cellular and molecular mechanisms of ion and fluid transport in these insects. Such information provides insight into the adaptive features of these mechanisms with regard to the ecology and evolution of this highly successful and medically important family of insects.

DEFINITION OF TERMS

In this review the term fresh water is used to refer to all waters more dilute than larval hemolymph ($\sim$300 mOsm). Such waters constitute the vast majority of habitats occupied by mosquito larvae. Saline waters are defined as those in which the total osmotic concentration exceeds 300 mOsm owing to the presence of salts of either marine or terrestrial origin. Mosquito larvae are often found in saline pools whose concentration exceeds that of seawater two to threefold as a result of evaporation. Brackish waters are a subset of saline waters. Brackish habitats are formed where fresh water and seawater mix, as in coastal estuaries and marshes. For the purpose of physiological studies we define brackish water as that having a concentration between the upper limit of fresh water (300 mOsm) and the concentration of seawater (1000 mOsm). This definition is not an arbitrary one; it describes aquatic habitats that are quite common in coastal areas and physiological limitations of a specific physiological type of mosquito larva.

OSMOREGULATION IN FRESHWATER LARVAE

Nearly 95% of all mosquito species have larvae that are restricted to fresh water (48). Mosquito larvae in fresh water face the problem of water uptake by osmosis through the cuticle and by ingestion. In addition, the ionic gradient between their hemolymph and the external water favors ion loss. Wigglesworth (72) showed that larvae of *Aedes aegypti* and *Culex pipiens* are able to overcome these problems and can osmoregulate and ionoregulate very effectively in essentially all media more dilute than their hemolymph.

The insects achieve this by reducing drinking to a minimum, producing a very dilute urine, and by actively transporting ions into the hemolymph from the external medium (55, 64). The organs responsible for water elimination and ion conservation and uptake are the midgut, Malpighian tubules, rectum, and anal papillae (Figure 1).

Food or fluid entering the mouth passes down the cuticle-lined esophagus and into the midgut, which is not lined by cuticle. The midgut cells are columnar and lined with microvilli, in keeping with their role in digestion and absorption (18, 73). The midgut is isosmotic to the hemolymph (55) under a variety of conditions. Following ingestion of food or water, osmotic equilibration of the midgut contents and the hemolymph occurs by diffusion of water. Thereafter, nutrients and ions enter the hemolymph from the midgut by active transport, with water following passively (55, 70).

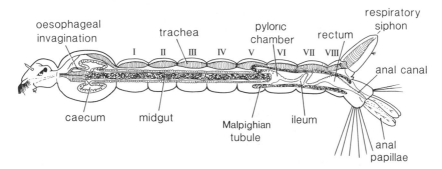

Figure 1   The anatomy of a freshwater larva. Modification of a drawing from Wigglesworth (70).

The primary urine in insects is produced by the Malpighian tubules. Five Malpighian tubules empty into the pyloric chamber at the posterior end of the midgut (18, 22, 32). The Malpighian tubules are blind-ended tubules, and their walls are a histologically simple epithelium. Two cell types are present, primary cells, which are large cuboidal cells containing round intracellular concretions, and stellate cells, which are squamous and lack crystals (18). The ultrastructure of the Malpighian tubules of freshwater mosquito larvae has not been described in detail. Work in my laboratory has shown, however, that Malpighian tubule cells in *A. aegypti* are essentially identical in type and ultrastructure to those in saline-water *Aedes*. The reader is therefore referred to the description of these cells in the next section.

Larval Malpighian tubules produce the primary urine by the active transport of ions, principally K^+ and Cl^- (55, 56). The Malpighian tubules are probably also the site of excretion of organic wastes such as uric acid (71). Production of the primary urine is of little direct use in osmoregulation, since the fluid is isosmotic to the hemolymph and rich in potassium and chloride (56).

Posterior to the midgut, the alimentary tract is again lined with cuticle (41). The narrow ileum possesses squamous cells with few basal and apical infoldings. The ileum is generally presumed to be unimportant in ion transport and to serve mainly in directing fluid and gut contents from the Malpighian tubules and midgut into the rectum (41).

The rectum of *A. aegypti* consists of a single segment with just one cell type (55, 41). Underlying the cuticle, the apical cell membranes are highly folded and studded on the cytoplasmic surface with particles or portasomes (29, 41). The report that the rectum of *A. aegypti* possesses apical microvilli and multiple cell types (24) is in error. The basal membranes are also highly folded, but in the form of invaginating flattened tubes. These ultrastructural features are consistent with an active role in ion transport.

Ramsay (55) showed that the primary urine from the Malpighian tubules is modified in the rectum of *A. aegypti*. He sampled intestinal contents derived from the midgut and Malpighian tubules and found them to be isosmotic to the hemolymph. The rectal contents, in contrast, were hyposmotic and reduced in chloride content relative to the hemolymph. Ramsay concluded that the primary urine is modified in the rectum by the resorption of ions, particularly Na^+, K^+, and Cl^-, prior to its elimination via the anus.

Fluid secreted by the Malpighian tubules flows into the pyloric chamber. From this region, the secreted fluid can either flow anteriorly into the midgut or posteriorly into the ileum and rectum. The ions contained in the Malpighian tubule secretions can thus be returned to the hemolymph either in the midgut or the rectum (55). Both Ramsay (55) and Stobbart (63) have suggested that the rate of production of larval urine and thus the volume of hemolymph is regulated by neuronal feedback from stretch receptors in the epidermis. If larval volume increases, gut peristalsis increases, passing more fluid back into the rectum. When urine flow is to be reduced, more of the Malpighian tubule secretions are passed anteriorly into the midgut. It is assumed that gut peristalsis is controlled directly by the nervous system. This does not necessarily preclude an additional system of hormonal control of secretion rate by the Malpighian tubules, as has been proposed for the tubules of saline-water larvae (37) and adults (49).

Freshwater mosquito larvae possess an additional extrarenal mechanism for maintaining the concentration of ions in the hemolymph. Surrounding the anus are four bulbous anal papillae (Figure 1). No investigator has described a cell border between nuclei in the papillae, and they are therefore assumed to be syncytial (20, 24, 25, 40, 41, 60). Such organization may make the anal papillae more impermeable by eliminating permeability pathways through extracellular junctional spaces. The papillae are covered externally by cuticle. The apical membranes are highly folded and studded on the cytoplasmic surface with portasomes. The basal membranes occur as deep infolds, which are similar to those found in the rectal cells. Regions of close association of the basal membranes and mitochondria, termed scalariform junctions, occur in all species examined (20, 24, 25, 40, 41, 60). The extensive folding of apical and basal membranes and the close association of these membranes with mitochondria are in keeping with the role of these tissues in ion transport.

Koch (34) was the first to report ion uptake by anal papillae, in this case active Cl^- uptake. Subsequent studies of isotopic uptake of Na^+ and Cl^- (62) and electrical potential measurements across the external body wall (64) have shown that both Na^+ and Cl^- are taken up against their respective electrochemical gradients. K^+, by contrast, may be taken up passively as a counterion for Cl^- (61, 64).

Experiments to demonstrate ion transport across anal papillae must overcome a number of technical difficulties. When using whole larvae, one must insure that net ion uptake into the hemolymph is not the result of entry via the mouth or anus. This is usually achieved by blocking the mouth and anus with wax, glue, or ligatures. In addition, it is necessary to show that the ions enter the hemolymph via the anal papillae and not some other portion of the external cuticle. The role of the general body wall is usually distinguished from that of the anal papillae by eliminating the papillae using caustic compounds such as NaOH or $AgNO_3$ (34, 69), which cause the papillae to drop off after 2–3 days. One can then compare ion uptake in blocked larvae with intact papillae to that in blocked larvae lacking papillae. The assumption is that any transport capacity lost following treatment must have resided in the anal papillae and not in the remaining cuticular surface, which was identically treated. This assumption may not be well founded, but data based on it are generally accepted because we lack any better technique for analyzing the function of anal papillae. Even measurements of the transepithelial electrical potential in the anal papillae must be viewed with considerable caution, since the potentials are reported to be highly unstable and of variable polarity (64). More disturbingly, inhibitors of cytochrome oxidase reduce Na^+ transport by about 85% (64) but have no effect on transepithelial potential (23). Methodological breakthroughs in the study of anal papillae would greatly increase our understanding of osmoregulation, not only in freshwater larvae but in brackish-water and saline-water forms as well.

The above description of osmoregulation in freshwater mosquito larvae is derived almost entirely from research on one species, *Aedes aegypti*. Wigglesworth (72) observed a similar pattern of osmoregulation and chloride regulation in *Culex pipiens*. Studies in my laboratory have confirmed that Na^+, K^+, Ca^{2+}, and Mg^{2+} are also well regulated in the hemolymph of *C. pipiens* in dilute media (T. J. Bradley, unpublished observations). No other species of freshwater larvae have been carefully examined for osmoregulatory patterns or mechanisms. Our studies on saline-tolerant mosquitoes from various genera (see below) illustrate the diversity of strategies that can evolve. It is particularly unfortunate that the physiological mechanisms of *Anopheles* larvae have not been examined, since *Anopheles* is in a separate subfamily from the other genera of mosquitoes and is in many respects quite different. Until additional species and genera have been examined, the generalization that the above mechanisms of osmoregulation apply to all freshwater mosquitoes must be viewed with some caution.

OSMOREGULATION IN SALINE-WATER LARVAE

Approximately 5% of all mosquito species have larvae that can survive and grow in saline waters (48). None of these species is truly marine, since for reasons that are not clear, from a physiological viewpoint at least, mosquitoes are never found in open seawater. However, some species are found in seawater that is physically separated from the sea, e.g. coastal splash pools [*Aedes togoi* (42)] or coastal marshes above high tide [*Aedes taeniorhynchus* (43)]. The most thoroughly studied of these saline-tolerant forms are in the genus *Aedes*. On the basis of anatomical features (41) and physiological characteristics (10), the larvae of these mosquitoes are characterized as saline-water mosquito larvae.

Saline-water mosquitoes must be able to withstand high and rapidly changing salinity in the external aquatic environment. Rainfall can cause the habitat to become rapidly diluted, perhaps even to concentrations approximating those of fresh water. Conversely, evaporation can cause the salinity of marshes or pools to rise slowly, but to very high levels (12, 43).

Saline-water mosquito larvae are osmoregulators. By hyperregulating in dilute media and hyporegulating in more concentrated media, they can maintain the osmotic concentration of their hemolymph within a narrow range in external media ranging from distilled water to water three times more saline than seawater (43). In dilute media, the larvae apparently osmoregulate using mechanisms identical to those of freshwater *Aedes* species. Midgut contents and Malpighian tubule secretions are isosmotic to the hemolymph (33, 53). Rectal contents of *Aedes detritus* and *A. taeniorhynchus* larvae reared in dilute media are hyposmotic to the hemolymph, which suggests that the rectum is the site of ion resorption from the primary urine (12, 55). Phillips & Meredith (54) showed that the anal papillae of *Aedes campestris* reared in dilute media were the site of ion uptake from the external medium.

In highly saline media (300 mOsm and above), saline-water mosquito larvae maintain the hemolymph at a concentration hyposmotic to the external medium. Therefore, although the cuticle in saline-water larvae is two orders of magnitude less permeable to water than that of freshwater species (45), saline-water larvae constantly lose water by osmosis across the cuticle (12). To counteract this water loss, saline-water species drink the external medium at substantial rates, ranging from 130% of body volume per day in *Aedes dorsalis* to 240% in *A. taeniorhynchus*. Drinking rates can, however, be quite variable. Asakura (2), in his study of *A. togoi*, found the rate was proportional to the external salinity to which the larvae were acclimated. Bradley & Phillips (12) found that drinking rates in *A. taeniorhynchus* were proportional to the larval surface area and/or metabolic rate, not to the external salinity. The larvae used had been starved for 24 hr, and these authors suggested that the larvae had increased the drinking rate to obtain dissolved nutrients. In all species examined, drinking rates are more than sufficient to replace the water lost across the cuticle by osmosis. In *A. campestris,* for example, drinking exceeds osmotic water loss 12-fold (12).

The water ingested during eating and drinking, and essentially all the ions in it, are completely absorbed across the midgut and into the hemolymph (33). This solves the mosquito's problem of volume regulation, but produces a substantial ion load in the hemolymph.

Beadle (4) was the first to examine the sites of osmoregulation in saline-water mosquito larvae. By placing ligatures at various locations on larvae of *A. detritus,* he demonstrated that regulation was dependent upon organs in the posterior portion of the animal. Using the same species, Ramsay (55) examined the osmotic concentration of the hemolymph and of fluid in the midgut, Malpighian tubules, and rectum. He found that the osmotic concentrations of all the fluids were identical, with the exception of the rectal contents, which were hyperosmotic to the hemolymph. Ramsay thus concluded that the rectum was the site of hyperosmotic urine formation. Unlike the rectum of the freshwater larva of *Aedes aegypti,* the rectum of *A. detritus* is divided into two histologically distinct regions, the anterior and posterior rectal segments (55). These segments have also been observed in *A. campestris* (41), *A. taeniorhynchus* (10), and *A. dorsalis* (68).

Meredith & Phillips (41) conducted ultrastructural studies on the saline-water larvae of *A. campestris* and found that each segment of the rectum is composed of cells of a single, morphologically distinct type. The cells of the anterior rectal segment resemble those comprising the rectal epithelium of *A. aegypti*. The cells have well-developed apical and basal membrane infoldings. Mitochondria are evenly distributed throughout the cytoplasm. The cells of the posterior rectal segment are approximately twice as thick as those of the anterior segment. The apical folds extend deeply into the cell cytoplasm. Most of the mitochondria in the cells of the posterior rectal segment are associated with the apical membrane infoldings. On the basis of Ramsay's (55) measurements of osmotic concentration and the ultrastructural characteristics of the larval rectum, Meredith & Phillips (41) proposed that saline-water larvae produce hyperosmotic urine by salt secretion rather than by water resorption. They proposed that hyperosmotic secretion occurs in the posterior rectal segment because (*a*) the ultrastructure is indicative of a very active role in ion transport and (*b*) the cells in this region are unlike any in the rectum of freshwater larvae.

Bradley & Phillips (10) demonstrated that hyperosmotic excretory fluid is indeed formed by fluid secretion in the rectum of saline-water larvae, and not by water resorption as occurs in terrestrial insects. Using both in vivo and in vitro rectal preparations (11, 12), they showed that Na^+, Mg^{2+}, Cl^-, and K^+ are all removed from the hemolymph by secretion of a hyperosmotic fluid (Figure 2). This was the first demonstration of a secretory function in any insect rectum. Measurements of intracellular and transepithelial electrical potentials, when coupled with measurements of chemical gradients, demonstrated that all of these ions were actively transported (13). The authors found that the concentrations of Na^+, Mg^{2+}, and Cl^- secreted into the rectal lumen were higher than those found in the hemolymph and tended to resemble the concentrations of these ions in the external medium to which the animal had been acclimated (11). The concentrations of Na^+ and Cl^- in the secretion were sufficient to account fully for the elimination of these ions from larvae in hyperosmotic media.

Some populations of saline-water mosquito larvae can survive in saline waters with an ionic makeup quite different from that of seawater. These waters are generally referred to as athalassohaline waters, derived from the Greek for "saline waters of nonmarine origin." Strange et al (67) conducted detailed studies of the mechanism of HCO_3^- regulation in larvae of *A. dorsalis* acclimated to hypersaline waters containing HCO_3^- as the principal anion (pH 10). These authors found lumen-to-hemolymph HCO_3^- and CO_3^{2-} gradients of

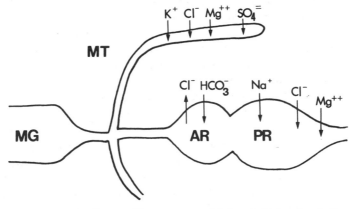

Figure 2   Demonstrated pathways of ion transport and their probable locations in the recta and Malpighian tubules of saline-water larvae.

21:1 and 241:1, respectively. The transepithelial potential under these conditions was -25 mV (lumen negative), demonstrating that these anions are actively transported.

In later studies (66) Strange & Phillips found that the rate of net CO_2 transport (essentially equivalent to net HCO_3^- and CO_3^{2-} transport) was unaffected by Na^+ or K^+ levels in either the lumen or hemolymph, but was quite sensitive to levels of Cl^- in the lumen. On the basis of this evidence and from studies with inhibitors of anion and cation transport, the authors proposed that HCO_3^- transport in the rectum of *A. dorsalis* involved a 1:1 exchange of HCO_3^- for Cl^- (see Figure 2). Electrophysiological studies coupled with studies in which inhibitors of anion transport were used provided support for a model of membrane transport in the anterior rectal segment. The authors proposed that Cl^- entry into the cell from the lumen occurred by a passive electrodiffusive movement through a Cl^- channel. HCO_3^- exit into the lumen was electrogenic but possibly passive. At the basal membrane, Cl^- exit from the cell and HCO_3^- entry occurred by an electrically silent 1:1 exchange. These mechanisms permit the larvae in HCO_3^--rich media to retrieve Cl^- from the contents of the rectum and to excrete HCO_3^-, thereby regulating hemolymph pH levels.

At present, the precise functions of individual rectal segments are unclear. Bradley & Phillips (13) examined segmental function in *A. taeniorhynchus* by tying ligatures between the anterior and posterior rectal segments in an isolated, in vitro rectal preparation. Each segment was incubated in artificial hemolymph for two hours and the recta were examined for swelling due to fluid secretion. The posterior segment swelled slightly with fluid that was hyperosmotic to the artificial hemolymph. No swelling or fluid secretion was apparent in the anterior rectal segments. The very small samples of fluid that could be removed from the lumina of the anterior segments were hyposmotic to the bathing medium. These results, in conjunction with the ultrastructural studies of Meredith & Phillips (41), led Bradley & Phillips (13) to hypothesize that the posterior rectal segment was the site of hyperosmotic fluid secretion, while the anterior rectum was involved in selective reabsorption of inorganic and organic solutes, particularly for larvae in fresh water.

Strange et al (68) examined this hypothesis using microcannulated rectal segments from A. dorsalis acclimated to HCO_3^--rich medium. They found that both rectal segments are capable of hyperosmotic fluid secretion. The posterior and anterior rectal segments accounted for 75% and 25%, respectively, of the fluid secretion observed in the whole rectum. Thus both studies (13, 68) support the conclusion that hyperosmotic fluid secretion occurs predominantly in the posterior rectal segment. It is clear that HCO_3^-/Cl^- exchange occurs in the anterior rectal segment.

Figure 2 is a summary of the results obtained to date on rectal and Malpighian tubule function in saline-water larvae (16). It shows the major sites of ion transport that have been identified. The rates of transport, but presumably not the locations, may differ from species to species, and depend quite strongly on the ionic makeup of the waters in which the larvae have been reared.

Aedes togoi, which is also capable of hyperosmotic urine formation, has two histologically distinct regions in the hindgut (1, 3). Unlike the recta of A. campestris, A. detritus, A. dorsalis, and A. taeniorhynchus, these regions in A. togoi are entirely separate and have been termed the rectum and anal canal (1). All information to date suggests that these regions are functionally equivalent to the anterior and posterior rectal segments, respectively, in the other species. It is noteworthy that A. togoi is placed taxonomically in the subgenus Finlaya, while all the above species possessing two-part recta are in the subgenus Ochlerotatus (19). The morphological differences noted in A. togoi may therefore reflect its taxonomic and perhaps evolutionary distance from the other saline-water aedine species.

Bradley et al (17) described the ultrastructure of the Malpighian tubules of A. taeniorhynchus. The tubules contain cells of two types, primary and stellate. Primary cells are characterized by their large size (70 μm $\times$ 70 μm $\times$ 10 μm) and an abundance of intracellular membrane-bound crystals. Two types of microvilli are found on the luminal surface of the primary cells: (a) small microvilli, which contain core microfilaments and extensions of endoplasmic reticulum, and (b) larger microvilli ($\sim$ 3 μm in length), which contain, in addition to the above components, a mitochondrion along their entire length. Both microvillar types have abundant portasomes lining the cytoplasmic surface of the microvillar membrane. The stellate cells are much smaller than the primary cells and lack intracellular crystals. Their microvilli are smaller as well (~ 0.6 μm in length) and contain no endoplasmic reticulum, mitochondria, or portasomes. The cell types found in the saline-water larvae of *A. taeniorhynchus* are identical to those in the freshwater larvae of *A. aegypti*. In fact, it is likely that the Malpighian tubules of larval and adult mosquitoes in all the genera are identical in cell type and ultrastructure to those described above in *A. taeniorhynchus* (57).

The Malpighian tubules produce a KCl-rich fluid isosmotic to the hemolymph (53). They therefore do not serve directly in osmoregulation; but their main function is to filter the hemolymph. In addition to serving as systems for the transport of monovalent ions, the Malpighian tubules of saline-water mosquito larvae exhibit interesting and unusual properties related to the transport of divalent cations. The tubules of *A. campestris* are able to transport SO_4^{2-} and Mg^{2+} against their respective electrochemical gradients (36, 53). Mg^{2+} is also secreted in the rectum, but the Malpighian tubules are the sole site for the excretion of SO_4^{2-}. In a later study using *A. taeniorhynchus*, Maddrell & Phillips found that sulfate transport in the tubules was inducible (37). Tubules from animals reared in sulfate-free artificial seawater showed very low levels of sulfate transport in vitro. Rates of sulfate transport increased in proportion to sulfate enrichment of the rearing medium. When larvae reared in sulfate-free seawater were placed in seawater augmented with 89 mM NaSO₄, increased transport of sulfate by the Malpighian tubules was demonstrable within 8 hr of transfer. Transport was augmented through an increase in V_{max} with no change in K_m. These results suggest that the presence of sulfate in the external medium results in the synthesis and insertion of additional transport "pumps" in the membranes of the Malpighian tubules (37).

Aedes taeniorhynchus and *Aedes sollicitans* are sibling species. Both species are found in coastal marshes. *Aedes campestris* and *Aedes dorsalis* are closely related species found in both coastal and inland saline waters. Sheplay & Bradley (59) compared larvae and eggs of *A. taeniorhynchus* and *A. dorsalis* for tolerance to high levels of MgSO₄. They reasoned that *A. taeniorhynchus* might be limited to coastal locations because of a relative intolerance to Mg^{2+} and SO_4^{2-} ions that are often found in high concentration in inland waters. However, they found no differences in MgSO₄ tolerance, in terms of egg-hatching success or larval survival, between the two species.

It has been shown that mechanisms of ion transport are inducible in both the Malpighian tubules (37) and the rectum (11). Bradley (7) found increasing saline tolerance over a two-year period in a laboratory colony of *A. taeniorhyncus* transferred from constant rearing in 10% seawater to rearing in 100% seawater. Potential therefore exists for both genetic drift in a population and physiological adaptation to varying saline environments in an individual.

As a result, saline-water larvae show tremendous capacities for acclimation to unusual and highly variable saline environments. Although population differences in saline tolerance have been observed in the field (12), to date it has not been demonstrated that different saline-water *Aedes* species use different types of ion pumps. Therefore these species are assumed to have similar mechanisms and capabilities.

Nicholson (44) showed that larvae of the species *Opifex fuscus* can osmoregulate in 0–200% seawater. *O. fuscus*, the only species in its genus, is thought to be closely related to the genus *Aedes* (35). Everything that we know about osmoregulation in *Opifex* suggests that it is a typical saline-water mosquito. Since we have no information on the histology or function of the Malpighian tubules and rectum in this species, this conclusion is still very tentative. *Opifex fuscus* may have evolved in isolation in New Zealand from an unknown saline-water species of *Aedes*.

OSMOREGULATION IN BRACKISH-WATER MOSQUITO LARVAE

Until 1984, the osmoregulatory mechanisms of saline-tolerant mosquito larvae had been studied only in larvae from the genera *Aedes* or *Opifex*. O'Meara (48), however, pointed out that saline-tolerant larvae are found in nine separate genera of mosquitoes. Garrett & Bradley (25, 26) therefore investigated larvae from some of these other genera, beginning with *Culex* and *Culiseta*.

It is impossible on the basis of present information to identify precisely the evolutionary pathway by which mosquito species with saline-tolerant larvae evolved from their presumably freshwater ancestors. Because the majority of mosquito species possess freshwater larvae, and because saline-water species of *Aedes* occur in closely related species clusters, it could be argued that saline-tolerant species in the genera *Aedes*, *Culex*, and *Culiseta* separately evolved the capacity to survive in saline waters. Larvae in the genera *Culiseta* and *Culex*, which use an osmoregulatory strategy entirely different from that described above for saline-water larvae, are classified as "brackish-water" larvae (25).

Pattern of Osmoregulation

Culiseta inornata is widely distributed in North America in both freshwater and saline-water habitats. In saline sites, the larvae of this species are generally not found in the highly saline pools in which *Aedes* larvae occur (25). Instead, they tend to occur in vegetated, tidal, and brackish portions of coastal marshes. The species is also found in some inland saline ponds.

Larvae of *Culiseta inornata* cannot survive or grow in concentrations of seawater above 800 mOsm (26). Larvae can survive for 24 hr at 1000 mOsm, but cannot undergo further development unless the salinity is again reduced (26). This mosquito's maximum saline tolerance is therefore well above that of freshwater species such as *Aedes aegypti* (500 mOsm), but well below that of saline-water species such as *Aedes taeniorhynchus* (3000 mOsm). Larvae of *C. inornata* therefore seem to be restricted, on physiological grounds, to brackish waters, i.e. waters no more concentrated than seawater.

Osmotic concentrations of the larval hemolymph, measured in various media, were quite different from those observed in saline-water larvae. At salinities below 400 mOsm, the larvae of *C. inornata* are fairly good osmoregulators. At salinities above 400 mOsm, however, they osmoconform, unlike any other mosquito species examined up to the time of Garrett & Bradley's studies (26). Although the hemolymph osmoconforms at higher salinities, hemolymph levels of Na^+, Cl^-, Ca^{2+}, and Mg^{2+} are well regulated. Garrett & Bradley (25, 26) coined the term "brackish-water mosquito" to describe species whose larvae (*a*) can survive in brackish but not highly saline waters, (*b*) exhibit the unique pattern of osmoconforming in water more concentrated than 400 mOsm, and (*c*) have a rectal ultrastructure with several unique aspects (see below).

Mechanisms of Osmoregulation in Brackish-Water Larvae

IONIC REGULATION   Osmoregulatory mechanisms in brackish-water mosquitoes have been under investigation for only two years. A great deal remains to be determined regarding the sites of ion transport and the mechanisms of ion regulation in these larvae. At present, it is not known at what rate brackish-water larvae drink the external medium. Theoretically at least, the larvae could regulate hemolymph volume without drinking. Since their hemolymph is isosmotic with the external medium at higher salinities, they presumably do not lose water osmotically under these conditions. If the number of osmotically active molecules in the hemolymph were increased by either inward transport of ions or biochemical synthesis of organics, osmotic movement of water across the cuticle would increase hemolymph volume. Similarly, if osmotically active compounds were excreted, hemolymph volume would be reduced. It is not known whether either of these scenarios occurs, and how they might be affected by external osmotic concentrations and changing salinities.

The Malpighian tubules of *Culiseta inornata* are identical in cell type and ultrastructure to those of the various species of *Aedes*. The physiological function of the tubules is also identical to that of all larvae examined to date; the tubules produce a secretion that is isosmotic to the hemolymph but rich in K^+ and Cl^- and low in Na^+. Therefore, although the Malpighian tubules are likely to have an important role in the excretion of wastes, the fluid they produce does not directly benefit osmoregulation or ionoregulation.

The rectum of *Culiseta inornata* consists of a single segment containing only one cell type. Thus the mechanisms of ionoregulation and osmoregulation in this species clearly must be different from those in saline-water mosquitoes, which involve the interaction of two rectal segments containing distinct cell types. Garrett & Bradley (26) analyzed the ions in rectal fluid of *C. inornata* using both in vivo and in vitro preparations. They found that the rectal fluid was isosmotic to the hemolymph and isotonic with regard to Na^+. If these findings represent the true capacity of the rectum in this species, then it would seem that its gut lacks the capacity to produce either a hyperosmotic or a hyperionic fluid. The same authors (25) proposed that the anal papillae of *C. inornata* may be the site of ionic transport, but their suggestion rests largely on the lack of any other plausible site.

OSMOTIC REGULATION   The above information on the mechanisms of ion regulation in brackish-water larvae, although slight, demonstrates that changes in hemolymph ion concentrations are not sufficient to account for the observed increases in hemolymph osmotic concentrations. M. Garrett & T. J. Bradley (manuscript submitted) therefore undertook a search for the organic compounds acting as osmolytes in the hemolymph of brackish-water larvae. The authors used *Culex tarsalis,* a species that has a saline tolerance identical to that of *Culiseta inornata*. The two species exhibit identical osmoregulation in dilute media, osmoconformation in concentrated media, ionoregulation in all media, rectal ultrastructure, and Malpighian tubule structure and function.

Having established the pattern of osmoregulation in *Culex tarsalis,* Garrett & Bradley (manuscript submitted) proceeded with studies to isolate and identify the organic compounds responsible for osmoconformation in the hemolymph of larvae at higher salinities. Pooled hemolymph from larvae acclimated to seawater diluted to 50 mOsm was compared with that from larvae acclimated to seawater at 600 mOsm. The total amino acid concentration in the hemolymph was five times higher in larvae adapted to high salinity than in those adapted to low salinity. Proline and serine were the major amino acids accounting for this increase. Proline concentration increased 15-fold, while serine levels increased 6-fold. Other solutes were measured using high-pressure liquid chromatography (HPLC) and an index-of-refraction detector. Trehalose levels increased 4.6-fold. Changes in the concentration of glucose, glycerol, and numerous organic acids were negligible.

The inorganic and organic compounds measured in the hemolymph of larvae adapted to 600 mOsm seawater accounted for 505 of the 595 mOsm of osmotic activity present in the hemolymph. The 90 mOsm unaccounted for consisted of compounds present in low concentrations (e.g. sugars other than trehalose, proteins, organic acids). Similar calculations conducted with the compounds identified in the hemolymph from larvae adapted to 50 mOsm seawater left 78 mOsm unaccounted for. On the basis of these findings, Garrett & Bradley (manuscript submitted) argued that similar unknown osmolytes were present in low concentrations in the hemolymph of larvae from both conditions.

Proline and trehalose are known to act as nontoxic osmolytes that accumulate in high concentration in bacteria, plants, and animals subjected to desiccation, freezing, or high salinity (21, 39, 71). In particular, the accumulation of amino acids in the tissues of marine molluscs, annelids, and crustaceans has been the subject of much study (28, 77). In these organisms, amino acids accumulate intracellularly during acclimation to increasing salinity, probably by means of osmotically activated enzymes (5, 30).

In contrast, the osmolytes in brackish-water larvae accumulate in the hemolymph. This strategy of extracellular accumulation of organic osmolytes is unlike that of any other aquatic animals except sharks and rays, which differ in using urea and trimethylamine oxides. Garrett & Bradley (manuscript submitted) have hypothesized that the intracellular compartments in the larvae may respond to increasing salinity by increasing the synthesis of organic solutes, much as those of marine invertebrates do. The difference may be that brackish-water mosquitoes always allow these osmolytes to leak out into the hemolymph. The additional step of active exclusion of ions allows osmoconformation of the hemolymph while the larvae maintain low extracellular salt levels. The authors pointed out that this strategy may be particularly useful for species that inhabit athalassohaline waters. By excluding ions from the hemolymph and using osmolytes of their own manufacture, the larvae isolate their hemolymph from the unusual and variable ionic ratios of terrestrial systems. This may explain why brackish-water osmoconforming larvae have evolved a strategy different from that of the marine invertebrates.

Many questions remain unanswered regarding the mechanisms of osmoregulation in brackish-water mosquitoes. Although we now recognize the organic osmolytes responsible for the increased osmotic concentration of the hemolymph of osmoconforming *Culex tarsalis* larvae, we do not know which cells are responsible for producing these compounds or which signals trigger their production. Osmoconforming larvae continue to regulate Na^+, Mg^{2+}, and Cl^-. We do not know which organs are responsible for this capacity. The role of the rectum and anal papillae remains largely unexplored. Perhaps most interestingly, we know that the larvae of *Culiseta inornata* and *Culex tarsalis* have very similar patterns of osmotic regulation, i.e. they both osmoconform by accumulating organic compounds in the hemolymph. This would seem to indicate that these rather distantly related insects either share a common brackish-water ancestor or have evolved a similar system by convergent evolution. Further comparative studies of species in these genera may shed some light on the mechanisms of osmolyte synthesis and accumulation, as well as on the evolutionary pathways by which these processes arose.

OSMOREGULATION IN PUPAE

During the pupal stage, larval mouthparts are lost and adult ones are formed (18). The larval midgut is shed into the gut lumen and is replaced by rapidly dividing cells, which form the adult midgut. It is therefore generally assumed that mosquito pupae cannot eat or drink. The Malpighian tubules do not undergo histolysis during pupation (19). In an electron microscopic study of the larval, pupal, and adult Malpighian tubules of *Aedes taeniorhynchus*, we have recently found that pupal tubules do undergo microvillar shrinkage and mitochondrial retraction from the apical microvilli. These pupal Malpighian tubules are incapable of fluid secretion, although the capacity returns within 24 hr of adult emergence (T. J. Bradley & C. Snyder, manuscript submitted). During the pupal period the larval rectum undergoes histolysis and the adult rectum forms from imaginal discs. It is not known precisely when the distinct functions of the larval rectum stop and those of the adult form begin. On the basis of these morphological observations, most workers have concluded that pupae do not drink or produce urine. Apart from our own observations on pupal Malpighian tubules, I am unaware of any direct studies of pupal osmoregulatory mechanisms.

How do the pupae survive and osmoregulate in a variety of fresh and saline waters when they lack the mechanisms of the larvae? The answer may be that the pupae are extremely impermeable and thus may survive by minimizing salt and water fluxes during the few days of the pupal stage. Sheplay & Bradley (59) examined the toxicity of high concentrations of $MgSO_4$ to the larvae and pupae of the saline-water mosquitoes *Aedes dorsalis* and *A. taeniorhynchus*. A solution of 1000 mM $MgSO_4$ in 50% seawater killed all the larvae of both species in 24 hr, but did not kill the pupae. The authors concluded that the pupae are very impermeable to $MgSO_4$ and perhaps also to water. Christophers (18) reported that while larval cuticle is thin and flexible, that of pupae is thick, heavily sclerotized, and covered by a waxy substance, which he was able to solubilize in chloroform.

Based on observations to date, pupal osmoregulation appears to be more dependent on passive resistance to salt and water fluxes than on active mechanisms of urine production and salt extrusion. However, I believe that some mechanisms for ion excretion and perhaps water uptake may yet be found in pupae, particularly in species with a protracted pupal stage.

OSMOREGULATION IN ADULTS

Anatomy of the Osmoregulatory Organs in Adults

During metamorphosis, the osmoregulatory organs are modified in form and function to adapt to the life-style of the adult, i.e. a terrestrial existence, flight, and the ingestion of liquid meals. Adults must conserve water to avoid desiccation during periods of nonfeeding, and must also possess mechanisms to rapidly jettison fluid following feeding on nectar or blood.

The foregut in adults is modified for the ingestion of large liquid meals (18, 19, 47). The crop, a diverticulum of the gut, stores part of the ingested meal. The midgut is also highly distensible. The crop is thought to be a highly impermeable storage organ (19), while the midgut is the site of digestion of the meal and absorption of nutrients, ions, and water.

Five Malpighian tubules empty into the pyloric chamber at the posterior end of the midgut. The adult Malpighian tubules of *Anopheles* (T. J. Bradley & J. K. Nayar, manuscript submitted) and of several species of *Aedes* (14, 38) have been shown to possess two cell types, primary and stellate. Malpighian tubules do not undergo histolysis during metamorphosis (19; T. J. Bradley & C. Snyder, manuscript submitted), and larval tubules from the genera *Aedes, Culex,* and *Culiseta* all possess these two cell types. It is therefore likely that the presence of the two cell types is widespread, if not universal, in the Malpighian tubules of adult mosquitoes.

Posterior to the pyloric chamber are the narrow ileum and the rectum. Hopkins (31) has described the ultrastructure of the rectum. The rectal wall is composed of three layers: an apical cuticle, underlying rectal cells, and a thick layer of circular and longitudinal muscles. Extending into the rectal sac are pear-shaped rectal papillae (four in males, six in females), which are contiguous with the rectal epithelium. The wall of the rectal sac is composed of simple, squamous cells with no elaborations of the apical or basal membranes. These cells are therefore not thought to perform any role in ion or water transport across the rectal wall.

The cells composing the rectal papillae have highly folded apical membranes extending a short distance into the cells. The basolateral membranes are very highly folded and extend throughout the depth of the cells. These basolateral membranes are coated with portasomes and are closely associated with mitochondria. The ultrastructure of the these cells is very similar to that of analogous cells in other terrestrial insects, particularly in the Diptera (51). Based on both current understanding of osmoregulation in adult mosquitoes (74, 75) and the extensive body of work on osmoregulation in insects with a similar rectal ultrastructure (51), we can assume that the rectal papillae are the site of ion and water resorption from the primary urine.

Physiological Aspects of Osmoregulation in Adults

In adult female mosquitoes, engorgement with blood is followed by the rapid production of copious urine (postprandial diuresis). The fluid in the urine derives indirectly from the plasma in the blood meal, in that the midgut epithelium transports fluid from the gut lumen to the hemolymph. The Malpighian tubules simultaneously remove fluid and ions from the hemolymph, thereby producing the primary urine. The urine produced during diuresis is more sodium-rich than that produced at other times. Diuresis rids the animal of 40% of the water, Na^+, and Cl^- in the ingested meal, and 20% of the ingested weight (65). Although blood-feeding often lasts less than 5 min, the initial drops of urine are sometimes expelled even before blood-feeding has ceased (65, 76). The rate and time course of diuresis has been examined in *Anopheles freeborni* (46) and *Aedes aegypti* (6, 65, 76).

In their study of *Anopheles freeborni*, Nijhout & Carrow (46) measured the rate of urine release over 10-min intervals. Short-term variations in secretion rate can therefore not be determined from their data. They found, however, that in this species diuresis starts almost immediately after cessation of feeding and continues for about 30 min at 24°C. Voiding of urine ceases almost altogether after about 50 min.

The same authors (46) were able to provide some insight into the control of diuresis in *Anopheles freeborni*. They found that decapitation of mosquitoes during diuresis led, with a delay of about 10 min, to a slowing of urine production. This was true whether the female was decapitated early or late in diuresis. Severance of the nerve cord in the first abdominal segment had the same effect, which suggests the involvement of either stretch receptors in the abdominal segments or neuronal control of urine release from the rectum. Hemolymph taken from mosquitoes undergoing diuresis did not stimulate tubules in vitro. Saline extracts of heads, thoracic ganglia, and (somewhat surprisingly) Malpighian tubules stimulated tubules in vitro, but only after the extracts had been boiled to reduce proteolysis.

In their study of *Aedes aegypti*, Williams et al (76) measured the rate at which individual drops of urine were produced (average drop volume 10 nl). On the basis of changing rates of fluid production, the researchers were able to resolve three distinct phases of diuresis: peak, postpeak, and late. Osmotic and ionic analyses of the urine produced in each of these periods revealed substantial differences. During peak diuresis, fluid flow was rapid, and the urine was rich in Na^+ and low in K^+. Hemolymph osmolality in mosquitoes is somewhat variable, especially during the period of rapid fluid and ion transport associated with diuresis. The range of hemolymph osmolalities measured by Williams et al (76) was 290–420 mOsm. The average osmotic concentration of the urine during peak diuresis was within this range. During the postpeak diuresis phase, when urine production slowed somewhat, Na^+ levels in the urine declined, and K^+ levels rose. Most significantly, the osmotic concentration of the urine declined below that of the hemolymph. Since the Malpighian tubules always produce an isosmotic fluid, Williams et al proposed that the secretions of the Malpighian tubules are modified during postpeak diuresis by transport processes in the rectum. During late diuresis the osmolality of the urine returns to the range of that of the hemolymph, Na^+ levels decline, and K^+ levels rise substantially (Table 1).

The three phases of diuresis that Williams et al defined obviously represent a physiological continuum involving complex interactions of regulatory and transport processes. It appears nonetheless that the concept of phases in diuresis may be very useful in directing research toward the mechanisms controlling temporal sequences of midgut, Malpighian tubule, and rectal function during the complex diuretic process.

Williams & Beyenbach (74) went on to examine isolated Malpighian tubules from *Aedes aegypti*. Using the Ramsay isolated-tubule assay they showed that while control tubules in insect Ringer solution secrete at 0.8 nl min^{-1}, tubules treated identically but stimulated either with dibutyryl cyclic AMP (1 mM) or a saline extract of mosquito heads secrete about 2.8 nl min^{-1}. The tubules treated with dibutyryl cyclic AMP showed increased Na^+ and decreased K^+ concentration in the secreted fluid. Sawyer & Beyenbach (58), in an elegant electrophysiological study of isolated tubules from the same species, showed that dibutyryl cyclic AMP increased the Na^+ con-

Table 1 Mean rates of urine excretion and urine composition during the three phases of diuresis in adult females of *Aedes aegypti*[a]

| Parameter | Phase of diuresis | | |
	Peak	Postpeak	Late
Time (min)[b]	0–10	11–50	50–120
Rate of urine excretion (nl min^{-1})	54	11	3
Osmolality of urine (mOsm kg^{-1})	309	217	298
Na^+ concentration of urine (mM)	175	132	106
K^+ concentration of urine (mM)	4	16	59
Cl^- concentration of urine (mM)	132	88	177

[a]Standard errors, sample sizes, and statistical analyses available in Reference 76.

[b]The time interval over which the phase occurs expressed as minutes since the initiation of blood-feeding.

ductance of the basolateral membranes of the primary cells in the tubules. They argued that this increased conductance was central to the rapid production of a sodium-rich urine following stimulation with cyclic AMP.

Recently, the question of the control of diuresis in adult *Aedes aegypti* has been approached from a slightly different angle. Using the same assays of in vitro fluid-secretion rates, electrophysiological changes, and ion concentrations in the secreted fluid, Petzel et al (49) have searched for factors contained in the heads of mosquitoes that might affect Malpighian tubule function. They examined a saline extract of heads using low pressure liquid chromatography on a Sep-pak® column followed by HPLC fractionation. The fractions were assayed using isolated Malpighian tubules monitored for changes in transepithelial voltage and/or fluid-secretion rates. The researchers thus isolated three peptide factors of 1900–2700 daltons, each with unique effects on adult Malpighian tubules.

Factor I reduces transepithelial resistance and transepithelial electrical potential. Significantly, application of Factor I alone does not increase fluid-secretion rates in isolated tubules.

Factor II depolarizes the transepithelial voltage of the tubule and increases fluid-secretion rates. Fluid secreted in response to stimulation with Factor II has elevated Na^+ concentration and lowered K^+ concentration relative to unstimulated controls.

Factor III hyperpolarizes transepithelial voltage, reduces transepithelial resistance, increases fluid secretion rates, increases Na^+ concentration, and decreases K^+ concentration of the secreted fluid. Factor III is therefore capable of stimulating the rapid secretion of an NaCl-rich fluid during the initial stages of diuresis. On the basis of these findings, Petzel et al (49, 50) proposed that Factor III be termed the mosquito natriuretic factor (MNF). The authors argued that MNF is likely to comprise at least part of the natural hormonal signal responsible for the production of postprandial natriuresis in adult mosquitoes.

Overall, then, the bulk of evidence is compatible with the view that postprandial diuresis in adult female mosquitoes is controlled by release of diuretic hormone(s). According to such a scheme the primary cells of the Malpighian tubules are the targets of the hormones. These cells respond by increasing the permeability to Na^+ of the basolateral cell membranes and by rapidly transporting an NaCl-rich fluid. Cyclic AMP is strongly implicated as an intracellular second messenger in this process.

It should be noted, however, that there is at present no direct evidence that compounds that stimulate the Malpighian tubules are actually present in the hemolymph or are released following blood-feeding. It is also presently unknown whether the various phases of diuresis represent a gradual decline in the level of stimulation or the sequential application of stimulation affecting separate organs (e.g. midgut, Malpighian tubules, and rectum). Stobbart (65) argued on the basis of ligation and surgical studies that neuronal control of peristaltic activity in the gut is also critical during diuresis in directing urine back to the rectum for excretion.

Despite a great deal of current activity in the field of adult mosquito osmoregulation, many interesting and important questions remain. Future studies are sure to result in fuller characterization of the compounds that influence Malpighian tubule function. Demonstration of the time and site of the release of these compounds will be important. In addition, work is needed on the role of the hindgut in modification and elimination of the primary urine produced by the Malpighian tubules. In particular, the coordination of the tubules and gut seems to be important during the various phases of diuresis.

ACKNOWLEDGMENTS

I thank Drs. Klaus Beyenbach, Jai Nayar, and Simon Maddrell for their helpful comments regarding an earlier version of this manuscript. The author's research reviewed in this article has been supported by grants from the National Institutes of Health and the National Science Foundation. I thank Lisa for her patience and assistance.

Literature Cited

1. Asakura, K. 1980. The anal portion as a salt-excreting organ in a seawater mosquito larva, *Aedes togoi* (Theobald). *J. Comp. Physiol.* 138:59–65
2. Asakura, K. 1982. A possible role of the gastric caecum in osmoregulation of the seawater mosquito larva, *Aedes togoi* (Theobald). *Annot. Zool. Jpn.* 55(1):1–8
3. Asakura, K. 1982. Ultrastructure and chloride cytochemistry of the hindgut epithelium of the larvae of the seawater mosquito, *Aedes togoi* (Theobald). *Arch. Histol. Jpn. Niigata Jpn.* 45 (2):167–80
4. Beadle, L. C. 1939. Regulation of the hemolymph in the saline water mosquito larva *Aedes detritus* Edw. *J. Exp. Biol.* 16:346–62
5. Bishop, S. H., Greenwalt, D. E., Burcham, J. M. 1981. Amino acid cycling in ribbed mussel tissues subjected to hyperosmotic shock. *J. Exp. Zool.* 215:277–87
6. Boorman, J. P. T. 1960. Observations on the feeding habits of the mosquito *Aedes (Stegomyia) aegypti* (L): the loss of fluid after a blood-meal and the amount of blood taken during feeding. *Ann. Trop. Med. Parasitol.* 54:8–14

7. Bradley, T. J. 1976. *The mechanism of hyperosmotic urine formation in the rectum of saline-water mosquito larvae.* PhD thesis. Univ. Br. Columbia, Vancouver. 170 pp.
8. Bradley, T. J. 1985. The excretory system: structure and function. In *Comprehensive Insect Physiology, Biochemistry and Pharmacology,* ed. G. A. Kerkut, L. I. Gilbert. London/New York: Pergamon
9. Deleted in proof
10. Bradley, T. J., Phillips, J. E. 1975. The secretion of hyperosmotic fluid by the rectum of a saline-water mosquito larva, *Aedes taeniorhynchus. J. Exp. Biol.* 63:331–42
11. Bradley, T. J., Phillips, J. E. 1977. Regulation of rectal secretion in saline-water mosquito larvae living in waters of diverse ionic composition. *J. Exp. Biol.* 66:83–96
12. Bradley, T. J., Phillips, J. E. 1977. The effect of external salinity on drinking rates and rectal secretion in the larvae of the saline-water mosquito larva, *Aedes taeniorhynchus. J. Exp. Biol.* 66:97–110
13. Bradley, T. J., Phillips, J. E. 1977. The

460 BRADLEY

location and mechanism of hyperosmotic fluid secretion in the rectum of the saline-water mosquito larva, *Aedes taeniorhynchus. J. Exp. Biol.* 66:111–26

14. Bradley, T. J., Sauerman, D. M. Jr., Nayar, J. K. 1984. Early cellular responses in the Malpighian tubules of the mosquito *Aedes taeniorhynchus* to infection with *Dirofilaria immitis* (dog heartworm). *J. Parasitol.* 70(1):82–88

15. Deleted in proof

16. Bradley, T. J., Strange, K., Phillips, J. E. 1984. Osmotic and ionic regulation in saline-water mosquito larvae. In *Osmoregulation in Estuarine and Marine Animals,* ed. A. Pequeux, R. Gilles, L. Bolis, pp. 35–50. Berlin: Springer-Verlag.

17. Bradley, T. J., Stuart, A., Satir, P. 1982. The ultrastructure of the larval Malpighian tubules of the saline-water mosquito, *Aedes taeniorhynchus. Tissue Cell* 14(4):759–73

18. Christophers, S. R. 1960. *Aedes aegypti (L.), The Yellow Fever Mosquito.* Cambridge: Cambridge Univ. Press

19. Clements, A. N. 1963. *The Physiology of Mosquitoes.* Oxford: Pergamon

20. Copeland, E. 1964. A mitochondrial pump in the cells of the anal papillae of mosquito larvae. *J. Cell Biol.* 23:253–64

21. Crowe, J. H., Crowe, L. M., Chapman, D. 1984. Preservation of membranes in anhydrobiotic organisms: the role of trehalose. *Science* 223:701–3

22. DeBoissezon, P. 1930. Contributions a l'étude de la biologie et de la histophysiologie de *Culex pipiens. Arch. Zool. Exp. Gen.* 70:281–431

23. Edwards, H. A. 1983. Electrophysiology of mosquito anal papillae. *J. Exp. Biol.* 102:343–46

24. Edwards, H. A., Harrison, J. B. 1983. An osmoregulatory syncytium and associated cells in a freshwater mosquito. *Tissue Cell* 15(2):271–80

25. Garrett, M. & Bradley, T. J. 1984. Ultrastructure of osmoregulatory organs in larvae of the brackish-water mosquito, *Culiseta inornata* (Williston). *J. Morphol.* 182:257–77

26. Garrett, M., Bradley, T. J. 1984. The pattern of osmotic regulation in larvae of the mosquito *Culiseta inornata. J. Exp. Biol.* 113:133–41

27. Deleted in proof

28. Gilles, R. 1979. Intracellular organic osmotic effectors. In *Mechanisms of Osmoregulation in Animals.* pp. 111–56. New York: Wiley

29. Harvey, W. R. 1980. Membrane physiology of insects. In *Insect Biology in the Future "VBW 80,"* ed. M. Locke, D. S. Smith, pp. 105–24. London/New York: Academic

30. Hillbish, T. J., Deaton, L. E., Koehn, R. K. 1982. Effect of allozyme polymorphism on regulation of cell volume. *Nature* 298:668–89

31. Hopkins, C. R. 1966. The finestructural changes observed in the rectal papillae of the mosquito *Aedes aegypti* (L) and their relation to the epithelial transport of water and inorganic ions. *J. R. Microsc. Soc.* 86:235–52

32. Jones, J. C. 1960. The anatomy and rhythmical activity of the alimentary canal of *Anopheles* larvae. *Ann. Entomol. Soc. Am.* 53:459–74

33. Kiceniuk, J. W., Phillips, J. E. 1974. Magnesium regulation in mosquito larvae, *Aedes campestris,* living in waters of high $MgSO_4$ content. *J. Exp. Biol.* 61:749–60

34. Koch, H. J. 1938. The absorption of chloride ions by the anal papillae of Diptera larvae. *J. Exp. Biol.* 15:152–60

35. Laird, M. 1956. Studies of mosquitoes and freshwater ecology in the South Pacific. *Bull. R. Soc. NZ* 6:1–213

36. Maddrell, S. H. P., Phillips, J. E. 1975. Active transport of sulphate ions by the Malpighian tubules of the larvae of the mosquito *Aedes campestris. J. Exp. Biol.* 62:367–78

37. Maddrell, S. H. P., Phillips, J. E. 1978. Induction of sulphate transport and hormonal control of fluid secretion by Malpighian tubules of the larvae of the mosquito *Aedes taeniorhynchus. J. Exp. Biol.* 72:181–202

38. Mathew, G., Rai, K. S. 1976. Fine structure of the Malpighian tubules in *Aedes aegypti. Ann. Entomol. Soc. Am.* 69(4):659–61

39. Measures, J. C. 1975. Role of amino acids in osmoregulation of non-halophilic bacteria. *Nature* 257:398–401

40. Meredith, J., Phillips, J. E. 1973. Ultrastructure of the anal papillae of saltwater mosquito larva, *Aedes campestris. J. Insect Physiol.* 19:1157–72

41. Meredith, J., Phillips, J. E. 1973. Rectal ultrastructure in salt- and freshwater mosquito larvae in relation to physiological state. *Z. Zellforsch. Mikrosk. Anat.* 138:1–22

42. Meredith, J., Phillips, J. E. 1973. Ultrastructure of the anal papillae from a sea-water mosquito larva (*Aedes togoi,* Theobald). *Can. J. Zool.* 51:349–53

43. Nayar, J. K., Sauerman, D. M. Jr. 1974. Osmoregulation in larvae of the

salt-marsh mosquito *Aedes taeniorhynchus. Entomol. Exp. Appl.* 17:367–80

44. Nicholson, S. W. 1972. Osmoregulation in larvae of the New Zealand salt-water mosquito *Opifex fuscus* (Hutton). *J. Entomol. Ser. A* 47:101–8

45. Nicholson, S. W., Leader, J. P. 1974. The permeability to water of the cuticle of the larva of *Opifex fuscus* (Hutton) (Diptera, Culicidae). *J. Exp. Biol.* 60:593–603

46. Nijhout, H. F., Carrow, G. M. 1978. Diuresis after a bloodmeal in female *Anopheles freeborni. J. Insect Physiol.* 24:293–98

47. Nuttall, G. H. F., Shipley, A. E. 1903. Studies in relation to malaria. II. The structure and biology of *Anopheles maculipennis. J. Hyg.* 2:58–84

48. O'Meara, G. F. 1976. Saltmarsh mosquitoes (Diptera: Culicidae). In *Marine Insects,* ed. L. Cheng, pp. 303–33. New York: North-Holland

49. Petzel, D. H., Hagedorn, H. H., Beyenbach, K. W. 1985. Preliminary isolation of mosquito natriuretic factor. *Am. J. Physiol.* 249:R379–86

50. Petzel, D. H., Hagedorn, H. H., Beyenbach, K. W. 1986. Peptide nature of two mosquito natriuretic factors. *Am. J. Physiol.* 19(3):R328–32

51. Phillips, J. E. 1981. Comparative physiology of insect renal function. *Am. J. Physiol.* 241:R241–57

52. Phillips, J. E., Bradley, T. J. 1977. Osmotic and ionic regulation in saline-water mosquito larvae. In *Transport of Ions and Water in Animals,* ed. B. L. Gupta, R. B. Moreton, J. L. Oschman, B. J. Wall, pp. 709–34. London: Academic

53. Phillips, J. E., Maddrell, S. H. P. 1974. Active transport of magnesium by the Malpighian tubules of the larvae of the mosquito, *Aedes campestris. J. Exp. Biol.* 61:761–71

54. Phillips, J. E., Meredith, J. 1969. Active sodium and chloride transport by anal papillae of a salt water mosquito larva *(Aedes campestris). Nature* 222:168–69

55. Ramsay, J. A. 1950. Osmotic regulation in mosquito larvae. *J. Exp. Biol.* 27:145–57

56. Ramsay, J. A. 1951. Osmotic regulation in mosquito larvae: the role of the Malpighian tubules. *J. Exp. Biol.* 28:62–73

57. Satmary, W., Bradley, T. J. 1984. The distribution of cell types in the Malpighian tubules of mosquitoes. *Int. J. Insect Morphol. Embryol.* 13(3):209–14

58. Sawyer, D. B., Beyenbach, K. W.

1985. Dibutyryl-cAMP increases basolateral sodium conductance of mosquito Malpighian tubules. *Am. J. Physiol.* 248:R339–45

59. Sheplay, A. W., Bradley, T. J. 1982. A comparative study of magnesium sulphate tolerance in saline-water mosquito larvae. *J. Insect Physiol.* 28(7):641–46

60. Sohal, R. S., Copeland, E. 1966. Ultrastructural variations in the anal papillae of *Aedes aegypti* (L) at different environmental salinities. *J. Insect Physiol.* 12:429–39

61. Stobbart, R. H. 1967. The effect of some anions and cations upon the fluxes and net uptake of chloride in the larvae of *Aedes aegypti* (L) and the nature of the uptake mechanisms for sodium and chloride. *J. Exp. Biol.* 47:35–57

62. Stobbart, R. H. 1971. Evidence for Na^+/H^+ and Cl^-/HCO_3^- exchange during independent sodium and chloride uptake by the larva of the mosquito *Aedes aegypti* (L). *J. Exp. Biol.* 54:19–27

63. Stobbart, R. H. 1971. Factors affecting the control of body volume in the larvae of the mosquitoes *Aedes aegypti* (L) and *Aedes detritus* (Edw.). *J. Exp. Biol.* 54:67–82

64. Stobbart, R. H. 1974. Electrical potential differences and ionic transport in the larva of the mosquito *Aedes aegypti* (L). *J. Exp. Biol.* 60:493–533

65. Stobbart, R. H. 1977. The control of diuresis following a bloodmeal in females of the yellow fever mosquito *Aedes aegypti* (L). *J. Exp. Biol.* 69:53–85

66. Strange, K., Phillips, J. E. 1984. Mechanisms of CO_2 transport in the microperfused rectal salt gland of *Aedes dorsalis.* I. Ionic requirements of CO_2 secretion. *Am. J. Physiol.* 15:R727–34

67. Strange, K., Phillips, J. E., Quamme, G. A. 1982. Active HCO_3^- secretion in the rectal salt gland of a mosquito larva inhabiting $NaHCO_3$-CO_3 lakes. *J. Exp. Biol.* 101:171–86

68. Strange, K., Phillips, J. E., Quamme, G. A. 1984. Mechanisms of CO_2 transport in the microperfused rectal salt gland of *Aedes dorsalis.* II. Site of $Cl^-/$ HCO_3^- exchange and function of anterior and posterior salt gland segments. *Am. J. Physiol.* 15:R735–40

69. Wigglesworth, V. B. 1933. The effect of salt on the anal gills of the mosquito larva. *J. Exp. Biol.* 10:1–15

70. Wigglesworth, V. B. 1933. The function of the anal gills of the mosquito larva. *J. Exp. Biol.* 10:16–26

71. Wigglesworth, V. B. 1933. The adaptation of mosquito larvae to salt water. *J. Exp. Biol.* 10:27–37

72. Wigglesworth, V. B. 1938. The regulation of osmotic pressure and chloride concentration in the hemolymph of mosquito larvae. *J. Exp. Biol.* 15:235–47

73. Wigglesworth, V. B. 1942. The storage of protein, fat, glycogen and uric acid in the fat-body and other tissues of mosquito larvae. *J. Exp. Biol.* 19:56–77

74. Williams, J. C., Beyenbach, K. W. 1983. Differential effects of secretagogues on Na$^+$ and K$^+$ secretion in Malpighian tubules of *Aedes aegypti* (L). *J. Comp. Physiol.* 149:511–17

75. Williams, J. C., Beyenbach, K. W. 1984. Differential effects of secretagogues on the electrophysiology of the Malpighian tubules of the yellow fever mosquito. *J. Comp. Physiol.* 154:301–9

76. Williams, J. C., Hagedorn, H. H., Beyenbach, K. W. 1983. Dynamic changes in the flow rate and composition of urine during the post-bloodmeal diuresis in *Aedes aegypti* (L). *J. Comp. Physiol.* 153:257–65

77. Yancey, P. H., Clark, M. E., Hank, S. C., Bowles, R. D., Somero, G. N. 1983. Living with water stress: evolution of osmolyte systems. *Science* 217:1214–22

Ann. Rev. Entomol. 1987. 32:463–78
Copyright © 1987 by Annual Reviews Inc. All rights reserved

ROLE OF SALIVA IN BLOOD-FEEDING BY ARTHROPODS

J. M. C. Ribeiro

Department of Tropical Public Health, Harvard School of Public Health, 665 Huntington Avenue, Boston, Massachusetts 02115

INTRODUCTION

Although various functions have been ascribed to the saliva of hematophagous arthropods, until recently no general role for saliva in hematophagy was apparent. Anticoagulants have been found in saliva or salivary gland homogenates of certain blood-sucking arthropods, but not in those of others (27). Hemagglutinins are present in salivary glands of some insects, but their role remains unknown (27). Saliva of some hard ticks contains prostaglandins that may help feeding by increasing host skin circulation at the site of the bite (7, 42). Whether saliva serves any purpose in hematophagy has been questioned, however, because insects are capable of taking blood even when they have been rendered incapable of salivating (37, 38, 45, 74).

Saliva functions in various ways unrelated to hematophagy. It is involved in dissolution of solid sugar (21) and in lubrication of the feeding stylets in mosquitoes (54). In hard ticks, the salivary apparatus contributes to ion and water metabolism by excreting the excess water acquired from the blood meal (7, 42). In addition, saliva of blood-sucking arthropods has been identified as the source of toxins or allergens; hence there is a vast literature on both the toxic and antigenic properties of these secretions (30, 90, 92). Allergic reactions to saliva are considered mainly as host functions that diminish the arthropod's feeding success (92). Because saliva has no apparent general function in hematophagy and has seemingly counterproductive allergenic properties, some scientists have theorized that salivary glands in many hematophagous arthropods may represent vestiges from plant-feeding ancestors (15).

463

During the past five years researchers have identified a role for saliva that appears to be applicable to all blood-sucking arthropods. This generalization derives from present approaches to vertebrate hemostasis that focus on the central role of platelets and deemphasize coagulation (53, 83). A number of unrelated arthropods seem to use common antiplatelet activities and other antihemostatic activities that facilitate blood location and prevent host hemostasis during feeding (60–63, 66).

This paper presents a summary of the problems faced by hematophagous arthropods in locating blood in their hosts, particularly hemostasis and inflammation. The summary is followed by a review of the salivary properties of various blood-sucking arthropods. These observations are then synthesized into a unified concept of the adaptive role of salivation in probing and feeding by blood-feeding arthropods.

PROBLEMS FACED BY ARTHROPODS WHEN LOCATING BLOOD

Blood Distribution in the Skin

Once in contact with their host's skin, hematophagous arthropods must locate blood before they can feed. The exploratory phase of feeding (probing) is characterized by introduction of the mouthparts into the skin followed by repetitive movements until blood is found. However, only a small fraction of the total skin volume is occupied by blood vessels. The superficial epidermis is avascular, while the dermis has a characteristic vessel distribution. A deeper plexus of venules and arterioles lying parallel to the skin serves capillary loops that project to the outer surface (75). During probing, the arthropod lacerates blood vessels and causes hemorrhages, thus increasing the relative volume of blood in the skin being probed. As a consequence, blood can be located inside blood vessels or inside hematomas, and the arthropod's chance of locating blood is increased. This laceration of blood vessels during probing is observed both in arthropods that feed from canulated vessels and in those that feed from hemorrhagic pools. Host hemostatic mechanisms are very effective, however, and small mechanical injuries to blood vessels are plugged within seconds by platelets (53, 83).

Platelet Aggregation and Vasoconstriction

Platelets have a central role in hemostasis (53, 83). When blood vessels are lacerated, platelets come into contact with collagen present in subendothelial structures. As a result, platelet phospholipases are activated, and they release free arachidonic acid. Other platelet enzymes convert arachidonic acid into thromboxane A_2 (TXA$_2$), a powerful platelet-aggregating, platelet-degranulating, and vasoconstricting substance. Platelet granules are rich in ADP,

ATP, and serotonin, which are also released during platelet-degranulating reactions such as those induced by TXA_2. ADP induces platelet aggregation and serotonin induces vasoconstriction. Blood contact with collagen therefore induces platelets to obstruct the injured vessel and causes the vessel to constrict. Both effects contribute to hemostasis (Figure 1).

ADP from sources other than platelets contributes similarly to hemostasis. Injured cells release their contents, particularly nucleotides that are disproportionally abundant in cytoplasm. ADP and ATP are present in millimolar concentrations intracellularly and at less than 10^{-7} M in plasma. Therefore, during hemostasis ADP derives from two sources: activated platelets and injured cells. This double source of ADP may explain why this substance has a key function in hemostasis (83).

Coagulation

Coagulation is initiated by a variety of mechanisms, each of which is set in motion by injury to blood vessels (97). Damaged cells release tissue thromboplastin, which leads to formation of thrombin via the extrinsic pathway of the coagulation cascade. The intrinsic pathway can be initiated by surface activa-

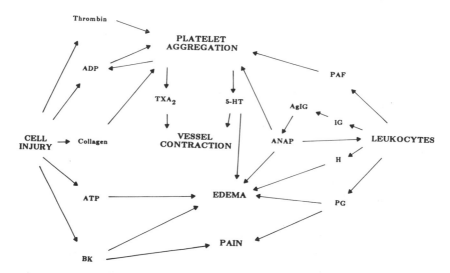

Figure 1 Highly schematic diagram of hemostasis and inflammation, oriented toward the relevant events at an arthropod's feeding site. The *left* portion indicates mediators released following cell injury. The *center* portion indicates the phenomena caused by such mediators. The *right* portion indicates leukocyte contributions to hemostasis and inflammation. Coagulation is not indicated because it has a minor role in repairing bleeding from small vessels. Abbreviations: *ANAP*, anaphylatoxin; *BK*, bradykinin; *H*, histamine; *5-HT*, serotonin; *IG*, immunoglobulins; *AgIG*, immunocomplexes; *PAF*, platelet-activating factor; *PG*, prostaglandins; *TXA₂*, thromboxane A_2.

tion of clotting factor XII following exposure of subendothelial components. This process is accelerated by factors released or exposed by activated platelets. Activation of the intrinsic pathway of blood coagulation produces thrombin, which not only acts on fibrinogen to form fibrin, the clot protein, but also stimulates platelet aggregation much as collagen does. The fibrin clot provides rigidity to the platelet plug, thereby preventing the secondary bleeding that could result from mechanical dislodgment of the aggregated mass of platelets. Hemostasis of large blood vessels is impaired when coagulation is impaired. However, in small blood vessels coagulation seems to have a negligible role; bleeding time of small skin lesions is not modified when coagulation is inhibited by heparin injection (53). Abolished platelet function, however, results in severe bleeding disorders, even from small skin lesions (53).

Inflammation and Hemostasis

During hemostasis, release of vasoactive substances such as serotonin induces increased vascular permeability and consequently edema (5). In addition to affecting coagulation, activated factor XII stimulates production of kalikrein from prekalikrein, leading to synthesis of bradykinin (56). Bradykinin causes vasodilation, increases capillary permeability, and promotes pain (56). In addition, ATP released by platelets and damaged cells enhances local inflammation by inducing mast-cell degranulation (13), which leads to release of thromboxane A_2, platelet-activating factor (PAF) (a platelet-aggregating phospholipid), other arachidonic acid derivatives with vasoactive properties, and large amounts of histamine or serotonin (5). ATP also contributes to inflammation by other mechanisms, which involve neutrophils. Indeed, ATP induces polymorphonuclear neutrophils to aggregate (22), and it produces an inflammatory exudate when injected intraperitoneally into rats (40). Because inflammatory reactions generally include erythema, edema, and pain (5, 14), hemostatic reactions are inherently linked to inflammatory reactions.

Inflammation and Immunity

After repeated exposure to salivary antigens, a host may produce antibodies that will further alter the site at which an arthropod feeds (90, 92). Antigen-antibody reactions may lead to cellular as well as humoral reactions. In cellular reactions, antigens stimulate platelet aggregation and degranulation of mast cells and basophils (Figure 1). In serum-mediated reactions, complement may be activated by certain classes of antibodies combined with antigen; this results in the production of anaphylatoxins, which are fragments of the proteolytic activation of components C3 and C5 of the complement cascade (39, 85). These substances stimulate platelet aggregation (31), induce mast-cell degranulation (41), and induce edema by increasing capillary permeability (85). Anaphylatoxins are also potent chemotactic substances that attract leukocytes to the site of complement activation (85). Other chemotactic substances include arachidonic acid (AA) derivatives released by activated platelets, mast cells, and leukocytes (5). Thus, by leading to the production of these substances, immune reactions in a skin lesion enhance the inflammatory process and change the composition of the cell population.

SALIVARY PROPERTIES OF HEMATOPHAGOUS ARTHROPODS

Both the volume of blood ingested and the time necessary for feeding to repletion vary greatly among blood-sucking arthropods. Mosquitoes may ingest a few microliters within a few minutes; blood-sucking bugs may ingest several milliliters in a half hour; and days may pass before a hard tick finishes its blood meal. Different skin sites may be chosen. Even the strategy of feeding varies; certain arthropods feed by cannulating a vessel, and others ingest blood flowing from lacerated tissue (43). Blood may be taken from superficial capillaries or from the deeper network of skin venules or arterioles (75). The mode and timing of feeding determine whether or not the arthropod encounters intense intradermal inflammatory reactions, which occur from minutes to several hours after stimulation, in addition to the more immediate hemostatic response (Figure 1). In the following discussion, the pharmacological properties of the saliva of diverse hematophagous arthropods are correlated with mode of blood-feeding. Suitably detailed information is available only for a few species of hematophagous arthropods.

Triatomine Bugs

Triatomine bugs locate blood vessels through intradermal probing with their flexible maxillae (44). Saliva is ejected during this probing phase (23, 44), and small hemorrhages form as blood vessels are ruptured by random thrusts of the maxillae. Eventually, the mouthparts come to rest within a venule or arteriole (not a capillary) in the network of vessels in the deeper skin, and engorgement begins. Probing and ingestion may take from a few minutes to nearly one hour, depending on the size of the meal. As described below, saliva injected during the probing phase facilitates location of blood vessels.

Adult *Rhodnius prolixus* whose salivary glands had been removed surgically probed hosts longer than intact bugs. Salivating bugs began to engorge within one minute of initial probing of the skin of a rabbit, but nonsalivating bugs failed to begin to engorge even after five minutes. When feeding on blood through a latex membrane, however, both groups of bugs fed equally well. Nonsalivating bugs eventually succeeded in their attempts to take blood from a living host if allowed sufficient time (60). These findings led to the conclusion that saliva facilitates blood-finding rather than blood-feeding.

Skin that was probed by salivarectomized bugs did not contain the minute hemorrhages that characterized skin lesions produced by bugs with intact salivary glands. This supports the suggestion that saliva has an antihemostatic effect (60). Indeed, *R. prolixus* saliva increased bleeding time of small cuts in the rat's tail (60) and prevented in vitro platelet aggregation (59, 67). One of the components inhibiting platelet aggregation in the saliva of these bugs is an apyrase enzyme (ATP-diphosphohydrolase) (58, 76, 78). This enzyme hydrolizes ATP and ADP to AMP and orthophosphate, thus preventing the effect of ADP on hemostasis. The enzyme exists in sufficient quantity to inhibit platelet aggregation under physiological conditions (58, 59). Saliva from *R. prolixus* also inhibited platelet aggregation induced by collagen (59); this effect may be explained by its potent antithromboxane activity (67). In addition to preventing TXA_2-induced platelet aggregation, the salivary antithromboxane activity also prevented the vessel-contracting effects of TXA_2 (67). Antiserotonin and antihistamine activities were also found in *Rhodnius* saliva (57); these activities help antagonize the vasoconstriction component of hemostasis. Additionally, this saliva contains an anticoagulant, apparently an antagonist of clotting factor VIII (34, 35), which delays thrombin formation. These products in *R. prolixus* saliva act synergistically to prevent host hemostasis.

Mosquitoes

Mosquitoes search for blood by repeatedly thrusting their mouthparts into the host's deep network of skin vessels (28, 29, 50, 66). Saliva is ejected during this intradermal probing period (29). Small hemorrhages accumulate during probing, until the mouthparts eventually come to rest inside a venule, arteriole, or hematoma. The whole process of probing and feeding usually takes less than 10 min. Like that of triatomine bugs, mosquito saliva serves to facilitate blood location.

Surgical procedures have been used in attempts to explore the role of saliva in blood-feeding by mosquitoes. In initial attempts to section the salivary ducts a few duct-ablated specimens survived and were capable of feeding from confined hosts (37, 38, 74). Mellink & Van Den Bovenkamp (51) observed that although nonsalivating *Aedes aegypti* could feed, duration of probing increased while engorgement time remained unchanged. This again suggested that saliva assists probing rather than blood ingestion. This idea has been confirmed, and saliva in mosquitoes has been correlated with anti–platelet aggregating activity both in salivary-gland homogenates and in oil-induced saliva of adult female mosquitoes (63). Thus, antihemostatic properties of saliva appear to induce formation of skin hematomas during intradermal probing, which ultimately increases the probability of blood location (63, 66).

Duration of probing depends on the degree of skin vascularization. Mosquitoes located blood faster in the well-vascularized skin of the ear of a guinea pig than on the poorly vascularized back. Indeed, when mosquitoes probed the homogeneous mass of blood beneath the membrane of an artificial feeder, ingestion began almost immediately (63, 66). Like that of bugs, saliva of mosquitoes reduces the period of blood-finding, thereby decreasing the duration of host contact. Decreased host contact seems to enhance survival, because many hosts kill mosquitoes or prevent them from feeding on blood (18, 19, 26).

The antiplatelet activity of *A. aegypti* saliva seems to derive from an apyrase (63, 68). Apyrase was abundant in the salivary glands of adult female *A. aegypti;* these glands contained the largest known specific activity for the enzyme in a crude organ homogenate (68). Apyrase was ejected by probing mosquitoes and was collected in oil-induced saliva; this demonstrated its secretory fate (63, 68). The salivary glands of diverse anopheline mosquitoes (*Anopheles freeborni, Anopheles stephensi,* and *Anopheles* sp. nr. *salbaii*) contain apyrase, but in differing quantities. Probing time for each species is inversely correlated with salivary apyrase, but is not correlated with salivary anticoagulants. Interestingly, the species that probed longest (and contained the least salivary enzyme) is an autogenous mosquito that seldom feeds on blood under natural conditions (65). Adult male mosquitoes, which do not feed on blood, have only 5% of the salivary apyrase of their female counterparts (72). These observations indicate that the secretory salivary apyrase is adapted to conditions favoring ingestion of blood in the shortest possible span of time.

No antihemostatic activities other than those of apyrase and anticoagulants have been ascribed to the saliva of mosquitoes. Although mosquitoes may have evolved other antihemostatic components, we could demonstrate no antithromboxane or antiserotonin activity in *A. aegypti* salivary glands (J. M. C. Ribeiro, unpublished). Perhaps mosquitoes depend exclusively on salivary apyrase to decrease duration of probing, as supported by the above-mentioned finding of an inverse correlation between stored salivary apyrase and duration of probing by *Anopheles* mosquitoes (65).

Sand Flies

Sand flies have very short mouthparts (less than 0.5 mm) that are unable to penetrate beyond the most superficial layers of skin (1, 46). The flies feed by lacerating the capillary loops and ingesting the blood pooling into the resulting hematomas. Adler & Theodor (1) demonstrated the presence of salivary anticoagulants in *Phlebotomus papatasi*. They also demonstrated that the sacular glands were emptied during the course of a blood meal, which indicates that saliva is ejected during probing and feeding.

The presence of antiplatelet activity, an apyrase, and an erythema-inducing substance (EIS) in salivary gland homogenates of *Lutzomyia longipalpis* was recently described (62). Within 2 min of initial probing, EIS induces erythema that lasts for 2–3 days in human skin. EIS is extremely active; intradermal injection of 2 ng of crude salivary protein on the back of a rabbit resulted in a well-delimited erythema that lasted for several hours. EIS-induced erythema does not result from a host allergic reaction; it was observed in people or rabbits not previously exposed to bites of sand flies, and the reaction is not accompanied by edema or itching (62). The EIS in *Lutzomyia* seems to be a protein, because it is destroyed by trypsin and can be precipitated by ethanol. Together, apyrase and EIS enhance the formation of hematomas and increase the blood flow to the superficial capillaries; this results in reduced duration of host contact (62). No data on salivary properties of other sand fly species have been published. Interestingly, salivary glands of the Old World sand fly *P. papatasi* contain antiplatelet and apyrase activities, but not EIS (J. M. C. Ribeiro, unpublished).

Tsetse

Tsetse thrust their mouthparts repeatedly into the host skin, causing hemorrhagic pools from which blood is rapidly ingested (28). Lester & Lloyd (45) successfully removed the salivary glands of tsetse and noted that although the flies could feed, probing was impaired. Eventually, some flies died with blood clotted in their feeding stylets. Anticoagulants, particularly an antithrombin, are present in their salivary glands (33, 48, 55). This antithrombin prevents thrombin-induced platelet aggregation (48). A salivary apyrase is also present (48). The activities of these substances may promote the formation of the hematomas from which these flies feed. The rapidity with which these flies locate blood and feed (28) suggests the presence of other antihemostatic salivary components.

Ixodid Ticks

Hard ticks insert their mouthparts into the host skin and remain attached for several days. The prolonged period of attachment provides ample time for development of various host inflammatory reactions, including infiltration of different leukocyte populations to the site of feeding (7, 42, 92). Ticks imbibe blood and other exudates from a pool that forms around the mouthparts rather than directly from a canulated vessel. Anticoagulants have been found in salivary gland homogenates of several ixodid ticks (7, 42), and prostaglandin E_2 (PGE_2) was demonstrated in saliva or salivary gland homogenates of at least three different species (17, 36, 61, 77). Because PGE_2 is a vasodilator, it may promote feeding by increasing the flow of blood to the tick mouthparts (7, 42). Salivary gland homogenates of *Ripicephalus sanguineus* contained antihistamine activity (12), and if such activity is excreted in saliva it could prevent some of the inflammatory reactions associated with host tick-rejection reactions. The literature on the immunological basis of host tick-rejection reactions is extensive, and focuses on the allergenic aspects of saliva, which are detrimental to the tick's feeding success (92).

More recently, antihemostatic, antiinflammatory, and immunosuppressive activities were demonstrated in the saliva of *Ixodes dammini* (61). The antihemostatic activity results from the synergistic action of a number of salivary components including an apyrase activity, PGE_2, and prostacyclin (J. M. C. Ribeiro, G. Makoul & D. Robinson, unpublished). In addition to deterring platelet aggregation (14, 98), prostacyclin and PGE_2 dilate host blood vessels, thus antagonizing the vasoconstrictor component of hemostasis. Increased vasodilation also facilitates feeding by increasing the supply of blood to the tick's mouthparts. Although prostacyclin has a brief half-life ($\sim$5 min at pH 7.4 and 37°C), hard ticks seem to salivate at regular 5–15-min intervals, alternating salivation with ingestion of fluid (42). A salivary anticoagulant, which acts on the intrinsic pathway of the coagulation cascade before clotting factor X is activated, has also been identified in *I. dammini* (61). Thus the combined effects of an apyrase, PGE_2, and prostacyclin effectively counteract hemostasis by preventing both platelet aggregation and vasoconstriction. Additionally, PGE_2 and prostacyclin inhibit cell-mediated processes such as mast-cell degranulation (5, 14), thus preventing release of other mediators of inflammation.

I. dammini saliva can rapidly destroy other inflammatory mediators such as bradykinin and anaphylatoxins (61, 69). It is possible that the same enzyme (a carboxypeptidase) destroys both activities (69). In guinea pigs, integrity of the complement system is important for expression of resistance to ticks (87, 91), and the anaphylatoxin inactivator of *I. dammini* saliva may prevent expression of complement-mediated inflammation. Whole tick homogenates contained abundant antiprotease activity, which prevents complement activation (84, 94, 95). Hosts exposed to such ticks develop precipitating antibodies that recognize purified antiprotease. It is possible that such complement-inhibiting activity is excreted in tick saliva, thus preventing complement activation. *I. dammini* saliva contains anticomplement activity that blocks human alternate-complement activation (J. M. C. Ribeiro, unpublished), but its mode of action is still unknown. It is likely that ticks possess redundant mechanisms for antagonizing host complement reactions.

I. dammini saliva prevents in vitro T-cell activation (61). This immunosuppressive activity could be mediated by salivary PGE_2, although other substances may be involved. Such activity may decrease host immune responses to salivary antigens, and may thus contribute to decreased local inflammation at the feeding site. Indeed, tick-infested animals demonstrate reduced lymphocyte responsiveness to mitogens (89, 93). Thus the saliva of *I. dammini* may prevent the inflammatory and hemostatic reactions of their natural hosts, thereby promoting successful blood-feeding.

THE FEEDING CAVITY AND TICK SALIVA   The site of feeding by ticks is characterized by a large hemorrhagic cavity (7, 42, 80). Initially it was speculated that saliva might have a proteolytic role, but attempts to find proteolytic activity failed (7). The feeding cavity was later discovered to be the result of neutrophils accumulating at the tick's feeding site and releasing their hydrolytic enzyme–rich granules. Dogs that were leukopenic showed neither the neutrophil accumulation nor the development of feeding cavities (81). Salivary gland homogenates of *Dermacentor* ticks, when incubated with human serum, developed a chemotactic substance indistinguishable from anaphylatoxin (6). Anaphylatoxin activity could be responsible for the neutrophil accumulation at the feeding site. This finding is paradoxical in the context of an antiinflammatory role of saliva, but an anaphylatoxin inactivator, if present, would reconcile the observations. Anaphylatoxin is usually inactivated by the removal of the terminal arginyl residue of the molecule (8), which halts its mast cell–degranulating and edema-promoting activities but does not abolish its chemotactic properties (69, 86). If ticks have this anaphylatoxin inactivator, then they could manipulate a host's leukocytes in their own interest. They could selectively attract them to the feeding site and cause them to dig a feeding cavity. This would increase the supply of blood for the tick's final rapid engorgement phase of feeding without causing the inflammatory reactions detrimental to feeding. Additionally, neutrophils may serve as food in the slow feeding phase of ixodid ticks and tungid fleas (44a, 80, 81).

TICK SALIVA AND HOST REJECTION REACTIONS   Trager originally established that certain vertebrates may become immune to particular ticks (82). After repeated exposure to *Ixodes* or *Dermacentor* ticks, guinea pigs developed a resistance that prevented attached ticks from obtaining a meal of blood or reduced the quantity of blood taken. However, another rodent *(Peromyscus leucopus)* did not develop such resistance. Similarly, *I. dammini* is rejected by guinea pigs but is tolerated by *Peromyscus leucopus* (P. Davidar & J. M. C. Ribeiro, unpublished). Related cattle species *(Bos indicus* and *Bos taurus)* differed in their ability to reject ticks (25, 70).

Basophil leukocytes appear to be important in tick-rejection reactions (2, 11, 92). Rejection may be mediated by histamine release stimulated by concentration of basophils, and *I. dammini* appears to lack histamine antagonists (J. M. C. Ribeiro, unpublished). Guinea pigs and rabbits contain abundant basophils (4), and their resistance to the tick *Ixodes ricinus* is abolished or suppressed by injected mepyramine, a H1 histamine antagonist (9). Antihistamines also inhibit the rejection of *Amblyomma* ticks by guinea pigs (88). Similarly, resistance of cattle to the bites of *Boophilus microplus* was correlated with skin histamine content (96). Other mediators released by basophils may also be involved in tick-rejection reactions (10). However, mice are devoid of basophils (4), and their cutaneous hypersensitivity reactions rely on mast cells and serum-derived mediators such as anaphylatoxins (16, 49, 85). A tick adapted to a particular host might evade rejection by countering against that host's inflammatory mechanism. Thus, effective host range may be defined by the properties of the salivary antiinflammatory battery of the tick (61). The antiinflammatory properties of the saliva of *I. dammini* may be a key that precisely fits the lock of the profile of the inflammatory mediators of *Peromyscus*. However, the tick *Ripicephalus sanguineus* does possess antihistamine salivary activity (12), which may permit its feeding on basophil-rich hosts. In contrast, a tick that possesses a powerful antihistamine but that is unable to counteract bradykinin and anaphylatoxin may be rejected by mice, whose inflammatory reactions rely heavily on these mediators. Thus, the ability to overcome a host's hemostatic as well as inflammatory defenses is a component of a tick's host specificity. This knowledge prepares the way for analyses of the mechanisms by which ticks evade the immune-inflammatory responses of particular hosts and the mechanisms by which other hosts come to reject these ectoparasites.

BLOOD-FINDING ROLE OF SALIVA IN HEMATOPHAGOUS ARTHROPODS

The antihemostatic properties of the saliva of such rapid-feeding arthropods as bugs and mosquitoes decrease the time required by these insects to locate the comparatively small masses of blood contained in the vessels of the skin (60, 63, 66). Formation of hematomas increases the probability of blood location by adding the mass of blood pooled in the skin to the total vascular volume present at that site. The diameter of a hematoma exceeds that of the originating blood vessel, thereby increasing the probability that randomly moving mouthparts will contact blood. If the insect sucks more blood from a hematoma than the lesion in the vessel provides, the hematoma wall (comprised of soft tissues) tends to collapse, drawing the vessel toward the mouthpart stylets. Therefore, whether the insect feeds from vessels or from pools is determined by the rate of feeding relative to the volume of the hematoma, and by its rate of refilling by the lacerated vessels.

COEVOLUTION OF HOSTS AND THEIR ECTOPARASITES

Hematophagy is polyphylectic in origin, but convergent evolution has equipped many blood-feeding arthropods similarly for locating blood. This convergent evolution has occurred not only among the arthropods. Saliva of the vampire bat *Desmodus rotundus* contains antiplatelet activity that prevents host hemostasis and allows continuous flow of blood from lesions produced by its sharp teeth (32). Similarly, extracts of various hookworms prevent platelet aggregation and may have a similar role in their feeding process (20, 79). Leeches have a vast array of anticoagulant and fibrinolytic activities (47, 52), and anti–platelet aggregating activities as well (24). Because saliva in hematophagous animals has an adaptative role in prevention of host hemostatic and inflammatory reactions, study of this subject not only reveals peculiar adaptations of parasites to their hosts, but may also increase our knowledge of vertebrate hemostasis and inflammation.

At least two sources of variation affect the evolution of salivary anti-hemostatic and antiinflammatory activities. Hemostasis and inflammation are complex phenomena with considerable variation among species, and parasites may evolve particular salivary activities that reflect such variation. Additionally, each blood-feeding species evolved to hematophagy starting with its own stock of evolutionary options. Therefore, the presence of a common salivary activity, such as apyrase, suggests both that ATP and ADP have general importance in vertebrate hemostasis and inflammation, and also that such an enzyme was available in the salivary glands of the species from which blood-feeding forms evolved (76). It may be that some salivary activities are common to most hematophagous arthropods and other activities are peculiar only to some.

Coevolution of Vector-Borne Parasites and their Vectors

Parasites use various mechanisms to manipulate their host's behavior to maximize their transmission to a new host. Parasite-induced salivary pathology may be one such mechanism. *Plasmodium gallinaceum* sporozoites reduced salivary apyrase in *A. aegypti* without affecting volume of salivary output (72). Duration of probing by infected mosquitoes was increased, and their biting rate on a host was doubled (73). *Trypanosoma rangeli*–infected *R. prolixus* demonstrated a similar behavior; their probing time was greatly enhanced, and the bugs were eventually unable to feed (3). Similarly, *A. triseriatus* infected with La Crosse virus tended to probe more and engorge less than uninfected siblings, and transmission rates increased as the level of probing increased (30a). Whether La Crosse virus and *T. rangeli* impair their vector's salivary function remains to be explored. Vector-borne parasites may facilitate vector feeding through the hemostatic disorders they commonly induce in vertebrate hosts (66). Mosquitoes located blood faster in Rift Valley fever virus–infected hamsters or *Plasmodium berghei*–infected mice than in uninfected animals (71). The parasites thus manipulate both the vertebrate host's hemostatic components and the vector's salivary antihemostatic properties to achieve maximum transmission.

CONCLUSION

Saliva of diverse blood-feeding arthropods has antihemostatic properties, particularly antiplatelet activity. Apyrase, an enzyme that hydrolyzes ATP and ADP to AMP and orthophosphate, is frequently abundant in the salivary glands. Because ADP is an important mediator of platelet aggregation, salivary apyrase is antihemostatic. Additional antihemostatic and antiinflammatory salivary substances counteract the host's redundant mechanisms of hemostasis and inflammation. Future studies of salivation may help explain how particular hematophagous arthropods evade their hosts' hemostatic and inflammatory reactions and how vector-borne parasites modify arthropod-vertebrate interactions to enhance parasite transmission.

ACKNOWLEDGMENTS

We are grateful to Drs. Andrew Spielman, Anthony James, Betz Halloran, Claudio Struchiner, and Mark Wilson for critical review of the manuscript. This work was supported in part by Contract DAMD-17-84-C-4003 from the United States Army Medical Research and Development Command, and by grants AI-15886 and AI-18694 from the National Institutes of Health.

Literature Cited

1. Adler, S., Theodor, O. 1926. The mouthparts, alimentary tract and salivary apparatus of the female *Phlebotomus papatasi*. *Ann. Trop. Med. Parasitol.* 20:109–42

2. Allen, J. R. 1973. Tick resistance: basophil in skin reactions of resistant guinea pigs. *Int. J. Parasitol.* 3:195–200

3. Anez, N., East, J. S. 1984. Studies on *Trypanosoma rangeli* Tejera, 1920. II. Its effects on the feeding behaviour of triatomine bugs. *Acta Trop.* 41:93–95

4. Askenase, P. W. 1977. Role of basophils, mast cells and vasoamines in hypersensitivity reactions with a delayed time course. *Prog. Allergy* 23:199–320

5. Bach, M. K. 1982. Mediators of anaphylaxis and inflammation. *Ann. Rev. Microbiol.* 36:371–413

6. Berenberg, J. L., Ward, P. A., Sonenshine, D. E. 1972. Tick-bite injury: Mediation by a complement-derived chemotictic chemotactic factor. *J. Immunol.* 109:451–56

7. Binnington, K. C., Kemp, D. H. 1980. Role of tick salivary glands in feeding and disease transmission. *Adv. Parasitol.* 18:316–40

8. Bokish, V. A., Muller-Eberhard, H. J. 1970. Anaphylatoxin inactivator of human plasma: its isolation and characterization as a carboxypeptidase. *J. Clin. Invest.* 49:2427–36

9. Brossard, M. 1982. Rabbits infested with adult *Ixodes ricinus* L. Effect of mepyramine on acquired resistance. *Experientia* 38:702–4

10. Brown, S. J., Askenase, P. W. 1985. Rejection of ticks from guinea pigs by antihapten-antibody-mediated degranulation of basophils at cutaneous basophil hypersensitivity sites: Role of mediators other than histamine. *J. Immunol.* 134:1160–65

11. Brown, S. J., Galli, S. J., Gleish, G. J., Askenase, P. W. 1982. Ablation of immunity to *Amblyomma americanum* by anti-basophil serum: Cooperation between basophils and eosinophils in expression of immunity to ectoparasites (ticks) in guinea pigs. *J. Immunol.* 129: 790–96

12. Chinery, W. A., Ayitey-Smith, E. 1977. Histamine blocking agent in the salivary gland homogenate of the tick *Rhipicephalus sanguineus sanguineus*. *Nature* 265:366–67

13. Coutts, A. A., Jorizzo, J. L., Eady, R. A. J., Greaves, M. W., Burnstock, G. 1981. Adenosine triphosphate-evoked vascular changes in human skin: Mechanism of action. *Eur. J. Pharmacol.* 76:391–401

14. Davies, P., Bailey, P. J., Goldenberg, M. M., Ford-Hutchinson, A. W. 1984. The role of arachidonic acid oxygenation products in pain and inflammation. *Ann. Rev. Immunol.* 2:335–58

15. De Meillon, B. 1949. The relationship between ectoparasites and host. IV. Host reaction to bites of arthropods. *Leech Johannesburg* Aug:43–46

16. Den Hollander, N., Allen, J. R. 1985. *Dermacentor variabilis*: Acquired resistance to ticks in BALB/C mice. *Exp. Parasitol.* 59:118–29

17. Dickinson, R. G., O'Hagan, J. E., Shotz, M., Binnington, K. C., Hegarty,

M. P. 1976. Prostaglandin in the saliva of the cattle tick *Boophilus microplus*. *Aust. J. Exp. Biol. Med. Sci.* 54:475–86

18. Edman, J., Kale, H. W. 1971. Host behavior: Its influence on the feeding success of mosquitoes. *Ann. Entomol. Soc. Am.* 64:513–16

19. Edman, J., Webber, L. A., Schmid, A. A. 1974. Effect of host defenses on the feeding pattern of *Culex nigripalpus* when offered a choice of blood sources. *J. Parasitol.* 60:874–83

20. Eiff, J. A. 1966. Nature of an anticoagulant from the cephalic glands of *Ancylostoma caninum*. *J. Parasitol.* 52: 833–43

21. Eliason, P. A. 1963. Feeding adult mosquitoes on solid sugars. *Nature* 200:289

22. Ford-Hutchinson, A. W. 1982. Aggregation of rat neutrophils by nucleotide triphosphates. *Br. J. Pharmacol.* 76 (3):367–71

23. Friend, W. G., Smith, J. J. B. 1971. Feeding in *Rhodnius prolixus*: mouthpart activity and salivation, and their correlation with changes of electrical resistance. *J. Insect Physiol.* 17:233–43

24. Gasic, G. J., Viner, E. D., Budzynski, A. Z., Gasic, G. P. 1983. Inhibition of lung tumor colonization by leech salivary gland extracts from *Aementeria ghilianii*. *Cancer Res.* 43:1633–36

25. George, J. E., Osburn, R. L., Wikel, S. K. 1985. Acquisition and expression of resistance by *Bos indicus* × *Bos taurus* calves to *Amblyomma americanum* infestation. *J. Parasitol.* 71:174–82

26. Gillett, J. D. 1967. Natural selection and feeding speed in a blood sucking insect. *Proc. R. Soc. London Ser. B* 167:316–29

27. Gooding, R. H. 1972. Digestive process of haematophagous insects. I. A literature review. *Quaest. Entomol.* 8:5–60

28. Gordon, R. M., Crewe, W. 1948. The mechanism by which mosquitoes and tsetse flies obtain their blood meal, the histology of the lesions produced, and the subsequent reactions of the mammalian host, together with some observations on the feeding of Chrisops and Cimex. *Ann. Trop. Med. Parasitol.* 42: 334–56

29. Gordon, R. M., Lunsden, W. H. R. 1939. A study of the behaviour of the mouthparts of mosquitoes when taking up blood from living tissue; together with some observations on the ingestion of microfilariae. *Ann. Trop. Med. Parasitol.* 33:259–78

30. Gothe, R., Kunze, K., Hoogstraal, H. 1979. The mechanism of pathogenicity in the tick paralyses. *J. Med. Entomol.* 16:357–69

30a. Grimstad, P. R., Ross, Q. E., Craig, G. B. Jr. 1980. *Aedes triseriatus* (Diptera: Culicidae) and La Crosse virus. II. Modification of mosquito feeding behavior by virus infection. *J. Med. Entomol.* 17:1–7

31. Grossklaus, C. B., Damerau, B., Lemgo, E., Vogt, W. 1976. Induction of platelet aggregation by the complement derived peptides C3a and C5a. *Naunyn-Schmiedebergs Arch. Pharmakol.* 295: 71–76

32. Hawkey, C. 1967. Inhibitor of platelet aggregation present in saliva of the vampire bat *Desmodus rotundus*. *Br. J. Haematol.* 13:1014–20

33. Hawkins, R. I. 1966. Factors affecting blood clotting from salivary glands and crops of *Glossina austeni*. *Nature* 212: 738–39

34. Hellmann, K., Hawkins, R. I. 1964. Anticoagulant and fibrinolytic activities from *Rhodnius prolixus* Stahl. *Nature* 201:1008–9

35. Hellmann, K., Hawkins, R. I. 1965. Prolixin-S and Prolixin-G: two anticoagulants from *Rhodnius prolixus* Stahl. *Nature* 207:265–67

36. Higgs, G. A., Vane, J. R., Hart, R. J., Porter, C., Wilson, R. G. 1976. Prostaglandins in the saliva of the cattle tick, *Boophilus microplus* (Canestrini) (Acarina, Ixodidae). *Bull. Entomol. Res.* 66: 665–70

37. Hudson, A. 1964. Some functions of the salivary glands of mosquitoes and other blood-sucking arthropods. *Can. J. Zool.* 42:113–20

38. Hudson, A., Bowman, L., Orr, C. W. M. 1960. Effects of absence of saliva on blood feeding by mosquitoes. *Science* 131:1730

39. Hugli, T. E., Muller-Eberhard, H. J. 1978. Anaphylatoxins: C3a and C5a. *Adv. Immunol.* 26:1–47

40. Ischikawa, A., Hayashi, H., Minami, M., Kenkichi, T. 1972. An acute inflammation induced by inorganic pyrophosphate and adenosine triphosphate, and its inhibition by cyclic 3'-5' adenosine monophosphate. *Biochem. Pharmacol.* 21:317–31

41. Johnson, A. R., Hugli, T. E., Muller-Eberhard, H. J. 1975. Release of histamine from rat mast cells by the complement peptides C3a and C5a. *Immunology* 28:1067–80

42. Kemp, D. H., Stone, B. F., Binnington, K. C. 1982. Tick attachment and feeding: Role of mouthparts, feeding apparatus, salivary gland secretions and the host response. In *Physiology of Ticks,* ed. F. D. Obenchain, R. Galun, pp. 119–68. Oxford: Pergamon

43. Lavoipierre, M. M. J. 1965. Feeding mechanisms of blood-sucking arthropods. *Nature* 208:302–3

44. Lavoipierre, M. M. J., Dickerson, G., Gordon, R. M. 1959. Studies on the methods of feeding of blood sucking arthropods. I. The manner in which triatomine bugs obtain their blood meal, as observed in the tissues of the living rodent, with some remarks on the effects of the bite on human volunteers. *Ann. Trop. Med. Parasitol.* 53:235–50

44a. Lavoipierre, M. M. J., Radovsky, F. J., Budwiser, P. D. 1979. The feeding process of a tungid flea, *Tunga monositus* (Siphonaptera: Tungidae), and its relationship to the host inflammatory and repair response. *J. Med. Entomol.* 15:187–217

45. Lester, H. M. O., Lloyd, L. 1928/1929. Notes on the process of digestion in tsetse flies. *Bull. Entomol. Res.* 19:39–60

46. Lewis, D. J. 1975. Functional morphology of the mouthparts in New World phlebotomine sandflies (Diptera: Psychodidae). *Trans. R. Entomol. Soc. London* 126:497–532

47. Malinconico, S. M., Katz, J. B., Budzynski, A. Z. 1984. Hementin: anticoagulant protease from the salivary gland of the leech *Haementeria ghilianii. J. Lab. Clin. Med.* 103:44–58

48. Mant, M. J., Parker, K. R. 1981. Two platelet aggregation inhibitors in tsetse (*Glossina*) saliva with studies of roles of thrombin and citrate in in vitro platelet aggregation. *Br. J. Haematol.* 48:601–8

49. Matsuda, H., Fukui, K., Kiso, Y., Kitamura, Y. 1985. Inability of genetically mast cell–deficient *W/W^v* mice to acquire resistance against larval *Haemaphysalis longicornis* ticks. *J. Parasitol.* 71:443–48

50. Mellink, J. J., Poppe, D. M. C., van Duin, G. J. T. 1982. Factors affecting the bloodfeeding process of a laboratory strain of *Aedes aegypti* on rodents. *Entomol. Exp. Appl.* 31:229–38

51. Mellink, J. J., Van Den Bovenkamp, W. 1981. Functional aspects of mosquito salivation in blood feeding in *Aedes aegypti. Mosq. News* 41:115–19

52. Muser, E. H., James, H. L., Budzynski, A. Z., Malinconico, S. M., Gasic, G. J. 1984. Protease inhibitors in *Haementeria* leech species. *Thromb. Haemostasis* 51:24–26

53. Mustard, J. F., Packham, M. A. 1977. Normal and abnormal haemostasis. *Br. Med. Bull.* 33:187–91

54. Orr, C. W. M., Hudson, A., West, A. S. 1961. The salivary glands of *Aedes aegypti*. Histological-histochemical studies. *Can. J. Zool.* 39:265–72

55. Parker, K. R., Mant, M. J. 1979. Effects of tsetse salivary gland homogenate on coagulation and fibrinolysis. *Thromb. Haemostasis* 42:743–51

56. Regoli, D., Barabe, J. 1980. Pharmacology of bradykinin and related kinins. *Pharmacol. Rev.* 32:1–46

57. Ribeiro, J. M. C. 1982. The antiserotonin and antihistamine activities of salivary secretion of *Rhodnius prolixus. J. Insect Physiol.* 28:69–75

58. Ribeiro, J. M. C., Garcia, E. S. 1980. The salivary and crop apyrase activity of *Rhodnius prolixus. J. Insect Physiol.* 26:303–7

59. Ribeiro, J. M. C., Garcia, E. S. 1981. Platelet antiaggregating activity in the salivary secretion of the blood sucking bug *Rhodnius prolixus. Experientia* 37:384–85

60. Ribeiro, J. M. C., Garcia, E. S. 1981. The role of saliva in feeding in *Rhodnius prolixus. J. Exp. Biol.* 94:219–30

61. Ribeiro, J. M. C., Makoul, G., Levine, J., Robinson, D., Spielman, A. 1985. Antihemostatic, antiinflammatory and immunosuppressive properties of the saliva of a tick, *Ixodes dammini. J. Exp. Med.* 161:332–44

62. Ribeiro, J. M. C., Rossignol, P. A., Spielman, A. 1986. Blood finding strategy of a capillary feeding sandfly, *Lutzomyia longipalpis. Comp. Biochem. Physiol.* 83A:683–86

63. Ribeiro, J. M. C., Rossignol, P. A., Spielman, A. 1984. Role of mosquito saliva in blood vessel location. *J. Exp. Biol.* 108:1–7

64. Deleted in proof

65. Ribeiro, J. M. C., Rossignol, P. A., Spielman, A. 1985. Salivary gland apyrase determines probing time in anopheline mosquitoes. *J. Insect Physiol.* 31(9):689–92

66. Ribeiro, J. M. C., Rossignol, P. A., Spielman, A. 1985. *Aedes aegypti:* Model for blood finding behavior and prediction of parasite manipulation. *Exp. Parasitol.* 60:118–32

67. Ribeiro, J. M. C., Sarkis, J. J. F. 1982. Anti-thromboxane activity in *Rhodnius prolixus* salivary secretion. *J. Insect Physiol.* 28:655–60

68. Ribeiro, J. M. C., Sarkis, J. J. F., Rossignol, P. A., Spielman, A. 1984. The salivary apyrase of *Aedes aegypti:* Characterization and secretory fate. *Comp. Biochem. Physiol.* 79B:81–86

69. Ribeiro, J. M. C., Spielman, A. 1986. *Ixodes dammini:* Salivary anaphylatoxin inactivating activity. *Exp. Parasitol.* In press

70. Riek, R. F. 1962. Studies on the reac-

tion of animals to infestation with ticks. VI. Resistance of cattle to infestation with the tick *Boophilus microplus* (Canestrini). *Aust. J. Agric. Res.* 13: 532–50

71. Rossignol, P. A., Ribeiro, J. M. C., Jungery, M., Turell, M. J., Spielman, A., Bailey, C. L. 1985. Enhanced mosquito blood-finding success on parasitemic hosts: Evidence for vector-parasite mutualism. *Proc. Natl. Acad. Sci. USA* 82:7725–27

72. Rossignol, P. A., Ribeiro, J. M. C., Spielman, A. 1984. Increased intradermal probing time in sporozoite-infected mosquitoes. *Am. J. Trop. Med. Hyg.* 33:17–20

73. Rossignol, P. A., Ribeiro, J. M. C., Spielman, A. 1986. Increased biting rate and reduced fertility in sporozoite-infected mosquitoes. *Am. J. Trop. Med. Hyg.* 35:277–79

74. Rossignol, P. A., Spielman, A. 1982. Fluid transport across the ducts of the salivary glands of a mosquito. *J. Insect Physiol.* 28:579–83

75. Ryan, T. J. 1976. The blood vessels of the skin. *J. Invest. Dermatol.* 67:110–18

76. Sarkis, J. J. F., Guimaraes, J. A., Ribeiro, J. M. C. 1986. Salivary apyrase of *Rhodnius prolixus:* kinetics and purification. *Biochem. J.* 233:885–91

77. Shemesh, M., Hadani, A., Shklar, A., Shore, L. S., Meleguir, F. 1979. Prostaglandins in the salivary glands and reproductive organs of *Hyalomma anatolicum* Koch (Acari: Ixodidae). *Bull. Entomol. Res.* 69:381–85

78. Smith, J. J. B., Cornish, R. A., Wilkes, J. 1980. Properties of a calcium-dependent apyrase in the saliva of the blood-feeding bug, *Rhodnius prolixus*. *Experientia* 36:898–900

79. Spellman, G. G., Nossell, H. L. 1971. Anti-coagulant activity of dog hookworm. *Am. J. Physiol.* 220:922–27

80. Tatchell, R. J., Moorhouse, D. E. 1968. The feeding processes of the cattle tick *Boophilus microplus* (Canestrini). Part II. The sequence of host-tissue changes. *Parasitology* 58:441–59

81. Tatchell, R. J., Moorhouse, D. E. 1970. Neutrophils: their role in the formation of a tick feeding lesion. *Science* 167:1002–3

82. Trager, W. 1939. Acquired immunity to ticks. *J. Parasitol.* 25:57–81

83. Vargaftig, B. B., Chignard, M., Benveniste, J. 1981. Present concepts on the mechanism of platelet aggregation. *Biochem. Pharmacol.* 30:263–71

84. Vermeulen, N. M. J., Neitz, A. W. H., Potgieter, D. J. J., Bezuidenhout, J. D. 1984. Anti-protease from *Amblyomma hebraeum. Insect Biochem.* 14:705–11

85. Vogt, W. 1974. Activation, activities and pharmacologically active products of complement. *Pharmacol. Rev.* 26: 125–69

86. Webster, R. O., Hong, S. R., Johnston, R. B. Jr., Henson, P. M. 1980. Biological effects of the human complement fragments C5a and C5a desarg on neutrophil function. *Immunopharmacology* 2:201–19

87. Wikel, S. K. 1979. Acquired resistance to ticks: Expression of resistance by C4 deficient guinea-pigs. *Am. J. Trop. Med. Hyg.* 28:586–90

88. Wikel, S. K. 1982. Histamine content of tick attachment sites and the effects of H1 and H2 histamine antagonists on the expression of resistance. *Ann. Trop. Med. Parasitol.* 76:179–85

89. Wikel, S. K. 1982. Influence of *Dermacentor andersoni* infestations on lymphocyte responsiveness to mitogens. *Ann. Trop. Med. Parasitol.* 76:627–32

90. Wikel, S. K. 1982. Immune responses to arthropods and their products. *Ann. Rev. Entomol.* 27:21–48

91. Wikel, S. K., Allen, J. R. 1977. Acquired resistance to ticks. III. Cobra venom factor and the resistance response. *Immunology* 32:457–65

92. Wikel, S. K., Allen, J. R. 1982. Immunological basis of host resistance to ticks. See Ref. 42, pp. 169–96

93. Wikel, S. K., Osburn, R. I. 1982. Immune responsiveness of the bovine host to repeated low-level infestation with *Dermacentor andersoni. Ann. Trop. Med. Parasitol.* 76:405–13

94. Willadsen, P., Riding, G. A. 1979. Characterization of a proteolytic-enzyme inhibitor with allergenic activity. Multiple functions of a parasite-derived protein. *Biochem. J.* 177:41–47

95. Willadsen, P., Riding, G. A. 1980. On the biological role of a proteolytic enzyme inhibitor from the ectoparasitic tick *Boophilus microplus. Biochem. J.* 189:295–303

96. Willadsen, P., Wood, G. M., Riding, G. A. 1979. The relation between skin histamine concentration, histamine sensitivity, and the resistance of cattle to the tick *Boophilus microplus. Z. Parasitenkd.* 59:87–93

97. Williams, J. W. 1972. Sequence of coagulation reactions. In *Hematology,* ed. W. J. Williams, E. Beutler, A. J. Erslev, R. W. Rundes, pp. 1085–91. New York: McGraw-Hill

98. Williams, T. J., Peck, M. J. 1977. Role of prostaglandin-mediated vasodilation in inflammation. *Nature* 270:530–32

Ann. Rev. Entomol. 1987. 32:479–505

ADVANCES IN MOSQUITO-BORNE ARBOVIRUS/VECTOR RESEARCH

G. R. DeFoliart

Department of Entomology, University of Wisconsin, Madison, Wisconsin 53706

P. R. Grimstad

Department of Biology, University of Notre Dame, Notre Dame, Indiana 46556

D. M. Watts

Department of Pathogenesis and Immunology, Division of Disease Assessment, United States Army Medical Research Institute of Infectious Diseases, Ft. Detrick, Frederick, Maryland 21701

PERSPECTIVES AND OVERVIEW

As defined in 1985 by the World Health Organization (WHO) (187), arboviruses are "maintained in nature principally, or to an important extent, through biological transmission between susceptible vertebrate hosts by haematophagous arthropods or through transovarian and possibly venereal transmission in arthropods; the viruses multiply and produce viremia in the vertebrates, multiply in the tissues of arthropods, and are passed on to new vertebrates by the bites of arthropods after a period of extrinsic incubation." Except for the addition of transovarian and venereal transmission, this definition is identical to the one issued by WHO in 1967. Despite the need for few modifications in definition, a tremendous amount of research on arboviruses has been published in the past 20 years. Of the approximately 500 arboviruses currently recognized (83), more than 200 are known or suspected to be mosquito-borne, among them the etiological agents of such extensively studied human diseases as yellow fever, dengue, Japanese and St. Louis encephalitides, and eastern equine, western equine, and Venezuelan equine encephalomyelitides. The

---

[1]The US Government has the right to retain a nonexclusive, royalty-free license in and to any copyright covering this paper.

majority of these viruses are zoonoses, and humans become infected only tangentially, with no primary involvement in the basic transmission cycles that perpetuate the viruses.

The concepts elaborated by Chamberlain & Sudia in their classic 1961 review (19) still form the basis of mosquito-borne arbovirology today, with surprisingly few modifications considering that 25 years have elapsed. Many advances have resulted simply from the steady accumulation of new knowledge about specific arbovirus survival systems; e.g. previously unknown vector or vertebrate host components have been discovered. We have summarized such changes for La Crosse, Jamestown Canyon, and dengue viruses in a companion review (32). In addition, however, notable progress has occurred in recent years in the development of sophisticated molecular techniques allowing more rapid serologic diagnosis, virus detection, and identification (95). These are providing a better understanding of viral structure, improved classification, and better tools for epidemiological studies. New modes of transmission have been discovered as noted in the WHO definition (164, 180). Researchers have gained new insights about the importance of genetically based intraspecific population differences (154), and about the diversity of complexly interacting ecological factors that affect vector populations, frequently in different ways in different geographical regions (98). Also, importantly, scientists are increasingly wary about data obtained from studies using long-colonized mosquitoes and/or high-passage virus strains, and are becoming more concerned that laboratory studies have far outstripped field studies as the basis of concepts that should be founded on a better balance between the two.

This review is focused on those aspects of arbovirology that have the greatest direct relevance to vector incrimination and biology. "Vector competence" is a term that has come into wide use in recent years. Intrinsic vector competence includes both internal physiological factors that govern infection of a vector and its ability to transmit the virus, if given the opportunity, and innate behavioral traits such as host preferences and probing activity. In addition to these intrinsic factors, virus transmission and epidemiological patterns are influenced by numerous extrinsic factors such as temperature, rainfall, and vertebrate host density. Four basic criteria have been used to incriminate the mosquito vector(s) of an arbovirus: (*a*) isolation of virus from naturally infected mosquitoes; (*b*) laboratory demonstration of the ability of the mosquitoes to become infected by feeding on a viremic host; (*c*) laboratory demonstration of the ability of these infected mosquitoes to transmit the virus during blood-feeding; and (*d*) evidence of blood-feeding contact between the suspected mosquito vector and suspected vertebrate hosts under natural conditions. We agree with Mitchell (110) that in evaluations of vector competence there has been a tendency to emphasize criteria *b* and *c* disproportionately compared to *a* and *d*. We further suggest that there has been an overemphasis on infection rates at the expense of transmission rates. Epidemiologically, the population transmission rate (i.e. the percentage of vectors that transmit virus of the total sample drawn and tested) is the vital statistic by which internal vector competence of a population can be judged.

Fine (40) suggested that arbovirologists need to define vector terminology more precisely. For example, malariologists have defined "vectorial capacity" as "the average number of potentially infective bites that will ultimately be delivered by all the vectors feeding upon a single host in one day." This crisp definition indicates that beyond incrimination on the basis of the four criteria above, a careful quantitative analysis and integration of data on almost every aspect of vector biology is required. It is not easy to quantify the dynamics of vector-host interactions when nonhuman vertebrates are involved in transmission cycles. Vertebrates introduce other competence variables (112): (*a*) abundance and dispersal, (*b*) seasonal breeding patterns, (*c*) attractiveness to mosquito vectors, (*d*) response to virus infection, (*e*) activity patterns, and (*f*) immune status. It has become increasingly clear over the past two decades that behavioral components of vector and vertebrate competence and local ecology are of critical importance as epidemiological determinants. *Aedes aegypti, Aedes simpsoni* complex (74), and *Aedes africanus,* the most studied of the African vectors of yellow fever virus, exhibit regional differences in anthropophily vs zoophily, endophily vs exophily, domestic vs peridomestic vs sylvan habitat, and vertical stratification (147). Such mosaics of regional behavior underline the importance of detecting the existence of species complexes and intraspecific differences in population behavior. Intraspecific population differences in internal vector competence have been similarly documented for the vectors of a number of arboviruses (56). Possibly the most important concept emerging from arbovirus research of the past decade is that there are no simple stories.

Recent major advances toward an understanding of the physicochemical, morphological, and genetic characteristics of arboviruses have demonstrated, not unexpectedly, that classification based on these features closely parallels classification formerly based on antigenic characteristics (83). Therefore, antigenic properties can be integrated with the physicochemical features that now serve as the essential criteria for classifying arboviruses within a three-taxon hierarchy that includes family, genus, and species; some genera are subdivided into serogroups (103, 185). Most mosquito-borne arboviruses have been placed in one of three genera, each in a different family; 26 are in the genus *Alphavirus,* family Togaviridae (83, 183); 31 in *Flavivirus,* Flaviviridae (83, 184), and more than 100 in *Bunyavirus,* Bunyaviridae (14, 83). Nearly 100 others are placed in other genera of Bunyaviridae or other families, or remain taxonomically unclassified.

Abbreviations listed in the *International Catalogue of Arboviruses* (83) are used for virus names that appear more than once in this article, as follows: chikungunya (CHIK), dengue (DEN or DEN 1, 2, 3, or 4), eastern equine encephalomyelitis (EEE), Jamestown Canyon (JC), Japanese encephalitis (JBE), La Crosse (LAC), Rift Valley fever (RVF), St. Louis encephalitis (SLE), San Angelo (SA), Venezuelan equine encephalomyelitis (VEE), western equine encephalomyelitis (WEE), West Nile (WN), and yellow fever (YF). LAC virus is additionally abbreviated in this article to LACV.

ADVANCES IN DETECTION, IDENTIFICATION, AND SEROLOGIC DIAGNOSIS OF ARBOVIRUSES

For years arbovirologists have utilized a variety of animal species for virus isolation; the newborn albino mouse was first used to isolate many arboviruses (83). In addition, a variety of cultured avian and mammalian cells (primary Pekin duck, and Vero, BHK, and LLC-MK2 lines) and mosquito cells (*Aedes aegypti,* C6/36-*Aedes albopictus*) have proven to be susceptible to viral infection and thus are very useful, especially in view of both the rising costs of animal husbandry and the international emphasis on the use of nonvertebrate systems whenever possible. A particularly useful and sensitive alternative is the propagation of viruses in live mosquitoes that can be laboratory-infected via intrathoracic inoculation (141). This procedure has increasingly been employed in dengue research, and has led to the isolation of DEN viruses in the field, including at localities where a functional cell culture laboratory was not readily available. It is limited in use, however, as most tick-, *Phlebotomus-,* and *Culicoides*-transmitted arboviruses apparently do not replicate in adult mosquito tissues (59).

Characteristic illness, paralysis, and/or death of the inoculated laboratory animal(s), specific cytopathic effects observed in cell cultures, or formation of distinct plaques in cell cultures signal the possible presence of an arboviral isolate. Reisolation is essential to rule out laboratory contamination or presence of a murine virus (149). Intrathoracic inoculation of mosquitoes gives rise to nonlethal infections, and additional procedures are necessary to detect the presence and identity of viral isolates (59, 149).

Arbovirologists have historically relied on the standard complement-fixation test, the neutralization test (18), and to a lesser extent the hemagglutination-inhibition test, in which extracted viral antigens are used to identify viral isolates (23). Isolates are tested routinely against a battery of mouse hyperimmune ascitic fluids prepared against heterologous reference viral strains and the homologous isolate (16, 166). These studies demonstrate that an isolate is identical to or related to a recognized virus, or may establish the isolate as a newly recognized agent. An excellent review of arbovirus typing was presented by Calisher et al (16). In a number of viral complexes, notably the California (CAL) serogroup, it is necessary to perform neutralization tests in cell culture to subtype the closely antigenically related agents. Calisher et al (16) suggested that the current test of choice for the CAL serogroup is the serum dilution–plaque reduction neutralization test with virus-specific immune hamster sera.

Recent advances in the use of monoclonal antibodies (54, 117), radioimmunoassay (42), and enzyme immunoassay (e.g. enzyme-linked immunosorbent assay or ELISA) (43, 91) have provided diagnostic tools for rapid virus detection and identification. Antigen-capture ELISA procedures for the CAL serogroup viruses (3), EEE (73), RVF (126), YF (115), and other arboviruses are now performed routinely in major public-health/research laboratories and are rapidly becoming more widely used. Combination of monoclonal antibodies and indirect immunofluorescent assay is also useful for rapid viral identification (16, 72). Immune electron microscopy can readily be used to visualize virions in infected host tissues (129). The most recent advanced approach is the utilization of molecular probes (DNA, RNA) to detect the presence of viral nucleic acid in infected cells. A limitation to use of radioimmunoassay (42) and DNA/RNA probes has been the need for radioactive reagents, including I^{125}; however, use of biotinylated DNA probes eliminates the need for isotopes (92).

Electrophoretic analyses of structural polypeptides and oligonucleotide fingerprint analyses of DEN (71, 139, 175), SLE (167), LAC (84), YF (34), and other arboviruses have led to a better understanding of viral structure and epitope composition, geographic strain variation, pathogenicity, epidemiology, and vector competence.

Serologic diagnosis has also evolved toward more sophisticated molecular techniques. For years the complement-fixation, hemagglutination-inhibition, and less frequently the neutralization test, all with known virus antigens, have been used to diagnose human and animal infection or have been used in serologic surveys for detecting past infection. More recently antibody-capture immunoassays, both solid-phase radioimmunoassay and ELISA, have been developed; some are becoming the tests of choice for diagnosis of DEN, JBE (77), RVF (127), SLE (116, 186), and other arboviral infections. ELISA procedures utilizing biotin-labeled antigens have also been developed (20). Calisher has suggested that because of its specificity and sensitivity, the IgM-capture ELISA (MAC-ELISA) is the test of choice for the rapid diagnosis of EEE and WEE viral infections in humans and WEE infections in equines (C. H. Calisher, personal communication). A MAC-ELISA is now available for diagnosis of La Crosse encephalitis (17).

Serologic confirmation of a case is often delayed the several weeks until convalescent serum can be obtained and tested in parallel by complement-fixation, hemagglutination-inhibition, or neutralization to determine if the antibody count has risen fourfold or more over that of serum. Diagnosis is often precluded when the convalescent serum cannot be obtained or when the acute serum is drawn too late. Because of failure to document clinical cases of arboviral diseases, the true public health importance of their agents has been greatly underestimated. Thus a major advantage of the IgM-capture ELISA procedure is the high probability of accurate diagnosis of an arboviral infection when the test is performed only with acute serum obtained while the patient is still clinically ill. Currently, these new procedures are used in only a few laboratories, but within a few years most diagnostic laboratories will undoubtedly use them. However, for some groups, notably the CAL serogroup, we agree with Calisher et al (16) that "serologic differentiation remains the most practical and expeditious approach."

INTRASPECIFIC GEOGRAPHIC VARIATION IN VECTOR BEHAVIOR AND COMPETENCE

Mosquito species vary geographically in their morphological characteristics (8, 99) and biological traits. Some of these have been shown to be heritable, such as host preference and endophily in *Aedes aegypti* (121, 168), diapause in *Aedes triseriatus* (151, 152), and photoperiodic response, diapause frequency, and development time in *Wyeomyia smithi* (78). Intraspecific geographic variation in internal vector competence has been shown among vector populations of WEE, CHIK, WN, DEN, YF, SLE, and LAC viruses. For some vectors this variation has been shown to be genetically based, although modes of inheritance apparently differ in different virus-vector systems (56, 68, 158, 176, and references therein). Selection experiments have produced strains with both increased oral susceptibility, e.g. WN virus in *Culex tritaeniorhynchus* (68), and decreased susceptibility, e.g. WEE virus in *Culex tarsalis* (61). Vector competence of natural populations should tend to be higher where sympatric viral and vector strains are paired. However, with WEE virus and *C. tarsalis* in California, wide differences in oral susceptibility have been shown between mosquito populations in fairly close geographic proximity (63). Further, it has been postulated that populations that vector LACV in the main endemic region have evolved toward reduced transmission capability (50).

Genetic change during adaptation of mosquitoes to the laboratory environment can be drastic and can occur quickly, resulting in reduced heterozygosity and number of alleles after only one generation of colonization (27, 122). Changes over time have been observed in behavioral and life history characteristics such as flight ability, autogeny, insemination and oviposition rates, and adult longevity of colonized mosquitoes (24, 70, 100). Similarly, changes in internal vector competence have been reported for *A. aegypti* and YF virus (97), *Culex pipiens* and RVF virus (45), and *A. triseriatus* and LACV. The oral transmission rate of a laboratory colony of *A. triseriatus* was reduced significantly during 22 months of colonization compared to that of females in the natural population from which the colony had been initiated (D. M. Watts, unpublished data). In a study involving several geographic populations of *A. triseriatus*, colonization had unpredictable effects on infection and transmission capability, "as might be expected with small populations and genetic drift" (50). Although no apparent alteration of susceptibility to virus has accompanied mosquito colonization in some cases (182), Lorenz et al (97) offered excellent advice: "Unless one examines the original population sample directly it is . . . best to avoid profound interpretations based solely on colony material."

"Conspecific" virus strains also differ in their characteristics, including virulence, RNA oligonucleotide fingerprints, capacity to produce viremia, and infectivity for mosquitoes (84, 102, 110, 113). For example, a low-passage Barsi strain of CHIK virus was over 100 times more infective for *Aedes albopictus* than a high-passage Ross strain of this virus (162). However, as in most studies, the separate possible effects of different passage history and different geographic origin (India and Tanzania, respectively) could not be distinguished with certainty.

Increasingly, isozyme variants are providing clues for vector biologists seeking to delimit the geographic extent of important behavioral, ecological, and vector competence traits; electrophoretic analyses identify intraspecific geographic enzyme groupings and genotypes unique to populations with distinctive characteristics, e.g. *A. aegypti* (134, 156–158, 177), *C. pipiens* (21), and *A. triseriatus* (122, 123). Powell et al (135) and others have emphasized the importance of using, whenever possible, field-collected mosquitoes or their F_1 progeny in population genetics studies; this is now obviously equally important for other experimental studies on vector competence.

HORIZONTAL TRANSMISSION BY INTERNALLY COMPETENT MOSQUITOES

Early studies demonstrated the existence of a "gut barrier" in mosquitoes, i.e. an infection threshold that must be overcome before ingested virus can initiate an infection and disseminate to the salivary glands. The studies revealed that the concentration of a given virus necessary to overcome this mesenteronal barrier varies between species (19, 102). Numerous studies have shown that in a given cohort of mosquitoes the infection rate will be higher, sometimes much higher, than the transmission rate. In general, the higher the concentration of virus ingested, the higher the ensuing infection and transmission rates

(62). To explain these phenomena, Hardy et al (62) proposed that a series of barriers prevent or reduce further dissemination of virus at various times between its appearance in the lumen of the mesenteron and its eventual shedding in the saliva when the mosquito again blood-feeds. The first of these is the mesenteronal infection barrier. The lumen of the mosquito mesenteron is separated from the hemocoel by a single layer of epithelial cells surrounded by a porous multilayered membrane, the basal lamina. The first proliferation of virus occurs in the epithelial cells if a sufficient dose of virus is ingested. However, infection is not established, or may be established only at very high virus dosages, if a mosquito is refractory to a given virus.

The second barrier is posed by the basal lamina. Again, as with the epithelial cells, the mechanism(s) of penetration are not fully understood. This barrier, termed the "mesenteronal escape" barrier, and a "salivary gland infection" barrier were demonstrated experimentally by Kramer et al (89) with WEE virus in *C. tarsalis*. Both barriers were found to be dose-dependent. In 20–30% of infected *C. tarsalis* females, virus failed to multiply to normal levels in the mesenteron and did not escape into the hemolymph regardless of the length of the extrinsic incubation period. In another 20–45% of infected females, virus multiplied normally and escaped into the hemolymph, but failed to infect the salivary glands. The salivary gland infection barrier was overcome in some of these females, however, after an extended extrinsic incubation period; this shows that this barrier is time as well as dose de-pendent. In addition, virus titers tended to stabilize by 8 days post-infection, which was consistent with the equilibrium state of infection described earlier by Murphy (124). Although the precise mechanism is unknown, Hardy et al (62) suggested that the ability of some female mosquitoes to modulate viral titers if not initially overwhelmed by a high dosage of virus best explains the dissemination barriers found in some *C. tarsalis* females. Weaver et al (182), working with VEE virus, obtained results consistent with those of Kramer et al and discussed the epidemiological significance of the mesenteronal infec-tion and escape barriers relative to transmission of Middle American enzootic and epizootic strains by the enzootic vector, *Culex (Melanoconion) taeniopus*. A fourth barrier, the "salivary gland escape" barrier hypothesized by Hardy et al (62) to explain the lack of transmission by mosquitoes with infected salivary glands, has since been demonstrated in *Aedes hendersoni* infected with LACV (52). The evidence suggested that the salivary gland escape barrier prevents *A. hendersoni* from being a highly competent vector of LACV.

Three values derivable from horizontal transmission experiments are of prime interest to epidemiologists and vector biologists: (*a*) the population transmission rate by orally infected mosquitoes, (*b*) the infection threshold, and (*c*) the length of the extrinsic incubation period. A host of variables can affect the outcome of experiments. These include virus dose, route of infection, and incubation temperature, in addition to several factors discussed earlier, i.e. geographic origin and colonization history of mosquitoes and geographic origin, animal source of original isolation, and passage history of virus strains. The transmission rate is often expressed as a percentage of the number of mosquitoes infected, with infection rate in turn expressed as a percentage of the number exposed to infection. Epidemiologically, however, the most meaningful expression of the transmission rate is as a percentage of the total sample exposed to infection. Grimstad et al (52) termed these the "modified" and "population" transmission rates, respectively; the population rate is based on the number of mosquitoes that survive long enough to refeed and thus to be tested for transmission. Infection rates not backed by transmission rates must be interpreted with caution because the infection rate is often much higher than the transmission rate in a given cohort of mosquitoes. In a recent comparison of the internal vector competence of seven strains of *A. hendersoni* with three strains of *A. triseriatus,* a known vector of LACV, the average disseminated infection rate (salivary gland infection rate) for the former species was 85% compared to 70% for the latter (52). The population transmission rates for these closely related species, however, were 9% and 59%, respectively. As all dissected *A. hendersoni* exhibited intense specific fluorescence within the salivary gland tissues, a salivary gland escape barrier was apparently responsible for their greatly reduced transmission. Similarly, only 9% of *Culex theileri* transmitted Sindbis virus despite a salivary gland infection rate of 39% (82). Thus, except with well-proven natural virus-vector combinations, high disseminated infection rates cannot be considered assurance of high transmission rates even though salivary glands may be heavily infected.

Population transmission rates are most useful when determined and interpreted in relation to conditions presumably encountered by a particular vector or suspected vector population in nature. Jupp (80), for example, interpreted the effect of virus dosage and temperature on laboratory transmission of WN virus by *Culex univittatus* in relation to wild bird viremias and the relatively cool seasonal temperatures found on the highveld of South Africa. Patrican et al (132, 133) determined that chipmunks, a major amplifying host of LACV, averaged only one day of viremia sufficiently high [3.5 $\log_{10}$ suckling mouse intracerebral lethal dose$_{50}$ (SMICLD$_{50}$)/0.025 ml] to ensure transmission by a high proportion (91%) of engorging *A. triseriatus*. A much lower percentage (33% or less) of mosquitoes engorging on chipmunks with viremias below this effective threshold subsequently transmitted virus. Oral infection thresholds as expressed by the infective dose$_{50}$ (ID$_{50}$) have been shown to be quite low in many cases, equivalent to about $10^{1.5}$ or $10^{2.5}$ SMICLD$_{50}$ or lower (67). Inasmuch as the epidemiological value of data on

infection thresholds lies in their relation to viremia levels exhibited by vertebrate hosts in nature [and it is well established that the higher the concentration of virus ingested, the higher (usually) will be the ensuing infection and
transmission rates], the feeding of concentrations of virus in the laboratory
that are higher than those encountered in nature, and that are far higher than
the mosquito ID_{50}, can lead to an overestimate of natural transmission rates. It
can similarly result in an underestimate of the true magnitude of differences in
internal vector competence between species and between populations of the
same species (2, 55), as well as an underestimate of the extrinsic incubation
period required for transmission in nature. Although modulation of virus titers
occurs even when the mosquito is initially overwhelmed by a large virus dose,
it occurs too late to prevent salivary gland infection (62).

Laboratory infection and transmission rates are also affected by the method
of exposure to virus. It has been shown that mosquitoes are generally more
susceptible to infection when fed on viremic animal donors than when fed on
blood/virus suspensions through membranes, in droplets, or in soaked cotton
pledgets (69, 81, 104, 113, 114, 133). The ID_{50} of SLE virus was at least
10,000-fold lower for *C. quinquefasciatus* fed on viremic chicks than for
those fed on virus-soaked pledgets (104). Parenteral inoculation of viruses,
while useful for some research objectives, bypasses the mosquito's first two
barriers to infection and therefore yields little valuable information for determining and comparing the internal vector competence of populations.

Females of most vector mosquito species have a daily life expectancy
within the range of 75–90% (146); this means that anything that happens in
the life of most mosquitoes happens within about three weeks or less after
emergence. Simulation models of CHIK and Turlock viruses indicate that
mosquito survival is the most important factor affecting population infection
rates (33, 145; also see 35 and 153). This suggests that experimental data on
the extrinsic incubation period should be interpreted in relation to mosquito
longevity, duration of ovarian cycle, and readiness to refeed under the same
temperature regimen at which extrinsic incubation data are obtained. Data on
extrinsic incubation period must ordinarily be interpreted cautiously, as they
are almost always obtained under constant temperature and humidity conditions rather than under the fluctuating conditions to which mosquitoes are
exposed in nature. In general, alphaviruses appear to require a shorter extrinsic incubation period, 2–9 days, than flaviviruses and bunyaviruses, which
require 6–14 days for release from the mesenteron and dissemination to
extramesenteronal organs, including salivary glands (143, 144). In studies on
EEE virus, salivary gland infection was detected within 2–3 days in *Culiseta
melanura* fed on viremic chicks. Viral titers in whole mosquitoes reached a
peak on the seventh day, after which they slowly decreased (143, 144).
Again, however, the mere fact of salivary gland infection within 2–3

days does not necessarily signal the beginning of significant transmission within that time. Similarly rapid infection of the salivary glands has been shown with another alphavirus, WEE, in *Culex tarsalis* fed on viremic chicks, as evidenced by oral transmission on the fourth day after feeding (7). The transmission rate was only 10% on the fourth day, however; it then steadily increased, reaching 84% after the thirteenth day. LACV (179) and other arboviruses (19) show similar initially ascending patterns of transmission.

In general, the rate of virus multiplication and the rapidity of salivary gland infection increase as the temperature of extrinsic incubation increases (19, 62, 75, 80, 88, 173, and references therein). In a recent study, however, Kramer et al (88) found that infection rates for WEE virus decreased in *C. tarsalis* held for more than six days at high temperature (32°C), but increased in those held at lower temperatures. High temperature enhanced the expression of virus modulation in some females, and high temperature during larval rearing selected for females with this ability.

VERTICAL (TRANSOVARIAL AND VENEREAL) TRANSMISSION AND VIRUS SURVIVAL DURING ADVERSE CLIMATIC CONDITIONS

Reports in the earlier literature indicated the possibility of transovarial transmission of YF, JBE, and SLE viruses (summarized in 161). As subsequent investigators were unable to confirm the reports, transovarial transmission of arboviruses by mosquitoes was generally discounted. Interest was renewed in 1973 when, following two isolations of LACV from naturally infected *A. triseriatus* larvae (130), Watts et al (180, 181) experimentally demonstrated transovarial transmission of LACV by *A. triseriatus* and showed that infection of the diapausing eggs of the mosquito provided an overwintering mechanism for the virus. The rapid confirmation of these findings for LACV and their extension to several other members of the CAL serogroup suggested the potential importance of transovarial transmission in the maintenance of these bunyaviruses. In addition to LACV, six other CAL serogroup viruses have been isolated from naturally infected *Aedes* larvae or from adults reared from them, and transovarial transmission has been demonstrated experimentally for four of the viruses in various *Aedes* species (161).

In addition to the temperate-zone CAL serogroup viruses, several tropical bunyaviruses belonging to other serogroups or other genera have displayed evidence of transovarial transmission. Ecological studies in Kenya implicated *Aedes lineatopennis* as an enzootic vector of RVF virus (genus *Phlebovirus*) (96). This mosquito survives the dry seasons by depositing eggs in shallow depressions known as dambos. Following seasonal rains, the mosquitoes emerge from temporary pools of water retained in the dambos. Multiple isolations of RVF virus obtained from males and females reared from field-collected larvae suggested that this virus is transmitted vertically. This needs confirmation, however, as only once was RVF virus reisolated from the virus-positive specimens despite a relatively high virus titer ($10^{5.5}/0.1$ ml) in the mosquito pool from which virus was initially isolated. Further, RVF virus was not isolated from larvae or pupae collected at the same time and location. There is evidence that another tropical bunyavirus, Gamboa virus, is transmitted vertically by *Aedeomyia squamipennis,* a species that breeds continuously in association with aquatic plants along the edges of the Chagres River in Panama (47). The virus has been principally associated with wild birds, the preferred hosts of the mosquito. Minimum field infection rates were 1:65 for adult male and female mosquitoes, 1:1530 for larvae, and 1:32 for adults reared from larvae.

Thus far, there is little evidence that transovarial transmission is of significance for survival among alphaviruses, while its significance among flaviviruses remains uncertain (161). In New York the alphavirus EEE was prevalent in female *Culiseta melanura,* the primary vector, during the summers. The minimum field infection rate was 1:256 for female mosquitoes (119), but the virus was not isolated from 1047 larvae, 8919 males, or 2140 recently emerged females. In Maryland, EEE virus was prevalent at a minimum field infection rate of 1:417 in female mosquitoes (D. M. Watts, unpublished data), but attempts to obtain virus from 4504 larvae, 5689 males, and 14,153 recently emerged females were unsuccessful. Further evidence against transovarial transmission was the failure to demonstrate EEE virus in the ovaries of infected *C. melanura* (144). There is also evidence against transovarial transmission of VEE virus (142) and several other alphaviruses (161). Secondary involvement of *Aedes* mosquitoes in alphavirus transmission, however, keeps alive the possibility of transovarial transmission in this group of viruses (see 49 for references about northern distribution and seasonal occurrence of WEE virus).

Recent experimental studies (44, 65) have shown that the flavivirus SLE can be transmitted vertically by *Culex pipiens* and *Culex tarsalis,* as well as by *Culex quinquefasciatus, Aedes epactius,* and *Aedes atropalpus.* Filial infection rates varied depending on species, ovarian cycle, and experimental conditions, but never exceeded 4%. Similarly low rates have been found for DEN viruses (32). Although experimentally obtained infection rates in progeny have been low in the Flaviviridae (usually less than 1%), increasing evidence for the existence of a variety of sylvan virus cycles involving *Aedes* mosquitoes are of interest, as are occasional virus isolations from field-collected larvae and males of these mosquitoes. In Africa three isolates of YF virus were reported from field-collected male mosquitoes of the *Aedes furcifer/Aedes taylori* complex (26), and a DEN-2 isolate was reported from males of the same mosquito complex (25). Two flaviviruses, YF and SLE, have been isolated from ticks in the Central African Republic and the United States, respectively (101, 161). Two recent findings of broad interest in relation to transovarial transmission of arboviruses are that larval rearing temperature can affect progeny infection rates (64, 65), and that virus titers in larvae, when assayed in cell cultures, can be reduced to undetectable levels depending on the number of larvae pooled for the assay (90).

Fine (39) defined vertical transmission as the "direct transfer of infection from a parent organism to his, her, or its progeny," and expressed the efficiency of vertical transmission as a function of four parameters (40): (a) the maternal vertical transmission rate, i.e. the proportion infected of progeny from infected females; (b) the paternal vertical transmission rate, i.e. the proportion infected of progeny from previously uninfected females mated to transovarially infected males; (c) the relative fertility of infected adults; and (d) the relative survival to reproductive age of infected young. Research has focused mainly on determining whether transovarial transmission occurs in different virus-vector associations and on determining maternal vertical transmission rates. Thompson & Beaty (164, 165), however, demonstrated in the laboratory that transovarially infected males of A. triseriatus can transmit LACV to uninfected females during copulation. Venereal (paternal) transmission is potentially important, but experimentally derived rates of oral and transovarial transmission by venereally infected females have been low (32). Furthermore, virus has not so far been detected in first-ovarian-cycle eggs, which comprised more than 80% of eggs deposited in a four-year field study of an A. triseriatus population in Wisconsin (S. V. Landry, unpublished data). This indicates that unless populations with higher maternal vertical transmission rates are found, or at least a small proportion of first-ovarian-cycle eggs are shown to become infected, venereal transmission can only be interpreted to partially offset the erosion of virus prevalence that occurs during maternal vertical transmission. Although by Fine's definition (39) venereal transmission is vertical, it has the effect of horizontal transmission via vertebrates, inasmuch as virus is passed to previously uninfected females of the concurrent vector generation.

Transovarially acquired LACV appears to have no adverse effect on A. triseriatus. No significant differences were detected between transovarially infected and uninfected groups in duration of larval stage, sex ratio, hatching success, time to ovarian maturation, fecundity, or adult survival through the second oviposition (131). Although increased probing and reduced blood-feeding success have been observed for A. triseriatus orally infected with LACV (53), these effects were not observed in A. triseriatus with transovarially acquired LACV (133). Transovarially acquired infections may be less harmful to their mosquito hosts, as virus replicates to lower titers than when orally or parenterally acquired (163). Also, some observed adverse effects from arbovirus infection may be the result of unnatural mosquito-virus combinations, e.g. Semliki Forest virus in *A. aegypti* (109) and SA virus in *A. albopictus* (160), or of virus in a mosquito species that may be primarily an epidemic vector, e.g. RVF virus in *C. pipiens* (170).

Reported maternal vertical transmission rates of CAL serogroup viruses have been well below 100% (13, 22, 107, 108, 163, 172); the highest yet reported was 70% for LACV in a population of *A. triseriatus* in Wisconsin (107). Filial infection rates determined on progeny of one to five females from several other Wisconsin localities suggested that natural populations with maternal vertical transmission rates higher than 70% probably exist. Lines of *Aedes dorsalis* and *A. albopictus* have been selected with filial infection rates of more than 90% for California encephalitis (171) and SA viruses (163), respectively, but in the latter these rates progressively declined when selection pressure for high maternal vertical transmission rates was relaxed (150).

As with horizontal transmission, intraspecific differences in transovarial transmission competence have been shown to occur among vector species (105, 108, 163, 172). Miller et al (105) reported that a population of *A. triseriatus* from Wisconsin inoculated intrathoracically with LACV exhibited a 50% filial infection rate, whereas populations from Connecticut and Massachussetts had rates of 26% and 13%, respectively. This suggests that evolution of populations of *A. triseriatus* highly competent in transovarial transmission in the north central states may have led to the high endemicity of LACV in this region and may have contributed to the evolution of reduced oral transmission capability observed there (50).

As maternal vertical transmission rates are less than 100%, it has been postulated that the erosion of virus prevalence during transovarial transmission is offset by horizontal amplification via vertebrate hosts during the summer (41, 93, 107). However, the efficiency of horizontal transmission and its adequacy in fully offsetting vertical erosion has been questioned in the case of LACV (31, 32). It has been suggested (161, 163, 171) that arboviruses can enter germinal cells (oogonia) of mosquitoes and establish stabilized infections similar to those of sigma virus in *Drosophila melanogaster*. Females with stabilized infections transmit sigma virus to nearly all of their progeny, and female progeny also have oogonial infections. With or without oogonial infection, however, transovarial transmission rates of less than 100% for all ovarian cycles would require periodic recruitment of new transovarially infected females via horizontal amplification in vertebrates, venereal transmission, some as yet unrecognized mechanism, or some combination of such mechanisms. Even with a vertical transmission rate of 90% and no deleterious effects from viral infection, virus prevalence without some form of amplification would erode by more than 50% within seven vector generations.

Aside from increased knowledge about transovarial transmission, little more is known about virus persistence during adverse climatic periods than when Reeves reviewed the subject in 1974 (137). One notable exception is that new information is now available on SLE virus, which is transmitted mainly by *C. pipiens* complex mosquitoes. During the winter, *C. pipiens* females hibernate as diapaused adults. Repeated isolations of SLE virus were obtained from adult females collected in Maryland during the winter seasons of 1977–1978 and 1978–1979 (4, 5). In the near absence of experimental or field evidence for vertical transmission of SLE virus, Eldridge (37) postulated that the most likely source of infection was viremic birds. Diapaused *C. pipiens* will blood-feed when transferred to warmer temperatures; if they are returned to reduced temperature before diapause is reversed, the ovaries remain in a diapaused state and the blood meal is utilized for development of the hypertrophic fat necessary for successful hibernation (38). A subsequent field study confirmed that diapaused *C. pipiens* blood-feed and that a high percentage survive in an overwintering hibernaculum (5). These findings suggest that contrary to previous belief, this mosquito, after feeding on a viremic host, can serve as an overwintering reservoir for SLE virus.

ECOLOGICAL FACTORS

During any sustained investigation of the epidemiology of an arbovirus, it becomes evident that information is required on almost every aspect of vector ecology and behavior. The vast expansion of ecological knowledge and methodology in recent years has made it clear that the factors involved are numerous, complexly interrelated, and dynamic (98, 146). Only a few examples of direct relevance to virus transmission can be mentioned here.

It has been found with *A. triseriatus* and LACV (32, 51, 133) and *C. tritaeniorhynchus* and JBE (159) and WN viruses (6) that smaller females that were nutritionally stressed as larvae have greater internal vector competence than larger females. While this would suggest that a high proportion of small females would enhance the vector competence of a population, the reverse appears to be true. Larval stress, resulting primarily from overcrowding and food deprivation, not only produces small adults, but can adversely affect larval survival, rate of larval development, adult fitness (including fecundity, ability to withstand dessication, expression of autogeny, longevity, and probability of completing the first gonotrophic cycle), sex ratio, net reproductive rate, and intrinsic rate of increase (6, 35, 60, 66, 133, 138, and papers by Bradshaw & Holzapfel, Chambers, Barr, and Service in 98). Size has as much implication for experimental work as for epidemiology; large, well-fed laboratory mosquitoes that may bear little resemblance to their undersized counterparts in nature will introduce bias into experiments. Effects of larval density on size and survival also have implications for vector control; Agudelo-Silva & Spielman (1), as well as several of the investigators cited above, have shown that partially effective larval control measures can be counterproductive because they may increase the number and fitness of adults that emerge.

As shown with LACV (32) and other viruses, vector-vertebrate host contact is a critical determinant of epidemiological patterns, but it is extremely difficult to quantify in the field. Contact is influenced by innate host preferences (125, 178), host availability, which may vary geographically and seasonally (36, 125, 178), the interplay of mosquito probing (140), refeeding behavior and host antimosquito behavior (30, 85, 86), and the effect of pathogen infection on both vertebrate (29, 46, 169) and vector (53, 133). As shown with *Anopheles,* even closely related species differ in their probing behavior (140). Ribeiro et al (140) found that duration of probing correlated inversely with salivary apyrase content, and that apyrase largely determines ability to locate blood. Day & Edman (30) found that mosquitoes fed most successfully on vertebrates that were inactive during the mosquitoes' primary flight periods, while vertebrates active at those times inflicted high mortality. Colonized *Culex nigripalpus* were more likely than field-caught or P_1 mosquitoes to be caught and eaten by active host animals. Probing without ingestion of sufficient blood to initiate completion of ovarian development increases the incidence of multiple feedings (85, 86); such probing can establish infection (53, 111), and such multiple feedings can therefore result in virus transmission within a single gonotrophic cycle (111). Some arboviruses cause elevated temperatures in viremic animals, and mosquitoes can discern a temperature difference of only 1–2°C (46). The preponderance of recent evidence, however, indicates that if mosquitoes in nature differentially feed on infected animals, they do so not because they distinguish between hyperthermia and hypothermia, but rather because of the reduced antimosquito behavior of sick animals (29, 169).

Because they blood-feed on a variety of hosts in nature, mosquitoes are exposed to the possibility of multiple infections with arboviruses, an arbovirus and homologous or heterologous antibody, or an arbovirus and other infective organisms. However, little is known of the interactions that occur in whole mosquitoes. Interference phenomena that result from ingestion of a virus and homologous or heterologous antibody, as may happen with LACV and JC virus (32, 133), might help to explain certain arbovirus distribution patterns. Although the data are preliminary, *A. triseriatus* that imbibed deer blood–LACV mixtures containing JC virus antibody became infected but did not transmit LACV (133). Similar interference has been observed in *A. triseriatus, C. quinquefasciatus,* and *A. aegypti* exposed experimentally to EEE, SLE, and WN viruses and their homologous antibody, respectively, and, in the case of WN virus, to certain heterologous antisera as well (19, 79). Similarly, *Culex annulirostris* infected intrathoracically with a temperature-sensitive, small plaque mutant of Semliki Forest virus isolated from cultured *A. albopictus* cells could not be infected 24 hr later with wild-type virus, although both viruses replicated normally in mosquitoes infected simultaneously (28). In the bunyaviruses, which have a tripartite genome containing large, medium, and small RNA segments, segment reassortment can occur in mosquitoes and such reassortant virus can be transmitted (10–12). Again, interference to superinfection, depending on sequential timing of exposure to the viruses, has been demonstrated (10, 12). The significance of reassortant virus transmission in nature, however, remains undetermined (12). As pointed out by Igarashi (76), arboviruses in nature are subject to selective pressures imposed by their alternating between two phylogenetically remote host systems (arthropod and vertebrate) and by different host components in different geographic areas; thus the range of variation in survivors of the evolutionary process is restricted.

Coinfection of mosquitoes by arboviruses and other infective agents may also be important. Turell et al (174) reported that concurrent ingestion of RVF virus and microfilariae of *Brugia malayi* by *Aedes taeniorhynchus* resulted in increased viral infection, rapidity of dissemination, and transmission compared to those of mosquitoes that ingested only virus. Gregarine protozoans apparently do not enhance arboviral infection (106, 120). *Plasmodium relictum* did not affect the vector competence of *C. tarsalis* for WEE virus (7), but there is evidence that *C. tarsalis* infected with the microsporidian *Amblyospora californica* is refractory to oral infection with WEE virus (62, 155).

DISEASE CONTROL: THE NEED TO APPLY EXISTING KNOWLEDGE

Despite the great proliferation of published research in the past 15–20 years, we appear to have gained little in our ability to prevent or control arboviral diseases through vector control. Biological control agents will not offer panaceas (94, 148). Because many arboviruses are widely dispersed, disease prevention and control have traditionally relied mainly on good virus surveillance through the use of indicator animals such as birds, susceptible equines, and mosquitoes, and on the ability to mobilize vector control equipment and procedures when and where indicated. Great progress is being made in new diagnostic and virus detection methods, which provide public health benefits by increasing the efficiency of virus monitoring and by enhancing research.

On the other hand, there remains a dearth of information on what constitutes a "safe" vector population density, i.e. what density is low enough that transmission is interrupted. A notable contribution is the *Culex tarsalis* light trap index developed in California by Reeves and colleagues (136). They found that St. Louis encephalitis disappeared as a clinical disease when the seasonal light trap index (females caught per light trap night) averaged between 10 and 20; clinical cases of western equine encephalomyelitis disappeared when the index fell below 10, although bird-to-bird transmission of WEE virus continued unless the index was maintained below 1.0. On the basis of a 20-year analysis of encephalitis cases in California, Olson et al (128) translated the rural light trap index of Reeves to a lower urban index suitable for areas with greater human population density. A *C. tarsalis* index based on sentinel chicken–flock trap collections, which show a rise in population one to two weeks earlier than light traps, may prove useful for predicting WEE epidemic activity in Manitoba (15). Similarly defined vector-population and virus-transmission thresholds are needed in other geographical areas as well as for other arboviruses.

Of all arboviral diseases, dengue presents the most frustrating of paradoxes. Although it is seemingly vulnerable to control by relatively simple source reduction methods, dengue has surged and resurged in both hemispheres; more than 850,000 human cases were reported from 1976 through 1982 (48). Because of difficulties in accurate diagnosis, the actual number of cases was probably a multiple of that total. In regard to the 1977 epidemic in Puerto Rico, Morens et al (118) stated that even with a good health system, high awareness of dengue, and efficient arbovirus response capability, there can be a minimum lag time of 20–35 days between the onset of epidemic activity and implementation of control responses. They suggested after the Puerto Rico experience that more emphasis on prevention, through incorporation of public education and assistance in source reduction at the household level, might be a better strategy.

If disease control is the real goal, many who are experienced in dealing with vector-borne diseases in tropical and semitropical latitudes are calling for similar prescriptions (9, 48, 57, 58, 87, 118). According to Gratz (48), large "vertical" programs, such as those aimed in the past at malaria and *A. aegypti* eradication, are no longer affordable. The future focus in developing countries will be on implementation of health care infrastructures that include not only primary health care but also disease-prevention programs involving participation by "individuals and families in and around their homes, by communities and by the health services at all levels. . . . [One of the best] targets for such individual and community involvement is *Aedes aegypti*." Gratz emphasized that the most critical need is not for more knowledge or better vector control methodology, but for more and better-trained vector control specialists: "Such trained personnel could then judge how best to apply the wealth of knowledge already available, ensure that it is adapted to local needs and train their own communities how to apply them. The core professional groups must also carry out the field research and entomological and epidemiological evaluation that can only be done in the endemic area where the control activities are being undertaken."

Halstead (58), pointing to the success of the Brazilian anti–*A. aegypti* campaign of half a century ago, emphasized source reduction as the cornerstone of *A. aegypti* and dengue control, and stressed that entomologically competent personnel, careful *A. aegypti* monitoring, and community participation, with the imposition of reasonable sanctions against citizens who fail to exert rigorous sanitary measures, are essential to success. He further attributed earlier success to a commitment "to discipline, to organization, and beyond all else, to the will to succeed." These words echo Gusmao's (57) description of the relentless attention to detail in the 1939–1940 program that eradicated the dangerous introduced malaria vector, *Anopheles gambiae,* from Brazil: "This spectacular win over a deadly vector, needless to say, was achieved through very hard work and flawless organization, long before the advent of DDT and the modern highly effective anti-malarial drugs. It was won with guts and fingernails."

ACKNOWLEDGMENTS

Partial support of research conducted in the laboratories of G. R. DeFoliart and P. R. Grimstad, and support during the writing of the manuscript, was provided by NIH grants AI 07453 and AI 19679, respectively. We thank Dr. Gary G. Clark, United States Army Medical Research Institute of Infectious Diseases, Frederick, Maryland, and Dr. Thomas M. Yuill, School of Veterinary Medicine, University of Wisconsin—Madison, for critically reading the manuscript, and also Dr. Ronald A. Ward, Walter Reed Army Institute of Research, Washington, DC, for suggestions that were helpful. We wish to thank Kim Viney for her expert secretarial assistance.

Literature Cited

1. Agudelo-Silva, F., Spielman, A. 1984. Paradoxical effects of simulated larviciding on production of adult mosquitoes. *Am. J. Trop. Med. Hyg.* 33:1267–69

2. Aitken, T. H. G., Downs, W. G., Shope, R. E. 1977. *Aedes aegypti* strain fitness for yellow fever virus transmission. *Am. J. Trop. Med. Hyg.* 26:985–89

3. Artsob, H., Spence, L. P., Th'ng, C. 1984. Enzyme-linked immunosorbent assay typing of California serogroup viruses in Canada. *J. Clin. Microbiol.* 20:276–80

4. Bailey, C. L., Eldridge, B. F., Hayes, D. E., Watts, D. M., Tammariello, R. F., et al. 1978. Isolation of St. Louis encephalitis virus from overwintering *Culex pipiens* mosquitoes. *Science* 199:1346–49

5. Bailey, C. L., Faran, M. E., Gargan, T. P. II, Hayes, D. E. 1982. Winter survival of blood-fed *Culex pipiens* L. *Am. J. Trop. Med. Hyg.* 31:1054–61

6. Baqar, S., Hayes, C. G., Ahmed, T. 1980. The effect of larval rearing conditions and adult age on the susceptibility of *Culex tritaeniorhynchus* to infection with West Nile virus. *Mosq. News* 40:165–71

7. Barnett, H. C. 1956. The transmission of western equine encephalitis virus by the mosquito *Culex tarsalis* Coq. *Am. J. Trop. Med. Hyg.* 5:86–98

8. Barr, A. R. 1982. The *Culex pipiens* complex. See Ref. 154, pp. 551–72

9. Beams, B. F. 1985. Analysis of mosquito control agency public education programs in the United States. *J. Am. Mosq. Control Assoc.* 1:212–19

10. Beaty, B. J., Bishop, D. H. L., Gay, M., Fuller, F. 1983. Interference between bunyaviruses in *Aedes triseriatus* mosquitoes. *Virology* 127:83–90

11. Beaty, B. J., Fuller, F., Bishop, D. H. L. 1983. Bunyavirus gene structure-function relationships and potential for RNA segment reassortment in the vector: La Crosse and snowshoe hare reas-

sortant viruses in mosquitoes. In *California Serogroup Viruses*, ed. C. H. Calisher, W. H. Thompson, pp. 119–28. New York: Liss
12. Beaty, B. J., Sundin, D. R., Chandler, L. J., Bishop, D. H. L. 1985. Evolution of bunyaviruses by genome reassortment in dually infected mosquitoes *(Aedes triseriatus)*. *Science* 230:548–50
13. Beaty, B. J., Thompson, W. H. 1976. Delineation of La Crosse virus in developmental stages of transovarially infected *Aedes triseriatus*. *Am. J. Trop. Med. Hyg.* 25:505–12
14. Bishop, D. H. L., Calisher, C. H., Casals, J., Chumakov, M. P., Gaidamovich, S. Ya., et al. 1980. Bunyaviridae. *Intervirology* 14:125–43
15. Brust, R. A. 1982. Population dynamics of *Culex tarsalis* Coquillett in Manitoba. In *Western Equine Encephalitis in Manitoba*, ed. L. Selka, pp. 21–30. Winnipeg: Manit. Minist. Health. 296 pp.
16. Calisher, C. H., Monath, T. P., Karabatsos, N., Trent, D. W. 1981. Arbovirus subtyping. *Am. J. Epidemiol.* 114:619–31
17. Calisher, C. H., Pretzman, C. I., Muth, D. J., Parsons, M. A., Peterson, E. D. 1986. Serodiagnosis of La Crosse virus infections in humans by detection of immunoglobulin M class antibodies. *J. Clin. Microbiol.* 23:667–71
18. Casey, H. L. 1965. Standardized diagnostic complement-fixation method and adaptation to microtest. II. Adaptation of LBCF method to microtechnique. *Public Health Serv. Monogr.* 74:31–34
19. Chamberlain, R. W., Sudia, W. D. 1961. Mechanisms of transmission of viruses by mosquitoes. *Ann. Rev. Entomol.* 6:371–90
20. Chang, H. C., Takashima, I., Arikawa, J., Hashimoto, N. 1984. Biotin-labeled antigen sandwich enzyme-linked immunosorbent assay (BLA-S-ELISA) for the detection of Japanese encephalitis antibody in human and a variety of animal sera. *J. Immunol. Methods* 72:401–9
21. Cheng, M. L., Hacker, C. S., Pryor, S. C., Ferrell, R. E., Kitto, G. B. 1982. The ecological genetics of the *Culex pipiens* complex in North America. See Ref. 154, pp. 581–627
22. Christensen, B. M., Rowley, W. A., Wong, Y. W., Dorsey, D. C., Hausler, W. J. Jr. 1978. Laboratory studies of transovarial transmission of trivittatus virus by *Aedes trivittatus*. *Am. J. Trop. Med. Hyg.* 27:184–86
23. Clarke, D. H., Casals, J. 1958. Techniques for hemagglutination and hemagglutination-inhibition with arthropod-borne viruses. *Am. J. Trop. Med. Hyg.* 7:561–73
24. Clarke, J. L. III, Rowley, W. A., Asman, M. 1983. The effect of colonization on the laboratory flight ability of *Culex tarsalis* (Diptera: Culicidae). *J. Fla. Anti-Mosq. Assoc.* 54:23–26
25. Cordellier, R., Bouchité, B., Roche, J. C., Monteny, N., Diaco, B., et al. 1983. Sylvatic circulation of the dengue 2 virus in 1980, in the sub-Sudanian savannas of the Ivory Coast. *Cahiers O.R.S.T.O.M. Ser. Entomol. Med. Parasitol.* 21:165–79 (In French)
26. Cornet, M., Robin, Y., Hème, G., Adam, C., Renaudet, J., et al. 1979. Une poussée épizootique de fièvre jaune selvatique au Senegal oriental. Isolement du virus de lots de moustiques adultes males et femelles. *Méd. Mal. Infect.* 9:63–66
27. Craig, G. B. Jr. 1964. Applications of genetic technology to mosquito rearing. *Bull. WHO* 31:469–73
28. Davey, M. W., Mahon, R. J., Gibbs, A. J. 1979. Togavirus interference in *Culex annulirostris* mosquitoes. *J. Gen. Virol.* 42:641–43
29. Day, J. F., Edman, J. D. 1984. The importance of disease induced changes in mammalian body temperature to mosquito blood feeding. *Comp. Biochem. Physiol.* 77:447–52
30. Day, J. F., Edman, J. D. 1984. Mosquito engorgement on normally defensive hosts depends on host activity patterns. *J. Med. Entomol.* 21:732–40
31. DeFoliart, G. R. 1983. *Aedes triseriatus*: Vector biology in relationship to the persistence of La Crosse virus in endemic foci. See Ref. 11, pp. 89–104
32. DeFoliart, G. R., Watts, D. M., Grimstad, P. R. 1986. Changing patterns in mosquito-borne arboviruses. *J. Am. Mosq. Control Assoc.* 2: In press
33. de Moor, P. P., Steffens, F. E. 1970. A computer-simulation model of an arthropod-borne virus transmission cycle, with special reference to chikungunya virus. *Trans. R. Soc. Trop. Med. Hyg.* 64:927–34
34. Deubel, V., Pailliez, J. P., Cornet, M., Schlesinger, J. J., Diop, M., et al. 1985. Homogeneity among Senegalese strains of yellow fever virus. *Am. J. Trop. Med. Hyg.* 34:976–83
35. Dye, C. 1984. Models for the population dynamics of the yellow fever mosquito, *Aedes aegypti*. *J. Anim. Ecol.* 53:247–68
36. Edman, J. D. 1971. Host-feeding pat-

terns of Florida mosquitoes. I. *Aedes, Anopheles, Coquillettidia, Mansonia* and *Psorophora. J. Med. Entomol.* 8: 687–95

37. Eldridge, B. F. 1981. Vector maintenance of pathogens in adverse environments (with special reference to mosquito maintenance of arboviruses). In *Vectors of Disease Agents*, ed. J. J. McKelvey, Jr., B. F. Eldridge, R. Maromorosch, pp. 141–57. New York: Praeger

38. Eldridge, B. F., Bailey, C. L. 1979. Experimental hibernation studies in *Culex pipiens* (Diptera: Culicidae): Reactivation of ovarian development and blood-feeding in prehibernatory females. *J. Med. Entomol.* 15:462–67

39. Fine, P. E. M. 1975. Vectors and vertical transmission: An epidemiologic perspective. *Ann. NY Acad. Sci.* 266: 173–94

40. Fine, P. E. M. 1981. Epidemiological principles of vector-mediated transmission. See Ref. 37, pp. 77–91

41. Fine, P. E. M., Le Duc, J. W. 1978. Towards a quantitative understanding of the epidemiology of Keystone virus in the eastern United States. *Am. J. Trop. Med. Hyg.* 27:322–38

42. Forghani, B. 1985. Radioimmunoassay systems. See Ref. 95, pp. 93–103

43. Forghani, B. 1985. Enzyme immunoassay systems. See Ref. 95, pp. 105–18

44. Francy, D. B., Rush, W. A., Montoya, M., Inglish, D. S., Bolin, R. A. 1981. Transovarial transmission of St. Louis encephalitis virus by *Culex pipiens* complex mosquitoes. *Am. J. Trop. Med. Hyg.* 30:699–705

45. Gargan, T. P., Bailey, C. L., Higbee, G. A., Gad, A., El Said, S. 1983. The effect of laboratory colonization on the vector-pathogen interactions of Egyptian *Culex pipiens* and Rift Valley fever virus. *Am. J. Trop. Med. Hyg.* 32:1154–63

46. Gillett, J. D., Connor, J. 1976. Host temperature and the transmission of arboviruses by mosquitoes. *Mosq. News* 36:472–77

47. Gorgas Memorial Laboratory. 1979. *Annual Report Fiscal Year 1978*, pp. 23–24. Washington, DC: US Gov. Print. Off. 65 pp.

48. Gratz, N. R. 1985. The future of vector biology and control in the World Health Organization. *J. Am. Mosq. Control Assoc.* 1:273–78

49. Grimstad, P. R. 1983. Mosquitoes and the incidence of encephalitis. *Adv. Virus Res.* 28:357–438

50. Grimstad, P. R., Craig, G. B. Jr., Ross, Q. E., Yuill, T. M. 1977. *Aedes triseriatus* and La Crosse virus: Geographic variation in vector susceptibility and ability to transmit. *Am. J. Trop. Med. Hyg.* 26:990–96

51. Grimstad, P. R., Haramis, L. D. 1984. *Aedes triseriatus* (Diptera: Culicidae) and La Crosse virus. III. Enhanced oral transmission by nutrition-deprived mosquitoes. *J. Med. Entomol.* 21:249–56

52. Grimstad, P. R., Paulson, S. L., Craig, G. B. Jr. 1985. Vector competence of *Aedes hendersoni* (Diptera: Culicidae) for La Crosse virus and evidence of a salivary-gland escape barrier. *J. Med. Entomol.* 22:447–53

53. Grimstad, P. R., Ross, Q. E., Craig, G. B. Jr. 1980. *Aedes triseriatus* (Diptera: Culicidae) and La Crosse virus. II. Modification of mosquito feeding behavior by virus infection. *J. Med. Entomol.* 17:1–7

54. Gubler, D. J., Kuno, G., Sather, G. E., Velez, M., Oliver, A. 1984. Mosquito cell cultures and specific monoclonal antibodies in surveillance for dengue viruses. *Am. J. Trop. Med. Hyg.* 33: 158–65

55. Gubler, D. J., Nalim, S., Tan, R., Saipan, H., Saroso, J. S. 1979. Variation in susceptibility to oral infection with dengue viruses among geographic strains of *Aedes aegypti. Am. J. Trop. Med. Hyg.* 28:1045–52

56. Gubler, D. J., Novak, R., Mitchell, C. J. 1982. Arthropod vector competence: Epidemiological, genetic, and biological considerations. See Ref. 154, pp. 343–78

57. Gusmao, H. H. 1982. Fighting disease-bearing mosquitoes through relentless field leadership. *Am. J. Trop. Med. Hyg.* 31:705–10

58. Halstead, S. B. 1984. Selective primary health care: Strategies for control of disease in the developing world. XI. Dengue. *Rev. Infect. Dis.* 6:251–64

59. Halstead, S. B. 1985. Arboviruses. See Ref. 95, pp. 147–69

60. Haramis, L. D. 1985. Larval nutrition, adult body size, and the biology of *Aedes triseriatus*. See Ref. 98, pp. 431–37

61. Hardy, J. L., Apperson, G., Asman, S. M., Reeves, W. C. 1978. Selection of a strain of *Culex tarsalis* highly resistant to infection following ingestion of western equine encephalomyelitis virus. *Am. J. Trop. Med. Hyg.* 27:313–21

62. Hardy, J. L., Houk, E. J., Kramer, L. D., Reeves, W. C. 1983. Intrinsic factors affecting vector competence of mos-

quitoes for arboviruses. *Ann. Rev. Entomol.* 28:229–62

63. Hardy, J. L., Reeves, W. C., Sjogren, R. D. 1976. Variations in the susceptibility of field and laboratory populations of *Culex tarsalis* to experimental infection with western equine encephalomyelitis virus. *Am. J. Epidemiol.* 103:498–505

64. Hardy, J. L., Rosen, L., Kramer, L. D., Presser, S. B., Shroyer, D. A., et al. 1980. Effect of rearing temperature on transovarial transmission of St. Louis encephalitis virus in mosquitoes. *Am. J. Trop. Med. Hyg.* 29:963–68

65. Hardy, J. L., Rosen, L., Reeves, W. C., Scrivani, R. P., Presser, S. B. 1984. Experimental transovarial transmission of St. Louis encephalitis virus by *Culex* and *Aedes* mosquitoes. *Am. J. Trop. Med. Hyg.* 33:166–75

66. Hawley, W. A. 1985. The effect of larval density on adult longevity of a mosquito, *Aedes sierriensis:* Epidemiological consequences. *J. Anim. Ecol.* 54:955–64

67. Hayes, C. G. 1979. Vector competence of colonized *Culiseta melanura* (Diptera: Culicidae) for western equine encephalomyelitis virus. *J. Med. Entomol.* 15:253–58

68. Hayes, C. G., Baker, R. H., Baqar, S., Ahmed, T. 1984. Genetic variation for West Nile virus susceptibility in *Culex tritaeniorhynchus. Am. J. Trop. Med. Hyg.* 33:715–24

69. Hayes, C. G., Basit, A., Baqar, S., Akhter, R. 1980. Vector competence of *Culex tritaeniorhynchus* (Diptera: Culicidae) for West Nile Virus. *J. Med. Entomol.* 17:172–77

70. Hayes, R. O., Montoya, M., Smith, G. C., Francy, D. B., Jakob, W. L. 1974. An analysis of *Culex tarsalis* Coquillett laboratory rearing productivity. *Mosq. News* 34:462–66

71. Henchal, E. A., McCown, J. M., Burke, D. S., Seguin, M. C., Brandt, W. E. 1985. Epitopic analysis of antigenic determinants on the surface of dengue-2 virions using monoclonal antibodies. *Am. J. Trop. Med. Hyg.* 34:162–69

72. Henchal, E. A., McCown, J. M., Seguin, M. C., Gentry, M. K., Brandt, W. E. 1983. Rapid identification of dengue virus isolates by using monoclonal antibodies in an indirect immunofluorescence assay. *Am. J. Trop. Med. Hyg.* 32:164–69

73. Hildreth, S. W., Beaty, B. J. 1984. Detection of eastern equine encephalomyelitis virus and Highlands J virus antigens within mosquito pools by enzyme immunoassay. *Am. J. Trop. Med. Hyg.* 33:965–72

74. Huang, Y.-M. 1986. *Aedes (Stegomyia) bromeliae* (Diptera: Culicidae), the yellow fever virus vector in East Africa. *J. Med. Entomol.* 23:196–200

75. Hurlbut, H. S. 1973. The effect of environmental temperature upon the transmission of St. Louis encephalitis virus by *Culex pipiens quinquefasciatus. J. Med. Entomol.* 10:1–12

76. Igarashi, A. 1984. A hypothesis on the geographical distribution of arboviruses. *Trop. Med.* 26:173–79

77. Igarashi, A., Bundo, K., Matsuo, S., Makino, Y., Lin, W.-J. 1981. Enzyme-linked immunosorbent assay for Japanese encephalitis virus: 1. Basic conditions for the assay of human immunoglobulin. *Trop. Med.* 23:49–60

78. Istock, C. A. 1985. Pattern and process in life history genetics. See Ref. 98, pp. 319–25

79. Johnson, B. K., Varma, M. G. R. 1975. Infection of the mosquito *Aedes aegypti* with infectious West Nile virus–antibody complexes. *Trans. R. Soc. Trop. Med. Hyg.* 69:336–41

80. Jupp, P. G. 1974. Laboratory studies on the transmission of West Nile virus by *Culex (Culex) univittatus* Theobald; factors influencing the transmission rate. *J. Med. Entomol.* 11:455–58

81. Jupp, P. G. 1976. The susceptibility of four South African species of *Culex* to West Nile and Sindbis viruses by two different infecting methods. *Mosq. News* 36:166–73

82. Jupp, P. G. 1985. *Culex theileri* and Sindbis virus; salivary glands infection in relation to transmission. *J. Am. Mosq. Control Assoc.* 1:374–76

83. Karabatsos, N., ed. 1985. *International Catalogue of Arboviruses Including Certain Other Viruses of Vertebrates.* San Antonio, Tex: Am. Soc. Trop. Med. Hyg. 1147 pp. 3rd ed.

84. Klimas, R. A., Thompson, W. H., Calisher, C. H., Clark, G. G., Grimstad, P. R., et al. 1981. Genotypic varieties of La Crosse virus isolated from different geographic regions of the continental United States and evidence for a naturally occurring intertypic recombinant La Crosse virus. *Am. J. Epidemiol.* 114:112–31

85. Klowden, M. J., Lea, A. O. 1978. Blood meal size as a factor affecting continued host-seeking by *Aedes aegypti* (L.). *Am. J. Trop. Med. Hyg.* 27:827–31

86. Klowden, M. J., Lea, A. O. 1979.

Effect of defensive host behavior on the blood meal size and feeding success of natural populations of mosquitoes (Diptera: Culicidae). *J. Med. Entomol.* 15:514–17

87. Knudsen, A. B. 1983. *Aedes aegypti* and dengue in the Caribbean. *Mosq. News* 43:269–75

88. Kramer, L. D., Hardy, J. L., Presser, S. B. 1983. Effect of temperature of extrinsic incubation on the vector competence of *Culex tarsalis* for western equine encephalomyelitis virus. *Am. J. Trop. Med. Hyg.* 32:1130–39

89. Kramer, L. D., Hardy, J. L., Presser, S. B., Houk, E. G. 1981. Dissemination barriers for western equine encephalomyelitis virus in *Culex tarsalis* infected after ingestion of low viral doses. *Am. J. Trop. Med. Hyg.* 30:190–97

90. Ksiazek, T. G., Hardy, J. L., Reeves, W. C. 1985. Effect of normal mosquito extracts upon arbovirus recoveries from mosquito pools. *Am. J. Trop. Med. Hyg.* 34:578–85

91. Kurstak, E., Tijssen, P., Kurstak, C. 1984. Enzyme immunoassays applied in virology: Reagents preparation and interpretation. In *Applied Virology,* ed. E. Kurstak, W. Al-Nakib, C. Kurstak, 29:479–503. Orlando/New York: Academic. 518 pp.

92. Leary, J. J., Brigati, D. J., Ward, D. C. 1983. Rapid and sensitive colorimetric method for visualizing biotin-labeled DNA probes hybridized to DNA or RNA immobilized on nitrocellulose: Bioblots. *Proc. Natl. Acad. Sci. USA* 80:4045–49

93. Le Duc, J. W. 1979. The ecology of California group viruses. *J. Med. Entomol.* 16:1–17

94. Legner, E. F., Sjogren, R. D. 1984. Biological mosquito control furthered by advances in technology and research. *Mosq. News* 44:449–56

95. Lennette, E. H., ed. 1985. *Laboratory Diagnosis of Viral Infections.* New York/Basel: Dekker. 506 pp.

96. Linthicum, K. J., Davies, F. G., Kairo, A., Bailey, C. L. 1985. Rift Valley fever virus (family Bunyaviridae, genus *Phlebovirus*). Isolations from Diptera during an inter-epizootic period in Kenya. *J. Hyg.* 95:197–209

97. Lorenz, L., Beaty, B. J., Aitken, T. H. G., Wallis, G. P., Tabachnick, W. J. 1984. The effect of colonization upon *Aedes aegypti* susceptibility to oral infection with yellow fever virus. *Am. J. Trop. Med. Hyg.* 33:690–94

98. Lounibos, L. P., Rey, J. R., Frank, J. H., eds. 1985. *Ecology of Mosquitoes.*

Vero Beach, Fla: Fla. Med. Entomol. Lab. 579 pp.

99. McClelland, G. A. H. 1974. A worldwide survey of variation in scale pattern of the abdominal tergum of *Aedes aegypti* (L.) (Diptera: Culicidae). *Trans. R. Entomol. Soc. London* 126:239–59

100. McDonald, P. T., Hanley, M., Wrensch, M. 1979. Comparison of reproductive characteristics of laboratory and field-collected *Culex tarsalis* in laboratory cages. *Mosq. News* 39:258–62

101. McLean, R. G., Francy, D. B., Monath, T. P., Calisher, C. H., Trent, D. W. 1985. Isolation of St. Louis encephalitis virus from adult *Dermacentor variabilis* (Acari: Ixodidae). *J. Med. Entomol.* 22:232–33

102. McLintock, J. 1978. Mosquito-virus relationships of American encephalitides. *Ann. Rev. Entomol.* 23:17–37

103. Melnick, J. L. 1974. Classification and nomenclature of viruses. *Prog. Med. Virol.* 17:290–94

104. Meyer, R. P., Hardy, J. L., Presser, S. B. 1983. Comparative vector competence of *Culex tarsalis* and *Culex quinquefasciatus* from the Coachella, Imperial, and San Joaquin Valleys of California for St. Louis encephalitis virus. *Am. J. Trop. Med. Hyg.* 32:305–11

105. Miller, B. R., Beaty, B. J., Lorenz, L. H. 1982. Variation of La Crosse virus filial infection rates in geographic strains of *Aedes triseriatus* (Diptera: Culicidae). *J. Med. Entomol.* 19:213–14

106. Miller, B. R., DeFoliart, G. R. 1979. Infection rates of *Ascocystis*-infected *Aedes triseriatus* following ingestion of La Crosse virus by the larvae. *Am. J. Trop. Med. Hyg.* 28:1064–66

107. Miller, B. R., DeFoliart, G. R., Yuill, T. M. 1977. Vertical transmission of La Crosse virus (California encephalitis group): Transovarial and filial infection rates in *Aedes triseriatus* (Diptera: Culicidae). *J. Med. Entomol.* 14:437–40

108. Miller, B. R., DeFoliart, G. R., Yuill, T. M. 1979. *Aedes triseriatus* and La Crosse virus: Lack of infection in eggs from the first ovarian cycle following oral infection of females. *Am. J. Trop. Med. Hyg.* 28:897–901

109. Mims, C. A., Day, M. F., Marshall, I. D. 1966. Cytopathic effect of Semliki Forest virus in the mosquito *Aedes aegypti. Am. J. Trop. Med. Hyg.* 15:775–84

110. Mitchell, C. J. 1983. Mosquito vector competence and arboviruses. In *Current Topics in Vector Research,* Vol. I, ed.

K. F. Harris, pp. 63–92. New York: Praeger

111. Mitchell, C. J., Bowen, G. S., Monath, T. P., Cropp, C. B., Kerschner, J. 1979. St. Louis encephalitis virus transmission following multiple feeding of *Culex pipiens pipiens* (Diptera: Culicidae) during a single gonotrophic cycle. *J. Med. Entomol.* 16:254–58

112. Mitchell, C. J., Francy, D. B., Monath, T. P. 1980. Arthropod vectors. In *St. Louis Encephalitis,* ed. T. P. Monath, pp. 313–79. Washington, DC: Am. Public Health Assoc. 680 pp.

113. Mitchell, C. J., Gubler, D. J., Monath, T. P. 1983. Variation in infectivity of Saint Louis encephalitis virus strains for *Culex pipiens quinquefasciatus* (Diptera: Culicidae). *J. Med. Entomol.* 20:526–33

114. Mitchell, C. J., Monath, T. P., Cropp, C. B. 1981. Experimental transmission of Rocio virus by mosquitoes. *Am. J. Trop. Med. Hyg.* 30:465–72

115. Monath, T. P., Nystrom, R. R. 1984. Detection of yellow fever virus in serum by enzyme immunoassay. *Am. J. Trop. Med. Hyg.* 33:151–57

116. Monath, T. P., Nystrom, R. R., Bailey, R. E., Calisher, C. H., Muth, D. J. 1984. Immunoglobulin M antibody capture enzyme-linked immunosorbent assay for diagnosis of St. Louis encephalitis. *J. Clin. Microbiol.* 20:784–90

117. Monath, T. P., Schlesinger, J. J., Brandriss, M. W., Cropp, C. B., Prange, W. C. 1984. Yellow fever monoclonal antibodies: Type-specific and cross-reactive determinants identified by immunofluorescence. *Am. J. Trop. Med. Hyg.* 33:695–98

118. Morens, D. M., Rigau-Peréz, J. G., López-Correa, R. H., Moore, C. G., Ruiz-Tibén, E. E., et al. 1986. Dengue in Puerto Rico, 1977: Public health response to characterize and control an epidemic of multiple serotypes. *Am. J. Trop. Med. Hyg.* 35:197–211

119. Morris, C. D., Srihongse, S. 1978. An evaluation of the hypothesis of transovarial transmission of eastern equine encephalomyelitis virus by *Culiseta melanura. Am. J. Trop. Med. Hyg.* 27:1246–50

120. Mourya, D. T., Soman, R. S. 1985. Effect of gregarine parasite, *Ascogregarina culicis* and tetracycline on the susceptibility of *Culex bitaeniorhynchus* to JE virus. *Indian J. Med. Res.* 81:247–50

121. Mukwaya, L. G. 1977. Genetic control of feeding preference in the mosquitoes *Aedes (Stegomyia) simpsoni* and *aegypti. Physiol. Entomol.* 2:133–45

122. Munstermann, L. E. 1985. Geographic patterns of genetic variation in the treehole mosquito *Aedes triseriatus.* See Ref. 98, pp. 327–43

123. Munstermann, L. E., Taylor, D. B., Matthews, T. C. 1982. Population genetics and speciation in the *Aedes triseriatus* group. See Ref. 154, pp. 433–53

124. Murphy, F. A. 1975. Cellular resistance to arbovirus infection. *Ann. NY Acad. Sci.* 266:197–203

125. Nasci, R. S. 1985. Behavioral ecology of variation in blood-feeding and its effect on mosquito-borne diseases. See Ref. 98, pp. 293–303

126. Niklasson, B. S., Gargan, T. P. II. 1985. Enzyme-linked immunosorbent assay for detection of Rift Valley fever virus antigen in mosquitoes. *Am. J. Trop. Med. Hyg.* 34:400–5

127. Niklasson, B., Peters, C. J., Grandien, M., Wood, O. 1984. Detection of human immunoglobulins G and M antibodies to Rift Valley fever virus by enzyme-linked immunosorbent assay. *J. Clin. Microbiol.* 19:225–29

128. Olson, J. G., Reeves, W. C., Emmons, R. W., Milby, M. M. 1979. Correlation of *Culex tarsalis* population indices with the incidence of St. Louis encephalitis and western equine encephalomyelitis in California. *Am. J. Trop. Med. Hyg.* 28:335–43

129. Oshiro, L. A. 1985. Application of electron microscopy to the diagnosis of viral infections. See Ref. 95, pp. 55–72

130. Pantuwatana, S., Thompson, W. H., Watts, D. M., Yuill, T. M., Hanson, R. P. 1974. Isolation of La Crosse virus from field collected *Aedes triseriatus* larvae. *Am. J. Trop. Med. Hyg.* 23:246–50

131. Patrican, L. A., DeFoliart, G. R. 1985. Lack of adverse effect of transovarially acquired La Crosse virus infection on the reproductive capacity of *Aedes triseriatus* (Diptera: Culicidae). *J. Med. Entomol.* 22:602–9

132. Patrican, L. A., DeFoliart, G. R., Yuill, T. M. 1985. La Crosse viremias in juvenile, subadult and adult chipmunks *(Tamias striatus)* following feeding by transovarially-infected *Aedes triseriatus. Am. J. Trop. Med. Hyg.* 34:596–602

133. Patrican, L. A., DeFoliart, G. R., Yuill, T. M. 1985. Oral infection and transmission of La Crosse virus by an enzootic strain of *Aedes triseriatus* feeding on chipmunks with a range of viremia levels. *Am. J. Trop. Med. Hyg.* 34:992–98

134. Powell, J. R., Tabachnick, W. J.,

Arnold, J. 1980. Genetics and the origin of a vector population: *Aedes aegypti*, a case study. *Science* 208:1385–87

135. Powell, J. R., Tabachnick, W. J., Wallis, G. P. 1982. *Aedes aegypti* as a model of the usefulness of population genetics of vectors. See Ref. 154, pp. 396–412

136. Reeves, W. C. 1971. Mosquito vector and vertebrate host interaction: The key to maintenance of certain arboviruses. In *Ecology and Physiology of Parasites*, ed. A. M. Fallis, pp. 223–30. Toronto: Univ. Toronto Press. 258 pp.

137. Reeves, W. C. 1974. Overwintering of arboviruses. *Prog. Med. Virol.* 17:193–220

138. Reisen, W. K., Milby, M. M., Bock, M. E. 1984. The effects of immature stress on selected events in the life history of *Culex tarsalis*. *Mosq. News* 44:385–95

139. Repik, P. M., Dalrymple, J. M., Brandt, W. E., McCown, J. M., Russell, P. K. 1983. RNA fingerprinting as a method for distinguishing dengue 1 virus strains. *Am. J. Trop. Med. Hyg.* 32:577–89

140. Ribeiro, J. M. C., Rossignol, P. A., Spielman, A. 1985. Salivary gland apyrase determines probing time in anopheline mosquitoes. *J. Insect Physiol.* 31:689–92

141. Rosen, L. 1981. The use of *Toxorhynchites* mosquitoes to detect and propagate dengue and other arboviruses. *Am. J. Trop. Med. Hyg.* 30:177–83

142. Scherer, W. F., Weaver, S. C., Taylor, C. A., Cupp, E. W. 1986. Vector competency: Its implication in the disappearance of epizootic Venezuelan equine encephalomyelitis virus from Middle America. *J. Med. Entomol.* 23:23–29

143. Scott, T. W., Burrage, T. G. 1984. Rapid infection of salivary glands in *Culiseta melanura* with eastern equine encephalitis virus: An electron microscopic study. *Am. J. Trop. Med. Hyg.* 33:961–64

144. Scott, T. W., Hildreth, S. W., Beaty, B. J. 1984. The distribution and development of eastern equine encephalitis virus in its enzootic mosquito vector, *Culiseta melanura*. *Am. J. Trop. Med. Hyg.* 33:300–10

145. Scott, T. W., McLean, R. G., Francy, D. B., Card, C. S. 1983. A simulation model for the vector-host transmission system of a mosquito-borne avian virus, Turlock (Bunyaviridae). *J. Med. Entomol.* 20:625–40

146. Service, M. W. 1976. *Mosquito Ecology: Field Sampling Methods*. New York: Wiley. 583 pp.

147. Service, M. W. 1982. Importance of vector ecology in vector disease control in Africa. *Bull. Soc. Vector Ecol.* 7:1–13

148. Service, M. W. 1983. Biological control of mosquitoes—has it a future? *Mosq. News* 43:113–20

149. Shope, R. E. 1985. Arboviruses. In *Manual of Clinical Microbiology*, ed. E. H. Lennette, A. Balows, W. J. Hausler, Jr., H. J. Shadomy, 77:785–89. Washington, DC: Am. Soc. Microbiol. 1149 pp. 4th ed.

150. Shroyer, D. A. 1986. Transovarial maintenance of San Angelo virus in sequential generations of *Aedes albopictus*. *Am. J. Trop. Med. Hyg.* 35:408–17

151. Shroyer, D. A., Craig, G. B. Jr. 1983. Egg diapause in *Aedes triseriatus* (Diptera: Culicidae): Geographic variation in photoperiodic response and factors influencing diapause termination. *J. Med. Entomol.* 20:601–7

152. Sims, S. R. 1985. Embryonic and larval diapause in *Aedes triseriatus:* Phenotypic correlation and ecological consequences of the induction response. See Ref. 98, pp. 359–69

153. Smith, C. E. G. 1975. The significance of mosquito longevity and blood-feeding behaviour in the dynamics of arbovirus infections. *Med. Biol.* 53:288–94

154. Steiner, W. W. M., Tabachnick, W. J., Rai, K. S., Narang, S., eds. 1982. *Recent Developments in the Genetics of Insect Disease Vectors*. Champaign, Ill: Stipes. 665 pp.

155. Stoddard, P., Asman, S. M. 1981. Possible integration of microbial and genetic control of *Culex tarsalis*. *Proc. Calif. Mosq. Vector Control Assoc.* 49:108–11

156. Tabachnick, W. J., Powell, J. R. 1979. A world-wide survey of genetic variation in the yellow fever mosquito, *Aedes aegypti*. *Genet. Res.* 34:215–29

157. Tabachnick, W. J., Wallis, G. P. 1985. Population genetic structure of the yellow fever mosquito *Aedes aegypti* in the Caribbean: Ecological considerations. See Ref. 98, pp. 371–81

158. Tabachnick, W. J., Wallis, G. P., Aitken, T. H. G., Miller, B. R., Amato, G. D., et al. 1985. Oral infection of *Aedes aegypti* with yellow fever virus: Geographic variation and genetic considerations. *Am. J. Trop. Med. Hyg.* 34:1219–24

159. Takahashi, M. 1976. The effects of environmental and physiological conditions of *Culex tritaeniorhynchus* on the

pattern of transmission of Japanese encephalitis virus. *J. Med. Entomol.* 13: 275–84

160. Tesh, R. B. 1980. Experimental studies on the transovarial transmission of Kunjin and San Angelo viruses in mosquitoes. *Am. J. Trop. Med. Hyg.* 29: 657–66

161. Tesh, R. B. 1984. Transovarial transmission of arboviruses in their invertebrate vectors. In *Current Topics in Vector Res.*, Vol. 2, ed. K. F. Harris, pp. 57–76. New York: Praeger

162. Tesh, R. B., Gubler, D. J., Rosen, L. 1976. Variation among geographic strains of *Aedes albopictus* in susceptibility to infection with chikungunya virus. *Am. J. Trop. Med. Hyg.* 25:326–35

163. Tesh, R. B., Shroyer, D. A. 1980. The mechanism of arbovirus transovarial transmission in mosquitoes: San Angelo virus in *Aedes albopictus. Am. J. Trop. Med. Hyg.* 29:1394–404

164. Thompson, W. H., Beaty, B. J. 1977. Venereal transmission of La Crosse (California encephalitis) arbovirus in *Aedes triseriatus* mosquitoes. *Science* 196:530–31

165. Thompson, W. H., Beaty, B. J. 1978. Venereal transmission of La Crosse virus from male to female *Aedes triseriatus. Am. J. Trop. Med. Hyg.* 27:187–96

166. Tikasingh, E. S., Spence, L., Downs, W. G. 1966. The use of adjuvant and sarcoma 180 cells in the production of mouse hyperimmune ascitic fluids to arboviruses. *Am. J. Trop. Med. Hyg.* 15:219–26

167. Trent, D. W., Naeve, C. W. 1980. Biochemistry and replication. See Ref. 112, pp. 159–99

168. Trpis, M., Hausermann, W. 1978. Genetics of house-entering behavior in East African populations of *Aedes aegypti* (L) (Diptera: Culicidae) and its relevance to speciation. *Bull. Entomol. Res.* 68:521–32

169. Turell, M. J., Bailey, C. L., Rossi, C. A. 1984. Increased mosquito feeding on Rift Valley fever virus–infected lambs. *Am. J. Trop. Med. Hyg.* 33:1232–38

170. Turell, M. J., Gargan, T. P. II, Bailey, C. L. 1985. *Culex pipiens* (Diptera: Culicidae) morbidity and mortality associated with Rift Valley fever virus infection. *J. Med. Entomol.* 22:332–37

171. Turell, M. J., Hardy, J. L., Reeves, W. C. 1982. Stabilized infection of California encephalitis virus in *Aedes dorsalis,*

and its implication for viral maintenance in nature. *Am. J. Trop. Med. Hyg.* 31:1252–59

172. Turell, M. J., Reeves, W. C., Hardy, J. L. 1982. Evaluation of the efficiency of transovarial transmission of California encephalitis viral strains in *Aedes dorsalis* and *Aedes melanimon. Am. J. Trop. Med. Hyg.* 31:382–88

173. Turell, M. J., Rossi, C. A., Bailey, C. L. 1985. Effect of extrinsic incubation temperature on the ability of *Aedes taeniorhynchus* and *Culex pipiens* to transmit Rift Valley fever virus. *Am. J. Trop. Med. Hyg.* 34:1211–18

174. Turell, M. J., Rossignol, P. A., Spielman, A., Rossi, C. A., Bailey, C. L. 1984. Enhanced arboviral transmission by mosquitoes that concurrently ingested microfilariae. *Science* 225:1039–41

175. Vezza, A. C., Rosen, L., Repik, P., Dalrymple, J., Bishop, D. H. L. 1980. Characterization of the viral RNA species of prototype dengue viruses. *Am. J. Trop. Med. Hyg.* 29:643–52

176. Wallis, G. P., Aitken, T. H. G., Beaty, B. J., Lorenz, L., Amato, G. D., et al. 1985. Selection for susceptibility and refractoriness of *Aedes aegypti* to oral infection with yellow fever virus. *Am. J. Trop. Med. Hyg.* 34:1225–31

177. Wallis, G. P., Tabachnick, W. J., Powell, J. R. 1984. Genetic heterogeneity among Caribbean populations of *Aedes aegypti. Am. J. Trop. Med. Hyg.* 33:492–98

178. Washino, R. K., Tempelis, C. H. 1983. Mosquito host bloodmeal identification: Methodology and data analysis. *Ann. Rev. Entomol.* 28:179–201

179. Watts, D. M., Morris, C. D., Wright, R. E., DeFoliart, G. R., Hanson, R. P. 1972. Transmission of La Crosse virus (California encephalitis group) by the mosquito *Aedes triseriatus. J. Med. Entomol.* 9:125–27

180. Watts, D. M., Pantuwatana, S., DeFoliart, G. R., Yuill, T. M., Thompson, W. H. 1973. Transovarial transmission of La Crosse virus (California encephalitis group) in the mosquito, *Aedes triseriatus. Science* 182:1140–41

181. Watts, D. M., Thompson, W. H., Yuill, T. M., DeFoliart, G. R., Hanson, R. P. 1974. Overwintering of La Crosse virus in *Aedes triseriatus. Am. J. Trop. Med. Hyg.* 23:694–700

182. Weaver, S. C., Scherer, W. F., Cupp, E. W., Castello, D. A. 1984. Barriers to dissemination of Venezuelan encephalitis viruses in the Middle American enzootic vector mosquito, *Culex*

(*Melanoconion*) *taeniopus*. *Am. J. Trop. Med. Hyg.* 33:953–60

183. Westaway, E. G., Brinton, M. A., Gaidamovich, S. Ya., Horzinek, M. C., Igarashi, A., et al. 1985. Togaviridae. *Intervirology* 24:125–39

184. Westaway, E. G., Brinton, M. A., Gaidamovich, S. Ya., Horzinek, M. C., Igarashi, A., et al. 1985. Flaviviridae. *Intervirology* 24:183–92

185. Wildy, P. 1971. Classification and nomenclature of viruses. *Monogr. Virol.* 5:27–74

186. Wolff, K. L., Muth, D. J., Hudson, B. W., Trent, D. W. 1981. Evaluation of the solid-phase radioimmunoassay for diagnosis of St. Louis encephalitis infections in humans. *J. Clin. Microbiol.* 14:135–40

187. World Health Organization. 1985. *Arthropod-Borne and Rodent-Borne Viral Diseases. WHO Tech. Rep. Ser. 719.* Geneva: WHO. 114 pp.

Ann. Rev. Entomol. 1987. 32:507–38

ECOLOGY AND MANAGEMENT OF SOYBEAN ARTHROPODS

Marcos Kogan

Office of Agricultural Entomology, University of Illinois and Illinois Natural History Survey, Champaign, Illinois 61820

Samuel G. Turnipseed

Department of Entomology, Clemson University, Edisto Experiment Station, Blackville, South Carolina 29817

INTRODUCTORY PERSPECTIVE

Soybean entomology was in its infancy when we wrote our first review ten years ago (209). That review included 264 references from the SIRIC[1] database of about 700 papers on soybean entomology at the end of 1975. SIRIC is again our primary bibliographic resource; in 1986 it has provided a database of nearly 5000 references (83). The growth of the literature has indeed followed the expansion of the crop throughout the world (Figure 1).

Multidisciplinary, multi-institutional collaborative programs, e.g. the Huffaker project (66) and the Consortium for Integrated Pest Management (CIPM) (35), and United States Department of Agriculture regional projects supported by the agricultural experiment stations of land-grant universities have contributed directly to the intensification of the research effort on soybean entomology in the United States. In Brazil, research on various aspects of soybean entomology gained momentum following the creation in 1973 of the National Soybean Research Center in Londrina, Paraná, a unit of the Empresa Brasileira de Pesquisas Agropecuárias (EMBRAPA) (141).

---

[1]The Soybean Insect Research Information Center (SIRIC) is the computerized database for soybean entomology of the College of Agriculture, University of Illinois, and the Illinois Natural History Survey.

0066-4170/87/0101-0507$02.00

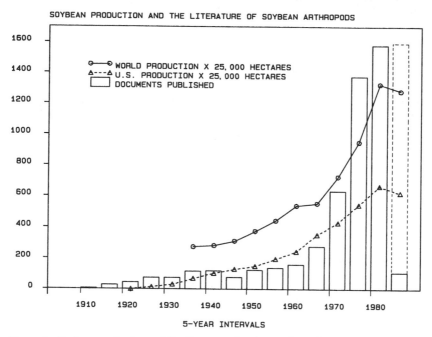

SOYBEAN PRODUCTION AND THE LITERATURE OF SOYBEAN ARTHROPODS

o——o WORLD PRODUCTION X 25,000 HECTARES
△---△ U.S. PRODUCTION X 25,000 HECTARES
□ DOCUMENTS PUBLISHED

5-YEAR INTERVALS

Figure 1   Soybean production and the literature of soybean arthropods.

Public and private support also led to expanded research efforts and a con-sequent increase in output of literature in China, Japan, India, Mexico, Argentina, and Australia. In addition, three international centers, the Interna-tional Institute of Tropical Agriculture (IITA) in Nigeria, the Asian Vegetable Research and Development Center (AVRDC) in Taiwan, and the In-ternational Rice Research Institute (IRRI) in the Philippines, have included soybean research in their programs (see the annual reports and research highlights of those international centers). The present maturity of research on soybean entomology is reflected in the number of major reviews, com-prehensive articles, symposia proceedings, and monographs on specific topics.

Regional assessments of soybean pest problems and control procedures have been published for the United States (31, 107, 139, 152, 210), Brazil (39, 141, 147, 206, 214), Japan (60, 79), and India (10, 11). Reports summarizing these problems in various countries have been presented in workshops and symposia and in the proceedings of three world soybean research conferences (27, 42, 59, 117, 171, 176, 178, 179). A book on sampling methods for soybean entomology (92) has provided extensive in-formation on geographic range, host plants, and biology of the most important pests, natural enemies, and diseases of soybean insects, in addition to a discussion of sampling techniques and definition of spatial patterns of arthropod populations for research and integrated pest management (IPM). Biological control of soybean insect pests in the United States has been thoroughly reviewed in a cooperative publication (154). Guidelines, extension brochures, and summaries of practical recommendations for the implementation of IPM programs have been developed for many states in the United States, e.g. Illinois (94), Virginia (121), and Louisiana (212); for the southern regions of Brazil (13, 39); and for Australia (202), Argentina (12, 67), Mexico (177), China (112, 218, 224), and Japan (133, 134). Finally, a book summarizing progress in soybean entomology in the United States, primarily based on the achievements of the Huffaker and CIPM projects, is in preparation (137).

Because of the large number of recent publications, we cannot provide a comprehensive account of the literature accumulated since our 1976 review. Instead, we choose to comment on the current status of ecological research on soybean-associated arthropods, focusing on the dynamics of communities and niche occupancy based on a tabulation of pest species in six major production regions of the world. Four case histories illustrate contemporary changes in the pest status of some species as production and control practices change. Summaries of advances in control tactics are followed by analyses of IPM programs in the United States, Brazil, and the People's Republic of China; these programs illustrate diverse strategies adapted to the agroecological conditions of the crop and the socioeconomic characteristics of the producers. We conclude with an assessment of current research directions following new trends in soybean production and advances in the science and technology of crop protection.

ECOLOGY AND ECONOMIC IMPACT OF SOYBEAN-ASSOCIATED ARTHROPODS

Arthropod communities in areas planted to soybean *(Glycine max)* undergo constant dynamic change as a consequence of changing crop management practices and the adaptive patterns of native colonizers (88). Worldwide, the area planted to soybean, currently about 52 million ha, continues to expand. The presumptive center of origin of the cultivated soybean is located in northeastern China between parallels 40° and 50° N. About 83% of the crop is now planted outside this area. The cultivated soybean is typically a short-season, temperate-zone plant; 77% of the soybean hectareage in China (156) and 81% in the United States (6) is found north of latitude 35° N, i.e. in the temperate zone. Much of the recent expansion, however, has been to lower latitudes north and south of the equator. Virtually all of the soybean production in South America is north of latitude 35° S and falls in subtropical zones; the more recent expansion in Brazil has been in tropical areas of the northeast.

Although soybean has adapted well to these lower latitudes and long-season conditions, crops grown under these conditions are much more vulnerable to insect pests. In 1976, for example, southern regions of the United States used 89% of the insecticides applied nationwide on soybean, but these regions contained only 37% of the total soybean area. The subtropical fauna has a rich reservoir of potential soybean colonizers, and this fauna adapts rapidly to the introduced crop (88). As a rule, however, new regions of cultivation lack key pests, and none of the best-adapted soybean specialists of East Asia have immigrated into new production regions.

Ecological Studies

The extensive literature accumulated in the last decade on the major species, species complexes, and guilds has concentrated on life histories, population dynamics, phenology, dispersal, and herbivore/plant and predator/prey interactions as bases for the development of pest management programs. Objects of the most intensive studies are the major worldwide pests of soybean, *Heliothis* spp. (159, 191–193) and *Nezara viridula* and other pentatomids (50, 146, 148, 198). Some comprehensive ecological studies have been conducted on species of regional importance, e.g. the velvetbean caterpillar, *Anticarsia gemmatalis,* in the southern United States (57); the green cloverworm, *Plathypena scabra,* in Iowa (18, 151, 152); the bean leaf beetle, *Cerotoma trifurcata,* in Illinois, Louisiana, and North Carolina (98, 113–115); and lepidopterous pod borers in China and Japan (64, 82, 157). Although these autoecological studies have supported the development of preliminary computer simulation models (166), they cannot presently be reliably used to forecast the time and size of population peaks because of gaps in the ecological database. Areas in need of research are short-range dispersal, overwintering behavior, colonization patterns, and age-specific mortality, including inter- and intraspecific competition.

Although research on soybean-associated arthropods continues to emphasize the ecology of individual species, community-level studies have been conducted. Fairly comprehensive faunal lists are available for many areas including Argentina (162, 200), Australia (33a, 175), Brazil (108, 116, 147), Canada (219), Central America (169), Colombia (47), Egypt (170), India (9, 10), Indonesia (69, 184), Japan (54, 79, 80, 81, 199), Kenya (77), Korea (25, 99), Malaysia (223), New Zealand (19a), China (96, 109), Taiwan (195, 217), Thailand (180), Turkey (201), and far eastern territories of the Soviet Union (100, 101). Surveys conducted in the United States prior to 1975 were cited in our previous review (209), and to those references two should be added (31, 220). The accuracy of identifications in some reports is questionable, but faunal lists are valuable in the analysis of community structure and development.

In addition to these general surveys there are reports on specific components of the faunal complex, e.g. the noncolonizing transient aphid fauna in Illinois soybean fields (46), the Plusiinae of Louisiana (3), the thrips on soybean in the midwestern United States (70), and the soil arthropods of Iowa soybean agroecosystems (111). Many surveys of natural enemies of specific pests have been conducted, and Pitre (154) provided references for surveys through 1982. These surveys have advanced our understanding of community-level processes in single fields (119, 155) and in wide areas such as the United States (88) and the Philippines (23).

Nonsaturation of Soybean Feeding Niches

Comparative analyses of faunal surveys suggest that some feeding niches remain either unoccupied or ineffectively occupied in soybean outside East Asia (China, Korea, Japan), and wide areas remain open for new colonizers. Table 1 compares major niche occupation by phytophagous species of some economic importance. North America, which accounts for about half of the total area planted to soybean worldwide, has no important upper stem and stalk borers, and pods and green seeds are attacked only by generalist feeders such as stink bugs and *Heliothis zea*. In East Asia a complex of pod borers, including the highly specialized *Leguminivora glycinivorella,* represents some of the most serious pests of soybean. Entire guilds are missing from the Western Hemisphere; foremost among them are such soybean-colonizing aphids as the soybean specialist *Aphis glycines,* a common pest in East Asia (68).

Native legume and nonlegume feeders are adapting to the expanding crop, and new colonizers are intermittently recorded. For example, the soybean nodule fly, *Rivellia quadrifasciata,* was observed in Louisiana in 1975 (32) and has since been recorded in several other states (82a); the hispid leafminers *Odontota horni* (93) and *Sumitrosis rosea* (17, 167) are rather recent colonizers of soybean in the midwestern United States; and the stalk borer *Dectes texanus texanus,* previously known only as a composite weed borer, now occurs on soybean throughout the central and eastern United States (20). The curculionid girdler and stem borer *Sternechus subsignatus* first recorded on soybean in southern Brazil in 1974, is causing increasing damage and is expanding its range northward (140). Yukawa et al (222) suggested that the soybean pod gall midge, *Asphondylia* sp., in Japan may have originated on a non-Leguminosae wild host. These observations suggest that even in East Asia host shifts continue to contribute to the expansion of the soybean fauna.

In summary, niche occupancy is a dynamic process. The nonsaturation of feeding niches suggests that new phytophagous species may adapt to the crop. The most serious threat to soybean in North, Central, and South America remains the potential immigration of the coadapted soybean oli-

Table 1 Principal soybean-associated arthropods in major agroecological regions

Plant parts or stage of growth	North America	Pest impact rating[a]	Central & South America	Pest impact rating	China, Korea, Japan	Pest impact rating
Pods and seeds on plant	Heliothis zea	1	—		Heliothis viriplaca	3
	—		Etiella zinckenella	3	Etiella zinckenella	1
	—		Maruca testulalis	2	Leguminivora glycinivorella	1
	Nezara viridula	1	Nezara viridula	1	Nezara viridula	1
	—		Piezodorus guildinii	1	Nezara antennata	
	Euschistus servus	3	Euschistus heros	3	Piezodorus hybneri	1
	Acrosternum hilare	2	Acrosternum sp.	?	Dolycoris baccarum	2
	—		—		Riptortus clavatus	1
	Cerotoma trifurcata (A)[b]	2	?		—	
	—		—		Asphondylia sp.	1
Blossoms	—		Laspeyresia fabivora	?	—	
	Frankliniella tritici	4	—		—	
	—		—		—	
	—		—		—	
Stalks and upper stems	—		Epinotia aporema	2	Matsumuraeses phaseoli	1
	—		Hedylepta indicata	3	Hedylepta indicata	2
	—		Maruca testulalis	?	Sylepta ruralis	?
	—		Laspeyresia fabivora	?	—	
	—		—		Melanagromyza sojae	2
	—		—		Melanagromyza shibatzuji	3
	—		Sternechus subsignatus	?	—	
Leaf blades	Heliothis zea	1	—		Heliothis viriplaca	3
	Heliothis virescens	?	—		—	
	Pseudoplusia includens	1	Pseudoplusia includens	2	Plusia agnata	3
	Trichoplusia ni	?	—		—	
	Anticarsia gemmatalis	1	Anticarsia gemmatalis	1	—	
	Spodoptera exigua	3	Spodoptera latifascia	2	Spodoptera litura	3
	Spodoptera eridania	3	Spodoptera eridania	2	—	
	Plathypena scabra	3	—		—	
	Spilosoma virginica	4	—		—	
	Vanessa cardui	4	Urbanus proteus	4	Ascotis selenaria	3
	Cerotoma trifurcata (A)	2	Cerotoma arcuata (A)	3	Paraluperodes nigrobilineatus	3
	—		Cerotoma ruficornis (A)	3	—	
	Colaspis brunnea (A)	4	Maecolaspis aeruginosus (A)	4	—	
	Diabrotica balteata (A)	3	Diabrotica speciosa	3	—	
	Epicauta vittata	4	Epicauta atomaria	4	Epicauta gorhami	3
	Epilachna varivestis	2	—		—	
	Popilia japonica	4	—		Popilia japonica	3
	—		—		Anomala rufocuprea (A)	2
	Melanoplus spp.	3	—		Attractomorpha bedeli	
	Sericothrips variabilis	4	Caliothrips spp.	?	—	
	—		—		Aphis glycines	1
	—		—		Aphis craccivora	2
	—		—		Aulacorthum solani	2
	—		Bemisia tabaci	3	—	
	—		—		Chauliops fallax	3
	Tetranychus urticae	3	—		—	

Plant parts or stage of growth	India	Pest impact rating	Australia & New Zealand	Pest impact rating	Africa	Pest impact rating
Pods and seeds on plant	*Heliothis armigera*	3	*Heliothis armigera*	1	*Heliothis armigera*	?
	—		*Heliothis puntigera*	1	*Cydia ptychora*	?
	—		*Etiella hebrii*	?	*Maruca testulalis*	?
	Nezara viridula	2	*Nezara viridula*	1	*Nezara viridula*	1
	—		*Piezodorus hybneri*	?	*Piezodorus punctiventris*	2
	—		—		*Aspavia armigera*	1
	—		*Riptortus serripes*	?	*Riptortus dentipes*	1
	—		—		—	
	—		*Melanagromyza pseudograta*	?	—	
Blossoms	—		—		—	
	—		—		*Taeniothrips sjostedti*	?
	—		—		*Caliothrips impurus*	?
	—		—		*Sericothrips occipitalis*	?
Stalks and upper stems	—		—		—	
	Hedylepta indicata	3	—		*Hedylepta indicata*	?
	—		—		—	
	Melanagromyza sojae	1	*Ophiomyia phaseoli*	?	*Ophiomyia phaseoli*	?
	—		—		—	
	Obereopsis brevis	2	*Zygrita diva*	?	—	
Leaf blades	*Heliothis armigera*	3	*Heliothis armigera*	1	—	
			Heliothis punctigera	1	—	
	Plusia orichalcea	3	*Plusia orichalcea*	?	*Plusia orichalcea*	?
	—		*Chrysodeixis eriosoma*	?	—	
	—		—		—	
	Spodoptera litura	3	*Spodoptera litura*	?	*Spodoptera littoralis*	?
	Spodoptera exigua	3	—		—	
	Spilosoma obliqua	1	*Utetheisa pulchelloides*	?	*Spilosoma aurantica*	?
	—		—		—	
	Monolepta sp.	4	*Monolepta australis*	?	*Gonocephalus simplex*	?
	—		*Aulacophora abdominalis*	?	*Aulacophora africana*	
	—		—		—	
	Epicauta mannerheim	4	—		*Lagria villosa*	?
	—		—		—	
	Popilia gemma	4	—		—	
	—		—		—	
	Attractomorpha crenulata	4	*Attractomorpha similis*	?	*Zonocerus variegatus*	?
	—		—		—	
	—		—		*Aphis craccivora*	?
	—		—		*Aphis gossypii*	?
	Bemisia tabaci	1	—		*Bemisia tabaci*	?
	Chauliops fallax	2	—		—	
	—		—		—	

Table 1 *(Continued)*

Plant parts or stage of growth	North America	Pest impact rating[a]	Central & South America	Pest impact rating	China, Korea, Japan	Pest impact rating
Lower stems	*Elasmopalpus lig-nosellus*	2	*Elasmopalpus lignosellus*	2	—	
	Spissistilus festinus	2	—		—	
	Dectes texanus	4	—		—	
Roots and nodules	*Cerotoma trifurcata* (L)	?	*Cerotoma arcuata* (L)	?	—	
	Rivellia quadrifasciata	?	—		*Melanagromyza dolicho-stigma*	2
Seedlings and seed in soil	*Delia platura*	3	—		*Delia platura*	2
	Agrotis ypsilon	3	—		—	

[a]1 = major, frequent pest, 2 = intermediate pest, 3 = minor, occasional pest, 4 = incidental pest. Ratings are based on published estimates (see 78, 87) and relative volume of literature.
[b]A = adults, L = larvae

gophages of East Asia. In addition, the relative pest status of the established soybean arthropod fauna can change. Study of such changes provides insight into the evolution of pest/crop interactions.

Changes in Pest Status

Changes in pest status can result from one or a combination of the following factors: (*a*) changes in cultural practices (cropping patterns, rotations, planting dates, row spacing, variety maturation date, land preparation patterns, cultivation systems); (*b*) changes in efficiency of natural or manipulated entomopathogens, predators, and parasitoids; (*c*) intentional or unintentional effects of pesticide applications; (*d*) changes in the assessment of the economic impact of pests with the consequent upward or downward adjustment of economic injury levels; and (*e*) changes in host association and consequent improved fitness on soybean. We review four examples of changes in pest status that appear to be related to one or more of these factors.

THE CORN EARWORM IN THE ATLANTIC COASTAL PLAINS OF THE UNITED STATES   The corn earworm, *Heliothis zea,* has had a greater economic impact on soybean production in the United States than any other insect pest (84, 87). Major damage is from pod feeding, but damage may also occur on leaves, blooms, and growing tips of plants.

In the early 1970s most defoliation by *H. zea* occurred in August on

Table 1   (*Continued*)

Plant parts or stage of growth	India	Pest impact rating[a]	Australia & New Zealand	Pest impact rating	Africa	Pest impact rating
Lower stems	—		—		—	
	Melanagromyza sojae	2	—		—	
	Obereopsis brevis	2	—		—	
Roots and nodules	—		—		—	
			Rivellia nr. *basilaris*	?	—	
Seedlings and seed in soil	*Delia platura*	2	—		*Barombiella humeralis*	?
	Melanagromyza sojae	2	—		—	
	Agrotis ypsilon	2	—		—	

late-maturing varieties, prior to seed enlargement (203). Soybean bloomed in early August on the coastal plains of North and South Carolina, and *H. zea* numbers increased during the pod-set stage in late August and early September (22, 191). Extensive studies in North Carolina (15, 186, 191) suggested, therefore, that for management of *H. zea* early-planted, early-maturing soybean should be grown.

Presently, however, the only crops at high risk from *H. zea* in South Carolina are early-planted, early-maturing group V varieties; this situation is the opposite of that described for the coastal plains region a decade earlier (208, 210). Peak *H. zea* moth flights occur in South Carolina in mid-July and mid-August, but larval populations capable of inflicting damage of economic significance seldom occur after the later flight.

Changes in maturity of host crops combined with increases in mortality due to entomopathogens may have contributed to this shift (208). During the early 1970s, growers began changing from the sole production of late-maturing, late-blooming soybean (groups VII–VIII) to production of a mix that included substantial hectareage of early-blooming varieties (groups V and VI), which are attractive to ovipositing moths of the July flight. At the same time corn growers shifted from long-season to early-maturing corn hybrids that are not attractive to ovipositing moths in mid-July. Thus, various soybean fields are now attractive to moths of both the July and August flights, and corn is attractive to neither. These relationships explain the increase in *H. zea* infestations in July. However, the decrease of infestations in August seems to be related to increased mortality caused by entomopathogens.

Entomopathogens are ineffective in reducing economic losses in late July and early August following the first moth flight into soybean. The only disease that appears in early August in epizootic proportions is the *Heliothis* nuclear polyhedrosis virus (HNPV), which was first detected in 1976 and is now an annually recurring mortality factor. While the epizootic is generally too late to prevent economic loss caused by first-generation larvae, HNPV can cause significant mortality in the second generation in soybean, particularly in combination with the fungi *Nomuraea rileyi* and *Erynia* sp. *N. rileyi* is almost cosmopolitan and its epizootics have been recorded for decades. However, it is a significant mortality factor of *H. zea* in South Carolina only after mid-August. An *Erynia* sp. that kills first and second instars of *H. zea* was detected in late August of 1981 (J. W. Chapin & G. R. Carner, unpublished). This disease often completely eliminates small larvae after heavy moth flights and high egg deposition. Combined larval mortality from these diseases often reaches 95–100% after mid-August; thus the chances for widespread economic losses after the second moth flight are greatly reduced.

Interestingly, in 1977, early-planted, early-maturing soybean in North Carolina sustained greater damage than later-planted, later-maturing soybean (15, 191). This damage was attributed to hot, dry weather during which corn matured rapidly and *H. zea* developed about two weeks early. However, the possibility that the population dynamics of *H. zea* in North Carolina is undergoing a shift similar to that in South Carolina also deserves consideration.

THREECORNERED ALFALFA HOPPER IN THE UNITED STATES   Until recently, feeding damage by the threecornered alfalfa hopper, *Spissistilus festinus*, was considered unimportant in soybean production; although girdling of stems of young plants by adults and nymphs was known to sometimes result in lodging of plants later in the season, compensation for lodging was believed to be almost complete, with little or no yield loss (209). The economic consequences of upper main stem and leaf petiole girdling were unknown (131), but main stem girdling was generally considered of little economic significance (132).

Recent research in Louisiana, however, has changed our perception of the economic consequences of *S. festinus* feeding (58, 103, 125, 126, 185). Feeding on petioles and racemes during the pod-setting and pod-filling stages may cause extensive yield losses (125). The girdled regions are characterized by necrotic epidermal tissue and disorganized vascular bundles, which function as nutrient sinks that enhance the nutritional quality of sap for subsequent feeding. Plants with girdled stems and petioles exhibit restricted nutrient flow, reduced leaf area, decreased nodule growth, and reductions in nodule number and nitrogen fixation (58). Experiments in which insecticides were used to regulate populations of *S. festinus* showed that early-season damage caused no yield loss; late-season populations, however, had a significant negative correlation with yield (185). As a consequence of this study, economic thresholds in Louisiana were established at one adult *S. festinus* per sweep after pod set (212). According to extension recommendations in Arkansas and Louisiana in the early 1970s, this species was too unimportant to warrant the establishment of a threshold; however, earlier investigators may have considered the lodging associated with early stem girdling to be the sole injury capable of causing economic loss, and may therefore have neglected to evaluate later upper stem girdling.

WHITEFLIES IN BRAZIL   Between 1962 and 1972 soybean hectareage in Brazil increased tenfold. In late 1972 and early 1973, explosive outbreaks of sweetpotato whiteflies, *Bemisia tabaci*, were recorded in northern Paraná and southwestern São Paulo (30). These unprecedented outbreaks on soybean were ascribed to the presence of the abundant, suitable new host, to late plantings of soybean in the regions affected, and to unusually favorable climatic conditions for whitefly development (30). Whiteflies were regarded at that time as a serious threat to soybean, common bean, and cotton (29); the crops were subject not only to direct damage caused by large populations, but also to various phytopathogenic viruses of which *B. tabaci* was an efficient vector.

Whether climatic conditions favorable to the whiteflies did not persist or whether there was a tendency in later years for earlier planting of soybean, *B. tabaci* populations dwindled in those regions in subsequent years. Current guides for identification and control of soybean pests in Brazil (e.g. 37) do not list *B. tabaci* even as an occasional pest, although endemic populations still occur throughout the soybean-growing regions. Monitoring of populations on various hosts during periods of low incidence may provide valuable baseline information to help explain future outbreaks should they occur.

PIEZODORUS GUILDINII IN BRAZIL   Worldwide, pentatomid stink bugs are among the most serious soybean pests; 54 species are currently associated with soybean in North and South America (148). *Nezara viridula*, a cosmopolitan species considered a major pest in most warmer soybean growing regions, was the dominant stink bug on soybean in Argentina (215) and in southern Brazil during the period of great expansion of the crop (86).

Since the late 1970s *Piezodorus guildinii* has begun to supplant *N. viridula* in some areas of Brazil (146–148). A smaller and more mobile pentatomid, *P. guildinii* seems to be better adapted to warmer climates than *N. viridula*, and also seems to be more capable of colonizing soybean in Brazil early in the season (146). The reasons for this apparent replacement are not clear. In addition to mobility and early colonization of the crop, such factors as lower rates of parasitization and differential susceptibility to the most common insecticides may be involved.

ADVANCES IN CONTROL TACTICS

The literature on control methods for soybean insect pests was never totally dominated by reports on chemical controls. Research on soybean entomology began to flourish during the era of IPM, and researchers were inclined toward a balanced approach to pest control. The following section summarizes recent trends in the four major control tactics used in soybean IPM.

Chemical Control

Conventional chemical insecticides have generally provided effective and economical suppression of soybean insect pests that exceeded economic injury levels (211). Strategies involving chemical insecticides have changed little since our 1976 review (209). At that time, application of certain insecticides at minimum effective rates when economic thresholds were reached controlled most pest species and allowed survival of beneficial species. Low rates of selective insecticides are now commonly recommended for use in soybean IPM (e.g. 40), and the adoption of these recommendations should be credited for the preservation of natural enemies in soybean ecosystems in the United States and most of South America (207).

New insecticides have been incorporated in established management strategies, sometimes with unexpected side benefits. The pyrethroid permethrin in a single midseason application for control of *H. zea* on soybean in South Carolina gave extended late-season control of the velvetbean caterpillar (J. W. Chapin & M. J. Sullivan, unpublished data). As this control method becomes more generally adopted, it will be essential to monitor development of resistance in *H. zea* populations owing to the extensive use of pyrethroids on both cotton and soybean (211).

Biological Control

Since the mid-1970s considerable progress has been made in studies of endemic and exotic natural enemies of soybean pests. The literature on major native and introduced natural enemies has recently been reviewed (154, 174). Predators are major causes of pest mortality in soybean ecosystems (139, 207). Researchers have recently attempted to quantify predation rate as an essential mortality component for population dynamics models (33, 158) and to evaluate predator/prey relationships (e.g. 43, 181).

Two species of exotic parasitoids have been imported and successfully used in biocontrol programs (75, 174). *Pediobius foveolatus* was imported from

India, and through annual releases it controls the Mexican bean beetle on the Atlantic coastal plain (34, 188). *Euplectrus puttleri,* imported from South America, has become established in southern Florida, where it parasitizes *Anticarsia gemmatalis* (75). Despite research on various aspects of parasitoid/ host interactions (e.g. 8, 26, 28, 120, 143, 150) and on diversity and incidence of indigenous parasitoids (48, 91, 187), the *Pediobius*/Mexican bean beetle program remains the sole example of intentional incorporation of biocontrol into a soybean IPM system by inoculative releases of a parasitoid (75).

Pathogens have been successfully incorporated into IPM systems for soybean through intentional manipulations and use of industrially produced pathogens (21, 129, 210). *N. rileyi* remains the most important fungal disease of soybean lepidopterous pests, and its epidemiology and pathogen/host interactions have been intensively studied (14, 36, 197). The natural spread and the successful manipulative use of various nuclear polyhedrosis viruses have raised concern about the possible interference of the viruses with *N. rileyi* (129, 130, 161). *Heliothis* NPV has been produced commercially, but it recurs annually as a natural mortality factor in certain regions (21). Its persistence in fields one to three years after application suggests that regional introductions may be possible by spot treatments (161).

Host Plant Resistance

Perhaps the most vigorous effort in the development of a noninsecticidal method of soybean pest control has been in the area of host plant resistance (HPR). Most institutional soybean entomology programs throughout the world have HPR components, and the private sector in the United States is becoming increasingly involved in breeding to develop resistant varieties. Despite these efforts, no variety has been released for which adequate levels of resistance to defoliators are claimed. "Shore," a variety released in Maryland (183), is almost an exception, but it has low levels of resistance to the Mexican bean beetle and its yields are consistently lower than those of commercial varieties. Several Chinese varieties have been reported resistant to the soybean pod borer, *L. glycinivorella,* but not yet to satisfactory levels (45, 63, 96).

Sources of resistance against at least 13 insect pests or pest complexes are being sought in world germplasm collections in some 12 countries. The most active programs have been searches for sources of resistance against stink bugs in Brazil (123, 149, 164) and in the United States (41, 73, 74). Through screening in the United States PI 171444 has been identified as a possible source of resistance to stink bugs. However, the polyphagous nature of stink bugs and their feeding behavior suggest that levels of resistance should be evaluated with great caution. Successes obtained with PIs 171451, 227687, and 229358, identified as resistant to defoliators, have been reviewed previously (194, 209). Breeding programs based on those three PIs have released germplasm lines in the United States for maturity groups III–IX (see 52, 194). These lines have moderate to high levels of resistance and are available for further improvement by private and public breeders.

Parallel to the activities of the screening and breeding programs has been an effort toward identification of mechanisms of resistance against these various pests. Physical factors, mainly pubescence, have long been known to provide resistance to small arthropods (205), particularly the potato leafhopper, *Empoasca fabae*. The barrier posed by pubescence affects both feeding behavior and the ability of the potato leafhopper to attach to the leaf surface (104). However, pubescence seems to facilitate oviposition by *H. zea* (144) and pod borers (225). Structural characteristics of the stems of *Glycine soja* are correlated with resistance to bean flies (24).

Recent reviews (89, 182) have dealt with the chemical bases of soybean resistance to leaf-feeding species. One novel mechanism of resistance with potential practical applications is the antiherbivory effect of phytoalexins, particularly glyceollins, which accumulate in stressed soybean tissue (51). These phytoalexins are known to be associated in soybean with resistance against pathogens. A better understanding of the mechanisms and the inheritance of resistance will accelerate the production of resistant varieties.

Cultural Control

Changes in cropping practices directly affect the dynamics of soybean/arthropod associations. Switches from multiple to single cropping, wide to narrow row spacing, and conventional to reduced tillage, as well as variations in planting dates and changes in pesticide type or patterns of pesticide usage in soybean or adjacent crops, alter the ecological conditions or change the quality of the plant as a host (56). These alterations contribute to fluctuations in pest status and the relative dominance of species in the community. Although cultural practices seldom are intentionally directed to control insect pests, they can be considered cultural control when they have an impact on the behavior and population dynamics of both herbivores and natural enemies.

Soybean producers in the United States are shifting from wide to narrower row spacing, and adoption of reduced tillage is increasing. Narrower row spacing results in rapid canopy closure. The effect of canopy structure on populations of some phytophagous insects and their natural enemies has been investigated, but results are inconclusive. Although the population levels of some species increase with narrow rows, the levels of others decrease; no effect has been observed on the majority of species (19, 118). The effects of row spacing are obscured because of such other variables as planting date, stand density, cultivar maturity, and presence of weeds (5, 118). Variations in sampling procedures and methods used to assess arthropod population density make interpretation of results even more difficult.

Effects of tillage systems are often considered in conjunction with crop rotations because of potential influences of residues from the previous crop. The seedcorn maggot, *Delia platura,* seems to be more abundant in minimum-tillage fields with freshly decomposing, partially buried crop residues (49). Herbivory and leaf nitrogen content are higher in conventional than in reduced tillage systems, but no-tillage systems support a larger and more diverse arthropod community than conventionally tilled fields (61).

The effects of planting dates have been illustrated in the earlier discussion of *Heliothis* on the coastal plains of the southeastern United States (see 76). Late planting was reported to increase agromyzid bean fly infestations of soybean in Korea (102).

Cultural practices are usually dictated by agronomic and economic considerations. Unless insect pests are an obvious production-limiting factor, growers are unlikely to change cultural practices merely because of pest-control recommendations. One specific cultural control method targeted for soybean insect pests is trap cropping. In trap cropping as originally conceived, a border strip of soybean is intentionally planted one to two weeks earlier than the rest of the field; pest populations that aggregate on the trap crop are then eliminated with insecticides (138). The system is effective for species closely bound to crop phenology, and best results have been obtained with the Mexican bean beetle (168), the bean leaf beetle, and stink bugs (122). Use of traps of the same crop species or of a more attractive plant, cowpea, have greatly improved stink bug control and final soybean yield in Nigeria (72). In experiments in Brazil, earlier plantings of soybean strips on contour terraces, with properly timed insecticide applications to the traps, have resulted in substantial reduction of stink bug populations on the main crop (145).

Planting of trap crops is a potentially useful tactic in an IPM program. However, on a regional basis farmers tend to stagger plantings; thus the fields planted earliest become natural traps, attracting the early colonizers. Regional IPM systems, therefore, must take the total crop mosaic with its natural traps into consideration for appropriate modeling (193).

Interactions Among Control Tactics

Under normal field conditions, complexes of insects, nematodes, weeds, and plant pathogens co-occur in soybean ecosystems. Control methods directed at the target pest(s) within a group often impinge on other members of the same or another group (55, 207, 221). Although the multiple effects of different types of pesticides and the interactions among control tactics are difficult to investigate experimentally, information on such effects is essential for IPM (135, 136).

PESTICIDE INTERACTIONS   Insecticides may interact with herbicides to reduce plant growth and seed yield. The herbicide metribuzin, for example, is phytotoxic to soybean if applied with organophosphate insecticides or nematicides, but not if applied with the carbamates aldicarb and carbofuran (221). However, metribuzin may interact negatively with carbamates under certain environmental conditions, particularly unfavorable weather early in the season (106). Cold stress prior to seedling emergence reduces metribuzin uptake, and thus reduces phytotoxicity when metribuzin and phorate are used (16).

PESTICIDE/BIOCONTROL INTERACTIONS   If there were not endemic natural enemies of soybean insect pests, multiple applications of insecticides would probably be required for insect control. Insecticide applications are infrequent in many soybean ecosystems in the United States because careful consideration is given to the role of natural enemies in chemical control recommendations for the crop (207, 211). Resurgences of pest species may occur when insecticides are improperly used. Following application of methyl parathion + methomyl, populations of *P. includens, A. gemmatalis,* and *H. zea* increased three- to fivefold over those in untreated plots. The number of predators was substantially reduced in treated plots, as was mortality caused by *N. rileyi.* (173). *H. zea* populations increased almost eightfold after treatment with methyl parathion or nematicidal rates of aldicarb (128). In another study (105) carbamates at nematicidal rates applied at planting had no effect on midseason pest or predator populations; however, the narrow plot width, only 2–4 rows, might have facilitated reinvasions by predators. Antagonism between the fungicide benomyl and natural control agents has been shown to affect the survival of *Cotesia marginiventris,* a parasitoid of various lepidopterous larvae (196), and the development of an *Entomophthora* species epizootic in a *P. includens* population (110). A recent review (213) has provided additional examples of the effects of pesticides on natural enemies of soybean insect pests.

EFFECTS OF HERBICIDES ON ARTHROPODS   The effects of herbicides on herbivores have been tested because growth regulators and other plant-stressing chemicals are known to induce changes in plants, often making them less susceptible to herbivores (95). The effects detected with many herbicides on field populations of the Mexican bean beetle have been modest, but some herbicides enhanced fitness of the herbivore (2). Toxaphene had a detrimental effect when used as an herbicide to control sicklepod *(Cassia obtusifolia)* (65, 71, 216), but toxaphene is also used directly as an insecticide. Most currently used herbicides do not seem to affect herbivore populations directly, but they do so indirectly by removing weed hosts of the prey of many important natural enemies (4, 5, 172).

INTERACTIONS OF PLANT RESISTANCE WITH OTHER CONTROL TACTICS Plant resistance has been considered highly compatible with other methods of pest control and therefore ideally suited for IPM. At one time HPR was thought to potentiate the effect of insecticides on herbivores (78), but this was not confirmed by later work (S. G. Turnipseed & M. J. Sullivan, unpublished). Although HPR is generally compatible with biological control (142), antagonistic effects have recently been detected between resistant PI 171444, *N. viridula,* and the egg parasitoid *Telenomus chloropus* (143); and between PIs 171451 and 229358, soybean looper larvae, and the predator *Geocoris punctipes* (163).

A recent review of interactions of HPR and cultural control with biocontrol (55) revealed that combined effects are more often negative than positive. The complexity of these interactions is only beginning to be appreciated, however, and results are preliminary and sometimes contradictory.

IPM STRATEGIES AND THEIR IMPLEMENTATION

Table 1 lists the major soybean arthropod pests in six diverse production regions of the world. Most pests have been assigned a rating of 1–4 based on an assessment of their economic impact on the crop. Impact ratings for the United States and Japan are based on published assessments (78, 87); those for other regions are inferred from the number of research papers devoted to a given species in a given region.

Below we describe strategies developed for the major pests (those rated 1 and 2) of North and South America and the system developed in the People's Republic of China for control of the soybean pod borer. These examples illustrate integration of tactical procedures and decision rules adjusted to actual ecological, economic, and cultural conditions. Programs that relied on the best available technologies that could be understood, accepted, and incorporated by farmers into existing practices with the least amount of disruption or risk have been most successful.

IPM Systems in the United States

Pest pressure in the United States differs sharply from North to South. Soybean producers in the subtropical latitudes of the far south usually face high risk of outbreaks of potentially yield-limiting insect pests; in the temperate Midwest, however, outbreaks of insect pests are sporadic in time and space. Between the extremes of rare incidence in Minnesota and nearly inevitable incidence in northern Florida lies a wide range of insect pressures exerted by diverse sets of species (87). The general principles of IPM for soybean in the United States have been summarized in several recent papers (139, 152, 210). These systems are targeted toward the control of two major pest complexes: (*a*) lepidopterous and coleopterous defoliators and (*b*) hemipterous and lepidopterous pod feeders (see Table 1). The main features of these systems include: (*a*) systematic scouting during periods of risk to monitor crop growth, level of extant damage, insect development, population density, and natural enemy activity, especially the onset of pathogen epizootics (92, 154); (*b*) the use of action-decision rules based on economic injury levels (85, 153); and (*c*) the application of carefully selected insecticides at minimum rates that are effective against the target pest but have the least adverse effects on natural enemies (207, 211).

Variants of this basic scheme have been adapted to meet specific pest conditions in the various production subregions. In Illinois, where insect pest problems are sporadic, systematic scouting may be unnecessary in most years. Thus efforts are being directed toward improving the ability to forecast the risk of green cloverworm, bean leaf beetle, grasshopper, and spider mite outbreaks; these pests combined account for 93% of the total insecticide use in the state (90, 127). In Maryland, where the Mexican bean beetle is a dominant species, inoculative releases of the imported parasitoid *Pediobius foveolatus* on early-nurse crops of common bean planted adjacent to soybean fields produce consistently high parasitization rates (60–90%) on soybean throughout the season (75, 188). The trap-crop technique for control of bean leaf beetle and stink bugs developed in Louisiana (122, 138) has already been discussed. As these examples illustrate, the basic IPM scheme has been expanded by the addition of biological and cultural control components. As high-yielding varieties with resistance to defoliating pests are developed they will rapidly be incorporated into these IPM schemes to fill a present void (194).

Economic analyses of simulated or real IPM programs for soybean in various states consistently present a highly positive balance in IPM programs (44, 53, 160). The most important development in soybean IPM in the United States since our 1976 review is the rate at which growers have adopted key tactical components. In some states over 90% of the producers follow established economic thresholds to initiate insecticide applications. Growers are increasingly aware of the importance of conserving natural enemies, and they use recommended insecticides at minimum dosages (1).

IPM Systems in Brazil

Table 1 reveals the similarities between the major foliage and pod pests (rated 1) in North and South America. The major defoliators in Brazil and the southern United States are *A. gemmatalis* and *P. includens* (less important in Brazil), and the major pod feeders are stink bugs. *H. zea* feeds on both foliage and pods in the southern United States, but not in Brazil (206). Similarities are also found between the major complexes of natural enemies of defoliators (129). Although several other pests in Brazil required special treatment in certain areas (e.g. *Epinotia aporema, Maruca testulalis,* and *Elasmopalpus lignosellus*), similarities in the major pest problems of these two countries suggested that insect pest management systems used in the southern United States might be suitable for use in southern and south-central Brazil (204). The programs that followed provided one of the most spectacular success stories of IPM implementation for a major annual crop over a wide area.

In 1975, paired fields, 10–30 ha each, were selected on each of nine farms (three in Rio Grande do Sul and six in the state of Paraná). One field in each pair was managed according to an IPM scheme; the other was managed according to the farmer's established methods. The IPM scheme consisted of (*a*) weekly scouting of pest populations and incidence of *N. rileyi* on *A. gemmatalis* larvae; (*b*) an assessment of levels of defoliation and plant growth; (*c*) the establishment of economic thresholds based on defoliation level, pest density, and stage of plant growth; and (*d*) the immediate application of minimum effective rates of insecticides when populations and levels of injury exceeded the threshold.

In the first year of the program the number of sprays in the IPM fields was 78% lower than that in farmer-treated fields. No differential effect on yields was found (97). The results of this large-scale experiment demonstrated that the system employed in the southern United States could be adopted in Brazil with only minor modifications.

In 1976 the program was expanded in the state of Paraná with participation by representatives from local production cooperatives and field personnel from a major agrichemical company. Results in 1976 confirmed those obtained in 1975, and adoption of the program by the large producer cooperatives guaranteed its success. The program was officially adopted for the state of Paraná (214). To facilitate the training of farm personnel in scouting and decision-making procedures, close links were established with the Extension Service, a public organization parallel to but independent of the research establishment in Brazil. In 1977, the program was adopted by the National Soybean Research Center (CNPSoja) of the federally supported EMBRAPA. Programs were greatly expanded through intensive training sessions and demonstration plots throughout most soybean-producing states in Brazil (38).

As the benefits of the IPM program became evident, the following plan was proposed so these benefits could reach broader segments of the producer community (86):

1. A network of cooperating farmers and extension agronomists would be established in sites that represented the major agroecological areas of the state (Paraná at the initial stage).
2. Pest and damage levels would be monitored on a weekly basis at each site

and these data would be reported to a central coordinating station located at the CNPSoja headquarters at Londrina.

3. These data would be quickly compiled and mapped so that information could be transmitted through mass communication media (newspapers, radio, television) to the various regions of the state.

4. Warnings would be released to initiate scouting of fields in regions at risk when pest populations were reported to be approaching the economic threshold; spray decisions would be left to individual farmers.

Furthermore, it was recommended that farmers be made aware that since the monitoring network was based on a random sample of the region, localized outbreaks might well go undetected.

The vigor with which the program was organized and deployed by the CNPSoja entomologists resulted in extraordinary success. By 1980, the program was reaching 200,000 farmers over an area of about 2.5 million hectares (D. L. Gazzoni, personal communication). The program resulted in 60% average reduction of insecticide use and a savings of about $10.00/ha on chemicals alone (141). Recent improvements in the program include the refinement of thresholds for local conditions, the use of light traps as an early alert for *A. gemmatalis,* and the spraying of *A. gemmatalis* nuclear polyhedrosis virus (AgNPV) suspensions for control of the velvetbean caterpillar (38).

Exceptional control of *A. gemmatalis* has been obtained through the application of AgNPV, which is endemic to Brazil (129, 130). Freshly killed larvae are collected by farmers in virus-treated areas, frozen, and kept until the following year, when they are used to treat larger areas. One application of the virus (50 larval equivalents/ha) has proved comparable to two insecticide applications per cropping year. An estimated 11,000 ha of soybean were treated with AgNPV (130) in 1983–1984, and the area was expected to increase to 300,000 ha in 1984–1985. In 1984 (38), an estimated 30% of soybean farmers in Brazil had adopted the IPM techniques, and increasing numbers were adopting the AgNPV spray method. The success of the Brazilian IPM program rests on five major factors: (*a*) governmental support, (*b*) simplicity and ease of adoption with minimal risks, (*c*) adequacy with regard to socioeconomic conditions, (*d*) close coordination between research and extension, and (*e*) strong and dedicated leadership.

IPM Systems for Soybean Pod Borers in the People's Republic of China

The soybean pod borer, *L. glycinivorella,* is a soybean specialist feeder that is very closely adapted to the phenology of the crop in East Asia (64). It is considered to be the principal soybean pest in northern China, Korea, and some areas of Japan (82, 157). The following account is based on numerous meetings between the authors and Chinese entomologists in the provinces of Heilongjiang, Jilin, and Liaoning (96).

The soybean pod borer has one generation per year. Adult moths begin to emerge from pupae in the soil in mid-July and fly to soybean fields. They are not strong fliers and rarely disperse far from the areas where they emerged. Adult populations peak in mid-August in Jilin Province, when soybean is in the pod-fill stage. Females lay single eggs on pods. Larvae emerge in 8–10 days and penetrate the pods, where they feed preferentially on the germ, destroying the seed. Mature larvae drop to the soil and overwinter within cocoons about 3 cm below the soil surface. Larvae and pupae spend about 10 months in the soil and are very susceptible to low temperatures. The host range is limited to *Glycine max, Glycine soja,* and leguminous shrubs in the genera *Sophora* and *Pueraria.* Losses caused by this pest in untreated fields average 20%, but losses of 40–50% are not uncommon (64).

The complete IPM system for *L. glycinivorella* has the following components (S. G. Gu, personal communication):

1. Use of a resistant cultivar. Many varieties are cited as having good levels of resistance, based on antixenosis to oviposition in glabrous types and hardiness of pod walls to deter larval penetration (45, 63).
2. Rotation of soybean with spring wheat and plowing under the wheat stubble soon after harvest. The moths, which are weak fliers, do not disperse, and larvae and pupae in the soil are destroyed. This practice may increase larval mortality in soil by 25%.
3. Cultivation of nonrotated fields in mid-July to destroy larvae and pupae in the soil.
4. Release of *Trichogramma dendrolinae* for control of eggs.
5. Application of *Beauveria bassiana* to the soil surface to infect larvae as they drop from pods to overwinter (62).
6. Attachment of dichlorvos-impregnated sorghum stems to soybean plants to kill adult moths. (No information on the efficacy of this tactic is available to us.)
7. Chemical control (usually dimethoate) applied at an economic injury level of 1 adult/m of row (sampled by flushing).

Although we have no information on the rates of adoption of the entire program, at least some of these control methods were applied to most fields. This IPM system exemplifies a diversified approach to the control of an important pest and demonstrates why a thorough understanding of pest biology and behavior is necessary to develop effective control systems.

CONCLUDING REMARKS

IPM can be conceptualized at three ascending levels of integration: (*a*) integration of control tactics into operational strategies for single pest classes, (*b*) integration of the multiple impacts of pest complexes on the crop, and (*c*) systems integration, in which control strategies for pest complexes are meshed within the total crop system. The theoretical basis of the first is population, of the second is community, and of the third is ecosystems ecology. Current IPM programs for soybean, however, remain at the first level of integration, probably because of difficulties in the research needed to support the second and third levels. One of the main requirements for further progress in soybean IPM is the study of pest complexes (136), i.e. of the interactions among the various pest classes and the diverse tactics used in their control.

Systems at the first level of integration will certainly benefit from the process of acquiring the information needed to advance soybean IPM systems to higher levels. Current IPM programs rely on field scouting, but lack of time and the cost of scouting often discourage farmers from using IPM, and they resort instead to scheduled or insurance spraying of fields. The Brazilian solution of regional networks of sampling sites was an attempt to solve the problem; but if the goal is to develop wide-area monitoring systems, much basic research remains to be done.

The next generation of IPM systems will require reliable forecasting of pest outbreaks. Although computer simulation models have not yet been used to solve real field situations in soybean IPM, a solid base has been built from which predictive models applicable to practical field programs can be developed. The SOYGRO model (124), which was developed to simulate soybean growth and phenology, represents a significant advance toward the practical application of simulation models in soybean IPM. The systems approach is useful in evaluating interactions of tactics (189). Population models for several important soybean pests (165) and natural enemies (190) already exist. Advances in microcomputing and the increasing accessibility of microcomputers to extension services and to individual farmers may stimulate the rapid development of decision-support and expert systems. These will provide on-line diagnoses of pest problems, guides for decision making, real time weather, and market and pesticide-cost information. Dynamic economic injury levels supported by on-line data will become the backbone of decision making; however, models must be based on reliable experimental information (7, 153).

Advances in control tactics will continue to help improve IPM systems in soybean. Just as it was ten years ago, the most anticipated event in soybean IPM is the release of good agronomic, high-yielding varieties with multiple pest resistance. Advances in our understanding of the mechanisms of resistance may accelerate variety development. As we enter the era of genetic engineering, an entirely new range of possibilities—not only in plant resistance but also in the development of more virulent entomopathogens and cold-hardy, pesticide-resistant biological control agents—is within reach. But none of these technological advances will mean much if programs do not rest on solid ecological information. To develop IPM programs for soybean at higher levels of integration, researchers must continue to gather basic information in population, community, and ecosystems ecology.

ACKNOWLEDGMENTS

We are grateful to Cathy Eastman, Charles Helm, and Audrey Hodgins for technical and editorial review of the manuscript; to Jenny Kogan and Ellen Brewer, SIRIC, for assistance with the bibliography; and to Jo Ann Auble and Joan Traub, Illinois Natural History Survey, for word processing. This work was supported in part by grants from the American Soybean Association and the USDA/CRGO (to MK), and by Regional Projects S-157, S-107, and the Consortium for Integrated Pest Management through Texas A & M University. The opinions expressed in this review are those of the authors and not necessarily those of the supporting institutions or agencies.

Literature Cited

1. Adkisson, P. L., Frisbie, R. E., Thomas, J. G., McWhorter, G. M. 1986. Impact of IPM on production of major crops. See Ref. 35, In press
2. Agnello, A. M., Bradley, J. R. Jr., Van Duyn, J. W. 1986. Plant-mediated effects of postemergence herbicides on *Epilachna varivestis* (Coleoptera: Coccinellidae). *Environ. Entomol.* 15:216–20
3. Alford, A. R., Hammond, A. M. Jr. 1982. Plusiinae (Lepidoptera: Noctuidae) populations in Louisiana soybean ecosystems as determined with looplure-baited traps. *J. Econ. Entomol.* 75:647–50
4. Altieri, M. A., Todd, J. W. 1981. Some influences of vegetational diversity on insect communities of Georgia soybean fields. *Prot. Ecol.* 3:333–38
5. Altieri, M. A., Todd, J. W., Hauser, E. W., Patterson, M., Buchanan, G. A., Walker, R. H. 1981. Some effects of weed management and row spacing on insect abundance in soybean fields. *Prot. Ecol.* 3:339–43
6. American Soybean Association. 1982. *Soya Bluebook.* St. Louis, Mo: ASA. 212 pp.
7. Andow, D. A., Kiritani, K. 1983. The economic injury level and the control threshold. *Jpn. Pestic. Inf.* 43:3–9
8. Beach, R. M., Todd, J. W. 1985. Parasitoids and pathogens of the soybean looper, *Pseudoplusia includens* (Walker), in south Georgia soybean. *J. Entomol. Sci.* 20:318–23

9. Bhattacharya, A. K., Chaudhary, R. R. P., Ram, S. 1976. Survey of insect pests of soybean during various stages of crop growth. *Govind Ballabh Pant Univ. Agric. Technol. Annu. Rep. Res. 1974–75*, pp. 76–77. Pantnagar, India: Govind Ballabh Pant Univ. Agric. Technol. 164 pp.
10. Bhattacharya, A. K., Rathore, Y. S. 1977. *Survey and Study of the Bionomics of Major Soybean Insects and their Chemical Control. Res. Bull. 107.* Pantnagar, India: Govind Ballabh Pant Univ. Agric. Technol. 324 pp.
11. Bhattacharya, A. K., Rathore, Y. S. 1980. Soybean insect problems in India. See Ref. 27, pp. 291–301
12. Bimboni, H. G. 1982. Manejo integrado de plagas en soja. *INTA Bol. Divulg. Tec. 2,* S. Pedro, B. A., Argentina. 6 pp.
13. Borgo, A., Bisotto, V., Soares, J. B., Donaduzzi, J. N., Schweig, A., et al. 1978. Manejo de pragas da cultura da soja no Rio Grande do Sul. *Trigo Soja Bol. Tec.* 36:3–7
14. Boucias, D. G., Bradford, D. L., Barfield, C. S. 1984. Susceptibility of the velvetbean caterpillar and soybean looper (Lepidoptera: Noctuidae) to *Nomuraea rileyi:* Effects of pathotype, dosage, temperature, and host age. *J. Econ. Entomol.* 77:247–53
15. Bradley, J. R. Jr., Van Duyn, J. W. 1980. Insect pest management in North Carolina soybeans. See Ref. 27, pp. 343–54

16. Britton, D. W., Corbin, F. T., Schmitt, D. P., Bradley, J. R. Jr., Van Duyn, J. W., Coble, H. D. 1982. Influence of preemergence temperature on the interaction of metribuzin with organophosphate insecticide-nematicides in soybeans. *Proc. South. Weed Sci. Soc.* 35:367

17. Buntin, G. D., Pedigo, L. P. 1982. Foliage consumption and damage potential of *Odontota hornii* and *Baliosus nervosus* (Coleoptera: Chrysomelidae) on soybean. *J. Econ. Entomol.* 75:1034–37

18. Buntin, G. D., Pedigo, L. P. 1983. Seasonality of green cloverworm (Lepidoptera: Noctuidae) adults and an expanded hypothesis of population dynamics in Iowa. *Environ. Entomol.* 12:1551–58

19. Buschman, L. L., Pitre, H. N., Hodges, H. F. 1984. Soybean cultural practices: Effects on populations of geocorids, nabids, and other soybean arthropods. *Environ. Entomol.* 13:305–17

19a. Cameron, P. J., Allan, D. J., Walker, G. P. 1986. Insect pests and crop damage in soybeans. *NZ J. Agric. Res.* 29:93–100

20. Campbell, W. V. 1980. Sampling coleopterous stem borers in soybean. See Ref. 92, pp. 357–73

21. Carner, G. R. 1980. Sampling pathogens of soybean insect pests. See Ref. 92, pp. 559–74

22. Carner, G. R., Shepard, M., Turnipseed, S. G. 1974. Seasonal abundance of insect pests of soybeans. *J. Econ. Entomol.* 67:487–93

23. Cervancia, C. R., Rejesus, R. S. 1984. Soybean insect pest abundance and succession in different cropping combinations. *Philipp. Entomol.* 6:47–64

24. Chiang, H. S., Norris, D. M. 1983. Physiological and anatomical stem parameters of soybean resistance to agromyzid beanflies. *Entomol. Exp. Appl.* 33:203–12

25. Choi, K. M., Hwang, C. Y. 1975. Survey of crop loss and insect fauna in soybean cultivation systems. *Annu. Rep. Inst. Agric. Sci. Off. Rural Dev.*, pp. 271–75. Suweon, Korea: ORD (In Korean)

26. Cobb, C. H., Grant, J. F., Shepard, M. 1985. Effect of parasitism by *Microplitis demolitor* (Hymenoptera: Braconidae) on foliage consumption by *Heliothis zea* (Lepidoptera: Noctuidae) larvae. *Fla. Entomol.* 68:490–92

27. Corbin, F. T., ed. 1980. *World Soybean Research Conference II: Proceedings.* Boulder, Colo: Westview. 897 pp.

28. Correa Ferreira, B. S., Oliveira, E. B. de. 1982. Utilização de parasitas no controle de percevejos. In *Result. Pesqui. Soja 1981–1982*, pp. 286–89. Londrina, Brazil: EMBRAPA/CNPSoja. 377 pp.

29. Costa, A. S. 1975. Increase in the populational density of *Bemisia tabaci*, a threat of widespread virus infection of legume crops in Brazil. In *Tropical Diseases of Legumes*, ed. J. Bird, K. Maramorosch, pp. 27–49. New York: Academic. 171 pp.

30. Costa, A. S., Costa, C. L., Sauer, H. F. G. 1973. Surto de mosca branca em culturas do Paraná e São Paulo. *An. Soc. Entomol. Bras.* 2:20–30

31. Deitz, L. L., Van Duyn, J. W., Bradley, J. R. Jr., Rabb, R. L., Brooks, W. M., Stinner, R. E. 1976. A guide to the identification and biology of soybean arthropods in North Carolina. *NC Agric Exp. Stn. Tech. Bull.* 238:1–264

32. Eastman, C. E., Wuensche, A. L. 1977. A new insect damaging nodules of soybean: *Rivellia quadrifasciata* (Macquart). *J. Ga. Entomol. Soc.* 12:190–99

33. Elvin, M. K., Stimac, J. L., Whitcomb, W. H. 1983. Estimating rates of arthropod predation on velvetbean caterpillar larvae in soybeans. *Fla. Entomol.* 66:319–30

33a. Evans, M. L. 1985. Arthropod species in soybeans in southeast Queensland. *J. Aust. Entomol. Soc.* 24:169–77

34. Flanders, R. V. 1985. Biological control of the Mexican bean beetle: Potentials for and problems of inoculative releases of *Pediobius foveolatus*. See Ref. 176, pp. 685–94

35. Frisbie, R., Adkisson, P. L., eds. 1986. *Integrated Pest Management on Major Agricultural Systems. Proceedings.* College Station, Texas: Texas A & M Univ. In press

36. Fuxa, J. R. 1984. Dispersion and spread of the entomopathogenic fungus *Nomuraea rileyi* (Moniliales: Moniliaceae) in a soybean field. *Environ. Entomol.* 13:252–58

37. Gazzoni, D. L. 1979. *Soja: Como Reconhecer e Identificar Suas Pragas.* São Paulo: Bayer Brasil. 20 pp.

38. Gazzoni, D. L., Oliveira, E. B. de. 1984. Soybean insect pest management in Brazil: I. Research effort; II. Program implementation. See Ref. 117, pp. 312–25

39. Gazzoni, D. L., Oliveira, E. B. de, Corso, I. C., Correa Ferreira, B. S., Villas Boas, G. L., et al. 1981. Manejo de Pragas da Soja. *EMBRAPA/CNPSoja Circ. Tec. 5.* 44 pp.

40. Gazzoni, D. L., Oliveira, E. B. de, Corso, I. C., Villas Boas, G. L., Correa

Ferreira, B. S., et al. 1982. Recomendações de inseticidas para utilização no programa de manejo de pragas da soja safra 1982/83 nos estados de Mato Grosso do Sul, Paranà e São Paulo. *EMBRAPA/CNPSoja Comun. Tec. 17.* 8 pp.

41. Gilman, D. F., McPherson, R. M., Newsom, L. D., Herzog, D. C., Williams, C. 1982. Resistance in soybeans to the southern green stink bug. *Crop Sci.* 22:573–76

42. Goodman, R. M., ed. 1976. *Expanding the Use of Soybeans. Proc. Conf. for Asia and Oceania, Chiang Mai, Thailand.* Urbana-Champaign: Univ. Ill./ INTSOY. 260 pp.

43. Grant, J. F., Shepard, M. 1985. Techniques for evaluating predators for control of insect pests. *J. Agric. Entomol* 2:99–116

44. Greene, C. R., Rajotte, E. G., Norton, G. W., Kramer, R. A., McPherson, R. M. 1985. Revenue and risk analysis of soybean pest management options in Virginia. *J. Econ. Entomol.* 78:10–18

45. Guo, S., Feng, Z. 1984. Studies on resistance of soybean varieties to soybean pod borer. In *Proc. 2nd US-China Soybean Symp., Changchun, PRC, 1983,* pp. 345–47. Washington, DC: USDA Off. Int. Coop. Dev. 464 pp.

46. Halbert, S. E., Irwin, M. E., Goodman, R. M. 1981. Alate aphid (Homoptera: Aphididae) species and their relative importance as field vectors of soybean mosaic virus. *Ann. Appl. Biol.* 97:1–9

47. Hallman, G. 1983. Artropodos asociados con la soya en el Tolima. *Rev. Colomb. Entomol.* 9:55–59

48. Hammond, R. B. 1983. Parasites of the green cloverworm (Lepidoptera: Noctuidae) on soybeans in Ohio. *Environ. Entomol* 12:171–73

49. Hammond, R. B., Funderburk, J. E. 1983. Influence of tillage practices on soil-insect population dynamics in soybean. See Ref. 176, pp. 659–66

50. Harris, V. E., Todd, J. W., Mullinix, B. G. 1984. Color change as an indicator of adult diapause in the southern green stink bug, *Nezara viridula. J. Agric. Entomol.* 1:82–91

51. Hart, S. V., Kogan, M., Paxton, J. D. 1983. Effect of soybean phytoalexins on the herbivorous insects Mexican bean beetle and soybean looper. *J. Chem. Ecol.* 9:657–72

52. Hartwig, E. E., Turnipseed, S. G., Kilen, T. C. 1984. Registration of soybean germplasm line D75-10169. *Crop Sci.* 24:214–15

53. Hatcher, J. E., Wetzstein, M. E., Douce, G. K. 1984. An economic evaluation of integrated pest management for cotton, peanuts, and soybeans in Georgia. *Ga. Agric. Exp. Stn. Res. Bull.* 318:1–28

54. Hayashi, H., Nakazawa, K., Umeda, K. 1981. Insects and other pests of soybean in Hiroshima Prefecture. *Bull. Hiroshima Prefect. Agric. Exp. Stn.* 44:45–52 (In Japanese)

55. Herzog, D. C., Funderburk, J. E. 1985. Plant resistance and cultural practice interactions with biological control. In *Biological Control in Agricultural IPM Systems,* ed. M. A. Hoy, D. C. Herzog, pp. 67–88. New York: Academic. 589 pp.

56. Herzog, D. C., Funderburk, J. E. 1986. Ecological basis for habitat management and pest control. In *Ecological Theory and Integrated Pest Management Practice,* ed. M. Kogan, pp. 217–50. New York: Wiley.

57. Herzog, D. C., Todd, J. W. 1980. Sampling velvetbean caterpillar on soybean. See Ref. 92, pp. 107–40

58. Hicks, P. M., Mitchell, P. L., Dunigan, E. P., Newsom, L. D., Bollich, P. K. 1984. Effect of threecornered alfalfa hopper (Homoptera: Membracidae) feeding on translocation and nitrogen fixation in soybeans. *J. Econ. Entomol.* 77:1275–77

59. Hill, L. D., ed. 1976. *World Soybean Research. Proc. World Soybean Res. Conf.* Danville, Ill: Interstate Print. 1073 pp.

60. Hirashima, Y. 1977. Soybean entomology in Japan. *Bull. Inst. Trop. Agric. Kyushu Univ.* 2:104–15

61. House, G. J., Stinner, B. R. 1983. Arthropods in no-tillage soybean agroecosystems: Community composition and ecosystem interactions. *Environ. Manage.* 7:23–28

62. Hsu, C. F., Feng, C., Ma, S. L. 1959. A preliminary study on the utilization of the fungus *Beauveria bassiana* to control the soybean pod borer (*Grapholitha glycinivorella* Mats). *Acta Entomol. Sin.* 9:203–17 (In Chinese)

63. Hsu, C. F., Kuo, S. K., Han, Y. M., Chang, J., Li, M. C. 1965. A preliminary study on the resistance of soybean varieties to the soybean pod borer (*Leguminivora glycinivorella*). *Acta Phytophylacica Sin.* 4:111–18 (In Chinese)

64. Hsu, C. F., Kuo, S. K., Han, Y. M., Feng, C., Chang, Y., Lee, Y. C. 1965. A study on the soybean pod borer (*Leguminivora glycinivorella* (Mats) Obrazt-

sov). *Acta. Entomol. Sin.* 14:461–79 (In Chinese)

65. Huckaba, R. M., Bradley, J. R., Van Duyn, J. W. 1983. Effects of herbicidal applications of toxaphene on the soybean thrips, certain predators and corn earworm in soybeans. *J. Ga. Entomol. Soc.* 18:200–7
66. Huffaker, C., ed. 1980. *New Technology in Pest Control.* New York: Wiley. 500 pp.
67. Instituto Nacional de Tecnologia Agropecuaria. 1983. Como Manejar las Plagas de la Soja. San Pedro, Argentina: Secr. Agric. Ganad., INTA, Estac. Exp. Agric. 14 pp.
68. Irwin, M. E. 1980. Sampling aphids in soybean fields. See Ref. 92, pp. 239–59
69. Irwin, M. E. 1980. Soybean insect pests in Indonesia. *INTSOY Staff Rep. 80–29,* Univ. Ill., Urbana-Champaign. 7 pp.
70. Irwin, M. E., Yeargan, K. V., Marston, N. L. 1979. Spatial and seasonal patterns of phytophagous thrips in soybean fields with comments on sampling techniques. *Environ. Entomol.* 8:131–40
71. Isenhour, D. J., Todd, J. W., Hauser, E. W. 1985. The impact of toxaphene applied as a postemergence herbicide for control of sicklepod, *Cassia obtusifolia* L., on arthropods associated with soybean. *Crop Prot.* 4:434–45
72. Jackai, L. E. N. 1984. Using trap plants in the control of insect pests of tropical legumes. See Ref. 117, pp. 101–12
73. Jones, W. A. Jr., Sullivan, M. J. 1978. Susceptibility of certain soybean cultivars to damage by stink bugs. *J. Econ. Entomol.* 71:534–36
74. Jones, W. A. Jr., Sullivan, M. J. 1979. Soybean resistance to the southern green stink bug *Nezara viridula. J. Econ. Entomol.* 72:628–32
75. Jones, W. A. Jr., Young, S. Y., Shepard, M., Whitcomb, W. H. 1983. Use of imported natural enemies against insect pests of soybean. See Ref. 154, pp. 63–77
76. Joshi, J. M. 1980. Effect of planting dates and soybean cultivars on pod damage by corn earworm. *Crop Sci.* 20:59–63
77. Kamau, A. W. 1979. A survey of insects associated with soybean, *Glycine max* L. in Kenya. *Kenya Entomol. News* 9:9–11
78. Kea, W. C., Turnipseed, S. G., Carner, G. R. 1978. Influence of resistant soybeans on the susceptibility of lepidopterous pests to insecticides. *J. Econ. Entomol.* 71:58–60
79. Kobayashi, T. 1977. Insect pests of soybeans in Japan. *Tech. Bull. ASPAC Food Fert. Technol. Cent.* 36:1–24
80. Kobayashi, T. 1981. Insect pests of soybeans in Japan. *Tohoku Natl. Agric. Exp. Stn. Misc. Publ.* 2:1–39
81. Kobayashi, T., Oku, T. 1976. Studies on the distribution and abundance of the invertebrate soybean pests in Tohoku district with special reference to the insect pests infesting the seeds. *Tohoku Natl. Agric. Exp. Stn. Bull.* 52:49–106
82. Kobayashi, T., Oku, T. 1980. Sampling lepidopterous pod borers on soybean. See Ref. 92, pp. 422–37
82a. Koethe, R. W., Van Duyn, J. W. 1984. Aspects of larva/host relations of the soybean nodule fly, *Rivellia quadrifasciata* (Diptera: Platystomatidae). *Environ. Entomol.* 13:945–47
83. Kogan, J., Kogan, M., Brewer, E., Helm, C. G. 1987. *The Literature of Arthropods Associated with Soybean. VI. A World Bibliography.* Urbana-Champaign, Ill: Univ. Ill. Soybean Insect Res. Inf. Cent. (SIRIC) In press
84. Kogan, J., Sell, D. K., Stinner, R. E., Bradley, J. R. Jr., Kogan, M. 1978. *The Literature of Arthropods Associated with Soybean. V. A Bibliography of* Heliothis zea *(Boddie), and* H. virescens *(F.) (Lepidoptera: Noctuidae).* Urbana, Ill: INTSOY. 242 pp.
85. Kogan, M. 1976. Evaluation of economic injury levels for soybean insect pests. See Ref. 59, pp. 515–33
86. Kogan, M. 1977. *Soybean Entomology Program. Consultant's Report.* Londrina, Paraná, Brazil: EMBRAPA, CNPSoja. 10 pp.
87. Kogan, M. 1980. Insect problems of soybeans in the United States. See Ref. 27, pp. 303–25
88. Kogan, M. 1981. Dynamics of insect adaptations to soybean: Impact of integrated pest management. *Environ. Entomol.* 10:363–71
89. Kogan, M. 1986. Natural chemicals in plant resistance to insects. *Iowa State J. Res.* 60:501–27
90. Kogan, M., Helm, C. G. 1984. Soybean insects in Illinois: Lessons of the 1983 season. *10th Annu. Ill. Crop Prot. Workshop,* pp. 110–19. Urbana, Ill: Coop. Ext. Serv. Univ. Illinois
91. Kogan, M., Helm, C., Kogan, J., Brewer, E. 1987. Distribution and economic importance of *Heliothis* spp. in North and South America including a listing and assessment of importance of their natural enemies and host plants. In *Biological Control of* Heliothis: *Increasing the Effectiveness of Natural Ene-*

mies. New Delhi, India: Conf. Biol. Control *Heliothis.* In press

92. Kogan, M., Herzog, D. C., eds. 1980. *Sampling Methods in Soybean Entomology.* New York: Springer-Verlag. 587 pp.

93. Kogan, M., Kogan, D. D. 1979. *Odontota horni,* a hispine leaf miner adapted to soybean feeding in Illinois. *Ann. Entomol. Soc. Am.* 72:456–61

94. Kogan, M., Kuhlman, D. E. 1982. Soybean insects: Identification and management in Illinois. *Ill. Agric. Exp. Stn. Bull.* 773:1–58

95. Kogan, M., Paxton, J. 1983. Natural inducers of plant resistance to insects. In *Plant Resistance to Insects. ACS Symp. Ser.,* Vol. 208, ed. P. A. Hedin, pp. 153–71. Washington, DC: Am. Chem. Soc. 375 pp.

96. Kogan, M., Turnipseed, S. G. 1982. Soybean entomology and biological control. In *Soybean Cultivation and Research in the People's Republic of China. Reports of the Study Tour Delegation to the People's Republic of China,* ed. R. D. Riggs, pp. 51–79. Washington, DC: USDA Off. Int. Coop. Dev. 80 pp.

97. Kogan, M., Turnipseed, S., Shepard, M., Oliveira, E. B. de, Borgo, A. 1977. Pilot insect pest management program for soybean in southern Brazil. *J. Econ. Entomol.* 70:659–63

98. Kogan, M., Waldbauer, G. P., Boiteau, G., Eastman, C. E. 1980. Sampling bean leaf beetles on soybean. See Ref. 92, pp. 201–36

99. Korean Society of Plant Protection. 1972. *A List of Plant Diseases, Insect Pests, and Weeds in Korea.* Suweon, Korea: Korean Soc. Plant Prot. 424 pp.

100. Kulikova, L. S. 1968. Pests of soybean in the Primorye territory. In *Fauna and Ecology of Insects of the Soviet Far East,* ed. A. I. Kurentsov, Z. A. Konovalova, pp. 108–19. Vladivostok, USSR: Acad. Sci. USSR Far East. Branch Inst. Biol. Pedol. 170 pp. (In Russian)

101. Kulikova, L. S. 1981. Formation of the entomofauna of soybean in cleared forest regions in the maritime territory. In *Noveishie Dostizheniya Sel' skokhozyaistvennoi Entomologii,* pp. 104–7. Vilnius, USSR: Vses. Entomol. Ovo. 224 pp. (In Russian)

102. Kwon, S. H., Chung, K. H., Lee, Y. I., Kim, J. R., Ryu, J. 1980. Responses of soybean varieties and planting dates to beanfly (*Melanagromyza* sp.) damage. *Korean J. Breed.* 12:30–34 (In Korean)

103. Layton, M. A. B. 1983. *Effects of the threecornered alfalfa hopper,* Spissistilus festinus *(Say), on yields of soybean, as determined by manipulation of populations by the use of insecticides.* PhD thesis. Louisiana State Univ., Baton Rouge. 71 pp.

104. Lee, Y. I., Kogan, M., Larsen, J. R. Jr. 1986. Attachment of the potato leafhopper to soybean plant surfaces as affected by morphology of the pretarsus. *Entomol. Exp. Appl.* In press

105. Lentz, G. L., Chambers, A. Y., Hayes, R. M. 1983. Effects of systemic insecticide-nematicides on midseason pests and predator populations in soybean. *J. Econ. Entomol.* 76:836–40

106. Lentz, G. L., Hayes, R. M., Chambers, A. Y. 1985. Interactions of insecticidenematicides, metribuzin, and environment on soybean injury and yield. *J. Econ. Entomol.* 78:1217–21

107. Lincoln, C., Boyer, W. P., Miner, F. D. 1975. The evolution of insect pest management in cotton and soybeans: Past experience, present status, and future outlook in Arkansas. *Environ. Entomol.* 4:1–7

108. Link, D. 1980. Insetos-pragas da soja no Brasil. *An. Congr. Brasil. Entomol., Campinas,* pp. 215–44. São Paulo, Brazil

109. Liu, T. R., Xin, H. P., Li, Q. X., eds. 1979. *Soybean Pests (Insects, Diseases and Weeds).* Beijing, China: Agric. Press. 248 pp.

110. Livingston, J. M., Yearian, W. C., Young, S. Y., Stacey, A. L. 1981. Effect of benomyl on an *Entomophthora* epizootic in a *Pseudoplusia includens* population. *J. Ga. Entomol. Soc.* 16:511–14

111. Loureiro, M. C. 1976. *Synecology of edaphic Arthropoda in Iowa agroecosystems.* PhD thesis. Iowa State Univ., Ames. 130 pp.

112. Ma, Z., Zhang, J. 1984. The strategy for controlling soybean pests in northern Shandong Province. See Ref. 45, pp. 372–74

113. Marrone, P. G., Stinner, R. E. 1983. Effects of soil moisture and texture on oviposition preference of the bean leaf beetle, *Cerotoma trifurcata* (Forster) (Coleoptera: Coccinellidae). *Environ. Entomol.* 12:426–28

114. Marrone, P. G., Stinner, R. E. 1983. Effects of soil physical factors on egg survival of the bean leaf beetle, *Cerotoma trifurcata* (Forster) (Coleoptera: Chrysomelidae). *Environ. Entomol.* 12: 673–79

115. Marrone, P. G., Stinner, R. E. 1984. Influence of soil physical factors on survival and development of the larvae and pupae of the bean leaf beetle, *Cerotoma trifurcata* (Coleoptera: Chrysomelidae). *Can. Entomol.* 116:1015–23

116. Massariol, A. A., Ramiro, Z. A., Calcagnolo, G. 1979. Insetos observados na cultura da soja no estado de São Paulo. *Biológico* 45:83–87

117. Matteson, P. C., ed. 1984. *Proc. Int. Workshop Integrated Pest Control Grain Legumes, Goiania, Goias, Brasil, 1983.* Brasilia, Brasil: Dep. Difus. Tecnol., EMBRAPA. 417 pp.

118. Mayse, M. A. 1984. Soybean row-spacing: Effects on arthropod population patterns and sampling considerations. *Environ. Manage.* 8:325–32

119. Mayse, M. A., Price, P. W. 1978. Seasonal development of soybean arthropod communities in East Central Illinois. *Agroecosystems* 4:387–405

120. McCutcheon, G. S., Turnipseed, S. G. 1981. Parasites of lepidopterous larvae in insect resistant and susceptible soybeans in South Carolina. *Environ. Entomol.* 10:69–74

121. McPherson, R. M., Allen, W. A., Smith, J. C. 1981. *Soybean Insect Pest Management for Virginia. Va. Agric. Ext. Serv. Pest Manage. Guide 32.* 8 pp.

122. McPherson, R. M., Newsom, L. D. 1984. Trap crops for control of stink bugs in soybean. *J. Ga. Entomol. Soc.* 19:470–80

123. Miranda, M. A. C. de, Rossetto, C. J., Rossetto, D., Braga, N. R., Mascarenhas, H. A. A., et al. 1979. Resistência de soja a *Nezara viridula* e *Piezodorus guildinii* em condições de campo. *Bragantia* 38:181–88

124. Mishoe, J. W., Jones, J. W., Swaney, D. P., Wilkerson, G. G. 1984. Using crop and pest models for management applications. *Agric. Syst.* 15:153–70

125. Mitchell, P. L., Newsom, L. D. 1984. Seasonal history of the threecornered alfalfa hopper (Homoptera: Membracidae) in Louisiana. *J. Econ. Entomol.* 77:906–14

126. Mitchell, P. L., Newsom, L. D. 1984. Histological and behavioral studies of threecornered alfalfa hopper (Homoptera: Membracidae) feeding on soybean. *Ann. Entomol. Soc. Am.* 77:174–81

127. Moffitt, L. J., Farnsworth, R. L., Zavaleta, L. R., Kogan, M. 1986. Economic impact of public pest information: Soybean insect forecasts in Illinois. *Am. J. Agric. Econ.* 68:274–79

128. Morrison, D. E., Bradley, J. R. Jr., Van Duyn, J. W. 1979. Populations of corn earworm and associated predators after applications of certain soil-applied pesticides to soybeans. *J. Econ. Entomol.* 72:97–100

129. Moscardi, F. 1984. Microbial control of insect pests in grain legume crops. See Ref. 117, pp. 189–223

130. Moscardi, F., Correa Ferreira, B. S. 1985. Biological control of soybean caterpillars. See Ref. 176, pp. 703–11

131. Mueller, A. J. 1985. Impact and economics of threecornered alfalfa hopper feeding on soybean. See Ref. 176, pp. 635–40

132. Mueller, A. J., Jones, J. W. 1983. Effects of main-stem girdling of early vegetative stages of soybean plants by threecornered alfalfa hoppers (Homoptera: Membracidae). *J. Econ. Entomol.* 76:920–22

133. Murakami, M. 1981. Pest control in soybeans. *Kongetsu No Noyaku* 25:25–33 (In Japanese)

134. Nagano, M. 1980. Ecology and control of insect pests of soybean cultivated in rotational paddy fields. 1. Seasonal prevalence in 1979 and chemical control. *Proc. Assoc. Plant Prot. Kyushu* 26:133–35 (In Japanese)

135. Newsom, L. D. 1976. Selective use of agricultural pesticides. See Ref. 59, pp. 539–48

136. Newsom, L. D. 1980. Interactions of control tactics in soybeans. See Ref. 27, pp. 261–73

137. Newsom, L. D., ed. 1987. *Integrated Pest Management Systems for Soybean.* New York: Wiley. In press

138. Newsom, L. D., Herzog, D. C. 1977. Trap crops for control of soybean pests. *La. Agric.* 20:14–15

139. Newsom, L. D., Kogan, M., Miner, F. D., Rabb, R. L., Turnipseed, S. G., Whitcomb, W. H. 1980. General accomplishments toward better pest control in soybean. See Ref. 66, pp. 51–98

140. Oliveira, E. B. de, Campo, C. B. H. 1984. Ocorrência e controle químico de *Sternechus subsignatus* Boheman, 1836 em soja no Paraná. In *III Sem: Nacl. Pesqui. Soja, Resumos*, pp. 20–24. Londrina, Brazil: EMBRAPA/CNPSoja. 136 pp.

141. Oliveira, F. T. G., Silva, J. B., Gazzoni, D. L., Roessing, A. C. 1980. Manejo de pragas na cultura da soja: Um caso de sucesso da pesquisa. *EMBRAPA/CNPSoja Dep. Difus. Tec. Doc. 01,* Brasilia, Brasil. 23 pp.

142. Orr, D. B., Boethel, D. J. 1985. Com-

parative development of *Copidosoma truncatellum* (Hymenoptera: Encyrtidae) and its host, *Pseudoplusia includens* (Lepidoptera: Noctuidae), on resistant and susceptible soybean genotypes. *Environ. Entomol.* 14:612–16

143. Orr, D. B., Boethel, D. J., Jones, W. A. 1985. Biology of *Telenomus chloropus* (Hymenoptera: Scelionidae) from eggs of *Nezara viridula* (Hemiptera: Pentatomidae) reared on resistant and susceptible soybean genotypes. *Can. Entomol.* 117:1137–42

144. Panda, N., Daugherty, D. M. 1978. Ovipositional preference of *Heliothis zea* (Hubn.) on glabrous and dense soybean genotypes. *Madras Agric. J.* 63:227–30

145. Panizzi, A. R. 1980. Uso de cultivar armadilha no controle de percevejos em soja. *Trigo Soja Bol. Tec.* 47:11–14

146. Panizzi, A. R. 1985. Dynamics of phytophagous pentatomids associated with soybean in Brazil. See Ref. 176, pp. 674–80

147. Panizzi, A. R., Correa, B. S., Gazzoni, D. L., Oliveira, E. B., Newman, G. C., Turnipseed, S. G. 1977. Insetos da soja no Brasil. *EMBRAPA/CNPSoja Bol. Tec.* 1:4–20

148. Panizzi, A. R., Slansky, F. Jr., 1985. Review of phytophagous pentatomids (Hemiptera: Pentatomidae) associated with soybean in the Americas. *Fla. Entomol.* 68:184–214

149. Panizzi, M. C. C., Bays, I. A., Kiihl, R. A. S., Porto, M. P. 1981. Identificação de genótipos fontes de resistência a percevejos-pragas da soja. *Pesqui. Agropecu. Bras.* 16:33–37

150. Parkman, P., Jones, W. A. Jr., Turnipseed, S. G. 1983. Biology of *Pediobius* sp. near *facialis* (Hymenoptera: Eulophidae), an imported pupal parasitoid of *Pseudoplusia includens* and *Trichoplusia ni* (Lepidoptera: Noctuidae). *Environ. Entomol.* 12:1669–72

151. Pedigo, L. P. 1980. Sampling green cloverworm on soybean. See Ref. 92, pp. 169–86

152. Pedigo, L. P., Higgins, R. A., Hammond, R. B., Bechinski, E. J. 1981. Soybean pest management. In *CRC Handbook of Pest Management in Agriculture,* ed. D. Pimentel, 3:427–537. Boca Raton, Fla: CRC. 656 pp.

153. Pedigo, L. P., Hutchins, S. H., Higley, L. G. 1986. Economic injury levels in theory and practice. *Ann. Rev. Entomol.* 31:341–68

154. Pitre, H. N., ed. 1983. Natural enemies of arthropod pests in soybean. *South.*

Coop. Ser. Bull. 285, Mississippi State Univ., Mississippi State. 90 pp.

155. Price, P. W. 1976. Colonization of crops by arthropods: Nonequilibrium communities in soybean fields. *Environ. Entomol.* 5:605–11

156. Pu, M., Pan, T. 1984. A study on the regionalization of the soybean producing area in China. See Ref. 45, pp. 45–53

157. Qu, Y., Kogan, J. 1984. *A Bibliography of Three Lepidopterous Pod Borers—Etiella zinckenella, Leguminivora glycinivorella and Matsumuraeses phaseoli—Associated with Soybean and Other Legumes.* Urbana-Champaign, Ill. Univ. Ill. Soybean Insect Res. Inf. Cent. (SIRIC). 81 pp.

158. Reed, T., Shepard, M., Turnipseed, S. G. 1984. Assessment of the impact of arthropod predators on noctuid larvae in cages in soybean fields. *Environ. Entomol.* 13:954–61

159. Reed, W., Kumble, V., eds. 1982. *Proc. Int. Workshop* Heliothis *Manage.* Patancheru, India: ICRISAT. 418 pp.

160. Reichelderfer, K. H., Bender, F. E. 1979. Application of a simulative approach to evaluating alternative methods for the control of agricultural pests. *Am. J. Agric. Econ,* 61:258–67

161. Richter, A. R., Fuxa, J. R. 1984. Timing, formulation, and persistence of a nuclear polyhedrosis virus and a microsporidium for control of the velvetbean caterpillar (Lepidoptera: Noctuidae) in soybeans. *J. Econ. Entomol.* 77:1299–306

162. Rizzo, H. F., Losada, A. D. 1975. Insectos encontrados en cultivos de soja (*Glycine max* (L) Merrill) en la zona Yraizoz (Provincia de Buenos Aires, Argentina). *Fitotec. Latinoam.* 11:3–8

163. Rogers, D. J., Sullivan, M. J. 1986. Nymphal performance of *Geocoris punctipes* (Hemiptera: Lygaeidae) on pest resistant soybeans. *Environ. Entomol.* In press

164. Rossetto, C. J., Lourenção, A. L., Igue, T., de Miranda, M. A. C. 1981. Picadas de alimentação de *Nezara viridula* em cultivares e linhagens de soja de diferentes graus de suscetibilidade. *Bragantia* 40:109–14

165. Rudd, W. G. 1980. Simulation of insect damage to soybeans. See Ref. 27, pp. 547–55

166. Rudd, W. G., Ruesink, W. G., Newsom, L. D., Herzog, D. C., Jensen, R. L., Marsolan, N. F. 1980. The systems approach to research and decision making for soybean pest control. See Ref. 66, pp. 99–122

167. Ruesink, W. G. 1984. Soybean as a host for the leafminer *Sumitrosis rosea* (Coleoptera: Chrysomelidae). *J. Econ. Entomol.* 77:108–9

168. Rust, R. W. 1977. Evaluation of trap crop procedures for control of Mexican bean beetle in soybeans and lima beans. *J. Econ. Entomol.* 70:630–32

169. Saunders, J. L., King, A. B. S., Vargas S., C. L. 1983. Plagas de cultivos en America Central: Una lista de referencia. *CATIE Ser. Tec. Bol. Tec. 9., Cent. Agron. Trop. Invest. Enseñ.*, Turrialba, Costa Rica. 90 pp.

170. Shaheen, A. H. 1977. Survey of pests attacking soybean plants in Egypt with some ecological notes. *Agric. Res. Rev.* 55:59–65

171. Shanmugasundaram, S., Sulzberger, E. W., eds. 1985. *Soybean in Tropical and Subtropical Cropping Systems*. Shanhua, Taiwan: Asian Veg. Res. Dev. Cent. 471 pp.

172. Shelton, M. D., Edwards, C. R. 1983. Effects of weeds on the diversity and abundance of insects in soybeans. *Environ. Entomol.* 12:296–98

173. Shepard, M., Carner, G. R., Turnipseed, S. G. 1977. Colonization and resurgence of insect pests of soybean in response to insecticides and field isolation. *Environ. Entomol.* 6:501–6

174. Shepard, M., Herzog, D. C. 1985. Soybean: Status and current limits to biological control in the southeastern U.S. See Ref. 55, pp. 557–74

175. Shepard, M., Lawn, R. J., Schneider, M. A. 1983. *Insects on Grain Legumes in Northern Australia; A Survey of Potential Pests and Their Enemies.* St. Lucia, Australia: Univ. Queensland Press. 89 pp.

176. Shibles, R. 1985. *World Soybean Research Conference III: Proceedings*. Boulder, Colo: Westview. 1262 pp.

177. Sifuentes, J. A. 1978. Plagas de la soya y su control en Mexico. *Mex. Inst. Nacl. Invest. Agric. Foll. Divulg. 70.* 24 pp.

178. Sinclair, J. B., Jackobs, J. A., eds. 1981. *Soybean Seed Quality and Stand Establishment*. Urbana-Champaign, Ill: Univ. Ill., INTSOY. 206 pp.

179. Singh, S. R., Van Emden, H. F., Taylor, T. A., eds. 1978. *Pests of Grain Legumes: Ecology and Control.* London: Academic. 454 pp.

180. Sirisingh, S., Sitchawat, T., Alapach, N. 1971. Soybean insect pests in northeast Thailand and their control. *Thailand Northeast Agric. Cent. Res. Bull.* 4:1–47 (In Thai)

181. Sloderbeck, P. E., Yeargan, K. V. 1983. Comparison of *Nabis americoferus* and *Nabis roseipennis* (Hemiptera: Nabidae) as predators of the green cloverworm (Lepidoptera: Noctuidae). *Environ. Entomol.* 12:161–65

182. Smith, C. M. 1985. Expression, mechanisms and chemistry of resistance in soybean, *Glycine max* L. (Merr.) to the soybean looper, *Pseudoplusia includens* (Walker). *Insect Sci. Appl.* 6:243–48

183. Smith, T. J., Camper, H. M. Jr., Smith, J. C., Alexander, M. W. 1979. The Shore soybean. *Va. Polytech. Inst. State Univ. Res. Div. Bull.* 225:1–3

184. Soekarna, D., Tengkano, W. 1979. The diversity and pests succession of soybean. *Kongr. Entomol.* 1:1–11 (In Indonesian)

185. Sparks, A. N. Jr., Newsom, L. D. 1984. Evaluation of the pest status of the threecornered alfalfa hopper (Homoptera: Membracidae) on soybean in Louisiana. *J. Econ. Entomol.* 77:1553–58

186. Sprenkel, R. K., Brooks, W. M., Van Duyn, J. W., Deitz, L. L. 1979. The effects of three cultural variables on the incidence of *Nomuraea rileyi*, phytophagous Lepidoptera, and their predators on soybeans. *Environ Entomol.* 8:334–39

187. Stadelbacher, E. A., Powell, J. E., King, E. G. 1984. Parasitism of *Heliothis zea* and *H. virescens* (Lepidoptera: Noctuidae) larvae in wild and cultivated host plants in the delta of Mississippi. *Environ. Entomol.* 13:1167–72

188. Stevens, L. M., Steinhauer, A. L., Coulson, J. R. 1975. Suppression of Mexican bean beetle on soybeans with annual inoculative releases of *Pediobius faveolatus*. *Environ. Entomol.* 4:947–52

189. Stimac, J. L., Barfield, C. S. 1980. Systems approach to pest management in soybean. See Ref. 27, pp. 249–59

190. Stimac, J. L., O'Neil, R. J. 1983. Modeling the impact of natural enemies on insect pests of soybean. See Ref. 154, pp. 78–87

191. Stinner, R. E., Bradley, J. R. Jr., Van Duyn, J. W. 1980. Sampling *Heliothis* spp. on soybean. See Ref. 92, pp. 407–21

192. Stinner, R. E., Rabb, R. L., Bradley, J. R. Jr. 1977. Natural factors operating in the population dynamics of *Heliothis zea* in North Carolina. *Proc. 15th Int. Congr. Entomol., Washington, DC*, pp. 622–742. Washington, DC: Entomol. Soc. Am.

193. Stinner, R. E., Regniere, J., Wilson, K. 1982. Differential effects of agroecosystem structure on dynamics of three soybean herbivores. *Environ. Entomol.* 11:538–43

194. Sullivan, M. J. 1985. Resistance to insect defoliators. See Ref. 176, pp. 400–5

195. Talekar, N. S., Chen, B. S. 1983. Seasonality of insect pests of soybean and mungbean in Taiwan. *J. Econ. Entomol.* 76:34–37

196. Teague, T. G., Horton, D. L., Yearian, W. C., Phillips, J. R. 1985. Benomyl inhibition of *Cotesia* (= *Apanteles*) *marginiventris* survival in four lepidopterous hosts. *J. Entomol. Sci.* 20:76–81

197. Thorvilson, H. G., Pedigo, L. P. 1984. Epidemiology of *Nomuraea rileyi* (Fungi: Deuteromycotina) in *Plathypena scabra* (Lepidoptera: Noctuidae) populations from Iowa soybeans. *Environ. Entomol.* 13:1491–97

198. Todd, J. W., Herzog, D. C. 1980. Sampling phytophagous Pentatomidae on soybean. See Ref. 92, pp. 438–78

199. Togashi, I. 1980. Preliminary report on the soybean pests and their natural enemies in Ishikawa Prefecture. *Bull. Ishikawa Agric. Coll.* 10:15–21

200. Trujillo, M. R. 1973. Insectos colectados en soja durante la campaña 1971/72 al norte de Corrientes. *IDIA* 306–8:115–18

201. Turhan, N., Tunc, A., Belli, A., Kismir, A., Kisakurek, N. 1983. Cukurova'da soya (*Glycina max* L.) 'da bocek ve akar faunasinin tesbiti uzerinde calismalar. *Bitki Koruma Bul.* 23:148–69

202. Turner, J. W. 1978. Pests of grain legumes and their control in Australia. See Ref. 179, pp. 73–81

203. Turnipseed, S. G. 1973. Insects. In *Soybeans: Improvement, Production, and Uses,* ed. B. E. Caldwell, pp. 545–72. Madison, Wis: Am. Soc. Agron. 681 pp.

204. Turnipseed, S. G. 1975. Manejo das pragas da soja no sul do Brasil. *Trigo Soja Bol. Tec.* 1:4–7

205. Turnipseed, S. G. 1977. Influence of trichome variations on populations of small phytophagous insects in soybean. *Environ. Entomol.* 6:815–17

206. Turnipseed, S. G. 1980. Soybean insects and their natural enemies in Brazil—A comparison with the southern United States. See Ref. 27, pp. 285–90

207. Turnipseed, S. G. 1984. Insecticide use and selectivity in soybean. See Ref. 117, pp. 227–37

208. Turnipseed, S. G., Herzog, D. C., Chapin, J. W. 1986. Evaluation of biological agents for pest management in soybean. See Ref. 35, In press

209. Turnipseed, S. G., Kogan, M. 1976. Soybean entomology. *Ann. Rev. Entomol.* 21:247–82

210. Turnipseed, S. G., Kogan, M. 1986. Integrated control of insect pests. In *Soybeans: Improvement, Production and Uses,* ed. J. R. Wilcox. Madison, Wis: Am. Soc. Agron. 2nd ed. In press

211. Turnipseed, S. G., Newsom, L. D. 1987. Chemical control. See Ref. 137. In press

212. Tynes, J. S., Boethel, D. J. 1986. *Control Soybean Insects. La. Coop. Ext. Serv. Publ. 2211 (Revised)*

213. Van Duyn, J. W., McLeod, P. J. 1983. Pesticide effects on natural enemies. See Ref. 154, pp. 56–62

214. Villacorta, A., Oliveira, E. B. de, Kogan, M. 1976. Pragas e seu controle. In *Manual Agropecuário para o Paraná,* pp. 213–27. Londrina, Brazil: Inst. Agron. Paraná

215. Vincentini, R., Jimenez, H. A. 1977. El vaneo de los frutos en soja. *Parana Estac. Exp. Reg. Agropecu. Ser. Tec.* 47:1–30

216. Walker, R. H., Patterson, M. G., Hauser, E., Isenhour, D. J., Todd, J. W., Buchanan, G. A. 1984. Effects of insecticide, weed-free period, and row spacing on soybean (*Glycine max*) and sicklepod (*Cassia obtusifolia*) growth. *Weed Sci.* 32:702–6

217. Wang, C. L. 1980. Soybean insects occurring at podding stage in Taichung, Taiwan. *J. Agric. Res. China* 29:283–86 (In Chinese)

218. Wang, Y. S. 1979. Integrated control of major soybean insect pests. In *Integrated Control of Major Insect Pests in China, Inst. Zool. Acad. Sin.,* pp. 213–21. Beijing, China: Sci. Press. 467 pp.

219. Whitfield, G. H., Ellis, C. R. 1976. The pest status of foliar insects on soybeans and white beans in Ontario. *Proc. Entomol. Soc. Ont.* 107:47–55

220. Witkowski, J. F. 1982. A multi-county survey of insects and mites found on Nebraska soybeans, 1980–1981. *Nebr. Agric. Exp. Stn. Bull.* 549:1–23

221. Yeargan, K. V. 1985. Pesticide compatibility in soybean pest management. See Ref. 176, pp. 695–702

222. Yukawa, J., Ohiani, Y., Yazawa, Y. 1983. Host-change experiments from wild plants to soybean in *Asphondylia* species (Diptera: Cecidomyiidae). *Proc.*

Assoc. Plant Prot. Kyushu 29:115–17 (In Japanese)

223. Yunus, A., Ho, T. H. 1980. List of economic pests, host plants, parasites and predators in West Malaysia (1920–1978). *Malays. Ministr. Agric. Bull.* 153:1–538

224. Zhang, G. R., Jin, Q. F., Wan, L. 1981. Soybean root fly and its integrated control. *Zhongguo Youliao* 2:63–67 (In Chinese)

225. Zhang, Z. J., Fu, C. Q. 1983. Inheritance of resistance to soybean pod borer in soybeans. See Ref. 45, pp. 74–79

SUBJECT INDEX

A

Abacomorphus, 24
Abbella subflava, 353
Ablattaria laevigata, 28
Ablerus, 60
Acanthomyops claviger, 393
Acaricide resistance, 361-80
Accessory glands, 395
Aceratogallia calcaris, 352
Acetylcholinesterase, 153
Acilius sulcatus, 397
Acraea, 402
Action potential, 5
Action thresholds for entomo-
    pathogens, 242-43
Activity cycles
    of fruit flies, 120-21
    of scorpions, 286-89
Activity patterns
    of aeolian fauna, 165
    of biting flies, 298
    of nival arthropods, 174-75
Acyrthosiphum pisum, 52, 56-
    57, 146
Adalia, 35
Adelium, 36
Adelphoparasitism, 60-61
Adipokinetic hormone, 6
ADP, 464-65, 468, 474
Aedeomyia squamipennis, 490
Aedes, 450, 455
    aegyptii, 8, 157, 299, 439-
        46, 449, 457-58, 468-69,
        474, 481-85, 492-95,
        497
    africanus, 481
    albopictus, 482, 485, 492,
        494
    atropalpus, 490
    campestris, 440, 445-49
    detritus, 440, 445, 448
    dorsalis, 440, 445-46, 449,
        454
    furcifer, 490
    hendersoni, 486-87
    lineatopennis, 489
    simpsoni, 481
    sollicitans, 449
    taeniorhynchus, 299, 440,
        444-47, 449, 451, 453
    taylori, 490
    togoi, 440, 444-45, 448
    triseriatus, 299, 474, 484-85,
        487, 489, 491-94
    vexans, 300
Aeolian faunas, 163-79
Aggregation pheromones, 391,
    394, 396-97

Aggregations
    and entomopathogens, 241
    and fruit flies, 127
    and population dynamics,
        332-33
Agrotis ipsilon, 344
Alarm pheromones, 27, 383,
    393
Aldicarb, 351, 353, 522
Aldrin, 151
Aleochara curtula, 29
Algon grandicollis, 30
Alianta, 29
Alkaloids and chemical de-
    fenses, 34-36
Allergens and saliva, 463
Allomones, 17-48, 381, 396,
    397-400
Alloxysta, 50, 52-53, 55-57, 59,
    61
    ancylocera, 55
    quateri, 59
    victrix, 53-54, 56, 58-59
Alphaviruses, 488, 490
Alpine aeolian systems, 163-
    79
Altitudinal zonation, 164
Amathes c-nigrum, 344
Amarygmus tristis, 37
Amblyomma, 472
    americanum, 400
    maculatum, 400
Amino acids and osmoregula-
    tion, 452-53
AMP, 468
Ampumixis, 264
Anal papillae, 441, 444, 453
Anaphylatoxins, 466-67
Anastrepha, 115
Anchytarsus, 254
Ancyronyx, 262, 264
Ancyrophorus, 31
Androctonus australis, 289
Aneristus, 60
Anisomorpha buprestoides, 393
Anommatelmis, 260
Anopheles, 368-69, 371, 444,
    455, 494
    albimanus, 149-50, 154
    arabiensis, 148, 371
    calicifacies, 368, 371
    freeborni, 456, 469
    gambiae, 150, 300, 368, 371,
        497
    salbaii, 469
    stephensi, 148, 371, 469
Anthonomus grandis, 231, 394
Antibiotics and chitin, 80-82
Antibodies and arboviruses, 494

Anticarsia gemmatalis, 228,
    237-38, 242, 510, 519,
    522, 524-26
Anticoagulants, 463, 470
Antigens and viruses, 482-84
Antimicrobics and chemical de-
    fenses, 25-26, 29
Ants
    and aeolian faunas, 169-70
    and beetles, 29, 35
    and chemical defenses, 33,
        40-41
    and cultural entomology, 186,
        188-89
    and exocrine compounds,
        384-85, 390, 405
    and fruit flies, 128
Anuroctonus phaiodactylus, 281-
    82, 285-88
Aphelinoidea plutella 353
Aphelinus, 59
Aphidecta obliterata, 35
Aphidencyrtus, 52-55, 57, 61
    aphidivorus, 54-55, 59
Aphidius, 52, 56-57, 59
    smithi, 52, 56
Aphids
    and aeolian faunas, 167, 169,
        171-72
    and alarm pheromone, 351,
        354
    and hyperparasitoids, 50, 52-
        61
    and insecticide resistance,
        148-49, 156
    and *Liriomyza*, 211
    and sugar beet, 342-51, 354
Aphis
    craccivora, 351
    fabae, 342, 344-47, 349,
        351
    glycines, 511
    pomi, 59
Apis, 186
    mellifera, 383, 433
Aploderus, 31
Apolysis, 79
Aposematic
    coloration, 33, 41
    patterns, 264
Apsena, 20
    pubescens, 37
Apyrase and blood-feeding, 469-
    70, 474, 494
Arboviruses and mosquitoes,
    479-505
Archanara geminipuncta, 325
Argoporis, 20, 37
Argyrotaenia

citrana, 386
velutinana, 386-87
Armyworms, 343-44
Aromia moschata, 38
Arphia pseudonietana, 5
Art and insects, 185-87
Arthropod fallout, 167-72
Artificial feeding and arboviruses, 488
Artificial intelligence, 428, 433
Artystona, 37
Asaphes, 52-56, 59, 61
californicus, 53, 56
lucens, 53-54
Asemobius, 32
Asphondylia, 511
Atholus, 27
ATP, 465-66, 468, 474
Augmentation of entomopathogens, 226
Austrelmis, 260, 264
consors, 260
Austrolimnius, 261, 267
Autogeny, 484, 493
Autoparasitism, 61
Azotus, 60

B

Bacillus
popilliae, 226, 231, 236
sphaericus, 231
thuringiensis, 226, 231, 238-39
Bacteria
and fruit flies, 126, 128, 131
and IPM, 231-32
Bacterial symbiotes of fruit flies, 118-19
Bactrocera, 118
Baculoviruses, 227-28, 236, 241, 244
Baits
and entomopathogens, 235-36
and fruit flies, 116, 126
Bark beetles
and aeolian faunas, 172
and exocrine compounds, 393-94
and nematodes, 232
Beauveria bassiana, 226, 229-30, 527
Bedbugs and cultural entomology, 190
Bees
and cultural entomology, 184-91
and exocrine compounds, 405
Beet
leafhopper, 344, 351-52
mosaic virus, 345
yellows virus, 345

Beetles and chemical defenses, 17-48
Bembidion, 166, 175
Bemisia tabaci, 517
Benomyl, 522
Benzimidazoles and chitin, 84
Benzoylphenyl ureas, 75, 79, 85-87
Binodoxys, 59
Biogeography
of Mexican entomofauna, 95-114
of riffle beetles, 266-67
Biological control
and hyperparasitism, 60-63
of soybean insects, 509
Biology
of Dryopidae, 257
of Elmidae, 254-56
of *Liriomyza*, 201-24
of Lutrochidae, 257
of Psephenidae, 257-58
Biomedical research, 1-16
Biosynthesis
of acetogenins, 385-91
of alkaloids, 402-4
of carboxylic acids, 383-85
of chemical defenses, 21-22
of exocrine compounds, 381-413
of hydrocarbons, 382
of peptides, 405-7
of pheromones, 382, 385-90, 393, 397
of quinones, 391-93
of sterols, 397
of terpenes, 393-97
Biotypes of aphids, 345, 355
Biting flies, 297-316
Black flies and vision, 301-3
Black widow spiders, 290
Blaps, 38
Blapstinus, 344
Blattella germanica, 368
Bledius, 31
Blister beetles and chemical defenses, 35
Blood-feeding
and osmoregulation, 456-59
and saliva, 463-78
Blow flies, 364, 369-73
Bolitochara, 29
Bombadier beetles, 23, 392
Bombykol, 388
Bombyx mori, 8, 10, 84-85, 388, 390
Booklungs, 287
Boophilus microplus, 371, 472
Brachinus, 392
Brackish-water mosquito larvae, 450-54
Brain, 9-12, 118

Brevicoryne brassicae, 55
Brood pouch, 9
Bruchids, 326
Bryothinusa, 29
Bumble bees and cultural entomology, 184
Bunyaviruses, 488-90, 495
Buprofezin and chitin, 84
Buthotus minax, 277
Buthus
hottentotta, 288
occitanus, 277
Butterflies and cultural entomology, 186-87, 189, 191
Byrsotria fumigata, 5

C

Callantra, 115, 117
Calliphora erythrocephala, 80, 82
Callistenus, 24
Calosoma, 24
Calpodes ethlius, 74, 79
Cambala annulata, 393
Campalita, 24
Camponotus, 170, 384
Cannibalism, 265
in scorpions, 283
Cantharidin, 35, 394-96
Captan, 83-84
Carabids
and aeolian faunas, 166-67, 169, 173-75
and chemical defenses, 23-25
and population dynamics, 330
Carabus
taedatus, 383
yaconinus, 24
Carbamates, 522
resistance to, 149, 151
Carbaryl resistance, 365, 371
Carbon dioxide and biting flies, 302-3, 305, 309
Cardenolides, 41-42
Cardiac glycosides, 39
Carrion beetles and chemical defenses, 27
Cell cultures and viruses, 482
Centipedes and exocrine compounds, 400
Centruroides
exilicauda, 282, 284, 286
gracilis, 277-78, 286
insulanus, 279, 281
sculpturatus, 278, 281, 288
thorelli, 284
vittatus, 279, 281
Ceratitis, 115
capitata, 133
Ceromae, 117
Cerotoma trifurcata, 510

Charips
  ancylocera, 55
  victrix, 56
Chauliognathus puchellus, 34
Chemical defenses of beetles,
  17-48, 264
Chemosystematics, 17-48
Chikungunya, 482
Chilo suppressalis, 82-83, 390
Chironomids and aeolian faunas,
  165-66, 170
Chitin
  biochemistry, 71-93
  polymerization, 72-75, 79-80,
    82, 84, 86
  synthesis, 71-76, 78
  synthetase, 74-76
Chitinases, 72
Chlaenius, 25
Chlorpyrifos resistance, 371-72
Chlosyne harrisii, 323
Cholesterol, 397
Cholinesterases, 149
Choristoneura, 169, 171, 330
  fumiferana, 388-89
Chromaphis juglandicola, 59
Chrysalids and cultural entomol-
  ogy, 189
Chrysochus cobaltinus, 41
Chrysolina brunsvicensis, 41
Chrysomela, 40
  tremulae, 399
Chrysomelids and aeolian
  faunas, 170
Chrysopeplus, 38
Chrysops
  atlanticus, 304
  fuliginosus, 304
Cicadas
  and cultural entomology, 189,
    191
  and fungal disease, 329
Cicadellids and aeolian faunas,
  171
Circadian rhythms
  in biting flies, 298-99
  in fruit flies, 120-21
  in scorpions, 288-89
Circadian systems, 11-12
Circulifer
  opacipennis, 351
  tenellus, 344, 351-52
Cladistics, 18-22
Clotting and saliva, 466
Coadaptation and resistance,
  372-74
Coagulation and saliva, 464-66
Coccinella
  septempunctata, 403-4
  undecimpunctata, 35
Coccinellids and aeolian faunas,
  170-71

Coccophagoides, 60
  similis, 61
  utilis, 61
Coccophagus, 60-61
Cochineal, 187
Cockroaches and cultural
  entomology, 187, 190
Codling moth, 326
Cold-hardiness, 172
Coleomegila maculata, 147
Collembola and aeolian faunas,
  166-67
Colorado potato beetle, 361
Color sensitivity, 304, 306-7
Colymbetes fuscus, 27
Competition
  and fruit flies, 132-33
  and *Liriomyza*, 211
Complement-fixation test, 482-
  84
Computers and decision-making,
  415-37
Conophthorus, 326
Control
  of *Liriomyza*, 216
  of soybean pests, 218-27
Copelatus, 26
Coprophilus, 30-31
Coremata, 404
Corpus allatum, 4-6, 8, 10
Corpus cardiacum, 3-4, 6, 10-
  11
Coruna, 52-53
Corydalus cornutus, 263
Cosmobaris americana, 344
Cost/benefit analysis, 421
Cotesia marginiventris, 522
Courtship by scorpions, 278,
  288
Creatonotus, 404
Cremastocheilus, 32
Creophilus, 20, 30
Crepitation mechanisms, 23
Crickets and cultural entomolo-
  gy, 184, 190
Cross-resistance, 366
Cryptolestes ferrugineus, 391,
  396-97
Crystallization of chitin, 76-78
Culex, 450, 455
  annulirostris, 494
  nigripalpus, 300, 494
  pipiens, 148-49, 439-41, 444,
    485, 490, 492-93
  quinquefasciatus, 153, 300,
    371-72, 488, 490, 494
  tarsalis, 440, 452-53, 484,
    486, 489-90, 495-96
  taeniopus, 486
  theileri, 487
  tritaeniorhynchus, 484, 493
  univittatus, 487

Culicoides, 482
Culiseta, 455
  melanura, 488, 490, 300
  inornata, 440, 450-53
Cultural
  control of soybean pests, 520-
    21
  entomology, 181-99
Curly top disease of sugar beet,
  351
Cuticle
  and chitin, 71-93
  and osmoregulation, 439,
    441-43, 445, 451, 454-
    55
  penetration, 31
Cutworms, 343
Cyanogenic
  compounds, 23, 40, 400-2
  glands, 399
Cybister, 26
Cyclic AMP, 457-58
Cyclodiene resistance, 155
Cylloepus, 264
Cytoplasmic polyhedrosis
  viruses, 228

D

Dacine fruit flies, 115-44
Dacnusa sibirica, 217
Dacus
  aglaia, 118
  amphoratus, 127
  aquilonus, 125
  binotatus, 127
  brevis, 127
  butianus, 127
  cacuminatus, 118-19, 125,
    131-35
  ciliatus, 122, 127, 132
  cucumis, 118-19, 125
  cucurbitae, 115-19, 122-27,
    129, 133
  diversus, 127
  dorsalis, 115-16, 118, 122-
    27, 129-32, 133
  expandens, 123
  frontalis, 127
  hageni, 127
  halfordiae, 118, 125, 132
  jarvisi, 119-20, 123, 133
  kraussi, 118
  latifrons, 122, 125, 133-34
  musae, 118-19, 132, 134
  neohumeralis, 116, 118-19,
    123, 125
  newmani, 131
  occipitalis, 125
  oleae, 115-24, 126-35
  opiliae, 117, 125, 127, 131-
    35

scutellaris, 127
serratus, 127
tenuifascia, 123, 131
tryoni, 115-27, 129-31
tsuneonsis, 123
umbeluzinus, 127
vertebratus, 125, 127
visendus, 118
vivittatus, 127
zonatus, 115-16, 125-27, 130, 132
Danaus
  chrysippus, 403-4
  gilippus, 403
Dasychira abietis, 320
DDT resistance, 150, 154, 371
Decision-making in entomology, 415-37, 528
Decision support systems, 418-19, 426-28
Dectes texanus, 511
Deer flies and vision, 303-6
Defenses of beetles, 17-48
Defensive
  glands, 399, 401
  substances, 17-48
Defoliation and population dynamics, 325-26
Deladenus siricidicola, 232
Deleaster, 20, 30-31
Delia
  antiqua, 241
  platura, 521
Demography of fruit flies, 133-34
Dendrocerus, 50, 52-55, 59, 61
  carpenteri, 54
  frontalis, 428, 433
Dendrolasin, 396
Dendrolimus superans, 322
Dengue, 482, 496-97
Density dependence and entomopathogens, 239-41
Density-dependent competition, 370
Dermacentor, 472
Detection of insecticide resistance, 145-62
Development rates of fruit flies, 129-31
Diabrotica, 41, 344
Diaeretiella rapae, 55
Diamesa, 165
Diamphidia, 42
Diapause, 11
  and arboviruses, 484, 496
  in fruit flies, 117, 134
  of *Liriomyza*, 212
Diapause hormone, 10
Diaulota, 29
Diazinon resistance, 364, 369, 371-73

Didymocentrus comondae, 286
Dieldrin resistance, 368-69, 371
Diflubenzuron, 85-87
Dimethoate, 148
Dimilin, 85
Dineutus, 26
Dinoponera, 188
Dioryctria, 326
Diplocentrus
  bigbendensis, 281
  spitzeri, 281, 285, 288
Diploptera punctata, 5, 391
Diprion pini, 328
Diseases and population dynamics, 329
Dispersal
  and insecticide resistance, 37-75
  of entomopathogens, 233-34
  of fruit flies, 126-27, 135
  of *Liriomyza*, 213
  of mosquitoes, 299
  of mountain entomofauna, 103
  of riffle beetles, 255
  of scorpions, 277, 286
Distribution of entomopathogens, 235
Diuresis, 456-58
DNA probes, 155, 157, 483
Dragonflies and cultural entomology, 191
Drifting by riffle beetles, 255
Drosophila, 8, 148, 155-57
  melanogaster, 73, 148, 157, 366, 390, 492
  virilis, 156
Dryops, 266-67
Dubiraphia, 260, 262-64
  brunnescens, 260
Dufour's gland, 383, 390
Dung beetles and cultural entomology, 186, 191
Dysaphis plantaginea, 59
Dyschirius, 20, 24
  wilsoni, 24
Dytiscus, 26

E

Ecdysis, 11
Ecdysone, 8, 382
Ecdysterone, 78
Eclosion hormone, 11
Ecological genetics, 361-80
Ecology
  of riffle beetles, 261-63
  of soybean arthropods, 509-18
Economic
  injury levels
    and entomopathogens, 233, 242

and soybean pests, 518
  thresholds, 517-18, 524
Ecosystems and population dynamics, 320-24
Ectopria, 259, 262
  thoracica, 260
Edema, 466, 470
Egg diapause, 10
Eggs of *Liriomyza*, 206
Elasmopalpus lignosellus, 525
Eleodes, 38
  longicollis, 392
  obsoleta, 37
ELISA tests, 149-51, 155, 157, 483
Elmis, 261
  condimentarius, 264
Elmoparnus, 253, 257, 267
Emergence of *Liriomyza*, 202
Empoasca, 342, 344, 347
  abrupta, 352
  arida, 352
  fabae, 352, 520
  mexara, 352
  solana, 352
Encapsulation, 62, 64
Encarsia, 60
Encephalitis, 482-83, 496
Encephalomyelitis, 482, 495
Endocrine control, 1-16
Endrin resistance, 368
Enhydrus, 25
Entomopathogens
  in IPM, 225-51
  of soybean pests, 514-16
Entomophthora, 522
  muscae, 229, 241
  sphaerosperma, 229
Environment
  and entomopathogens, 243-44
  and population dynamics, 319-21
Enzyme immunoassay, 483
Enzymes and insecticide resistance, 147-56
Ephedrus, 59
Epicauta, 344
Epidinocarsis lopezi, 62
Epilachna varivestis, 80
Epinotia aporema, 525
Epizootics, 225-51, 516
Equine encephalomyelitis, 482
Eradication of fruit flies, 116
Erynia, 226, 229
Erythema, 466, 470
Estigmene acrea, 344
Ethnoentomology, 191
Eubria, 259
Eubrianax, 258-59, 267
Euceraphis, 169
Euophrys omnisuperstes, 165
Euplectrus puttleri, 519

Euplognatha ovata, 170
Euryurus maculatus, 399
Euscorpius flavicaudis, 280
Eusphalerum, 28
Euxanthellus, 60
Euxesta notata, 146
Evolution of resistance, 367
Excretion
  and blood-feeding, 440
  and saliva, 463
Exocrine compounds, 381-413
Expert systems, 419, 422-23,
  428-35, 528

F

Fall webworm, 328
Farnesene, 351
Fat body, 8, 11
Fecundity
  of *Liriomyza*, 204
  of scorpions, 280-81
Feeding
  biology of scorpions, 281-83
  by fruitflies, 121, 126
  punctures of *Liriomyza*, 204
Feltia subterranea, 344
Fenitrothion resistance, 154
Fenvalerate, 149, 155
Fibrillogenesis of chitin, 76-78
Film and insects, 184-85, 188
Finlaya, 448
Fiorinia externa, 326
Fireflies,
  and toxic steroids, 34
  and cultural entomology, 190
Fitness modifiers, 372-73
Flaviviruses, 488, 490
Fleas and cultural entomology,
  190-91
Flies and cultural entomology,
  192
Flight
  ability, 484
  activity, 299-300
  of aphids, 349
  of biting flies, 299
  of fruit flies, 124
  of riffle beetles, 255-57
Folklore and insects, 188-90
Food webs and models, 61-64
Foraging by aeolian faunas, 174
Formica, 170, 384
  schaufussi, 390
Formic acid, 24, 384-85
Forest tent caterpillar, 320
Fossil riffle beetles, 266
Freshwater mosquito larvae,
  441-44
Fruit flies, 115-44
Fulgora, 190

Fungi
  and chitin, 71, 78, 82
  and entomopathogens, 236
  and IPM, 226, 228-30, 236,
    238
  and parasites, 245
  and riffle beetles, 266
Fungicides and chemical defenses, 25-26

G

GABA receptor, 155
Gabonia, 41
Galleria mellonella, 390
Gastrolina depressa, 40
Gastrophysa, 41
Gene
  cloning, 154
  duplications, 374
  flow, 366-67
  regulation, 155-56
Genes for resistance, 154-57
Genetic
  drift, 499, 485
  engineering, 239, 243
  markers, 375
  variability, 330-31
Genetics of resistance, 361-80
Geocoris punctipes, 523
Geodessus, 26
Gestation in scorpions, 279
Gills of riffle beetles, 261
Gilpinia
  hercyniae, 226, 243
  verticalis, 331
Glands and chemical defenses,
  17-19, 23-42
Glomeris marginata, 402-3
Glossina
  brevipalpis, 308
  morsitans, 307-8, 382
  pallidipes, 308
Glycoproteins, 72
Gonielmis, 264
Gonioctena viminalis, 40
Grandlure, 394
Granulosis viruses, 227-28, 239,
  329
Grasshoppers
  and cultural entomology, 184,
    186
  and fungi, 229
Gregarines and riffle beetles,
  266
Grylloblatta, 173
Grylloblattids and aeolian
  faunas, 166-67, 173-74
Gymnodacus, 117
Gypsy moth, 325-26, 329, 331,
  333

Gyrinus, 26
Gyrophaena, 29

H

Hadogenes 283
Hadrurus
  arizonensis, 279, 281, 283,
    286-87
  hirsutus, 283
  spadix, 283
Haemagogus, 299
Haematobia irritans, 307
Haliplus, 25
Halobates, 164
Halomaeusa, 29
Harpalus capito, 24
Harpaphe haydeniana, 400-1
Helichus, 253, 255, 257, 259,
  263-68
  fastigiatus, 255
  lithophilus, 257
  productus, 257
  suturalis, 257
Heliconius, 402
Heliothis, 226, 241, 243, 510-
  11, 514
  armigera, 147
  virescens, 146-47, 151, 373,
    389-90
  zea, 241, 389, 515, 518,
    520, 522, 524
Hemagglutination, 463, 482-84
Hematomas, 468, 470, 473
Hematophagy, 297-316
Hemlock scale, 328
Hemolymph
  and chemical defenses, 17
  and hormones, 4, 11
  and osmoregulation, 439-47,
    449, 451-53, 456-58
Hemostasis, 464-66, 468, 471,
  473-75
Henosepilachna
  pustulosa, 323
  vigintioctopunctata, 323
Herbicides, 522
Heterelmis, 262-63
Heterlimnius, 264
Heterometrus fulvipes, 289
Heterosis, 371
Hexacylloepus, 262
Hibernation by mosquitoes,
  300-1
Hieroglyphs, 184
Hintonelmis, 262
Hippodamia, 35
  convergens, 147, 171
Hirsutella thompsonii, 226, 229
Hispaniolara, 261
Hodegia apatela, 167
Homopterus arrowi, 23

Honey, 187, 190
Honey bees and cultural
    entomology, 184-87
Honeydew
    and biting flies, 299
    and fruit flies, 126
Horizontal transmission
    of entomopathogens, 234
    by mosquitoes, 485-89
Hormonal signals, 2
Hormones
    and chitin, 78-79
    and osmoregulation, 458
Horn flies, 307
Hornia, 36
Horse flies and vision
Host
    location by biting flies, 300,
        305
    races in *Liriomyza*, 214
    range of entomopathogens,
        236
    specificity, 55-59
Host preferences
    of fruit flies, 122
    of mosquitoes, 484
Hot springs and riffle beetles,
    260-61
House flies
    and chitin, 73
    and insecticide resistance,
        148-51, 371-72
Hulstia undulatella, 344
Hyalophora cecropia, 78, 82-83
Hybomitra schineri, 304
Hydaticus, 26
Hydora, 267
Hydrovatus, 26
Hygrobia tarda, 26
Hylemya platura, 344
Hypera postica, 226, 229
Hyperparasitism, 49-70

I

Ilybius, 20, 26-27
    fenestratus, 27
Imaginal disks, 73-74, 76, 454
Immunity
    to entomopathogens, 238-39
    and saliva, 466
Immunofluorescent assay, 483
Incubation period for arbovir-
    uses, 488
Infection by arboviruses, 485-
    89, 91-92
Inflammation and saliva, 466
Ingestion by scorpions, 282
Inheritance of resistance, 362-66
Inhibition of chitin formation,
    79-81
Insecticide resistance, 145-62,
    361-80

Insecticides
    and *Liriomyza*, 216-18
    and soybean pests, 522
Insect symbols, 186
Integrated pest management
    and decision making, 415,
        419-23, 428, 433-35
    and entomopathogens, 225-51
    and soybeans, 509, 519-28
    and sugar beet, 343, 347-49
    and resistance, 362
International centers, 508
Introduction of entomo-
    pathogens, 225
Ion transport, 442-43, 446-47,
    449, 451
Ipsenol, 394
Ips paraconfusus, 394-95
Ismarus, 50
Isometrus maculatus, 277
Ixodes
    dammini, 471-73
    ricinus, 472

J

Juglone, 40
Juvenile hormone, 4, 8-9, 11
    esterase, 156

K

Katydids and cultural entomolo-
    gy, 184

L

Labidomera clivicollis, 41-42
Laccophilus, 26
Laccornis, 26
Lac insects, 186
Lacon murinus, 33
Lady beetles and chemical de-
    fenses, 334
Lampyris, 34
Lara avara, 256
Larch bud moth, 329, 333
Larch casebearer, 328
Lasius fuliginosus, 383, 396
Laspeyresia, 326
    pomonella, 228
Leafhoppers
    and insecticide resistance,
        148
    and sugar beet, 342-44, 347,
        351-52
Leafminers, 201-24
Lebistina, 42
Leguminivora glycinivorella,
    511, 519
Leiopilio glaber, 173
Lekking by fruitflies, 123
Lepthyphantes tenuis, 170

Leptinotarsa decemlineata, 7,
    42, 229, 361
Leptoglossus phyllopus, 390,
    399
Leucophaea maderae, 3, 5-10
Lice and cultural entomology,
    190
Life cycle of fruit flies, 116
Life history of scorpions, 276-
    78
Life tables
    and fruit flies, 133
    and scorpions, 276
    and decision making, 418
Light
    intensity and activity, 299
    traps
        and arboviruses, 495-96
        and soybean pests, 526
Ligurotettix coquilletti, 320-
    21
Limonius, 344
Lindane, 155, 368-69, 374
Liocheles australasiae, 278
Liparocephalus, 29
Liriomyza, 344
    brassicae, 206, 213-14
    bryoni, 216
    congesta, 206-9
    huidobrensis, 203, 206, 217-
        18
    pictella, 215
    sativae, 203, 207-10, 213-16
    trifolii, 202-3, 205-13, 215-
        17
    urophorina, 215
Literature and insects, 182-84
Locusta migratoria, 10
Locusts
    and chitin biochemistry, 73
    and cultural entomology, 188,
        191
Lomechusa, 29
Lounsburyia, 60
Loxostege sticticalis, 344
Lucilia, 22
    cuprina, 74, 80-81, 83, 86,
        364, 369
Lures for fruit flies, 116, 125
Lutrochus, 253-54, 257, 261-62,
    266-68
    arizonicus, 257
    laticeps, 257
    luteus, 257
Lutzomyia longipalpis, 470
Lygus
    elisus, 344
    hesperus, 146, 344
Lymantria, 2
    dispar, 226, 391
Lytoxysta, 50, 52-53
Lytta, 344
    vesicatoria, 394

M

Machilanus swani, 165
Machilids and aeolian faunas,
  165
Macrelmis, 264
Macronychus, 255, 262, 264-
  65, 267-68
  glabratus, 255
  quadrituberculatus, 265
Macrosiphum euphorbiae, 59,
  346
Malacosoma disstria, 228, 330
Malaria, 154, 495-97
Malathion, 147-48, 153-54,
  371-72
Malpighian tubules, 441-59
Mamestra
  brassicae, 76, 87
  configurata, 397-99
Management systems, 424-26
Mandibular glands, 38, 405
Manduca sexta, 11, 86
Manna, 187
Mantids and cultural entomolo-
  gy, 188, 190-91
Marietta, 60
Maruca testulalis, 525
Massospora cicadina, 329
Maternal care in scorpions, 280
Mating
  of black flies, 301
  of fruit flies, 121, 123-25,
    130, 135
  of *Liriomyza*, 202-3
  of scorpions, 278, 288
  of stable flies, 306
  of tabanids, 303-4
Megacephala, 23
Megacormus gertschi, 279, 281
Melanism, 367
Melanoplus sanguinipes, 80
Melanotus, 344
Melitaea harrisii, 326
Melittin, 405-7
Metamorphosis
  and cultural entomolgy, 189
  and osmoregulation, 455
Metarhizium anisopliae, 229
Metaseiulus occidentalis, 365,
  372
Metasystox, 350
Methyl eugenol, 125
Metribuzin, 522
Metriorrhynchus, 20, 33, 39
  rhipidus, 33, 39
Mexican
  bean beetle, 521-22, 524
  entomofauna, 95-114
Microbial control, 225-51
Microcylloepus, 255, 260, 265
  thermarum, 255
Microfibrils of chitin, 71-72, 76

Microfilariae, 495
Microglotta, 29
Microsporidia
  and arboviruses, 495
  and entomopathogens, 230,
    236, 238, 244
  and IPM, 230, 238
  and population dynamics, 329
Migration and population dynamics, 323
Millipedes and exocrine compounds, 391, 393, 399-403
Mimicry, 17, 25, 33, 42
Mirids and aeolian faunas, 170
Mites
  and aeolian faunas, 167
  and chitin, 78, 80
  and insecticide resistance, 151
  and *Liriomyza*, 211
  and soybean, 524
  and sugar beet, 343
Mitochondria, 443, 446, 449,
  455
Models
  for biomedical research, 1-16
  and decision-making, 418
  and entomopathogens, 232-
    37, 240, 242
  and hyperparasitism, 61-64
  and insecticide resistance, 367
  for soybean pests, 528
Molecular
  biology, 157
  genetics, 151, 155
  probes, 483
  techniques and resistance,
    145-62
Molting hormone, 78-79
Monoclonal antibodies, 151,
  154, 483
Monoctonus, 59
Monogenic resistance, 364-66
Montane insects of Mexico, 95-
  114
Moonlight and biting flies, 299
Morpho, 185
Mosquitoes
  and aeolian faunas, 170
  and arboviruses, 479-505
  and blood-feeding, 463, 467-
    69
  and Bt, 231, 238
  and fungi, 229
  and insecticide resistance,
    147-49, 151, 154
  and IPM, 238
  and nematodes, 232
  and osmoregulation, 439-62
  and protozoa, 230
  and vision, 298-301
Mucopolysaccharides, 72
Musca domestica, 157, 371,
  382

Muscalure, 382
Muscid visual ecology, 306-7
Music and insects, 184
Mycetomes, 119
Myrrha 35
Myzus persicae, 148, 344-47,
  349, 351, 374

N

Nabis alternatus, 170
Nanobius, 32
Narceus
  annularis, 393
  gordanus, 393
Narpus, 263-64
Nasutitermes octopilis, 394, 396
Natural enemies
  and population dynamics,
    327-29
  of soybean pests, 524, 528
Nebria
  paradisi, 174
  vandykei, 173
Necrodes, 20, 28
  surinamensis, 28
Nectar
  and biting flies, 299, 302-4,
    310
  and fruit flies, 126
Nematicides and soybean pests,
  522
Nematodes
  and arboviruses, 495
  and entomopathogens, 241
  and IPM, 232, 238
Neoaplectana carpocapsae, 226-
  27
Neodiprion sertifer, 331
Neogregarines
  and entomopathogens, 241
  and IPM, 230
Neuroendocrine
  research, 1-16
  systems, 118
Neuropeptides, 5
Neurosecretory neurons, 3
Nezara viridula, 383, 390, 510,
  517
Nicotine, 155
Nikkomycins and chitin, 80-82
Nilaparvata lugens, 84
Nomuraea rileyi, 226, 229, 236,
  244, 519, 522, 525
Nosema
  heliothidis, 241
  locustae, 226, 230, 236
  pyrausta, 231, 244
Notocelia roborana, 326
Nuclear polyhedrosis viruses,
  226-28, 236-38, 241-42,
  329, 516, 519
  and soybean pests, 526

Nucleic acid probes, 154-55
Nutrition of fruit flies, 128
Nysius
    ericae, 170, 344
    raphanus, 171-72
    wekiuicola, 167

O

Ocelli of scorpions, 289
Ochlerotatus, 448
Odontota horni, 511
Odors and fruit flies, 117
Oeceoptoma, 28
Olfaction by fruit flies, 121-22
Oligota, 29
Oncopeltus fasciatus, 86
Ontholestes, 20, 30
Onychiurus, 344
Oocytes, 9
Oogonial infection, 492
Ootheca, 9
Opifex fuscus, 440, 450
Opisthophthalmus, 285
    carinatus, 283
Opius oophilus, 132
Optioservus, 261, 264
Orgyia pseudotsugata, 226, 228
Oreina, 41
Orientation of biting flies, 299,
    302
Oryctes, 228-29, 241
    rhinoceros, 228, 241
Oryzaephilus surinamensis, 371
Oscinella frit, 171
Osmeteria, 29, 383, 396
Osmoderma, 33
Osmoregulation in mosquitoes,
    439-62
Ostrinia nubilalis, 229, 231
Oulimnius, 262, 265, 267
Ovarian maturation, 129
Ovaries of fruit flies, 118, 129
Ovary, 8-9
Oviposition
    by fruit flies, 122-23, 127-28
    by *Liriomyza*, 203-4, 212-13
    by mosquitoes, 301
Oxidus gracilis, 399-400
Oxygen consumption of scor-
    pions, 287, 289
Oxytelus, 20, 31-32
    piceus, 32
    sculpturatus, 32

P

Pacemakers, 11-12
Pachyneuron, 52-53, 56, 59-60
    concolor, 60
Paederus, 20, 28
    fuscipes, 385

Pagelmis, 260-61
Palamnaeus longimanus, 280
Paltothyreus tarsatus, 405
Pandinus
    imperator, 288
    pallidus, 283
Papilio
    aegus, 383
    helenus, 396
    protenor, 396
Parabuthus villosus, 283
Paraoxon, 148-49, 151
Parasites
    and *Liriomyza*, 216-17
    and population dynamics,
        328-29
    of fruit flies, 131-33
    of leafhoppers, 353
    of riffle beetles, 265
Parasitoids, 49-70
Parathion, 147, 366, 522
Pardia tripunctata, 326
Pardosa mackenziana, 166
Paropsis atomaria, 40
Parthenogenesis in scorpions,
    278
Parturition in scorpions, 279-
    80
Paruroctonus, 279-81
    becki, 287
    boreus, 276, 281, 284, 286-
        88
    mesaensis, 276-88
    utahensis, 287-88
Pathogens of fruit flies, 132
Pederin, 28, 385
Pediobius, 519
    foveolatus, 518, 524
Pegomya, 344
Pelonomus, 266
Pemphigus populivenae, 344
Peptide hormones, 4
Peptidergic neurons, 5
Periplaneta americana, 7, 12
Peritrophic membrane, 71-72,
    76, 79-80, 82, 85
Permethrin, 518
Persistence of entomopathogens,
    236-37
Phaenoglyphis, 50, 52-53, 59
Phalangids and aeolian faunas,
    166-67, 173-74
Phanocerus, 266
Phausis, 34
Phenacoccus manihoti, 61-62
Phenyl carbamates and chitin,
    84-85
Pheromones
    and chemical defenses, 29
    and fruit flies, 117-18, 121,
        123-25, 135
    and traps, 146

and tsetse flies, 309
of aphids, 351
Phlebotomus papatasi, 469-70,
    482
Phloeopara, 29
Phoracantha, 20, 38
    onyma, 38-39
    semipunctata, 39
    synonyma, 39
Phorate, 351, 353, 522
Phorodon humili, 374
Phosphuga, 28
Photinus, 33-34
Photoperiod, 5-8, 10-12
    and vectors, 484
Photuris, 34
Phratora vitellinae, 399
Phrenapates, 37
Phthorimaea operculella, 239
Phylogeny of Coleoptera, 19
Physcus, 60
Phytoalexins, 520
Phytosus, 29
Picrotoxinin, 368
Pieris
    brassicae, 81
    rapae, 228
Piezodorus guildinii, 517
Pirimicarb, 351
Plant resistance, 519-20, 523,
    527-28
Plasmodium gallinaceum, 474
Plastron respiration, 260-61
Platambus, 20, 26-27
Platelets and saliva, 464-74
Plathypena scabra, 510
Platynota stultana, 344
Platynus dorsalis, 24
Platystethus, 31
Plodia interpunctella, 74, 76,
    148
Plumbagin and chitin, 85
Poison glands, 384-85
Pollen
    and biting flies, 304
    and fruit flies, 126
Polygenic resistance, 364-65
Polymorphism and population
    dynamics, 332
Polyoxins and chitin, 80-82, 84
Popillia japonica, 226
Population
    dynamics, 317-40
    of ants, 325
    of aphids, 325
    of bark beetles, 323, 327
    of grasshoppers, 320-21
    of gypsy moths, 323, 325,
        328
    of lady beetles, 323
    of leafminers, 325
    of locusts, 323

of pine sawflies, 323
of psyllids, 323, 325
of sawflies, 324
of tent caterpillars, 324-
  25
ecology
  of fruit flies, 129-35
  genetics, 146, 362, 367
  stability, 318-19
Potamodytes, 261
Potamophilops, 261
Predation by scorpions, 281-83
Predators and population dyna-
  mics, 327-28
Prococcophagus, 60
Progesterone, 397
Proline, 452-53
Promoresia, 255-56, 264
  elegans, 255-56
Propylaea, 35
Prostaglandins, 463, 470-71
Prothoracic
  defensive glands, 27, 36,
    38
  gland, 4, 11, 397
Protocerebrum, 4
Protolin, 7
Protoparnus, 267
Protozoa and IPM, 230-31
Proxenus mindara, 344
Psephenivorus, 265
Psephenoides, 259, 261, 268
Psephenus, 257-68
  herricki, 257, 259
  marlieri, 259
  murvoshi, 261
Pseudoplusia includens, 237
Pseudopolydesmus erasus, 399
Pseudopsis, 32
Psilopa leucostoma, 344
Psorophora columbiae, 300
Psyllids and aeolian faunas,
  167, 171
Pteromalus macronychivorus,
  265
Pterostichus lucublandus, 25
Pullus, 35
Pupation hormone, 2-3
Pygidial glands, 23-25, 33
Pyrausta nubilalis, 244
Pyrethroids, 146-47, 351
Pyrochroa, 36
Pyrrhalta luteola, 41

Q

Quedius, 30
Queen bees, 405
Quinones and chemical de-
  fenses, 19, 21, 23-25, 29,
  31, 37

R

Radioimmunoassay, 483
Receptors of scorpions, 281
Rectal papillae, 455
Reflex bleeding, 17, 34-35, 39,
  41
Resistance
  management, 369
  of mites, 361, 365, 372
  of mosquitoes, 368-71, 373
  of ticks, 371
  to aphids, 349
  to entomopathogens, 238-39
  to insecticides, 145-62, 216,
    361-80
Respiration of riffle beetles,
  260-61
Rhagoletis, 115, 120
Rhinocricus holomelanus, 393
Rhizobius, 35
Rhodnius prolixus, 467-68, 474
Rhopalosiphum insertum, 59
Rhythms in fruit flies, 128
Riffle beetles, 253-73
Riolus, 260
Ripicephalus sanguineus, 470
Rivellia quadrifasciata, 511
RNA
  and arboviruses, 495
  probes, 483
Romanomermis culicivorax, 226
Rotenone, 361
Rove beetles and chemical de-
  fenses, 28-30

S

Sand flies and blood-feeding,
  469-70
Saissetia oleae, 326
Salicin, 40
Saline-water mosquito larvae,
  444-50
Saliva
  and arboviruses, 486
  and blood-feeding, 463-78
Salivary
  antigens, 466, 472
  glands
    and blood-feeding, 463,
      467, 369, 472, 475
    and arboviruses, 485-89
Scarabs and cultural entomolo-
  gy, 182, 186, 188, 191
Schistocerca gregaria, 73
Schizotus, 36
Scolytids and aeolian faunas,
  170
Scorpions
  and bionomics, 275-95

and cultural entomology, 185,
  188-91
Scotogramma trifolii, 343-44
Scutigerella immaculata, 344
Secaptin, 407
Seleniphera lunigera, 320
Sensilla
  of fruit flies, 117
  of scorpions, 281
Sequential sampling, 216
Sequestration, 35, 39-41
Serology and arboviruses, 482-83
Serotonin, 465-66
Serpentine leafminer, 201-24
Serradigitus, 279, 281
  deserticola, 284
Sex
  pheromones, 123-24
  ratio, 493
Shellac, 187
Sigma virus, 492
Silk
  and cultural entomology, 187
  moths and chitin, 78, 82-83
Silpha, 20, 27-28
  americana, 28
Silphids and aeolian faunas, 166
Simulation modeling, 418-23,
  427, 431-35, 488, 528
Simuliidae, 301-3
Simulium euryadminiculum, 303
Sirex noctilio, 232
Sitophilus
  oryzae, 374-75
  zeamais, 375
Snowfield arthropods, 163-79
Soybean arthropods, 507-38
Spastonyx, 36
Speciation, 102-3
Species complexes, 481
Spectral sensitivity, 121
Spermatophore, 278
Sperm transfer, 278
Spiders
  and aeolian faunas, 165-69,
    174-75
  and cutural entomology, 183,
    189-91
Spiroplasma tenella, 352
Spissistilus festinus, 516-17
Spodoptera
  exigua, 343-44
  praefica, 344
Spruce budworm, 169, 171,
  320, 323, 328, 330, 332
Spruce sawfly, 328
Stable flies, 306-7
Staphylinids and aeolian faunas,
  166-67, 169, 174
Staphylinus, 30
Stegomyia, 299
Steinernema feltiae, 226-27, 238

Stenechus subsignatus, 511
Stenelmis, 256, 260, 263, 265, 267
  crenata, 256
  quadrimaculata, 260
Stenhelmoides, 260-61
Stenocentrus ostricilla, 38
Stenopelmatus, 290
Stenus comma, 29
Steroids and chemical defenses, 27, 34, 41
Stomoxys calcitrans, 74-75, 82-83, 86, 306
Stresses and population dynamics, 321, 324
Stretch receptors, 443, 456
Stridulation, 38
  by dacines, 117, 123-24
  by *Liriomyza*, 203
Suction traps, 300
Sugar beet pests, 341-60
Sumitrosis rosea, 511
Supercooling, 172
Superstitionia donensis, 281
Swarming
  of biting flies, 298-99, 301-2, 309
  of riffle beetles, 256
Syllites grammicus, 38
Symbiotic bacteria, 118-19
Synthesis of chitin, 71-93
Syntomium, 32
Syntropis, 279
Syrphids and aeolian faunas, 169-70
Syrphus torvus, 169
Systematics of scorpions, 276

T

Tabanidae, 303-6
Tabanus
  bromius, 304
  nigrovittatus, 304-5
Telenomus chloropus, 523
Temephos resistance, 371
Tenebrio, 37
Tent caterpillar, 328, 331-32
Tergal glands, 29-32
Termites
  and beetles, 29
  and exocrine compounds, 394, 396
Tetanops myopaeformis, 344
Tetranychus, 344
  urticae, 80, 372
Tetraopes
  basilis, 39
  femoratus, 39
  oregonensis, 39
  tetrophthalmus, 39

Tetrastichus, 52-53
Thanatosis, 27, 33-34, 38
Thectura, 29
Theridion bimaculatum, 170
Therioaphis trifolii, 57
Thermal pools and riffle beetles, 260
Thermoregulation, 306
Thrombin, 465-66
Thyreocephalus, 30
Ticks
  and arboviruses, 482, 491
  and blood-feeding, 463, 467, 470-73
  and exocrine compounds, 400
Tiger beetles, 20
Tityus
  bahiensis, 277-78
  serrulatus, 278
  trivittatus, 277
Toxaphene, 522
Toxins and saliva, 463
Toxorhynchites, 299
Tracheal gills, 261
Transmission
  of arboviruses, 485-92, 494
  of beet viruses, 351
  of entomopathogens, 233-34
Transovarial transmission of arboviruses, 489, 491-92
Trap cropping, 521
Traps
  for biting flies, 302-4, 306, 308-9
  for fruit flies, 116
  for mosquitoes, 301
  for sugar beet pests, 349
  for riffle beetles, 255, 261
Trehalose, 72, 452-53
Triatoma infestans, 75
Triatomine bugs and blood-feeding, 467-68
Tribolium castaneum, 38, 74-76, 81-85, 148, 372, 391
Trichogramma dendrolinae, 527
Trichoplusia ni, 81-83, 149, 228, 387
Trichopria, 265
Trimerotropis pallidipennis, 320-21
Trioxys pallidus, 59
Trogoderma, 241
Trogophloeus, 31
Trypanosomes and blood-feeding, 474
Tsetse flies
  and blood-feeding, 470
  and vision, 307-9
Tussock moth, 328
Tyrophagus similis, 344

U

Udea
  profundalis, 344
  rubigalis, 344
Uloma tenebrionides, 37
Uranotaenia sapphirina, 300
Uric acid, 442
Urine and osmoregulation, 439, 441-43, 445-46, 448-49, 454-59
Uroctonus, 279-81, 285
  mordax, 276-78, 281-82, 284, 287-88
Urodacus, 289
  abruptus, 276-81, 285
  hoplurus, 285
  yaschenkoi, 276-77, 285
Uroplata, 321
UV
  detection, 300, 302-3, 305, 307-8
  receptors, 299-300

V

Vaejovis, 279-81, 283, 286
  confusus, 283-84, 286-88
  gertschi, 276, 282-84, 287-88
  hirsuticauda, 287
  janssi, 284
  littoralis, 284, 288
  spinigerus, 279, 281, 283-84, 286-88
  wupatkiensis, 287
Vairimorpha necatrix, 226-27, 231
Vector
  behavior, 493
  competence, 480-85, 88, 95
  control, 495
  ecology, 493
Vectors of sugar beet viruses, 345-50
Vejovoidus, 281
  longiunguis, 285
Venereal transmission of arboviruses, 491
Venom
  peptides, 405
  of scorpions, 282
Vertical transmission, 234, 489-93
Verticillium lecanii, 226
Virulence of entomopathogens, 234-35
Viruses
  and IPM, 226-28
  and soybean pests, 519
  and sugar beet, 342-55
  identification of, 483
  transmission of, 355, 480

Vision and biting flies, 297-316
Visual ecology, 297-316
Vitellogenesis, 8, 126

W

Wasps and cultural entomology, 187
Water beetles and chemical defenses, 25-27
Wax, 186
Webworms, 344
Weeds and sugar beet, 342-43, 345-46, 349
Whiteflies and soybeans, 517

Wind-borne nutrients, 164
Wing dimorphism, 255
Winter moth, 327
Wyeomyia smithi, 484

X

Xantholinus, 20, 30
Xenelmis, 262

Y

Yellow fever virus, 481
Yolk deposition, 8-9

Z

Zaitzevia, 254-55, 260, 263
    thermae, 255
Zalobius, 32
Zonocerus, 61
Zophobas, 38
    rugipes, 392
Zygaena trifolii, 402
Zyras
    comes, 29
    humeralis, 29
    japonicus, 29

CUMULATIVE INDEXES

CONTRIBUTING AUTHORS, VOLUMES 23–32

A

Akre, R. D., 23:215–38
Allan, S. A., 32:297–316
Alstad, D. N., 27:369–84
Altieri, M. A., 29:383–402
Altner, H., 30:273–95
Ananthakrishnan, T. N., 24:159–83
Andersen, S. O., 24:29–61
Anderson, N. H., 24:351–77
Asman, S. M., 26:289–318

B

Baker, H. G., 28:407–53
Baker, R. R., 28:65–89
Balashov, Yu. S., 29:137–56
Barfield, C. S., 28:319–35
Baron, R. L., 26:29–48
Beck, S. D., 28:91–108
Beckage, N. E., 30:371–413
Bedford, G. O., 23:125–49; 25:309–39
Beeman, R. W., 27:253–81
Bellotti, A., 23:39–67
Bentzien, M., 26:233–58
Berg, C. O., 23:239–58
Berlocher, S. H., 29:403–33
Berry, S. J., 27:205–27
Blomquist, G. J., 27:149–72
Blum, M. S., 32:381–413
Bownes, M., 31:507–31
Brader, L., 24:255–54
Bradley, T. J., 32:439–62
Brittain, J. E., 27:119–47
Brogdon, W. G., 32:145–62
Bronson, L., 26:345–71
Brown, A. W. A., 25:xi–xxvii
Brown, H. P., 32:253–73
Brown, K. S. Jr., 26:427–56
Brown, T. M., 32:145–62
Burkholder, W. E., 30:257–72
Bush, G. L., 29:471–504
Byers, G. W., 28:203–28

C

Caltagirone, L. E., 26:213–32
Carlson, S. D., 24:379–416
Catts, E. P., 27:313–38
Chapman, R. F., 31:479–505

Chen, P. S., 29:233–55
Cheng, L., 30:111–35
Chi, C., 24:379–416
Chiang, H. C., 23:101–23
Claridge, M. F., 30:297–317
Cochran, D. G., 30:29–49
Cohen, E., 32:71–93
Corbet, P. S., 25:189–217
Coulson, R. N., 24:417–47; 32:415–37
Crossley, D. A. Jr., 31:177–94

D

Daly, H. V., 30:415–38
Daoust, R. A., 31:95–119
Davies, J. E., 23:353–66
Davis, H. G., 23:215–38
Day, J. F., 32:297–316
DeFoliart, G. R., 32:479–505
De Jong, D., 27:229–52
de Kort, C. A. D., 26:1–28
Delcomyn, F., 30:239–56
Denlinger, D. L., 31:239–64
Dettner, K., 32:17–48
Diehl, S. R., 29:471–504
Dixon, A. F. G., 30:155–74
Dohse, L., 28:319–35
Druk, A. Ya., 31:533–45
Duffey, S. S., 25:447–77
Dumser, J. B., 25:341–69
Dunn, P. E., 31:321–39

E

Edman, J. D., 32:297–316
Edmunds, G. F. Jr., 27:369–84
Edwards, J. S., 32:163–79
Ehler, L. E., 23:367–87
Eickwort, G. C., 25:421–46; 27:229–52
Elliott, M., 23:443–59
Evenhuis, N. L., 26:159–81

F

Fahmy, M. A. H., 31:221–37
Farber, P. L., 23:91–99
Ferron, P., 23:409–42
Fletcher, B. S., 32:115–44
Fletcher, D. J. C., 23:151–71; 30:319–43

Frankie, G. W., 23:367–87
Freed, V., 23:353–66
Friedman, S., 23:389–407
Friend, J. A., 31:25–48
Futuyma, D. J., 30:217–38
Fuxa, J. R., 32:225–51

G

Gage, S. H., 26:259–87
Gagné, W. C., 29:383–402
Gamboa, G. J., 31:431–54
Getz, W. M., 27:447–66
Ginsberg, H. S., 25:421–46
Gonçalves, L. S., 23:197–213
Gorham, J. R., 24:209–24
Granger, N. A., 26:1–28
Grégoire, J.-C., 28:263–89
Griffiths, G. W., 25:161–87
Grimstad, P. R., 32:479–505
Gut, L. J., 31:455–78
Gutierrez, A. P., 27:447–66

H

Hackman, R. H., 27:75–95
Hagedorn, H. H., 24:475–505
Halffter, G., 32:95–114
Hardy, J. L., 28:229–62
Hargrove, W. W., 31:177–96
Harpaz, I., 29:1–23
Harris, M. K., 28:291–318
Hassell, M. P., 29:89–114
Haynes, D. L., 26:259–87
Henry, J. E., 26:49–73
Higley, L. G., 31:341–68
Hogue, C. L., 32:181–99
Hoogstraal, H., 26:75–99
Horie, Y., 25:49–71
Houk, E. J., 25:161–87; 28:229–62
Howard, R. W., 27:149–72
Howarth, F. G., 28:365–89
Hoy, M. A., 30:345–70
Huddleston, E. W., 27:283–311
Hutchins, S. H., 31:341–68

I

Ikeda, T., 29:115–35
Illies, J., 28:391–406
Iwantsch, G. F. 25:397–419

J

Jackai, L. E. N., 31:95–119
Janes, N. F., 23:443–69
Jay, S. C., 31:49–65
Jeanne, R. L., 25:371–96

K

Kaneshiro, K. Y., 28:161–78
Keeley, L. L., 23:329–52
Keh, B., 30:137–54
Kevan, P. G., 28:407–53
Kiritani, K., 24:279–312
Kirschbaum, J. B., 30:51–70
Knutson, L., 23:329–58
Kobayashi, F., 29:115–35
Kogan, M., 32:507–38
Kosztarab, M., 24:1–27
Kramer, L. D., 28:229–62
Krantz, G. W., 24:121–58
Kristensen, N. P., 26:135–57
Krivolutsky, D. A., 31:533–45
Kunkel, J. G., 24:475–505

L

Labeyrie, V., 23:69–89
Lacey, L. A., 31:265–96
Lange, W. H., 26:345–71;
    32:341–60
Langley, P. A., 23:283–307
Larsen-Rapport, E. W., 31:145–
    75
Laverty, T. M., 29:175–99
Lawton, J. H., 28:23–39
Levine, J. F., 30:439–60
Levins, R., 25:287–308
Lindquist, E. E., 24:121–58
Liss, W. J., 31:455–78
Lloyd, J. E., 28:131–60
Lockley, T., 29:299–320
Loftus, R., 30:273–95

M

Ma, M., 30:257–72
Mackay, R. J., 24:185–208
Marks, E. P., 25:73–101
Masaki, S., 25:1–25
Matteson, P. C., 29:383–402
May, M. L., 24:313–49
McCaffery, A. R., 31:479–505
McDonald, P. T., 26:289–318
McKenzie, J. A., 32:361–80
McLintock, J., 23:17–37
Merritt, R. W., 25:103–32
Metcalf, R. L., 25:219–56
Miller, D. R., 24:1–27
Mitchell, R., 26:373–96
Morse, R. A., 27:229–52
Mumford, J. D., 29:157–74

N

Norton, G. A., 29:157–74

O

OConnor, B. M., 27:385–409
Opler, P., 26:233–58
Owens, E. D., 28:337–64
Owens, J. C., 27:283–311

P

Page, R. E. Jr., 31:297–320
Page, W. W., 31:479–505
Parrella, M. P., 32:201–24
Parker, G. A., 23:173–96
Pasteels, J. M., 28:263–89
Pedigo, L. P., 31:341–68
Petersen, C. E., 28:455–86
Peterson, S. C., 30:217–38
Pfennig, D. W., 31:431–54
Piesman, J., 30:439–60
Pinder, L. C. V., 31:1–23
Plowright, R. C., 29:175–99
Potter, C., 23:443–69
Powell, J. A., 25:133–59
Prestwich, G. D., 29:201–32
Pritchard, G., 28:1–22
Prout, T., 26:289–318
Pyle, R., 26:233–58

R

Rabinovich, J. E. 26:101–33
Radcliffe, E. B., 27:173–204
Randolph, S. E., 30:197–216
Reeve, H. K., 31:431–54
Reeves, W. C., 28:229–62
Ribeiro, J. M. C., 32:463–78
Richards, A. G., 23:309–28
Richardson, A. M. M., 31:25–
    48
Riechert, S. E., 29:299–320
Robinson, M. H., 27:1–20
Rogers, D. J., 30:197–216
Rose, D. J. W., 23:259–82
Ross, K. G., 30:319–43
Rothfels, K. H., 24:507–39
Roush, R. T., 32:361–80
Rowell-Rahier, M., 28:263–89
Russell, L. M., 23:1–15

S

Saunders, M. C., 32:415–37
Scharrer, B., 32:1–16
Schmidt, J. O., 27:339–68
Schowalter, T. D., 31:177–96
Schuh, R. T., 31:67–93
Scriber, J. M., 26:183–211
Seabrook, W. D., 23:471–85

Seastedt, T. R., 29:25–46
Sedell, J. R., 24:351–77
Sehnal, F., 30:89–109
Singh, S. R., 24:255–78
Slansky, F. Jr., 26:183–211
Smith, R. F., 23:353–66
Sōgawa, K., 27:49–73
Sonenshine, D. E., 30:1–28
Southgate, B. J., 24:449–73
Spielman, A., 30:439–60
Staal, G. B., 31:391–429
Stanford, J. A., 27:97–117
Stark, R. W., 27:479–509
Steffan, W. A., 26:159–81
Stimac, J. L., 28:319–35
Stort, A. C., 23:197–213
Strong, D. R. Jr., 24:89–119
Sullivan, D. J., 32:49–70
Sylvester, E. S., 25:257–86;
    30:71–88

T

Tallamy, D. W., 31:369–90
Taylor, L. R., 29:321–57
Tempelis, C. H., 28:179–201
Terriere, L. C., 29:71–88
Thompson, S. N., 31:197–219
Thornhill, R., 28:203–28
Thornton, I. W. B., 30:175–96
Tinsley, T. W., 24:63–87
Tobe, S. S., 23:283–307
Turnipseed, S. G., 32:507–38

U

Undeen, A. H., 31:265–96

V

van Emden, H. F., 24:255–78
Van Schoonhoven, A., 23:39–
    67
Viggiani, G., 29:257–76
Visser, J. H., 31:121–44

W

Waage, J. K., 29:89–114
Wallace, J. B., 25:103–32
Wallner, W. E., 32:317–40
Wallwork, J. A., 28:109–30
Ward, J. V., 27:97–117
Warren, C. E., 31:455–78
Washino, R. K., 28:179–201
Watanabe, H., 25:49–71
Waters, W. E., 25:479–509
Watts, D. M., 32:479–505
Watts, J. G., 27:283–311
Wehner, R., 29:277–98
Weinstein, L. H., 27:369–84

Welch, S. M., 29:359–81
Westigard, P. H., 31:455–78
Whalon, M. E., 29:435–70
Wharton, G. W., 23:309–28
Whitcomb, R. F., 26:397–425
Wiegert, R. G., 28:455–86
Wiggins, G. B., 24:185–208
Wikel, S. K., 27:21–48

Wille, A., 28:41–64
Williams, S. C., 32:275–95
Wilson, M. L., 25:287–308;
    30:439–60
Wirtz, R. A., 29:47–69
Wood, D. L., 27:411–46
Wood, T. K., 31:369–90
Wootton, R. J., 26:319–44

Y

Yamane, A., 29:115–35

Z

Zacharuk, R. Y., 25:27–47
Zeledón, R., 26:101–33

CHAPTER TITLES, VOLUMES 23–32

ACARINES, ARACHNIDS, AND OTHER ARTHROPODS

Evolution of Phytophagous Mites (Acari)	G. W. Krantz, E. E. Lindquist	24:121–58
Courtship and Mating Behavior in Spiders	M. H. Robinson	27:1–20
Mite Pests of Honey Bees	D. De Jong, R. A. Morse, G. C. Eickwort	27:229–52
Evolutionary Ecology of Astigmatid Mites	B. M. OConnor	27:385–409
Oribatids in Forest Ecosystems	J. A. Wallwork	28:109–30
Pheromones and Other Semiochemicals of the Acari	D. E. Sonenshine	30:1–28
Recent Advances in Genetics and Genetic Improvement of the Phytoseiidae	M. A. Hoy	30:345–70
Biology of Terrestrial Amphipods	J. A. Friend, A. M. M. Richardson	31:25–48
Scorpion Bionomics	S. C. Williams	32:275–95

AGRICULTURAL ENTOMOLOGY

Mite and Insect Pests of Cassava	A. Bellotti, A. van Schoonhoven	23:39–67
Pest Management in Corn	H. C. Chiang	23:101–23
Agromedical Approach to Pesticide Management	J. E. Davies, R. F. Smith, V. Freed	23:353–66
Integrated Pest Control in the Developing World	L. Brader	24:225–54
Insect Pests of Grain Legumes	S. R. Singh, H. F. van Emden	24:255–78
Pest Management in Rice	K. Kiritani	24:279–312
Changing Role of Insecticides in Crop Protection	R. L. Metcalf	25:219–56
The Cereal Leaf Beetle in North America	D. L. Haynes, S. H. Gage	26:259–87
Insect Pests of Tomatoes	W. H. Lange, L. Bronson	26:345–71
Insect Pests of Potato	E. B. Radcliffe	27:173–204
Rangeland Entomology	J. G. Watts, E. W. Huddleston, J. C. Owens	27:283–311
Integrated Pest Management of Pecans	M. K. Harris	28:291–318
Economics of Decision Making in Pest Management	J. D. Mumford, G. A. Norton	29:157–74
Developments in Computer-Based IPM Extension Delivery Systems	S. M. Welch	29:359–81
Modification of Small Farmer Practices for Better Pest Management	P. C. Matteson, M. A. Altieri, W. C. Gagné	29:383–402
Apple IPM Implementation in North America	M. E. Whalon, B. A. Croft	29:435–70
Insect Pests of Cowpeas	L. E. N. Jackai, R. A. Daoust	31:95–119
Economic Injury Levels in Theory and Practice	L. P. Pedigo, S. H. Hutchins, L. G. Higley	31:341–68
Perspectives on Arthropod Community Structure, Organization, and Development in Agricultural Crops	W. J. Liss, L. J. Gut, P. H. Westigard, C. E. Warren	31:455–78
Improved Detection of Insecticide Resistance Through Conventional and Molecular Techniques	T. M. Brown, W. G. Brogdon	32:145–62
Insect Pests of Sugar Beet	W. H. Lange	32:341–60

Computer-Assisted Decision-Making as
    Applied to Entomology   R. N. Coulson, M. C. Saunders   32:415–37
Ecology and Management of Soybean
    Arthropods   M. Kogan, S. G. Turnipseed   32:507–38

APICULTURE AND POLLINATION
The African Bee, *Apis mellifera adansonii,* in
    Africa   D. J. C. Fletcher   23:151–71
Honey Bee Improvement Through Behavioral
    Genetics   L. S. Gonçalves, A. C. Stort   23:197–213
Foraging and Mating Behavior in Apoidea G. E. Eickwort, H. S. Ginsberg   25:421–46
Mite Pests of Honey Bees   D. De Jong, R. A. Morse, G. C.
                                           Eickwort   27:229–52
Insects As Flower Visitors and Pollinators P. G. Kevan, H. G. Baker   28:407–53
Spatial Management of Honey Bees on Crops S. C. Jay   31:49–65

BEHAVIOR
Evolution of Competitive Mate Searching G. A. Parker   23:173–96
Evolution of Social Behavior in the Vespidae R. L. Jeanne   25:371–96
Foraging and Mating Behavior in Apoidea G. C. Eickwort, H. S. Ginsberg   25:421–46
Courtship and Mating Behavior in Spiders M. H. Robinson   27:1–20
Bioluminescence and Communication in
    Insects   J. E. Lloyd   28:131–60
Visual Detection of Plants by Herbivorous
    Insects   R. J. Prokopy, E. D. Owens   28:337–64
Defense Mechanisms of Termites   G. D. Prestwich   29:201–32
Astronavigation in Insects   R. Wehner   29:277–98
Pheromones and Other Semiochemicals of the
    Acari   D. E. Sonenshine   30:1–28
Factors Regulating Insect Walking   F. Delcomyn   30:239–56
Pheromones for Monitoring and Control of
    Stored-Product Insects   W. E. Burkholder, M. Ma   30:257–72
Acoustic Signals in the Homoptera: Behavior,
    Taxonomy, and Evolution   M. F. Claridge   30:297–317
Host Odor Perception in Phytophagous Insects J. H. Visser   31:121–44
Convergence Patterns in Subsocial Insects D. W. Tallamy, T. K. Wood   31:369–90
The Evolution and Ontogeny of Nestmate
    Recognition in Social Wasps   G. J. Gamboa, H. K. Reeve, D. W.
                                           Pfennig   31:431–54
Visual Ecology of Biting Flies   S. A. Allan, J. F. Day, J. D.
                                           Edman   32:297–316

BIOGEOGRAPHY
See SYSTEMATICS, EVOLUTION, AND BIOGEOGRAPHY

BIOLOGICAL CONTROL
Biological Control of Insect Pests by
    Entomogenous Fungi   P. Ferron   23:409–42
Integrated Pest Control in the Developing
    World   L. Brader   24:255–54
Host Suitability for Insect Parasitoids S. B. Vinson, G. F. Iwantsch   25:397–419
Natural and Applied Control of Insects by
    Protozoa   J. E. Henry   26:49–73
Landmark Examples in Classical Biological
    Control   L. E. Caltagirone   26:213–32
The Chemical Ecology of Defense in
    Arthropods   J. M. Pasteels, J.-C. Grégoire, M.
                                           Rowell-Rahier   28:263–89
Spiders as Biological Control Agents   S. E. Riechert, T. Lockley   29:299–320
Nutrition and In Vitro Culture of Insect
    Parasitoids   S. N. Thompson   31:197–219
Insect Hyperparasitism   D. J. Sullivan   32:49–70

BIONOMICS
See also ECOLOGY

Biology and Ecology of the Phasmatodea	G. O. Bedford	23:125–49
Biology and Pest Status of Venomous Wasps	R. D. Akre, H. G. Davis	23:215–38
Biology of the Bruchidae	B. J. Southgate	24:449–73
Biology of Odonata	P. S. Corbet	25:189–217
Biology, Ecology, and Control of Palm Rhinoceros Beetles	G. O. Bedford	25:309–39
Biology of *Toxorhynchites*	W. A. Steffan, N. L. Evenhuis	26:159–81
The Rice Brown Planthopper: Feeding Physiology and Host Plant Interactions	K. Sōgawa	27:49–73
Biology of Mayflies	J. E. Brittain	27:119–47
Biology of New World Bot Flies: Cuterebridae	E. P. Catts	27:313–38
Biology of Tipulidae	G. Pritchard	28:1–22
Biology of the Stingless Bees	A. Wille	28:41–64
Biology of the Mecoptera	G. W. Byers, R. Thornhill	28:203–28
The Ecology and Sociobiology of Bumble Bees	R. C. Plowright, T. M. Laverty	29:175–99
Bionomics of the Aphelinidae	G. Viggiani	29:257–76
Population Ecology of Tsetse	D. J. Rogers, S. E. Randolph	30:197–216
Bionomics of the Variegated Grasshopper (*Zonocerus variegatus*) in West and Central Africa	R. F. Chapman, W. W. Page	31:479–505
The Biology of Dacine Fruit Flies	B. S. Fletcher	32:115–44
Biology of *Liriomyza*	M. P. Parrella	32:201–24

ECOLOGY
See also BIONOMICS; BEHAVIOR

The Significance of the Environment in the Control of Insect Fecundity	V. Labeyrie	23:69–89
Ecology of Insects in Urban Environments	G. W. Frankie, L. E. Ehler	23:367–87
Recent Advances in the Study of Scale Insects	D. R. Miller, M. Kosztarab	24:1–27
Ecological Diversity in Trichoptera	R. J. Mackay, G. B. Wiggins	24:185–208
Detritus Processing by Macroinvertebrates in Stream Ecosystems	N. H. Anderson, J. R. Sedell	24:351–77
Filter-Feeding Ecology of Aquatic Insects	J. B. Wallace, R. W. Merritt	25:103–32
Biology of Odonata	P. S. Corbet	25:189–217
Ecological Theory and Pest Management	R. Levins, M. Wilson	25:287–308
Biology, Ecology, and Control of Palm Rhinoceros Beetles	G. O. Bedford	25:309–39
Foraging and Mating Behavior in Apoidea	G. C. Eickwort, H. S. Ginsberg	25:421–46
The Nutritional Ecology of Immature Insects	J. M. Scriber, F. Slansky, Jr.	26:183–211
Insect Conservation	R. Pyle, M. Bentzien, P. Opler	26:233–58
Insect Behavior, Resource Exploitation, and Fitness	R. Mitchell	26:373–96
Thermal Responses in the Evolutionary Ecology of Aquatic Insects	J. V. Ward, J. A. Stanford	27:97–117
Effects of Air Pollutants on Insect Populations	D. N. Alstad, G. F. Edmunds, Jr., L. H. Weinstein	27:369–84
A Perspective on Systems Analysis in Crop Production and Insect Pest Management	W. M. Getz, A. P. Gutierrez	27:447–66
Plant Architecture and the Diversity of Phytophagous Insects	J. H. Lawton	28:23–39
Insect Territoriality	R. R. Baker	28:65–89
Dispersal and Movement of Insect Pests	R. E. Stinner, C. S. Barfield, J. L. Stimac, L. Dohse	28:319–35
Ecology of Cave Arthropods	F. G. Howarth	28:365–89
Energy Transfer In Insects	R. G. Wiegert, C. E. Petersen	28:455–86

The Role of Microarthropods in
  Decomposition and Mineralization
  Processes   T. R. Seastedt   29:25–46
Host-Parasitoid Population Interactions   M. P. Hassell, J. K. Waage   29:89–114
Biology of *Halobates* (Heteroptera: Gerridae) L. Cheng   30:111–35
Structure of Aphid Populations   A. F. G. Dixon   30:155–74
Genetic Variation in the Use of Resources by
  Insects   D. J. Futuyma, S. C. Peterson   30:217–38
Pheromones for Monitoring and Control of
  Stored-Product Insects   W. E. Burkholder, M. Ma   30:257–72
Biology of Freshwater Chironomidae   L. C. V. Pinder   31:1–23
Herbivory in Forested Ecosystems   T. D. Schowalter, W. W. Hargrove,
                                                 D. A. Crossley, Jr.   31:177–96
Dormancy in Tropical Insects   D. L. Denlinger   31:239–64
The Biology of Dacine Fruit Flies   B. S. Fletcher   32:115–44
Arthropods of Alpine Aeolian Ecosystems   J. S. Edwards   32:163–79
Biology of Riffle Beetles   H. P. Brown   32:253–73
Factors Affecting Insect Population Dynamics:
  Differences Between Outbreak and
  Non-Outbreak Species   W. E. Wallner   32:317–40

EVOLUTION
See SYSTEMATICS, EVOLUTION, AND BIOGEOGRAPHY

FOREST ENTOMOLOGY
Population Dynamics of Bark Beetles   R. N. Coulson   24:417–47
Forest Pest Management: Concept and Reality   W. E. Waters, R. W. Stark   25:479–509
The Role of Pheromones, Kairomones, and
  Allomones in the Host Selection and
  Colonization Behavior of Bark Beetles   D. L. Wood   27:411–46
The Japanese Pine Sawyer Beetle as the
  Vector of Pine Wilt Disease   F. Kobayashi, A. Yamane, T. Ikeda   29:115–35

GENETICS
Honey Bee Improvement Through Behavioral
  Genetics   L. S. Gonçalves, A. C. Stort   23:197–213
Recent Advances in the Study of Scale
  Insects   D. R. Miller, M. Kosztarab   24:1–17
Cytotaxonomy of Black Flies (Simuliidae)   K. H. Rothfels   24:507–39
Field Studies of Genetic Control Systems for
  Mosquitoes   S. M. Asman, P. T. McDonald, T.
                                                 Prout   26:289–318
Sexual Selection and Direction of Evolution
  in the Biosystematics of Hawaiian
  Drosophilidae   K. Y. Kaneshiro   28:161–78
Potential Implication of Genetic Engineering
  and Other Biotechnologies to Insect Control J. B. Kirschbaum   30:51–70
Recent Advances in Genetics and Genetic
  Improvement of the Phytoseiidae   M. A. Hoy   30:345–70
Imaginal Disc Determination: Molecular and
  Cellular Correlates   E. W. Larsen-Rapport   31:145–75
Expression of the Genes Coding for
  Vitellogenin (Yolk Protein)   M. Bownes   31:507–31
Ecological Genetics of Insecticide and
  Acaricide Resistance   R. T. Roush, J. A. McKenzie   32:361–80

HISTORICAL
Leland Ossian Howard: A Historical Review   L. M. Russell   23:1–15
A Historical Perspective on the Impact of the
  Type Concept on Insect Systematics   P. Farber   23:91–99
The First Twenty-Five Years of the Annual
  Review of Entomology: An Overview   A. W. A. Brown   25:xi–xxvii

Frederick Simon Bodenheimer (1897–1959):
  Idealist, Scholar, Scientist   I. Harpaz   29:1–23
  Cultural Entomology   C. L. Hogue   32:181–99

INSECTICIDES AND TOXICOLOGY
  The Future of Pyrethroids in Insect Control   M. Elliot, N. F. James, C. Potter   24:443–69
  Changing Role of Insecticides in Crop
    Protection   R. L. Metcalf   25:219–56
  Delayed Neurotoxicity and Other
    Consequences of Organophosphate Esters   R. L. Baron   26:29–48
  Recent Advances in Mode of Action of
    Insecticides   R. W. Beeman   27:253–81
  Induction of Detoxication Enzymes in Insects   L. C. Terriere   29:71–88
  Derivatization Techniques in the Development
    and Utilization of Pesticides   M. A. H. Fahmy   31:221–37
  Chitin Biochemistry: Synthesis and Inhibition   E. Cohen   32:71–93
  Improved Detection of Insecticide Resistance
    Through Conventional and Molecular
    Techniques   T. M. Brown, W. G. Brogdon   32:145–62
  Ecological Genetics of Insecticide and
    Acaricide Resistance   R. T. Roush, J. A. McKenzie   32:361–80

MEDICAL AND VETERINARY ENTOMOLOGY
  Mosquito-Virus Relationships of American
    Encephalitides   J. McLintock   23:17–37
  The Significance for Human Health of Insects
    in Food   J. R. Gorham   24:209–24
  Changing Patterns of Tickborne Diseases in
    Modern Society   H. Hoogstraal   27:75–99
  Chagas' Disease: An Ecological Appraisal
    With Special Emphasis on Its Insect
    Vectors   R. Zeledón, J. E. Rabinovich   26:101–33
  Immune Responses to Arthropods and Their
    Products   S. K. Wikel   27:21–48
  Biology of New World Bot Flies:
    Cuterebridae   E. P. Catts   27:313–38
  Mosquito Host Bloodmeal Identification:
    Methodology and Data Analysis   R. K. Washino, C. H. Tempelis   28:179–201
  Intrinsic Factors Affecting Vector Competence
    of Mosquitoes for Arboviruses   J. L. Hardy, E. J. Houk, L. D.
                                                    Kramer, W. C. Reeves   28:229–62
  Allergic and Toxic Reactions to Non-Stinging
    Arthropods   R. A. Wirtz   29:47–69
  Interaction Between Blood-Sucking
    Arthropods and Their Hosts, and its
    Influence on Vector Potential   Yu. S. Balashov   29:137–56
  Scope and Applications of Forensic
    Entomology   B. Keh   30:137–54
  Ecology of Ixodes dammini-borne Human
    Babesiosis and Lyme Disease   A. Spielman, M. L. Wilson, J. F.
                                                    Levine, J. Piesman   30:439–60
  Microbial Control of Black Flies and
    Mosquitoes   L. A. Lacey, A. H. Undeen   31:265–96
  Role of Saliva in Blood-Feeding by
    Arthropods   J. M. C. Ribeiro   32:463–78
  Advances in Mosquito-Borne
    Arbovirus/Vector Research   G. R. DeFoliart, P. R. Grimstad,
                                                    D. M. Watts   32:479–505

MORPHOLOGY
  The Functional Morphology of the Insect
    Photoreceptor   S. D. Carlson, C. Chi   24:379–416

Ultrastructure and Function of Insect
  Chemosensilla   R. Y. Zacharuk   25:27–47
Intracellular Symbiotes of the Homoptera   E. J. Houk, G. W. Griffiths   25:161–87
Structure and Function in Tick Cuticle   R. H. Hackman   27:75–95
The Functional Morphology and Biochemistry
  of Insect Male Accessory Glands and Their
  Secretions   P. S. Chen   29:233–55
Morphology of Insect Development   F. Sehnal   30:89–109
Ultrastructure and Function of Insect Thermo-
  and Hygroreceptors   H. Altner, R. Loftus   30:273–95

PATHOLOGY
The Potential of Insect Pathogenic Viruses as
  Pesticidal Agents   T. W. Tinsley   24:63–87
Potential Implication of Genetic Engineering
  and Other Biotechnologies to Insect Control   J. B. Kirschbaum   30:51–70
Microbial Control of Black Flies and
  Mosquitoes   L. A. Lacey, A. H. Undeen   31:265–96
Ecological Considerations for the Use of
  Entomopathogens in IPM   J. R. Fuxa   32:225–51

PHYSIOLOGY AND BIOCHEMISTRY
Reproductive Physiology of Glossima   S. S. Tobe, P. A. Langley   23:283–307
Water Vapor Exchange Kinetics in Insects
  and Acarines   G. W. Wharton, A. G. Richards   23:309–28
Endocrine Regulation of Fat Body
  Development and Function   L. L. Keeley   23:329–52
Trehalose Regulation, One Aspect of
  Metabolic Homeostasis   S. Friedman   23:389–407
Neurobiological Contributions to
  Understanding Insect Pheromone Systems   W. D. Seabrook   23:471–85
Biochemistry of Insect Cuticle   S. O. Andersen   24:29–61
Insect Thermoregulation   M. L. May   24:313–49
Vitellogenin and Vitellin in Insects   H. H. Hagedorn, J. G. Kunkel   24:475–505
Summer Diapause   S. Masaki   25:1–25
Insect Tissue Culture: An Overview
  1971–1978   E. P. Marks   25:73–101
Intracellular Symbiotes of the Homoptera   E. J. Houk, G. W. Griffiths   25:161–87
The Regulation of Spermatogenesis in Insects   J. B. Dumser   25:341–69
Host Suitability for Insect Parasitoids   S. B. Vinson, G. F. Iwantsch   25:397–419
Sequestration of Plant Natural Products by
  Insects   S. S. Duffey   25:447–77
Regulation of the Juvenile Hormone Titer   C. A. D. de Kort, N. A. Granger   26:1–28
The Rice Brown Planthopper: Feeding
  Physiology and Host Plant Interactions   K. Sōgawa   27:49–73
Structure and Function in Tick Cuticle   R. H. Hackman   27:75–95
Chemical Ecology and Biochemistry of Insect
  Hydrocarbons   R. W. Howard, G. J. Bloomquist   27:149–72
Maternal Direction of Oogenesis and Early
  Embryogenesis in Insects   S. J. Berry   27:205–27
Biochemistry of Insect Venoms   J. O. Schmidt   27:339–68
Insect Thermoperiodism   S. D. Beck   28:91–108
Nitrogen Excretion in Cockroaches   D. G. Cochran   30:29–49
Regulation of Reproduction in Eusocial
  Hymenoptera   D. J. C. Fletcher, K. G. Ross   30:319–43
Endocrine Interactions Between Endoparasitic
  Insects and Their Hosts   N. E. Beckage   30:371–413
Imaginal Disc Determination: Molecular and
  Cellular Correlates   E. W. Larsen-Rapport   31:145–75
Nutrition and In Vitro Culture of Parasitoids   S. N. Thompson   31:197–219
Sperm Utilization in Social Insects   R. E. Page, Jr.   31:297–320
Biochemical Aspects of Insect Immunology   P. E. Dunn   31:321–39

Anti Juvenile Hormone Agents   G. B. Staal   31:391–429
Expression of the Genes Coding for
  Vitellogenin (Yolk Protein)   M. Bownes   31:507–31
Insects as Models in Neuroendocrine Research   B. Scharrer   32:1–16
Chitin Biochemistry: Synthesis and Inhibition   E. Cohen   32:71–93
Biosynthesis of Arthropod Exocrine
  Compounds   M. S. Blum   32:381–413
Physiology of Osmoregulation in Mosquitoes   T. J. Bradley   32:439–62

POPULATION ECOLOGY
Assessing and Interpreting the Spatial
  Distributions of Insect Populations   L. R. Taylor   29:321–57

SERICULTURE
Recent Advances in Sericulture   Y. Horie, H. Watanabe   25:49–71

SYSTEMATICS, EVOLUTION, AND BIOGEOGRAPHY
Biology and Systematics of the Sciomyzidae   C. O. Berg, L. Knutson   23:239–58
Recent Advances in the Study of Scale
  Insects   D. R. Miller, M. Kosztarab   24:1–27
Biogeographic Dynamics of Insect-Host Plant
  Communities   D. R. Strong Jr.   24:89–119
Evolution of Phytophagous Mites (Acari)   G. W. Krantz, E. E. Lindquist   24:121–58
Biosystematics of Thysanoptera   T. N. Ananthakrishnan   24:159–83
Evolution of Larval Food Preferences in
  Microlepidoptera   J. A. Powell   25:133–59
Phylogeny of Insect Orders   N. P. Kristensen   26:135–57
Palaeozoic Insects   R. J. Wootton   26:319–44
The Biology of *Heliconius* and Related
  Genera   K. S. Brown, Jr.   26:427–56
Changing Concepts in Biogeography   J. Illies   28:391–406
Insect Molecular Systematics   S. H. Berlocher   29:403–33
An Evolutionary and Applied Perspective of
  Insect Biotypes   S. R. Diehl, G. L. Bush   29:471–504
The Geographical and Ecological Distribution
  of Arboreal Psocoptera   I. W. B. Thornton   30:175–96
Insect Morphometrics   H. V. Daly   30:415–38
The Influence of Cladistics on Heteropteran
  Classification   R. T. Schuh   31:67–93
The Evolution and Ontogeny of Nestmate
  Recognition in Social Wasps   G. J. Gamboa, H. K. Reeve, D. W.
    Pfennig   31:431–54
Fossil Oribatid Mites   D. A. Krivolutsky, A. Ya. Druk   31:533–45
Chemosystematics and Evolution of Beetle
  Chemical Defenses   K. Dettner   32:17–48
Biogeography of the Montane Entomofauna of
  Mexico and Central America   G. Halffter   32:95–114

VECTORS OF PLANT PATHOGENS
Epidemiology of Maize Streak Disease   D. J. W. Rose   23:259–82
Circulative and Propagative Virus
  Transmission by Aphids   E. S. Sylvester   25:257–86
The Biology of Spiroplasmas   R. F. Whitcomb   26:397–425
Multiple Acquisition of Viruses and
  Vector-Dependent Prokaryotes:
  Consequences on Transmission   E. S. Sylvester   30:71–88

Annual Reviews Inc.

A NONPROFIT SCIENTIFIC PUBLISHER

AR 4139 El Camino Way
P.O. Box 10139
Palo Alto, CA 94303-0897 • USA

Annual Reviews Inc. publications may be ordered directly from our office by mail or use our Toll Free Telephone line (for orders paid by credit card or purchase order, and customer service calls only); through booksellers and subscription agents, worldwide; and through participating professional societies. Prices subject to change without notice.   ARI Federal I.D. #94-1156476

- **Individuals:** Prepayment required on new accounts by check or money order (in U.S. dollars, check drawn on U.S. bank) or charge to credit card — American Express, VISA, MasterCard.
- **Institutional buyers:** Please include purchase order number.
- **Students:** $10.00 discount from retail price, per volume. Prepayment required. Proof of student status must be provided (photocopy of student I.D. or signature of department secretary is acceptable). Students must send orders direct to Annual Reviews. Orders received through bookstores and institutions requesting student rates will be returned.
- **Professional Society Members:** Members of professional societies that have a contractual arrangement with Annual Reviews may order books through their society at a reduced rate. Check with your society for information.
- **Toll Free Telephone orders:** Call 1-800-523-8635 (except from California) for orders paid by credit card or purchase order and customer service calls only. California customers and all other business calls use 415-493-4400 (not toll free). Hours: 8:00 AM to 4:00 PM, Monday-Friday, Pacific Time.

Regular orders: Please list the volumes you wish to order by volume number.
Standing orders: New volume in the series will be sent to you automatically each year upon publication. Cancellation may be made at any time. Please indicate volume number to begin standing order.
Prepublication orders: Volumes not yet published will be shipped in month and year indicated.
California orders: Add applicable sales tax.
Postage paid (4th class bookrate/surface mail) **by Annual Reviews Inc.** Airmail postage or UPS, extra.

ANNUAL REVIEWS SERIES		Prices Postpaid per volume USA/elsewhere	Regular Order Please send:	Standing Order Begin with:
			Vol. number	Vol. number
Annual Review of ANTHROPOLOGY				
Vols. 1-14	(1972-1985)	$27.00/$30.00		
Vol. 15	(1986)	$31.00/$34.00		
Vol. 16	(avail. Oct. 1987)	$31.00/$34.00	Vol(s). _____	Vol. _____
Annual Review of ASTRONOMY AND ASTROPHYSICS				
Vols. 1-2, 4-20	(1963-1964; 1966-1982)	$27.00/$30.00		
Vols. 21-24	(1983-1986)	$44.00/$47.00		
Vol. 25	(avail. Sept. 1987)	$44.00/$47.00	Vol(s). _____	Vol. _____
Annual Review of BIOCHEMISTRY				
Vols. 30-34, 36-54	(1961-1965; 1967-1985)	$29.00/$32.00		
Vol. 55	(1986)	$33.00/$36.00		
Vol. 56	(avail. July 1987)	$33.00/$36.00	Vol(s). _____	Vol. _____
Annual Review of BIOPHYSICS AND BIOPHYSICAL CHEMISTRY				
Vols. 1-11	(1972-1982)	$27.00/$30.00		
Vols. 12-15	(1983-1986)	$47.00/$50.00		
Vol. 16	(avail. June 1987)	$47.00/$50.00	Vol(s). _____	Vol. _____
Annual Review of CELL BIOLOGY				
Vol. 1	(1985)	$27.00/$30.00		
Vol. 2	(1986)	$31.00/$34.00		
Vol. 3	(avail. Nov. 1987)	$31.00/$34.00	Vol(s). _____	Vol. _____

ANNUAL REVIEWS SERIES	Prices Postpaid per volume USA/elsewhere	Regular Order Please send:	Standing Order Begin with:

		Vol. number	Vol. number

Annual Review of COMPUTER SCIENCE
Vol. 1   (1986) . . . . **$39.00/$42.00**
Vol. 2   (avail. Nov. 1987) . . . . **$39.00/$42.00**   Vol(s). _____   Vol. _____

Annual Review of EARTH AND PLANETARY SCIENCES
Vols. 1-10   (1973-1982) . . . . **$27.00/$30.00**
Vols. 11-14   (1983-1986) . . . . **$44.00/$47.00**
Vol. 15   (avail. May 1987) . . . . **$44.00/$47.00**   Vol(s). _____   Vol. _____

Annual Review of ECOLOGY AND SYSTEMATICS
Vols. 1-16   (1970-1985) . . . . **$27.00/$30.00**
Vol. 17   (1986) . . . . **$31.00/$34.00**
Vol. 18   (avail. Nov. 1987) . . . . **$31.00/$34.00**   Vol(s). _____   Vol. _____

Annual Review of ENERGY
Vols. 1-7   (1976-1982) . . . . **$27.00/$30.00**
Vols. 8-11   (1983-1986) . . . . **$56.00/$59.00**
Vol. 12   (avail. Oct. 1987) . . . . **$56.00/$59.00**   Vol(s). _____   Vol. _____

Annual Review of ENTOMOLOGY
Vols. 10-16, 18-30   (1965-1971, 1973-1985) . . . . **$27.00/$30.00**
Vol. 31   (1986) . . . . **$31.00/$34.00**
Vol. 32   (avail. Jan. 1987) . . . . **$31.00/$34.00**   Vol(s). _____   Vol. _____

Annual Review of FLUID MECHANICS
Vols. 1-4, 7-17   (1969-1972, 1975-1985) . . . . **$28.00/$31.00**
Vol. 18   (1986) . . . . **$32.00/$35.00**
Vol. 19   (avail. Jan. 1987) . . . . **$32.00/$35.00**   Vol(s). _____   Vol. _____

Annual Review of GENETICS
Vols. 1-19   (1967-1985) . . . . **$27.00/$30.00**
Vol. 20   (1986) . . . . **$31.00/$34.00**
Vol. 21   (avail. Dec. 1987) . . . . **$31.00/$34.00**   Vol(s). _____   Vol. _____

Annual Review of IMMUNOLOGY
Vols. 1-3   (1983-1985) . . . . **$27.00/$30.00**
Vol. 4   (1986) . . . . **$31.00/$34.00**
Vol. 5   (avail. April 1987) . . . . **$31.00/$34.00**   Vol(s). _____   Vol. _____

Annual Review of MATERIALS SCIENCE
Vols. 1, 3-12   (1971, 1973-1982) . . . . **$27.00/$30.00**
Vols. 13-16   (1983-1986) . . . . **$64.00/$67.00**
Vol. 17   (avail. August 1987) . . . . **$64.00/$67.00**   Vol(s). _____   Vol. _____

Annual Review of MEDICINE
Vols. 1-3, 6, 8-9   (1950-1952, 1955, 1957-1958)
11-15, 17-36   (1960-1964, 1966-1985) . . . . **$27.00/$30.00**
Vol. 37   (1986) . . . . **$31.00/$34.00**
Vol. 38   (avail. April 1987) . . . . **$31.00/$34.00**   Vol(s). _____   Vol. _____

Annual Review of MICROBIOLOGY
Vols. 18-39   (1964-1985) . . . . **$27.00/$30.00**
Vol. 40   (1986) . . . . **$31.00/$34.00**
Vol. 41   (avail. Oct. 1987) . . . . **$31.00/$34.00**   Vol(s). _____   Vol. _____